WILEY SERIES IN PROBABILITY AND STATISTICS

Established by WALTER A. SHEWHART and SAMUEL S. WILKS



A complete list of the titles in this series appears at the end of this volume.

# Models for Probability and Statistical Inference

## The Wiley Bicentennial–Knowledge for Generations

Each generation has its unique needs and aspirations. When Charles Wiley first opened his small printing shop in lower Manhattan in 1807, it was a generation of boundless potential searching for an identity. And we were there, helping to define a new American literary tradition. Over half a century later, in the midst of the Second Industrial Revolution, it was a generation focused on building the future. Once again, we were there, supplying the critical scientific, technical, and engineering knowledge that helped frame the world. Throughout the 20th Century, and into the new millennium, nations began to reach out beyond their own borders and a new international community was born. Wiley was there, expanding its operations around the world to enable a global exchange of ideas, opinions, and know-how.

For 200 years, Wiley has been an integral part of each generation's journey, enabling the flow of information and understanding necessary to meet their needs and fulfill their aspirations. Today, bold new technologies are changing the way we live and learn. Wiley will be there, providing you the must-have knowledge you need to imagine new worlds, new possibilities, and new opportunities.

Generations come and go, but you can always count on Wiley to provide you the knowledge you need, when and where you need it!

William J. Pesce
President and Chief Executive Officer

Peter Booth Wiley
Chairman of the Board

# Models for Probability and Statistical Inference

## Theory and Applications

JAMES H. STAPLETON

Michigan State University
Department of Statistics and Probability
East Lansing, Michigan

WILEY-INTERSCIENCE

A JOHN WILEY & SONS, INC., PUBLICATION

Published by John Wiley & Sons, Inc., Hoboken, New Jersey.
Published simultaneously in Canada.



***Library of Congress Cataloging-in-Publication Data:***

Stapleton, James H., 1931–
Models for probability and statistical inference: theory and applications/James H. Stapleton.
p. cm.
ISBN 978-0-470-07372-8 (cloth)
1. Probabilities—Mathematical models. 2. Probabilities—Industrial applications. I. Title.
QA273.S7415 2008
519.2—dc22 2007013726

10 9 8 7 6 5 4 3 2 1

To Alicia, who has made my first home so pleasant
for almost 44 years.
To Michigan State University and its Department of Statistics
and Probability, my second home for almost 49 years,
to which I will always be grateful.

# Contents

# Preface

This book was written over a five to six-year period to serve as a text for the two-semester sequence on probability and statistical inference, STT 861–2, at Michigan State University. These courses are offered for master's degree students in statistics at the beginning of their study, although only one-half of the students are working for that degree. All students have completed a minimum of two semesters of calculus and one course in linear algebra, although students are encouraged to take a course in analysis so that they have a good understanding of limits. A few exceptional undergraduates have taken the sequence. The goal of the courses, and therefore of the book, is to produce students who have a fundamental understanding of statistical inference. Such students usually follow these courses with specialized courses on sampling, linear models, design of experiments, statistical computing, multivariate analysis, and time series analysis.

For the entire book, simulations and graphs, produced by the statistical package S-Plus, are included to build the intuition of students. For example, Section 1.1 begins with a list of the results of 400 consecutive rolls of a die. Instructors are encouraged to use either S-Plus or R for their courses. Methods for the computer simulation of observations from specified distributions are discussed.

Each section is followed by a selection of problems, from simple to more complex. Answers are provided for many of the problems.

Almost all statements are backed up with proofs, with the exception of the continuity theorem for moment generating functions, and asymptotic theory for logistic and log-linear models. Simulations are provided to show that the asymptotic theory provides good approximations.

The first six chapters are concerned with probability, the last seven with statistical inference. If a few topics covered in the first six chapters were to be omitted, there would be enough time in the first semester to cover at least the first few sections of Chapter Seven, on estimation. There is a bit too much material included on statistical inference for one semester, so that an instructor will need to make judicious choices of sections. For example, this instructor has omitted Section 7.8, on Fisher information, the Cramér–Rao bound, and asymptotic normality of MLEs, perhaps the most difficult material in the book. Section 7.9, on sufficiency, could be omitted.

Chapter One is concerned with discrete models and random variables. In Chapter Two we discuss discrete distributions that are important enough to have names: the binomial, hypergeometric, geometric, negative binomial, and Poisson, and the Poisson process is described. In Chapter Three we introduce continuous distributions, expected values, variances, transformation, and joint densities.

Chapter Four concerns the normal and gamma distributions. The beta distribution is introduced in Problem 4.3.5. Chapter Five, devoted to conditional distributions, could be omitted without much negative effect on statistical inference. Markov chains are discussed briefly in Chapter Five. Chapter Six, on limit theory, is usually the most difficult for students. Modes of convergence of sequences of random variables, with special attention to convergence in distribution, particularly the central limit theorem for independent random variables, are discussed thoroughly.

Statistical inference begins in Chapter Seven with point estimation: first methods of evaluating estimators, then methods of finding estimators: the method of moments and maximum likelihood. The topics of consistency and the $\delta$-method are usually a bit more difficult for students because they are often still struggling with limit arguments. Section 7.7, on confidence intervals, is one of the most important topics of the last seven chapters and deserves extra time. The author often asks students to explain the meaning of confidence intervals so that "your mother [or father] would understand." Students usually fail to produce an adequate explanation the first time. As stated earlier, Section 7.8 is the most difficult and might be omitted. The same could be said for Section 7.9, on sufficiency, although the beauty of the subject should cause instructors to think twice before doing that.

Chapter Eight, on testing hypotheses, is clearly one of the most important chapters. We hope that sufficient time will be devoted to it to "master" the material, since the remaining chapters rely heavily on an understanding of these ideas and those of Section 7.7, on confidence intervals.

Chapter Nine is organized around the distributions defined in terms of the normal: multivariate normal, chi-square, $t$, and $F$ (central and noncentral). The usefulness of each of the latter three distributions is shown immediately by the development of confidence intervals and testing methods for "normal models." Some of "Student's" data from the 1908 paper introducing the $t$-distribution is used to illustrate the methodology.

Chapter Ten contains descriptions of the two- and one-sample Wilcoxon tests, together with methods of estimation based on these. The Kolmogorov–Smirnov one- and two-sample tests are also discussed.

Chapter Eleven, on linear models, takes the linear space-projection approach. The geometric intuition it provides for multiple regression and the analysis of variance, by which sums of squares are simply squared lengths of vectors, is quite valuable. Examples of S-Plus and SAS printouts are provided.

Chapter Twelve begins with logistic regression. Although the distribution theory is quite different than the linear model theory discussed in Chapter Eleven and is asymptotic, the intuition provided by the vector-space approach carries over to logistic regression. Proofs are omitted in general in the interests of time and the students' level of understanding. Two-way frequency tables are discussed for models which suppose that the logs of expected frequencies satisfy a linear model.

Finally, Chapter Thirteen has sections on survival analysis, including the Kaplan–Meier estimator of the cumulative distribution function, bootstrapping, Bayesian statistics, and sampling. Each is quite brief. Instructors will probably wish to select from among these four topics.

I thank the many excellent students in my Statistics 861–2 classes over the last seven years, who provided many corrections to the manuscript as it was being developed. They have been very patient.

JIM STAPLETON

*March 7, 2007*

CHAPTER ONE

# Discrete Probability Models

## 1.1 INTRODUCTION

The mathematical study of probability can be traced to the seventeenth-century correspondence between Blaise Pascal and Pierre de Fermat, French mathematicians of lasting fame. Chevalier de Mere had posed questions to Pascal concerning gambling, which led to Pascal's correspondence with Fermat. One question was this: Is a gambler equally likely to succeed in the two games: (1) at least one 6 in four throws of one six-sided die, and (2) at least one double-6 (6–6) in 24 throws of two six-sided dice? At that time it seemed to many that the answer was yes. Some believe that de Mere had empirical evidence that the first event was more likely to occur than the second, although we should be skeptical of that, since the probabilities turn out to be 0.5178 and 0.4914, quite close. After students have studied Chapter One they should be able to verify these, then, after Chapter Six, be able to determine how many times de Mere would have to play these games in order to distinguish between the probabilities.

In the eighteenth century, probability theory was applied to astronomy and to the study of errors of measurement in general. In the nineteenth and twentieth centuries, applications were extended to biology, the social sciences, medicine, engineering—to almost every discipline. Applications to genetics, for example, continue to grow rapidly, as probabilistic models are developed to handle the masses of data being collected. Large banks, credit companies, and insurance and marketing firms are all using probability and statistics to help them determine operating rules.

We begin with discrete probability theory, for which the events of interest often concern count data. Although many of the examples used to illustrate the theory involve gambling games, students should remember that the theory and methods are applicable to many disciplines.

## 1.2 SAMPLE SPACES, EVENTS, AND PROBABILITY MEASURES

We begin our study of probability by considering the results of 400 consecutive throws of a *fair die*, a six-sided cube for which each of the numbers 1, 2, ..., 6 is equally likely to be the number showing when the die is thrown.

61635 52244 21641 36536 52114 64452 33132 26324 62624 63134
36426 33552 65554 64623 56111 32256 36435 64146 53514 56364
52624 12534 15362 65261 43445 13223 66126 53623 63265 21564
21524 13552 65253 21225 42234 32361 62454 54561 15125 36555
45215 66442 42635 52522 13242 15434 16336 63241 13111 54343
32261 63155 55235 13611 54346 56323 41666 31221 53233 52414
53366 62336 11265 55136 56524 64215 44221 14222 15145 31662
55241 54223 25156 56155 43324 36566 23466 51123 11414 24653

The frequencies are:

| 1 | 2 | 3 | 4 | 5 | 6 |
|---|---|---|---|---|---|
| 60 | 73 | 65 | 58 | 74 | 70 |

We use these data to motivate the definitions and theory to be presented. Consider, for example, the following question: What is the probability that the five numbers appearing in five throws of a die are all different? Among the 80 consecutive sequences of five numbers above, in only four cases were all five numbers different, a relative frequency of $5/80 = 0.0625$. In another experiment, with 2000 sequences of five throws each, all were different 183 times, a relative frequency of 0.0915. Is there a way to determine the long-run relative frequency? Put another way, what could we expect the relative frequency to be in 1 million throws of five dice?

It should seem reasonable that all possible sequences of five consecutive integers from 1 to 6 are equally likely. For example, prior to the 400-throw experiment, each of the first two sequences, 61635 and 52244, were equally likely. For this example, such five-digit sequences will be called *outcomes* or *sample points*. The collection of all possible such five-digit sequences will be denoted by $S$, the sample space. In more mathematical language, $S$ is the Cartesian product of the set $A = \{1, 2, 3, 4, 5, 6\}$ with itself five times. This collection of sequences is often written as $A^{(5)}$. Thus, $S = A^{(5)} = A \times A \times A \times A \times A$. The number of outcomes (or sample points) in $S$ is $6^5 = 7776$. It should seem reasonable to suppose that all outcomes (five-digit sequences) have probability $1/6^5$.

We have already defined a *probability model* for this experiment. As we will see, it is enough in cases in which the sample space is discrete (finite or countably infinite) to assign probabilities, nonnegative numbers summing to 1, to each outcome in the sample space $S$. A discrete probability model has been defined for an experiment when (1) a finite or countably infinite sample space has been defined, with each possible result of the experiment corresponding to exactly one outcome; and (2) probabilities, nonnegative numbers, have been assigned to the outcomes in such a way that they sum to 1. It is not necessary that the probabilities assigned all be the same as they are for this example, although that is often realistic and convenient.

We are interested in the *event* $A$ that all five digits in an outcome are different. Notice that this event $A$ is a subset of the sample space $S$. We say that an event $A$ has *occurred* if the outcome is a member of $A$. In this case event $A$ did not occur for any of the eight outcomes in the first row above.

We define the *probability* of the event $A$, denoted $P(A)$, to be the sum of the probabilities of the outcomes in $A$. By defining the probability of an event in this way, we assure that the probability measure $P$, defined for all subsets (events, in probability language) of $S$, obeys certain axioms for probability measures (to be stated later). Because our probability measure $P$ has assigned all probabilities of outcomes to be equally likely, to find $P(A)$ it is enough for us to determine the number of outcomes $N(A)$ in $A$, for then $P(A) = N(A)[1/N(S)] = N(A)/N(S)$. Of course, this is the case only because we assigned equal probabilities to all outcomes.

To determine $N(A)$, we can apply the multiplication principle. $A$ is the collection of 5-tuples with all components different. Each outcome in $A$ corresponds to a way of filling in the boxes of the following cells:

|   |   |   |   |   |
|---|---|---|---|---|

The first cell can hold any of the six numbers. Given the number in the first cell, and given that the outcome must be in $A$, the second cell can be any of five numbers, all different from the number in the first cell. Similarly, given the numbers in the first two cells, the third cell can contain any of four different numbers. Continuing in this way, we find that $N(A) = (6)(5)(4)(3)(2) = 720$ and that $P(A) = 720/7776 = 0.0926$, close to the value obtained for 2000 experiments. The number $N(A) = 720$ is the number of permutations of six things taken five at a time, indicated by $P(6, 5)$.

***Example 1.2.1*** Consider the following discrete probability model, with sample space $S = \{a, b, c, d, e, f\}$.

| Outcome $\omega$ | $a$ | $b$ | $c$ | $d$ | $e$ | $f$ |
|---|---|---|---|---|---|---|
| $P(\omega)$ | 0.30 | 0.20 | 0.25 | 0.10 | 0.10 | 0.05 |

Let $A = \{a, b, d\}$ and $B = \{b, d, e\}$. Then $A \cup B = \{a, b, d, e\}$ and $P(A \cup B) = 0.3 + 0.2 + 0.1 + 0.1 = 0.7$. In addition, $A \cap B = \{b, d\}$, so that $P(A \cap B) = 0.2 + 0.1 = 0.3$. Notice that $P(A \cup B) = P(A) + P(B) - P(A \cap B)$. (Why must this be true?). The *complement* of an event $D$, denoted by $D^c$, is the collection of outcomes in $S$ that are not in $D$. Thus, $P(A^c) = P(\{c, e, f\}) = 0.15 + 0.15 + 0.10 = 0.40$. Notice that $P(A^c) = 1 - P(A)$. Why must this be true? □

Let us consider one more example before more formally stating the definitions we have already introduced.

***Example 1.2.2*** A penny and a dime are tossed. We are to observe the number $X$ of heads that occur and determine $P(X = k)$ for $k = 0, 1, 2$. The symbol $X$, used here for its convenience in defining the events $[X = 0]$, $[X = 1]$, and $[X = 2]$, will be called a *random variable* (rv). $P(X = k)$ is shorthand for $P([X = k])$. We delay a more formal discussion of random variables.

Let $S_1 = \{\text{HH, HT, TH, TT}\} = \{\text{H, T}\}^{(2)}$, where the result for the penny and dime are indicated in this order, with H denoting head and T denoting tail. It should seem reasonable to assign equal probabilities 1/4 to each of the four outcomes. Denote the resulting probability measure by $P_1$. Thus, for $A =$ [event that the coins give the same result] $= \{\text{HH, TT}\}$, $P_1(A) = 1/4 + 1/4 = 1/2$. □

The 400 throws of a die can be used to simulate 400 throws of a coin, and therefore 200 throws of two coins, by considering 1, 2, and 3 as heads and 4, 5, and 6 as tails. For example, using the first 10 throws, proceeding across the first row, we get TH, TH, TT, HH, TT. For all 400 die throws, we get 50 cases of HH, 55 of HT, 47 of TH, and 48 of TT, with corresponding relative proportions 0.250, 0.275, 0.235, and 0.240. For the experiment with 10,000 throws, simulating 5000 pairs of coin tosses, we obtain 1288 HH's, 1215 HT's, 1232 TH's, and 1265 TT's, with relative frequencies 0.2576, 0.2430, 0.2464, and 0.2530. Our model $(S_1, P_1)$ seems to fit well.

For this model we get $P_1(X = 0) = 1/4$, $P_1(X = 1) = 1/4 + 1/4 = 1/2$, and $P_1(X = 2) = 1/4$. If we are interested only in $X$, we might consider a slightly smaller model, with sample space $S_2 = \{0, 1, 2\}$, where these three outcomes represent the numbers of heads occurring. Although it is tempting to make the model simpler by assigning equal probabilities 1/3, 1/3, 1/3 to these outcomes, it should be obvious that the empirical results of our experiments with 400 and 10,000 tosses are not consistent with such a model. It should seem reasonable, instead, to assign probabilities 1/4, 1/2, 1/4, thus defining a probability measure $P_2$ on $S_2$. The model $(S_2, P_2)$ is a *recoding* or *reduction* of the model $(S_1, P_1)$, with the outcomes HT and TH of $S_1$ corresponding to the single outcome $X = 1$ of $S_2$, with corresponding probability determined by adding the probabilities 1/4 and 1/4 of HT and TH.

The model $(S_2, P_2)$ is simpler than the model $(S_1, P_1)$ in the sense that it has fewer outcomes. On the other hand, it is more complex in the sense that the probabilities are unequal. In choosing appropriate probability models, we often have two or more possible models. The choice of a model will depend on its approximation of experimental evidence, consistency with fundamental principles, and mathematical convenience.

Let us stop now to define more formally some of the terms already introduced.

**Definition 1.2.1** A *sample space* is a collection $S$ of all possible results, called *outcomes*, of an experiment. Each possible result of the experiment must correspond to one and only one outcome in $S$. A sample space is *discrete* if it has a finite or countably infinite number of outcomes. (A set is *countably infinite* if it can be put into one-to-one correspondence with the positive integers.) □

**Definition 1.2.2** An *event* is a subset of a sample space. An event $A$ is said to *occur* if the outcome of an experiment is a member of $A$. □

**Definition 1.2.3** A *probability measure* $P$ on a discrete sample space $S$ is a function defined on the subsets of $S$ such that:

**(a)** $P(\{\omega\}) \geq 0$ for all points $\omega \in S$.
**(b)** $P(A) = \sum_{\omega \in A} P(\omega)$ for all subsets $A$ of $S$.
**(c)** $P(S) = 1$.

For simplicity, we write $P(\{\omega\})$ as $P(\omega)$. □

**Definition 1.2.4** A *probability model* is a pair $(S, P)$, where $P$ is a probability measure on $S$. □

In writing $P(\{\omega\})$ as $P(\omega)$, we are abusing notation slightly by using the symbol $P$ to denote both a function on $S$ and a function on the subsets of $S$. We assume that students are familiar with the notation of set theory: *union*, $A \cup B$; *intersection* $A \cap B$; and *complement*, $A^c$. Thus, for events $A$ and $B$, the event $A \cup B$ is said to occur if the outcome is a member of $A$ or $B$ (by "or" we include the case that the outcome is in both $A$ and $B$). The event $A \cap B$ is said to occur if both $A$ and $B$ occur. $A^c$, called a complement, is said to occur if $A$ does not occur. For convenience we sometimes write $A \cap B$ as $AB$.

We also assume that the student is familiar with the notation for relationships among sets, $A \subset B$ and $A \supset B$. Thus, if $A \subset B$, the occurrence of event $A$ implies that $B$ must occur. We sometimes use the language "event $A$ implies event $B$." For the preceding two-coin-toss example, the event $[X = 1]$ implies the event $[X \geq 1]$.

Let $\emptyset$ denote the *empty event*, the subset of $S$ consisting of no outcomes. Thus, $A \cap A^c = \emptyset$. We say that two events $A$ and $B$ are *mutually exclusive* if their intersection is empty. That is, $A \cap B = \emptyset$. Thus, if $A$ and $B$ are mutually exclusive, the occurrence of one of them implies that the other cannot occur. In set-theoretic language we say that $A$ and $B$ are *disjoint*. *DeMorgan's laws* give relationships among intersection, union, and complement:

$$(1) \quad (A \cap B)^c = A^c \cup B^c \quad \text{and} \quad (2) \quad (A \cup B)^c = A^c \cap B^c.$$

These can be verified from a *Venn diagram* or by showing that any element in the set on the left is a member of the set on the right, and vice versa (see Figure 1.2.1).

### Properties of a Probability Measure $P$ on a Sample Space $S$

1. $P(\emptyset) = 0$.
2. $P(S) = 1$.
3. For any event $A$, $P(A^c) = 1 - P(A)$.

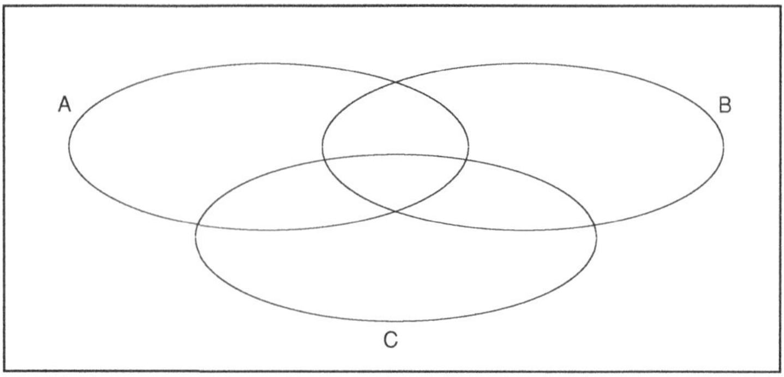


**FIGURE 1.2.1** Venn diagram for three events.

4. For any events $A$ and $B$, $P(A \cup B) = P(A) + P(B) - P(A \cap B)$. For three events $A, B, C$, $P(A \cup B \cup C) = P(A) + P(B) + P(C) - P(A \cap B) - P(A \cap C) - P(B \cap C) + P(A \cap B \cap C)$. This follows from repeated use of the identity for two events. An almost obvious similar result holds for the probability of the union of $n$ events, with $2^n - 1$ terms on the right.
5. For events $A$ and $B$ with $A \cap B = \emptyset$, $P(A \cup B) = P(A) + P(B)$. More generally, if $A_1, A_2, \ldots,$ are disjoint (mutually exclusive) events, $P(\cup_{k=1}^{\infty} A_k) = \sum_{k=1}^{\infty} P(A_k)$. This property of $P$ is called *countable additivity*. Since $A_k$ for $k > n$ could be $\emptyset$, the same equality holds when $\infty$ is replaced by any integer $n > 0$.

Let us make use of some of these properties in a few examples.

***Example 1.2.3*** Smith and Jones each throw three coins. Let $X$ denote the number of heads for Smith. Let $Y$ denote the number of heads for Jones. Find $P(X = Y)$.

We can simulate this experiment using the 400-die-tossing example, again letting 1, 2, 3 correspond to heads, 4, 5, 6 correspond to tails. Let the first three tosses be for Smith, the next three for Jones, so that the first six tosses determine one trial of the experiment. Repeating, going across rows, we get 36 trials of the experiment. Among these 36 trials, 10 resulted in $X = Y$, suggesting that $P(X = Y)$ may be approximately $10/36 = 0.278$. For the experiment with 9996 tosses, 525 among 1666 trials gave $X = Y$, suggesting that $P(X = Y)$ is close to $525/1666 = 0.3151$. Let us now try to find the probability by mathematical methods.

Let $S_1 = \{\text{H, T}\}^{(3)}$, the collection of 3-tuples of heads and tails. $S_1$ is the collection of outcomes for Smith. Also, let $S_2 = \{\text{H, T}\}^{(3)} = S_1$, the collection of outcomes for Jones. Let $S = S_1 \times S_2$. This Çartesian product can serve as the sample space for the experiment in which Smith and Jones both toss three coins. One outcome in $S$, for example (using shorthand notation), is (HTH, TTH), so that $X = 2$, $Y = 1$. The event $[X = Y]$ did not occur. Since $N(S_1) = N(S_2) = 2^3 = 8$, $N(S) = 64$. Define the probability measure $P$ on $S$ by assigning probability $1/64$ to each outcome. The pair $(S, P)$ constitutes a probability model for the experiment.

Let $A_k = [X = Y = k]$ for $k = 0, 1, 2, 3$. By this bracket notation we mean the collection of outcomes in $S$ for which $X$ and $Y$ are both $k$. We might also have

**TABLE 1.2.1 Box Diagram**

| | $R_2$ | $R_2^c$ | |
|---|---|---|---|
| $R_1$ | | 0.2 | 0.6 |
| $R_1^c$ | | | |
| | 0.5 | | 1.0 |

written $A_k = [X = k, Y = k]$. The events $A_0, A_1, A_2, A_3$ are mutually exclusive, and $[X = Y] = A_0 \cup A_1 \cup A_2 \cup A_3$. It follows from property 5 above that $P(X = Y) = P(A_0) + P(A_1) + P(A_2) + P(A_3)$. Since $N(A_0) = 1$, $N(A_1) = 3^2$, $N(A_2) = 3^2$, $N(A_3) = 1$, and $P(A_k) = N(A_k)/64$, we find that $P(X = Y) = 20/64 = 5/16 = 0.3125$, relatively close to the proportions obtained by experimentation. □

***Example 1.2.4*** Suppose that a probability model for the weather for two days has been defined in such a way that $R_1 =$ [rain on day 1], $R_2 =$ [rain on day 2], $P(R_1) = 0.6$, $P(R_2) = 0.5$, and $P(R_1 R_2^c) = 0.2$. Find $P(R_1 R_2)$, $P(R_1^c R_2)$, and $P(R_1 \cup R_2)$.

Although a Venn diagram can be used, a *box diagram* (Table 1.2.1) makes things clearer. From the three probabilities given, the other cells may be determined by subtraction. Thus, $P(R_1^c) = 0.4$, $P(R_2^c) = 0.5$, $P(R_1 R_2) = 0.4$, $P(R_1^c R_2) = 0.1$, $P(R_1^c R_2^c) = 0.3$, $P(R_1^c \cup R_2^c) = 0.6$. Similar tables can be constructed for the three events. □

***Example 1.2.5*** A jury of six is to be chosen randomly from a panel of eight men and seven women. Let $X$ denote the number of women chosen. Let us find $P(X = k)$ for $k = 0, 1, \ldots, 6$.

For convenience, name the members of the panel $1, 2, \ldots, 15$, with the first eight being men. Let $D = \{1, 2, \ldots, 15\}$. Since the events in which we are interested do not depend on the order in which the people are drawn, the outcomes can be chosen to be subsets of $D$ of size 6. That is, $S = \{B \mid B \subset D, N(B) = 6\}$. We interpret "randomly" to mean that all the outcomes in $S$ should have equal probability. We need to determine $N(S)$. Such subsets are often called *combinations*. The number of combinations of size $k$ of a set of size $n$ is denoted by $\binom{n}{k}$. Thus, $N(S) = \binom{15}{6}$.

The number of *permutations* (6-tuples of different people) of 15 people six at a time, is $P(15, 6) = (15)(14)(13)(12)(11)(10) = 15!/9!$. The number of ways of ordering six people is $P(6, 6) = 6!$. Since (number of subsets of $D$ of size 6) × (number of ways of ordering six people) $= P(15, 6)$, we find that $N(S) = \binom{15}{6} = P(15, 6)/6! = 15!/[9!6!] = 5005$. Each outcome is assigned probability $1/5005$.

Consider the event $[X = 2]$. An outcome in $[X = 2]$ must include exactly two females and therefore four males. There are $\binom{7}{2} = (7)(6)/(2)(1) = 21$ such combinations. There are $\binom{8}{4}$ combinations of four males. There are

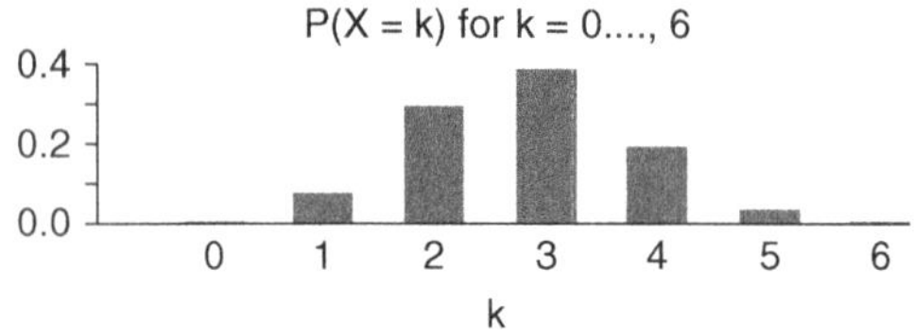


**FIGURE 1.2.2** Probability mass function for $X$.

therefore $\binom{7}{2}\binom{8}{4} = (21)(70) = 1470$ outcomes in the event $[X = 2]$. Therefore, $P(X = 2) = \binom{7}{2}\binom{8}{4}\Big/\binom{15}{6} = 1470/5005 = 0.2937.$

Similarly, we find $P(X = 3) = \binom{7}{3}\binom{8}{3}\Big/N(S) = (35)(56)/5005 = 1960/5005 = 0.3916$, $P(X = 0) = 28/5005 = 0.0056$, $P(X = 1) = 392/5005 = 0.0783$, $P(X = 4) = 980/5005 = 0.1691$, $P(X = 5) = 168/5005 = 0.0336$, $P(X = 6) = 7/5005 = 0.0014$. Figure 1.2.2 shows that these probabilities go "uphill," then "downhill," with the maximum at 3. □

***Example 1.2.6*** (Poker) The cards in a 52-card deck are classified in two ways: by 13 ranks and by four suits. The 13 ranks and four suits are indicated by the column and row headings in Table 1.2.2. In the game of poker, five cards are chosen randomly without replacement, so that all possible subsets (called *hands*) are equally likely. Hands are classified as follows, with decreasing worth: straight flush, 4-of-a-kind, full house, flush, straight, 3-of-a-kind, two pairs, one pair, and "bad." The category "bad" was chosen by the author so that all hands fall in one of the categories. *k-of-a-kind* means that $k$ cards are of one rank but the other $5 - k$ cards are of differing ranks. A straight consists of five cards with five consecutive ranks. For this purpose the ace is counted as either high or low, so that ace–2–3–4–5 and 10–J–Q–K–ace both constitute straights. A flush consists of cards that are all of the same suit. So that a hand falls in exactly one of these categories, it is always classified in the higher category if it satisfies the definition. Thus, a hand that is both a straight and a flush is classified as a straight flush but not as a straight or as a flush. A full house has three cards of one rank, two of another. Such a hand is not counted as 3-of-a-kind or as 2-of-a-kind.

**TABLE 1.2.2 52-Card Deck**

| | Ace | 2 | 3 | 4 | 5 | 6 | 7 | 8 | 9 | 10 | Jack | Queen | King |
|---|---|---|---|---|---|---|---|---|---|---|---|---|---|
| Spades | | | | | | | | | | | | | |
| Hearts | | | | | | | | | | | | | |
| Diamonds | | | | | | | | | | | | | |
| Clubs | | | | | | | | | | | | | |

Let $D$ be the set of 52 cards. Let $S$ be the collection of five-card hands, subsets of five cards. Thus, $S = \{B \mid B \subset D,\ N(B) = 5\}$. "Randomly without replacement" means that all $N(S) = \binom{52}{5} = 2{,}598{,}960$ outcomes are equally likely. Thus, we have defined a probability model.

Let $F$ be the event [full house]. The rank having three cards can be chosen in 13 ways. For each of these the rank having two cards can be chosen in 12 ways. For each choice of the two ranks there are $\binom{4}{3}\binom{4}{2} = (4)(6)$ choices for the suits. Thus, $N(F) = (13)(12)(6)(4) = 3744$, and $P(F) = 3744/2,598,960 = 0.0001439 = 1/694$. Similarly, $P(\text{straight}) = 10[4^5 - 4]/N(S) = 10{,}200/N(S) = 0.003925 = 1/255$, and $P(2\,\text{pairs}) = \binom{13}{2}\binom{4}{2}\binom{4}{2}(44)/N(S) = 123{,}552/N(S) = 0.04754 = 1/21.035$. In general, as the value of a hand increases, the probability of the corresponding category decreases (see Problem 1.2.3). □

***Example 1.2.7*** (The Birthday Problem) A class has $n$ students. What is the probability that at least one pair of students have the same birthday, not necessarily the same birth year?

So that we can think a bit more clearly about the problem, let the days be numbered 1, ..., 365, and suppose that $n = 20$. Birth dates were randomly chosen using the function "sample" in S-Plus, a statistical computer language.

| | | | | | | | | | | | | | | | | | | | | |
|---|---|---|---|---|---|---|---|---|---|---|---|---|---|---|---|---|---|---|---|---|
| (1) | 52 | 283 | 327 | 15 | 110 | 214 | 141* | 276 | 16 | 43 | 130 | 219 | 337 | 234 | 64 | 262 | 141* | 336 | 220 | 10 |
| (2) | 331 | 106 | 364 | 219 | 209 | 70 | 11 | 54 | 192 | 360 | 75 | 228 | 132 | 172 | 30 | 5 | 166 | 15 | 143 | 173 |
| (3) | 199 | 361* | 211 | 48 | 86 | 129 | 39 | 202 | 339 | 347 | 22 | 361* | 208 | 276 | 75 | 115 | 65 | 291 | 57 | 318 |
| (4) | 300 | 252 | 274 | 135 | 118 | 199 | 254 | 316 | 133 | 192 | 238 | 189 | 94 | 167 | 182 | 5 | 235 | 363 | 160 | 214 |
| (5) | 110 | 187 | 107 | 47 | 250 | 341 | 49 | 341 | 258 | 273 | 290 | 225 | 31 | 108 | 334 | 118 | 214 | 87 | 315 | 282 |
| (6) | 195 | 270^ | 24 | 204# | 69 | 233 | 38% | 204# | 12* | 358 | 38% | 138 | 149 | 76 | 71 | 186 | 106 | 270^ | 12* | 87 |
| (7) | 105 | 354 | 259 | 10 | 244 | 22 | 70 | 28 | 278 | 127 | 320 | 238 | 60 | 8 | 165 | 339 | 119 | 346 | 295 | 92 |
| (8) | 359# | 289 | 112 | 299 | 201 | 36 | 94 | 75 | 269 | 359# | 122 | 288 | 310 | 329 | 133 | 117 | 291 | 61* | 61* | 336 |
| (9) | 300 | 346 | 72 | 296 | 221 | 176 | 109 | 189 | 3 | 114 | 83 | 222 | 292 | 318 | 238 | 215 | 246 | 183 | 220 | 236 |
| (10) | 337 | 98 | 17 | 357 | 75 | 32 | 138 | 255 | 150 | 12 | 88 | 133 | 135 | 5 | 319 | 198 | 119 | 288 | 183 | 359 |

Duplicates are indicated by *'s, #'s, ^'s, and %'s. Notice that these 10 trials had 1, 0, 0, 3, 0, 2, 0, 0 duplicates. Based on these trials, we estimate the probability of at least one duplicate to be 4/10. This would seem to be a good estimate, since 2000 trials produced 846 cases with at least one duplicate, producing the estimate 0.423. Let us determine the probability mathematically.

Notice the similarity of this example to the die-throw example at the beginning of the chapter. In this case let $D = \{1, \ldots, 365\}$, the "dates" of the year. Let $S = D^{(n)}$, the $n$-fold Cartesian product of $D$ with itself. Assign probability $1/N(S) = 1/365^n$ to each outcome. We now have a probability model. □

Let $A$ be the event of at least one duplicate. As with most "at least one" events, it is easier to determine $N(A^c)$ than $N(A)$ directly. In fact, $N(A^c) = P(365, n) = 365(364)\cdots(365 - n + 1)$. Let $G(n) = P(A^c)$. It follows that $G(n) = N(A^c)/N(S) = \prod_{k=1}^{n}[(365 - k + 1)/365] = \prod_{k=1}^{n}[1 - (k - 1)/365]$. We can find

**TABLE 1.2.3 Probabilities of Coincident Birthdays**

| | $n$ | | | | | | | | |
|---|---|---|---|---|---|---|---|---|---|
| | 10 | 20 | 22 | 23 | 30 | 40 | 50 | 60 | 70 |
| $P(A)$ | 0.1169 | 0.4114 | 0.4757 | 0.5073 | 0.7063 | 0.8912 | 0.9704 | 0.9941 | 0.9991 |
| $h(n)$ | 0.1160 | 0.4058 | 0.4689 | 0.5000 | 0.6963 | 0.8820 | 0.9651 | 0.9922 | 0.9987 |

a good approximation by taking logarithms and converting the product to a sum. $\ln\ G(n) = \sum_{k=1}^{n} \ln\,[1 - (k-1)/365]$. For $x$ close to zero, $\ln(1-x)$ is close to $-x$, the Taylor series linear approximation. It follows that for $(n-1)/365$ small, $\ln(G(n))$ is approximately $-\sum_{k=1}^{n}[(k-1)/365] = -[n(n-1)/2]/365 = -n(n-1)/730$. Hence, a good approximation for $P(A) = 1 - G(n)$ is $h(n) = 1 - e^{-n(n-1)/730}$. Table 1.2.3 compares $P(A)$ to its approximation $h(n)$ for various $n$. Notice that the relative error in the approximation of $P(A^c)$ by $1 - h(n)$ increases as $n$ increases.

### Pascal's Triangle: An Interesting Identity for Combinatorics

Consider a set of five elements $A = \{a_1, a_2, \ldots, a_5\}$. $A$ has $\binom{5}{3} = 10$ subsets of size 3. These are of two types: those that contain element $a_1$ and those that do not. The number that contains $a_1$ is the number of subsets $\binom{4}{2} = 6$ of $\{a_2, \ldots, a_5\}$ of size 2. The number of subsets of $A$ that do not contain $a_1$ is $\binom{4}{3} = 4$. Thus, $\binom{5}{3} = \binom{4}{2} + \binom{4}{3}$.

More generally, if $A$ has $n$ elements $\{a_1, a_2, \ldots, a_n\}$, $A$ has $\binom{n}{k}$ subsets of size $k$ for $0 < k \le n$. These subsets are of two types, those that contain $a_1$ and those that do not. It follows by the same reasoning that $\binom{n}{k} = \binom{n-1}{k-1} + \binom{n-1}{k}$. The same equality can be proved by manipulation of factorials. Pascal, in the mid-seventeenth century, represented this in the famous *Pascal triangle* (Figure 1.2.3). Each row begins and ends with 1, and each interior value is the sum of the two immediately above. The $n$th row for $n = 0, 1, \ldots$ has $\binom{n}{k}$ in the $k$th place for $k = 0, 1, \ldots, n$. Row $n = 4$ has elements 1, 4, 6, 4, 1. Notice that these sum to $16 = 2^4$.

### The Equality $\binom{n}{0} + \binom{n}{1} + \cdots + \binom{n}{n} = 2^k$

The collection $B$ of subsets of a set with $n$ elements is in one-to-one correspondence to the set $C = \{0, 1\}^{(n)}$. For example, for the set $A = \{a_1, a_2, a_3, a_4\}$, the point (0, 1, 1, 0) in $C$ corresponds to the subset $\{a_2, a_3\}$, and $\{1, 0, 1, 1\}$ corresponds to the subset $\{a_1, a_3, a_4\}$. Thus, $N(B) = N(C) = 2^k$. But we can count the elements in $B$ another way. There are those with no elements, those with one, those with 2, and so on. The

$$\binom{0}{0} = 1$$

$$\binom{1}{0} \quad \binom{1}{1}$$

$$\binom{2}{0} \quad \binom{2}{1} \quad \binom{2}{2}$$

$$\binom{3}{0} \quad \binom{3}{1} \quad \binom{3}{2} \quad \binom{3}{3}$$

.........

**FIGURE 1.2.3** Pascal's triangle.

equality above follows. For example, $2^5 = 1 + 5 + 10 + 10 + 5 + 1$. The sum of the numbers in the row of Pascal's triangle labeled $n$ is $2^n$.

### Relationship Between Two Probability Models

***Example 1.2.8*** Suppose that a coin is tossed twice. This experiment may be reasonably modeled by $(S_1, P_1)$, where $S_1 = \{\text{HH, HT, TH, TT}\}$ and $P_1$ assigns probability $1/4$ to each outcome, and by $(S_2, P_2)$, where $\{X = 0, X = 1, X = 2\}$, where $X =$ (no. heads), and $P_2$ assigns probabilities $1/4, 1/2, 1/4$. In this case we have essentially "glued" together the outcomes HT and TH in $S_1$ to create one outcome $(X = 1)$ in $S_2$. We have also added the probabilities $1/4$ and $1/4$ in $(S_1, P_1)$ to get $P_2(X = 1) = 1/2$. The model $(S_2, P_2)$ is simpler than $(S_1, P_1)$ in the sense that it has fewer outcomes, but it is more complex in the sense that the probabilities aren't equal. We will say that the probability model $(S_2, P_2)$ is a *reduction* of the probability model $(S_1, P_1)$, and that $(S_1, P_1)$ is an *expansion* of the probability model $(S_2, P_2)$.

The model $(S_1, P_1)$ can be used to determine the probability of the event $H_1 =$ [first toss is heads]. The probability model $(S_2, P_2)$ cannot be used to determine the probability of $H_1$. The event $H_1$ is not "measurable" with respect to the model $(S_2, P_2)$. Such questions on measurability are considered as part of the subject of measure theory. We say very little about it here. □

In order to consider a general definition of reduction and expansion of probability models, we need to recall that for any function $g$: $S_1 \rightarrow S_2$,

$$g^{-1}(A) = \{\omega \mid g(\omega) \in A\} \quad \text{for any subset } A \text{ of } S_2.$$

**Definition 1.2.5** Let $(S_1, P_1)$ and $(S_2, P_2)$ be two discrete probability models. Then $(S_2, P_2)$ is said to be a *reduction* of $(S_1, P_1)$ if there exists a function $g$ from $S_1$ to

$S_2$ such that $P_1(g^{-1}(A)) = P_2(A)$ for any subset $A$ of $S_2$. $(S_1, P_1)$ is said to be an *expansion* of $(S_2, P_2)$. □

If $S_1$ is finite or countably infinite, $P_1(g^{-1}(A)) = P_2(A)$ is assured if it holds whenever $A$ is a one-point set. This follows from the fact that $g^{-1}(A) = \{g^{-1}(\omega)\,|\,\omega \in A\}$ is a union of mutually exclusive events.

If $X$ is a discrete random variable defined on $(S_1, P_1)$, let $S_2 = \{k\,|\,P(X = k) > 0\}$. Let $P_2(k) = P_1(X = k)$. Then $X$ plays the role of $g$ in the definition so that $(S_2, P_2)$ is a reduction of $(S_1, P_1)$. This is the most common way to reduce a probability model. More generally, if $X_1, \ldots, X_n$ are random variables defined on $(S_1, P_1)$, $\mathbf{X} = (X_1, \ldots, X_n)$, then we can take $S_2 = \{\mathbf{x} \in R_n\,|\,P(\mathbf{X} = \mathbf{x}) > 0\}$ and assign $P_2(\mathbf{x}) = P_1(\mathbf{X} = \mathbf{x})$.

***Example 1.2.9*** A husband and wife and two other couples are seated at random around a round table with six seats. What is the probability that the husband and wife in a particular couple, say $C_1$, are seated in adjacent seats?

Let the people be $a, b, \ldots, g$, let the seats be numbered $1, \ldots, 6$, reading clockwise around the table, and let $(x_1, \ldots, x_6)$, where each $x_i$ is one of the these letters, all different, correspond to the outcome in which person $x_i$ is seated in seat $i$, $i = 1, 2, \ldots, 6$. Let $S_1$ be the collection of such arrangements. Let $P_1$ assign probability $1/6!$ to each outcome. Let $A$ be the collection of outcomes for which $f$ and $g$ are adjacent. If $f$ and $g$ are the husband and wife in $C_1$, then $fg$ in this order may be in seats $12, 23, \ldots, 56, 61$. They may also be in the order of $21, 32, \ldots, 16$. For each of these the other four people may be seated in 4! ways. Thus, $N(A) = (2)(6)(4!) = 288$ and $P(A) = 2/5$.

We may instead let an outcome designate only the seats given to the husband and wife in $C_1$, and let $S_2$ be the set of pairs $(x, y)$, $x \neq y$. We have combined all 4! seating arrangements in $S_1$ which lead to the same seats for the husband and wife in $C_1$. Thus, $N(S_2) = (6)(5)$. Let $P_2$ assign equal probability $1/[(5)(6)]$ to each outcome, $4! = 24$ times as large as for the outcomes in $S_1$. Let $B =$ [husband and wife in $C_1$ are seated together] $= \{12, 23, \ldots, 61, 21, \ldots, 16\}$, a subset of $S_2$. Then $P(B) = (2)(6)/(6)(5) = 2/5$, as before. Of course, if we were asked the probability of the event $D$ that all three couples are seated together, each wife next to her husband. we could not answer the question using $(S_2, P_2)$, although we could using the model $(S_1, P_1)$. $P_1(D) = (2)(3!)(2^3)/6! = 96/720 = 2/15$. (Why?) $D$ is an event with respect to $S_1$ (a subset of $S_1$), but there is no corresponding subset of $S_2$. □

## Problems for Section 1.2

**1.2.1** Consider the sample space $S = \{a, b, c, d, e, f\}$. Let $A = \{a, b, c\}$, $B = \{b, c, d\}$, and $C = \{a, f\}$. For each outcome $x$ in $S$, let $P(\{x\}) = p(x)$, where $p(a) = 0.20$, $p(b) = 0.15$, $p(c) = 0.20$, $p(d) = 0.10$, $p(e) = 0.30$. Find $p(f)$, $P(A)$, $P(B)$, $P(A \cup B)$, $P(A \cup B^c)$, $P(A \cup B^c \cup C)$. Verify that $P(A \cup B) = P(A) + P(B) - P(A \cap B)$.

**1.2.2** Box 1 has four red and three white balls. Box 2 has three red and two white balls. A ball is drawn randomly from each box. Let $(x, y)$ denote an outcome for which the ball drawn from box 1 has color $x$ and the ball drawn from box 2 has color $y$. Let $C = \{\text{red, white}\}$ and let $S = C \times C$.

**(a)** Assign probabilities to the outcomes in $S$ in a reasonable way. [In 1000 trials of this experiment the outcome (red, white) occurred 241 times.]

**(b)** Let $X =$ (no. red balls drawn). Find $P(X = k)$ for $k = 0, 1, 2$. (In 1000 trials the events $[X = 0]$, $[X = 1]$, and $[X = 2]$ occurred 190, 473, and 337 times.)

**1.2.3** For the game of poker, find the probabilities of the events [straight flush], [4-of-a-kind], [flush], [3-of-a-kind], [one pair].

**1.2.4** Find the elements in the rows labeled $n = 6$ and $n = 7$ in Pascal's triangle. Verify that their sums are $2^6$ and $2^7$.

**1.2.5** A coin is tossed five times.

**(a)** Give a probability model so that $S$ is a Cartesian product.

**(b)** Let $X =$ (no. heads). Determine $P(X = 2)$.

**(c)** Use the die-toss data at the beginning of the chapter to simulate this experiment and verify that the relative frequency of cases for which the event $[X = 2]$ occurs is close to $P(X = 2)$ for your model.

**1.2.6** For a Venn diagram with three events $A, B, C$, indicate the following events by darkening the corresponding region:

**(a)** $A \cup B^c \cup C$.

**(b)** $A^c \cup (B^c \cap C)$.

**(c)** $(A \cup B^c) \cap (B^c \cup C)$.

**(d)** $(A \cup B^c \cap C)^c$.

**1.2.7** Two six-sided fair dice are thrown.

**(a)** Let $X =$ (total for the two dice). State a reasonable model and determine $P(X = k)$ for $k = 2, 3, \ldots, 12$. (Different reasonable people may have sample spaces with different numbers of outcomes, but their answers should be the same.)

**(b)** Let $Y =$ (maximum for the two dice). Find $P(Y = j)$ for $j = 1, 2, \ldots, 6$.

**1.2.8** **(a)** What is the (approximate) probability that at least two among five nonrelated people celebrate their birthdays in the same month? State a model first. In 100,000 simulations the event occurred 61,547 times.

**(b)** What is the probability that at least two of five cards chosen randomly without replacement from the deck of 48 cards formed by omitting the aces are of the same rank? Intuitively, should the probability be larger or smaller than the answer to part (a)? Why? In 100,000 simulations the event occurred 52,572 times.

**1.2.9** A small town has six houses on three blocks, $B_1 = \{a, b, c\}$, $B_2 = \{d, e\}$, $B_3 = \{f\}$. A random sample of two houses is to be chosen according to two different methods. Under method 1, all possible pairs of houses are written on slips of paper, the slips are thoroughly mixed, and one slip is chosen. Under method 2, two of the three blocks are chosen randomly without replacement, then one house is chosen randomly from each of the blocks chosen. For each of these two methods state a probability model, then use it to determine the probabilities of the events [house $a$ is chosen], [house $d$ is chosen], [house $f$ is chosen], and [at least one of houses $a$, $d$ is chosen].

**1.2.10** Four married couples attend a dance. For the first dance the partners for the women are randomly assigned among the men. What is the probability that at least one woman must dance with her husband?

**1.2.11** From among nine men and seven women a jury of six is chosen randomly. What is the probability that two or fewer of those chosen are men?

**1.2.12** A six-sided die is thrown three times.

**(a)** What is the probability that the numbers appearing are in increasing order? *Hint:* There is a one-to one correspondence between subsets of size 3 and increasing sequences from $\{1, 2, \ldots, 6\}$. In 10,000 simulations the event occurred 934 times.

**(b)** What is the probability that the three numbers are in nondecreasing order? (2, 4, 4) is not in increasing order, but *is* in nondecreasing order. Use the first 60 throws given at the beginning of the chapter to simulate the experiment 20 times. For the 10,000 simulations, the event occurred 2608 times.

**1.2.13** Let $(S, P)$ be a probability model and let $A, B, C$ be three events such that $P(A) = 0.55$, $P(B) = 0.60$, $P(C) = 0.45$, $P(A \cap B) = 0.25$, $P(A \cap C) = 0.20$, $P(B^c \cap C) = 0.15$, and $P(A \cap B \cap C) = 0.10$.

**(a)** Present a box diagram with $2^3 = 8$ cells giving the probabilities of all events of the form $A^* \cap B^* \cap C^*$, where $A^*$ is either $A$ or $A^c$, and $B^*$ and $C^*$ are defined similarly.

**(b)** Draw a Venn diagram indicating the same probabilities.

**(c)** Find $P(A^c \cap B \cap C^c)$ and $P(A \cup B^c \cup C)$. *Hint:* Use one of DeMorgan's laws for the case of three events.

**1.2.14** (*The Matching Problem*)

**(a)** Let $A_1, \ldots, A_n$ be $n$ events, subsets of the sample space $S$. Let $S_k$ be the sum of the probabilities of the intersections of all $\binom{n}{k}$ choices of these $n$ events, taken $k$ at a time. For example, for $n = 4$,

$S_3 = P(A_1A_2A_3) + P(A_1A_2A_4) + P(A_1A_3A_4) + P(A_2A_3A_4)$. Prove that $P(A_1 \cup A_2 \cup \cdots \cup A_n) = S_1 - S_2 + \cdots + (-1)^{n+1}S_n$.

**(b)** Let $X = (X_1, \ldots, X_n)$ be a random permutation of the integers $1, \ldots, n$. Let $A_i = [X_i = i]$. Thus, $A_i$ is the event of a *match* in the $i$th place. Express the probability of at least one match as a sum of $n$ terms, and then use this to find an approximation for large $n$. For 1000 simulations with $n = 10$, the frequencies $f(k)$ of $k$ matches were as follows: $f(0) = 351$, $f(1) = 372$, $f(2) = 195$, $f(3) = 60$, $f(4) = 14$, $f(5) = 8$, for an average of 1.038 matches per experiment. The probabilities for the numbers of matches for $n = 3$, are $f(0) = 1/3$, $f(1) = 3/6$, $f(3) = 1/6$. Later we will be able to show that the "expected number" of matches per experiment is 1.0.

**(c)** Apply the formulas obtained in part (b) to answer Problem 1.2.8.

**1.2.15** Give two models $(S_i, P_i)$ for $i = 1, 2$ for the two tosses of a six-sided die, so that $(S_2, P_2)$ is a reduction of $(S_1, P_1)$. Both should enable you to determine the probability that the sum of the numbers appearing exceeds 10, while $(S_1, P_1)$ allows the determination of the probability that the first toss results in 6, but $(S_2, P_2)$ does not.

## 1.3 CONDITIONAL PROBABILITY AND INDEPENDENCE

### Conditional Probability

Suppose that two six-sided fair dice are tossed and you learn that at least one of the two dice had resulted in 6. What is the probability now that the total of the numbers on the two dice is at least 10? Obviously, a revised probability should be larger than it was before you learned of the 6.

To answer this, consider the sample space $S = D \times D$, where $D = \{1, 2, \ldots, 6\}$, with the assignment of equal probabilities $1/36$ to each outcome (see Table 1.3.1). The event $A =$ [at least one 6] has 11 outcomes, and since you know that $A$ has occurred, can serve as a new sample space, again with equal probabilities. However, these probabilities must be $1/11$ rather than $1/36$, in order to sum to 1. Let us refer

**TABLE 1.3.1 Sample Space for Throw of Two Dice**

| | Second Die | | | | | |
|---|---|---|---|---|---|---|
| First Die | 1 | 2 | 3 | 4 | 5 | 6 |
| 1 | | | | | | $a$ |
| 2 | | | | | | $a$ |
| 3 | | | | | | $a$ |
| 4 | | | | | | $ab$ |
| 5 | | | | | $b$ | $ab$ |
| 6 | $a$ | $a$ | $a$ | $ab$ | $ab$ | $ab$ |

to the new model as the *conditional model*, given $A$, and write the new probability of an event $B$ as $P(B|A)$. For $B$ = [total of 10 or more] = $\{(4,6), (5,5), (5,6), (6,4), (6,5), (6,6)\}$, we get $P(B \mid A) = 5/11 = (5/36)/(11/36) = P(A \cap B)/P(A)$. This is larger than $P(B) = 6/36 = 1/6$.

Let $a$ and $b$ denote outcomes in $A$ and $B$, respectively. Consider Example 1.2.1 with sample space $S = \{a, b, c, d, e, f\}$, with probabilities 0.30, 0.20, 0.25, 0.10, 0.10, 0.05. As before, let $A = \{a, b, d\}$ and $B = \{b, d, e\}$. If $A$ is known to have occurred, then since $P(A) = 0.60$, we can form a new probability model with sample space $A$ and revised probabilities $0.30/0.60 = 1/2$, $0.20/0.60 = 1/3$, and $0.10/0.60 = 1/6$. Since $A \cap B = \{b, d\}$, we find $P(B \mid A) = 1/3 + 1/6 = 1/2 = P(A \cap B)/P(A)$. Since $P(B) = 0.40$, the occurrence of $A$ has again increased the probability of $B$.

We can avoid the need to define a new probability model by simply defining $P(B \mid A)$ for events $A$, $B$ as follows.

**Definition 1.3.1** For a given probability model $(S, P)$, let $A$ and $B$ be two events with $P(A) > 0$. The *conditional probability* of $B$, given $A$, is $P(B \mid A) = P(A \cap B)/P(A)$. □

Consider Example 1.2.4. Since $P(R_1) = 0.6$, $P(R_1 \cap R_2) = 0.4$, we find that $P(R_2 \mid R_1) = 2/3$, while $P(R_2) = 0.5$. Rain on the first day makes it more likely that it will rain the second day. Similarly, $P(R_1 \mid R_2) = P(R_1 \cap R_2)/P(R_2) = 4/5 > P(R_1)$.

The definition $P(B \mid A) = P(A \cap B)/P(A)$ is useful in the form $P(A \cap B) = P(B \mid A)P(A)$, since in many cases conditional probabilities can be determined more easily from the fundamentals of the experiment than can the probabilities of intersections. In this form, conditional probabilities can be used to build probability models.

***Example 1.3.1*** Suppose that a sexual disease is present in 0.6% of 18- to 24-year-old men in a large city. A blood test for the disease is good, but not perfect, in the following way. The probability that a man with the disease is positive on the test is 0.98 (the *sensitivity* of the test). The probability that a man who does not have the disease is positive for the test is 0.01. (The *specificity* of the test is therefore 0.99.) What are:

**(a)** The probability that a man of that age selected randomly will be positive for the test?

**(b)** Given that such a man is positive for the test, what is the probability that he actually has the disease? The answer to this question may surprise you.

Let $S = \{n, d\} \times \{+, -\} = \{(d, +), (d, -), (n, +), (n, -)\}$, where $d$ means that the man has the disease, $n$ means that he does not, $+$ indicates that the test is positive, and $-$ indicates that the test is negative. Let $D$ = [man has disease] $= \{(d, +), (d, -)\}$, and $V$ = [test is positive] $= \{(d, +), (n, +)\}$. We are given

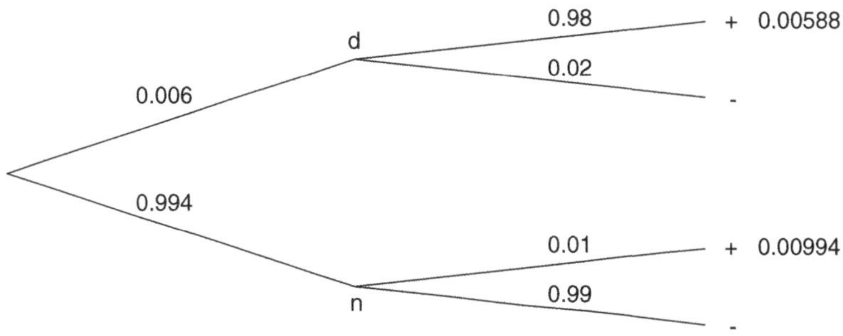


**FIGURE 1.3.1** Tree diagram for sexual disease.

$P(D) = 0.006$, $P(V \mid D) = 0.98$, $P(V \mid D^c) = 0.01$. We want to determine $P(V)$ and $P(D \mid V)$.

Figure 1.3.1 presents these and other relevant unconditional and conditional probabilities. For example, $P((d, +)) = P(D \cap V) = P(D)P(V \mid D) = (0.006)(0.98) = 0.00588$. Similarly, $P((n, +)) = P(D^c \cap V) = P(D^c)P(V \mid D^c) = (0.994)(0.01) = 0.00994$. Since $V = (D \cap V) \cup (D^c \cap V)$ and the two events within parentheses are mutually exclusive, $P(V) = P(D \cap V) + P(D^c \cap V) = 0.00588 + 0.00994 = 0.01582$. Then $P(D|V) = P(D \cap V)/P(V) = 0.00588/0.01582 = 0.3717$. Only 37% of all those testing positive actually have the disease! This certainly suggests the retesting of those whose first test is positive. □

***Example 1.3.2*** A box contains $r$ red and $w$ white balls. Let $N = r + w$. Two balls are drawn consecutively and randomly without replacement. What is the probability that the second ball drawn is red?

We use two different approaches to answer the question. With luck the answers will be the same. Since the question concerns the ball chosen second, we choose a sample space in which the outcomes indicate the order in which the balls are chosen. Let $R$ be the set of red balls and let $W$ be the set of white balls. Let $B = R \cup W$. Let $S = \{(b_1, b_2) \mid b_1 \in B, b_2 \in B, b_1 \neq b_2\}$. Assign equal probability $1/N(S) = 1/[N(N-1)]$ to each outcome. Let $R_2 =$ [red on the second ball chosen]. Let us determine $N(R_2)$. For each possible choice of a red ball for the second ball chosen, there are $(N-1)$ choices for the first ball. Therefore, $N(R_2) = r(N-1)$ and $P(R_2) = [r(N-1)]/N(N-1) = r/N$, the proportion of red balls in the box. This is, of course, also the probability of the event $R_1$ that the first ball chosen is red.

Now consider the problem using conditional probability. Then $P(R_2) = P(R_1 R_2) + P(R_1^c R_2) = P(R_1)P(R_2 \mid R_1) + P(R_1^c)P(R_2 \mid R_1^c) = (r/N)[(r-1)/(N-1)] + (w/N)[r/(N-1)] = [r/N(N-1)][(r-1) + w] = r/N = P(R_1)$.
□

The problem many students have when first confronted with this type of example is caused by their difficulty in distinguishing between conditional and unconditional probabilities.

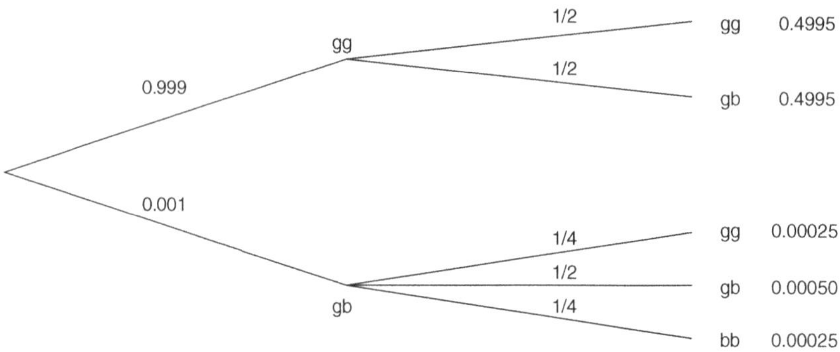


**FIGURE 1.3.2** Tree diagram for gene analysis.

Let $A_1, A_2, \ldots, A_n$ be events. Then, applying the definition of conditional probability and canceling factors in the numerator and denominator, we get

$$P(A_1)P(A_2 \mid A_1)P(A_3 \mid A_1A_2) \cdots P(A_n \mid A_1A_2 \cdots A_{n-1}) = P(A_1A_2 \cdots A_n),$$

assuming that these conditional probabilities are all defined. For example, in consecutive random-without-replacement draws of four balls from a box with six red and seven white balls, the probability of the order (red–white–red–white) is $(6/13)(7/12)(5/11)(6/10)$.

***Example 1.3.3*** Suppose that a gene has two forms, $g$ for good and $b$ for bad. Every person carries a pair of genes, so there are three possible genotypes, *gg, gb*, and *bb*. Persons with genotype *bb* suffer from the disease and always die young, so they never have children. Persons of genotype *gb* do not suffer from the disease but are carriers in the sense that their offspring (children) may acquire the bad gene $b$ from them. Suppose now that the proportions of people in the population of adults of genotypes *gg* and *gb* are 0.999 and 0.001. A female of known genotype *gb* has a child with a male drawn randomly from the adult male population (see Figure 1.3.2).

**(a)** What are the probabilities that a child of such a mating is of each of genotypes *gg, gb, bb*?
**(b)** Given that the child is of genotype *gb*, what is the probability that the male is of genotype *gb*?

If the male is of genotype *gg*, the child is equally likely to be *gg* or *gb*. If the male is of genotype *gb*, the child has probabilities $1/4$, $1/2$, $1/4$ of being *gg, gb*, or *bb*.

From Figure 1.3.2, the answers to part (a) are 0.49975, 0.50000, 0.00025. More generally, if $p$ is the proportion of genotype *gb* in the population of adult males, the probabilities are $(1 - p)/2 + p/4 = 1/2 - p/4$, $(1 - p)/2 + p/2 = 1/2$, and $p/4$. In general, in answer to part (b), $P(\text{male is } gb \mid \text{offspring is } gb) = (p/2)/(1/2) = p$, so the conditional probability that the male is *gb* is the same as the probability that the child is *gb*. □

The examples for which we have used a tree diagram suggest useful identities. Suppose that a sample space can be partitioned into $k$ disjoint subsets $A_1, \ldots, A_k$ whose probabilities, often called *prior probabilities,* are known and are positive. Suppose also that $B$ is another event and that $P(B \mid A_i)$ is known for each $i$. Then:

**Theorem 1.3.1** $P(B) = \sum_{i=1}^{k} P(A_i)P(B \mid A_i)$.

*Proof:* Since $B = BA_1 \cup \cdots \cup BA_k$, these events are disjoint, and $P(BA_i) = P(A_i)P(B \mid A_i)$, the identity follows from the additivity of $P$. □

The identity of Theorem 1.3.1 is sometimes called the *total probability formula.* From this formula and the definition of conditional probability, we get:

**Theorem 1.3.2 (Bayes' Formula)**

$$P(A_j \mid B) = \frac{P(B \mid A_j)P(A_j)}{\sum_{i=1}^{k} P(A_i)P(B \mid A_i)} \qquad \text{for } j = 1, \ldots, k.$$

The probabilities $P(A_j \mid B)$ are sometimes called *posterior probabilities,* since they are the revised probabilities [from the prior probabilities $P(A_j)$] of the events $A_j$ given the occurrence of the event $B$.

***Example 1.3.4*** Boxes 1, 2, and 3 each have four balls, each ball being red or white. Box $i$ has $i$ red balls, $i = 1, 2, 3$. A six-sided fair die is tossed. If a 1 occurs, a ball is drawn from box 1. If a 2 or 3 occurs, a ball is drawn from box 2. If a 4, 5, or 6 occurs, a ball is drawn from box 3. What are (a) the probability that the ball drawn is red, and (b) the conditional probability that the ball was drawn from box $j$ given that it was red?

Let $A_i =$ [ball is drawn from box $i$] for $i = 1, 2, 3$. Let $B =$ [ball drawn is red]. Then $P(A_j) = j/6$ and $P(B \mid A_j) = j/4$ for $j = 1, 2, 3$. Therefore, from Theorem 1.3.1, $P(B) = (1/6)(1/4) + (2/6)(2/4) + (3/6)(3/4) = 14/24$. From Bayes' theorem, $P(A_1 \mid B) = (1/24)/(14/24) = 1/14$, $P(A_2 \mid B) = (4/24)/(14/24) = 4/14$, and $P(A_3 \mid B) = (9/24)/(14/24) = 9/14$. The posterior probability $P(A_3 \mid B) = 9/14$ that the ball was drawn from box 3 given that it was red is larger than the prior probability $P(A_3) = 3/6$ that the ball would be drawn from box 3. □

***Example 1.3.5*** Your friend Zeke has been reasonably honest in the past, so that your prior evaluation of the probability that he is telling the truth when he claims to be tossing a fair coin rather than his two-headed coin is 0.9. The prior probability that he is tossing the two-headed coin is therefore 0.1. Zeke then tosses the coin $n$ times and gets a head on every toss. What is the posterior probability that he tossed the fair coin?

Let $F =$ [coin tossed is fair] and let $B =$ [all tosses result in heads]. Then $P(B) = P(F)P(B|F) + P(F^c)P(B \mid F^c) = (0.9)(1/2^n) + (0.1)(1) = 0.9/2^n + 0.1$. Therefore, $P(F \mid B) = (0.9/2^n)/[0.9/2^n + 0.1] = 1/[1 + 2^n/9]$. As $n$ becomes larger, the posterior probability that he is telling the truth goes rapidly to zero. Students can draw their own conclusions about friends who tell them often of low-probability events that have occurred in their lives. □

### Simpson's Paradox

In a paper appearing in the journal *Science*, Bickel et al. (1975) studied the rates of admission to graduate school by gender and department at the University of California at Berkeley. To make their point they invented the following data for the departments of "Machismatics" and "Social Warfare." For the combined departments their data were:

| | Deny | Admit | Percentage |
|---|---|---|---|
| Men | 300 | 250 | 45.5 |
| Women | 400 | 250 | 38.5 |

Assuming relatively equal ability among men and women, there seems to be discrimination against women. But the frequencies for the separate departments were:

| | Machismatics | | | Social Warfare | | |
|---|---|---|---|---|---|---|
| | Admit | Deny | Percentage | Admit | Deny | Percentage |
| Men | 200 | 200 | 50.0 | 50 | 100 | 33.3 |
| Women | 100 | 100 | 50.0 | 150 | 300 | 33.3 |

Assigning equal probabilities 1/1200 to each applicant and using obvious notation, with $D_1$ and $D_2$ denoting the events that the student applied to the Department of Machismatics and the Department of Social Warfare, we have $P(A \mid M) = 0.455$, $P(A \mid W) = 0.385$, $P(A \mid M \cap D_1) = 0.50$, $P(A \mid W \cap D_2) = 0.50$. When drawing conclusions about the relationships between variables, the tendency to "collapse" (combine) tables in this way leads to what is called *Simpson's paradox* (from a paper by E. H. Simpson, 1951, *not* named after the famous O. J. Simpson. In this case the department variable is called a *lurking variable*. Failure to take it into consideration leads to the wrong conclusion.

### Independence

Consider the experiment in which a fair six-sided die is thrown twice. Let $D = \{1, \ldots, 6\}$, $S = D \times D$, and assign probability 1/36 to each outcome in $S$. Let $A$ be the event that the number appearing on the first throw is 5 or 6, and let $B$ be the event that the number appearing on the second throw is at least 3. Then $P(A) = 12/36 = 1/3$, $P(B) = 24/36 = 2/3$, $P(AB) = 8/36 = 2/9$, and $P(B \mid A) = (2/9)/(1/3) = 2/3 = P(B)$. Thus, the occurrence of the event $A$ does not affect the probability of the event $B$. This, of course, should seem intuitively clear, unless one believes that dice have memories.

We take this as a starting point in developing a definition of *independence* of events. Suppose that for two events $A$ and $B$ with $P(A) > 0$, $P(B \mid A) = P(B)$. Then

$$P(A \cap B) = P(B \mid A)P(A) = P(B)P(A),$$

and if $P(B) > 0$ so that if $P(A \mid B)$ is defined, $P(A \mid B) = P(A)$. Since

$$P(AB) = P(A)P(B) \tag{1.3.1}$$

implies both $P(B \mid A) = P(B)$ and $P(A \mid B) = P(A)$, and since (1.3.1) is symmetric in $A$ and $B$ and does not require either $P(A) > 0$ or $P(B) > 0$, we take (1.3.1) as the definition of the independence of two events.

**Definition 1.3.2** Two events $A$ and $B$ are *independent* if $P(AB) = P(A)P(B)$. □

WARNING: Do not confuse the statement that two events $A$ and $B$ are independent with the statement that they are mutually exclusive (disjoint). In fact, if $A$ and $B$ are mutually exclusive, then $P(AB) = 0$, so that they cannot be independent unless at least one of them has probability zero.

It is easy to show that independence of $A$ and $B$ implies independence of the following pairs of events: $(A, B^c)$, $(A^c, B)$, $(A^c, B^c)$. For example, $P(A^c B) = P(B) - P(AB) = P(B) - P(A)P(B) = P(B)[1 - P(A)] = P(B)P(A^c)$. In fact, independence is best thought of as a property of the probability measure on the sample space as applied to the partitioning of the sample space into four parts produced by the two events $A$ and $B$. The fundamental idea of independence of events is used very often to build probability models.

Suppose that two experiments are to be performed, with corresponding probability models $(S_1, P_1)$ and $(S_2, P_2)$. Suppose also that it is reasonable to believe that the outcome of either experiment should not change the probability of any event in the other experiment. We can then produce a probability model for the combination of experiments as follows. Let $S = S_1 \times S_2$, and for $(s_1, s_2) \in S$, let $P((s_1, s_2)) = P_1(s_1)P_2(s_2)$. In this way we have defined a probability measure on $S$. To see this, note that for any events,

$$\begin{aligned} A_1 \subset S_1, A_2 \subset S_2, P(A_1 \times A_2) &= \sum_{s_1 \in A_1, s_2 \in A_2} P((s_1, s_2)) = \sum_{s_1 \in A_1} P_1(s_1) \sum_{s_2 \in A_2} P_2(s_2) \\ &= P_1(A_1)P_2(A_2). \end{aligned}$$

In particular, $P(S) = P(S_1 \times S_2) = P_1(S_1)P_2(S_2) = (1)(1) = 1$. Let $B_1 = A_1 \times S_2$ and $B_2 = S_1 \times A_2$, where $A_1 \subset S_1$ and $A_2 \subset S_2$. In the language of set theory, $B_1$ and $B_2$ are *cylinder sets*. $B_1$ is defined entirely in terms of the first experiment, $B_2$ entirely in terms of the second experiment. For the model $(S, P)$, $B_1$

**TABLE 1.3.2 Product Model**

| | $e$ | $f$ | $g$ | sum |
|---|---|---|---|---|
| $a$ | 0.08 | 0.12 | 0.20 | 0.40 |
| $b$ | 0.06 | 0.09 | 0.15 | 0.30 |
| $c$ | 0.04 | 0.06 | 0.10 | 0.20 |
| $d$ | 0.02 | 0.03 | 0.05 | 0.10 |
| sum | 0.20 | 0.30 | 0.50 | |

and $B_2$ are independent. Since $B_1 \cap B_2 = A_1 \times A_2$, $P(B_1 \cap B_2) = P(A_1 \times A_2) = P_1(A_1)P_2(A_2) = P(A_1 \times S_2)P(S_1 \times A_2) = P(B_1)P(B_2)$. (Pay attention to the subscripts or lack of subscripts on $P$!)

***Example 1.3.6*** Let $S_1 = \{a, b, c, d\}$ and let $P_1$ assign probabilities 0.4, 0.3, 0.2, and 0.1 to its outcomes. Let $S_2 = \{e, f, g\}$ and let $P_2$ assign probabilities 0.2, 0.3, and 0.5 to its outcomes. Then the outcomes of $S = S_1 \times S_2$ and the probability assignments under the independence model for the two experiments correspond to the 12 cells of Table 1.3.2.

Let $A_1 = \{b, c\}$, $A_2 = \{e, g\}$. Then the event $A_1 \times S_2$, the set of outcomes in the rows headed by $b$ and $c$, has probability 0.50. The event $S_1 \times A_2$, the set of outcomes in the columns headed by $e$ and $f$, has probability 0.70. The event $A_1 \times A_2$, the set of outcomes in the rectangle formed from the rows headed by $b$ and $c$ and the columns headed by $e$ and $f$, has probability 0.35, which is, of course, $P(A_1 \times S_2)P(S_1 \times A_2) = (0.50)(0.70)$. □

In generalizing the property of independence for two events to that of three or more events $A_1, \ldots, A_n$, we want to be able to use the multiplication rule,

$$P(A_1^* \cap \cdots \cap A_n^*) = P(A_1^*) \cdots P(A_n^*), \tag{1.3.2}$$

where each $A_i^*$ is either $A_i$ or $A_i^c$. To assure this, it is not enough to require that these events be independent in pairs. It is equivalent that for any integer $k$, $1 \le k \le n$, and indices $1 \le i_1 < \cdots < i_k \le n$,

$$P(A_{i_1} \cap \cdots \cap A_{i_k}) = P(A_{i_1}) \cdots P(A_{i_k}). \tag{1.3.3}$$

For example, for $n = 3$, we need $P(A_1 \cap A_2 \cap A_3) = P(A_1)P(A_2)P(A_3)$, $P(A_1 \cap A_2) = P(A_1)P(A_2)$, $P(A_1 \cap A_3) = P(A_1)P(A_3)$, and $P(A_2 \cap A_3) = P(A_2)P(A_3)$.

An inductive proof of the equivalence of (1.3.2) and (1.3.3) can be constructed, but we will avoid the messy details and instead show that (1.3.3) implies (1.3.2) for the special case of $A_1$, $A_2$, $A_3^c$. Since $(A_1 \cap A_2 \cap A_3^c) \cup (A_1 \cap A_2 \cap A_3) = A_1 \cap A_2$, $P(A_1 \cap A_2 \cap A_3^c) = P(A_1 \cap A_2) - P(A_1 \cap A_2 \cap A_3)$. From (1.3.3) this is $P(A_1)P(A_2)[1 - P(A_3)] = P(A_1)P(A_2)P(A_3^c)$.

**Definition 1.3.3** Events $A_1, \ldots, A_n$ are *independent* if for any integer $k$, $1 \le k \le n$, and indices $1 \le i_1 < \cdots < i_k \le n$, (1.3.3) holds. □

Given two models $(S_1, P_1)$ and $(S_2, P_2)$ for two experiments and independence of these experiments, we constructed a product model $(S, P)$, where $S = S_1 \times S_2$ and $P(A_1 \times A_2)$ for $A_1 \subset S_1, A_2 \subset S_2$. We can generalize this to the case of $n$ models $(S_1, P_1), \ldots, (S_n, P_n)$. We let $S = S_1 \times \cdots \times S_n$, and for any outcomes $s_1 \in S_1, \ldots, s_n \in S_n$, assign probability $P_1(s_1) \ldots P_n(s_n)$ to $(s_1, \cdots, s_n) \in S$. Again we find that events which are defined in terms of nonoverlapping indices are independent. For example, for $n = 4$, independence of the events $A_1, \ldots, A_4$ implies the independence of $A_1 \cup A_3^c$, $A_2$, and $A_4^c$.

***Example 1.3.7*** Three new car salespeople, Abe, Betty, and Carl, are to be assigned to the next three customers. The three have differing sales skills: Abe makes a sale with probability 0.3, Betty with probability 0.2, and Carl with probability 0.1. If the three salespeople do or do not make sales independently, what is the probability that the three make a total of at least one sale?

Let $A$ be the event that Abe makes a sale, and define $B$ and $C$ similarly for Betty and Carl. Then, since $P(\text{at least one sale}) = P(A \cup B \cup C) = 1 - P((A \cup B \cup C)^c) = 1 - P(A^c B^c C^c) = 1 - P(A^c)P(B^c)P(C^c) = 1 - (0.7)(0.8)(0.9) = 1 - 0.504 = 0.496$. □

***Example 1.3.8*** During the seventeenth century the French nobleman Antoine Gombauld, the Chevalier de Mere, a gambler, wrote to the mathematician Blaise Pascal concerning his experience throwing dice. He had been able to win regularly by betting that at least one 6 would appear in four rolls of a die. On the other hand, he was losing money when he bet that at least one double-6 would occur in 24 throws of two dice. It seemed to de Mere that he should have about the same chance of winning on each of these two bets.

It seemed "obvious" that the probability of at least one 6 should be 2/6 for two throws of a die, 3/6 for three throws, and so on. This reasoning seems to go bad for seven throws, however, so perhaps we need to think a bit more carefully. Using independence and DeMorgan's law, similarly to Example 1.3.5, for $n$ throws of one die we get $P(\text{at least one } 6) = 1 - P(\text{no } 6\text{'s}) = 1 - (5/6)^4 = 1 - 0.4823 = 0.5177 > 0.5$, so de Mere should have been a winner, although he had to be patient. On the other hand, for $n = 24$ throws of two dice, $P(\text{at least one } 6\text{–}6) = 1 - (35/36)^{24} = 1 - 0.5086 = 0.4914 < 0.5$, so that de Mere should have expected to lose, although slowly. To determine the difference in success rates between the two games experimentally, de Mere must have played very often and must have kept very good records. We have reason to be skeptical about de Mere's story. □

***Example 1.3.9*** Consider a system consisting of three components, 1, 2, and 3. Current (in the case that this is a wiring diagram with resistors 1, 2, and 3) or traffic (in the case that this is a system of highways with bridges 1, 2, and 3) must travel

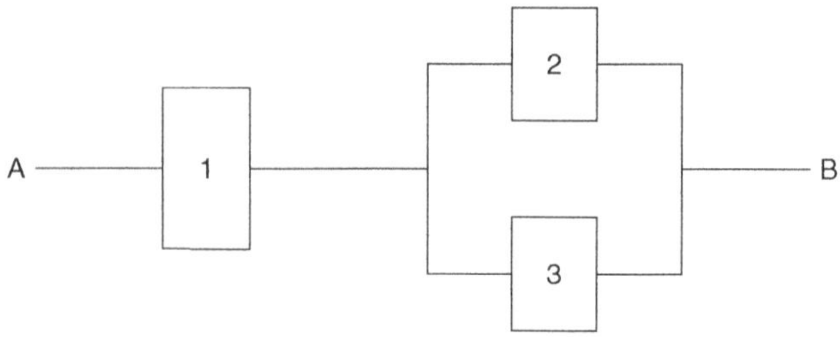


**FIGURE 1.3.3** Reliability diagram.

from point $A$ to point $B$. The system works if component 1 works and either of 2 or 3 works. Suppose that the three components have reliabilities (probabilities of working) 0.9, 0.8, and 0.7. Suppose also that the events that the three components function as they should are independent. What is the reliability of the system? That is, what is the probability that the system works? (See Figure 1.3.3.)

Let $W_i$ for $i = 1, 2, 3$ be the probability that component $i$ works. Then the reliability of the system $= P(W_1 \cap (W_2 \cup W_3)) = P(W_1)[1 - P(W_2^c)P(W_3^c)] = (0.9)[1 - (0.2)(0.3)] = 0.846$. □

***Example 1.3.10*** Our football team will play games against teams 1, 2, and 3. It has probabilities 0.7, 0.6, and 0.4 of winning each game, and it is reasonable to believe that the events of winning each game are independent. Given that the team wins at least one of the first two games, what is the conditional probability that it wins at least one of the last two?

Let $W_i =$ [our team wins the game with team $i$] for $i = 1, 2, 3$. We need to find $P(W_2 \cup W_3 \mid W_1 \cup W_2)$. A Venn diagram is helpful. Note that $(W_1 \cup W_2)(W_2 \cup W_3) = W_2 \cup (W_1 W_2^c W_3)$, so that using the independence of $W_1, W_2, W_3$, we get $P((W_1 \cup W_2)(W_2 \cup W_3)) = 0.6 + (0.7)(0.4)(0.4) = 0.712$. Since $P(W_1 \cup W_2) = 0.7 + 0.6 - (0.7)(0.6) = 0.88$, we get $P(W_2 \cup W_3 \mid W_1 \cup W_2) = 0.712/0.880 = 89/110$. What is the conditional probability that the team wins at least two games given that it wins at least one of the first two? □

## Problems for Section 1.3

**1.3.1** Let $S = \{a, b, c, d, e\}$ and let $P$ assign probabilities 0.2, 0.3, 0.1, 0.3, and 0.1, respectively. Let $A = \{a, b, c\}$ and $B = \{b, c, d\}$. Find $P(A)$, $P(B)$, $P(A \mid B)$, $P(B \mid A)$, and $P(A^c \mid B^c)$.

**1.3.2** A fair coin is tossed three times. Let $A =$ [at least one of the first two tosses is a head], $B =$ [same result on tosses 1 and 3], $C =$ [no heads], $D =$ [same result on tosses 1 and 2]. Among these four events there are six pairs. Which of these pairs are independent? Which are mutually exclusive?

**1.3.3** A bowl contains five balls numbered 1, 2, 3, 4, 5. One ball is drawn randomly, that ball is replaced, balls with larger numbers are withdrawn, then a second ball is drawn randomly.

(a) Given that the second ball drawn has the number $k$, what is the probability that the first ball drawn had the number $j$, for $j = 1, \ldots, 5$, and $k = 1, 2, \ldots, 5$?

(b) Repeat part (a) for the case of $N$ balls numbered $1, \ldots, N$.

**1.3.4** For the experiment consisting of three coin tosses with eight equally likely outcomes, define three events $A_1, A_2, A_3$ which are pairwise independent but are not independent as a collection of three events.

**1.3.5** How many tosses of two dice must be made to have probability at least 0.90 that at least one double-six occurs?

**1.3.6** For Figure 1.3.3 we can say that the single component 1 is in series with the subsystem consisting of (2, 3) in parallel. Consider a similar diagram except that there are four components, with the subsystem of components (1, 2) in parallel, the subsystem of components (3, 4) in parallel, and the subsystems (1, 2) and (3, 4) in series.

(a) Suppose that component $k$ has reliability $p_k = 0.5 + 0.1k$ for $k = 1, 2, 3, 4$. Find the reliability of the system.

(b) If the four components with these four reliabilities could be placed arbitrarily in the four positions, how should they be placed to maximize the reliability of the system?

(c) Answer part (b) for any four reliabilities $0 \leq p_1 < p_2 < p_3 < p_4 \leq 1$.

**1.3.7** You have found that your friend is quite honest, telling the truth with probability 0.99. One day she tells you that she threw a coin $n$ times and that she observed heads every time. When she lies she always says that she tossed the coin $n$ times and got heads each time. What is the conditional probability that she was telling the truth? How large must $n$ be before you begin to believe that she was more likely to be lying?

**1.3.8** A terrible new virus, XXX, has been discovered. It has infected already 0.03% of the population of the island of Oceanana. Fortunately, a test has been developed that is positive with probability 0.96 for someone who has XXX, and is positive for someone who does not have XXX with probability 0.07. We say that the test has sensitivity 0.96 and specificity 0.93. A person is chosen randomly from the population.

(a) What is the conditional probability that the person has XXX given that the test is positive?

(b) Suppose that whenever the test is positive, the test is given again, with the outcomes of the two tests being independent, given that the person has XXX and also when the person does not. Given that the test is positive both times it is given, what is the conditional probability that the person has XXX?

**(c)** If a person is diagnosed to have the virus only if every one of $k$ tests are positive, what must $k$ be before the conditional probability that the person has XXX given that all $k$ tests are positive is at least $1/2$?

**(d)** For $k$ tests as determined in part (c), what is the probability that a person with XXX will be diagnosed correctly?

**1.3.9** Suppose that whenever we toss a die, we say that a success has occurred if it results in 5 or 6. Note that among the results of the tosses of five dice 80 times as presented at the beginning of the chapter, the number $X$ of successes were as indicated below.

| | | | | | | |
|---|---|---|---|---|---|---|
| 3113 | 1201 | 2122 | 4222 | 2223 | 2123 | 1032 |
| 3212 | 3101 | 2324 | 2222 | 0121 | 0113 | 3123 |
| 3011 | 3223 | 3200 | 2221 | 3404 | 2122 | |

**(a)** Find the probability of exactly two successes when five dice are thrown. Are the results of these 80 trials consistent with that? Exactly two successes actually occurred 33 times. Plot both $P(X = k)$ and the relative frequency $r(k) = (\text{no. of trials in which } X = k)/80$ for $k = 0, 1, \ldots, 5$.

**(b)** Among the groups of four as presented above, let $Y = (\text{no. of 0's})$. Thus, the $Y$'s were: 010 000 101 001 101 020 10. Compare $P(Y = k)$ with $s(k) = (\text{no. } Y\text{'s} = k)/20$ for $k = 0, 1, 2, 3, 4$.

**1.3.10** **(a)** Chao and Francois each throw two dice. What is the probability that they get the same total?

**(b)** If Bernardo also throws two dice, what is the probability that all three get the same total?

**1.3.11** Patrick and Hildegarde, who do not know one another, were both born in July 1989. That year July 1 fell on Saturday. State a reasonable model and then determine the probability that they were born on the same day of the week: both on Sunday, both on Monday, and so on.

**1.3.12** (*The Secretary or Marriage Problem*) Originally, this problem was described as a problem concerning the choice of the best of $n$ candidates for a secretarial position. However, it seems to be more interesting, particularly to the young people likely to be reading this book, when described as the marriage problem. Suppose that you, at 18 or so, will meet $n > 0$ possible marriage partners in your lifetime, and being, as you know, very attractive, so that whomever you choose will be suitably thrilled by the opportunity, you wish to devise a suitable strategy that will maximize the probability of choosing the best of the $n$. The only rule is that the candidates appear one at a time in random order, and that for each candidate you must either choose him or her, or discard him or her, before considering the next candidate. That is, there is no possibility of going back to someone who was discarded earlier. You must therefore choose one of the strategies $S_k$, $k = 0, \ldots, n-1$. Under $S_k$ you will judge each of the first $k$ but will never choose one of those. If, for example, $n = 6$, $k = 3$, and the ranks (1 = best) appearing are 3, 5, 4, 6, 1, 2, then you would discard those with ranks 3, 5, 4 and successfully choose the best. However, 3, 5, 4, 2, 1, 6 and 3, 5, 4, 2, 6, 1 would lead to the choice of 2, a failure.

**(a)** Express $p(k;n) = P(S_k$ leads to success) as a function of $k$. For example, counting, for $n = 4$, shows that $p(0;4) = 1/4$, $p(1;4) = 11/24$, $p(2;4) = 10/24$, $p(3;4) = 1/4$. *Hint*: Let $A_k =$ [strategy $k$ is successful]. Let $B_j =$ [best is in $j$th position] and $C_j = [j$th is chosen]. Then $A_k = \cup_{j=k+1}^{n-1} B_j \cap C_j$ and $P(A_k) = \Sigma_{j=k+1}^{n-1} P(B_j) P(C_j \mid B_j)$.

**(b)** Let $\eta_n = (1 + 1/2 + \cdots + 1/n) - \log(n)$. Euler showed in the eighteenth century that $\lim_{n\to\infty} \eta_n = \eta$, *Euler's constant*, approximately equal to 0.5772. Use this to find an approximation of $P(A_k)$, then use this approximation to find $k$ to maximize $p(k;n) = P(A_k)$. For this choice of $k$, say $k_{\max}$, give an approximation of $p(k_{\max};n)$.

**1.3.13** The dice game "craps" is played as follows. Two six-sided dice are thrown by a player. If the player get a total of 7 or 11, he wins. If he (we use "he" rather than "she" because most people who are foolish enough to play the game are male) gets 2, 3, or 12, he loses. Otherwise (4, 5, 6, 8, 9, or 10), the total is called the *point*, and the player continues to throw until he either throws the point or a 7. He wins if the point occurs before the 7. He loses if the 7 occurs before the point.

**(a)** Suppose that an experiment is repeated independently until one of two mutually exclusive events $A_1$ and $A_2$ occurs. Let $p_1 = P(A_1)$ and $p_2 = P(A_2)$ for one experiment. Prove that $P(A_1$ occurs before $A_2) = p_1/(p_1 + p_2)$. *Hint*: Let the sample space $S$ consist of all finite sequences of the form $(B, B, \ldots, B, A_i)$ for $i = 1$ or 2.

**(b)** Use a conditional probability argument to find the probability that the player wins a game.

## 1.4 RANDOM VARIABLES

Suppose that two dice are thrown. Let $M$ denote the maximum of the numbers appearing on the two dice. For example, for the outcome (4, 5), $M$ is 5, and for the outcome (3, 3) $M$ is 3. For the model with $S = \{1, 2, \ldots, 6\}^{(2)}$ with probabilities 1/36 for each outcome, $[M = k] = \{(x, y) \mid \max(x, y) = k\}$. Let $f(k) = P([M = k])$. Counting then shows that $f(k) = (2k - 1)/36$ for $k = 1, 2, \ldots, 6$. Or $f(k)$ can be determined as follows: $f(k) = P(M \le k) - P(M \le k - 1) = (k/6)^2 - [(k - 1)/6]^2 = (2k - 1)/36$.

$M$ is an example of a discrete *random variable*. Formally, a random variable will be defined as a function. However, to a probabalist or statistician it can be thought of in two fundamental ways: as a list of possible values, together with probabilities of those values, and as a *potential number*, which will become a number only *after* an experiment having random outcomes is completed.

**Definition 1.4.1** Let $S$ be a finite or countably infinite sample space. A discrete *random variable* is a real-valued function on $S$. □

Although formally a random variable $X$ is a function that assigns a real number $X(\omega)$ to each point $\omega \in S$, we usually think of a random variable somewhat differently. For any subset $A$ of the real line, we write $[X \in A]$ to denote $\{\omega \in S | X(\omega) \in A\}$, and $P(X \in A)$ rather than $P([X \in A])$. Thus, for $M$ as defined above and $A = \{2, 3\}$, we write $[M \in A] = [M = 2] \cup [M = 3] = \{(x, y) \in S \mid \max(x, y) \in A\}$ and $P(M \in A) = P(M = 2 \text{ or } 3) = P(M = 2) + P(M = 3)$. Thus, a random variable $X$ is a potential number. After the experiment has been performed and an outcome $\omega$ determined, the random variable $X$ takes the value $X(\omega)$. We will be interested in determining $P(X \in A)$ for various sample spaces $S$, random variables $X$, subsets $A$ of the real line, and probability measures $P$. Since discrete random variables can take at most a countably infinite number of values, we will be particularly interested in the *probability mass function* $f(k) = P(X = k)$, defined for every real number $k$. For the random variable $M$, $f(k) = (2k - 1)/36$ for $k = 1, \ldots, 6$ and $f(k) = 0$ for other $k$.

**Definition 1.4.2** Let $X$ be a random variable defined on a discrete sample space $S$ and let $P$ be a probability measure on $S$. The *probability mass function* for $X$ is the function $f$ defined on the real line by $f(k) = P(X = k)$ for each real number $k$. □

Note that $P$ is a set function, defined for subsets of $S$, while $f$ is a point function. In some discussions when more than one random variable is defined on $S$ it will be necessary to indicate which mass function we are referring to, with subscript notation. For example, if $X_1$ and $X_2$ are both random variables defined on $S$, we might write $f_1$ and $f_2$ to denote the probability mass functions for $X_1$ and $X_2$. Or if the random variables are $X$ and $Y$, we might indicate their mass functions by $f_X$ and $f_Y$ or by $f$ and $g$. We sometimes shorten *probability mass function* to *probability function* or *mass function*.

***Example 1.4.1*** Two balls are drawn randomly with replacement from a box containing four balls, numbered 1, 2, 2, 3. Let $M$ denote the maximum of the numbers on the two balls chosen, and let $T$ denote their total. $M$ takes the values 1, 2, 3. That is, $M$ has the *range* $R_M = \{1, 2, 3\}$. $T$ has the range $\{2, 3, 4, 5, 6\}$. The probability mass function for $M$ is $f$, where $f(1) = 1/16$, $f(2) = 8/16$, and $f(3) = 7/16$, and

$f(j) = 0$ for $j \notin R_M$. It is left to the student to find the probability mass function $g$ for $T$. *Hint*: $g(4) = 3/8$. □

Discrete random variables may take an infinite number of possible values. Consider the experiment in which a die is tossed consecutively until a 6 has occurred. Then we can take $S$ to be the collection of $k$-tuples of $(n, n, \dots, n, 6)$, where $n$ denotes non-6 and 6 denotes the occurrence of a 6. For any point $\omega \in S$, let $X(\omega)$ be the length of $\omega$. Thus, $X$ is the number of tosses necessary to get a 6. From independence of the tosses, we can assign $P(\omega) = (5/6)^{k-1}(1/6)$, where $k = X(\omega)$ is the number of tosses. Let $f$ be the probability mass function for $X$. Thus, $f(k) = (5/6)^{k-1}(1/6)$ for $k = 1, 2, \dots$. Note that $\sum_{k=1}^{\infty} f(k) = (1/6)\frac{1}{1-(5/6)} = 1$.

Of course, every probability mass function must sum to 1 over its domain. In fact, if $f$ is any nonnegative function on the real line that is positive for a finite or countably infinite number of values, we can take $S$ to be the real line $R$, assign $P(A) = \sum_{k \in A} f(k)$ to each subset $A$ of $S$, and define $X(\omega) = \omega$ for each $\omega \in S$. Then $X$ has probability mass function $f$.

**Definition 1.4.3** A *probability mass function* is a nonnegative function on the realline $R$, positive for at most a countably infinite number of values, with positive probabilities summing to 1. □

For example, for $0 < p < 1$ the function $f(k) = (1-p)^{k-1}p$ for $k = 1, 2, \dots$, and 0 otherwise is a probability mass function. The random variable $X$ = (no. tosses of a die necessary to get a 6) has this probability function for $p = 1/6$. This probability function is useful enough to have a name, the *geometric probability function*. A random variable having this probability function is said to have the *geometric distribution*.

WARNING ON NOTATION: Random variables are almost always denoted by capital letters, usually from the end of the alphabet. Recall that for a subset $A$ of the real line $[X \in A] = \{\omega \in S \mid X(\omega) \in A\}$, so that $[X = x] = \{\omega \in S \mid X(\omega) = x\}$, and that for brevity we write $P(X \in A)$ rather than $P([X \in A])$ and $P(X = x)$ rather than $P([X = x])$. The arguments of probability mass functions are often taken to be $i, j, k$ for the case that the random variable takes integer values. More generally, the lowercase letter corresponding to the uppercase letter denoting the random variable is used as the argument of the mass function. Thus, if $X$ is the number of heads in two tosses of a coin, $X$ has the mass function $f(x) = 1/4$ for $x = 0$ and 2 and $f(x) = 1/2$ for $x = 1$, $f(x) = 0$ for other $x$. *Do not use a capital letter as the argument of a mass function.* The author has learned from long experience that some students write $X$ and $x$ so that they are indistinguishable. *Do not make this mistake.* If you now write $X$ and $x$ in the same way, practice until they are distinguishable. Similarly, considering the example with which we began this chapter, we say that the random variable $M$ has mass function $f(m) = (2m-1)/36$ for $m = 1, 2, \dots, 6$. We could equivalently write $f(x) = (2x-1)/36$ for $x = 1, 2, \dots, 6$. Do *not* write $f(M) = (2M-1)/36$.

**TABLE 1.4.1 Joint Probability Mass Function** $f(r, w)$

| | $w$ | | | | |
|---|---|---|---|---|---|
| $r$ | 0 | 1 | 2 | 3 | $P(R = r) = f_R(r)$ |
| 0 | 0 | 3/84 | 6/84 | 1/84 | 10/84 |
| 1 | 4/84 | 24/84 | 12/84 | 0 | 40/84 |
| 2 | 12/84 | 18/84 | 0 | 0 | 30/84 |
| 3 | 4/84 | 0 | 0 | 0 | 4/84 |
| $P(W = w) = f_W(w)$ | 20/84 | 45/84 | 18/84 | 1/84 | 1 |

***Example 1.4.2*** From a class of six men and five women, a random sample of 4 is chosen without replacement. Let $X$ be the number of men chosen. Let us find the probability mass function $f$ for $X$. Let $A = \{m_1, m_2, \ldots, m_6, w_1, \ldots, w_5\}$, with the $m$'s and $w$'s denoting the men and women, and let $S = \{B \subset A \,|\, N(B) = 4\}$ be the sample space. Assign probability $1/\binom{11}{4} = 1/330$ to each outcome in $S$. Then $X$ has mass function $f$, with, for example, $f(3) = P(X = 3) = \binom{6}{3}\binom{5}{1}/\binom{11}{4} = 100/330$. Similarly, $f(k) = \binom{6}{k}\binom{5}{4-k}/\binom{11}{4}$ for $k = 0, 1, \ldots, 4$, so that $f(0) = 5/330$, $f(1) = 60/330$, $f(2) = 150/330$, $f(3) = 100/330$, $f(4) = 15/330$. The random variable $X$ is said to have the *hypergeometric distribution.* □

Later we collect together more formal definitions of random variables which have special names. At this point the student is advised to become acquainted with them through these examples.

We will need methods that deal with more than one random variable defined on the same sample space. Let us begin with an example.

***Example 1.4.3*** A box contains 12 balls, of which four are red, three are white, and two are blue. Since these are the colors of the American flag, we will call this the "American box." A random sample of three balls is chosen without replacement. Let $R$ and $W$ be the numbers of red and white balls in the sample. Then, for example, $P([R = 2] \cap [W = 1]) = P(R = 2, W = 1) = \binom{4}{2}\binom{3}{1}\binom{2}{0}/\binom{9}{3} = 18/84$. More generally, define $f(r, w) = P(R = r, W = w) = \binom{4}{r}\binom{3}{w}\binom{2}{3-r-w}/\binom{9}{3}$, for nonnegative integers $r, w$ with $0 \le r + w \le 3$. We can study the function $f$ more easily in a two-way table such as Table 1.4.1. The marginal probability functions $f_R$ and $f_W$ are given on the margins of the table. For example, $P(R = 2) = f_R(2) = 30/84$ and $P(W = 1) = f_W(1) = 45/84$. Notice that in this case $f(2, 1)$ is *not* equal to $f_R(2) f_W(1)$. That is, $P(R = 2, W = 1) \ne P(R = 2)P(W = 1)$. □

In general, the joint probability mass function $f$ for two random variables $X$ and $Y$ has the properties $\sum_k f(j, k) = f_X(k)$, for each real $j$, where the sum is over all $k$

for which $f(j,k)$ is positive. This follows by writing $[X = j] = \bigcup_k [X = j, Y = k]$. Similarly, $\sum_j f(j,k) = f_Y(k)$ for each $k$, where the sum is taken over all $j$ for which $f(j,k)$ is positive. The mass functions $f_X$ and $f_Y$ are called the *marginal probability mass functions for f.*

More generally, the joint probability mass function $f$ for random variables $X_1, X_2, \ldots, X_n$ is defined by $f(k_1, \ldots, k_n) = P(X_1 = k_1, \ldots, X_n = k_n)$ for all points $(k_1, \ldots, k_n)$ in $R_n$.

***Example 1.4.4*** Suppose that $X_1, X_2, X_3$ have joint probability mass function $f$ defined by $f(k_1, k_2, k_3) = (k_1 + k_2 + k_3)/144$ for $k_1 = 1, 2, k_2 = 1, 2, 3,$ and $k_3 = 1, 2, 3, 4$. We can obtain the marginal joint mass function for $(X_1, X_3)$ by summing over $k_2$. We obtain $f_{13}(k_1, k_3) = (3k_1 + 6 + 3k_3)/144$ for $k_1 = 1, 2$ and $k_3 = 1, 2, 3, 4$. The marginal mass function for $X_3$ may then be obtained by summing $f_{13}$ over $k_1$. We obtain $f_3(k_3) = (9 + 12 + 6k_3)/144 = (7 + 2k_3)/48$ for $k_3 = 1, 2, 3, 4$.

Let $Y = X_1 + X_2 - X_3$. There are (2)(3)(4) = 24 combinations of values for $(X_1, X_2, X_3)$. By listing these, determining $Y$ and $f$ for each, and some careful arithmetic, we find that $Y$ takes the values $-2, -1, 0, 1, 2, 3, 4$ with probabilities $6/144, 19/144, 32/144, 36/144, 28/144, 17/144, 6/144$. □

We now consider one of the most important concepts in probability and statistics, that of *independent random variables.*

**Definition 1.4.4** Two random variables $X$ and $Y$ are *independent* if the events $[X \in A]$ and $[Y \in B]$ are independent for all subsets $A$ and $B$ of the real line. □

If $X$ and $Y$ are independent, it follows immediately by taking $A = \{j\}$ and $B = \{k\}$ that the joint probability mass function $f$ has $(X, Y)$ factors; that is;

$$f(j,k) = f_X(j) f_Y(k) \qquad \text{for all } (j,k) \in R_2. \tag{1.4.1}$$

That (1.4.1) implies that $X$ and $Y$ are independent follows from the fact that if (1.4.1) holds,

$$\begin{aligned} P(X \in A, Y \in B) &= \sum_{j \in A,\, k \in B} f(j,k) = \sum_{j \in A} \sum_{k \in B} f(j,k) = \sum_{j \in A} \sum_{k \in B} f_X(j) f_Y(k) \\ &= \sum_{j \in A} f_X(j) \sum_{j \in A} f_Y(k) = P(X \in A) P(Y \in B). \end{aligned}$$

If $(S_1, P_1)$ and $(S_2, P_2)$ are two probability models, $(S = S_1 \times S_2, P = P_1 \times P_2)$ is their product model, and $X_1$ is completely defined by the outcome in $S_1$ and $X_2$ by the outcome in $S_2$, then $X_1$ and $X_2$ are independent. More formally, if $X_i(s_1, s_2)$ depends only on $s_i$ for $i = 1, 2$, then $X_1, X_2$ are independent. To prove this, note that $\{(s_1, s_2) \mid X_1(s_1, s_2) \in A\} \cap \{(s_1, s_2) \mid X_2(s_1, s_2) \in B\} = D_1 \times D_2$, where $D_1 = \{s_1 \mid X_1(s_1, s_2) \in A \text{ for all } s_2\}$ and $D_2 = \{s_2 \mid X_2(s_1, s_2) \in B \text{ for all } s_1\}$. Since

$P(X_1 \in A) = P_1(D_1)$ and $P(X_2 \in B) = P_2(D_2)$, we conclude that $P(X_1 \in A, X_2 \in B) = P(D_1 \times D_2) = P_1(D_1)P_2(D_2) = P(X_1 \in A)P(X_2 \in B)$.

***Example 1.4.5*** Box 1 has four balls with numbers 1, 2, 2, 3. Box 2 has three balls with numbers 0, 1, 1. One ball is drawn from each box at random. Let $X_i$ be the number on the ball drawn from box $i$, $i = 1, 2$. Let $S_1 = \{1, 2, 2^*, 3\}$ and $S_2 = \{0, 1, 1^*\}$. Let $S = S_1 \times S_2$, and assign probability 1/12 to each outcome. Then $X_1$ and $X_2$ are independent with probability mass functions $f_1$ and $f_2$, where $f_1(1) = 1/4$, $f_1(2) = 1/2$, $f_1(3) = 1/4$, $f_2(0) = 1/3$, $f_2(1) = 2/3$. If $Y = X_1 + X_2$, then $Y$ has the probability function $g$, where $g(1) = 1/12$, $g(2) = f_1(1)f_2(1) + f_1(2)f_2(0) = 1/3$, $g(3) = 5/12$, $g(4) = 1/6$. Actually, it is easier to determine the probability function for $Y$ directly from the probability model $(S, P)$. □

Although we discuss these random variables later more formally, we will now informally introduce *binomial* random variables. Each of these arises so often in applications that they certainly warrant special study. We begin with examples.

***Example 1.4.6*** Suppose that a husband and wife each is of genotype $aA$, where the gene $a$ is recessive while $A$ is dominant. Children who are of genotype $aa$ have six toes, whereas those of genotypes $aA$ or $AA$ have five. Since each parent will give the genes $A$ and $a$ with equal probabilities to each of their offspring, the probability that an offspring is of genotype $aa$, and therefore will have six toes, is 1/4. Suppose now that these parents have five children, none being identical twins, triplets, and so on. Let $X$ be the number who are of genotype $aa$.

The random variable $X$ is said to have the *binomial distribution* because (1) $X$ is the number of "successes" ($aa$ children in this case) of a fixed number five of "trials", (2) the trials are independent, and (3) the probability of success on each trial is the same, in this case 1/4.

Let us try to determine the probability function for $X$. $X$ takes the values $0, 1, \ldots, 5$. First consider the event $[X = 2]$. Let $s$ denote success, and $f$, failure. Then *fssff*, for example, can denote the outcome (failure, success, success, failure, failure) for the five children in the order of birth. Thus, $[X = 2] = \{$*ssfff*, *sfsff*, *sffsf*, *sfffs*, *fssff*, *fsfsf*, *fsffs*, *ffssf*, *ffsfs*, *fffss*$\}$. Because of properties 2 and 3 above, each of these 10 outcomes has probability $(1/4)^2\ (3/4)^3$. Therefore, $P(X = 2) = (10)(1/4)^2(3/4)^3 = 270/1024$. Is there an easier way to determine the number of outcomes in $[X = 2]$ than simply counting? Yes, each such outcome corresponds to a subset of two positions for the successes taken from the five positions, and also to a subset of three positions for the three failures taken from the five positions. Therefore, the number of such outcomes is $\binom{5}{2} = \binom{5}{3} = 10$. A similar argument shows that $P(X = 1) = \binom{5}{1}(1/4)^1(3/4)^4 = 405/1024$, $P(X = 0) = (3/4)^5 = 243/1024$, $P(X = 3) = \binom{5}{3}(1/4)^3(3/4)^2 = 90/1024$, $P(X = 4) = \binom{5}{4}(1/4)^4(3/4) = 15/1024$, and $P(X = 5) = (1/4)^5 = 1/1024$. We are ready for some formal definitions. □

**Definition 1.4.5** Suppose that $n$ independent experiments are performed. For each experiment (called a *trial*) the probability of success is $p$, where $0 < p < 1$. Let $X$ be the number of successes. Then $X$ is said to have the *binomial distribution* with parameters $n$ and $p$. If $n = 1$, then $X$ is said to have the *Bernoulli distribution* with parameter $p$. □

Example 1.4.5 has essentially proved the following very useful theorem.

**Theorem 1.4.1** Let $X$ have the binomial distribution with parameters $n$ and $p$. Then $X$ has the probability mass function $f(k) = P(X = k) = \binom{n}{k} p^k (1-p)^{n-k}$ for $k = 0, 1, \ldots, n$.

The *indicator random variable* for an event $A$ is the random variable $I_A$ defined by $I_A(s) = 1$ for $s \in A$ and $I_A(s) = 0$ for $s \notin A$. The indicator random variable $I_A$ has the Bernoulli distribution with parameter $p = P(A)$. Note that $I_{A^c} = I_S - I_A = 1 - I_A$. $I_{A \cap B} = I_A I_B$ and $I_{A \cup B} = I_A + I_B - I_A I_B$. (To say that two random variables are equal means that the random variables are equal as functions on the sample space, and the same for all outcomes in $S$.)

We need to consider the independence of more than two random variables.

**Definition 1.4.6** Discrete random variables $X_1, \ldots, X_n$ are said to be *independent* if for every choice of subsets $A_1, \ldots, A_n$ of the real line the events $[X_1 \in A_1], \ldots, [X_n \in A_n]$ are independent. □

It follows by a slight generalization of the argument we made for the case $n = 2$ that independence of $X_1, \cdots, X_n$ is equivalent to $f = f_1 \cdots f_n$, where $f$ is the joint probability function $X_1, \ldots, X_n$ and $f_1, \ldots, f_n$ are the marginal probability functions of $X_1, \ldots, X_n$.. That is, independence of $X_1, \ldots, X_n$ is equivalent to $f(k_1, \ldots, k_n) = f_1(k_1) f_2(k_2) \cdots f_n(k_n)$ for every $(k_1, \ldots, k_n) \in R_n$. In the same way, if $(S_i, P_i)$ is a probability model for experiment $E_i$ for $i = 1, \ldots, n$, $(S, P)$ is their product model, and for each $i$, $X_i$ depends only on the outcome of $E_i$ (only on the outcome $s_i \in S_i$), then $X_1, \ldots, X_n$ are independent.

***Example 1.4.7*** Box $k$ has $k$ balls, numbered $1, \ldots, k$, for $k = 2, 3, 4$. Balls are chosen independently from these boxes, one per box. Let $X_k$ be the number on the ball chosen from box $k$. Then a product model seems reasonable, so that $X_2, X_3, X_4$ are independent and $X_k$ has the uniform distribution on $\{1, \ldots, k\}$. That is, $X_k$ has the probability function $f_k(i) = 1/k$ for $i = 1, \ldots, k$. The random vector $\mathbf{X} = (X_2, X_3, X_4)$ takes each of 24 possible values with probability 1/24. Let $Y = X_2 + X_3 + X_4$. Then, for example, $[Y = 5]$ consists of five outcomes (one is $\mathbf{X} = (1, 3, 1)$, for example), so that $Y$ has the probability function $g$, with $g(5) = 5/24$. Similarly, $g(3) = 1/24$, $g(4) = 3/24$, $g(6) = 6/24$, $g(7) = 5/24$, $g(8) = 3/24$, $g(9) = 1/24$. Similarly, let $M = \max(X_1, X_2, X_3)$ have probability function h. Then $h(1) = 1/24$, $h(2) = P(M \le 2) - P(M \le 1) = 1/3 - 1/24 = 7/24$, $h(3) = P(M \le 3) - P(M \le 2) = 3/4 - 1/3 = 10/24$, $h(4) = 1 - P(M \le 3) = 1 - 3/4 = 6/24$.

Similarly, if there are $B$ boxes each with $N$ balls numbered $1, \ldots, N$, one ball is drawn from each, and $M$ is the maximum number drawn, then $P(M \le m) = (m/N)^B$, for $m = 1, \ldots, N$. From these probabilities the probability function for $M$ can easily be determined. □

A binomial random variable is the sum of the indicators of success on the individual trials. That is, if $X$ is the number of successes in $n$ independent trials, each having probability $p$ of success, then $X = I_1 + \cdots + I_n$, where $I_i$ is 1 when the $i$th trial results in success, zero otherwise. Because of the independence of the trials, $I_1, \ldots, I_n$ are independent. As we will see in the next section, the ability to express a random variable as a sum of random variables, especially as a sum of independent random variables, is very useful.

### The Multinomial and Generalized Bernoulli Distributions

***Example 1.4.8*** Suppose that three coins are tossed and the number of heads is observed 10 times. Let $X_j$ be the number of times among these 10 tosses for which the number of heads is $j$, for $j = 0, 1, 2, 3$. It should be clear that $X_j$ has the binomial distribution with parameters $n = 10$ and $p_j$, with $p_0 = 1/8$, $p_1 = 3/8$, $p_2 = 3/8$, $p_3 = 1/8$. The experiment was repeated independently 12 times, with outcomes as follows:

| $X_0$ | $X_1$ | $X_2$ | $X_3$ | $X_0$ | $X_1$ | $X_2$ | $X_3$ | $X_0$ | $X_1$ | $X_2$ | $X_3$ |
|---|---|---|---|---|---|---|---|---|---|---|---|
| 1 | 5 | 3 | 1 | 0 | 7 | 3 | 0 | 2 | 3 | 2 | 3 |
| 2 | 5 | 2 | 1 | 0 | 3 | 4 | 3 | 2 | 3 | 3 | 2 |
| 1 | 4 | 4 | 1 | 0 | 4 | 5 | 1 | 1 | 4 | 4 | 1 |
| 0 | 2 | 5 | 3 | 2 | 2 | 5 | 1 | 0 | 7 | 3 | 0 |

□

**Definition 1.4.7** Suppose that $n$ independent experiments are performed and that for each there are possible outcomes $B_1, \ldots, B_k$ with corresponding probabilities $p_1, \ldots, p_k$, with $\Sigma_{j=1}^k p_j = 1$. Let $X_i$ be the number of occurrences of outcome $B_i$. Then $\mathbf{X} = (X_1, \ldots, X_k)$ is said to have the *multinomial distribution* with parameters $n$ and $\mathbf{p} = (p_1, \ldots, p_k)$. In the case that $n = 1$, $\mathbf{X}$ is also said to have the *generalized Bernoulli distribution* with parameter $\mathbf{p}$. □

COMMENTS: A generalized Bernoulli random vector takes the $i$th unit vector, having 1 as its $i$th component, zeros elsewhere (the indicator of the $i$th outcome) with probability $p_i$. The $i$th component $X_i$ of the multinomial random vector $\mathbf{X}$ has the binomial distribution with parameters $n$ and $p_i$. If $\mathbf{Y}_1, \ldots, \mathbf{Y}_n$ are independent, each with the generalized Bernoulli distribution with parameter $\mathbf{p} = (p_1, \ldots, p_k)$, then $\mathbf{Y} = \sum_{i=1}^n \mathbf{Y}_i$ has the multinomial distribution with parameters $n$ and $\mathbf{p}$.

If $\mathbf{X}$ is as in Example 1.4.7, then, for example, $P(X_0 = 1, X_1 = 3, X_2 = 4, X_3 = 2) = \binom{10}{1\ 3\ 4\ 2}(1/8)^1(3/8)^3(3/8)^4(1/8)^2 = \frac{10!}{1!\,3!\,4!\,2!}(1/8)^1(3/8)^3(3/8)^4(1/8)^2$. More generally, if $\mathbf{X} = (X_1, \ldots, X_k)$ has the multinomial distribution with parameters $n$ and $\mathbf{p} = (p_1, \ldots, p_k)$, then $\mathbf{X}$ has probability function

$$f(x_1, \ldots, x_k) = \binom{n}{x_1 \ \cdots \ x_k} p_1^{x_1} \cdots p^{x_k} \qquad (1.4.2)$$

for nonnegative integers $x_1, \ldots, x_k$ with $\Sigma_{i=1}^k x_i = n$.

### Reduction of a Model $(S_1, P_1)$

If $X$ is a discrete random variable defined on $(S_1, P_1)$, let $S_2 = \{k \mid P(X = k) > 0\}$. Let $P_2(k) = P_1(X = k)$. Then $X$ plays the role of $g$ in the definition so that $(S_2, P_2)$ is a *reduction* of $(S_1, P_1)$. This is the most common way to reduce a probability model. More generally, if $X_1, \ldots, X_n$ are random variables defined on $(S_1, P_1)$, $\mathbf{X} = (X_1, \ldots, X_n)$, we can take $S_2 = \{\mathbf{x} \in R_n \mid P(\mathbf{X} = \mathbf{x}) > 0\}$ and assign $P_2(\mathbf{x}) = P_1(\mathbf{X} = \mathbf{x})$.

**Theorem 1.4.2** Let $Y_i = g_i(X_i)$ for $i = 1, \ldots, n$, be random variables, where $X_1, \ldots, X_n$ are independent. Then $Y_1, \ldots, Y_n$ are independent.

*Proof:* That the events $[Y_i \in B_i] = [X_i \in g_i^{-1}(B_i)]$ are independent follows from the definition. □

### Problems for Section 1.4

**1.4.1** Each of two dice has 1 on three sides, 2 on two sides, and 3 on one side.

**(a)** Let $X$ be the total when the two dice are thrown. Find the probability function for $X$.

**(b)** Let $M$ be the maximum of the two numbers appearing. Find the probability function for $M$.

**1.4.2** Let $I_1, I_2, I_3$ be the indicators of success on three independent projects having success probabilities 0.2, 0.3, 0.4. Find the probability function for the total number of successes $X = I_1 + I_2 + I_3$.

**1.4.3** Suppose that Larry and Ling have probabilities $p_1 > 0$ and $p_2 > 0$ of success on any trial. Each performs the trials independently until a success occurs. Let $X_1$ and $X_2$ be the numbers of trials they must use. $X_1$ and $X_2$ are said to have geometric distributions with parameters $p_1$ and $p_2$.

**(a)** Show that $P(X_1 = X_2) = P(\text{both succeed on trial } 1 \mid \text{at least one of them succeeds on trial } 1)$. *Hint*: Express each in terms of $p_1$, $q_1 = 1 - p_1$, $p_2$, and $q_2 = 1 - p_2$.

**(b)** Let $W = \min(X_1, X_2)$. Show that the event $A \equiv [X_1 = X_2]$ and the random variable $W$ are independent. That is, $P(A \cap [W = w]) = P(A)P(W = w)$ for $w = 1, 2, \ldots$.

**(c)** Present a simpler expression for $P(A)$ for the case $p_1 = p_2 = p$, and evaluate it for the case that the trials are coin tosses with success = heads and for the case that the trials are tosses of a six-sided die, with success = [6 occurs].

**(d)** For the case $p_1 = p_2 = p$, find the probability function for $Y = X_1 + X_2$. ($Y$ is said to have the negative binomial distribution.)

**1.4.4** A "deck" of nine cards has four spades, three hearts, and two diamonds. Five cards are chosen randomly without replacement. Let $X$ = (no. spades chosen) and $Y$ = (no. hearts chosen).

**(a)** Present a matrix whose entries are values of the joint probability function for $X$ and $Y$.

**(b)** Show that $X$ and $Y$ are dependent random variables.

**(c)** Use this matrix to determine the marginal probability functions for $X$ and for $Y$. Verify your answers using the fact that $X$ and $Y$ have hypergeometric distributions.

**1.4.5** George throws two coins 10 times. What is the probability that he gets two heads at least three times?

**1.4.6** A coin is tossed twice. Let $I_j$ be the indicator of [heads on toss $j$] for $j = 1, 2$ and let $I_3 = [I_1 = I_2]$ = [same outcome on both tosses]. Show that $I_1, I_2, I_3$ are pairwise independent, but that as a triple of random variables, they are not independent.

**1.4.7** Let $M$ be the maximum of the numbers appearing when four fair six-sided dice are thrown. Find the mass function for $M$.

**1.4.8** Mary and Bob are automobile salespeople. Mary tries to make sales to three customers, each with success probability 0.4. Bob tries to makes sales to two customers, each with success probability 0.3. The five trials for Mary and Bob are independent.

**(a)** Find the probability mass function for $T$ = total number of sales that Bob and Mary make.

**(b)** Find the probability that Bob and Mary make the same number of sales.

**1.4.9** Let $A_1, A_2, A_3$ be events, let $I_1, I_2$, and $I_3$ be their indicators, note that $I_S$ is identically 1, and let $B = A_1 \cup A_2 \cup A_3$. Use the fact that $B = (A_1^c \cap A_2^c \cap A_3^c)^c$ to show that the indicator of $B$ is $I_B = 1 - (1 - I_1)(1 - I_2)(1 - I_3) = I_1 + I_2 + I_3 - I_1 I_2 - I_1 I_3 - I_2 I_3 + I_1 I_2 I_3$.

**1.4.10** Two six-sided fair dice are thrown $n$ times.

(**a**) For $n = 24$, find the probability of at least one double-6 (6–6). (This is the answer to one of the problems Chevalier de Mere posed, as described at the beginning of the chapter.)

(**b**) How large must $n$ be in order that $P$(at least one 6–6) is at least 0.99?

(**c**) Show that for large $n$, $P$(at least one 6–6 in $n$ tosses) $\doteq 1 - e^{-d_n}$, where $d_n = n/36$. Check your answer to part (a) using this approximation. *Hint:* $\lim_{k\to\infty}(1 - a/k)^k = e^{-a}$.

**1.4.11** A box has five balls, numbered $1, 2, \ldots, 5$. A ball is chosen randomly. Let $X_1$ be the number on the ball. Then another ball is drawn randomly from among the balls numbered $1, \ldots, X_1$.

(**a**) Find the joint probability mass function for the pair $(X_1, X_2)$.

(**b**) Find the probability mass function for $X_2$.

(**c**) Find $P(X_1 = 4 \mid X_2 = 2)$.

(**d**) Find the probability mass function for $Y = X_1 + X_2$.

**1.4.12** Suppose that three fair coins are tossed 10 times. Let $X_i$ be the number of times for which $i$ heads occur.

(**a**) Find $P(X_0 = 2, X_1 = 3)$.

(**b**) What is the probability function for $\mathbf{X} = (X_0, X_1, X_2, X_3)$?

(**c**) What is the distribution of $X_2 + X_3$?

**1.4.13** Use the identity $(a_1 + \ldots + a_k)^m = \sum_{x_1+\cdots+x_k=m} \binom{m}{x_1 \;\cdots\; x_k} a_1^{x_1} \cdots a_k^{x_k}$ and equation (1.4.2) to show that in Definition 1.4.7, $X_1 \sim$ Binomial$(n, p_1)$.

**1.4.14** Let $\mathbf{Y}_1$ and $\mathbf{Y}_2$ be independent, each with the multinomial distribution, with parameters $n_1$ and $n_2$, each with the same parameter vector $\mathbf{p} = (p_1, \ldots, p_k)$. What is the distribution of $T = Y_1 + Y_2$? Prove it.

## 1.5 EXPECTATION

***Example 1.5.1*** A gambling jar has 10 marbles, of which five are red, three are white, and one is blue. The jar is used to make bets as follows. You are to pay $x$ dollars each time you play the game. You will then draw one marble at random. If it is red, you are given nothing. If it is white, you receive \$1. If it is blue, you receive \$3. If you wish to at least break even, how large can $x$ be? Put another way, what will be your long-run average return from many independent draws from the jar?

Let $S = \{\text{red, white, blue}\}$, let $P$ assign probabilities $5/10, 4/10$, and $1/10$ to these three outcomes, and let $X(\text{red}) = 0$, $X(\text{white}) = 1$, $X(\text{blue}) = 3$. What will the long-run average value of $X$ be?

Here are the results of 100 simulations:

| | | | | |
|---|---|---|---|---|
| 0 1 0 0 1 | 1 1 1 0 1 | 1 1 0 0 1 | 1 0 0 1 | 0 |
| 0 3 1 3 0 | 0 0 0 1 1 | 0 1 0 1 0 | 0 0 1 1 | 0 |
| 1 3 0 0 0 | 0 1 0 0 0 | 0 3 1 1 0 | 0 3 1 0 | 0 |
| 3 3 0 1 0 | 1 0 1 3 3 | 0 0 1 1 1 | 1 0 1 1 | 0 |
| 1 0 0 0 1 | 0 0 1 0 1 | 0 3 0 0 0 | 3 0 0 1 | 0 |

There were 53 0's, 37 1's, and 11 3's, for an arithmetic mean of 0.69. Let us try to predict the mean for 10,000 simulations. We should expect the total to be approximately $[10{,}000(5/10)(0) + 10{,}000(4/10)(1) + 10{,}000(1/10)(3)] = [(5/10)(0) + (4/10)(1) + (1/10)(3)](10{,}000)$, so that the mean should be approximately $[(5/10)(0) + (4/10)(1) + (1/10)(3)] = 0.7$. We will refer to 0.7 as the *expected value* of $X$ and write $E(X) = 0.7$. Actually, for these 10,000 simulations we obtained 5031 0's, 3973 1's, and 996 3's, for a mean of 0.6961. For 100,000 simulations the mean was 0.70298. □

Figure 1.5.1 shows the results of independent simulations on $X$. In the top two graphs the consecutive values of the sample mean of the first $n$ X's is given for $n = 101, \ldots, 10{,}000$. Notice that the mean of the first 10,000 is in each case quite close to $E(X) = 0.7$. The *strong law of large numbers* states that with probability 1 each of these sample paths will converge to $E(X)$. The lower two graphs are histograms of 10,000 values of sample means of $n = 100$ and of $n = 400$ $X$'s. Notice that the spread of these sample means is about one-half as much for 400 as for 100.

The *weak law of large numbers*, proved later in this chapter, states that the probability that the sample mean will lie within a fixed distance, say ε, of $E(X)$ converges to 1 as the sample size $n$ goes to infinity. Another interesting feature of the histograms is that their "shape" is roughly the same for $n = 100$ as for $n = 400$, the shape of

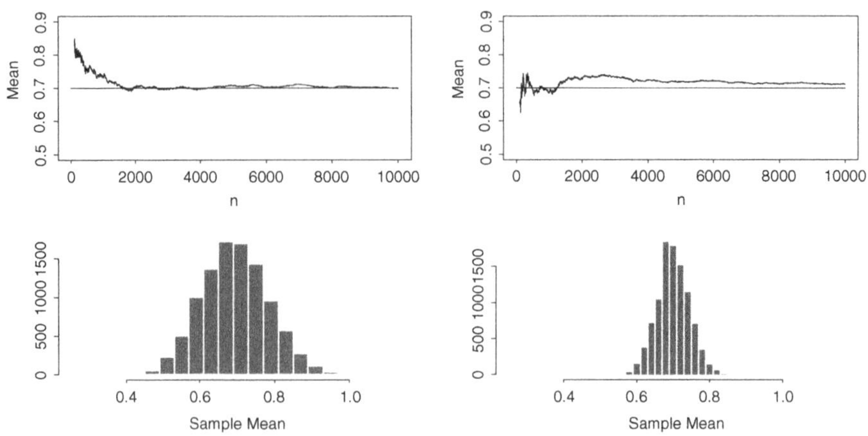


**FIGURE 1.5.1** Sample means.

the normal or Gaussian density. This will be made explicit in Chapter Six when we discuss the central limit theorem.

We are ready for the definition. To accommodate the case that the sample space $S$ is infinite, we require a special condition.

**Definition 1.5.1** Let $X$ be a discrete random variable defined on a probability space $(S, P)$. If $\sum_{\omega \in S} |X(\omega)| P(\omega)$ converges, the *expected value* of $X$ is said to exist, with value

$$E(X) = \sum_{\omega \in S} X(\omega) P(\omega).$$

$E(X)$ is also called the *mean* of $X$. □

***Example 1.5.2*** Suppose that a coin is tossed until the first head occurs. Let $S$ be the set of positive integers, where an outcome $\omega$ is the number of tosses necessary. Assign $P(\omega) = 1/2^{\omega}$. Define $X(\omega) = 2^{\omega}$. Then $\sum_{\omega \in S} |X(\omega)|\, P(\omega) = 1 + 1 + \cdots$ does not converge, so "the expected value of $X$ does not exist." On the other hand, if $Y(\omega) = \omega$, then $\sum_{\omega \in S} |Y(\omega)| P(\omega)) = (1)(1/2) + (2)(1/2)^2 + (3)(1/2)^3 + \cdots$. To evaluate this, consider that for $|t\,| < 1$, $g(t) \equiv 1 + t + t^2 + \cdots = 1/(1 - t)$, and $\frac{d}{dt} g(t) = 0 + 1 + 2t + 3t^2 + \cdots = (1 - t)^{-2}$, so that $E(Y) = (1/2)(1 - 1/2^{-2})$ $= 2$. □

As the definition is given, the summation is over the entire sample space $S$. Assuming that the expectation exists, we can sometimes compute its value by using the probability function of $X$.

**Theorem 1.5.1** Suppose that $X$ is a discrete random variable defined on a probability space $(S, P)$. Suppose that $E(X)$ exists. Then $E(X) = \sum_k k P(X = k)$, where the sum is taken over all $k$ for which $P(X = k) > 0$.

*Proof:* Since $E(X)$ exists, the $\sum_{\omega \in S} X_{(\omega)} P(\omega)$ converges absolutely, and therefore converges for any rearrangement of the terms. Thus,

$$E(X) = \sum_k \sum_{\{\omega | X(\omega) = k\}} X(\omega) P(\omega) = \sum_k \sum_{\{\omega | X(\omega) = k\}} k P(\omega) = \sum_k k P(X = k).$$

□

***Example 1.5.3*** Let $I_A$ be the indicator for the event $A$. Then $E(I_A) = 0 \cdot P(A^c) + 1 \cdot P(A) = P(A)$. □

***Example 1.5.4*** Let $\mathbf{x} = (x_1, \ldots, x_n)$ be a sequence of $n$ real numbers, a point in $R_n$. Let $X$ be the random variable that takes each of the values $x_k$ with probability $1/n$. Then $E(X) = (1/n)x_1 + \cdots + (1/n)x_n = [\sum_{k=1}^{n} x_k]/n$, the arithmetic mean of the $x_k$ (usually called simply the *mean* $x$), which is often denoted by $\bar{x}$, read as "$x$-bar." □

***Example 1.5.5*** Suppose that independent Bernoulli trials are performed until the first success, each trial having probability $0 < p < 1$ of success. If $X$ is the number of trials necessary, then $X$ has probability function $f$, where $f(k) = q^{k-1}p$, for $k = 1, 2, \ldots$, where $q = 1 - p$. $X$ has the geometric distribution. Then $E(X) = \sum_{k=1}^{\infty} kq^{k-1}\ p = p[1 + 2q + 3q^2 + 4p^3 + \cdots] = p/(1-q)^{-2} = 1/p$, where the second-to-last equality follows from Example 1.5.2. □

***Example 1.5.6*** Let $S = \{a, b, c, d, e, f\}$, and let $P$, $X$, and $Y$ be defined as follows:

| $\omega$ | $a$ | $b$ | $c$ | $d$ | $e$ | $f$ |
|---|---|---|---|---|---|---|
| $P(\omega)$ | 0.20 | 0.10 | 0.15 | 0.25 | 0.20 | 0.10 |
| $X(\omega)$ | −2 | −2 | −1 | 0 | 1 | 2 |
| $Y(\omega) = X^2(\omega)$ | 4 | 4 | 1 | 0 | 1 | 4 |

Then $E(Y)$ may be computed in three ways. By the definition $E(Y) = (-2)^2(0.20) + (-2)^2(0.10) + (-1)^2(0.15) + (0)^2(0.25) + (1)^2(0.2) + (2)^2(0.10) = 1.95$. Or, using the probability function of $Y$, $E(Y) = (4)(0.40) + (1)(0.35) + (0)(0.25) = 1.95$. Finally, by grouping together only outcomes having the same $X$-value, $E(Y) = E(X^2) = (-2)^2(0.30) + (-1)^2(0.15) + (0)^2(0.25) + (1)^2(0.20) + (2)^2(0.10) = 1.95$. □

That this last method is always valid for discrete random variables is summarized in Theorem 1.5.2.

**Theorem 1.5.2** Let $X$ be a discrete random variable with probability function $f_X$, let $g$ be a real-valued function defined on the range of $X$, let $Y = g(X)$, and suppose that $E(Y)$ exists. Then $E(Y) = \sum_k g(k) f_X(k)$, where the sum is taken over all $k$ for which $f_X(k) > 0$.

Before proving this, some comments are in order. Since $Y(\omega) = g(X(\omega))$, $Y$ is a random variable. It follows that $E(Y)$ is defined to be $\sum_{\omega \in S} Y(\omega)P(\omega)$. Then by Theorem 1.5.1, $E(Y) = \sum_j j f_Y(j)$, where $f_Y$ is the probability function for $Y$ and the sum is over all $j$ for which $f_Y(j) > 0$. Theorem 1.5.2 states that it is not necessary to determine the probability function for $Y$. Instead, we may work only with the probability function for $X$.

*Proof of Theorem 1.5.2:* By definition, $E(Y) = \sum_{\omega \in s} g(X(\omega))P(\omega)$. Recall that for each real $k$, $[X = k] = \{\omega \mid g(X(\omega)) = k\}$. Then

$$E(Y) = \sum_{\omega \in S} Y(\omega)P(\omega) = \sum_k \sum_{\omega \in [X=k]} g(X(\omega))P(\omega) = \sum_k g(k) \sum_{\omega \in [X=k]} P(\omega)$$

$$= \sum_k g(k)P(X = k). \qquad \square$$

***Example 1.5.7*** ***(The Jelly Donut Problem)*** A bakery has demand $D$ for packages of jelly donuts each day, where $D$ is a random variable with probability function $f$ and $f$ is given in the following table.

| $D$ | 1 | 2 | 3 | 4 | 5 |
|---|---|---|---|---|---|
| $P(D=d)$ | 0.1 | 0.2 | 0.3 | 0.25 | 0.15 |
| $g(d;3)$ | −3 | 0 | 3 | 3 | 3 |
| $g(d;4)$ | −5 | −2 | 1 | 4 | 4 |

The cost to the bakery for each package it makes is \$2. The packages sell for \$3. How many packages should the bakery make each day if it must discard every package not sold that day?

If the bakery bakes three packages each day, its profit is $g(D;3) = 3D - 2(3)$ for $D \le 3$, and 3 for $D > 3$. Then $E(g(D;3)) = (-3)(0.1) + (0)(0.2) + (3)(0.3) + (3)(0.25) + (3)(0.15) = 1.80$. If $g(D;b)$ is the profit when $b$ packages are baked and $D$ are "demanded," $E(g(D;4)) = 1.0$. Similarly, $E(g(D;2)) = 1.7$, $E(g(D;1)) = 1.0$, $E(g(D;5)) = -0.55$, so that the expected profit is largest when $b = 3$ are baked. Later we try to find a solution for the case that $D$ has any probability distribution. □

***Example 1.5.8*** Let $X$ have the geometric distribution with parameter $p > 0$. Let $q = 1 - p$ and $Y = a^X$, where $a \ge 0$. Then $E(Y) = \sum_{k=1}^{\infty} a^k q^{k-1} p = a\sum_{k=1}^{\infty}(aq)^{k-1} = a/(1-aq)$ for $aq < 1$. For example, for $p = 1/2$, $E(Y) = a/(2-a)$ exists for $a < 2$. More generally, if $\mathbf{X} = (X_1, \ldots, X_n)$ is a random vector with joint probability function $f$ and $Y = g(X)$, we can evaluate $E(Y)$ using $f$. □

**Theorem 1.5.3** $E[g(X_1, \ldots, X_n)] = \sum_{\mathbf{x}\in R_n} g(\mathbf{x})f(\mathbf{x})$.

*Proof:* Simply replace $[X = k]$ by $[\mathbf{X} = \mathbf{x}]$ in the proof of Theorem 1.5.2. □

## Properties of Expectation

The *expectation operator* $E$ takes random variables $X$ with finite expectation into points $E(X)$ on the real line. We now list some properties of $E$. For this discussion, $b$ and $c$ will denote constants, while $X$, with and without subscripts, will always denote a random variable. When $E(X)$ is written, the assumption is made that it exists.

1. $E(c) = c$. *Proof*: $E(c) = \sum_{\omega\in S} cP(\omega) = c\sum_{\omega\in S} P(\omega) = c(1) = c$.
2. $E(cX) = cE(X)$. *Proof*: Similar to property 1.
3. $E(X_1 + X_2) = E(X_1) + E(X_2)$. *Proof*: $E(X_1 + X_2) = \sum_{\omega\in S}[X_1(\omega) + X_2(\omega)]P(\omega) = \sum_{\omega\in S} X_1(\omega)P(\omega) + \sum_{\omega\in S} X_2(\omega)P(\omega) = E(X_1) + E(X_2)$.

More generally, by induction,

$$E\left(\sum_{k=1}^{n} X_k\right) = \sum_{k=1}^{n} E(X_k).$$

**4.** If $X_1$ and $X_2$ are independent, then $E(g_1(X_1)g_2(X_2)) = E[g_1(X_1)]E[g_2(X_2)]$ for any functions $g_1$, $g_2$ defined on the ranges of $X_1$ and $X_2$. *Proof*: By Theorem 1.5.3, $E(g_1(X_1)g_2(X_2)) = \sum_{(x_1,x_2)} g_1(x_1)g_2(x_2)f(x_1, x_2)$, where $f$ is the joint probability function for $X_1$ and $X_2$. Since $X_1$ and $X_2$ are independent, $f = f_1 f_2$, where $f_1$ and $f_2$ are the marginal probability functions. Thus, $E(X_1X_2) = \sum_{x_1} g_1(x_1)f_1(x_1)\sum_{x_2} g_2(x_2)f_2(x_2) = E(g_1(X_1))E(g_2(X_2))$. By induction it follows that independence of $X_1, X_2, \ldots, X_n$ implies that

$$E\left[\prod_{i=1}^{n} g_i(X_i)\right] = \prod_{i=1}^{n} E[g_i(X_i)].$$

**5.** Suppose that $E(X)$ exists. If $P(X \geq 0) = 1$, so that $X$ is a "nonnegative" random variable, then $E(X) \geq 0$. If also $P(X > 0) > 0$, then $E(X) > 0$. If $X \geq Y$ for all outcomes, then $E(X) \geq E(Y)$. *Proof*: $E(X) = \sum_k kP(X = k) = \sum_{k>0} kP(X = k) \geq 0$. This is greater than 0 if $P(X = k) > 0$ for any $k > 0$. If $X \geq Y$ for all outcomes, $E(X - Y)$ and $E(Y)$ must exist and $X - Y \geq 0$ for all outcomes, so that $E(X - Y) \geq 0$, implying that $E(X) \geq E(Y)$. It follows from property 5 that if $P(X \leq B) = 1$ for a constant $B$ and $E(X)$ exists, then since $Y = B - X$ is a nonnegative random variable, $E(Y) = B - E(X) \geq 0$, so that $E(X) \leq B$. Similarly, if $P(X \geq A) = 1$, then $P(X - A \geq 0) = 1$, so that $E(X) \geq A$.

Suppose that $m$ balls are distributed randomly into $n$ cells with replacement. Let $Y$ be the number of empty cells. Let us try to determine $E(Y)$. For example, for the case of $n = 4$, $m = 3$, 100 simulations produced the following $Y$ values:

```
22212  12113  21322  21121  22221  12122  12221
12122  12212  21232  22221  12222  22121  12122
12222  11322  12322  12121  21122  13221
```

There were 35 1's, 59 2's, and 6 3's, so the mean was 1.71. Similarly, for 10,000 simulations the frequencies were 3750, 5673, and 577, and the mean was 1.6827. Careful counting shows that $P(Y = 1) = 24/64$, $P(Y = 2) = 36/64$, and $P(Y = 3) = 4/64$, so $E(Y) = 108/64 = 1.6875$.

The probability function for $Y$ is not easy to find in the general case, but by expressing $Y$ as the sum of random variables, we can use property 3 of the expectation operator $E$ to find $E(Y)$. Let $I_i$ be the indicator of the event that cell $i$ is empty for $i = 1, \ldots, n$. Then $E(I_i) = P(\text{cell } i \text{ is empty}) = [(n-1)/n]^m$. Then, since $Y = I_1 + \cdots + I_n$, $E(Y) = nE(I_1) = n[(n-1)/n]^m = n[1 - 1/n]^m$. For

$n = 4, m = 3$, we get $4[(3/4)^3] = 108/64 = 1.675$. Notice that these indicator random variables are *not* independent. It is interesting to note that if for each positive integer $n, m_n$ is also a positive integer so that $\lim_{n\to\infty} m_n/n = \alpha$, then for $Y_n = Y, \lim_{n\to\infty} E(Y_n)/n = e^{-\alpha}$, so that $E(Y)$ is approximately $n\exp(-m/n)$ for large $n$ and $m$. For $m = 50, n = 25$, 10,000 simulations produced the mean 3.247, while $E(Y) = 25(24/25)^{50} = 3.247145$ and its approximation is $n\exp(-m/n) = 3.383382$.

***Example 1.5.9*** Let $X$ have the binomial distribution with parameters $n$ and $p$. Then $E(X) = \sum_{k=0}^{n} \binom{n}{k} kp^k(1-p)^{n-k}$. It is possible to find a simple expression for this by using a careful argument involving factorials (see Problem 1.5.5). However, there is a much simpler argument using indicator random variables. If $n$ independent Bernoulli trials are performed, each with probability $p$ of success, the number $X$ of successes may be written in the form $X = I_1 + I_2 + \cdots + I_n$, where $I_j$ is the indicator of success on trial $j$. $E(I_j) = P(j\text{th trial is a success}) = p$, so that $E(X) = p + \cdots + p = np$. □

***Example 1.5.10*** **(The Coupon Problem)** In an effort to inspire possible customers to purchase the breakfast cereal Sugar Clumps, the Clump Company puts numbered coupons in the bottom of each box. The numbers are chosen randomly with replacement from $1, 2, \ldots, 20$. Any customer who accumulates all 20 numbers receives a shipment of 89 packages of Sugar Clumps free of charge. Mrs. Johnson's children love Sugar Clumps, so she is determined to buy packages until she has all 20 numbers. How many packages should she expect to have to purchase?

Let $X_k$ be the number of additional boxes she must purchase after having $k-1$ different numbers in order to get the $k$th different number. For example, if the numbers on the consecutive coupons are $13, 7, 8, 9, 7, 18, 9, 13, 15, \ldots$, then $X_1 = 1, X_2 = 1, X_3 = 1, X_4 = 1, X_5 = 2, X_6 = 3$. Then the total of packages she will need to buy is $T = X_1 + \cdots + X_{20}$. $X_1$ is identically 1. For $k > 1$, $X_k$ has the geometric distribution with $p = p_k = (20-k)/20$. From Example 1.5.2, $E(X_k) = 1/p_k$. Therefore, $E(T) = 1/p_1 + \cdots + 1/p_{20} = 20[1/20 + 1/19 + 1/18 + \cdots + 1/2 + 1] = 71.955$. For the case of six coupon numbers rather than 20, which may be simulated by die throws as at the beginning of Section 1.2, we get $6[1/6 + 1/5 + \cdots + 1/2 + 1] = 294/20 = 14.70$.

Mrs. Johnson's experiment was simulated 100 times, resulting in the following values of $T$:

48 46 86 91 90 61 64 84 97 67 96 60 61 75 59 83 102 57 139 69
27 85 48 99 105 45 66 44 61 50 67 85 73 90 67 64 57 91 48 103
47 51 51 72 42 106 57 90 68 33 76 95 67 157 48 63 95 82 59 88
56 74 55 82 87 66 53 68 44 60 102 77 59 92 73 47 148 87 123 57
42 48 70 56 36 74 90 115 89 48 53 51 70 64 95 68 103 72 55 56,

for an arithmetic mean of 72.22. The experiment was repeated 1000 times with the mean 72.091. A histogram of the 1000 values of $T$ is presented in Figure 1.5.2. For

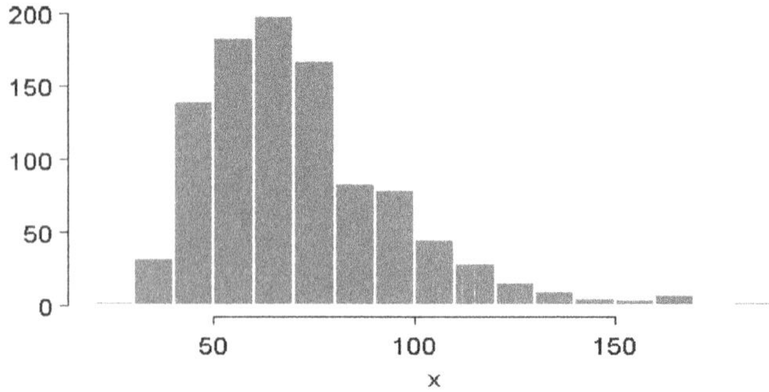


**FIGURE 1.5.2** Coupon problem.

the general case of $N$ coupon numbers, we can take advantage of Euler's approximation of $\eta_n = (1 + 1/2 + \cdots + 1/n) - \log(n)$. Euler showed that $\lim_{n\to\infty} \eta_n = \eta$, *Euler's constant*, which is approximately 0.5772 (see Problem 1.3.12). Thus, $E(T) = n[1 + \cdots + 1/n]$ is approximately $n[\log(n) + 0.5772]$. For $n = 6, 20$, and 100, we get $E(T) = 14.70, 71.95, 518.74$ and the approximations $14.21, 71.46, 518.24$.

Notice that the histogram of $T$ values does not seem to have the "shape" of the normal density. Although $T$ is the sum of independent random variables, their distributions vary and the value taken by $T$ depends more heavily on the $X_k$'s for $k$ near 20. □

We will often be interested in the probability that a nonnegative random variable is large. The following inequality provides an upper bound on the probability that a nonnegative random variable $X$ exceeds a constant in terms of $E(X)$.

**The Markov Inequality:** Let $X$ be a random variable such that $P(X < 0) = 0$. Then for any constant $C > 0$, $P(X \geq C) \leq E(X)/C$.

*Proof:* Let $I_C = [X \geq C]$. Then by the inequalities $X \geq XI_C \geq CI_C$, holding for all outcomes, and property 5 of an expectation operator, $E(X) \geq E(XI_C) \geq E(CI_C) = CP(X \geq C)$. □

COMMENTS: The Markov inequality is "sharp" in the following sense. Let $C > 1$, and $X = C$ with probability $1/C, 0$ otherwise. Then $E(X) = 1$ and $P(X \geq C) = 1/C = E(X)/C$.

## Problems for Section 1.5

**1.5.1** Let $S = \{a, b, c, d, e\}$ with $P(a) = 0.2$, $P(b) = 0.3$, $P(c) = 0.1$, $P(d) = 0.3$, $P(e) = 0.1$. Define $X(a) = 1$, $X(b) = 2$, $X(c) = 2$, $X(d) = 3$, $X(e) = 4$. Let $g(x) = |x - 2|$ for $x \in R$. Let $Y = g(X)$.

**(a)** Find the probability functions $f_X$ for $X$ and $f_Y$ for $Y = g(X)$.

**(b)** Determine $E(X)$ by (1) using the definition, and (2) using $f_X$.

**(c)** Determine $E(Y)$ by (1) using the definition, (2) using $f_X$, and (3) using $f_Y$.

**(d)** Find $E(XY)$.

**1.5.2** Two dice are thrown. Let $S = \{1, 2, \ldots, 6\}^{(2)}$, and let $P(\omega) = 1/36$ for all $\omega = (\omega_1, \omega_2) \in S$. Let $M = \max(\omega_1, \omega_2)$.

**(a)** Find $E(M)$ using the definition.

**(b)** Find the probability function $f$ for $M$ and use this to find $E(M)$.

**1.5.3** A box has three red, two white, and one blue ball. A random sample of two is chosen without replacement. Let $R$, $W$, and $B$ be the numbers of red, white, and blue balls chosen.

**(a)** Find the joint probability function for $R$ and $W$ and the marginal probability functions $f_R$ and $f_W$.

**(b)** Find $E(R)$, $E(W)$, and $E(RW)$.

**(c)** Express $B$ as a linear function of $R$ and $W$ and use this to determine $E(B)$. Also find the probability function for $B$ and use this to determine $E(B)$.

**(d)** Find $E(M)$, where $M = \max(R, W, B)$.

**1.5.4** Let $X$ and $Y$ have the joint probability function $f(x, y) = Kxy$, for $x = 1, 2, 3$, $y = 1, 2, 3$, and $x + y \le 4$, $f(x, y) = 0$ otherwise.

**(a)** Find $K$.

**(b)** Find $E(X)$, $E(Y)$, and $E(XY)$. Are $X$ and $Y$ independent? Does $E(XY) = E(X)E(Y)$?

**(c)** Let $W = X + Y$. Find the probability function for $W$, and verify that $E(W) = E(X) + E(Y)$.

**1.5.5** Let $X$ have the binomial distribution with parameters $n$ and $p$. Show that $E(X) = np$ by using the summation given in Example 1.5.9.

**1.5.6** Henry takes a multiple-choice exam on the history of Albasta, 1357–1461, with two parts. Part I has 50 true–false questions. Part II has 40 multiple-choice questions, each having four possible answers, only one of which is correct. Henry has failed to study, so he uses a coin to choose his answer to each question, so that probability $1/2$ of getting the true–false questions correct, probability $1/4$ of getting the others correct. Define the notation, state a model, and use the model to determine the expected number of correct answers Henry will have.

**1.5.7** Five balls are placed randomly and independently in four cells. Let $X_k$ be the number of cells with exactly $k$ balls. Find $E(X_k)$ for $k = 0, 1, \ldots, 5$. *Hint:* Express $X_k$ as the sum of indicators. What should $\sum_{k=0}^{5} E(X_k)$ be?

**1.5.8** Suppose that the demand $D$ for packages of jelly donuts at your bakery shop takes the values $0, 1, \ldots, 5$ with probabilities 0.10, 0.15, 0.20, 0.25, 0.20, 0.10. Each package sells for \$4 but costs \$2 to make, and unsold packages must be destroyed. How many packages should your shop bake if you wish to maximize $E(X)$, where $X$ is the profit? What is $E(X)$ for this choice?

**1.5.9** Your task, should you decide to accept it, is to throw two dice consecutively until you have gotten 6–6 five times. How many times should you expect to have to throw the two dice? Define notation and give a careful explanation that a classmate would understand, assuming that the classmate is struggling to get a B. How many times must you throw the two dice to make the probability at least 0.60 that a 6–6 will occur? [This is the 0.60-quantile of the rv $X$ =(no. tosses necessary).]

**1.5.10** Let $A_k$ be an event and $I_k$ its indicator for $k = 1, 2, 3, 4$. Let $B = \bigcup_{k=1}^{4} A_k$. Express the indicator $I_B$ of $B$ in terms of the $I_k$ and use the linearity of the expectation operator to express $P(B)$ in terms of the probabilities of the $A_k$ and their intersections. Verify your formula for the case that the $A_k$ are independent with $P(A_k) = k/5$.

**1.5.11** Let $X_1, X_2, X_3$ be independent Bernoulli random variables with parameters $0.8, 0.7, 0.6$. Let $T = X_1 + X_2 + X_3$.

**(a)** Find the probability function for $T$.

**(b)** Compare the probability function given in part (a) to the binomial probability function with $n = 3$, $p = 0.7$.

**(c)** Verify that the expectations are the same for both probability functions in part (b). Would this be true if $0.8, 0.7, 0.6$ were replaced by $p_1, p_2, p_3$ and $p$ by $(p_1 + p_2 + p_3)/3$? If so, prove it. If not, give an example.

**1.5.12** **(a)** Use the fact that $1 + 1/2^2 + 1/3^2 + \cdots = \pi^2/6$ to construct a probability function for which the mean does not exit. *Hint:* The sequence $\{a_n\}$ for $a_n = 1/1 + 1/2 + 1/3 + \cdots + 1/n$ does not converge.

**(b)** The infinite series $1 + 1/2^k + 1/3^k + \cdots$ converges to a constant, say $C_k$, if and only if $k > 1$. Let $M > 0$. Give an example of a distribution for $X$ for which the moments $E(X^m)$ exist for $m < M$, but not for $m \geq M$.

**1.5.13** Prove that for any random variable $X$ for which $E(X^2)$ exists, $E(X^2) \geq [E(X)]^2$. *Hint:* Let $\mu = E(X)$ and consider $E[(X - \mu)^2]$.

**1.5.14** Let balls numbered $1, 2, \ldots, n$ be placed randomly without replacement into cells numbered $1, 2, \ldots, n$. Find the expected number of matches, the number of cells $k$ that receive ball $k$, $1 \leq k \leq n$. *Hint:* Use indicator rv's.

**1.5.15** Let $X$ be a discrete rv, taking only nonnegative integer values. Prove that $E(X) = \sum_k P(X \geq k)$. Verify it for the case that $X$ has the geometric distribution.

## 1.6 THE VARIANCE

In an important sense, probability and statistics are concerned with variability. We have studied random variables and their expectations. $E(X)$ is a numerical description of the "location" of the probability distribution of $X$. However, probability distributions having the same mean can differ greatly in their variation. The distribution assigning probability $1/3$ to each of 49, 50, 51 has the same mean as the distribution assigning probability $1/3$ to each of 0, 50, 100 or to the distribution assigning $1/101$ to each of $0, 1, \ldots, 100$. We need a measure of variation.

If $X$ is a random variable with mean $\mu$, we would like to describe the variation of the deviations $X - \mu$. One such measure is $E(|X - \mu|)$, called the *mean deviation* of $X$. However, the absolute value function is often mathematically awkward, and it turns out that the variance is a much more useful measure of variability of a random variable.

**Definition 1.6.1** Let $X$ be a random variable and let $\mu = E(X)$. The *variance* of $X$ is $\text{Var}(X) = E[(X - \mu)^2]$, provided that the expectation exists. The *standard deviation* of $X$ is $\text{SD}(X) = \text{Var}(X)^{1/2}$. □

Often, the symbols $\sigma^2$ and $\sigma$ are used to denote the variance and standard deviation. To distinguish among the variances of different random variables, we may use subscripts, so that, for example, $\sigma_X^2 = \text{Var}(X)$.

***Example 1.6.1*** Let $X$ have probability function $f(k) = k/10$ for $k = 1, 2, 3, 4$. Then $\mu = 30/10 = 3$ and $\text{Var}(X) = (1-3)^2(1/10) + (2-3)^2(2/10) + (3-3)^2(3/10) + (4-3)^2(4/10) = 10/10 = 1$. If $Y$ has the probability function $g(k) = k/15$ for $k = 1, \ldots, 5$, a similar computation shows that $\text{Var}(Y) = 14/9$. □

***Example 1.6.2*** Let $x_1, \ldots x_n$ be real numbers and let $X$ take each of these values with probability $1/n$. Then $\mu = E(X) = (1/n)\Sigma_{i=1}^n x_i$ and $\text{Var}(X) = \sigma_X^2 = (1/n)\sum_{i=1}^n (x_i - \mu)^2$. Often, the symbol $\bar{x}$ is used rather than $\mu$, depending on the application. For reasons to be explained later, $S_X^2 = \Sigma_{i=1}^n (x_i - \bar{x})^2/(n-1)$, rather than $\sigma_X^2$, is used as a measure of variation, depending on the applications. Roughly speaking, since the standard deviation is a certain type of average deviation from the mean (the *root mean square*), we can usually expect 50 to 80% of the $x_i$'s to lie within one standard deviation of the mean. For example, for the data set $1, 2, \ldots, 20$, $\mu = 10.5$ and $\sigma_X = [399/12]^{1/2} = 5.77$, so 12 of the 20 $x_i$'s are within one standard deviation of $\mu$. □

### Properties of the Variance

For the purpose of this discussion, $c$ will denote a constant, and $X$, with or without a subscript, will denote a random variable. Let $\mu = E(X)$.

1. $\text{Var}(c) = 0$. *Proof:* In a sense we are abusing notation a bit here by using $c$ to denote a real number as well as the random variable that assigns the number $c$ to all outcomes. No harm is usually done by such an abuse. Since $E(c) = c$, it follows that the deviations $(c - c)$ are 0 with probability 1, so $\text{Var}(c) = E(c - c)^2 = 0$.
2. $\text{Var}(cX) = c^2 \ \text{Var}(X)$. *Proof:* Let $\mu = E(X)$. $E(cX) = c\mu$, so $\text{Var}(cX) = E[(cX - c\mu)^2] = E[c^2(X - \mu)^2] = c^2 \ \text{Var}(X)$.
3. $\text{Var}(X + c) = \text{Var}(X)$. *Proof:* Since $E(X + c) = \mu + c$, $\text{Var}(X + c) = E[(X + c) - (\mu + c)]^2 = E(X - \mu)^2 = \text{Var}(X)$.
4. Let $X_1$ and $X_2$ be independent. Then $\text{Var}(X_1 + X_2) = \text{Var}(X_1) + \text{Var}(X_2)$. *Proof:* Let $\mu_1 = E(X_1)$ and $\mu_2 = E(X_2)$. Let $D_1 = X_1 - \mu_1$ and $D_2 = X_2 - \mu_2$. Then $\text{Var}(X_1 + X_2) = E[(X_1 + X_2) - (\mu_1 + \mu_2)]^2 = E[D_1 + D_2]^2 = E[D_1^2 + D_2^2 + 2D_1D_2] = E(D_1^2) + E(D_2^2) + 2E(D_1)E(D_2)$, since in the last step, independence implies that $E(D_1D_2) = E(D_1)E(D_2)$. But $E(D_1) = E(D_2) = 0$.
5. Let $X_1, \ldots, X_n$ be independent and define $T_n = X_1 + \cdots + X_n$. Then $\text{Var}(T_n) = \text{Var}(X_1) + \cdots + \text{Var}(X_n)$. In the special case that $X_1, \ldots, X_n$ are independent, all with the same variance $\sigma^2$, with $T_n = X_1 + \cdots + X_n$, $\text{Var}(T_n) = n\sigma^2$, and therefore $\text{SD}(T_n) = \sqrt{n}\sigma$. For $\bar{X} = T_n/n$, $\text{Var}(\bar{X}) = \text{Var}(T_n)/n^2 = n\sigma^2/n^2 = \sigma^2/n$. This last formula,

   $$\text{Var}(\bar{X}) = \sigma^2/n, \text{ for independent random variables,}$$

   is one of the most important in probability and statistics. *Proof:* This follows from property 4 by induction.

***Example 1.6.3*** Suppose that $n$ independent Bernoulli trials are performed, each having probability $p$ of success, for $0 < p < 1$. Let $I_k$ be the indicator of success on trial $k$, and let $X = I_1 + \cdots + I_k$ be the total number of successes. Then, of course, $X$ has the binomial distribution with parameters $n$ and $p$. As shown earlier, since $E(I_k) = p$, $E(X) = np$. Using property 5 we can now determine $\text{Var}(X)$ easily. $\text{Var}(I_k) = (0 - p)^2(1 - p) + (1 - p)^2p = p(1 - p)[p + (1 - p)] = p(1 - p)$, so $\text{Var}(X) = np(1 - p)$. If, for example, $X$ is the number of heads in 16 tosses of a coin, then $\text{Var}(X) = 16(1/2)(1 - 1/2) = 4$, so $\text{SD}(X) = 2$, while for $64 = (4)(16)$ tosses, $\text{SD}(X) = 4$. For the four binomial probability functions presented in Figure 1.6.1, the probabilities that $X$ will fall within one standard deviation of its expected values are $7/8, 0.790, 0.740, 0.712$. The corresponding probabilities for two standard deviations are $1, 0.989, 0.967, 0.961$. The central limit theorem will show that these probabilities converge to those given by the area under the standard normal density, yet to be defined. □

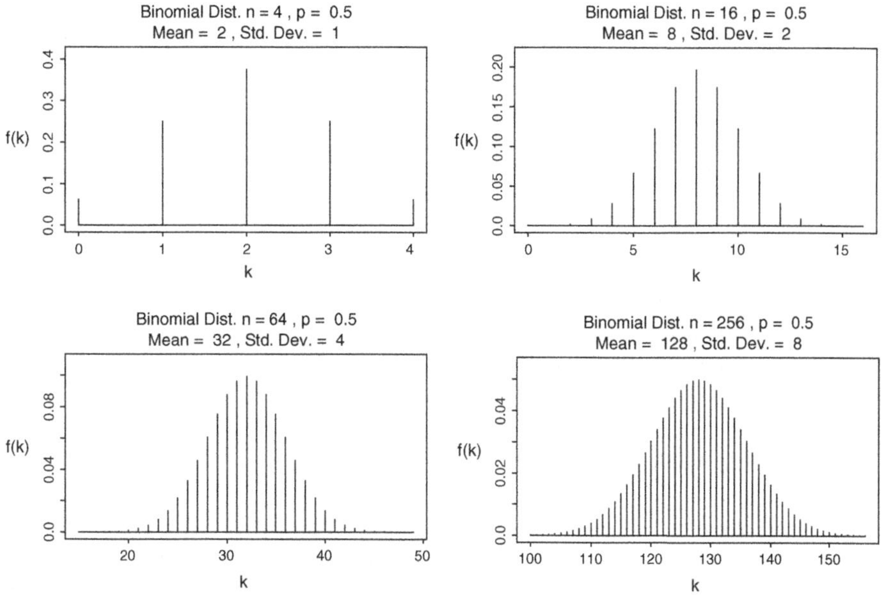


**FIGURE 1.6.1** Binomial probability mass functions.

***Example 1.6.4*** Let $X$ take the values $0, 1, 2$ with probabilities $0.3, 0.5, 0.2$. Let $Y$ take the values $1, 2, 3$ with probabilities $0.3, 0.4, 0.3$. Suppose that $X$ and $Y$ are independent. Let $T = X + Y$. Some work with the $3 \times 3$ matrix of probabilities shows that $T$ takes the values $1, 2, 3, 4, 5$ with probabilities $0.09, 0.27, 0.35, 0.23, 0.06$. More computation shows that $E(X) = 0.9$, $E(Y) = 2$, $E(T) = 2.9$, $\text{Var}(X) = 0.49$, $\text{Var}(Y) = 0.60$, $\text{Var}(T) = 1.09$. Let $W = 3X - 2Y$. Then $E(W) = 3(0.9) - 2(2.0) = -1.3$, and $\text{Var}(W) = \text{Var}(3X) + \text{Var}(-2Y) = (3^2)(0.49) + (-2)^2(0.60) = 6.81$. (A common error for neophytes to probability is to forget to square the coefficients, therefore sometimes getting negative variances! Of course, readers of this book would never do that.) □

## Computational Formulas

The variance and standard deviation of a random variable $X$ are measures of the variability of $X$. From the point of view of physics, $\text{Var}(X)$ is the *moment of inertia* about the center of mass, $\mu = E(X)$. The following formula expresses the moment of inertia about an arbitrary point $c$ as the sum of the variance and the squared distance of $c$ from the mean:

$$E(X - c)^2 = \text{Var}(X) + (c - \mu)^2. \tag{1.6.1}$$

*Proof:* $E(X - c)^2 = E[(X - \mu) + (\mu - c)]^2 = E(X - \mu)^2 + (c - \mu)^2 + 2(\mu - c)E(X - \mu)$. Equation (1.6.1) follows from the fact that $E(X - \mu) = 0$.

□

Formula (1.6.1) is sometimes useful in the computation of Var($X$), since

$$\mathrm{Var}(X) = E(X - c)^2 - (c - \mu)^2. \tag{1.6.2}$$

$c$ can be chosen arbitrarily to make computations as simple as possible. For example, if $c$ is chosen to be 0, then

$$\mathrm{Var}(X) = E(X^2) - \mu^2 = E(X^2) - [E(X)]^2. \tag{1.6.3}$$

***Example 1.6.5*** Let $X$ take the values 173.5, 173.6, 173.7 with probabilities 0.3, 0.5, 0.2. The random variable $Y = X - 173.5$ takes the values 0, 0.1, 0.2 with the same probabilities. The random variable $W = 10Y$ takes the values 0, 1, 2 with these probabilities. From (1.6.3), $\mathrm{Var}(W) = E(W^2) - [E(W)]^2 = [0^2(0.3) + 1^2(0.5) + 2^2(0.2)] - [0.9]^2 = 1.30 - 0.81 = 0.49$. Therefore, $\mathrm{Var}(X) = \mathrm{Var}(Y) = 0.49/10^2 = 0.0049$. □

***Example 1.6.6*** (The Discrete Uniform Distribution) Let $X$ have the uniform distribution on $(1, 2, \ldots, N)$. Then, since $\sum_{k=1}^{N} k = N(N+1)/2$ and $\sum_{k=1}^{N} k^2 = N(N+1)(2N+1)/6$, $E(X) = (N+1)/2$, $E(X^2) = (N+1)(2N+1)/6$, and $\mathrm{Var}(X) = E(X^2) - [E(X)]^2 = (N^2 - 1)/12$. □

Suppose that we have "observations" $X_1, \ldots, X_n$ from a probability distribution with mean $\mu$ and variance $\sigma^2$. That is, $X_1, \ldots, X_n$ are independent, each with the distribution of a random variable $X$, where $E(X) = \mu$ and $\mathrm{Var}(X) = \sigma^2$. If $\mu$ and $\sigma^2$ are unknown, we may wish to use the data $X_1, \ldots, X_n$ to estimate these unknown *parameters.* This is a problem in statistical inference. Consider the *estimators*

$$\bar{X} = \frac{\sum_{k=1}^{n} X_k}{n} \quad \text{and} \quad \hat{\sigma}^2 = \frac{\sum_{k=1}^{n}(X_k - \bar{X})^2}{n}.$$

Then $E(\bar{X}) = n\mu/n = \mu$, so $\bar{X}$ is an *unbiased estimator* of $\mu$. To determine $E(\hat{\sigma}^2)$, first consider any numbers $c, x_1, \ldots, x_n$. Let $\bar{x} = (\sum_{k=1}^{n} x_k)/n$. Then $\left[\sum_{k=1}^{n}(x - c)^2\right]/n = \sum_{k=1}^{n}\left[(x_k - \bar{x}) + (\bar{x} - c)\right]^2/n = (1/n)\sum_{k=1}^{n}(x_k - \bar{x})^2 + (\bar{x} - c)^2$. The middle term drops out because $\sum_{k=1}^{n}(x_k - \bar{x}) = 0$. This numerical identity is a special case of (1.6.1), where $X$ takes each of the values $x_1, \ldots, x_n$ with probability $1/n$. From the identity we get

$$\sum_{k=1}^{n}(x_k - \bar{x})^2 = \sum_{k=1}^{n}(x_k - c)^2 - n(\bar{x} - c)^2, \tag{1.6.4}$$

which, of course, simplifies further by taking $c = 0$. Now, to determine $E(\hat{\sigma}^2)$, replace $x_k$ by $X_k$, $\bar{x}$ by $\bar{X}$, and $c$ by $\mu$. We get

$$\sum_{k=1}^{n}(X_k - \bar{X})^2 = \sum_{k=1}^{n}(X_k - \mu)^2 - n(\bar{X} - \mu)^2.$$

Taking expectation on the right, we get $n\sigma^2 - n \text{ Var}(\bar{X}) = n\sigma^2 - n(\sigma^2/n) = (n-1)\sigma^2$. Therefore, $E(\hat{\sigma}^2) = (n-1)\sigma^2/n = \sigma^2[(n-1)/n]$. Thus, $\hat{\sigma}^2$ is a biased estimator of $\sigma^2$. For that reason, people usually use $S^2 = \sum_{k=1}^{n} (X_k - \bar{X})^2/(n-1)$ as an estimator of $\sigma^2$, since $S^2$ is an unbiased estimator of $\sigma^2$.

***Example 1.6.7*** Consider the die-throw data at the beginning of the chapter. The uniform distribution on $1, \ldots, 6$ has mean $(6+1)/2 = 3.5$ and variance $(6^2 - 1)/12 = 35/12 = 2.917$. The frequency table for the data was

| 1 | 2 | 3 | 4 | 5 | 6 |
|---|---|---|---|---|---|
| 60 | 73 | 65 | 58 | 74 | 70 |

There were $n = 400$ observations: $X_1, \ldots, X_{400}$. Then $\sum_{k=1}^{n} X_k = 60(1) + 73(2) + \cdots + 70(6) = 1423$, and $\sum_{k=1}^{n} X_k^2 = 6235$. Hence, $\bar{X} = 1423/400 = 3.5575$, the estimate of $\mu = 3.5$, and $\hat{\sigma}^2 = [6235 - 400(3.5575)^2]/400 = 6235/400 - 3.5575^2 = 2.932$, the estimate of $\sigma^2 = 2.917$. The estimates were both reasonably close to the corresponding parameters. Since $\text{Var}(\bar{X}) = \sigma^2/400$ and $\text{SD}(\bar{X}) = \sigma/400^{1/2} = 0.0854$, we should not be surprised that $\bar{X}$ differed from $\mu$ by 0.0575. If, however, $\bar{X}$ had taken this value for $n = 10{,}000$, in which case $\text{SD}(\bar{X}) = 0.0171$, we would have been somewhat surprised, and perhaps suspected that we had an unbalanced die. □

### The Chebychev Inequality

Recall that the Markov inequality (Section 1.5) states that for any $C > 0$ and any random variable $X$ for which $P(X \geq 0) = 1$ and $E(X)$ exists, $P(X \geq C) \leq E(X)/C$. We can make use of this to relate the probability that a random variable differs from its mean by more than a prescribed amount to the variance of the random variable.

**The Chebychev Inequality** Let $Y$ be a random variable whose mean $\mu$ and variance $\sigma^2$ exists. Let $k > 0$ and $h > 0$ be constants. Then

$$P(|Y - \mu| \geq k) \leq \frac{\sigma^2}{k^2} \quad \text{and} \quad P(|Y - \mu| \geq h\sigma) \leq \frac{1}{h^2}.$$

*Proof:* Let $X = |Y - \mu|^2$. Then $E(X) = \sigma^2$ and $P(|Y - \mu| \geq k) = P(X \geq k^2) \leq \sigma^2/k^2$. The last step follows from the Markov inequality. The second inequality follows by taking $k = h\sigma$. Taking complements, we get the inequalities

$$P(|Y - \mu| < k) \geq 1 - \frac{\sigma^2}{k^2} \quad \text{and} \quad P(|Y - \mu| < h\sigma) \geq 1 - \frac{1}{h^2}. \qquad \square$$

COMMENTS: The Chebychev inequality is sharp is the following sense. For $k > 0$, let $Y$ take the values $-k, 0, k$ with probabilities $1/(2k)$, $1 - 1/k$, $1/(2k)$. Then $E(Y) = 0$, $\sigma^2 = \text{Var}(Y) = k$, and $P(|Y| \geq k) = 1/k = \sigma^2/k^2$.

***Example 1.6.8*** Let $Y$ be the number of heads in 100 tosses of a coin. Then $Y$ has the binomial distribution with $n = 100$, $p = 1/2$. Therefore, $\mu = E(Y) = np = 50$, $\sigma^2 = np(1-p) = 25$, $\sigma = 5$. By the Chebychev inequality, $P(|Y-50| \geq 20) = P(|Y-50| \geq 4(5)) \leq 1/4^2 = 1/16$. It follows that $P(31 \leq Y \leq 70) \geq 1 - 1/16 = 15/16$. Actually, $P(|Y-50| \geq 20) = 0.0000322$, much smaller than $1/16$. We needed to know the distribution of $Y$, not just the mean and variance, to determine this. □

As indicated just a little earlier, we often wish to use $\bar{X}$, a sample mean, to estimate a population mean $\mu$. The Chebychev inequality gives us a way to relate the population variance $\sigma^2$ and the sample size $n$ to the probability that $\bar{X}_n = \bar{X}$ will differ by a prescribed amount from $\mu$. To be more precise, let $X_1, \ldots, X_n$ be independent, each with mean $\mu$ and variance $\sigma^2$. Let $k > 0$ be a constant. Then

$$P(|\bar{X}_n - \mu| \leq k) \leq \frac{\mathrm{Var}(\bar{X}_n)}{k^2} = \frac{\sigma^2/n}{k^2} = \frac{\sigma^2}{nk^2}. \tag{1.6.5}$$

By letting $n \to \infty$, we find that $P(|\bar{X}_n - \mu| \geq k) \to 0$ as $n \to \infty$. This is an important result, making precise what should be intuitive. For the case that the $X_k$ are independent and identically distributed, we indicate the importance of this result with the name the *weak law of large numbers* (WLLN).

**The Weak Law of Large Numbers** Let $X_1, X_2, \ldots$ be independent and identically distributed with mean $\mu$. Let $\bar{X}_n = (X_1 + \cdots + X_n)/n$. Then for any $\varepsilon > 0$,

$$\lim_{n\to\infty} P(|\bar{X}_n - \mu| \geq \varepsilon) = 0 \quad \text{and} \quad \lim_{n\to\infty} P(|\bar{X}_n - \mu| < \varepsilon) = 1.$$

This statement of the WLLN requires the $X_k$ to be identically distributed, whereas the argument in (1.6.5) requires only that $\mathrm{Var}(\bar{X}_n)$ converges to zero as $n \to \infty$. On the other hand, the WLLN does not require the existence of the variance, as (1.6.5) does. We do not present a proof of this form of the WLLN here but refer the reader to books by Feller (1950) and Durrett( ). The idea behind the proof is to make use of the Markov inequality by truncating the random variables: that is, by considering the random variables (for the case $\mu = 0$, as we can assume without loss of generality) $X_{kn}^* = X_k I[|X_k| \leq n]$ and showing that the sums of the $X_{kn}^*$ and $X_k$ do not differ too much. The term *weak* is used for contrast with the strong law of large numbers, which states that $P(\lim_{n\to\infty} \bar{X}_n = \mu) = 1$.

The upper bound on $P(|\bar{X} - \mu| \geq k)$ given by (1.6.5) is not a good approximation. The central limit theorem, given in Section 6.3, provides an approximation that is quite good, even for moderate $n$. Let $T_n = X_1 + \cdots + X_n$, $\bar{X}_n = T_n/n$, $Z_n = (T_n - \mu)/\sqrt{n\sigma^2} = (\bar{X}_n - \mu)/\sqrt{\sigma^2/n}$ and let $G_n(k) = P(|Z_n| \leq k)$. Then $Z_n$ is the *standardized version* of both $T_n$ and $\bar{X}_n$, since $E(Z_n) = 0$ and $\mathrm{Var}(Z_n) = 1$. Then $\lim_{n\to\infty} G_n(1) = 0.6827$, $\lim_{n\to\infty} G_n(2) = 0.9545$, $\lim_{n\to\infty} G_n(3) = 0.9973$. For example, for large $n$ the probability that $\bar{X}_n$ will lie within two of its standard deviations of $\mu$ is approximately 0.9545. We simulated the throws of 100 dice 100,000 times, each time determining $\bar{X}_{100}$. Since $\sigma^2 = 35/12$, $\text{Var}(\bar{X}_{100}) = 35/1200 = 0.029133$, and $\text{SD}(\bar{X}_{100}) = 0.17078$. In 100,000 simulations, 69,354, 95,556, and 99,765 of the $\bar{X}_{100}$'s were within 0.17078, 2(0.17078), and 3(0.17078) of 3.50. Similarly, since $\text{SD}(T_{100}) = 17.078$, these were the frequencies within 17.078, 2(17.078), and 3(17.078) of $E(T_{100}) = 100\mu = 350$.

If $X$ is the number of heads in 400 tosses of a coin, since $X$ has the binomial distribution, $E(X) = np = 200$ and $\text{Var}(X) = np(1-p) = 100$, so $\text{SD}(X) = 10$. It follows that $P(|X - 200|/10 \le 2) = P(180 \le X \le 220)$ should be approximately 0.9545. The actual probability, determined by summing binomial probabilities using S-Plus, is 0.9598. The lower bound provided by the Chebychev inequality is $1 - 1/2^2 = 3/4$.

## Problems for Section 1.6

**1.6.1** Let $X$ take the values 3, 4, 8 with probabilities 0.4, 0.5, 0.1.

**(a)** Find $\mu = E(X)$ and $\text{Var}(X)$ using Definition 1.6.1.

**(b)** Find $\text{Var}(X)$ using (1.6.2) for $c = 3$ and for $c = 0$.

**1.6.2** Show that $G(a) = E(X - a)^2$ is minimum for $a = E(X)$:

**(a)** Using calculus.

**(b)** Using (1.6.2).

**1.6.3** Let $X$ have the geometric distribution with parameter $p, 0 < p < 1$. Show that $\text{Var}(X) = q/p^2$, where $q = 1 - p$. *Hint:* Differentiate twice with respect to $q$ on both sides of $1/(1-q) = 1 + q + q^2 + \cdots$ to find $E[X(X-1)]$, then use (1.6.3).

**1.6.4** Find the sample mean $\bar{X}$ and variance $S^2$ for the data 1, 3, 3, 5, 6, 12.

**1.6.5** Let $x_1, \ldots, x_n$ be real numbers for $n > 1$. Let $\bar{u}_{n-1}$ and $\bar{u}_n$ be the arithmetic means of the first $n-1$ and the first $n$. Prove that $= \sum_{i=1}^{n}(x_i - \bar{u}_n)^2 = \sum_{n-1}^{n}(x_i - \bar{u}_{n-1})^2 + (\bar{u}_{n-1} - x_n)^2(n-1)/n$. (This is an "update formula" for the sum of squares of deviations from the mean.) Verify the formula for the data 1, 3, 5, 7.

**1.6.6** Our nonstudying student friend (see Problem 1.5.6) must take another exam, knowing nothing about the history of Albasta. The exam has 50 true–false questions, 40 multiple-choice questions with four possible answers, and 30 multiple-choice questions with five possible answers. The correct answer for each question is equally likely to be any of the possible answers, independently. The total score $T$ is the number of points earned, where $k$ points are earned on a question if the answer is correct and the question has $k$ possible answers. Define a model, and use it to find $E(T)$ and $\text{SD}(T)$.

**1.6.7** Let $X$ and $Y$ be independent. Let their probability functions be $f_X$ and $f_Y$, where $f_X(k) = k/6$ for $k = 1, 2, 3$ and $f_Y(j) = (3 - j)/6$ for $j = 0, 1, 2$. Let $T = X + Y$.

**(a)** Find $E(X)$, $E(Y)$, $\text{Var}(X)$, $\text{Var}(Y)$.

**(b)** Find the probability function for $T$.

**(c)** Find $E(T)$, $\text{Var}(T)$, and verify that $E(T) = E(X) + E(Y)$ and $\text{Var}(T) = \text{Var}(X) + \text{Var}(Y)$.

**1.6.8** A six-sided die is thrown 100 times. Let $T$ be the total of the numbers showing.

**(a)** Use the Chebychev inequality to give a lower bound on $P(300 \le T \le 400)$. Also give an upper bound on $P(|\bar{X} - 3.5| \ge 0.5)$, where $\bar{X} = T/100$.

**(b)** Give an upper bound for $P(|\bar{X} - 3.5\,| \ge 0.5)$ for the case of 400 throws.

**1.6.9** Let $X$ be the number of Clump cereal boxes that Mrs. Johnson must buy to get all $N$ coupons (see Section 1.5). We showed that $E(X) = N[1/1 + 1/2 + \cdots + 1/N]$.

**(a)** Express $\text{Var}(X)$ as a function of $N$. For 1000 simulations with $N = 6$ the mean was 14.6 and the variance was 38.3. For 1000 simulations with $N = 20$ the mean was 72.2 and the variance was 582.6 (see Problem 1.6.3).

**(b)** Use the equality $\sum_{k=1}^{\infty} 1/k^2 = \pi^2/6$ to give an approximation for $\text{Var}(X)$, and determine its values for $N = 6$ and $N = 20$.

**1.6.10** Gamblers and mathematicians of the sixteenth century, including Girolamo Cardano, a distinguished mathematician of his time, considered the following sort of problem. Suppose that $A$ and $B$ play a game of chance, with each having probability 1/2 of winning. They play for a stake of \$1000, the first to win 10 games taking the entire "pot" of \$1000. After they play 11 games, $A$ has won seven games and $B$ has won four. At that point, $A$ must leave to join his wife, who is about to deliver triplets (or to attend a statistics class).

**(a)** How should $A$ and $B$ split the pot? It should seem reasonable that $A$ should receive an amount equal to his expected return. That is, the amount $x_A$ he should receive should be $p_A$ (\$1000), his expected profit, where $p_A$ is the probability that $A$ would win the \$1000 if the games had continued.

**(b)** Answer the question for the case that the pot is won when one player has won $N$ games, and they stop playing when $A$ has won $w_A$ and $B$ has won $w_B$ games. Also consider the case that $A$ wins any given game with probability $p$. Blaise Pascal, father of Pascal's triangle and Pierre Fermat (of Fermat's last theorem fame), in an exchange of letters in 1654, solved the problem (see Hald, 2003, pp. 54–63).

**(c)** In part (a) the phrase "it should seem reasonable" was used. Why is it reasonable?

**1.6.11** (*Bertrand's Paradox*) In his book (Bertrand, 1889) discussed the "paradox of the three jewelry boxes." Three such boxes each have two drawers, each with one coin. Box $A$ has one gold coin in each drawer. Box $B$ has a silver coin in each drawer. Box $C$ has one gold coin and one silver coin in the two drawers. A drawer is chosen at random with equal probabilities, then one of the two drawers is chosen with equal probabilities. The drawer chosen has a gold coin. What is the revised (conditional) probability that the drawer chosen was $C$. Since the drawer chosen must be $A$ or $C$, it seems that drawers $A$ and $C$ are equally likely, with probabilities $1/2$, $1/2$. Do you agree? Bertrand did not.

## 1.7 COVARIANCE AND CORRELATION

We need a numerical measure of the relationship between a pair of random variables. In general, a joint distribution may be hard to determine or even to estimate. And even when the joint distribution is known, it may be so complex that it is difficult to make interpretations. It turns out that the joint *linear* behavior of two random variables can be studied through the covariance function.

**Definition 1.7.1** Let $(X, Y)$ be a pair of random variables with means $\mu_X$, $\mu_Y$. The *covariance* of $X$ and $Y$, provided that the expectation exists, is

$$\text{Cov}(X, Y) = E[(X - \mu_X)(Y - \mu_Y)]. \qquad \square$$

COMMENTS: The covariance exists whenever the variances of $X$ and $Y$ exist, although that is not necessary. Note that $\text{cov}(X, Y)$ exists if and only if the three expectations $E(XY)$, $\mu_X$, and $\mu_Y$ exist, and in that case, $\text{Cov}(X, Y) = E(XY) - \mu_X\mu_Y$.

### Properties of the Covariance

Proofs are left to students:

1. For any constants $a, b, c, d$, $\text{Cov}(aX + b, cY + d) = ac\,\text{Cov}(X, Y)$.
2. For random variables $Y, X_1, \ldots X_k$ $\text{Cov}(X_1 + \cdots + X_k, Y) = \text{Cov}(X_1, Y) + \cdots + \text{Cov}(X_k, Y)$.
3. If $X$ and $Y$ are independent, then $\text{Cov}(X, Y) = 0$.

***Example 1.7.1*** Let $(X, Y)$ take the values $(-1, 0), (0, 1), (1, 0)$ with probabilities $0.3, 0.4, 0.3$. Then $E(X) = 0$ and $Z = XY$ is zero with probability 1, so that $\text{Cov}(X, Y) = 0$. However, $X$ and $Y$ are *not* independent. $\square$

***Example 1.7.2*** Consider Example 1.4.3. $E(R) = 4/3$, $E(W) = 1$, $E(RW) = 1$, so that $\text{Cov}(X, Y) = 1 - (4/3)(1) = -1/3$. $\square$

***Example 1.7.3*** Let $A$ and $B$ be events, and let $I_A$ and $I_B$ be their indicators. Thus, $I_A I_B = I_{A\cap B}$ is the indicator of $A \cap B$. Then $\text{Cov}(I_A, I_B) = E(I_{A\cap B}) - E(I_A)E(I_B) = P(A \cap B) - P(A)P(B)$. Thus, $A$ and $B$ are independent if and only if $\text{Cov}(I_A, I_B) = 0$. More generally, suppose that $X_1$ takes only the two values $a_1 < b_1$ and $X_2$ takes only the two values $a_2 < b_2$. Let $r_1 = b_1 - a_1, r_2 = b_2 - a_2, U_1 = (X_1 - a_1)/r_1$, and $U_2 = (X_2 - a_2)/r_2$. Then each $U_i$ is an indicator random variable, and $X_i = r_i U_i + a_i$. Therefore, $\text{Cov}(X_1, X_2) = r_1 r_2 \text{Cov}(U_1, U_2) = 0$ if and only if $U_1, U_2$, hence $X_1, X_2$, are independent. As shown in Example 1.7.1, independence and zero covariance are *not* equivalent if either of $X_1, X_2$ take more than two values with positive probabilities. $\square$

Recall the mathematical identifies you were so proud of learning in your first algebra course: $(a + b) = a^2 + 2ab + b^2$ and $(a + b + c)^2 = a^2 + b^2 + c^2 + 2ab + 2ac + 2bc$. If $X_1$ and $X_2$ are random variables with means $\mu_1, \mu_2$, variances $\sigma_1^2, \sigma_2^2$, covariance $\sigma_{12}$, we can easily express the variance of $Y = X_1 + X_2$ in terms of $\sigma_1^2, \sigma_2^2$, and $\sigma_{12} \equiv \text{Cov}(X, Y)$. Let $X_i^* = X_i - \mu_i$ for each $i$. Then $\text{Var}(Y) = E(X_1^* + X_2^*)^2 = E(X_1^*)^2 + E(X_2^*)^2 + 2E(X_1^* X_2^*) = \text{Var}(X_1) + \text{Var}(X_2) + 2\,\text{Cov}(X_1, X_2)$. This is important enough to emphasize it on its own line:

$$\text{Var}(Y) = \text{Var}(X_1) + \text{Var}(X_2) + 2\,\text{Cov}(X_1, X_2). \tag{1.7.1}$$

***Example 1.7.4*** Consider Example 1.4.3. Let $Z = R + W$. Then $Z$ has a hypergeometric distribution, taking the values 1, 2, 3 with probabilities $1/12, 6/12, 5/12$. Thus, $\text{Var}(Z) = E(Z^2) - [E(Z)]^2 = 35/6 - (7/3)^2 = 7/18$. But $\text{Var}(R) = 5/9$, $\text{Var}(W) = 1/2$, and $\text{Cov}(R, W) = -1/3$, so $\text{Var}(Z) = 5/9 + 1/2 + 2(-1/3) = 7/18$. $\square$

Induction allows us to extend (1.7.1): For random variables $X_1, \ldots, X_n$:

$$\text{Var}(X_1 + \cdots + X_n) = \sum_{i=1}^{n} \text{Var}(X_i) + 2\sum_{i<j} \text{Cov}(X_i, X_j). \tag{1.7.2}$$

A "trick" allows us to find a formula for the variance of the sum of the observations in a simple random sample (without replacement) from a population of $N$. Let the population of $N$ have measurements $y_1, \ldots, y_N$. Let $\mu = (1/N)\sum_{i=1}^N y_i$ and $\sigma^2 = (1/N)\sum_{i-1}^N (y_i - \mu)^2$. Let $X_1, \cdots, X_n$ be the $n$ observations. Let $T = X_1 + \cdots + X_n$ be their total. Of course, $\text{Var}(X_j) = \sigma^2$ for each $j$, but the $X_j$ are *not* independent. To determine the covariances, let $y_i^* = y_i - \mu$. Then $0 = (\sum_{i=1}^N y_i^*)^2 = \sum_{i=1}^N y_i^{*2} + \sum_{i\neq j} y_i^* y_j^*$. Thus, $\sum_{i\neq j} y_i^* y_j^* = -\sum_{i=1}^N y_i^{*2}$. Therefore, for $i \neq j$, $\text{Cov}(X_i, X_j) = \sum_{i\neq j} y_i^* y_j^* / [N(N-1)] = -\sigma^2/(N-1)$, and

$$\text{Var}(T) = n\sigma^2 - n(n-1)\sigma^2/(N-1) = n\sigma^2 \frac{N-n}{N-1}. \tag{1.7.3}$$

The fraction $(N - n)/(N - 1)$ is called the *finite population correction factor*. It follows that for $\bar{X} = T/n$,

$$\mathrm{Var}(\bar{X}) = \frac{N - n}{N - 1}\frac{\sigma^2}{n}. \tag{1.7.4}$$

Notice that the fpc is a decreasing function of $n$, that for $n = 1$ the fpc is 1, whereas for $N = n$ it is zero. (Why would you guess that it must be zero?) The fpc is approximately $1 - n/N$, so that if $n$ is small compared to $N$, as it is in many applications, it can safely be ignored in the formulas for $\mathrm{Var}(T)$ and $\mathrm{Var}(\bar{X})$.

***Example 1.7.5*** Consider a class of 12 men and eight women. A simple random sample of five is chosen. Let $X$ be the number of women chosen. What are $E(X)$ and $\mathrm{Var}(X)$? Since $X$ has a hypergeometric distribution, we could first determine $P(X = k)$ for $k = 0, 1, \ldots, 5$, then find $E(X)$ and $\mathrm{Var}(X)$. However, we can take advantage of Example 1.6.4 to make life much easier. Consider the population of measurements: 1 for females, zero for males. Then the mean $\mu = p = 0.4$ is the proportion of females, and the variance is $\sigma^2 = p - p^2 = p(1 - p)$. The number $X$ of women chosen is the sample total ($T$ of Example 1.7.4). It follows that $E(X) = np = 2$ and $\mathrm{Var}(X) = n\sigma^2[(N - n)/(N - 1)] = np(1 - p)[(N - n)/(N - 1)] = 5(0.24)(15/19) = 18/19$, just a bit smaller than the variance of the corresponding binomial distribution, $np(1 - p) = 1.20$. □

We can generalize: Let $X \sim$ hypergeometric with parameters $r$ and $b$ (see Section 1.2), sample size $n$. Let $p = r/(r + b)$. Then $E(X) = np$ and $\mathrm{Var}(X) = np(1 - p)[(N - n)/(N - 1)]$.

It is very useful to consider a measure of the relationship between $X$ and $Y$ which does not depend on their scales. Let $Z_X = (X - \mu_X)/\sigma_X$ and $Z_Y = (Y - \mu_Y)/\sigma_Y$, where $\mu$ and $\sigma$ with corresponding subscripts denote means and standard deviations. (We assume, of course, that $\sigma_X > 0$ and $\sigma_Y > 0$.) $Z_X$ and $Z_Y$ are the *standardized forms* of $X$ and $Y$. They have means zero and variances 1.

**Definition 1.7.2** The *correlation coefficient* between $X$ and $Y$ is $\rho = \rho(X, Y) = E[Z_X Z_Y]$. □

**Properties of $\rho = \rho(X, Y)$**

1. $\rho = \mathrm{Cov}(X, Y)/\sigma_X\sigma_Y$.
2. $\rho(aX + b, cY + d) = \rho(X, Y)\,\mathrm{sign}(ac)$ if $a \neq 0, c \neq 0$.
3. $-1 \leq \rho \leq 1$ with $\rho^2 = 1$ if and only if there exist constants $a, b \neq 0$ such that $P(Y = a + bX) = 1$. In this case, $\rho = 1$ if $b > 0$, $\rho = -1$ if $b < 0$. *Proof:* $0 \leq E(Z_Y - \rho Z_X)^2 = 1 - 2\rho E(Z_Y Z_X) + \rho^2 = 1 - \rho^2$. This proves the inequalities. $\rho^2 = 1$ if and only if $Z_Y - \rho Z_X = 0$ with probability 1, i.e., $P(Y = \mu_Y + \rho(\sigma_Y/\sigma_X)(X - \mu_X)) = 1$. This is equivalent to the existence

of $a, b \neq 0$ such that $P(Y = a + bX) = 1$, since in this case, $b = \sigma_Y/\sigma_X$ if $\rho = 1, b = -\sigma_Y/\sigma_X$ if $\rho = -1$.

## Linear Prediction

Consider first the prediction of $Z_Y$ by a linear predictor $\hat{Z}_Y \equiv g(Z_X) = bZ_X + a$. Consider the error $U = Z_Y - \hat{Z}_Y$. The mean squared error of prediction is $E(U)^2 = E(Z_Y - \hat{Z}_Y)^2 = E(Z_Y^2) + b^2 E(Z_X^2) - 2bE(Z_Y Z_X) + a^2 - 2aE(Z_Y) + 2abE(Z_X) = 1 + b^2 - 2b\rho + a^2 = (b - \rho)^2 + 1 - \rho^2 + a^2$. This is obviously minimum for $b = \rho, a = 0$, so that $\hat{Z}_Y = \rho Z_X = g(Z_X)$. In that case the mean squared error of prediction is $1 - \rho^2$. Thus, $\rho$ is the slope of the least squares prediction line for $Z_Y$ on $Z_X$. This was Galton's original definition of the correlation coefficient (Galton, 1890). Note that $\text{Cov}(U, Z_X) = \text{Cov}(Z_Y, Z_X) - \rho\ \text{cov}(Z_X, Z_X) = \rho - \rho = 0$. Thus, the error of the predictor is uncorrelated with $X$ and therefore with the predictor $\hat{Z}_Y = \rho Z_X$.

Now consider the more general linear predictor $\hat{Y} \equiv h(X) = cX + d$ of $Y$. Let $D = Y - \hat{Y} = \sigma_Y[Z_Y - bZ_X - a]$, where $b = c(\sigma_X/\sigma_Y)$ and $a = (\mu_Y - c\mu_X - d)/\sigma_Y$. It follows that $E(D^2) = \sigma_Y^2 E(U^2)$ is minimum for $b = \rho, c = \rho\sigma_Y/\sigma_X, a = 0, d = \mu_Y - c\mu_X$. Thus, we let $\hat{Y} = h(X) = cX + d = \mu_Y + \rho\sigma_Y Z_X$. The mean squared error of $\hat{Y} = h(X)$ as a predictor of $Y$ is therefore $E(D^2) = \sigma_Y^2(1 - \rho^2)$. From the paragraph above it follows that $\text{Cov}(D, X) = \text{Cov}(D, Z_X)\sigma_Y = 0$. The error is uncorrelated with $X$ and therefore with $\hat{Y}$. Of course, $E(\hat{Y}) = d + c\mu_X = \mu_Y$. $Y = \hat{Y} + (Y - \hat{Y})$ is a decomposition of $Y$ into two parts: the predictable part, $\hat{Y} = h(X)$, and the unpredictable part, $D = Y - \hat{Y}$. Since the two parts have covariance zero, $\text{Var}(Y) = \sigma_Y^2 = \text{Var}(\hat{Y}) + \text{Var}(Y - \hat{Y}) = \rho^2\sigma_Y^2 + (1 - \rho^2)\sigma_Y^2$. $\rho^2 = \text{Var}(\hat{Y})/\text{Var}(Y)$ is the proportion of variation in $Y$ which can be explained by the least squares linear predictor of $Y$.

***Example 1.7.6*** Consider Examples 1.4.3, 1.7.2, and 1.7.4. Since $\text{Var}(R) = 5/9$, $\text{Var}(W) = 1/2$, and $\text{Cov}(R, W) = -1/3$, $\rho = \rho(R, W) = -2/\sqrt{10} = -0.632$. Therefore, the least squares predictor of $W$ as a linear function of $R$ is $\hat{W} = (\mu_W - c\mu_R) + cR$, where $c = \rho\sigma_W/\sigma_R = \text{Cov}(R, W)/\sigma_R^2 = -3/5$. Since $\mu_R = 4/3$ and $\mu_w = 1$, $\hat{W} = (9 - 3R)/5$. $\text{Var}(\hat{W}) = 3^2\sigma_R^2/25 = 1/5 = -\rho^2\sigma_W^2$, and $\text{Var}(W - \hat{W}) = (1 - \rho^2)\sigma_W^2 = 3/10$. The proportion of variation in $W$ that can be explained by a linear function of $R$ is $\rho^2 = 4/10$. □

## Matrix Methods

When computing means for and covariances among several random variables, matrix methods can reduce labor and increase understanding. For that purpose we write a random vector as a row or column vector $\mathbf{X} = (X_1, \ldots, X_k)$. Similarly, we consider random $r \times s$ matrices $\mathbf{W} = (X_{ij})$. By $E(\mathbf{X})$ and $E(\mathbf{W})$ we mean the vector or matrix with the components of $\mathbf{X}$ or $\mathbf{W}$ replaced by their expectations. Thus, for an American box with four, three, and two red, white, and blue balls, with three balls drawn and

$\mathbf{X} = (R, W, B)$, $E(\mathbf{X}) = (4, 3, 2)(3/9) = (4/3, 1, 2/3)$. We can represent the collection of variances and covariances for a random vector conveniently as a matrix.

**Definition 1.7.3** Let $\mathbf{X} = (X_1, \ldots, X_m)$ and $\mathbf{Y} = (Y_1, \ldots, Y_n)$ be random vectors, written as row or column vectors. Suppose that the covariances $\sigma_{ij} = \text{Cov}(X_i, Y_j)$ exist for each $i$ and $j$. The *covariance matrix* for the pair $(\mathbf{X}, \mathbf{Y})$ is the $m \times n$ matrix $\sum$ with $(ij)$th element $\sigma_{ij}$. We write $\text{Cov}(\mathbf{X}, \mathbf{Y}) = \sum = (\sigma_{ij})$. The covariance matrix for a single random vector $\mathbf{X}$ is $\text{Cov}(\mathbf{X}) = \text{Cov}(\mathbf{X}, \mathbf{X})$. □

COMMENTS: In the determination of $\text{Cov}(X, Y)$, it makes no difference whether the random vectors are written as row or columns. Only the order is relevant in the definition of the covariance matrix. In general, $\text{Cov}(X, Y)$ need not be square, although $\text{Cov}(X) = \text{Cov}(X, X)$ must be square and symmetric.

***Example 1.7.7*** The covariance matrix $\text{Cov}(\mathbf{X}, \mathbf{X}) = \text{Cov}(\mathbf{X}) = \sum_{\mathbf{X}}$ for the random vector $\mathbf{X} = (R, W, B)$ determined by three draws from the American box is $(1/18)\begin{pmatrix} 10 & -6 & -4 \\ -6 & 9 & -3 \\ -4 & -3 & 7 \end{pmatrix}$. If $\mathbf{Y} = (X_1, X_3)$, then $\text{Cov}(\mathbf{X}, \mathbf{Y})$ is the $3 \times 2$ matrix consisting of all the rows and the first and third columns of $\sum_{\mathbf{X}}$. Notice that the row sums of $\sum_{\mathbf{X}}$ are each zero, so that $\sum_{\mathbf{X}}$ is singular. Why this must be the case will be evident from the properties of covariance matrices as described below. □

Consider the column vectors $\mathbf{X} = (X_1, \ldots, X_m)^{\text{T}}$ and $\mathbf{Y} = (Y_1, \ldots, Y_n)^{\text{T}}$ (the T indicates the transpose). Let $\mu = E(\mathbf{X})$ and $\nu = E(\mathbf{Y})$. Then the $m \times n$ matrix $(\mathbf{X} - \mu)(\mathbf{Y} - \nu)^{\text{T}}$ has the $(ij)$th element $(X_i - \mu_i)(Y_j - \nu_j)$, so that $\text{Cov}(\mathbf{X}, \mathbf{Y}) = E[(\mathbf{X} - \mu)(\mathbf{Y} - \nu)^{\text{T}}]$. Let $A = (a_{ij})$ and $B = (b_{ij})$ be $r \times m$ and $s \times n$ matrices of constants. Let $\mathbf{U} = A\mathbf{X}$ and $\mathbf{V} = B\mathbf{Y}$. Then from the linearity of expectations, $E(\mathbf{U}) = AE(\mathbf{X}) = A\mu$ and $E(\mathbf{V}) = B\nu$, and $\text{Cov}(\mathbf{X}, \mathbf{Y}) = E[(A\mathbf{X} - A\mu)(B\mathbf{Y} - B\nu)^{\text{T}}] = AE[(\mathbf{X} - \mu)(\mathbf{Y} - \nu)^{\text{T}}]B^{\text{T}}$, again using the linearity of expectations. Thus,

$$\text{Cov}(\mathbf{AX}, \mathbf{BY}) = \mathbf{A}\,\text{Cov}(\mathbf{X}, \mathbf{Y})\mathbf{B}^{\text{T}} \quad \text{and} \quad \text{Cov}(\mathbf{AX}) = \mathbf{A}\,\text{Cov}(\mathbf{X})\mathbf{A}^{T}. \tag{1.7.5}$$

In particular, for $W = \mathbf{aX}$, where $\mathbf{a}$ is an $m$-component row vector of constants, we get $0 \le \text{Var}(W) = \mathbf{a}\sum_{\mathbf{X}}\mathbf{a}^{\text{T}}$, so that every covariance matrix is nonnegative definite, positive definite if $\sum_{\mathbf{X}}$ is nonsingular.

**Definition 1.7.4** The *correlation matrix* for the random vectors $\mathbf{X} = (X_1, \ldots, X_m)$ and $\mathbf{Y} = (Y_1, \ldots, Y_n)$ is the $m \times n$ matrix $\text{Cor}(\mathbf{X}, \mathbf{Y})$ with $(ij)$th element $\rho_{ij} = \rho(X_i, Y_j)$. □

COMMENTS: We can easily determine the correlation matrix from the covariance matrices. Let $D_X$ and $D_Y$ be the diagonal $m \times m$ and $n \times n$ matrices with diagonals the reciprocals of the standard deviations of the components $\mathbf{X}$ and $\mathbf{Y}$. Thus, $D_X$ and $D_Y$ may be obtained from $\text{Cov}(\mathbf{X})$ and $\text{Cov}(\mathbf{Y})$ by replacing the diagonal terms by the reciprocals of the square roots and off-diagonal terms by zero. Then $\text{Cor}(\mathbf{X}, \mathbf{Y}) = D_X \text{Cov}(\mathbf{X}, \mathbf{Y}) D_Y$.

***Example 1.7.8*** Let $\mathbf{X} = (X_1, X_2, X_3, X_4)^{\text{T}}$ have equal variances $\sigma^2$ with zero covariances. Let $T_i = X_1 + \cdots + X_i$ for $i = 1, \ldots, 4$. Let $\mathbf{T} = (T_1, T_2, T_3, T_4)^{\text{T}}$. Then $\mathbf{T} = \mathbf{AX}$, where $A = \begin{pmatrix} 1 & 0 & 0 & 0 \\ 1 & 1 & 0 & 0 \\ 1 & 1 & 1 & 0 \\ 1 & 1 & 1 & 1 \end{pmatrix}$. Since $\text{Cov}(\mathbf{X}) = \sigma^2 I_4$, $\text{Cov}(\mathbf{T}) = \mathbf{A}(\sigma^2 I_4)\mathbf{A}^{\text{T}} = \sigma^2 \begin{pmatrix} 1 & 1 & 1 & 1 \\ 1 & 2 & 2 & 2 \\ 1 & 2 & 3 & 3 \\ 1 & 2 & 3 & 4 \end{pmatrix} = \sigma^2(\min(i, j))$. Since $\text{Var}(T_i) = i\sigma^2$, $\text{Cor}(\mathbf{T}) = (\rho_{ij})$, where $\rho_{ij} = \rho(T_i, T_j) = \sqrt{\min(i, j)/\max(i, j)}$. □

Let $\mathbf{X}_1$ and $\mathbf{X}_2$ be $n$-component column random vectors with mean vectors $\boldsymbol{\mu}_1, \boldsymbol{\mu}_2$. Let $\mathbf{X}_1^* = \mathbf{X}_1 - \boldsymbol{\mu}_1$, $\mathbf{X}_2^* = \mathbf{X}_2 - \boldsymbol{\mu}_2$. Then $\text{Cov}(\mathbf{X}_1 + \mathbf{X}_2) = E((\mathbf{X}_1^* + \mathbf{X}_2^*)(\mathbf{X}_1^* + \mathbf{X}_2^*)^T) = E(\mathbf{X}_1^*\mathbf{X}_1^{*\text{T}}) + E(\mathbf{X}_2^*\mathbf{X}_2^{*\text{T}}) + E(\mathbf{X}_1^*\mathbf{X}_2^{*\text{T}}) + E(\mathbf{X}_2^{*\text{T}}\mathbf{X}_1^{*\text{T}}) = \text{Cov}(\mathbf{X}_1) + \text{Cov}(\mathbf{X}_2) + \text{Cov}(\mathbf{X}_1, \mathbf{X}_2) + \text{Cov}(\mathbf{X}_2, \mathbf{X}_1)$. It is tempting to think, but not true in general, that these last two terms must be equal. Using induction, we conclude that

$$\text{Cov}(\mathbf{X}_1 + \cdots + \mathbf{X}_k) = \sum_{i,j} \text{Cov}(\mathbf{X}_i, \mathbf{X}_j) = \sum_i \text{Cov}(\mathbf{X}_i) + \sum_{i \neq j} \text{Cov}(\mathbf{X}_i, \mathbf{X}_j). \tag{1.7.6}$$

In particular, if $\mathbf{X}_1, \ldots, \mathbf{X}_k$ have zero covariances, as they do when they are independent,

$$\text{Cov}(\mathbf{X}_1 + \cdots + \mathbf{X}_k) = \sum_i \text{Cov}(\mathbf{X}_i), \tag{1.7.7}$$

and if each $\mathbf{X}_i$ has covariance matrix $\sum$,

$$\text{Cov}(X_1 + \cdots + X_k) = k \sum \quad \text{and} \quad \text{Cov}\left(\frac{\mathbf{X}_1 + \cdots + \mathbf{X}_k}{k}\right) = (1/k)\sum. \tag{1.7.8}$$

**Problems for Section 1.7**

**1.7.1** From a box with four tickets numbered 1, 2, 3, 4, two tickets are drawn randomly without replacement. Let $X$ and $Y$ be the minimum and maximum of the numbers on the tickets drawn. Find:

**(a)** The joint probability function for $(X, Y)$.

**(b)** $E(X)$, $E(Y)$, $\text{Var}(X)$, $\text{Var}(Y)$, $\text{Cov}(X, Y)$, $\rho(X, Y)$.

**(c)** The least squares predictor $\hat{Y}$ of $Y$.

**(d)** $\text{Var}(\hat{Y})$ and $\text{Var}(Y - \hat{Y})$. Show that $\text{Cov}(\hat{Y}, Y - \hat{Y}) = 0$ and that $\text{Var}(\hat{Y}) + \text{Var}(Y - \hat{Y}) = \text{Var}(Y)$.

**1.7.2** Consider the set $A = \{1, 3, 5, 7\}$. A simple random sample of $n = 2$ is chosen from the population $A$. Let $T$ and $\bar{X}$ be the sample total and mean.

**(a)** Determine the mean $\mu$ and variance $\sigma^2$ of $A$.

**(b)** Find the probability function for $T$ and use it to find $E(T)$, $\text{Var}(T)$, $E(\bar{X})$, and $\text{Var}(\bar{X})$.

**(c)** Verify that $E(T) = n\mu$, $E(\bar{X}) = \mu$, $\text{Var}(T) = n\sigma^2[(N - n)/(N - 1)]$, $\text{Var}(\bar{X}) = (\sigma^2/n)[(N - n)/(N - 1)]$.

**1.7.3** Let $X_1, X_2, \ldots, X_n$ be uncorrelated random variables, all with variance $\sigma^2$. For $1 \le k \le n$, let $T_k = X_1 + \cdots + X_k$. Find $\rho(T_k, T_n)$.

**1.7.4** Let $X$ and $Y$ be indicator random variables, with $P(X = 1) = P(Y = 1) = 1/2$ and $P(X = 1, Y = 1) = \theta$ for $0 \le \theta \le 1/2$. Express $\rho(X, Y)$ as a function of $\theta$. Plot it. Use the relationship between $\rho$ and $\theta$ to show that $\rho = 0$ if any only if $X$ and $Y$ are independent.

**1.7.5** Let $X_1, \ldots, X_n$ be random variables with the same variance $\sigma^2$ and equal correlations $\rho(X_i, X_j) = \rho$ for $i \ne j$. Let $T = X_1 + \cdots + X_n$.

**(a)** Express $\text{Var}(T)$ and $\text{Var}(T/n)$ as functions of $\sigma^2$ and $\rho$. Find $\lim_{n\to\infty} \text{Var}(T/n)$.

**(b)** Use the fact that $\text{Var}(T) \ge 0$ to give a lower bound on $\rho$.

**(c)** This model holds, for example, if the $X_i$ are $n$ scores on repeated attempts on an exam of the same type by a randomly chosen student, where $X_i = A + \varepsilon_i$ and $A, \varepsilon_1, \ldots, \varepsilon_n$ are uncorrelated random variables, $A$ is a measure of ability of the student, $\text{Var}(A) = \sigma_A^2$ and $\text{Var}(\varepsilon_i) = \sigma_\varepsilon^2$. Express $\rho(X_i, X_j)$ in terms of $\sigma_A^2$ and $\sigma_\varepsilon^2$.

**1.7.6** Let $(x_i, y_i)$ for $i = 1, \ldots, n$ be $n$ pairs of numbers. Let $P((X, Y) = (x_i, y_i)) = 1/n$ for $i = 1, \ldots, n$.

**(a)** Determine simple formulas for the least squares linear predictor $\hat{Y}$ of $Y$ and for $\rho(X, Y)$. For simplicity of notation, define $\bar{x}$, $\bar{y}$, the means, $S_{xx} = \sum_{i=1}^n (x_i - \bar{x})^2$, $S_{yy} = \sum_{i=1}^n (y_i - \bar{y})^2$, $S_{xy} = \sum_{i=1}^n (x_i - \bar{x})(y_i - \bar{y})$.

**(b)** Apply the formulas for the three pairs $(1, 5)$, $(2, 6)$, $(3, 1)$. Plot these points and the least squares line $y = g(x) = a + bx$. What is the proportion of variation in $y$ explained by linear regression on $x$? Determine $\rho(X, Y)$.

CHAPTER TWO

# Special Discrete Distributions

## 2.1 INTRODUCTION

In Chapter One we discussed the properties of discrete random variables in a rather general way, although we did define the binomial, geometric, and uniform random variables. Some random variables and their distributions arise so often that it becomes useful to refer to them by name, and we should try to summarize their properties. In addition, under certain conditions, limit arguments allow us to approximate some probability distributions by simpler distributions. We begin with what is almost certainly the most useful discrete probability distribution, the binomial.

## 2.2 THE BINOMIAL DISTRIBUTION

Definition 1.4.4 is important enough to repeat:

**Definition 2.2.1** Suppose that $n$ independent experiments are performed. For each experiment (called a *trial*) the probability of success is $p$, where $0 \le p \le 1$. Let $X$ be the number of successes. Then $X$ is said to have the *binomial distribution* with parameters $n$ and $p$. We will use the notation $\mathbf{X} \sim B(n, p)$ to indicate this. If $n = 1$, then $X$ is said to have the *Bernoulli distribution* with parameter $p$. □

In Section 1.4 we showed that if $X$ has the binomial distribution, its probability function is

$$b(k; n, p) = P(X = k) = \binom{n}{k} p^k (1-p)^{n-k} \qquad \text{for } k = 0, 1, \ldots, n.$$

In general, $b(k; n, p) = b(k)$ is an increasing function of $k$ for $k < M$, and a decreasing function for $k > M$, where $M$ is a point of maximum. As we will see, $M$ is "near" $E(X) = np$.

---

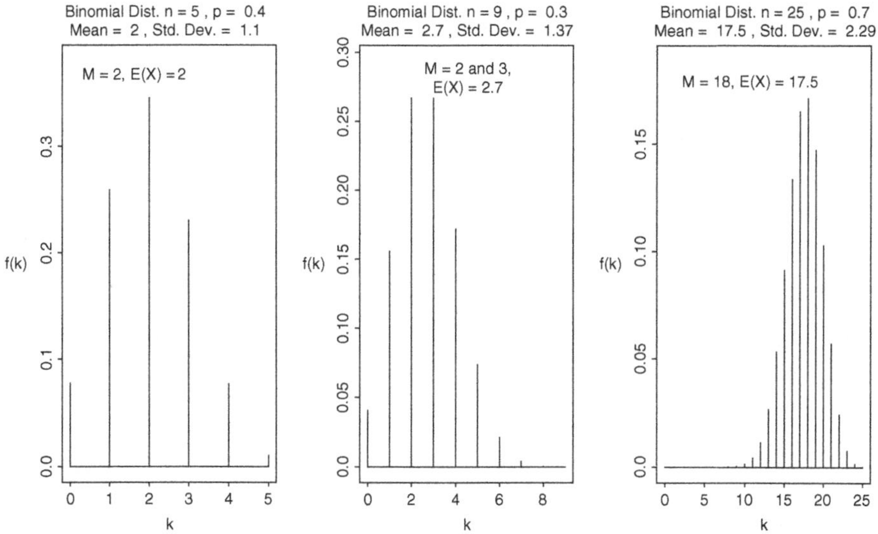


**FIGURE 2.2.1** Binomial mass functions.

To determine $M$, and for computational purposes, consider $r(k) = b(k)/b(k-1)$ for $k = 1, \ldots, n$. (Why can't we use calculus to find $M$?) Fortunately, $r(k)$ turns out to be a reasonably simple function:

$$r(k) = \frac{n!}{k!(n-k)!}\left[\frac{n!}{(k-1)!(n-k+1)!}\right]^{-1}\frac{p}{q} = \frac{(n-k+1)}{k}\frac{p}{q} \text{ where } q = 1-p. \tag{2.2.1}$$

Since $b(k) = b(k-1)r(k)$, a simple loop on $k$ can be used to determine $b(k)$ for all $k$. For $n = 5$, $p = 0.4$, for example, $b(0) = 0.6^5 = 0.0778$, $b(1) = b(0)[5/1](2/3) = 0.259$, $b(2) = b(1)[4/2](2/3) = 0.346$, $b(3) = b(2)[3/3](2/3) = 0.234$, $b(4) = b(3)(2/4)(2/3) = 0.077$, $b(5) = b(4)[1/5](2/3) = 0.4^5 = 0.010$. Notice that in this case, $M = 2 = np = E(X)$. From (2.2.1), $b(k) \geq b(k-1)$ if and only if $r(k) \geq 1$, equivalently, $k \leq (n+1)p$. Thus, $M$ is the largest integer less than or equal to $(n+1)p$. In the case that $(n+1)p$ is an integer $M$, the maximum is taken at both $M$ and $M - 1$ (see Figure 2.2.1).

Appendix Table 1 presents cumulative probabilities $B(k; n, p) = P(X \leq k) = \sum_{j=0}^{k} b(j; n, p)$ for $p = 0.05, 0.10, 0.20, \ldots, 0.90, 0.95$ and $n = 5, 10, 15, 20, 25$. Exact probabilities can be determined from $b(k; n, p) = B(k; n, p) - B(k-1; n, p)$. For integers $a < b$, $P(a < X \leq b) = B(b; n, p) - B(a; n, p)$. Taking $b = n$, we get $P(a < X) = 1 - B(a; n, p)$.

In Sections 1.5 and 1.6 we showed that when $\mathbf{X} \sim B(n, p)$ and the representation $X = I_1 + \cdots + I_n$, $E(X) = np$. Similarly, in Section 1.6 we showed that $\text{Var}(X) = np(1-p)$.

## Problems for Section 2.2

**2.2.1** A student tosses three coins five times. Find the probability that she obtains

**(a)** Exactly two heads exactly three of the five times.

**(b)** At least two heads exactly three of the five times.

**(c)** Exactly two heads at least three of the five times.

**(d)** At least two heads at least three of the five times.

**(e)** Let $X$ be the number of times for which exactly two heads appear. Find $k$ so that $P(X = k)$ is maximized.

**2.2.2** Let $M_p$ maximize $b(k; n, p)$. Find $M_p$ for the following $(n, p)$ pairs. If $M_p - 1$ is also a maximum, say so.

**(a)** $n = 5, p = 0.4$.

**(b)** $n = 5, p = 0.15$.

**(c)** $n = 9, p = 0.3$.

**(d)** $n = 100, p = 0.4$.

**(e)** For which values of $p$ does $k = 2$ maximize $b(k; 5, p)$?

**2.2.3** Let $X$ have the binomial distribution with parameters $n = 15$ and $p = 0.3$. Use Appendix Table 1 to find:

**(a)** $P(X \leq 5)$.

**(b)** $P(3 \leq X \leq 5)$.

**(c)** $P(X = k)$ for $k = 3, 4, 5$.

**(d)** $P(X \geq 7)$.

**2.2.4** **(a)** Prove that $\sum_{j=0}^{k} \binom{m}{j}\binom{n}{k-j} = \binom{m+n}{k}$ for $0 \leq k \leq m + n$. *Hint:* If sets $A$ and $B$ have $m$ and $n$ elements, each subset of size $k$ from $A \cup B$ must have $j$ elements from $A$, $k - j$ from $B$, for some $j$.

**(b)** Prove that $\sum_{j=0}^{m} b(j; m, p)b(k - j; n, p) = b(k; m + n, p)$ for positive integers $m$ and $n$, and $0 \leq p \leq 1$. Equivalently, if $X \sim B(m, p)$ and $Y \sim B(n, p)$ with $X, Y$ independent, then $X + Y \sim B(m + n, p)$. *Hint:* Use either part (a) or a probability argument.

**2.2.5** Let $X$ have the binomial distribution with parameters $n$ and $p$. Conditionally on $X = k$, let $Y$ have the binomial distribution with parameters $k$ and $r$. What is the marginal distribution of $Y$? *Hint:* Two approaches are possible. The intuitively more satisfying approach considers $n$ Bernoulli trials with a two-stage process on each trial. (That's all the hint we will provide.) The second involves working directly with factorials and summation signs.

**2.2.6** Let $X$ be the number of heads in 10 tosses of a coin. What is the conditional distribution of the number of heads $Y$ among the first six tosses given that

$X = k, 0 \le k \le n$? Generalize your result to the case that $X$ is the number of successes in $n$ independent Bernoulli trials, each with probability $p$ of success, $0 < p < 1$, and $Y$ is the number among the first $m < n$ trials. Find $P(Y = j|X = k)$ for $0 \le j \le k \le n$.

**2.2.7** Let $X_n \sim B(n, p)$ and $\hat{p}_n = X_n/n$, a sample proportion of successes, for $0 < p < 1$. Show that $\lim_{n\to\infty} P(|\hat{p}_n - p| \le \varepsilon) = 1$ for every $\varepsilon > 0$. That is, $\{\hat{p}_n\}$ converges in probability to $p$. *Hint:* Use the Chebychev inequality.

**2.2.8** Let $X_1 \sim B(n_1, p_1)$, $X_2 \sim B(n_2, p_2)$, with $X_1$ and $X_2$ independent. Let $N = n_1 + n_2$ and define $p = (n_1/N)p_1 + (n_2/N)p_2$. Let $T = X_1 + X_2$. Suppose that $Y \sim B(N, p)$.

**(a)** For $n_1 = 2, n_2 = 1, p_1 = 0.4, p_2 = 0.7$, compare the mass functions of $T$ and $Y$.

**(b)** Show that $E(T) = E(Y)$.

**(c)** Show that $\text{Var}(Y) = \text{Var}(T) + (n_1 n_2/N)(p_1 - p_2)^2$, so that the distribution of $T$ cannot be binomial unless $p_1 = p_2$. *Hint:* Express Var($Y$) and Var($T$) in terms of $n_1, n_2, p_1, p_2, q_1 = 1 - p_1$, and $q_2 = 1 - p_2$.

**2.2.9** Two teams $A$ and $B$ play consecutive games until one of the two teams wins $k$ games, for $k > 0$. That team is said to have won the *k-game playoff*. $A$ wins each game with probability $p$, $0 < p < 1$, and the outcomes are independent. Let $G_k(p)$ be the probability that $A$ wins a $k$-game playoff. Then, for example, $G_1(p) = p$, $G_2(p) = p^2 + 2p^2q$, where $q = 1 - p$. Express $G_3(p)$ and $G_4(p)$ in terms of $p$ and $q$, and plot them on a graph with $G_1(p)$ and $G_2(p)$. Check to see that $G_k(0) = 0$, $G_k(1/2) = 1/2$, and $G_k(1) = 1$. It should also be true that $G_k(p) > G_{k+1}(p)$ for $p < 1/2$ and $G_k(p) < G_{k+1}(p)$ for $p > 1/2$.

## 2.3 THE HYPERGEOMETRIC DISTRIBUTION

**Definition 2.3.1** Let $X$ be the number of red balls in a random sample of $n$ balls taken without replacement from a box of $R$ red and $B$ black balls. Then $X$ is said to have the *hypergeometric distribution* with parameters $R$, $B$, and $n$. □

Before we present a formula for the probability mass function for $X$ for the general case, let us determine it for a special case.

***Example 2.3.1*** A committee of five people is to be chosen at random from a club consisting of seven women and six men. Let $X$ denote the number of women chosen. Let the sample space $S$ consist of all subsets of size 5 from the set of 13 people. Then $N(S) = \binom{13}{5} = 1287$. The number of these subsets for which $X = 3$ is $\binom{7}{3}\binom{6}{2} = (35)(15) = 525$, so that $P(X = 3) = 0.4079$. Similarly, $P(X = 2) = \binom{7}{2}\binom{6}{3}\Big/\binom{13}{5} = (21)(20)/1287 = 0.3263$, and for $f(k) = P(X = k)$, $f(0) = 0.0040$, $f(1) = 0.0816$, $f(4) = 0.1632$, $f(5) = 0.0163$. □

More generally, for $X$ as given in Definition 2.3.1,

$$P(X = k) = \binom{R}{k}\binom{B}{n-k}\Big/\binom{R+B}{n} \qquad \text{for } k = 0, 1, \ldots, n. \tag{2.3.1}$$

Note that in the case that $k > R$ or $n - k > B$, the corresponding combinatoric in the numerator is zero. It is interesting to note that for $N = R + B$,

$$P(X = k) = \binom{n}{k}\binom{N-n}{R-k}\Big/\binom{N}{R} \qquad \text{for } k = 0, 1, \ldots, R. \tag{2.3.2}$$

That is, we can consider the $n$ sampled balls as fixed and the $R$ balls colored red chosen randomly. Equation (2.3.2) follows directly from (2.3.1) by expressing the right side in terms of factorials.

Letting $f(k) = P(X = k)$ and $r(k) = f(k)/f(k-1)$ for $k \geq 1$, $r(k) = [(R - k + 1)(n - k + 1)]/[k(B - n + k)]$. $r(k)$ is a decreasing function of $k$, and, since $r(k) \geq 1$ for $k \leq M \equiv$ {greatest integer $\leq [(n+1)(R+1)/(R+B+2)]$}, $f(k)$ is maximum for $k = M$ (see Problem 2.3.2 for details).

Letting $I_j$ be the indicator of the event [$j$th ball chosen is red], we can write $X$ as in the definition in the form $X = I_1 + I_2 + \cdots + I_n$. Since $E(I_j) = P$($j$th ball chosen is red) $= R/(R + B) \equiv p$, we get $E(X) = np = n[R/(R + B)]$. Since these indicator rv's are dependent, Var($X$) is not the sum of the variances of the $I_j$. From Section 1.7, $\text{Var}(X) = np(1 - p)[(N - n)/(N - 1)]$. Since the factor in brackets (the *finite correction factor*) is less than 1 for $n > 1$, Var($X$) is smaller than it would be for the case of sampling with replacement, in which case $X$ would have the binomial distribution with parameters $n$ and $p$. In many practical cases $N \gg n$, so the finite correction factor is close to 1. In fact, as we will show soon, the hypergeometric and binomial distributions are close for $N \gg n$. To convince the reader that hypergeometric distributions become closer and closer to a binomial as $p = R/(R + B)$ stays fixed but $R$ and $B$ increase, consider Table 2.3.1 for a sample size $n = 5$ taken from a box with $(R, B)$ numbers of red and black balls. In general, for smaller values of $R$ and $B$ the distribution of $X$ is less spread out, with smaller variance, as indicated by the finite correction factor. This should be expected intuitively, since an early excess of red balls, for example, makes it less likely that balls sampled later will be red.

**TABLE 2.3.1** $f(k)$ **for the Pair** $(R, B)$ **and** $b(k; 5, 0.6)$

| $k$ | (6,4) | (60,40) | (600,400) | (6000,4000) | $b(k; 5, 0.6)$ |
|---|---|---|---|---|---|
| 0 | 0.0000 | 0.0087 | 0.0101 | 0.0102 | 0.0102 |
| 1 | 0.0238 | 0.0728 | 0.0764 | 0.0768 | 0.0768 |
| 2 | 0.2381 | 0.2323 | 0.2306 | 0.2304 | 0.2304 |
| 3 | 0.4762 | 0.3545 | 0.3465 | 0.3457 | 0.3456 |
| 4 | 0.2381 | 0.2591 | 0.2592 | 0.2592 | 0.2592 |
| 5 | 0.0238 | 0.0725 | 0.0772 | 0.0777 | 0.0778 |

We now state and prove the convergence of the hypergeometric to the binomial. Let $(R_N, B_N = N - R_N)$ for $N = 1, 2, \ldots$ be a sequence of pairs of integers such that $\lim_{N\to\infty} R_N/N = p$, where $0 \le p \le 1$. Let $f(k; N) = \binom{R_N}{k}\binom{B_N}{n-k}\Big/\binom{N}{n}$.

Then

$$\lim_{N\to\infty} f(k; N) = b(k; n, p) \qquad \text{for all } 0 \le k \le n. \tag{2.3.3}$$

To show this, write $f(k; N)$ as the product of four factors: $F_{1N} = R_N(R_N - 1)\cdots(R_N - k + 1)/N^k$, $F_{2N} = B_N(B_N - 1)\cdots(B_N - n + k + 1)/N^{n-k}$, $F_{3N} = N(N - 1)\cdots N - n + 1)/N^n$, and $F_4 = \binom{n}{k}$. Then $\lim_{N\to\infty} F_{1N} = p^k$, $\lim_{N\to\infty} F_{2N} = (1 - p)^{n-k}$, $\lim_{N\to\infty} F_{3N} = 1$, and $\lim_{N\to\infty} F_4 = F_4$, so (2.3.3) follows.

## Problems for Section 2.3

**2.3.1** From a jury panel of nine men and six women, a sample of five jurors is selected randomly. Let $X$ be the number of men among the five chosen.

**(a)** Find the probability function for $X$ and compare it to the approximating binomial with $n = 5$ and the value of $p$ suggested by (2.3.3). (The binomial probabilities may be determined from the binomial tables.)

**(b)** Find $E(X)$ and $\text{Var}(X)$ using the result of part (a) and verify the formulas given for $E(X)$ and $\text{Var}(X)$.

**2.3.2** Let $X$ be as defined in Definition 2.3.1, let $f(k)$ be its probability mass function, and let $r(k) = f(k)/f(k-1)$.

**(a)** Verify the formula for $r(k)$ below equation (2.3.2) and use this to show that $f$ is maximized for $M$ as given there.

**(b)** Suppose that $N$ is known to be 400, but $R$ and $B$ are unknown. $X$ is observed. Define an estimator $\hat{R}$ of $R$ that has the property $E(\hat{R}) = R$, and determine $\text{Var}(\hat{R})$ as a function of $n$, $R$, and $B$. For $n = 100$, $X = 43$, estimate $R$ and its variance. Denote this estimate of $\text{Var}(\hat{R})$ by $\widehat{\text{Var}}(\hat{R})$. An approximate 95% *confidence interval* (in a sense to be given later) is given by $[\hat{R} \pm 1.96\sqrt{\widehat{\text{Var}}(\hat{R})}]$. Determine this interval.

**2.3.3** Demonstrate the convergence of the hypergeometric mass function to the binomial by considering $n = 3$, and the $(R, B)$ pairs (4, 6), (10, 15), and (40, 60).

## 2.4 THE GEOMETRIC AND NEGATIVE BINOMIAL DISTRIBUTIONS

**Definition 2.4.1** Let $r$ be a positive integer, and let $0 < p \leq 1$. Let $Y_r$ be the number of independent Bernoulli trials, each with probability $p$ of success, necessary to have $r$ successes. Then $Y_r$ is said to have the *negative binomial distribution* with parameters $r$ and $p$. $Y_1$ is said to have the *geometric distribution* with parameter $p$. □

*Note:* Some authors define a negative binomial random variable as the number $W_r$ of failures before the $r$th success. Thus, $Y_r = W_r + r$. Readers of other material should be careful to note which definition is employed. S-Plus uses this second definition. □

***Example 2.4.1*** Suppose that a six-sided fair die is tossed consecutively until three 6's have occurred. Let $Y_3$ be the number of tosses necessary. Then $Y_3$ has the negative binomial distribution with parameters $r = 3$ and $p = 1/6$. $Y_3$ was observed through computer simulation 10,000 times (see Figures 2.4.1 and 2.4.2). The sample mean and variance were 18.001 and 90.0024. The first 100 of these were

| | | | | | | | | | | | | | | | | | | | |
|---|---|---|---|---|---|---|---|---|---|---|---|---|---|---|---|---|---|---|---|
| 8 | 38 | 17 | 14 | 13 | 33 | 28 | 33 | 18 | 13 | 7 | 20 | 40 | 14 | 7 | 38 | 6 | 9 | 13 | 12 |
| 32 | 7 | 19 | 28 | 12 | 42 | 22 | 6 | 22 | 20 | 17 | 25 | 25 | 16 | 11 | 16 | 7 | 15 | 8 | 11 |
| 34 | 31 | 13 | 16 | 9 | 24 | 18 | 36 | 45 | 12 | 52 | 19 | 33 | 24 | 11 | 17 | 23 | 15 | 28 | 23 |
| 29 | 9 | 22 | 11 | 33 | 14 | 10 | 4 | 21 | 22 | 38 | 20 | 16 | 7 | 26 | 8 | 28 | 21 | 14 | 8 |
| 23 | 17 | 10 | 10 | 17 | 21 | 11 | 11 | 15 | 17 | 10 | 16 | 18 | 20 | 11 | 14 | 40 | 25 | 14 | 11 |

□

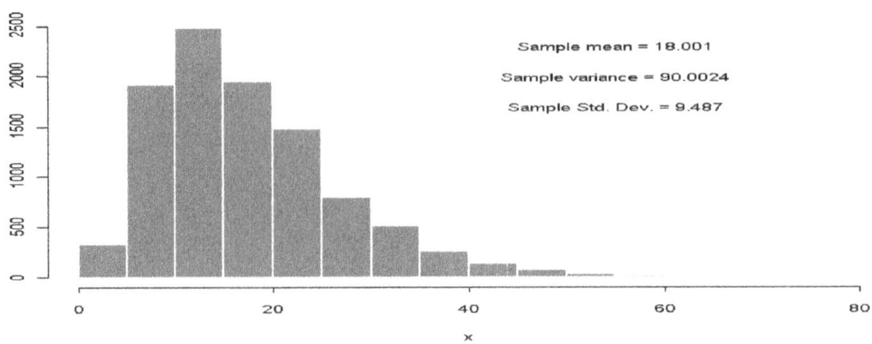


**FIGURE 2.4.1** Histogram for 10,000 observations from a negative binomial distribution.

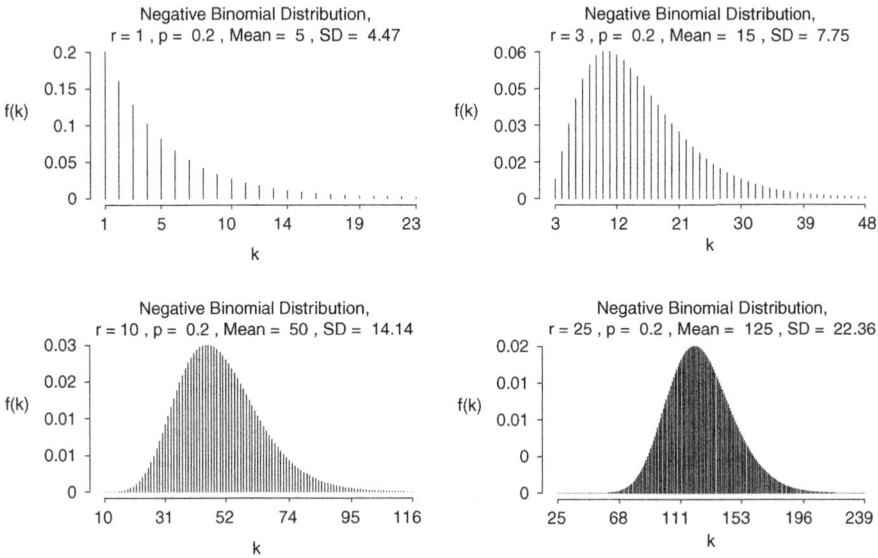


**FIGURE 2.4.2** Negative binomial mass functions.

$Y_r$ can be represented as a sum as follows. Let $Y_0$ be identically zero, and for $1 \le k \le r$ let $D_k = Y_k - Y_{k-1}$. Then $D_1, D_2, \ldots, D_r$ depend on disjoint subsets of trials, are therefore independent, and $Y_r = D_1 + \cdots + D_r$. Each $D_k$ has the geometric distribution with parameter $p$, so that from Example 1.5.3, $E(D_k) = 1/p$. Hence, $E(Y_r) = r/p$.

The representation of $Y_r$ as a sum enables us to find $\text{Var}(Y_r)$. First we need $\text{Var}(D)$, where $D$ has the geometric distribution with parameter $p$. Letting $q = 1 - p$. We begin with the infinite series $1 + q + q^2 + q^3 + \cdots = 1/(1-q)$. The series converges for $-1 < q < 1$, although in our case, $0 \le q < 1$. Differentiating with respect to $q$ on both sides twice, we get $2(1) + 3(2)q + 4(3)q^2 + \cdots = 2/(1-q)^3$. Hence, $E[D(D-1)] = 1(0)p + 2(1)qp + 3(2)q^2p + \cdots = 2qp/(1-q)^3 = 2q/p^2$. It follows that $E(D^2) = 2q/p^2 + E(D) = (2q+p)/p^2$ and $\text{Var}(D) = (2q + p - 1)/p^2 = q/p^2$. Finally, $\text{Var}(Y_r) = rq/p^2$. Notice that for $r = 3$, $p = 1/6$, we get $\text{Var}(Y_3) = 90$, consistent with the result of the simulation.

We can determine an explicit formula for the probability mass function for $Y_r$ as follows. For $k$ an integer, $k \ge r$, let $X$ be the number of successes among the first $(k-1)$ trials. Let $[Y_r = k] = [X = (k-1)] \cap [k\text{th trial is a success}]$. $X$ has the binomial distribution with parameters $(k-1)$ and $p$, so that the first event has probability $\binom{k-1}{r-1} p^{r-1} q^{k-r}$. The second has probability $p$, and the two events are independent. Thus,

$$P(Y_r = k) = \binom{k-1}{r-1} p^r q^{k-r} \qquad \text{for } k = r, r+1, \ldots \tag{2.4.1}$$

Of course, these probabilities must sum to 1. It is not obvious using the right side, $f(k)$, alone. We can do that by extending the definition of the combinatoric symbol to all real numbers:

$$\binom{x}{k} = [x(x-1)\cdots(x-k+1)/k! \qquad \text{for } k = 0, 1, \ldots.$$

For example, $\binom{-1}{3} = (-1)(-2)(-3)/3! = -1$ and $\binom{-1/2}{k} = (-1/2)(-3/2)$ $(-5/2)\cdots[-(2k-1)/2)]/k! = (-1)^k \binom{2k}{k}/2^{2k}$. The last equality is obtained by multiplying and dividing by $(2)(4)\cdots(2k) = 2^k k!$. Finally, we will need, for any $r > 0$, nonnegative integer $j$,

$$\binom{-r}{j} = (-1)^j \binom{r+j-1}{j}. \tag{2.4.2}$$

Exploiting this, we can write $P(W_r = j) = P(Y_r = j + r) = \binom{-r}{j}(-1)^j p^r q^j$ for $r > 0$ and $j = 0, 1, \ldots$. This allows the extension of the negative binomial distribution to the case that $r > 0$, not necessarily an integer. The distribution can then, for example, serve as a model for the number of tomato worms on tomato plants.

We are now in position to exploit *Newton's formula*, which we give without proof. $(1+a)^x = 1 + \binom{x}{1}a^1 + \binom{x}{2}a^2 + \cdots$ for any $x$ and $a$ for which the series converges. For a proof, read a good calculus book.

Summing the negative binomial probabilities $f(k)$, letting $j = k - r$, we get $f(r) + f(r+1) + \cdots = \sum_{j=0}^{\infty} \binom{j+r-1}{j} p^r q^j = p^r \sum_{j=0}^{\infty} \binom{-r}{j}(-1)^j q^j = p^r(1-q)^{-r} = p^r/p^r = 1$.

The word *negative* in the definition could be taken to refer to this way of expressing the combinatoric symbol. However, "negative" in this context has the meaning "opposite," in that we wait until $r$ successes have occurred, so that the number of successes is fixed rather than having a fixed number of trials as in the binomial, with the number of successes random.

Many interesting questions concerning $Y_r$ can be stated in terms of "tail" probabilities of the form $P(Y_r > k)$ or $P(Y_r \leq k)$. Except for the case $r = 1$, these probabilities can be expressed only as sums. For $r = 1$, $P(Y_1 > k) = P(\text{no successes in } k \text{ trials}) = q^k$, for $k = 0, 1, 2, \ldots$.

For $r$ an integer greater than 1, $P(Y_r > k) = 1 - \sum_{j=0}^{k} f(j)$, of course, but the computations involved may be heavy. For the case that $r$ is relatively small, we can take advantage of the relationship between the negative binomial and binomial tail probabilities. Let $X_k$ be the number of successes among the first $k$ trials. Then $[Y_r > k] = [X_k < r]$, so that $P(Y_r > k) = P(X_k < r)$, which may be determined by summing binomial probabilities, from tables, or using the normal approximation (discussed later).

***Example 2.4.2*** A telephone marketer is successful in getting a magazine sale in 10% of all cases in which telephone numbers are chosen randomly. What is the probability of the event $A$ that the third sale occurs before the twentieth call?

Let $Y_3$ be the number of calls necessary to make three sales. Then $Y_3$ has the negative binomial distribution with parameters $r = 3$, $p = 0.10$. Thus, $A = [Y_3 < 20]$ and $P(A) = \sum_{k=3}^{19} \binom{k-1}{r-1} p^r q^{k-r}$. However, $A = [X_{19} \geq 3]$, where $X_{19}$ is the number of successes among the first 19 trials, and since $X_{19} \sim B(19, 0.10)$, $P(A) = 1 - \sum_{k=0}^{3} b(k; 19, 0.10) = 1 - (0.135 + 0.285 + 0.285) = 1 - 0.705 = 0.295$. □

## Problems for Section 2.4

**2.4.1** Let $Y_3$ be the number of tosses of two coins necessary to get three H–H's.

**(a)** Simulate an observation on $Y_3$, more if you have more coins or a computer. Find:

**(b)** $P(Y_3 = 5)$.

**(c)** $P(Y_3 \leq 4)$ using the negative binomial probability mass function.

**(d)** $P(Y_3 \leq 4)$ using the binomial probability mass function.

**(e)** $E(Y_3)$ and $\text{Var}(Y_3)$.

**2.4.2** Let $Y_3$ and $Y_5$ be the numbers of trials necessary to get three, then five, 6's in consecutive tosses of a six-sided fair die.

**(a)** Find $P(Y_5 - Y_3 > 10)$. *Hint:* You need to add just two probabilities.

**(b)** Find $P(Y_5 = 12 \mid Y_3 = 5)$.

**2.4.3** Recall that $W_r = Y_r - r$ is the number of failures before the $r$th success. Use (2.4.2), Newton's formula, and differentiation of $(1 - q)^{-r}$ with respect to $q$ to show that $E(W_r) = rq/p$, $E(Y_r) = r/p$. Differentiate twice to get $E[W_r(W_r - 1)]$, then determine $\text{Var}(W_r) = \text{Var}(Y_r)$.

**2.4.4** Let $X$ and $Y$ be independent, each with a negative binomial distribution, with parameters $(r_1, p)$ for $X$ and $(r_2, p)$ for $Y$. Let $W = X + Y$.

**(a)** What is the distribution of $W$?

**(b)** Find $P(X = x \mid W = w)$ for $r_1 \leq x \leq w < \infty$.

**(c)** Let $M = \min(X, Y)$, with $r_1 = r_2 = 1$. What is the probability mass function for $M$? *Hint:* First find $P(M > m)$ for $m = 1, 2, \ldots$.

**2.4.5** Let $X$ and $Y$ be independent, with geometric distributions with parameters $p_1$ and $p_2$.

**(a)** Give a simple expression for $P(X = Y)$. In 1000 observations on the pair $(X, Y)$ for the case $p_1 = p_2 = 1/6$, the event $[X = Y]$ occurred 91 times. For $p_1 = 1/6$, $p_2 = 1/2$, the event occurred 141 times.

**(b)** Give a simple expression for $P(X > Y)$. Among 1000 independent observations of $(X, Y)$ for $p_1 = 1/6$, $p_2 = 1/6$, the event $[X > Y]$ occurred 482 times. For $p_1 = 1/6$, $p_2 = 1/2$, it occurred 724 times.

**2.4.6** Four hundred independent observations $X_i$ were taken from a negative binomial distribution, but $r$ and $p$ are unknown. (The author knows but won't tell you.) The first 50 observations were

23 15 22 11 6 24 22 7 10 12 14 31 17 27 17 4 9 15 25
7 13 21 16 23 18 26 26 8 3 12 23 10 7 27 15 18 3 7 11
13 16 35 12 14 20 31 23 6 22 7

For the 400 observations the sample mean and variance were 14.94 and 58.89. Estimate $r$ and $p$. *Hint:* Express $r$ and $p$ as functions of the mean $\mu$ and variance $\sigma^2$ of the $X_i$ . Solve for $r$ and $p$ in terms of $\mu$ and $\sigma^2$. Replace r and p by their estimates from the data to obtain a pair of estimates $(\hat{r}, \hat{p})$. The pair $(\hat{r}, \hat{p})$ is known as the *method of moments estimate of* $(r, p)$. If $r$ were known to be an integer, what would you choose $(\hat{r}, \hat{p})$ to be? [Don't tell others: $(r, p) = (3, 0.2)$.]

**2.4.7** Let $f(k; r, p)$ be the probability mass function for the negative binomial distribution with parameters $r$ and $p > 0$. For which value of $k$ is $f(k; r, p)$ maximum? Apply it for the pairs considered in Figure 2.4.2.

## 2.5 THE POISSON DISTRIBUTION

The definitions of binomial, hypergeometric, negative binomial, and geometric random variables and their distributions were stated in terms of experiments. These definitions are more useful than those given in terms of the corresponding probability mass functions because applications become more evident. Although an understanding of the experiment leads naturally to formulas for mass functions, the reverse is not necessarily true. It is true that we extended the definition of the negative binomial distribution to the case $r > 0$, not necessarily an integer, through the mass function, but the more fundamental definition (and the name) was defined in terms of an experiment.

The Poisson distribution, certainly one of the most useful models for an understanding of count data, will be given in terms of its mass function. The distribution was named for Simeon Denis Poisson (1781–1840), who mentioned it briefly on one page of a 1837 paper (see Stigler, 1986, p. 183). Although there are certainly many experiments in which the random variables observed are modeled well by the Poisson distribution, the fundamental justification is through its approximation of the binomial

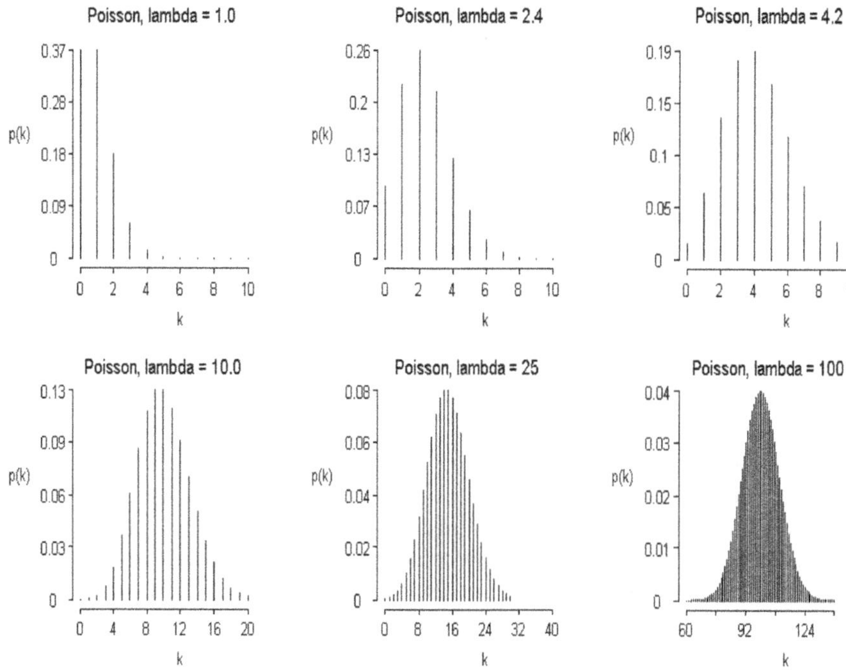


**FIGURE 2.5.1** Poisson probability mass functions.

for the case that $n$ is large and $p$ is small. It serves, for example, as a good approximation of the distribution of the numbers of deaths among the 60-year-old men insured by an insurance company, the numbers of highway accidents along a 20-mile stretch of freeway during nonbusy, reasonable weather days, and the numbers of deaths from breast cancer among women 50 to 55 during a one-year period in Lansing, Michigan.

**Definition 2.5.1** Let $\lambda > 0$. A random variable $X$ is said to have the *Poisson distribution* with parameter $\lambda$ if

$$P(X = k) = \frac{e^{-\lambda}\lambda^k}{k!} \qquad \text{for } k = 0, 1, 2, \ldots. \tag{2.5.1}$$

We will write $X \sim \text{Poisson}(\lambda)$. □

We should first verify that the function $f(k;\lambda)$ defined by (2.5.1) is a probability mass function (see Figure 2.5.1). But $\sum_{k=0}^{\infty} f(k;\lambda) = e^{-\lambda}[1 + \lambda/1! + \lambda^2/2! + \lambda^3/3! + \cdots]$. The term in brackets is the Taylor series expansion of $e^{\lambda}$, so the sum is 1.

Since $r(k) \equiv f(k;\lambda)/f(k-1;\lambda) = \lambda/k$, the maximum of $f(k;\lambda)$ is taken for $k = [\lambda]$, the greatest integer less than or equal to $\lambda$, with a tie for the maximum at $\lambda - 1$ and $\lambda$ when $\lambda$ is an integer. Consider a sample of 100 independent observations from $f(k; 3.5)$:

3 3 2 2 1 3 2 1 4 0 2 1 1 2 2 2 5 1 4 3 3 3 3 6 5 3 4 8 4 8 4 3 3 4 6 3
1 5 0 7 1 5 2 1 0 5 3 4 4 4 7 7 7 0 3 0 5 5 2 3 5 5 6 3 5 3 4 3 4 2 4 3
4 5 2 3 5 2 6 4 4 2 3 5 3 4 6 3 4 3 5 1 6 1 1 4 2 10 1 2

For 10,000 observations the frequencies were:

| $X$ | 0 | 1 | 2 | 3 | 4 | 5 | 6 | 7 | 8 | 9 | 10 | 11 | 12 |
|---|---|---|---|---|---|---|---|---|---|---|---|---|---|
| Freq. | 281 | 1886 | 1040 | 1886 | 2139 | 1889 | 1378 | 384 | 159 | 69 | 24 | 7 | 5 |

The sample mean, variance, and standard deviations were 3.507, 3.470, and 1.863.

It is relatively easy to determine the mean and variance of a Poisson rv: $E(X) = e^{-\lambda}[0(1) + 1(\lambda/1!) + 2(\lambda^2/2!) + \cdots] = e^{-\lambda}\lambda[0 + 1 + \lambda/1! + \lambda^2/2! + \cdots] = \lambda$. Similarly, $E[X(X-1)] = \lambda^2$, so that $E(X^2) = \lambda^2 + \lambda$ and $\text{Var}(X) = \lambda$. The mean and variance are both $\lambda$. Notice that the distribution becomes more and more like the "normal density" (yet to be defined) as $\lambda$ increases. See Chapter six for a proof. The Poisson distribution serves as an approximation of the binomial as $n$ increases and $p = p_n$ decreases in such a way that the product $\lambda_n = np_n$ approaches a limit.

**Theorem 2.5.1** Let $\{p_n\}$ be a sequence of probabilities, define $\lambda_n = np_n$, and suppose that $\lim_{n\to\infty} \lambda_n = \lambda > 0$. Then $\lim_{n\to\infty} b(k; n, p_n) = e^{-\lambda}\lambda^k/k!$ for $k = 0, 1, 2, \ldots$.

Before proving Theorem 2.5.1, consider the following example.

***Example 2.5.1*** The binomial and Poisson probabilities with $\lambda_n \equiv \lambda = 2.4$ are as follows: □

| | Binomial($n$, $p_n$) | | | | | | Poisson |
|---|---|---|---|---|---|---|---|
| $k$ | (10,0.24) | (20,0.12) | (40,0.06) | (80,0.03) | (160,0.015) | (320,0.0175) | $\lambda = 2.4$ |
| 0 | 0.0643 | 0.0776 | 0.0842 | 0.0874 | 0.0891 | 0.0899 | 0.0907 |
| 1 | 0.2030 | 0.2115 | 0.2149 | 0.2164 | 0.2171 | 0.2174 | 0.2177 |
| 2 | 0.2885 | 0.2740 | 0.2675 | 0.2643 | 0.2628 | 0.2620 | 0.2613 |
| 3 | 0.2429 | 0.2242 | 0.2162 | 0.2125 | 0.2108 | 0.2099 | 0.2090 |
| 4 | 0.1343 | 0.1299 | 0.1277 | 0.1265 | 0.1260 | 0.1257 | 0.1254 |
| 5 | 0.0509 | 0.0567 | 0.0587 | 0.0595 | 0.0599 | 0.0600 | 0.0602 |
| 6 | 0.0134 | 0.0193 | 0.0218 | 0.0230 | 0.0235 | 0.0238 | 0.0241 |
| 7 | 0.0024 | 0.0053 | 0.0068 | 0.0075 | 0.0079 | 0.0081 | 0.0083 |
| 8 | 0.0003 | 0.0012 | 0.0018 | 0.0021 | 0.0023 | 0.0024 | 0.0025 |
| 9 | 0.0000 | 0.0002 | 0.0004 | 0.0005 | 0.0006 | 0.0006 | 0.0007 |
| 10 | 0.0000 | 0.0000 | 0.0001 | 0.0001 | 0.0001 | 0.0001 | 0.0002 |

**Proof of Theorem 2.5.1** $b(k; n, p_n)$ can be written as the product of five factors: $F_{1n} = [n(n-1)\cdot(n-k+1)]/n^k$, $F_{2n} = 1/k!$, $F_{3n} = \lambda_n^k$, $F_{4n} = (1-\lambda_n/n)^n$, and $F_{5n} = (1-\lambda_n/n)^{-k}$. But $\lim_{n\to\infty} F_{1n} = 1$, $\lim_{n\to\infty} F_{2n} = 1/k!$, $\lim_{n\to\infty} F_{3n} = \lambda^k$, $\lim_{n\to\infty} F_{4n} = e^{-\lambda}$, and $\lim_{n\to\infty} F_{5n} = 1$. This proves the theorem. □

The approximation provided by Theorem 2.5.1 may be improved upon through a theorem of Lucien Le Cam.

**Theorem 2.5.2 (Le Cam)** Let $Y_1, \ldots, Y_n$ be independent, with $Y_i \sim B(1, p_i)$. Let $T = \sum_{i=0}^n Y_i$. Let $W$ have the Poisson distribution with parameter $\lambda = \sum_{i=1}^n p_i$. Then for any subset $A$ of the real line $G(A) \equiv |P(T \in A) - P(W \in A)| \le \sum_{i=1}^n p_i^2$.

We do not provide a proof, but instead, refer to a paper of T. W. Brown (1984). If, for example, one ball is drawn from urn $i$, which has $i$ red and $(50-i)$ black balls for $i = 1, 2, 3, 4$, with $Y_i =$ [ball drawn from urn $i$ is red], then $Y_i \sim$ Bernoulli($p_i = i/50$). Easy but tedious computation shows that the probability mass functions for $T$ and $W \sim$ Poisson($\lambda = 10$) , are:

| | $t$ | | | | |
|---|---|---|---|---|---|
| | 0 | 1 | 2 | 3 | 4 |
| $P(T=t)$ | 0.8136 | 0.1732 | 0.0128 | 0.0004 | 0.0000 |
| $P(W=t)$ | 0.8187 | 0.1637 | 0.0164 | 0.0011 | 0.0000 |

Taking $A = \{1, 2\}$, for example, we get $P(T \in A) = 0.1860$, while $P(W \in A) = 0.1801$, so that $|P(T \in A) - P(W \in A)| = 0.0059$. Since $\sum_{i-1}^n p_i^2 = 0.0120$, the inequality of Theorem 2.5.2 is satisfied. If we take $p_i \equiv p$, so that $T \sim B(n, p)$, the Le Cam upper bound becomes $G(A) \le np^2$. For $n = 80$, $p = 0.030$, for example, we get the upper bound 0.072, while from Example 2.5.1, for $A = \{2, 3, 4\}$ we get $|P(T \in A) - P(W \in A)| = 0.0076$.

### Sums of Independent Poisson RVs

Let $X_1$ and $X_2$ be independent, with Poisson distributions with parameters $\lambda_1$ and $\lambda_2$. Let $Y = X_1 + X_2$. Then, for any nonnegative integer $y$, $P(Y = y) = P(X_1 = 0, X_2 = y) + P(X_1 = 1, X_2 = y-1) + \cdots + P(X_1 = y, X_2 = 0) = \sum_{j=0}^y [e^{-\lambda_1}\lambda_1^j/j!]$ $[e^{-\lambda_2}\lambda_2^{y-j}/(y-j)!] = e^{-\lambda_1-\lambda_2}[\sum_{j=0}^y \binom{y}{j} p^j (1-p)^{y-j}](\lambda_2+\lambda_2)^y/y!$, where $p = \lambda_1/(\lambda_1+\lambda_2)$. The sum in brackets is 1. It follows that $Y$ has the Poisson distribution with parameter $\lambda_1 + \lambda_2$. By induction it follows that if $X_1, \ldots, X_n$ are independent, each with a Poisson distribution, with parameters $\lambda_1, \ldots, \lambda_n$, then $Y = X_1 + \cdots + X_n \sim$ Poisson($\lambda_1 + \cdots + \lambda_n$).

### The Poisson Process

Consider the births of babies at a hospital that averages 12 births per day. For simplicity let us ignore multiple births: twins, triplets, and so on. The probability of a

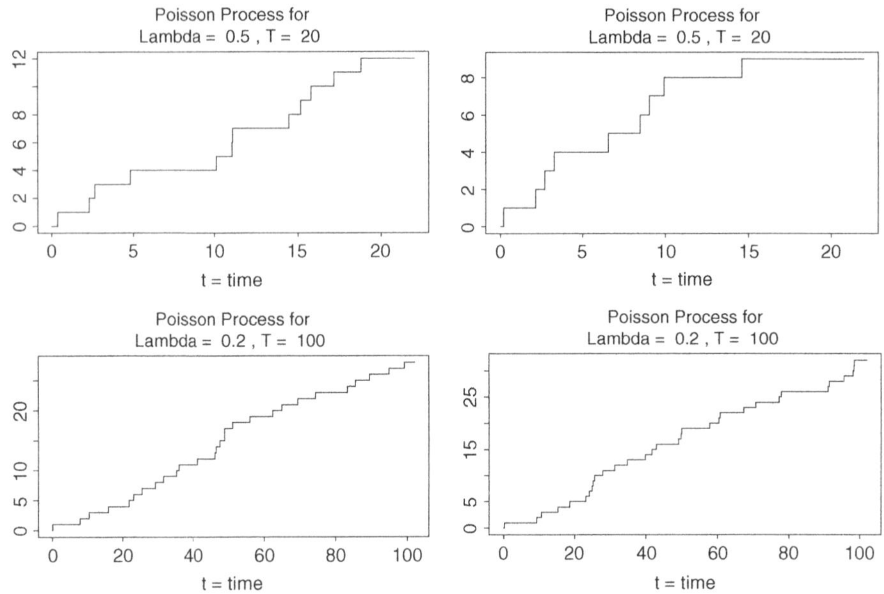


**FIGURE 2.5.2** Sample paths for Poisson processes.

birth in a short interval, say 1 minute, is approximately $12/[24(60)] = 1/120$. Occurrences of births in nonoverlapping intervals of time should be independent, at least in approximation. Therefore, the number of births in an interval of $t$ minutes should be, in good approximation, distributed as Poisson with parameter $\lambda = \lambda(t) = t/120$ (see Figure 2.5.2). Such a random function of time, a *stochastic process*, is said to be a *Poisson process* with parameter $1/120$ (when the time unit is 1 minute). The following formally defines a Poisson process. $X(t)$ should be considered to represent the number of occurrences of an event in the time interval $[0, t]$. The number of occurrences in the interval $(s, t]$ is therefore $X(t) - X(s)$.

**Definition 2.5.2** Let $X(t)$ be a random variable for each $t$, $0 \le t \le T$, where $T > 0$. The family of random variables (or stochastic process) $\{X(t) \mid 0 \le t \le T\}$ is a *Poisson process* with rate $\lambda$ if:

**(a)** $P(X(0) = 0) = 1$.
**(b)** There exists a constant $\lambda > 0$ such that for each pair $(s, t)$ with $0 \le s < t \le T$, $X(t) - X(s) \sim \text{Poisson}((t - s)\lambda)$.
**(c)** For any positive integer $k$ and any $0 \le a_1 < b_1 \le a_2 < b_2 < \cdots \le a_k < b_k < T$, the rv's $D_1 = X(b_1) - X(a_1), \ldots, D_k = X(b_k) - X(a_k)$ are independent. [Thus, $X(t)$ is an *independent increment process*.]

If (a) and (c) hold, (b) may be replaced by the statement that $X(t) \sim \text{Poisson}(\lambda t)$ (see Problem 2.5.7 for a proof).

If events occur at positive times in such a way that only one event can occur at any time, $X(t)$ is the number of occurrences in the time interval $[0, t]$ for $0 \leq t \leq T$, $X(t)$ has independent increments as in (c), and the distribution of $X(t) - X(s)$ for $0 \leq s < t \leq T$ depends only on $t - s$, it follows that (a) and (b) hold, so that $\{X(t) \mid 0 \leq t \leq T\}$ is a Poisson process (see Doob, 1953, p. 402, for a proof). We can extend the definition of the Poisson process by changing (b) so that there exists a function $\lambda$ on the positive real line so that $X(t) - X(s) \sim \text{Poisson}(\int_s^t \lambda(u)du)$. The resulting stochastic process with the function $\lambda$ varying is called a *nonhomogeneous Poisson process*. In this book the term *Poisson process* will imply a homogeneous process for which $\lambda$ is constant. □

Some examples of cases in which the Poisson process may be a reasonable model for count data:

1. Nonmultiple natural births occurring in a hospital
2. Traffic passing a point on a freeway at a nonbusy time
3. Alpha particles radiating from a disk
4. Cases of deaths among babies in their first year in a large city
5. Traffic accidents causing death on a freeway during June and July

In none of these cases would the model apply strictly. There might, for example, be a slight possibility that cars would pass a point on a freeway in small clusters, especially as traffic becomes heavier. Traffic volume varies with the time period, so that the model would serve as a reasonable approximation only over a limited interval of time. For case 1 the word *natural* was added because cesarean births may tend to occur during times more convenient to the hospital. In case 5 we consider traffic accidents causing death rather than deaths, because multiple deaths can occur, contrary to the Poisson process.

***Example 2.5.2*** Suppose that murders occur in a large city at a rate of 540 per year. Assuming the Poisson process model and a year of 360 days, find the probability of each of the following.

**(a)** Two or more murders in one day.
Let $X(t)$ be the number of murders in $t$ days. Then $X(t) \sim$ Poisson $(540t/360)$. $P(X(1) \geq 2) = 1 - P(X(1) \leq 1) = 1 - e^{-3/2}[1 + (3/2)] = 0.4422$.

**(b)** Two or more murders for each of three consecutive days,
$P(2 \text{ or more for 3 consecutive days}) = (0.4422)^3 = 0.0865$.

**(c)** No murders for five days
$P(\text{no murders in 5 days}) = P(X(5) = 0) = e^{-15/2} = 5.5 \times 10^{-4}$.

**(d)** Suppose that the rate is 1.2 murders per day during the workweek but 2.5 murders weekends. What is the probability of 10 or more murders in a one-week period?

Let $X$ and $Y$ be the numbers of murders during the workweek and on the weekend. Then $X$ and $Y$ are independent, $X \sim$ Poisson(6), $Y \sim$ Poisson(5), so that $X + Y \sim$ Poisson(11) and $P(X + Y \geq 10) = 1 - 0.3405 = 0.6595$. □

## Problems for Section 2.5

**2.5.1** Let $X_n \sim B(n, p_n)$, and $X_0 \sim$ Poisson(3). Find $P(X_n = 2)$ and $P(X_n = 3)$ for $(n, p_n)$ as follows: (30, 0.10), (150, 0.02), (750, 0.004), and compare these to $P(X_0 = 2)$ and $P(X_0 = 3)$.

**2.5.2** Show that in Definition 2.5.2, (a), (c), and the condition (b*) there exists $\lambda > 0$ such that $X(t) \sim$ Poisson$(t\lambda)$ for $0 \leq t \leq T$ implies (b).

**2.5.3** The occurrences of death among 70-year-old men insured by the Bigrock Insurance Company may be modeled by the Poisson process with a rate 0.8 per day. Find:

**(a)** The probability of three or more deaths in a two-day period.

**(b)** The probability that at least once among the four Saturday–Sunday weekends in February 2010, three or more die.

**(c)** The expected value and variance of $W$, the number dying in all of 2010.

**2.5.4** Let $\{X(t) \mid 0 \leq t \leq T\}$ be a Poisson process with rate $\lambda > 0$. Let $T_k = \inf\{t \mid X(t) \geq k\}$ for $k = 0, 1, 2, \ldots$. Let $D_k = T_k - T_{k-1}$ for $k = 1, 2, \ldots$. Then $P(D_k > d) = P(T_k - T_{k-1} > d) = P$(no occurrences in the time interval $(T_{k-1}, T_{k-1} + d]) = P$(no occurrences in the time interval $(0, d]$). Express this last probability as a function of $\lambda$ and $d$. The rv $D_k$ has a continuous distribution, the exponential distribution, the same for each $k = 1, 2, \ldots$.

**2.5.5** Let $X_1, X_2$ be independent with Poisson distributions with parameters $\lambda_1$ and $\lambda_2$. Let $N = X_1 + X_2$. Show that $P(X_1 = k \mid N = n) = b(k; n, p)$, where $p$ is a simple function of $\lambda_1$ and $\lambda_2$.

**2.5.6** Suppose that telephone calls are received at an emergency 911 number in a nonhomogeneous Poisson process, beginning at midnight, measuring time in hours, with $\lambda(t) = 0.4$ for $0 \leq t \leq 7$, 0.8 for $7 < t \leq 17$, and 1.5 for $17 < t \leq 24$. Find the probability of

**(a)** No calls between 6:00 and 10:00 a.m.

**(b)** At least three calls between 4:00 and 7:00 p.m.

**(c)** Let $W$ be the total number of calls received during a one-week period. What is the distribution of $W$?

**2.5.7** (More difficult than most problems) Let $X(t)$ satisfy (a) and (c) of Definition 2.5.2, and $X(t) \sim$ Poisson$(\lambda t)$ for $\lambda > 0$ for all $t > 0$. Prove that (b) holds;

that is, $D(s,t) \equiv X(t) - X(s) \sim$ Poisson($\lambda(t-s)$) for all $0 < s < t$. *Hint:* Prove first that $P(D(s,t) = 0) = e^{-\lambda(t-s)}$. Let $g(j) = P(D(s,t) = j)$. Then $P(X(t) = k) = \sum_{j=0}^{k} g(j)P(X(s) = k - j)$. To use induction on $k$, express $g(k)$ in terms of $P(X(t) = k)$ and $\sum_{j=0}^{k-1} g(j)P(X(s) = k - j)$, and take advantage of the fact that $P(X(s) = k - j)$ for $j = 1, \ldots, k$ and $P(X(t) = k)$ are Poisson probabilities.

**2.5.8** Let $X \sim B(1000, 0.002)$, and $Y \sim$ Poisson(2). Compare the probability functions of $X$ and $Y$ and determine the maximum value of $G(A)$ as defined for the Le Cam theorem. Show that this maximum satisfies the inequality given by the theorem.

CHAPTER THREE

# Continuous Random Variables

## 3.1 INTRODUCTION

To this point we have concentrated on random variables that take only a finite or countably infinite number of values. Since computers have the capacity to handle only a finite number of values, in one sense that should be enough. However, for mathematical convenience we need to be able to consider models in which the range of a random variable may be the unit interval, the interval $[0, \pi]$, the positive real line, or the entire real line, in the same way that the rational numbers alone are inadequate for the treatment of mathematical models for heights, weights, temperatures, and other physical measurements.

We must represent probability distributions for random variables taking noncountably infinite numbers of values in a different way than by assigning positive probabilities to points. Just as the length of an interval is not the sums of the lengths of the points in the interval, the probability that a continuous random variable $X$ will fall in an interval $[a, b]$ will *not* be the sum of probabilities $P(X = x)$ for $x \in [a, b]$. Instead, $P(X \in [a, b])$ will be expressed as the integral of a "density function for $X$." The student is warned that density values are *not probabilities*. In fact, for continuous random variables these probabilities $P(X = x)$ will be zero for *all* $x$.

## 3.2 CONTINUOUS RANDOM VARIABLES

**Definition 3.2.1** A random variable $X$ is said to be *continuous* if there exists a function $f(x)$ defined on the real line so that $f(x) \geq 0$ for all $x$, such that for each subset $A$ of the real line

$$P(X \in A) = \int_A f(x)\,dx. \tag{3.2.1}$$

The function $f$ is called the *probability density* for $X$. □

*Models for Probability and Statistical Inference: Theory and Applications,* By James H. Stapleton

*Notes:* The integral will be assumed to be the Riemann integral, the integral with which most students are familiar, although for a more rigorous theory, in which countable additivity is required, we would usually consider the Lebesque integral. In addition, for the purpose of rigor, the phrase "every subset $A$" needs to be modified to "every Borel subset or "every Lebesgue subset." Most students need not worry about these niceties. "Reasonable sets" $A$ will all satisfy (3.2.1). The function $f$ may be changed arbitrarily at a finite or even a countably infinite number of values without changing the value of the integral in (3.2.1), although we will never let a density $f$ take negative values. □

WARNING: In every class in which the author has introduced probability density functions there have been a few students who seem either unable or very reluctant to distinguish between $X$ and $x$. On the text page the difference is evident from size alone. But on the chalkboard and in a student's notes, size alone will not distinguish these symbols. A student who cannot distinguish between upper and lower case is inviting disaster. Although your class is, of course, smarter than most, it may happen in your class as well. Those who find that their $X$ and $x$ appear to be the same are advised to write each 100 times as they did when they were 15 or 20 years younger. Please trust the author and your instructor. It *is* important. Now test yourself—ask a friend if he or she can distinguish between your $x$ and your $X$. Similarly, $Y$ and $y$, $W$ and $w$, must be distinguishable.

***Example 3.2.1*** Let $f(x) = 2$ for $0 \le x \le 1/2$, 0 otherwise. If $X$ has this probability density function, then, for example, $P(0.2 \le X \le 0.4) = 2(0.4 - 0.2) = 0.4$, and $P(X = 0.3) = 0$, since an integral over a single point is zero. Notice that the density value 2 should make it clear that density values are *not* probabilities.

Following is a sample of 50 from the density $f$, values given to three decimal places:

0.175 0.215 0.063 0.083 0.216 0.063 0.014 0.491 0.206 0.273
0.446 0.257 0.089 0.317 0.267 0.364 0.180 0.040 0.451 0.166
0.142 0.356 0.020 0.078 0.310 0.355 0.253 0.222 0.249 0.130
0.339 0.388 0.227 0.079 0.436 0.289 0.024 0.485 0.172 0.346
0.095 0.211 0.465 0.493 0.226 0.289 0.050 0.398 0.453 0.176
Mean 0.243, std. dev. = 0.142

The frequency table for 1000 observations is

| Interval | 0–.05 | .05–.10 | .10–.15 | .15–.20 | .20–.25 | .25–.30 | .30–.35 | .35–.40 | .40–.45 | .45–.50 |
|---|---|---|---|---|---|---|---|---|---|---|
| Freq. | 95 | 110 | 101 | 113 | 99 | 89 | 100 | 87 | 105 | 101 |

Since probabilities are never negative and $P(-\infty < X < +\infty) = 1$, it follows that a density must integrate to 1. □

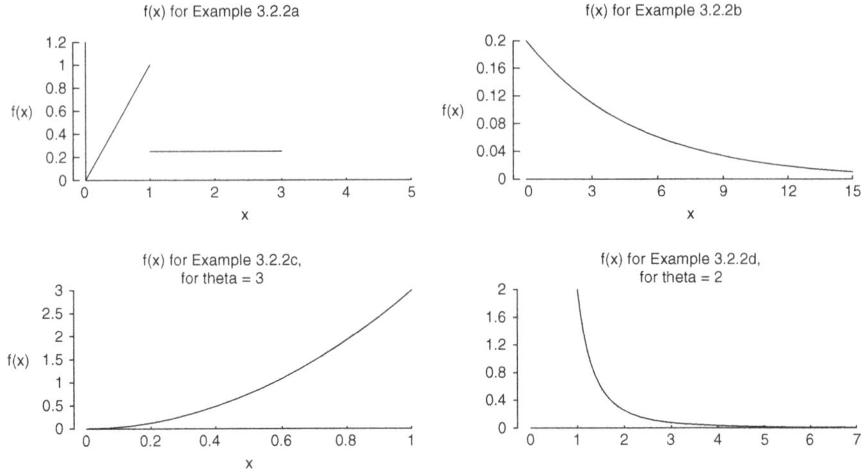


**FIGURE 3.2.1** Densities of Example 3.1.2.

**Definition 3.2.2** Let $f$ be a real-valued function on the real line satisfying:

**(a)** $f(x) \geq 0$ for all $x$.
**(b)** $\int_{-\infty}^{+\infty} f(x)\,dx = 1$.

Then $f$ is a *probability density* function. □

***Example 3.2.2*** The following are all density functions. For each, the set on which the density is positive is indicated (see Figure 3.2.1). The density is zero otherwise.

**(a)** $f(x) = x$ for $0 \leq x < 1$, $1/4$ for $1 \leq x \leq 3$.
**(b)** $f(x) = (1/5)e^{-x/5}$ for $x > 0$.
**(c)** $f(x) = \theta x^{\theta-1}$ for $0 \leq x \leq 1$, $\theta > 0$.
**(d)** $f(x) = \theta x^{-\theta-1}$ for $x \geq 1$, $\theta > 0$.

A sample of 50 from $f$ as in part (a):

0.971 1.903 0.863 2.444 2.057 1.223 0.884 0.340 0.792 2.918
0.645 1.558 1.226 0.868 0.983 2.054 2.266 2.673 2.196 0.874
2.797 0.911 0.792 0.934 0.866 0.404 1.247 2.099 1.648 0.055
0.528 1.747 2.319 2.439 1.307 2.929 1.324 0.746 0.866 0.724
0.740 2.259 0.755 0.962 0.941 1.829 1.928 1.723 0.921 1.639
Mean = 1.402, variance = 0.546

The frequency table for 10,000 observations from the density in Example 3.2.2(a) is

| Interval | 0–0.50 | 0.50–1.00 | 1.00–1.50 | 1.5–2.00 | 2.00–2.50 | 2.50–3.00 |
|---|---|---|---|---|---|---|
| Frequency | 1266 | 3709 | 1232 | 1271 | 1272 | 1250 |

For part (a), $P(0.50 \le X \le 1.00) = \int_{0.5}^{1} f(x)\,dx = (1/2)(1^2 - 0.50^2) = 3/8$, so that the frequency 3709 for the interval [0.50, 1.00] is reasonable. □

Since probabilities are determined for continuous random variables by integration of densities, it is sometimes useful to have available a point function, the cumulative distribution function, which makes it unnecessary to repeatedly evaluate integrals.

**Definition 3.2.3** The function $F(u) = P(X \le u)$, defined for all real $u$, is called the *cumulative distribution function* of $X$. □

*Notes:* The cumulative distribution function (cdf) is defined for every real number $u$. It is most useful for continuous random variables, primarily because, for $a < b$, $P(a < X \le b) = P(X \le b) - P(X \le a) = F(b) - F(a)$. If $X$ is the number of heads in two tosses of a fair coin, then

$$F(u) = \begin{cases} 0 & \text{for } u < 0 \\ 1/4 & \text{for } 0 \le u < 1 \\ 3/4 & \text{for } 1 \le u < 2 \\ 1 & \text{for } u \ge 2. \end{cases}$$

Since for $u < v$, $F(u) = P(X \le u) \le P(X \le v) = F(v)$, a cdf $F$ must be monotone nondecreasing. Let $\{u_n\}$ be a sequence of numbers converging to $+\infty$. Then $1 = P(X < +\infty) = \lim_{n\to\infty} P(X \le u_n)$. It follows that $\lim_{n\to\infty} F(u_n) = 1$ for any such sequence. Similarly, $\lim_{n\to\infty} F(v_n) = 0$ for any sequence $\{v_n\}$ converging to $-\infty$. We write these limits equivalently as $\lim_{u\to\infty} F(u) = 1$ and $\lim_{u\to-\infty} F(u) = 0$.

□

One more property that a cdf must possess is continuity on the right. To show this for a random variable $X$ and any $b$, let $B_n = [-\infty, b + h_n]$, where $\{h_n\}$ is a sequence of positive numbers converging monotonically to zero. Then $\bigcap_{n=1}^{\infty} B_n = (-\infty, b]$, so that $F(b) = P(X \le b) = P(X \in (-\infty, B]) = \lim_{n\to\infty} P(X \in B_n) = \lim_{n\to\infty} P(X \le b + h_n) = \lim_{n\to\infty} F(b + h_n)$. Thus, $FP(X \in [-\infty, B])$ is continuous on the right. The third equality follows from the countable additivity of all probability measures.

Although we have required more in order that the random variable $X$ be called continuous, it is true that the cdf $F$ of a continuous random variable is continuous. The mathematically curious may wish to try to prove that the cdf of $Y = \sum_{k=1}^{\infty} X_n \cdot 3^{-k}$, where the $X_n$ are 0 or 2 independently with probabilities ½, has a continuous cdf but does not possess a density. $Y$ is said to have the *Cantor distribution*. (*Advice*: Take a more advanced course before trying to prove it.)

***Example 3.2.2*** (*Continued*) The corresponding cdf's are:

$$\textbf{(a)}\ F(u) = \begin{cases} 0 & \text{for } u < 0 \\ u^2/2 & \text{for } 0 \le u < 1 \\ 1/2 + (u-1)/4 & \text{for } 1 \le u < 3 \\ 1 & \text{for } u \ge 3. \end{cases}$$

**(b)** $F(u) = \begin{cases} 0 & \text{for } u < 0 \\ 1 - e^{-u/5} & \text{for } u \geq 0. \end{cases}$

**(c)** For $\theta > 0$, $F(u) = \begin{cases} 0 & \text{for } u < 0 \\ u^\theta & \text{for } 0 \leq u < 1 \\ 1 & \text{for } u \geq 1. \end{cases}$

**(d)** For $\theta > 0$, $F(u) = \begin{cases} 0 & \text{for } u < 0 \\ 1 - u^{-\theta} & \text{for } u \geq 1. \end{cases}$ □

A common error in determining $F$ in part (a) is to neglect $P(X \leq 1)$ in determining the value of $F(u)$ for $1 \leq u < 3$. That sort of error can be avoiding by always checking the conditions that a function $F$ must satisfy in order to be a cdf:

Necessary and sufficient conditions that a real-valued function $G$ on the real line be a cdf:

**1.** $u < v$ implies that $G(u) \leq G(v)$ (monotone nondecreasing).
**2.** $\lim_{u\to\infty} G(u) = 1$ and $\lim_{u\to-\infty} G(u) = 0$.
**3.** $G$ is continuous on the right.

If a random variable $X$ has a density function $f$, then of course its cdf $F$ may be expressed as its integral: $F(u) = P(X \leq u) = \int_{-\infty}^{u} f(x)\,dx$. If a density $f$ is continuous at a point $x_0$, then $\frac{d}{dx}F(x)|_{x=x_0} = F'(x_0)$. Conversely, the *fundamental theorem of calculus* states that if $f(x)$ is continuous on the interval $a \leq x \leq b$ and $G$ is a function such that $\frac{d}{du}G(u) = f(u)$ for $a < u < b$, then

$$\int_a^b f(x)\,dx = G(b) - G(a).$$

If a cdf $F$ has a derivative $f$ on an interval $[a, b]$ that is continuous on $(a, b)$, it follows that $\int_a^b f(x)\,dx = F(b) - F(a)$ and if $a \leq u \leq b$, then $F(u) = F(a) + \int_a^u f(x)\,dx$. For Example 3.2.2(a), for $1 < u \leq 3$, $\frac{d}{du}F(u) = 1/4$, and $F(u) = F(1) + \int_1^u f(x)\,dx = 1/2 + (u-1)/4$. In (d), $F(1) = 0$, and for $u > 1$, $\frac{d}{du}F(u) = \theta u^{-\theta-1}$, $F(u) = 0 + \int_1^u f(x)\,dx = 1 - u^{-\theta}$ for $u > 0$. Of course, $f$ may be changed arbitrarily at a finite or even a countably infinite number of points without changing the corresponding cdf $F$. When we have a choice we usually choose the "version" that is as continuous as possible.

Also, from calculus, a cdf $F(u) = (\int_{-\infty}^{u} f(x)\,dx$ is continuous for all $u$. If $f$ is continuous at a point $u$, then the derivative of $F$ exists at $u$ and has the value f($u$). In Example 3.2.2(b), $F$ does not have a derivative at $u = 0$, although it does at $u = 1$, where the derivative is $f(1) = (1/5)e^{-1/5}$. However, if we change the definition of $f$ so that $f(1) = 17$ (one of the author's favorite numbers—the set of such numbers is infinite), then $F$ is unchanged, so that its derivative remains the same and is not 17.

A random variable may be partly continuous, partly discrete. For example, the distribution of the lengths of life $X$ of AAA batteries might be as in Example 3.2.2(a), with density $f$. However, if 5% are faulty, so that $X = 0$, then $P(X = 0) = 0.05$. Then we can write $P(X \in A) = 0.05I[0 \in A] + 0.95\int_A f(x)\,dx$. $I[0 \in A]$ is 1 if $0 \in A$, 0 otherwise. $X$ then has cdf $F(u) = 0.05I[0 \le u] + 0.95\int_{-\infty}^u f(x)\,dx$ for $u \ge 0$, 0 otherwise.

For most of our discussion we will consider either discrete or continuous random variables, although much will also remain true for mixtures. In general, a random variable $X$ is said to be a mixture of continuous and discrete distributions if there exist $0 < \alpha < 1$, a probability mass function $f_d$, and a probability density function $f_c$ such that

$$P(X \in A) = \alpha \sum_{x \in A} f_d(x) + (1 - \alpha) \int_A f_c(x)\,dx \tag{3.2.2}$$

for any (Borel) subset $A$ of the real line.

As we develop methods for continuous random variables, it will be convenient to have three simple families to enhance understanding. Discussion of other families will be postponed until after we have developed appropriate probabilistic tools.

**Definition 3.2.4** A random variable $X$ is said to have the *uniform distribution* on an interval $[a, b]$ if $X$ has density $f(x) = 1/(b - a)$ for $a \le x \le b$, 0 otherwise. We will use Unif$(a, b)$ to denote this distribution. □

Using geometry, it follows that $X$ has cdf $F_X(u) = (u - a)/(b - a)$ for $u \in [a, b]$. In particular, if $U \sim$ Unif(0, 1), $U$ has cdf $F_U(u) = u$ for $0 \le u \le 1$. Then $Y = a + (b - a)U \sim$ Unif$(a, b)$.

**Definition 3.2.5** A random variable $X$ is said to have the *exponential distribution* with rate parameter $\lambda > 0$ if $X$ has density $f(x; \lambda) = \lambda e^{-\lambda x}$ for $x > 0$. □

*Notes:* The corresponding cdf has a simple form: $F(u) = 1 - e^{-\lambda u}$ for $u > 0$. The "no memory property" is $P(X > u + h | X > u) = P(X > u + h)/P(X > u) = e^{-\lambda(u+h)}/e^{-\lambda u} = e^{-\lambda h}$ for all $u > 0, h > 0$. Thus, if $X$ is the waiting time until the occurrence of an event, the additional waiting time $X - u$, given $[X > u]$, is the same for all $u \ge 0$. This is the "memoryless property" of the exponential distribution, the same property possessed by the geometric distribution. □

**Definition 3.2.6** A random variable $X$ has the *double exponential* or *Laplace distribution* with scale parameter $\lambda > 0$ and location parameter $\theta$ if $X$ has density $f(x; \theta, \lambda) = (1/2\lambda)e^{-|x-\theta|/\lambda}$ for all $x$. We sometimes refer to the distribution of $X$ as the *Laplace*$(\theta, \lambda)$ distribution. □

*Notes:* If $X$ has this density, then $Y = (X - \theta)/\lambda$ has the Laplace (0, 1) distribution, with density $f_Y(y) = (1/2)e^{-|y|}$ for all $y$. The cdf for $Y$ is $F_Y(y) = \begin{cases} e^y/2 & \text{for } y < 0 \\ 1 - e^{-y}/2 & \text{for } y \geq 0 \end{cases}$ Then $F_X(u) = P(X \leq u) = P(Y \leq (u - \theta)/\lambda) = F_Y((u - \theta)/\lambda)$. □

## Symmetry

A random variable $X$ is said to have a symmetric distribution about 0 if $X$ and $-X$ have the same distribution. If $X$ has cdf $F$, then $Y = -X$ has cdf $G(u) = P(Y \leq u) = P(-X \leq u) = P(X \geq -u) = P(X = -u) + P(X > -u) = P(X = -u) + 1 - F(-u)$. In the continuous case with density function $f$, symmetry is equivalent to $f(x) = f(-x)$ for all but countably many $x$. In the discrete case with probability function $f$, symmetry about zero is equivalent to $f(-x) = f(x)$ for all $x$. $Y$ is said to be *symmetrically distributed* about a constant $b$ if $Y - b$ is symmetrically distributed about zero. Thus, $P(Y - b \leq u) = P(Y - b = -u) + 1 - P(Y - b \leq -u)$, so that $F_Y(u + b) = P(Y = b - u) + 1 - F_Y(b - u)$, and in the continuous case, $F_Y(b + u) = 1 - F_Y(b - u)$ for all $u$. The Laplace distributions of $X$ and $Y$ as defined under "Notes" above are symmetric about $\theta$ and 0.

## Quantiles, Percentiles, Deciles, Median, and Quartiles

**Definition 3.2.7** Let $X$ be a random variable. For any $\alpha, 0 < \alpha < 1$, an $\alpha$-th-*quantile* for $X$ is any number $x_\alpha$ satisfying $P(X \leq x_\alpha) \geq \alpha$ and $P(X \geq x_\alpha) \geq 1 - \alpha$. $x_\alpha$ is also called a $100\alpha$-th-*percentile* for $X$ or for $F$. If $\alpha = k/10$ for some $k = 1, 2, \ldots, 9$, then $x_\alpha$ is also called a $k$th-*decile*. If $\alpha = k(0.25)$ for $k = 1, 2, 3$, then $x_\alpha$ is called a $k$th-*quartile*. The *median* is $x_{0.5}$. □

A median may be called a 50th percentile, a fifth decile, or a second quartile. For some $\alpha$, $x_\alpha$ may not be unique. This explains why we used "a" rather than "the" in the definitions. For example, if $X \sim$ Binomial (2, 1/2), then $x_{0.75}$ need only satisfy $1 \leq x_{0.75} \leq 2$. In terms of the cdf $F$ of $X$, $x_\alpha$ needs to satisfy $\alpha \leq P(X \leq \alpha) = F(x_\alpha)$ and $P(X \geq x_\alpha) = \lim_{u \uparrow x_\alpha}[1 - F(u)] = 1 - \lim_{u \uparrow x_\alpha} F(u) \geq 1 - \alpha$; equivalently, $\lim_{u \uparrow x_\alpha} F(u) \equiv F(x_{\alpha^-}) \leq \alpha$.

***Example 3.2.3*** Let $X$ have cdf

$$F(u) = \begin{cases} 0 & \text{for } u < 0 \\ u^2/2 & \text{for } 0 \leq u < 1 \\ 3/4 & \text{for } 1 < u \leq 2 \\ 3/4 + (u - 2)/4 & \text{for } 2 \leq u < 3 \\ 1 & \text{for } u \geq 3. \end{cases}$$

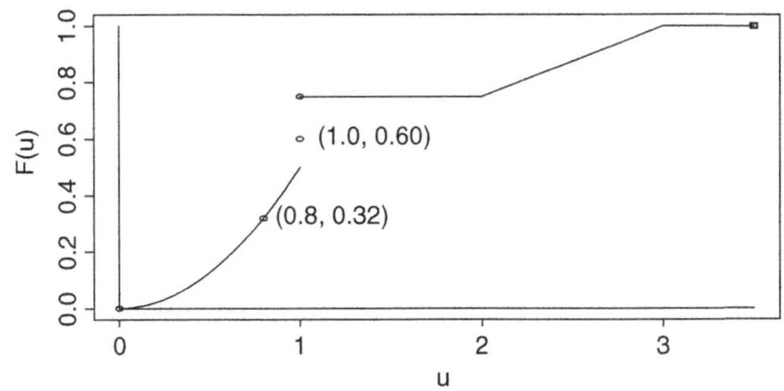


**FIGURE 3.2.2** CDF of Example 3.2.3.

Then $F(0.8) = 0.32$, so that $x_{0.32} = 0.8$, the 32nd percentile. Since $P(X \le 1) = F(1) = 0.75 > 0.6$, and $\lim_{u \uparrow 1} F(u) = 0.50 < 0.6$, $x_{0.6} = 1$, the 60th percentile (see Figure 3.2.2). Any $x$ in the interval [1, 2] qualifies as a 0.75-quantile or a 75th percentile. □

## Effects of Linear Transformations $Y = a + bX$

Let $X$ have density $f$, cdf $F$, and suppose that $Y = a + bX$, where $a$ and $b \neq 0$ are constants. What is the density of $Y$? In general, we will want to know the density of $Y = g(X)$, where $g$ is defined on the range of $X$. We postpone a more general treatment to Section 3.4. As a gentler introduction, we consider a linear transformation.

First consider $b > 0$. Let the density and cdf of $Y$ be $g$ and $G$. Then $G(u) = P(Y \le u) = P(X \le (u - a)/b) = F((u - a)/b)$. It follows that at points $x = (u - a)/b$ at which $f$ is continuous, $g(u) = \dfrac{d}{du}G(u) = f((u - a)/b)(1/b)$. Similarly, for $b < 0$, $G(u) = P(X \ge (u - a)/b) = 1 - F((u - a)/b)$, and at points $x = (u - a)/b$ at which $f$ is continuous, $g(u) = f((u - a)/b)\,(-1/b)$. In both cases $b > 0$ and $b < 0$, $g(u) = f((u - a)/b)(1/|b|)$ for any $u$ for which $f$ is continuous at $x = (u - a)/b$. For example, $X$ has density $f(x) = (1/2)e^{-|x|}$ for all $x$, then $Y = 5 + 3X$ has density $g(u) = (1/6)e^{-|u-5|/3}$ for all $u$. If $U \sim$ Unif(0, 1), then $W = 5U + 7 \sim$ Unif(7, 12), with density $g(t) = f_X((t - 7)/5)(1/5) = (1/5)$ for $0 \le (t - 7)/5 \le 1$ (i.e., $7 \le t \le 12$).

## Problems for Section 3.2

**3.2.1** Let $X$ have density $f(x) = C|x - 1|$ for $0 \le x \le 2$.

**(a)** Find $P(0.5 \le X < 1.5)$.

**(b)** Find the cdf $F(u)$ for $X$ and graph it. Use $F$ to verify your answer to (a).

**(c)** For which $u$ does $\dfrac{d}{du}F(u) = f(u)$?

**(d)** Define a density function $g$ that differs from $f$ at at least two points, so that the corresponding cdf is $F$. At what points $u$ does $\frac{d}{du}F(u)$ exist, and for which of these points does it equal $g(u)$?

**(e)** Find $x_{0.6}$, the 60th percentile for $X$.

**3.2.2** Let $W$ be a random variable with density $f(x) = x^2$ for $0 \le x \le 1$ and $C$ for $1 < x \le 4$.

**(a)** Find $C$ so that $f$ is a density and graph $f$.

**(b)** Find the cdf $F$ for $W$ and graph it.

**(c)** Express $P(|W - 2| < 1.5)$ in terms of $F$ and evaluate it.

**(d)** For which $u$ does the derivative of $F(u)$ exist? For those $u$ for which the derivative does exist, is its value $f(u)$?

**3.2.3** Let $U \sim \text{Unif}(0, 1)$. Find the cdf and density of $Y = U^2$. Graph both.

**3.2.4** Let $X$ have cdf $F(u) = 0.1I[u \ge 0] + 0.2I[u \ge 2] + C\int_0^u x\,dx$ for $0 \le u \le 4$, 0 for $u < 0$, and 1 for $u > 4$.

**(a)** Find $C$ so that $F$ is a cdf and graph it.

**(b)** Find $P(1 \le X < 3)$.

**(c)** For which $u$ does the derivative of $F(u)$ exist, and what are its values?

**(d)** Let $Y = X^2 + 1$. Find the cdf $G$ for $Y$. *Hint*: First express $F_Y$ in terms of $F_X$.

**(e)** Find the 0.09 and 0.80-quantiles for $X$.

## 3.3 EXPECTED VALUES AND VARIANCES FOR CONTINUOUS RANDOM VARIABLES

In the case that a random variable $X$ has a discrete distribution, $E(X)$ was expressed as a sum. When $X$ has a continuous distribution with density $f$, we will be unable to employ such a definition. However, if $a$ and $b$ are chosen such that $P(X < \boldsymbol{a})$ and $P(X > \boldsymbol{b})$ are "small," choose points $x_0 = \mathbf{a} < x_1 < x_2 < \cdots < x_n = \text{b}$ in such a way that each $x_i - x_{i-1} = \Delta_i$ is small, the random variable $Y$ that is **a** when $X \le \mathbf{a}$, $y_i = (x_{i-1} + x_i)/2$ when $x_{i-1} < X \le x_i$, and **b** when $X > \mathbf{b}$ should have approximately the same expectation as $Y$. Since $Y$ is discrete and $E(Y) = \mathbf{a}P(X \le a) + \sum_{i=1}^n y_i(F(x_i) - F(x_{i-1})) + \mathbf{b}P(X > b)$, that should suggest the following definition.

**Definition 3.3.1** Let $X$ be a continuous random variable with density $f$. Then the *expected value* of $X$ is $E(X) = \int_{-\infty}^{\infty} xf(x)\,dx$, provided that $\int_{-\infty}^{\infty} |x|f(x)\,dx$ exists. □

Our definition for the case that $X$ is a continuous random variable differs from that for discrete random variable in that it is defined in terms of its density function rather than in terms of a probability measure on the sample space as it was in the discrete case. In a more advanced and rigorous course built on measure theory, $E(X)$ would be defined as $\int_S X(\omega)\,dP(\omega)$, where $P$ is the probability measure on the sample space $S$. However, this would require that we define the integral on an abstract space $S$. Since its value is the same, we have chosen this definition instead.

***Example 3.3.1*** Consider Example 3.2.2. For (a), $E(X) = \int_0^1 x^2\,dx + \int_1^3 x(1/4)\,dx = 1/3 + 1 = 4/3$. Compare this with the sample mean 1.336 for 10,000 observations. For (b), replacing $x/5$ by $y$ and integrating by parts, $E(X) = \int_0^\infty (x/5)e^{-x/5}\,dx = -5ye^{-y}|\int_0^\infty + 5\int_0^\infty e^{-y}dy = 5$. In general, if 5 is replaced by $\theta > 0$, $E(X) = \theta$. For (c), $E(X) = \theta\int_0^1 x^\theta\,dx = \theta/(\theta+1)$ for $\theta > 0$. For (d), suppose that $\theta > 1$. Then $E(X) = \theta\int_1^\infty x^{-\theta}dx = \theta/(\theta-1)$. $E(X)$ does not exist for $\theta \le 1$. □

The linear properties of the expectation operator continue to hold in the continuous case: $Y = a + bX$ has density $g(y)$, $E(Y) = \int y g(y)\,dy = \int y f((y-a)/b)/b dy) = \int (a+bx)f(x)\,dx = a + bE(X)$, where we have used the change of variable $x = (y-a)/b$. Thus, it is not necessary to determine the density of $Y$ in order to determine its expectation. For example, if $U \sim \text{Unif}(0, 1)$ and $Y = a + (b-a)U$, then $E(Y) = a + (b-a)/2 = (a+b)/2$. We could also have determined this from the fact that $Y \sim \text{Unif}(a, b)$. We will be concerned with the distributions of functions $h(X)$ for the case that $X$ has density $f$. Since $h(X) = Y$ is itself a random variable, we would seem to need to find the density of $Y$ (if it has one) before attempting to determine its expected value. Section 3.4 is devoted to techniques for doing that. Fortunately, however, as indicated by Theorem 3.3.1, we will not need to do that. For a proof a student will have to read a more advanced book or take a more advanced course. Theorem 3.3.1 simply states that Theorem 1.5.2, stated there for discrete random variables, remains true for continuous random variables.

**Theorem 3.3.1** Let $X$ be a random variable with density $f$, and let $Y = g(X)$. Then $E(Y) = \int_{-\infty}^\infty g(x)f(x)\,dx$.

***Example 3.3.2*** Let $U \sim \text{Unif}(0, 1)$ and let $Y = U^2$. Then $Y$ has cdf $G(w) = P(U^2 \le w) = P(U \le w^{1/2}) = w^{1/2}$ for $0 \le w \le 1$, so that $Y$ has density $g(w) = (1/2)w^{-1/2}$ for $0 < w \le 1$. Hence, by definition, $E(Y) = \int_0^1 (1/2)ww^{-1/2}\,dw = (1/2)(2/3) = 1/3$. However, from Theorem 3.3.1, $E(Y) = \int_0^1 u^2 f_U(u)\,du = \int_0^1 u^2du = 1/3$, since $U$ has density $f(u) = 1$ on $[0, 1]$.

By the linearity of the integral it follows that $E(h_1(X) + h_2(X)) = E(h_1(X)) + E(h_2(X))$ whenever both expectations exist, so that, for example, if $U \sim \text{Unif}(0, 1)$, then $E(U^2 + U^3) = 1/3 + 1/4$. It is a bit difficult to determine the density of $Y = U^2 + U^3$ in order to compute $E(Y)$ directly from the definition. □

If $X$ has a distribution that is a mixture distribution of (3.2.1), with $P(X \in A)$ expressed as a convex combination of the integral of a density $f_c$ and the sum over a probability mass function $f_d$, it is natural to define

$$E(X) = \alpha \int x f_c(x)\,dx + (1-\alpha) \sum k f_d(k).$$

If $F$ is the corresponding cdf, then $E(X)$ is usually expressed as a Stieltjes integral $\int x\,dF(x)$.

## The Jelly Donut Problem

Suppose that the Jelly Donut Company must decide how many packages of jelly donuts to bake each day. Packages sell for $s$ dollars and cost $c$ dollars to make, $s > c$. The demand for packages is a random variable $D$, with density $f$, cdf $F$. For simplicity suppose that $f$ is continuous, and for $0 < F(d) < 1$, $f(d) > 0$. The demand must really be an integer, but again for simplicity, this model should provide a reasonably good approximation. This example has much greater applicability to business than is implied by "jelly donut," although, of course, real-world problems tend to be much more complex. We will pretend, for example, that there is no cost to the goodwill lost when customers learn that there are no jelly donuts left, especially after they have dreamed of having two or three jelly donuts for breakfast.

Suppose that the JDC bakes $B$ packages. Then its profit when the demand is $D$ is $Y = sX - cB$, where $X = g(D; B)$ is the number of packages sold and $g(D; B) = D$ for $D \le B$ and $B$ for $D > B$. $X$ has a mixture distribution, with density $f_c = f/F(B)$ for the continuous part, mass function $f_d$ assigning probability 1 to the point $B$, and mixture parameter $\alpha = F(B)$. Therefore, its expected profit is $H(B) \equiv E(Y) = sE(g(D; B)) - cB = sF(B) \int_0^B x f(x)\,dx / F(B) + sB[1 - F(B)] - cB = s\int_0^B x f(x)\,dx - sBF(B) + (s-c)B$. Then $\frac{d}{dB} H(B) = sBf(B) - sF(B) - sBf(B) + (s-c) = -sF(B) + (s-c)$. The first equality follows by the fundamental theorem of calculus. This derivative is zero if $B$ satisfies $F(B) = 1 - c/s$. Since the second derivative of $H(B)$ is negative whenever $f(B) > 0$, this $B$ does maximize $H(B)$. In Problem 3.3.5 students are asked to prove a corresponding result for the case that $D$ is discrete.

***Example 3.3.3*** Suppose that the Jelly Donut Company sells packages of jelly donuts for \$5 that cost \$3 to make. Demand $D$ has density $f(d) = (1/\theta)e^{-d/\theta}$ for $d > 0$ and cdf $F(u) = 1 - e^{-u/\theta}$ for $u > 0$; $H(B) = 5[1 - e^{-B/\theta}] - 3B$, and $\frac{d}{dB} H(B) = e^{-B/\theta} - c$ (see Figure 3.3.1). The expected profit is maximized for $F(B) = 1 - e^{-B/\theta} = 1 - c/s$, or $e^{-B/\theta} = c/s$, $B = \theta \log(s/c) = 0.5108\theta$, roughly one-half of $E(D) = \theta$. As the ratio $s/c$ increases from 1 to $\infty$, the optimum B increases from zero to $\infty$. Dropping the cost to nearly zero allows the optimum $B$

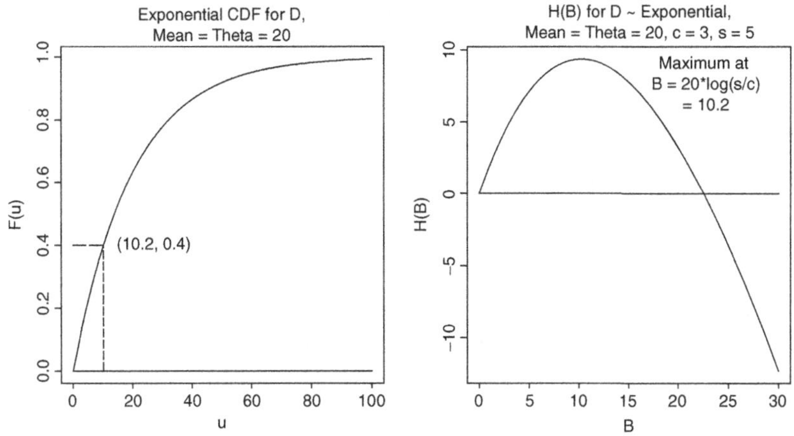


**FIGURE 3.3.1** Jelly donut problem.

to become large. It is tempting, of course, to increase $s$, but naive to think that the distribution of demand would not change as the price increases. We'll leave it to an economist to determine the optimum $s$. □

**Variance** As for discrete random variables, we define the variance of any random variable $X$, with mean $\mu = E(X)$ to be

$$\text{Var}(X) = E[(X - \mu)^2],$$

provided that the expectation exists.

***Example 3.3.4*** Let $U \sim \text{Unif}(0, 1)$. Then $\mu = 1/2$ and $\text{Var}(U) = \int_0^1 (x - 1/2)^2 \cdot 1\,dx = 1/12$. □

Since $Y = a + (b - a)U \sim \text{Unif}(a, b)$, it follows from $\text{Var}(Y) = (b - a)^2\,\text{Var}(U)$ that any random variable that has the Unif$(a, b)$ distribution has variance $(b - a)^2/12$. Compare that to the variance of the distribution which assigns probabilities 1/2 and 1/2 to $a$ and $b$: $(b - a)^2/4$ and the uniform distribution on the points 1, 2, ..., $N$: $(N^2 - 1)/12$ (see Example 1.6.6). It should not be surprising that as $N$ approaches infinity, the ratio of the variances of the discrete uniform distributions on $1, \ldots, N$ and the continuous uniform on $[1, N]$ approaches 1.

We remind the student now of the Markov inequality (Section 1.5): If $P(X \geq 0) = 1$, then $P(X \geq k) \leq E(X)/k$ for all $k > 0$, and the Chebychev inequality (Section 1.6): If Var$(X)$ exists and $\mu = E(X)$, then $P(|X - \mu| \geq k) \leq \text{Var}(X)/k^2$ for any $k > 0$. For the exponential distribution with parameter $\theta = 1$(mean 1), for example, the Markov inequality states that $P(X \geq k) \leq 1/k$, while $P(X \geq k) = e^{-k}$ for each $k > 0$. The Chebychev inequality states that $P(|X - 1| \geq k) \leq 1/k^2$, while $P(|X - 1| \geq k) = e^{-(1+k)}$ for k $\geq$ 1. Formulas (1.6.1), (1.6.2), and (1.6.3) hold as well, again because they depend only on properties of expectation. It is quite simple, for example, to show that when $U \sim \text{Unif}(0, 1)$, then $E(U) = 1/2$, $E(U^2) = 1/3$, so that $\text{Var}(U) = E(U^2) - [E(U)]^2 = 1/3 - (1/2)^2 = 1/12$.

### Problems for Section 3.3

**3.3.1** Let $X$ have density $f(x) = |x|$ for $-1 < x < 1$. Find $E(X)$ and $\text{Var}(X)$.

**3.3.2** Let $X$ have density $f(x) = x^2$ for $0 \le x < 1$ and $2/3$ for $1 \le x \le 2$. Find $E(X)$ and $\text{Var}(X)$.

**3.3.3** Suppose that $X$ has a density $f$ that is symmetric about $c$. That is, $f(c + h) = f(c - h)$ for all real $h$. Show that, if it exists, $E(X) = c$. *Hint*: Make the change of variable $h = x - c$.

**3.3.4** Show that the Markov inequality for continuous random variables $X$ with density $f$ is "sharp" in the following sense. For every $k > 0$ and every $\varepsilon > 0$ there exists $f$, depending on $\varepsilon$, so that $f(x) = 0$ for $x < 0$ and $P(X \ge k) > E(X)/k - \varepsilon$. Is there any pair $(k, f)$ with $k > 0$ and $f$ a density such that $f(x) = 0$ for $x < 0$ and $P(X \ge k) = E(X)/k$? *Hint*: Suppose that the density $f$ is continuous except perhaps on a countable set $B$ of points and use the fact that the integral of such a nonnegative function over an interval can be zero only if the function is zero for all points outside $B$.

**3.3.5** Let $X$ have density $f$, cdf $F$, and suppose that $F(0) = 0$ and that $E(X)$ exists.

**(a)** Prove that $E(X) = \int_0^\infty [1 - F(u)]\, du$ (see Problem 1.5.14). *Hint*: Integrate by parts.

**(b)** Let $X$ have cdf $F(u) = 1 - e^{-u/\theta}$ for $u > 0$, $\theta > 0$. Verify using part (a) that $E(X) = \theta$.

**(c)** Verify part (a) for the case that the random variable is $U \sim \text{Unif}(0, 1)$.

**3.3.6** The demand $D$ for packages of jelly donuts for the Jelly Donut Company has density $f(d) = (1 - d/100)(2/100)$ for the expected profit $0 < d < 100$. If $c = 2$ and $s = 5$, which choice for $B$ will maximize the profit? What is $H(B)$ for this $B$? What are $H(B + 10)$ and $H(B - 10)$? *Hints*: Define $D_1 = D/100$ and $B_1 = B/100$. $D_1$ has a triangular density on $[0, 1]$ and is therefore easier to work with. The profit is $g(D, B) = 100g(D_1, B_1)$. Find $B_1$ to maximize $H_1(B_1) \equiv E[g(D_1, B_1)] = H(B)/100$. Use geometry.

**3.3.7** The Old Donut Company guarantees that it will take all unsold packages from the Jelly Donut Company and will pay \$2 for each. Assume that the demand $D$ for new packages from the JDC will not be affected. Give formulas for the profit $G(D; B)$ and the expected profit $H(B)$.

## 3.4 TRANSFORMATIONS OF RANDOM VARIABLES

Suppose that $U \sim$ Unif(0, 1). What is the density of $X = U^2$? Or of $Y = \tan(2\pi U)$? Or $W = 1/X$? In general, if $X$ has density $f$ and $Y = h(X)$, can we write down a simple expression for the density of $Y$ in terms of $f$ and $h$? In the case of discrete random variables, finding the probability mass function for $Y$ is relatively easy, because we need only sum, for each possible value $y$ of $Y$, the probabilities $f(x)$ for which $h(x) = y$. For the continuous case, we cannot do that. The technique instead will be to express the cdf $G$ of $Y$ in terms of the cdf $F$ of $X$ and the function $h$. Let us begin with a simple example.

***Example 3.4.1*** Let $X$ have density $f(x) = -x/2$ for $-1 \le x < 0$ and $3x/2$ for $0 \le x \le 1$, and let $Y = h(X) = X^2$. We first notice that the range of $Y$ is the interval [0, 1]. It pays to graph $h(x)$ and note the set of values $x$ such that $y = h(x)$ with $f(x) > 0$. Let $Y$ have cdf $G$. Then for $v \ge 0$, $G(v) = P(h(X) \le v) = P(-v^{1/2} \le X \le v^{1/2}) = F(v^{1/2}) - F(-v^{1/2})$. If we have a convenient expression for $F$, we could now present an expression for $G$. Problem 3.4.2 asks us to do that. If we are satisfied to find the density of $Y$ without finding $G$, we can formally differentiate at this point. The density of $Y$ is then, for $v > 0$, $g(v) = f(v^{1/2})(v^{-1/2}/2) - f(-v^{-1/2})(-v^{-1/2}/2) = f(v^{1/2})(v^{-1/2}/2) + f(-v^{1/2})(v^{-1/2}/2) = (v^{-1/2}/2)[f(v^{1/2}) + f(-v^{1/2})]$.

Notice that to this point we have not needed to know $f$. The density $g$ is a weighted sum of the values of $f$ at $v^{1/2}$ and at $-v^{1/2}$ because these two values of $X$ produce the same $Y$ at the possible value $v$ of $Y$; that is, $\{x \mid h(x) = y\}$ has two members for each $y$ in the range of $Y$. Substituting $f$, we get $g(v) = (v^{-1/2}/2)[v^{1/2}/2 + 3v^{1/2}/2] = 1$ for $0 \le v \le 1$, 0 otherwise. Thus, $Y \sim$ Unif(0, 1) (see Figure 3.4.1). □

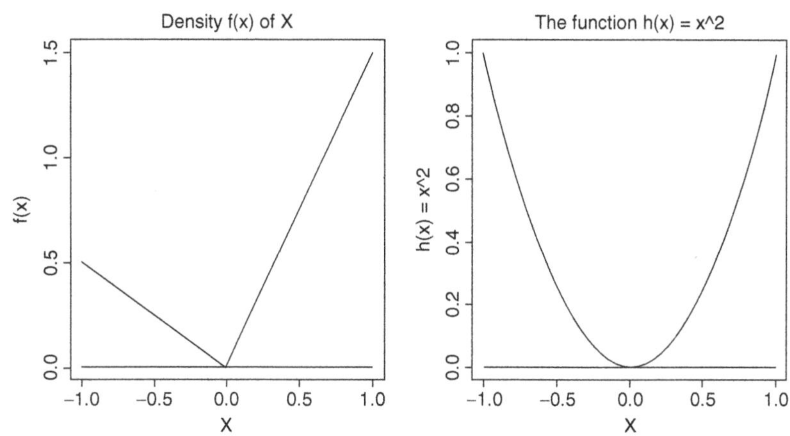


**FIGURE 3.4.1** Density $f$ and transformation $h$.

We can now determine $E(Y)$ in two ways. The definition gives $E(Y) = \int_0^1 yg(y)\,dy = 1/2$. On the other hand, Theorem 3.3.1 states that $E(Y) = \int_{-1}^1 x^2 f(x)\,dx = \int_{-1}^0 (-x^2/2)\,dx + \int_0^1 (3x^3/2)\,dx = 1/8 + 3/8 = 1/2$. The first way turned out to be easier because we had first determined the density of $Y$.

***Example 3.4.2*** Let $X \sim \text{Unif}(0, 1/2)$, and let $Y = 1/X$. Let us find the density $g$ of $Y$. $X$ has cdf $F(u) = 2u$ for $0 \le u \le 1/2$. Then $Y$ has range $[2, \infty)$, with cdf $G(v) = P(Y \le v) = P(X \ge 1/v) = 1 - F(1/v) = 1 - 2/v$ for $v \ge 2$, so that $Y$ has density $g(v) = 2/v^2$ for $v \ge 2$. The author has found in the past that some students ignore domains of $G$ and $g$, failing to state the set over which $g$ is positive. Of course, he knows that *you* would never do that. When $h$ is a monotone function on the range of $X$, we can determine a relatively simple formula for the density $g$ of $Y$. The author offers such a formula with some trepidation, because such formulas can easily be misused. In fact, he suggests that the method by which $G$ is first expressed be in terms of $F$. That was done for Example 3.4.1. That method should *always* be used when $h$ is not monotone on the set on which $f$ is positive. □

**Theorem 3.4.1** Let $X$ have continuous density $f$, cdf $F$. Let $Y = h(X)$, where $h$ is monotone (increasing or decreasing) on the set $A$ on which $f$ is positive, with derivative $h'$. Let $B = h(A) \equiv \{y \mid y = h(x), x \in A\}$. Let $k(y)$ satisfy $h(k(y)) = y$ for each $y \in B$. Then $Y$ has density

$$g(y) = f(k(y))|k'(y)| = f(k(y))/|h'(k(y))| \qquad \text{for all } y, \tag{3.4.1}$$

where $k'$ and $h'$ are derivatives of $k$ and $h$.

*Notes:* Students should work through the following example, drawing some graphs as they go. The absolute values defining the Jacobian, $|k'(y)|$, are needed for the case that $h$ is monotone decreasing. The function $k$ is usually called the *inverse* of $h$ and denoted by $h^{-1}$. Equation (3.4.1) requires only that we can find a "version" of $f$ (a density $f^*$ with the same $F$) for which the hypotheses hold. We delay a proof until after the following example. □

***Example 3.4.3*** Let $A = (0, 1]$ and let $X$ have density $f(x) = 2x$ for $x \in A$. Let $h(x) = 1/x^2$ for $x \in A$. Then $h$ is monotone decreasing on $A$, and $B = h(A) = [1, \infty)$. The inverse of $h$ is $k(y) = 1/y^{1/2}, k'(y) = -y^{-3/2}/2$, defined for $y \in B$. $h'(x) = -2x^{-3}$, and $h'(k(y)) = -2y^{-3/2}$ for $y \in B$. Thus, the density of $Y$ is $g(y) = (2/y^{1/2})|-y^{-3/2}/2| = y^{-2}$ for $y \in B = [1, \infty)$. $k'(y)$ can also be expressed as $1/h'(k(y)) = 1/(-2y^{-1/2})^{-3} = -y^{-3/2}/2$, as before. Notice that $f(0)$ could be redefined to be 1 without changing the probability distribution of $X$ or of $Y$, and that $h(0)$ is not defined. No harm done. Since $P(X = 0) = 0$, we can ignore $X = 0$. Similarly, if $h(1/2)$ is defined to be 17, $Y$ continues to have density $g$. □

*Proof of Theorem 3.4.1* If $h$ is monotone increasing, then $Y$ has cdf $G(y) = P(h(X) \leq y) = P(X \leq k(y)) = F(k(y))$ for all $y$. Differentiating with respect to $y$, we get (3.4.1). For the case that $h$ is monotone decreasing we have $G(y) = P(h(X) \leq y) = P(X \geq k(y)) = 1 - F(k(y))$, which has derivative $-f(k(y))k'(y)$, which is again (3.4.1). The second equality in (3.4.1) is a standard result on the derivatives of inverse functions. □

Two examples of transformations are of special interest. First, let $X$ have continuous cdf $F$ and let $Y = F(X)$. For $0< u < 1$, define $F^{-1}(u) = \inf\{x|F(x) \geq u\}$. For the case that $F$ is monotone increasing (not ever "flat"), $F^{-1}$ is the usual inverse function. The continuity of $F$ on the right guarantees that for each $u$ there is an $x$ such that $x = F^{-1}(u) = F^{-1}(u)$. Then $Y$ has cdf $G(y) = P(Y \leq y) = P(F(X) \leq y) = P(X \leq F^{-1}(y)) = F(F^{-1}(y)) = y$ for $0 < y < 1$. But this is the cdf of the Unif(0, 1) distribution, so that $Y = F(X)$ has the uniform distribution on [0, 1]. Notice that the continuity of $F$ is essential. If $X$ is discrete, so must $Y = F(X)$ be discrete. In a certain sense $Y$ is "approximately" distributed as Unif(0, 1). For example, if $X$ is the number of heads in three tosses of a fair coin, then $Y$ has cdf $G$ with $G(1/8) = 1/8$, $G(1/2) = 1/2$, $G(7/8) = 7/8$, $G(1) = 1$, but $G$ is flat between these jump points. Looking again at Example 3.4.3, we see that $F(x) = x^2$ on [0, 1]. Thus, $F^{-1}(y) = y^{1/2}$, so that $Y = X^{1/2} \sim$ Unif(0, 1).

Let $F$ be any cdf. For purposes of simulation it will be useful to be able to generate observations $X$ in a computer which act as if they are random variables having cdf $F$. As stated in Theorem 3.4.2, if $U \sim$ Unif(0, 1), then $X = F^{-1}(U)$ has cdf F. Most statistical and mathematical software packages have build-in commands that produce independent observations on $U$. Usually, these are based on some simple multiplicative algorithm of the type $a_{n+1} = (a_n m) \bmod M$, with some starting point $a_0$, where $M$ is a large positive integer, $m$ a smaller integer, relatively prime to $M$. For example, $M$ might be $2^{32}$ or $2^{32} - 1$ and $m$ could be $7^5$. If $M = 7$, $m = 3$, we might take $a_0 = 2$, so that the sequence $a_n$ becomes 2, 6, 4, 5, 1, 3, 2, so that the period is 6. Then, if $M$ and $m$ are chosen wisely, the sequence $\{U_n = a_n/M\}$ often behaves much like a sequence of independent Unif(0, 1) random variables. These may then be used to generate "pseudo" (not really random) random variables. Although the sequence generated is not truly random, it is true that the observations behave much like they are, which is important.

**Theorem 3.4.2** Let $U \sim$ Unif(0, 1), let $F$ be a cdf, and let $F^{-1}(u) = \inf(x|F(x) \geq u\}$ for each $u$, $0 < u < 1$. Then $X = F^{-1}(U)$ has cdf $F$.

*Notes:* We have written $F(x)$ rather than $F(u)$, to avoid confusion with the argument $u$ of $F^{-1}$. $F^{-1}(u)$ is the leftmost $u$th-quantile for $F$. If $F(x) = u$ for $x \in [F^{-1}(u), b)$ for some $b > F^{-1}(u)$, any such $x$ qualifies as a $u$th-quantile. □

*Proof:* $X$ has cdf $G(x) \equiv P(X \leq x) = P(F^{-1}(U) \leq x) = P(U \leq F(x)) = F(x)$ for all real numbers $x$. The next-to-last equality follows from the fact that for each pair $u$, $0 < u < 1$, and $x$, $F^{-1}(u) \leq x$ if and only if $u \leq F(x)$. □

***Example 3.4.4***

**(a)** Let $X$ be the discrete random variable taking the values 1, 2, 3 with probabilities 0.2, 0.5, 0.3. Then $F$ takes the values 0, 0.2, 0.7, 1.0 on the intervals $(-\infty, 1)$, $[1, 2)$, $[2, 3)$, $[3, \infty)$. Therefore, $X$ is 1 for $U < 0.2$, 2 for $0.2 \le U < 0.7$, and 3 for $0.7 \le U \le 1$, so $X$ has cdf $F$.

**(b)** Let $X$ have the density of Example 3.2.2: $f(x) = x$ on $[0, 1]$, $1/4$ on $(1, 3]$. Then, as indicated in Section 3.2, $F(x) = x^2/2$ on $[0, 1]$, $1/2 + (x-1)/4$ on $(1, 3]$, and $F^{-1}(u) = \sqrt{2u}$ for $0 \le u \le 1/2$, $4(u - 1/2) + 1 = 4u - 1$ for $1/2 < u \le 1$. Then $X = F^{-1}(U)$ has cdf $F$, density $f$.

**(c)** Let $F(x) = 1 - e^{-x/\theta}$ for $x > 0$, the cdf of the exponential distribution with mean $\theta > 0$. Then $F^{-1}(u) = -\theta \log(1 - u)$ for $u \in [0, 1)$, so that if $U \sim \text{Unif}(0, 1)$, then $X = F^{-1}(U) = -\theta \log(1 - U)$ has the exponential distribution with mean $\theta$, rate $1/\theta$. The distribution would remain the same if $1 - U$ were replaced by $U$.

**(d)** Suppose that we wish to generate random variables with density $f(x) = (1/\pi)e^{-x^2}$ for all $x$. (Please believe me—it is a density. Unfortunately, there is no closed-form expression for the corresponding cdf. It can be expressed as an integral or as an infinite series, and its inverse can be expressed an as infinite series. We will need to find another way to generate such random variables. □

We can increase our ability to simulate sampling from prescribed distributions a bit as follows. Let $X$ have cdf $F(x) = \sum_{i=1}^k p_i F_i(x)$, where $\sum_{i=1}^k p_i = 1$ and the $F_i$ are cdf's. $F$ is a *mixture* of the $F_i$. If $B$ takes the values $1, \ldots, k$ with probabilities $p_1, \ldots, p_k$ and $X_i \sim F_i$ for each $i$, and $B$ and $(X_1, \ldots, X_k)$ are independent, then $X = \sum_{i=1}^k I([B = i])X_i = X_j$ if $B = j$ for $j = 1, \ldots, k$ has cdf $F$. To see this, note that $P(X \le x) = \sum_{i=1}^k P(X \le x | B = i)P(B = i) = \sum_{i=1}^k p_i F_i(x)$.

***Example 3.4.5*** Consider the density $f$ of Example 3.2.2: $f(x) = x$ for $x \in [0, 1]$, $f(x) = 1/4$ for $x \in (1, 3]$. Then $f = (1/2)f_1 + (1/2)f_2$, where $f_1(x) = 2x$ on $[0, 1]$, and $f_2$ is the uniform density on $[1, 3]$. Let $U_1$, $U_2$ be independent, each Unif(0, 1). Let $B$ be 1 if $U_1 \le 1/2$, 2 otherwise. If $B = 1$, let $X = \sqrt{U_2}$. If $B = 2$, let $X = 1 + 2U_2$. Then $X$ has a density that is a $(1/2, 1/2)$ mixture of $f_1$ and $f_2$, as we wished. □

### Problems for Section 3.4

**3.4.1** Let $X \sim B(4, 1/2)$. Find the probability function for $Y = |X - 2|$.

**3.4.2** Let $X$ have density $f(x) = |x|$ for $-1 \le x \le 1$.

**(a)** Find the density $g$ of $Y = X^2$.

**(b)** Determine $E(X^2)$ using $g$ and also by using $f$.

**(c)** Let $U \sim \text{Unif}(0, 1)$. Give a function $H$ on $[0, 1]$ such that $X = H(U)$ has density $f$.

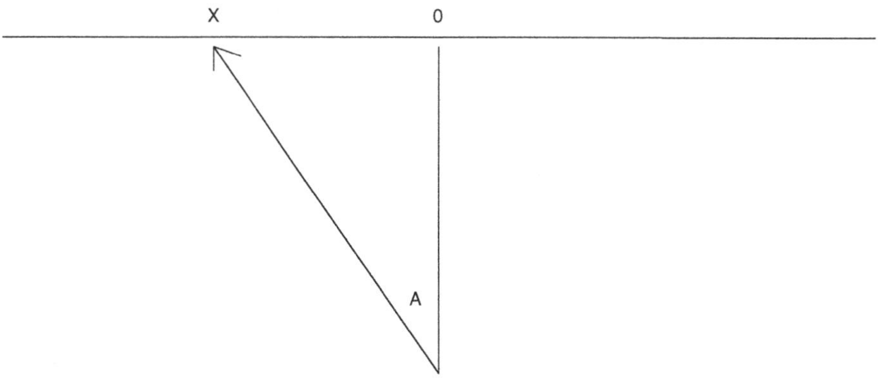


**FIGURE 3.4.5** Beam of light and the Cauchy distribution.

**3.4.3** Let $X$ have density $f(x) = 2x$ on $[0, 1]$. Find a function $H$ on $[0, 1]$ such that:

**(a)** $H(X) \sim \text{Unif}(0, 1)$.

**(b)** $H(X)$ has the density $g(w) = 1/w^2$ for $w > 1$.

**(c)** $H(X)$ has the density $g(w) = w$ for $0 < w < 1$, $1/8$ for $1 \le w < 5$.

**3.4.4** Let $F(x) = x/(1 + x)$ for $x \ge 0$.

**(a)** Let $X \sim F$. Find a function $H$ such that $H(X) \sim \text{Unif}(0, 1)$.

**(b)** Let $U \sim \text{Unif}(0, 1)$. Find a function $K$ such that $K(U) \sim F$.

**(c)** Let $X \sim F$. Find the density of $Y = 1/X$.

**3.4.5** A beam of light (see Figure 3.4.5) is sent from a point $d$ meters from a straight wall of infinite length at an angle between 0 to $\pi$ radians so that the beam always strikes the wall. (The Great Wall of China without bends would be an approximation.)

**(a)** If the angle $A$ is uniformly distributed between 0 and $\pi$, what is the density of $X$, the location of the point at which the beam strikes the wall, where the zero point is the point nearest the point of origin of the beam?

**(b)** $X$ is said to have the *Cauchy distribution*. Show that the expectation of $X$ does not exist.

**3.4.6** Let $X$ have the exponential distribution, with density $f(x) = e^{-x}$ for $x > 0$. Find the cdf and density of of $Y = \sqrt{X}$.

## 3.5 JOINT DENSITIES

We need to be able to describe the joint behavior of two or more random variables. For example, let $U_1, U_2, U_3$ be independent, each Unif(0, 1). Let $X = \min(U_1, U_2, U_3)$

and let $Y = \max(U_1, U_2, U_3)$. Intuitively, we should not expect $X$ and $Y$ to be independent since $P(X < Y) = 1$. If we wish to model the heights $X$ and $Y$ of fathers and sons, we should not expect a probability model under which $X$ and $Y$ are independent to provide an accurate description of the distribution of $(X, Y)$.

We begin with a discussion of the joint behavior of just two random variables, then generalize to the case $n \geq 2$.

**Definition 3.5.1** Two random variables $X$ and $Y$ are said to have *joint density* $f(x, y)$ if for every (Borel) subset $A$ of $R_2$ $P((X, Y) \in A) = \int_A f(x, y)\,dx\,dy$ (see Figure 3.5.1). □

*Notes:* A function $f$ can be a joint density on $R_2$ whenever $f(x, y) \geq 0$ for all $(x, y)$ and its integral over $R_2$ is 1. □

***Example 3.5.1*** Let $(X, Y)$ have joint density $f(x, y) = 8xy$ for $0 \leq y \leq x \leq 1$, 0 otherwise. Then $\int_0^1 \left(\int_0^x 8xy\,dy\right) dx = 1$, so that $f$ is a joint density. Let $T = X + Y$. Then $P(T > 1) = \int_{1/2}^1 (\int_{1-x}^x (8xy\,dy)\,dx = 4\int_{1/2}^1 x(-1 + 2x)\,dx = 5/6$. See Figure 3.5.2 to understand the limits of integration. □

If $A = A_1 \times R_1$, a *cylinder set* in $R_2$, then $P(X \in A_1) = P((X, Y) \in A)$ $= \int_{A_1} \int_{R_1} f(x, y)\,dy\,dx$. Since this is true for every (Borel) subset $A_1$, it follows that $\int_{R_1} f(x, y)\,dy$ is a density for $X$. That is, the *marginal density* of $X$ is $f_X(x) = \int_{R_1} f(x, y)\,dy$. Similarly, the marginal density of $Y$ is $f_Y(y) = \int_{R_1} f(x, y)\,dx$ (see Figure 3.5.3).

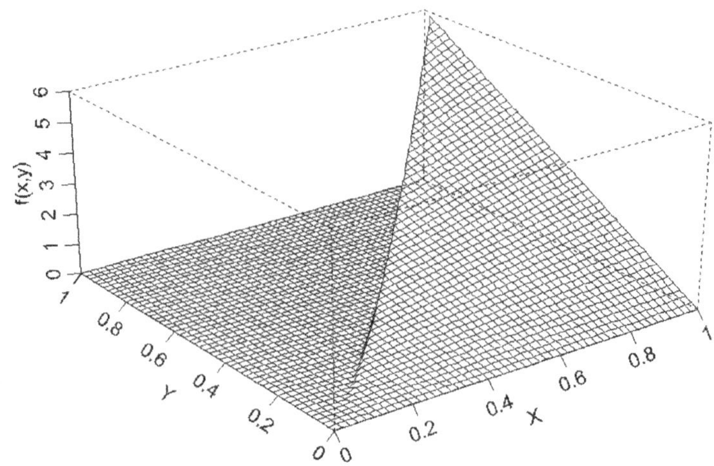


**FIGURE 3.5.1** Joint density $f(x, y)$.

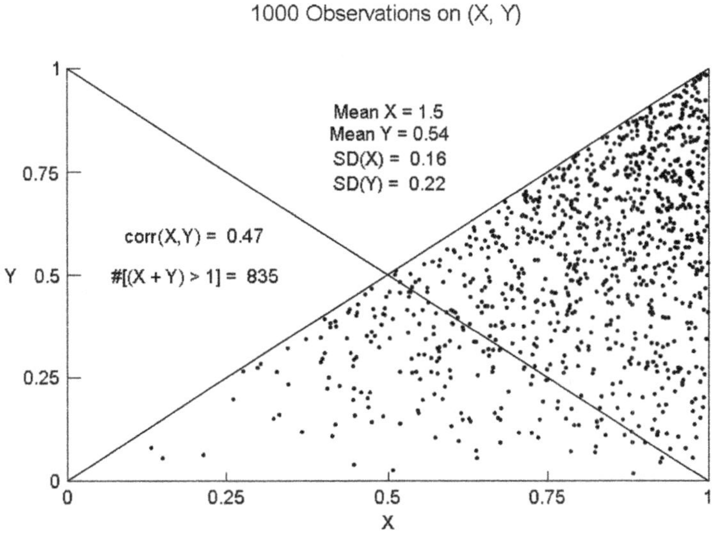


**FIGURE 3.5.2** Simulations on $(x, y)$.

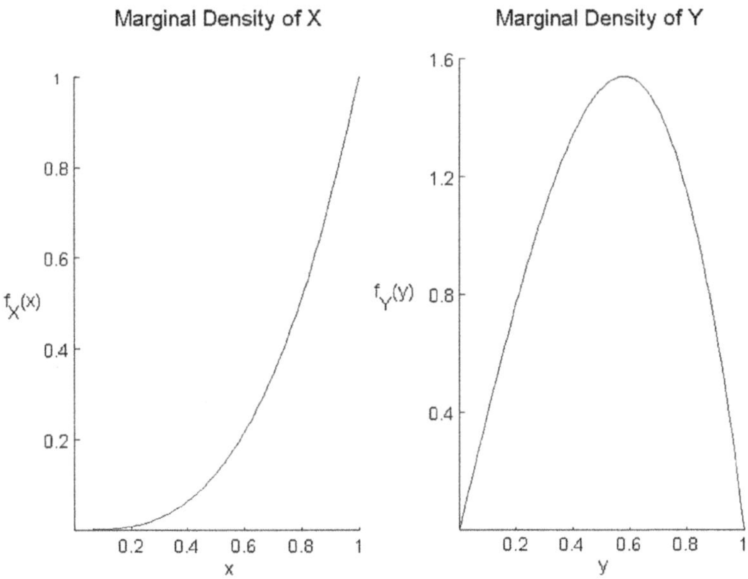


**FIGURE 3.5.3** Marginal densities of $X$ and $Y$.

***Example 3.5.1 Continued*** Integrating $f(x, y)$ with respect to $y$, we get $f_X(x) = \int_0^x (8xy)\,dy = 4x^3$ for $0 \leq x \leq 1$. Similarly, $f_Y(y) = \int_y^1 (8xy)\,dx = 4y[1 - y^2]$ for $0 \leq y \leq 1$. $\square$

***Example 3.5.2*** Suppose again that $(X, Y)$ has joint density $f(x, y) = 2e^{-x-y}$ for $0 \le y \le x < \infty$, as in Example 3.5.1. Then $X$ has marginal density $f_X(x) = \int_0^x f(x, y)\,dy = 2e^{-x}(1 - e^{-x})$ for $x \ge 0$ and $Y$ has density $f_Y(y) = \int_y^\infty f(x, y)\,dx = 2e^{-2y}$ for $y \ge 0$. Let $T = X + Y$. Let's try to find the cdf and density of $T$. For $t \ge 0$, $F_T(t) = P(T \le t) = \int_0^{t/2} \int_y^{t-y} f(x, y)\,dx\,dy = 1 - e^{-t}(1 + t)$. Thus, $T$ has density $f_T(t) = \frac{d}{dt} F_T(t) = te^{-t}$ for $t \ge 0$. □

If $X$ and $Y$ are independent with densities $f$ and $g$, then for any (Borel) subsets $A$ and $B$ of $R_1$, $P(X \in A, Y \in B) = P(X \in A)P(Y \in B) = \int_A f(x)\,dx \int_B g(y)\,dy = \int_A \int_B f(x)g(y)\,dx\,dy$. Since $P((X, Y) \in A \times B)$ is the integral of $f(x)g(y)$ over $A \times B$, it follows that for any (Borel) subset $C$ of $R_2$,

$$P((X, Y) \in C) = \int_C f(x)g(y)\,dx\,dy. \qquad (3.5.1)$$

A rigorous proof of (3.5.1) is beyond the scope of this course. (Take a course on real analysis or measure theory.) From (3.5.1) and Definition 3.5.1 it follows that the joint density of $(X, Y)$ is the product of $f$ and $g$.

If $(X, Y)$ has joint density $h(x, y) = h_1(x)h_2(y)$ for all $x, y$, where $h_1$ and $h_2$ are nonnegative functions, then for any (Borel) subsets $A, B$ of $R_1$, $P((X, Y) \in A \times B) = \int_{A\times B} h_1(x)h_2(y)\,dx\,dy = \int_A h_1(x)\,dx \int_B h_2(y)\,dy$. Let $C_1 = \int_{R_1} h_1(x)\,dx$ and $C_2 = \int_{R_1} h_2(y)\,dy$. Taking $A = R_1$, $B = R_1$, we see that $C_1C_2 = 1$. Taking $B = R_1$ we find that $P(X \in A) = \int_A h_1(x)C_2\,dx$, so $X$ has density $h_1(x)C_2 = h_1(x)/C_1$. Similarly, $Y$ has density $h_2(y)C_1 = h_2(y)/C_2$. Thus, $(X, Y)$ has a joint density that is the product of the densities of $X$ and $Y$, so $X$ and $Y$ are independent.

A pair of random variables $(X, Y)$ possesses a *joint cdf*. This point function on $R_2$ is sometimes useful, although for most applications it turns out to be somewhat unwieldy or can be expressed only as an integral.

**Definition 3.5.2** The pair of random variables $(X, Y)$ has *joint cumulative distribution function* $F(u, v) = P(X \le u, Y \le v)$, defined for all $(u, v) \in R_2$. □

*Notes:* Since $F(u, \infty) \equiv \lim_{v\to\infty} F(u, v) = P(X \le u) = F_X(u)$ for each $u$, and, similarly, $F(\infty, v) = F_Y(v)$ for each $v$, the marginal cdf's may be obtained from the joint cdf. Similarly to the one-dimensional case, if $(X, Y)$ has joint density $f(x, y)$, which is continuous at a point $(u, v)$, then $\frac{\delta^2}{\delta u \delta v} F(u, v) = f(u, v)$ at that point. □

***Example 3.5.3*** Let $(X, Y)$ have joint density $f(x, y) = 2$ for $0 \le y \le x \le 1$. Then $F(u, v) = v^2 + 2v(u - v)$ for $0 \le v \le u \le 1$ and $u^2$ for $0 \le u < v \le 1$. $X$ has marginal cdf $F_X(u) = F(u, \infty) = F(u, 1) = u^2$ for $0 \le u \le 1$ and $Y$ has marginal cdf, $F_Y(v) = F(\infty, v) = v^2 + 2v(1 - v) = 2v - v^2$ for $0 \le v \le 1$. □

In the rare case in which $F$ has a tractable form, we can express the probability that $(X, Y)$ will lie in a rectangle in terms of the joint cdf: $P(a_1 < X \le b_1,\ a_2 < Y \le b_2) = P(a_1 < X \le b_1,\ Y \le b_2) - P(a_1 < X \le b_1,\ Y \le a_2) = [P(X \le b_1,\ Y \le b_2) - P(X \le a_1,\ Y \le b_2)] - [P(X \le b_1,\ Y \le a_2) - P(X \le a_1,\ Y \le a_2)] = F(b_1, b_2) - F(a_2, b_2) - F(b_1, a_2) + F(a_1, a_2)$. If $(X, Y)$ has a joint density, we can afford to be sloppy about $\le$ and $<$ signs. However, if $(X, Y)$ is discrete, we must be careful.

We have discussed the two-dimensional case at some length because generalizations to higher dimensions are relatively obvious after the two-dimensional case is "conquered."

**Definition 3.5.3** The random variables $X_1, \ldots, X_n$ have joint density $f(x_1, \ldots, x_n)$ if for every (Borel) subset $A \subset R_n$,

$$P((X_1, \ldots, X_n) \in A) = \int_A f(x_1, \ldots, X_n)\, dx_1 \cdots dx_n. \qquad \square$$

*Notes:* Rather than repeat definitions, statements, and proofs of the theorems for $n > 2$ already presented for the case $n = 2$, we simply summarize the more important properties. Let $\mathbf{X} = (X_1, \ldots, X_n)$ and let $\mathbf{x} = (x_1, \ldots, x_n)$ be a point in $R_n$.

1. Suppose that $\mathbf{X}$, having $n$ components, has density $f$. If $D$ is a subset of the indices $1, 2, \ldots, n$, a joint density of $(X_i, i \in D)$ is the integral of $f$ over all $x_i$ with $i \notin D$. For example, if $n = 5$, a joint density of $(X_1, X_3, X_5)$ is the integral of $f$ over all $x_2, x_4$. We say *a* joint density rather than *the* joint density because there always exist an infinity of functions $f$ that produce the same integral values.
2. If $\mathbf{X}$ has density $f$ and $f(\mathbf{x}) = \prod_{i=1}^n f_i(x_i)$ for all $\mathbf{x}$, then $X_1, \ldots, X_n$ are independent. (The "for all $x$" can be modified by allowing a countable number of exceptions—(more rigorously, a set of exceptions of "measure" zero.)
3. If $X_1, \ldots, X_n$ are independent and $X_i$ has density $f_i$ for each $i$, $\mathbf{X}$ has a density $f(\mathbf{x}) \equiv \prod_{i=1}^n f_i(x_i)$ for all $\mathbf{x} \in R_n$. $\square$

The model under which $n$ random variables are independent, all with the same cdf $F$, or density or probability function $f$, is so useful that special shorthand language is commonly used.

**Definition 3.5.4** Let $X_1, \ldots, X_n$ be independent, each with the same distribution with cdf $F$ (or density or probability function $f$). Then $X_1, \ldots, X_n$ is said to be a *random sample* from $F$ (or from $f$). We also say that $X_1, \ldots, X_n$ are *independent and identically distributed* (iid). Probabalists usually say "iid" whereas statisticians say "random sample." $\square$

***Example 3.5.4***

**(a)** Suppose that $\mathbf{X} = (X_1, X_2, X_3)$, the $X_i$ are independent, and $X_i$ has density $f_i$ for $i = 1, 2, 3$. Then $\mathbf{X}$ has density $f(x_1, x_2, x_3) = f_1(x_1) f_2(x_2) f_3(x_3)$ for all $x_1, x_2, x_3$.

**(b)** Let $X_{(1)}, X_{(2)}, X_{(3)}$ be the ordered $X_1, X_2, X_3$, so that $0 < X_{(1)} < X_{(2)} < X_{(3)} < 1$ with probability 1. It follows that $\mathbf{X}_0 \equiv (X_{(1)}, X_{(2)}, X_{(3)})$ has density $f_0(y_1, y_2, y_3) = \sum_{(i_1,i_2,i_3)\in J} f_i(x_i)$, where $J$ is the set of 3! = 6 permutations of (1, 2, 3), for each $y_1 < y_2 < y_3$. If, for example, $X_i \sim$ Unif(0, 1) for each i, then $\mathbf{X}_0$ has joint density $f(y_1, y_2, y_3) = 6$ for $0 < y_1 < y_2 < y_3 < 1$. Then $(X_{(1)}, X_{(3)})$ has joint density $\int_{y_1}^{y_3} f(y_1, y_2, y_3)\,dy_2 = 6(y_3 - y_1)$ for $0 < y_1 < y_3 < 1$, $X_{(1)}$ has density $f_{(1)}(y_1) = \int_{y_1}^{1} 6(y_3 - y_1)\,dy_3 = 3(1 - y_1^2) - 6y_1(1 - y_1) = 3(1 - y_1)^2$ for $0 < y_1 < 1$, and similarly, $X_{(3)}$ has density $f_{(3)}(y_3) = 3y_3^2$ for $0 < y_3 < 1$. $X_{(1)}, X_{(2)}$, and $X_{(3)}$ are called the *order statistics* corresponding to $(X_1, X_2, X_3)$. □

## Maxima and Minima

In Example 3.4.4 we were able to determine the joint density of the minimum and the maximum of three independent random variables by integrating over $x_2$. With enough patience we could do the same for the case of $n > 3$ observations. However, it is instructive to determine the joint distribution and the marginal distributions by more direct means.

Suppose that $X_1, \ldots, X_n$ are independent and that for each $i$, $X_i \sim F_i$. Let $X_{(n)} = \max(X_1, \ldots, X_n)$ and $X_{(1)} = \min(X_1, \ldots, X_n)$. Let their cdf's be $G_{nn}$ and $G_{n1}$. Then $[X_{(n)} \le u] = \bigcap_{i=1}^n [X_i \le u]$. The events $[X_i \le u]$ are independent and $P(X_i \le u) = F_i(u)$ for each $i$. Therefore, the cdf of $X_{(n)}$ is $G_{nn}(u) \equiv \prod_{i=1}^n F_i(u)$. If the $F_i$ are all the same, say $F$, then $G_{nn}(u) = [F(u)]^n$ for each $u$. In the special case that $F(u) = u$ for $0 \le u \le 1$, i.e., each $X_i \sim$ Unif(0, 1), $G_{nn}(u) = u^n$. Easy computation shows that when each $X_i$ is Unif(0, 1) $X_{(n)}$ has median $x_{0.5} = 0.5^{1/n}$, $E(X_{(n)}) = n/(n+1)$, $E(X_{(n)}^2) = n/(n+2)$, and Var $(X_{(n)}) = n/[(n+1)^2(n+2)]$.

Similarly, the events $[X_i > u]$ are independent, $[X_{(1)} > u] = \bigcap_{i=1}^n [X_i > u]$ and $P(X_i > u) = 1 - F_i(u)$. It follows that $X_{(1)}$ has cdf $G_{n1}(u) = P(X_{(1)} \le u) = 1 - P(X_{(1)} > u) = 1 - \prod_{i=1}^n [1 - F_i(u)]$. If all $F_i = F$, then $G_{n1}(u) = P(X_{(1)} \le u) = 1 - [1 - F(u)]^n$, and if $F$ is the cdf of the Unif(0, 1) distribution, it follows that $X_{(1)}$ has cdf $G_{n1}(u) = 1 - (1-u)^n$ for $0 \le u \le 1$. If $F(u) = 1 - e^{-u/\theta}$ for $u > 0$ (i.e., each $X_i$ has the exponential distribution with mean $\theta$), then $G_{n1}(u) = 1 - e^{-nu/\theta}$, so that $X_{(1)}$ has the exponential distribution with mean $\theta/n$. These ideas may be exploited a bit more to give an explicit formulas for the joint cdf of $(X_{(1)}, X_{(n)})$. Consider any $u_1 \le u_2$ and let $A_i = [u_1 < X_i \le u_2]$. Then $B \equiv [u_1 < X_{(1)} \le X_{(n)} \le u_2] = \bigcap_{i=1}^n A_i$. These $A_i$ are independent, and $P(A_i) = F_i(u_2) - F_i(u_1)$, so that $P(B) = \prod_{i=1}^n [F_i(u_2) - F_i(u_1)]$. Therefore, $(X_{(1)}, X_{(n)})$ has joint cdf $G(u_1, u_2) = P(X_{(1)} \le u_1,\ X_{(n)} \le u_2) = P(X_{(n)} \le u_2) - P(X_{(1)} > u_1,\ X_{(n)} \le u_2) = G_{nn}(u_2) - P(B) = \prod_{i=1}^n F_i(u_2) - \prod_{i=1}^n [F_i(u_2) - F_i(u_1)]$. The marginal cdf's are $G_{n1}(u_1) = G(u_1, \infty)$ and $G_{nn}(u_2) = G(u_2, u_2)$, as given above. For the case that $F_i \equiv F$ for each $i$, $G$ becomes $G(u_1, u_2) = F(u_2)^n - [F(u_2) - F(u_1)]^n$. Specializing still further to the case that each $F$ is the Unif(0, 1) cdf, we get $G(u_1, u_2) = u_2^n - (u_2 - u_1)^n$ for $0 \le u_1 < u_2 \le 1$. It follows in this last case by taking partial derivatives that $(X_{(1)}, X_{(n)})$ has joint density $g(u_1, u_2) = n(n-1)(u_2 - u_1)^{n-2}$ for $0 \le u_1 < u_2 \le 1$.

The marginal densities, already given above, can be obtained by integrating with respect to $u_1$ or $u_2$.

### Expectations of Functions of RVs

Let $X$ and $Y$ be independent, each with the Unif(0, 1) distribution. Let $g(x, y) = x + y$ for $0 < x < y < 1$, 0 otherwise. Let $W = g(X, Y)$. Then $W$ is a 1/2, 1/2 mixture of the distribution with mass 1 at zero and the triangular distribution on [0, 2]. Therefore, $E(W) = 0(1/2) + (1)(1/2) = 1/2$. In this case we were able to find the distribution of a function of $(X, Y)$.

In general, since $W$ is a random variable, $E(W) = E(g(X, Y))$ has already been defined. However, the definition would seem to require that the distribution of $W$ be found. This is not always easy. Fortunately, the following theorem, a generalization of Theorem 3.3.1, again given without proof, often makes it much easier.

**Theorem 3.5.1** Let $\mathbf{X} = (X_1, \ldots, X_n)$ have density $f(\mathbf{x})$ on $R_n$. Let $g(\mathbf{x})$ be defined on a set $A$ such that $P(\mathbf{X} \in A) = 1$. Then

$$E(g(\mathbf{X})) = \int_A g(\mathbf{x}) f(\mathbf{x})\, d\mathbf{x} \equiv \int_A g(x_1, \ldots, x_n) f(x_1, \ldots, x_n)\, dx_1 \cdots dx_n.$$

In the example above, letting $A$ be the triangle $\{(x, y) \mid 0 < x < y < 1\}$, $E(g(X, Y)) = \int_{A^c} (0) f(x, y)\, dx\, dy + \int_A (x + y) f(x, y)\, dx\, dy = 0 + 1/2 = 1/2$. $E[g(X)^2]$ and $\mathrm{Var}(g(X))$ can be found by similar methods.

### Problems for Section 3.5

**3.5.1** Let $(X, Y)$ have joint density $f(x, y) = Ce^{-x}$ for $0 < y < x < \infty$.

(a) Find $C$. *Hint*: $\int_0^a xe^{-x} dx = 1 - (1 + a)e^{-a}$ for $a > 0$.

(b) Find the marginal densities of $X$ and of $Y$.

(c) Find $P(X > bY)$ for $b > 0$. Use this to find the density of $W = Y/X$.

(d) Find $E(XY)$.

**3.5.2** Let $(X, Y)$ have joint density $f(x, y) = x + y$ for $0 \le x \le 1, 0 \le y \le 1$.

(a) Find the marginal densities of $X$ and of $Y$.

(b) Find $P(X > 2Y)$.

**3.5.3** Let $X_1, \ldots, X_n$ be a random sample from the distribution with density $f(x) = e^{-x}$ for $x > 0$. Let $X_{(1)}$ and $X_{(n)}$ be the minimum and maximum of these $X_i$'s.

(a) Give the joint density of $(X_{(1)}, X_{(n)})$.

(b) Find the marginal density of $X_{(1)}$.

**3.5.4** Let $X_1, X_2, X_3$ be independent, and $X_k \sim \text{Unif}(0, k)$ for $k = 1, 2, 3$. Let $M = \max(X_1, X_2, X_3)$. Find the density of $M$.

**3.5.5** $f_M(m) = 2m^2$ for $m \in [0, 1]$, $m^2/6$ for m $\in$ (1, 2], and $m/3$ for m $\in$ (2, 3].

**3.5.6** Let $(X, Y)$ have density $f(x, y) = Cxy$ for $0 < y < x < 1$. Find:

**(a)** The marginal density of $X$.

**(b)** $C$.

**(c)** The marginal density of $Y$.

**(d)** $P(Y < X/2)$.

**(e)** $E(X)$, $E(Y)$, $\text{Var}(X)$, $\text{Var}(Y)$, $\text{Cov}(X, Y)$, the correlation coefficient $\rho(X, Y)$.

## 3.6 DISTRIBUTIONS OF FUNCTIONS OF CONTINUOUS RANDOM VARIABLES

Let $\mathbf{X} = (X_1, \ldots, X_n)$ be a random vector with range $A$, a subset of $R_n$. Let $h(\mathbf{x})$ for $\mathbf{x} = (x_1, \ldots, x_n)$ be defined on $A$, taking values in $R_k$. Given the distribution of $\mathbf{X}$ and the function $h$, we would like to be able to describe the distribution of $Y = h(\mathbf{X})$. If $\mathbf{X}$ has a discrete distribution with probability function $f$, then $P(Y = y) \equiv g(y) = \sum_{h(\mathbf{x})=y} f(\mathbf{x})$. The trick is to find a simple expression, if one exists, for $g$. For example, if $X_1, X_2$ are the results of two independent die tosses, then for $Y = X_1 + X_2$, simple counting yields $P(Y = y) = [6 - |y - 7|]/36$ for $y = 2, \ldots,$ 12. In Section 2.5 we showed that sums of independent Poisson random variables have Poisson distributions. For continuous random variables, the method must be a bit different.

Let's begin with a simple example.

***Example 3.6.1*** Suppose that $X_1$ and $X_2$ are independent, each with the exponential density, means both 1. Then $\mathbf{X} = (X_1, X_2)$ has density $f(x_1, x_2) = e^{-x_1 - x_2}$ for $x_1 > 0, x_2 > 0$. Let $Y = X_1 + X_2$. Our technique will be first, to find the cdf of $Y$, then differentiate to get the density of $Y$. The cdf of $Y$ is

$$\begin{aligned} G(y) = P(X_1 + X_2 \le y) &= \int_{-\infty}^{\infty} \int_{-\infty}^{y-x_1} f(x_1, x_2)\, dx_2\, dx_1 \\ &= \int_0^y \int_0^{y-x_1} f(x_1, x_2)\, dx_2\, dx_1 = \int_0^y e^{-x_1}(1 - e^{-y+x_1})\, dx_1 \\ &= 1 - e^{-y}(1 + y) \qquad \text{for } y > 0. \end{aligned} \tag{3.6.1}$$

Differentiating with respect to $y$, we find that $Y$ has density $g(y) = ye^{-y}$ for $y > 0$ (an example of the gamma distribution with shape parameter 2). □

Notice that the first three equalities of (3.6.1) hold for any joint density $f$. Differentiating under the first integral sign with respect to $y$ for the case that $f$ is continuous in $x_2$ for each $x_1$, using the fundamental theorem of calculus, we find that $Y$ has density $g(y) = \int_0^y f(x_1, y - x_1)\,dx_1$. This is the *convolution formula* for the density of the sum of two random variables. Note that in Example 3.6.1, if we fail to note that $f$ is 0 for either argument negative, we are led astray.

Consider another example in which we must be extra careful.

***Example 3.6.2*** Let $U_1, U_2$ be independent, each Unif(0, 1) [i.e., $(U_1, U_2)$ is distributed uniformly on the unit square]. Let $Y = U_2/U_1$. Let us find the cdf for $Y$, using geometry, then the density for $Y$. Consider two cases: $0 \le y \le 1$ and $y > 1$. Readers are urged to sketch graphs. For $0 \le y \le 1$, $G(y) \equiv P(Y \le y) = y/2$, and for $y > 1$, $G(y) = P(Y \le y) = 1 - 1/(2y)$. Thus, $Y$ has density $g(y) = 1/2$ for $0 \le y \le 1$ and $1/(2y^2)$ for $y > 1$ (see Figure 3.6.1). □

We can express the cdf of $Y = X_2/X_1$ in more general form:

$$G(y) \equiv P\left(X_2/X_1 \le y\right) = \int_{-\infty}^{\infty} \int_{-\infty}^{x_1 y} f(x_1, x_2)\,dx_2\,dx_2.$$

Assuming that $f$ is continuous in $x_2$ for each $x_1$, we can differentiate under the first integral to determine the density of $Y$: $g(y) = \int_{-\infty}^{\infty} f(x_1, x_1 y) x_1\,dx_1$. Similarly, we can show under these conditions that $Y = X_1X_2$ has density $g(y) = \int_{-\infty}^{\infty} f(x_1, y/x_1)(1/x_1)\,dx_1$. (see Problem 3.6.1).

Next we consider a somewhat more complex situation in which $X$ and its transform $Y = h(\mathbf{X})$ take values in $R_k$. For the case that $\mathbf{X}$ has density $f$ on $R_k$, we would like to have a convenient formula for the density of $\mathbf{Y}$. Such a formula exists under special

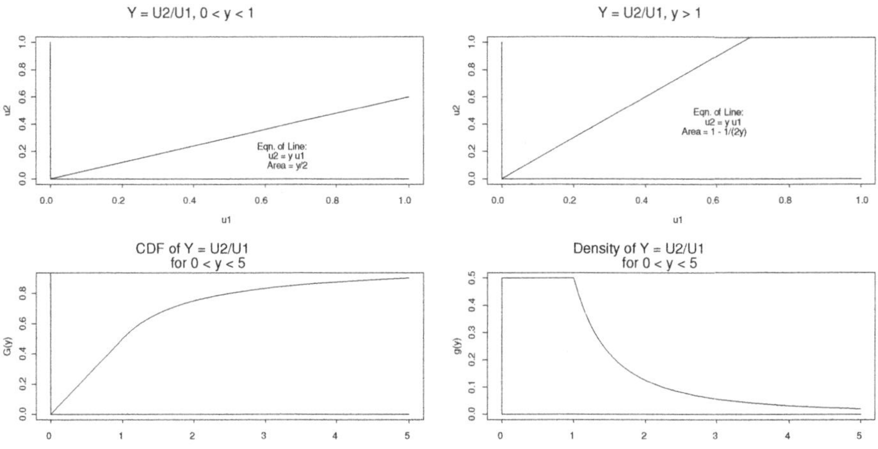


**FIGURE 3.6.1** CDF and density for $Y = U_2/U_1$.

conditions on $h$. Suppose that $\mathbf{X}$ has density $f$ on $R_k$ and takes values in an open subset $A$ of $R_k$. Suppose also that $h$ is a function from $R_k$ onto a subset $h(A) \equiv B$ of $R_k$ such that $h$ is one-to-one on $A$. Let $w(\mathbf{y}) \equiv (w_1(\mathbf{y}), \dots, w_k(\mathbf{y})) \equiv h^{-1}(\mathbf{y})$ be the inverse of $h$. Thus, $h(w(\mathbf{y})) = \mathbf{y}$ for every $\mathbf{y} \in B$. Let $J_w(\mathbf{y})$ be the Jacobian of $w$, evaluated at $\mathbf{y}$. That is, $J_w(\mathbf{y})$ is the determinate of the $n \times n$ matrix with $(ij)$th element $\frac{\delta}{\delta y_i} w_j(\mathbf{y})$. Suppose that these partial derivatives exist and are continuous for $y \in B$. Suppose also that $J_w(\mathbf{y})$ is never 0 for $y \in B$. Then it follows that for $y \in B$, $Y$ has density

$$g(\mathbf{y}) = f(w(\mathbf{y}))|J_w(\mathbf{y})|. \tag{3.6.2}$$

The Jacobian can also be expressed in terms of the Jacobian of its inverse $h$:

$$J_w(\mathbf{y}) = \frac{1}{J_h(w(\mathbf{y}))}. \tag{3.6.3}$$

***Example 3.6.3*** Let $X_1$ and $X_2$ be independent, each with the exponential distribution with mean 1. Thus, $f(x_1, x_2) = e^{-x_1 - x_2}$ for $x_1 > 0, x_2 > 0$. Thus, $A$ is the first quadrant in $R_2$. Let $Y_1 = X_1 + X_2$ and $Y_2 = X_1/X_2$. Let $h(x_1, x_2) = h(\mathbf{x}) = (x_1 + x_2, x_1/x_2)$. $B = h(A)$ is also the first quadrant of $R_2$. Then $w(\mathbf{y}) = (y_1 y_2/(1 + y_2), y_1/(1 + y_2))$. Some calculus shows that $J_w(\mathbf{y}) = -y_1(1 + y_2)$. Since $x_1 + x_2 = y_1$, we get $g(\mathbf{y}) = e^{-y_1} y_1 (1 + y_2)^{-2}$ for $y_1 > 0$, $y_2 > 0$. Since $g$ is the product of the two densities $g_1(y_1) = e^{-y_1} y_1$ for $y_1 > 0$ and $g_2(y_2) = (1 + y_2)^{-2}$ for $y_2 > 0$, we see that (surprisingly) the sum $Y_1$ and ratio $Y_2$ are independent random variables. We will learn later that in a more general case, $Y_1$ has a gamma distribution and that $Y_2$ has an $F$-distribution (after multiplication by a constant). As a check on (3.5.3), we determine $J_h(w(\mathbf{y})) = -(w_1 + w_2)/w_2^2 = -(1 + y_2)^2/y_1$, the reciprocal of $J_w(\mathbf{y})$. □

***Example 3.6.4*** Let $U_1, \dots, U_n$ be a random sample from the Unif(0, 1) distribution (see Table 3.6.1). Let $X_1, X_2, \dots, X_n$ be the corresponding order statistics. That is, $X_1$ is the smallest among the $U_i$'s, $X_n$ the largest and for $k = 2, \dots, n - 1$. $X_k$ is the $k$th in rank. Since $U \equiv (U_1, \dots, U_n)$ has density 1 on the unit cube and

**TABLE 3.6.1**

| Samples of Size 10 from Unif(0, 1) | $Y_1 = X_{(3)}$ | $Y_2 = X_{(7)}$ |
|---|---|---|
| 0.731 0.675 0.495 0.390 0.040 0.565 0.257 0.021 0.469 0.979 | 0.257 | 0.565 |
| 0.759 0.307 0.571 0.355 0.781 0.568 0.992 0.527 0.513 0.613 | 0.513 | 0.613 |
| 0.324 0.230 0.202 0.573 0.995 0.831 0.520 0.626 0.858 0.758 | 0.324 | 0.758 |
| 0.693 0.754 0.127 0.535 0.421 0.852 0.722 0.222 0.046 0.571 | 0.222 | 0.693 |
| 0.003 0.790 0.304 0.295 0.952 0.642 0.676 0.213 0.229 0.357 | 0.229 | 0.642 |

there are $n!$ values of $\mathbf{U}$ which produce the same $\mathbf{X} = (X_1, \ldots, X_n)$, $\mathbf{X}$ has density $f(x) = n!$ for any $\mathbf{x} = (x_1, \ldots, x_n)$ satisfying $0 < x_1 \cdots < x_n < 1$. Let $A$ be the set of such $x$. Let $r$ and $s$ be integers such that $1 \le r < s \le n$ and suppose that we wish to find the density of the pair $(X_r, X_s)$. For intuitive purposes consider the case $n = 10, r = 3, s = 7$. Let $Y_1 = X_1, Y_2 = X_2, Y_3 = X_3, Y_4 = X_4 - X_3, Y_5 = X_5 - X_3, Y_6 = X_6 - X_3, Y_7 = X_7, Y_8 = X_8 - X_7, Y_9 = X_9 - X_7, Y_{10} = X_{10} - X_7$. The more general case should be obvious. □

Let us write $\mathbf{X}$ and $\mathbf{Y}$ as column vectors. The transformation from $\mathbf{X}$ to $\mathbf{Y}$ is $\mathbf{Y} = h(\mathbf{X}) = \mathbf{C}\mathbf{X}$, where $\mathbf{C}$ is the $10 \times 10$ matrix with 1's on the diagonal, and zeros and –1's in the $(i+1, i)$ places, zeros otherwise. The determinants of the Jacobians (in both directions) are 1. Thus, $\mathbf{Y}$ has density $n! = 10!$ in the range of $\mathbf{Y}$, $h(A) = B = \{y \mid 0 < y_1 < y_2 < y_3 < y_7 < 1, 0 < y_4 < y_5 < y_6 < y_7 - y_3, 0 < y_8 < y_9 < y_{10} < 1 - y_7\}$. To get the density of $(Y_3 = X_3, Y_7 = X_7)$ we need only integrate out the other $y_i$ for fixed $y_3, y_7$. The integrals may be broken into three pieces: $(y_1, y_2)$, $(y_4, y_5, y_6)$, and $(y_8, y_9, y_{10})$. The middle three, for example, are $\int_0^{x_7 - x_3} \int_0^{x_4} \int_0^{x_5} dx_4\, dx_5\, dx_6 = (x_7 - x_3)^3/3!$ With a little care, we get the following: $(X_3, X_7)$ has joint density

$$g(x_3, x_7) = n!r(x_3, 3-1)r(x_7 - x_3, 7-3-1)r(1 - x_7, 11-7-1) \quad (3.6.4)$$

for $0 < x_3 < x_7 < 1$, where $r(u, k) = u^k/k!$ (see Figure 3.6.2). For the general case with $1 \le r < s \le n$, $(X_r, X_s)$ has density $f(x_r, x_s)$ determined by replacing 3 by $r$, 7

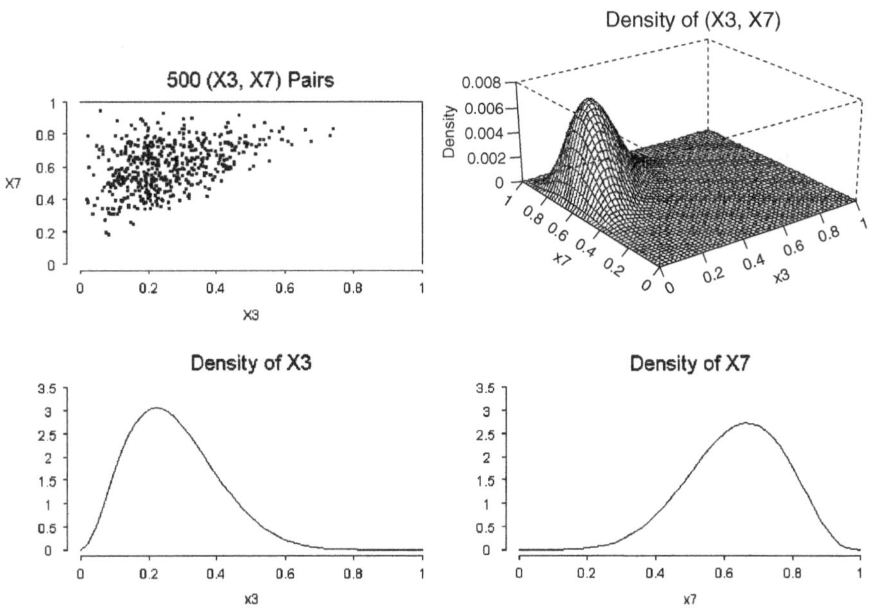


**FIGURE 3.6.2** Densities of $X_3$ and $X_7$.

by $s$ in (3.6.4). For example, if we take $r = 1$, $s = n$ in (3.6.4), we obtain the density $g(x_1, x_n) = n(n-1)(x_n - x_1)^{n-2}$ for $0 < x_1 < x_n < 1$, as obtained by a more direct method in Section 3.4.

Integrating with respect to $x_3$, we find that $X_7$ has density $g_7(x_7) = 10!\, r(x_7, 7-1)$ $r(1 - x_7, 11 - 7 - 1)$, for $0 < x_7 < 1$. For the density of the $s$th order statistic in the general case, we need only replace 10 by $n$, 7 by $s$. We find that $U_{(s)}$ has the Beta($\alpha$, $\beta$) distribution, with density $f(x; \alpha, \beta) = C(\alpha, \beta)x^{\alpha-1}(1-x)^{\beta-1}$ for $0 \le x \le 1$, where $C(\alpha, \beta) = \Gamma(\alpha + \beta)/\Gamma(\alpha)\Gamma(\beta)$ for $\alpha = s$, $\beta = n - s + 1$.

### Problems for Section 3.6

**3.6.1** Let $U_1, U_2$ be independent, each Unif(0, 1).

- **(a)** Find the density of $Y = U_1 U_2$. Verify your answer by determining each of $E(Y)$ and Var($Y$) with and without using the density of $Y$.
- **(b)** Repeat part (a) for the case that $Y = U_1 - U_2$.

**3.6.2** Let $X_1$ and $X_2$ be independent, each with density $f(x) = e^{-x}$ for $x > 0$. Let $X_{(1)}$ and $X_{(2)}$ be the corresponding order statistics, so that $X_{(1)} < X_{(2)}$ with probability 1.

- **(a)** Find the density of $(X_{(1)}, X_{(2)})$.
- **(b)** Find the density of $D = X_{(2)} - X_{(1)}$.
- **(c)** Let $Y_1 = X_{(1)}$ and $Y_2 = X_{(2)} - X_{(1)}$. Find the density of $(Y_1, Y_2)$. Use this to determine the marginal density of $Y_2$, verifying your answer to part (b).

**3.6.3** Let $(X_1, X_2)$ have density $f(x_1, x_2) = 3x_1$ for $0 \le x_2 \le x_1 \le 1$.

- **(a)** Let $Y_1 = X_1 + X_2$ and $Y_2 = X_1 - X_2$. Find the density of $(Y_1, Y_2)$. Use this to determine the density of $Y_1$.
- **(b)** Let $Y_1 = X_1 X_2$ and $Y_2 = X_1 / X_2$. Find the density of $(Y_1, Y_2)$. Use this to determine the density of $Y_1$.

**3.6.4** Let $X$ have the same density as $X_1$ in Problem 3.6.2. Let $B$ be $-1$ or $+1$ with probabilities, 1/2, 1/2, independent of $X$.

- **(a)** Show that $Y = BX$ has the Laplace (or double-exponential) distribution.
- **(b)** Give an algorithm that will use two independent Unif(0, 1) random variables to produce an observation $Y$ having the Laplace distribution.

**3.6.5** Let $U_1, U_2$ be as in Problem 3.6.1. Let $Y_1 = U_1 + U_2$ and $Y_2 = U_1 - U_2$. Find the density of $(Y_1, Y_2)$. Carefully define the domain on which the density is positive.

**3.6.6** Let $U_1, U_2$ be as in Problem 3.6.1. Consider the quadratic function $Q(x) = x^2 + U_1 x - U_2$.

- **(a)** Find the density of the largest root $R$ of $Q(x) = 0$.

**(b)** (For more ambitious students). Let $S$ be the smaller root. Find the density of $(S, R)$. The more difficult part of this is to find the region $B$ on which $(R, S)$ has positive density. *Hint*: What are the images of the line segments bounding the unit square under the transformation $(U_1, U_2) \to (S, R)$?

**3.6.7** (*Buffon's Needle Problem*) Suppose that a floor is marked with parallel lines running east to west a distance $L$ apart. A needle of length $d$ is thrown randomly on the floor. What is the probability that the needle crosses a line? *Hint*: We must interpret the meaning of "randomly." Suppose that the southernmost end of the needle is at a distance $X$ from the line that is closest in a southerly direction. Let the needle have angle $\theta$ with this line. Measure the angle in radians in a clockwise or counterclockwise direction so that $0 \le \theta \le \pi/2$. Suppose that $X \sim$ Unif$(0, L)$, that $\theta \sim$ Unif$(0, \pi/2)$, and that $X$ and $\theta$ are independent. First give conditions on $X$ and $\theta$ for which the needle crosses the line. Let $D = d/L$. Show that the probability is $p(D) = 2D/\pi$ for $D \le 1$ and $[(2D/\pi)(1 - \sqrt{1 - 1/D^2}) + 1 - (2\theta_0)/\pi$ for $D > 1$, where $\sin\theta_0 = 1/D$. In 100,000 simulations with $D = d/L = 0.5$, the needle crossed a line 32,008 times. For $D = d/L = 0.3$, it crossed 19,116 times. For $D = 2$ it crossed 83,768 times. Notice that $\lim_{D\to\infty} p(D) = 1$. For enough simulations we could estimate $\pi$ quite precisely. Let $\hat{p}(D)$ be the proportion of "successes." For $D \le 1$, we can estimate $\pi$ by $2/\hat{p}(D)$. For $D > 1$, since $p(D)$ is of the form $1 - g(D)/\pi$ for every $D$, we can estimate $\pi$ with $g(D)/(1 - \hat{p}(D))$. Find the estimates for the three frequencies given.

**3.6.8** Let $X_1$ and $X_2$ be independent, each with the exponential distribution with mean 1. Let $Y = X_1 - X_2$. Show that $Y$ has density $f_y(y) = (1/2)e^{-|y|}$ for all $y$. ($Y$ is said to have the *double-exponential* or *Laplace distribution.*)

**3.6.9** Let $U_1, U_2, U_3$ be a random sample from the Unif(0, 1) distribution. Let $M = \text{median}(U_1, U_2, U_3)$. Find the density and cdf for $M$. Use it to find $P(0.4 \le M \le 0.6)$ and $P(0.3 \le M \le 0.7)$.

CHAPTER FOUR

# Special Continuous Distributions

## 4.1 INTRODUCTION

As for discrete distributions, some continuous distributions are so useful and have applications that occur so frequently that they have names and deserve special attention. Actually, these are families of distributions, with certain parameters determining a unique member. They include the (1) uniform, (2) normal, (3) gamma, (4) beta, (5) Weibull, (6) chi-square, (7) $F$, and (8) Student's $t$ distributions. In this chapter we study systematically only the normal and gamma distributions. The beta and chi-square (a special case of the gamma) distributions are described briefly in Section 4.2, and the Weibull is mentioned only among the problems for that section. The last three are discussed in detail in Chapter Nine, after their use in statistical inference has been motivated more fully. We have already studied the uniform distribution.

All probabalists and statisticians would agree that among these distributions the normal distribution plays the most important role, both because it describes the distributions of random variables occurring naturally in nature and because it serves as an excellent approximation of the distribution of sums of independent random variables, as described in relation to the central limit theorem in Chapter Six. The gamma distribution serves as a model for the distribution of lifetimes. Since it is the distribution of the sum of independent exponential random variables, the waiting time until the $k$th occurrence of an event in a Poisson process has the gamma distribution with shape parameter $k$.

The beta distribution (or family) is the distribution of the $k$th-order statistic $X_{(k)}$ for a sample of $n$ from Unif(0, 1). The Weibull family often provides a good model for lifetime distributions. The chi-square distribution, of great importance in statistical inference, is a special case of the gamma distribution. The $F$ (named for Ronald A. Fisher) distribution arises in statistical inference in connection with regression analysis and the analysis of variance. Student's $t$-distribution is the distribution of a "standardized" sample mean, when the population standard deviation is replaced

*Models for Probability and Statistical Inference: Theory and Applications,* By James H. Stapleton

by the sample standard deviation. We begin with the most important: the normal distribution.

## 4.2 THE NORMAL DISTRIBUTION

We first discuss the standard normal distribution, then expand to the family of normal distributions through location and scale changes. To provide some intuition, we list 50 observations, generated using the S-Plus function "rnorm," then list the same observations in ordered form, the corresponding *order statistics*.

| | | | | | | | | | |
|---|---|---|---|---|---|---|---|---|---|
| −0.308 | 2.063 | 0.662 | 0.313 | 0.127 | 1.991 | 1.121 | 0.752 | −0.011 | −0.288 |
| −0.578 | −0.433 | −1.065 | 1.351 | 1.028 | −0.452 | −0.630 | 1.303 | 1.134 | 0.443 |
| 0.946 | 0.516 | 0.713 | −0.850 | −0.883 | 0.782 | 1.904 | 1.696 | −0.119 | −1.181 |
| 1.639 | −2.123 | 1.127 | 0.310 | −3.101 | −1.037 | −0.611 | 0.618 | −1.815 | 0.058 |
| −0.368 | −1.025 | −0.757 | −0.720 | 0.177 | 0.291 | 0.378 | −0.256 | 1.671 | 1.674 |

The corresponding order statistics:

| | | | | | | | | | |
|---|---|---|---|---|---|---|---|---|---|
| −3.101 | −2.123 | −1.815 | −1.181 | −1.065 | −1.037 | −1.025 | −0.883 | −0.850 | −0.757 |
| −0.720 | −0.630 | −0.611 | −0.578 | −0.452 | −0.433 | −0.368 | −0.308 | −0.288 | −0.256 |
| −0.119 | −0.011 | 0.058 | 0.127 | 0.177 | 0.291 | 0.310 | 0.313 | 0.378 | 0.443 |
| 0.516 | 0.618 | 0.662 | 0.713 | 0.752 | 0.782 | 0.946 | 1.028 | 1.121 | 1.127 |
| 1.134 | 1.303 | 1.351 | 1.639 | 1.671 | 1.674 | 1.696 | 1.904 | 1.991 | 2.063 |

If the original observations are denoted by $X_1, \ldots, X_{50}$, we denote the ordered observations as $X_{(1)}, \ldots, X_{(50)}$. For example, $X_{41} = -0.368$, while $X_{(41)} = 1.134$.

**Definition 4.2.1** The *standard normal density* is

$$\phi(x) = \frac{1}{\sqrt{2\pi}} e^{-x^2/2} \qquad \text{for } -\infty < x < +\infty.$$

The corresponding cdf is $\Phi(u) = \int_{-\infty}^{u} \phi(x)\, dx$ for $-\infty < u < +\infty$ (see Figure 4.2.1). □

We reserve the special symbols $\phi$ and $\Phi$ (lowercase and capital phi) for the standard normal density and cdf because, as we will show, this distribution plays such an important role in probability and statistics. It is not obvious at all that $\phi$ is a density. It requires a bit of a trick to show it. We prove it by showing that the square of the integral of $g(x) = e^{-x^2/2}$ is $2\pi$, using double integration. Let this integral be $I$. Then $I^2 = \int_{-\infty}^{+\infty} g(x)\, dx \int_{-\infty}^{+\infty} g(y)\, dy = \int\int e^{-(x^2+y^2)/2} dx\, dy$, where the double integration is over all of $R_2$. Letting $x = r\cos\theta$, $y = r\sin\theta$ (changing to polar coordinates) and remembering that the Jacobian is $r$, we get $I^2 = \int_0^{2\pi} \int_0^{+\infty} e^{-r^2/2} r\, d\theta\, dr = 2\pi$.

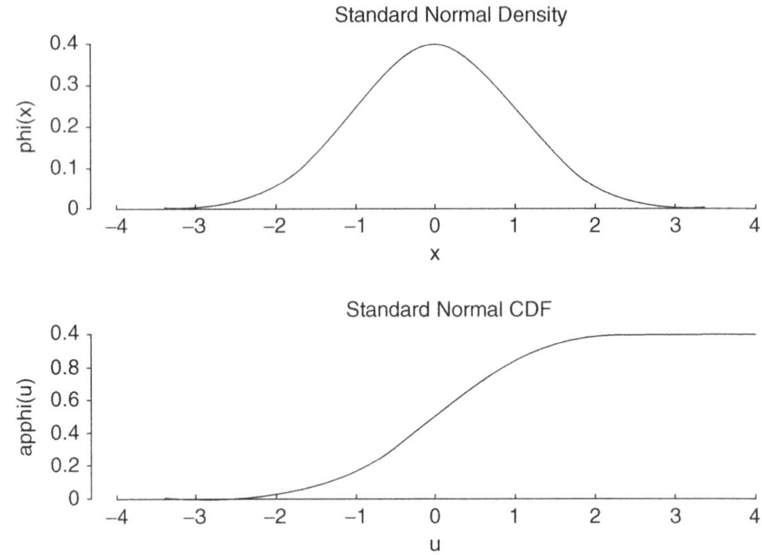


**FIGURE 4.2.1** Standard normal density and CDF.

$\Phi(u)$ does not have a closed-form representation except as an integral of $\phi$. An approximation for $u$ near 0 may be obtained by expanding $G(u) \equiv \log[\Phi(u) - 1/2]$ in a Taylor series, resulting in $\Phi(u) = 1/2 + \phi(u)(u + u^3/3 + u^5/(3 \cdot 5) + u^7/(3 \cdot 5 \cdot 7) + \cdots)$. Other approximations are available for $u$ near 0, and still others for $|u|$ large. Most books on probability and statistics have tables providing values of $\Phi(u)$ for $u$ between 0 and 3 or 4 in increments of 0.01. From the symmetry of $\phi$ about 0, it follows that $\Phi(-u) = 1 - \Phi(u)$ for all $u$, so that tables are necessary only for $u > 0$. Most statistical software packages and calculators provide values of $\Phi(u)$ for all $u$. A few values are so useful that experienced statisticians, even beginning students, know them by heart:

| $u$ | 0 | 0.841 | 1.000 | 1.282 | 1.645 | 1.960 | 2.000 | 2.326 | 2.580 | 3.000 |
|---|---|---|---|---|---|---|---|---|---|---|
| $\Phi(u)$ | 0.500 | 0.800 | 0.841 | 0.900 | 0.950 | 0.975 | 0.977 | 0.990 | 0.995 | 0.9987 |

If $Z$, the usual symbol employed, has the standard normal distribution, then, of course, for any $a < b$, $P(a < Z \le b) = \Phi(b) - \Phi(a)$. For example, $P(-1 \le X \le 2) = \Phi(2) - \Phi(-1) = \Phi(2) - [1 - \Phi(1)] = 0.977 - 0.159 = 0.818$, and for any $w > 0$, $P(-w < Z < w) = \Phi(w) - [1 - \Phi(w)] = 2\Phi(w) - 1$, and if we wish to find $w$ such that this is $\gamma$, we can solve for $w$, giving $w = \Phi^{-1}((1 + \gamma)/2)$. This forces us to read the standard normal table "inside out."

It is easy to show that $x = -1$ and $x = 1$ are points of inflection for $\phi$ (points at which the second derivative is zero). Since $\phi$ is symmetric about zero, all odd moments are zero. To find the second moment, we can integrate by parts with $u = x, dv = x\phi(x)$, to get $\int_{-\infty}^{\infty} x^2\phi(x)dx = -x\phi(x)|_{-\infty}^{\infty} + \int_{-\infty}^{\infty} \phi(x) = 1$. Thus, the standard normal distribution has variance 1.

We now introduce the family of normal distributions, all defined in terms of the "mother" of the family, the standard normal distribution.

**Definition 4.2.2** Let $Z$ have the standard normal distribution, and let $X = a + bZ$ for any constants $a$ and $b \neq 0$. Then $X$ is said to have the *normal distribution* with mean $\mu = a$ and variance $\sigma^2 = b^2$. We write this as $X \sim \mathrm{N}(\mu, \sigma^2)$. □

The density and cdf of $X$ are easy to express in terms of $\phi$ and $\Phi$. $X$ has cdf

$$F(u) = P(X \leq u) = P\left(Z \leq \frac{u-\mu}{\sigma}\right) = \Phi\left(\frac{u-\mu}{\sigma}\right)$$

and density

$$f(u) = \frac{\phi(u - \mu/\sigma)}{\sigma} = \frac{1}{\sqrt{2\pi}\sigma} e^{-[(u-\mu)/\sigma]^2/2} \qquad \text{for } -\infty < u < +\infty.$$

Figure 4.2.2 presents a scatterplot of 928 pairs of heights in inches of (midparent, son) pairs. These were taken from the Galton data, reproduced in a paper by Stigler (1989) on the correlation coefficient. Midparent heights were the means of the heights of the father and mother (see Figure 4.2.3), with the mother's height multiplied by 1.08. The heights were *jittered*, that is, independent Unif($-1/2$, $1/2$) were added to the heights, which were reported to the nearest 1/2 inch in order to smooth out the data, making the plot a more realistic portrayal of the actual heights. The normal density with the same means and standard deviation as for these data were drawn to show that the heights are approximately normally distributed. It is reasonable to suppose that the pairs have a bivariate normal distribution, as described in Section 5.3.

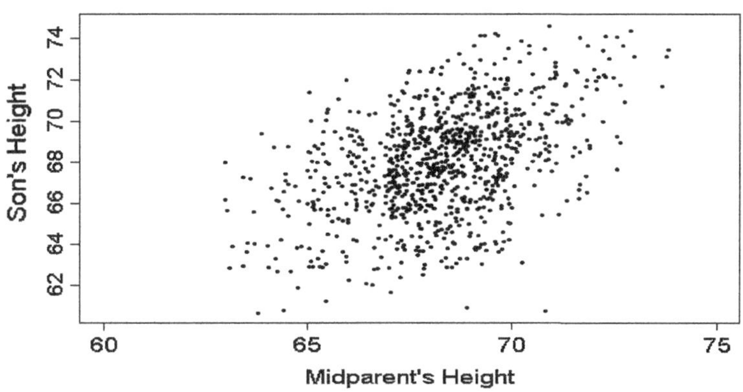


**FIGURE 4.2.2** Galton height data.

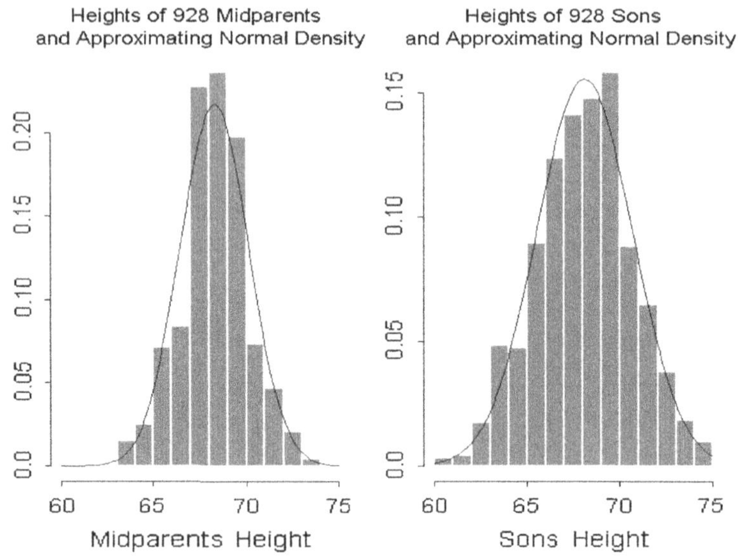


**FIGURE 4.2.3** Histograms of midparent and son heights.

If $X \sim \text{N}(\mu, \sigma^2)$, then $Z = (X - \mu_X)\sigma_X \sim \text{N}(0, 1)$, the standard normal distribution, and for any constants $c$ and $d \neq 0$, $Y = c + dX = (c + d\mu_X) + (d\sigma_X)Z$, so that $Y \sim \text{N}(c + d\mu_X, \sigma_Y^2 = d^2\sigma_X^2)$. Thus, a linear transform of a single normally distributed rv is normal if the constant multiplier is not zero. Like many other families of distributions (binomial, Poisson, negative binomial, gamma), the normal family is closed under the addition of independent members. To be precise, first consider two independent standard normal random variables $Z_1$ and $Z_2$, and let $c$ be a constant. We will show that

$$X = Z_1 + cZ_2 \sim \text{N}(0, 1 + c^2). \tag{4.2.1}$$

From this it will follow that if $X_1 \sim \text{N}(\mu_1, \sigma_1^2)$ and $X_2 \sim \text{N}(\mu_2, \sigma_2^2)$ and are independent, then for constants $a_1$ and $a_2$, $Y = a_1X_1 + a_2X_2 = a_1\sigma_1\{(X_1 - \mu_1)/\sigma_1 + [a_2\sigma_2/(a_1\sigma_1)](X_2 - \mu_2)/\sigma_2\} + a_1\mu_1 + a_2\mu_2 \sim \text{N}(a_1\mu_1 + a_2\mu_2, a_1^2\sigma_1^2 + a_2^2\sigma_2^2)$. The case of $n$ independent random variables follows by induction. The result is summarized in Theorem 4.2.1.

**Theorem 4.2.1** Let $X_1, \ldots, X_n$ be independent, with $X_i \sim \text{N}(\mu_i, \sigma_i^2)$ for $i = 1, \ldots, n$. Then $Y = X_1 + \cdots + X_n \sim \text{N}(\mu_1 + \cdots + \mu_n, \sigma_1^2 + \cdots + \sigma_n^2)$.

*Proof:* $X = Z_1 + cZ_2$ has density $f(x) = \int_{-\infty}^{\infty} \phi(u)\phi((x - u)/c)(1/c)du = \int_{-\infty}^{\infty} \phi(u - x/C)(\phi(x/C)/C)du)$, where $C = (1 + c^2)^{1/2}$. The term depending on $u$ is the density of the $\text{N}(x/C, 1)$ distribution, and therefore integrates to 1. $\phi(x/C)/C$ is the density of the $\text{N}(0, C^2)$ distribution. Thus, $X \sim \text{N}(0, C^2 = 1 + c^2)$. □

As we discuss in Chapter Six in a section on the *central limit theorem*, the conclusion of Theorem 4.2.1 holds in increasingly better approximation as $n$ increases when the $X_i$ are identically distributed with finite variance, independent, not necessarily normally distributed.

In the special case that $X_1, \ldots, X_n$ are independent, each with the $N(\mu, \sigma^2)$ distribution, $S_n = X_1 + \cdots + X_n$ has the $N(n\mu, n\sigma^2)$ distribution and $\bar{X}_n \equiv S_n/n$ has the $N(\mu, \sigma^2/n)$ distribution. It follows that $Z_n \equiv (S_n - n\mu)/(\sigma\sqrt{n}) = (\bar{X}_n - \mu)/(\sigma/\sqrt{n}) \sim N(0, 1)$.

***Example 4.2.1*** It is reasonable to believe that the weights in pounds of the men and women who use an elevator in a 50-story building are normally distributed, N(180, $20^2$) for men, N(130, $15^2$) for women, and that weights are independent. (a) If 16 men and 12 women get on an elevator, what is the probability that they exceed the nominal weight capacity of the elevator, 4600 pounds?

Let $X_i$ for $i = 1, \ldots, 16$ and $Y_j$ for $j = 1, \ldots, 12$ be the weights of the men and women, respectively. Let $T = \sum_{i=1}^{16} X_i + \sum_{j=1}^{12} Y_j$. Then, from Theorem 4.2.1, $T \sim N(16(180) + 12(130) = 4440, 16(20^2) + 12(15^2) = 9100)$. Then $P(T > 4600) = 1 - \Phi((4600 - 4440)/\sqrt{9100}) = 1 - \Phi(1.677) = 0.0467$. With one fewer man (perhaps you, in a moment of sanity), the probability drops to 0.00013. If you are a woman, you should see what the probability is in *your* moment of sanity.

(b) What is the probability that the mean weight of the 16 men is less than 170 pounds?

For the 16 men,

$$P(\bar{X}_{16} < 170) = P\left(\frac{\bar{X}_{16} - 180}{20/\sqrt{16}} < \frac{170 - 180}{5}\right) = \Phi(-2) = 0.0228.$$

(c) What is the probability that a randomly chosen man is heavier than a randomly chosen woman?

Let their weights be $M$ and $W$, so that $M \sim N(180, 20^2)$ and $W \sim N(130, 15^2)$. Since $M$ and $W$ are independent, $D \equiv M - W \sim N(50, 625)$, so that $P(M > W) = P(D > 0) = 1 - \Phi((0 - 50)/25) = \Phi(2) = 0.977$. □

Would the answer be the same if those chosen constitute a randomly chosen married couple? No, since $M$ and $W$ would not be independent in this case. People are more likely to marry someone of similar weight, light or heavy. We need to consider the joint distribution of $M$ and $W$ more carefully. We postpone that until we discuss the multivariate normal distribution in Section 5.2.

## Simulating Observations from the Normal Distribution

If $Z \sim N(0, 1)$, then $X = \sigma Z + \mu \sim N(\mu, \sigma^2)$, so it is enough to show how we can generate $Z$. Most statistical computer packages have built-in functions for the generation of such $Z$ in efficient (quick) ways. Given the speed of modern computers, it is not as important that algorithms be efficient, at least when the number to be generated is less than 50,000 or so. We present two methods, the polar coordinate method, which is exact but inefficient, and the sum of uniforms method, which is quick but not exact.

The *Polar coordinate method* proceeds as follows. Let $U_1$ and $U_2$ be independent, each Unif(0, 1). Let $W = -\log(U_1)$, so that $W$ has the exponential distribution with mean 1. Let $R = \sqrt{2W}$, $X = R\cos\ 2\pi U_2$, $Y = R\sin 2\pi U_2$. Then $(X, Y)$ are independent, each with the N(0, 1) distribution. Students are asked to prove this in Problem 4.2.6. The method is efficient in the sense that two uniform random variables are used to produce two independent N(0, 1) random variables, but the need to take two logs, a square root, a cosine, and a sine makes the method relatively slow.

The *sum of uniforms method* simply takes $X = U_1 + \cdots + U_{12}$, where the $U_i$ are independent, each Unif(0, 1). It is easy to show that $E(X) = 6$, $\text{Var}(X) = 12(1/12) = 1$, so that $Z = X - 6$ has the right mean and variance. From the central limit theorem of Chapter Six, sums of independent random variables have distributions that are (usually) close to normal for large $n$. In fact, for most purposes, $Z$, as defined here, has a distribution that is close enough to N(0, 1). For a better approximation we can add more uniform random variables (always standardizing the sum so that the mean is zero and the variance is 1).

### Problems for Section 4.2

**4.2.1** Let $Z \sim$ N(0, 1). Find:

- **(a)** $P(Z > 1.23)$.
- **(b)** $P(-1.87 < Z < 0.74)$.
- **(c)** $P(|Z| > 2.11)$.

**4.2.2** Let $X \sim$ N(50, 100). Find:

- **(a)** $P(X > 62.3)$.
- **(b)** $P(1.3 < X \leq 57.4)$.
- **(c)** $P(|X - 50| < 21.1)$.
- **(d)** The 80th percentile for $X$.
- **(e)** $d$ such that $P(X - 50| \leq d) = 0.95$.

**4.2.3** Let $Z \sim$ N(0, 1). Find the density of $Z^2$.

**4.2.4** Heights in centimeters of randomly selected adults are distributed as N(178, 50) for males and N(165, 42) for females.

- **(a)** What is the probability that a randomly selected male is taller than a randomly selected female?
- **(b)** What is probability that the mean height of three randomly selected males is greater than the mean height of three randomly selected females?

**4.2.5** Show that the standard normal density $\phi$ has points of inflection at $-1$ and $+1$. Use this to show that the N($\mu$, $\sigma^2$) density has points of inflection at $\mu \pm \sigma$.

**4.2.6** Let $U_1$ and $U_2$ be independent, each Unif(0, 1). Let $X = \sqrt{-2\log(U_1)}\cos(2\pi U_2)$ and $Y = \sqrt{-2\log(U_1)}\sin(2\pi U_2)$. Prove that $X$ and $Y$ are independent and that each has the N(0, 1) distribution. This is the polar coordinate method for the generation of N(0, 1) random variables.

**4.2.7** Let $X \sim N(\mu, \sigma^2)$ and let $Y = e^x$. $Y$ is said to have the *log normal distribution*. Find $E(Y)$ and $\mathrm{Var}(Y)$.

## 4.3 THE GAMMA DISTRIBUTION

We begin with the definition of the gamma function, defined first by Leonhard Euler in the eighteenth century. It is the continuous version of the factorial function.

**Definition 4.3.1** For each $\alpha > 0$, let $\Gamma(\alpha) = \int_0^\infty x^{\alpha-1}e^{-x}dx$. $\Gamma$ is called the *gamma function*. □

Although we will not need it, the domain of $\Gamma$ can be extended to the subset of the complex plane with positive real part. $\Gamma$ is continuous. Its most important property may be obtained by integrating by parts: $\Gamma(\alpha+1) = \alpha\Gamma(\alpha)$ for all $\alpha > 0$. Since $\Gamma(1) = 1$, it follows that for positive integers $n$, $\Gamma(n) = (n-1)!$

In general, $\Gamma(\alpha)$ may be obtained for noninteger $\alpha$ only by numerical integration. However, $\Gamma(3/2) = \int_0^\infty x^{1/2}e^{-x}dx$. Letting $x = y^2/2$, we obtain $\int_0^\infty ye^{-y2/2}ydy/2^{1/2} = (1/2)\sqrt{\pi}\int_{-\infty}^\infty y^2\phi(y)dy = \sqrt{\pi}/2$, since the variance of the standard normal distribution is 1. Since $\Gamma(3/2) = (1/2)\Gamma(1/2)$, it follows that $\Gamma(1/2) = \sqrt{\pi}$. Since $\Gamma(1+\eta) = \eta\Gamma(\eta)$ for all $\eta > 0$ and $\Gamma(1) = 1$, it follows that $\Gamma(\eta)$ approaches infinity as $\eta$ approaches zero through positive values.

Stirling's formula, discovered around 1730 by James Stirling for integer $\alpha$, was later extended to the gamma function (Stirling, 1730).

**Stirling's Formula** Let $G(\alpha) = (2\pi)^{1/2}\alpha^{\alpha+1/2}e^{-\alpha}$. Then $\lim_{\alpha\to\infty}\Gamma(\alpha+1)/G(\alpha) = 1$. Thus, we can approximate $\Gamma(\alpha+1)$ by $G(\alpha)$ for large $\alpha$ with small relative error. We write $\Gamma(\alpha+1) \sim (2\pi)^{1/2}\alpha^{\alpha+1/2}e^{-\alpha}$. See Table 4.3.1, for example.

Let $X$ be the number of heads in $2n$ tosses of a coin, so that $X \sim B(2n, 1/2)$. In 1730 and 1733, Abraham DeMoivre showed that $P(X = n) \sim 1/\sqrt{2\pi(2n/4)}$. This follows easily from Stirling's approximation. From this he showed that $P(X = n+k) \sim e^{-k^2/(2n/4)}/\sqrt{2\pi(2n/4)} \equiv H(k, n)$ for $k$ small compared to $n$. Although he

**TABLE 4.3.1 Binomial Probability Mass Functions**

| | α | | | | | |
|---|---|---|---|---|---|---|
| | 3 | 5 | 7 | 13.5 | 4.5 | 100 |
| $\Gamma(\alpha+1)$ | 6 | 120 | 5040 | $2.30923 \times 10^{10}$ | $1.19622 \times 10^{56}$ | $9.33262 \times 10^{157}$ |
| $G(\alpha)$ | 5.83621 | 118.0192 | 4980.396 | $2.29502 \times 10^{10}$ | $1.19401 \times 10^{56}$ | $9.32485 \times 10^{157}$ |
| $r(\alpha) = \Gamma(\alpha+1)/G(\alpha)$ | 1.0281 | 1.0167 | 1.0120 | 1.0062 | 1.0019 | 1.0008 |

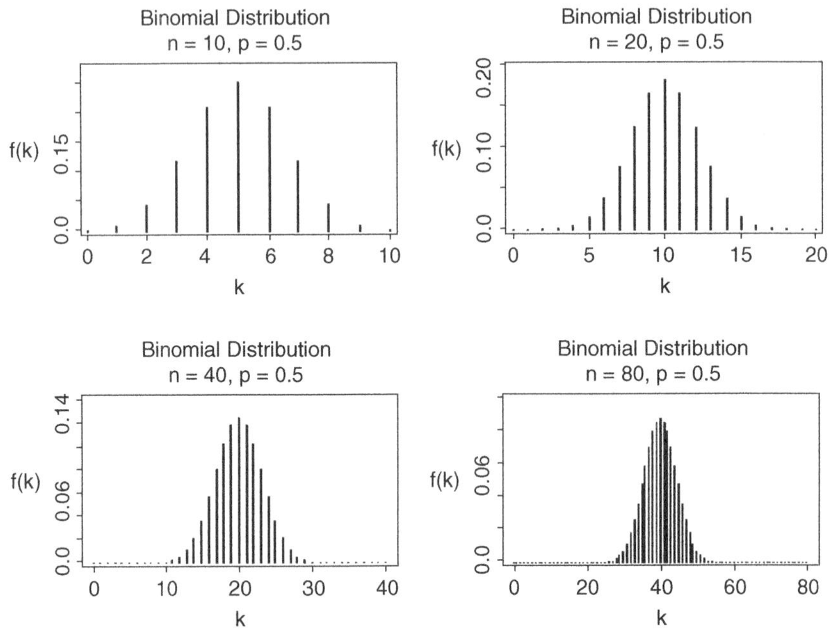


**FIGURE 4.3.1**

did not use this notation, we can see now (and it was known by the early nineteenth century) that $H(k, n) = \phi(k/\sigma_n)\sigma_n$ for $\sigma_n^2 = \text{Var}(X) = 2n/4$, the density of the N(0, $2n/4$) distribution at $k$ (see Figure 4.3.1). From this DeMoivre was able to find a normal approximation for $P(n + k_1 \le X \le n + k_2)$ for (somewhat) arbitrary $k_1 < k_2$. The eager student can turn now to Chapter Six for a discussion of the central limit theorem, describing the normal approximation to the distribution of sums of independent random variables.

We are now ready to define the family of gamma distributions.

**Definition 4.3.2** The *gamma distribution* with scale parameter $\theta = 1$ and shape parameter $\alpha > 0$ has density

$$f(x; \alpha, 1) = \frac{x^{\alpha-1}e^{-x}}{\Gamma(\alpha)} \qquad \text{for } x > 0. \tag{4.3.2}$$

If $X$ has this density, then $Y = \theta X$ is said to have the gamma distribution with scale parameter $\theta$ and shape parameter $\alpha$ (see Figure 4.3.2). We say that $Y$ has the $\Gamma(\alpha, \theta)$ distribution. □

*Notes:*

1. That $f$ as defined above is a density follows from the definition of the gamma function.

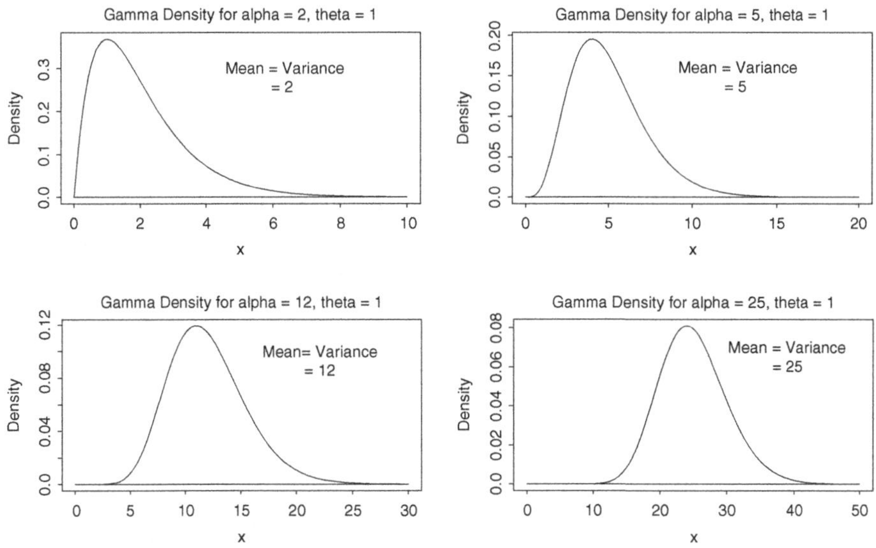


**FIGURE 4.3.2** Gamma densities.

2. The $k$th moment for $X$ is $\nu_k \equiv E(X^k) = \Gamma(\alpha + k)/\Gamma(\alpha)$, and therefore the $\Gamma(\alpha, \theta)$ distribution has $k$th moment $\theta^k \nu_k$. From this it follows that $E(X) = \alpha$, $\text{Var}(X) = E(X^2) - \alpha^2 = (\alpha + 1)\alpha - \alpha^2 = \alpha$, $E(Y) = \theta\alpha$, $\text{Var}(Y) = \theta^2\alpha$.
3. Let $U$ have the $\Gamma(\alpha, 1)$ distribution, let $V$ have the $\Gamma(\beta, 1)$ distribution, and let $U$ and $V$ be independent. Then $W = U + V$ has density $g(w) = \int_0^\infty f(w - v; \alpha, 1)\, f(v; \beta, 1)\; dv \;=\; \int_0^\infty (w - v)^{\alpha-1}\, e^{-(w-v)}\, (v^{\beta-1} e^{-v}) dv / [\Gamma(\alpha)\Gamma(\beta)]$. A change of variables, $s = v/w$, produces $g(w) = C w^{\alpha+\beta-1} e^{-w}$, where $C = \int_0^1 s^{\alpha-1}(1 - s)^{\beta-1} ds / [\Gamma(\alpha)\Gamma(\beta]$. Since $g$ is a constant multiple of the $\Gamma(\alpha + \beta, 1)$ density it must be the $\Gamma(\alpha + \beta, 1)$ density, so that $C = 1/\Gamma(\alpha + \beta)$. We have learned that $\int_0^1 s^{\beta-1}(1 - s)^{\alpha-1} dv = [\Gamma(\alpha)\Gamma(\beta)]/\Gamma(\alpha + \beta)$. By induction and multiplication by $\theta$ it follows that if $X_1, \ldots, X_n$ are independent, $X_j \sim \Gamma(\alpha_j, \theta)$, then $S_n = X_1 + \cdots + X_n \sim \Gamma(\alpha_1 + \cdots + \alpha_n, \theta)$. In particular, the sum of $n$ independent exponential random variables, each with the same scale parameter $\theta$, has the $\Gamma(n, \theta)$ distribution.
4. Recall that a Poisson process with rate $\lambda > 0$ has independent increments. That is, if the times of occurrence of events are $0 < X_1 < \cdots < X_k$ in a time interval $[0, T]$, the increments $D_1 = X_1, D_2 = X_2 - X_1, \ldots, D_k = X_k - X_{k-1}$ are independent. But $P(D_j > d) = P(\text{no occurrence in the interval } (X_{j-1}, X_{j-1} + d) = e^{-(\lambda)d}$, since the number of occurrences of the event in an interval of length $d$ has the Poisson distribution with mean $\lambda d$. Thus, each $D_j$ has cdf $F_D(d) = 1 - e^{-\lambda d}$ for $d > 0$, the cdf of the exponential distribution with rate $\lambda$, mean $1/\lambda$. From note 3 it follows that the *waiting time* until the $k$th occurrence has the $\Gamma(k, 1/\lambda)$ distribution.

5. If $D_1, D_2, \ldots$ are independent, each with the exponential distribution with mean 1, the sequence $X_1 = D_1, X_2 = X_1 + D_2, X_3 = X_2 + D_3, \ldots$ are the occurrence times of a Poisson process with rate 1. The number $Y$ of occurrences in time interval $[0, \lambda]$ has the Poisson distribution with mean $\lambda$. It follows that $P(X_r > \lambda) = P(Y < r)$, so that $\int_\lambda^\infty f(x; r, 1)\, dx = \sum_{k=0}^{r-1} p(k; \lambda)$, where $f$ is the gamma density and $p$ is the Poisson probability function. This equality may be established directly by integrating by parts on the left $r - 1$ times.
6. Recall also that the sum of independent geometric random variables, each with parameter $p, 0 < p < 1$, has the negative binomial distribution. That is, the waiting time until the $k$th success in a sequence of independent Bernoulli trials, each with probability $p$ of success, has the negative binomial distribution with parameters $k$ and $p$. Thus, from note 4, the gamma distribution is to the exponential as the negative binomial is to the geometric. We show in Chapter Six that in a certain sense the exponential distribution is the limit of a sequence of geometrics, and the gamma distribution is the limit of a sequence of negative binomials.
7. The $\Gamma(\nu, 2)$ distribution is called the *chi-square distribution* with $2\nu$ *degrees of freedom*, called the $\chi^2_{2\nu}$ *-distribution*. The distribution is quite useful in statistics. For example, if $X_1, \ldots, X_n$ is a random sample from the $N(\mu, \sigma^2)$ distribution and $S^2$ is the corresponding sample variance, then $(n-1)S^2/\sigma^2 = (1/\sigma^2)\sum_{i=1}^n (X_i - \bar{x})^2$ has the chi-square distribution with $n - 1$ degrees of freedom. This is proved in Chapter Nine. If $Y$ has the $\chi^2_\nu$ -distribution, the tables given in most statistics books present tables of quantiles $u_\gamma$, so that $P(Y \le u_\gamma) = \gamma$ for some $\gamma$. This can be exploited to give quantiles of the gamma distribution as follows. Suppose that $X_r \sim \Gamma(r, 1)$. Then $Y = 2X_r$ has the chi-square distribution with $2r$ degrees of freedom. Hence, if $u_\gamma$ is the $\gamma$-quantile for $Y$, the $\gamma$-quantile for $X_r$ is $u_\gamma/2$. □

### Estimation of α and θ

Suppose that we have observed the failure times of 400 different 100-watt light bulbs in our factory and feel that the gamma distribution may be a reasonable model for the observations but don't know the values of the parameters $\alpha$ and $\theta$. How can we use the data to estimate the pair $(\alpha, \theta)$? Let's act as nature may for a moment and generate a sample from, for example, the $\Gamma(3, 5)$ distribution (see Figure 4.3.3). Using S-Plus, the author obtained a sample of 400. The first 20 were:

| | | | | | | | | | |
|---|---|---|---|---|---|---|---|---|---|
| 54.08 | 19.77 | 19.12 | 13.64 | 12.25 | 12.26 | 12.89 | 9.56 | 7.11 | 2.11 |
| 26.38 | 21.10 | 59.75 | 19.98 | 4.92 | 17.55 | 2.70 | 12.31 | 14.75 | 8.17 |

The mean for the 400 was $\bar{X} = 14.582$ and the sample variance was $\hat{\sigma}^2 = (1/400)\sum_{i=1}^{400}(X_i - \bar{X})^2 = 76.787$. We can estimate the pair $(\alpha, \theta)$ by using *the method of moments.* Express the mean $\mu$ and variance $\sigma^2$ as functions of $(\alpha, \theta)$,

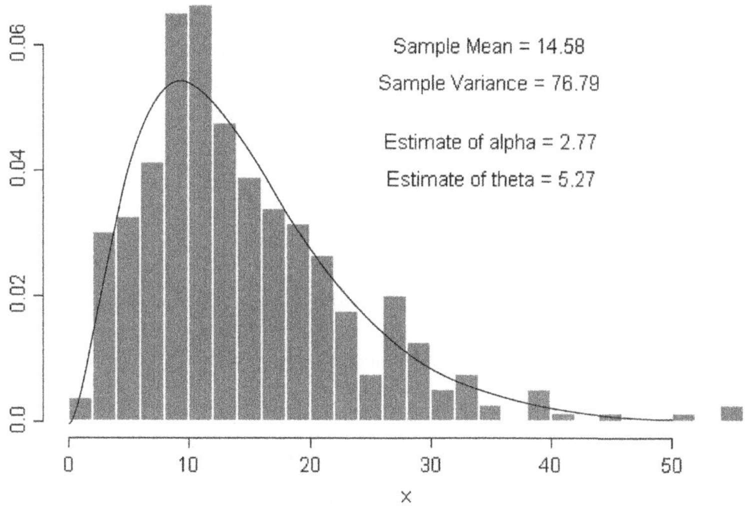


**FIGURE 4.3.3** Histogram and estimate of gamma density.

then express $\alpha$ and $\theta$ as functions of $\mu$ and $\sigma^2$. Since $\mu = \alpha\theta$, and $\sigma^2 = \alpha\theta^2$, we obtain $\theta = \sigma^2/\mu$ and $\alpha = \mu^2/\sigma^2$. An estimator of $(\alpha, \theta)$ is then obtained by replacing $\mu$ and $\sigma^2$ by their moment estimators $\bar{X}$ and $\hat{\sigma}^2$. Thus, we obtain the estimates $(\hat{\alpha} = \bar{X}^2/\hat{\sigma}^2 = 2.731, \hat{\theta} = 5.302)$. In Chapter Seven we discuss the properties of such estimators.

## Simulation of Observations from the $\Gamma(\alpha, \theta)$ Distribution

If $\alpha = k$ is an integer, the idea is simple. Recall that if $U \sim$ Unif(0, 1), then $X = -\log(U)$ has the exponential distribution with mean 1. It follows that if $U_1, \ldots, U_k$ are independent, each Unif(0, 1), and we let $X_j = -\log(U_j)$ then $Y = X_1 + \cdots + X_k \sim \Gamma(k, 1)$. Now consider the case $\alpha = k + d$, where $k = [\alpha]$ is the largest integer less than or equal to $\alpha$. Generate $Y$ as for the case $\alpha = k$ and let $U_{k+1}, U_{k+2}$ be independent, each Unif(0, 1), independent of the other $U_j$. Let $X = -\log(U_{k+1})$, so that $X$ has the exponential distribution with mean 1. Let $I = [U_{k+2} > d]$, an indicator random variable, so that $I$ has the Bernoulli distribution with parameter $d$. Let $W = Y + XI$. The random variable $W$ does not have an exact $\Gamma(\alpha, 1)$ distribution, but in good approximation, it does. It is easy to verify (see Problem 4.3.9) that $E(W) = \alpha$, as for the $\Gamma(\alpha, 1)$ distribution, although $\text{Var}(W) = \alpha - d^2$, slightly less than for the $\Gamma(\alpha, 1)$ distribution. Those seeking a more exact method are referred to Kennedy and Gentle's *Statistical Computing* (1980) (pp. 209–216).

## Simulation of Poisson Random Variables

Let $0 < X_1 < X_2 < \cdots$ be occurrence times of events in a Poisson process with rate 1. These may be generated as described in note 4. Let $Y$ be the largest $k$ such that

$X_k < \lambda$, 0 if $X_1 > \lambda$. Then as stated in note 5, $Y$ has the Poisson distribution with mean $\lambda$. Hence, to generate a Poisson random variable with mean $\lambda$, we need only generate consecutive $X_j = X_{j-1} + D_j$, where the $D_j$ are independent, exponential with mean 1. Of course, if $\lambda$ is large, this forces the generation of many $D_j$'s. Call this the *waiting-time method*.

## Problems for Section 4.3

**4.3.1** Give the density of the $\Gamma(\alpha, \theta)$ distribution.

**4.3.2** Find $r(\alpha) \equiv \Gamma(\alpha + 1)/G(\alpha)$, where $G(\alpha)$ is Stirling's approximation, for $\alpha = 2.5, 4.5, 6$.

**4.3.3** Show that for $\alpha > 1$ the density $f(x; \alpha, \theta)$ of the $\Gamma(\alpha, \theta)$ distribution has a mode (maximizing value of its argument) at $x = \theta(\alpha - 1)$. *Hint*: First find the $x$ that maximizes $\log f(x; \alpha, 1)$.

**4.3.4** Let $X \sim \Gamma(\alpha, 1)$ and $Y = 1/X$.

**(a)** For which $\alpha$ does $E(Y)$ exist? What is its value?

**(b)** Suppose that $f$ is a density function with the properties that $f(0) > 0$ and that $f$ is continuous on the right at 0 [as is the density of the $\Gamma(1, 1)$ distribution]. If $X$ has density $f$, may $E(1/X)$ exist?

**4.3.5** Let $X \sim \Gamma(\alpha, 1)$, $Y \sim \Gamma(\beta, 1)$, and suppose that $X$ and $Y$ are independent. Let $U = X + Y$, and $V = X/(X + Y)$.

**(a)** Show that $U$ and $V$ are independent and that $V$ has a density of the form $g(v) = C(\alpha, \beta)v^{\alpha-1}(1 - v)^{\beta-1}$ for $0 \leq v \leq 1$. $V$ has the *beta distribution* with parameters $\alpha$ and $\beta$.

**(b)** Show that $C(\alpha, \beta) = \Gamma(\alpha + \beta)/\Gamma(\alpha)\Gamma(\beta)$. What is the distribution of $U$?

**(c)** Let $\alpha = 1$, so that $X$ has the exponential distribution. Use parts (a) and (b) to find the cdf of $R = X/Y$, and use it to find $P(R \leq 10)$ for $\beta = 10$.

**4.3.6** Let $X$ have the chi-square distribution with $\nu$ degrees of freedom.

**(a)** What is the density of $X$? What are its mean and variance?

**(b)** Use Appendix Table 5 to find the 95th percentile of the $\Gamma(20, 10)$ distribution.

**4.3.7** Let $Z_1$ and $Z_2$ be independent, each with the N(0, 1) distribution. Show that $R = Z_1^2 + Z_2^2$ has the chi-square distribution with two degrees of freedom (exponential with mean 2, or $\Gamma(1, 2)$).

**4.3.8** See Problem 4.3.5. Let $X \sim \text{Beta}(\alpha, \beta)$.

(a) Show that $E(X) = \alpha/(\alpha + \beta)$ and that $\text{Var}(X) = \alpha\beta/[(\alpha + \beta)^2(\alpha + \beta + 1)]$.

(b) Show that $Y = 1 - X \sim \text{Beta}(\beta, \alpha)$.

(c) Find the 0.90-quantiles of the Beta(1, 1), Beta(1, 2), and Beta(2,1) distributions.

(d) Recall from Section 3.6 that the $s$th order statistic $U_{(s)}$ for a sample of $n$ from the Unif(0, 1) distribution has the Beta($s, n + 1 - s$) distribution. Find $E(U_{(s)})$ and $\text{Var}(U_{(s)})$ for $n = 10, s = 7$.

**4.3.9** Prove that $\int_0^\lambda f(x; r, 1)dx = \sum_{k=0}^{r-1} p(k; \lambda)$, where $f$ is the gamma density and $p$ is the Poisson probability function, by integrating by parts, as stated in note 5.

**4.3.10** See "Simulation of Observations from the $\Gamma(\alpha, \theta)$ Distribution" in this section. Show that $E(W) = \alpha$ and $\text{Var}(W) = \alpha - d^2$.

**4.3.11** Using your calculator and the waiting-time method, generate five independent Poisson random variables with $\lambda = 2$.

**4.3.12** Let $X$ have the exponential distribution with mean 1. Let $m > 0$, $\theta > 0$, and define $Y = \theta X^{1/m}$. Then $Y$ is said to have the *Weibull distribution* with power parameter $m$ and scale parameter $\theta$. The two-parameter Weibull family is often used to model lifetimes.

(a) Show that $Y$ has cdf $F(u) = 1 - e^{-(u/\theta)^m}$ for $u > 0$.

(b) Express the mean and variance of $Y$ in terms of $m$ and $\theta$.

(c) Find the 0.90-quantile of the Weibull distribution with $m = 3, \theta = 5$.

(d) Find a function $H(u; m, \theta)$, so that if $U \sim \text{Unif}(0, 1)$, then $Y = H(U; m, \theta)$ has the Weibull distribution with parameters $m$ and $\theta$.

**4.3.13** Let $Y_1, \ldots, Y_n$ be independent, each with the Weibull distribution with parameters $m$ and $\theta$.

(a) Let $W = \min(Y_1, \ldots, Y_n)$. Show that $W$ has a Weibull distribution. What are its parameters?

(b) Express the method of moments estimator of $(m, \theta)$ based on the observations $Y_1, \ldots, Y_n$ and the gamma function. *Hint*: See the answers to Problem 4.3.12(b).

(c) For a random sample of 100, the sample mean and variance were 2.571 and 2.070. Find the method of moments estimate $(\hat{m}, \hat{\theta})$ of $(m, \theta)$. The values of $m$ and $\theta$ used to generate the data were $m = 2, \theta = 3$. You will need to be able to compute $\Gamma(1/m + 1)$ and $\Gamma(2/m + 1)$.

**4.3.14** (More difficult than most problems) Let $f(x) = (\phi(x) + \phi(x - \mu))/2$ for all real $x$ (see Figure 4.3.14). $f$ is the density of $X = Z + \mu B$, where $B$ is 0

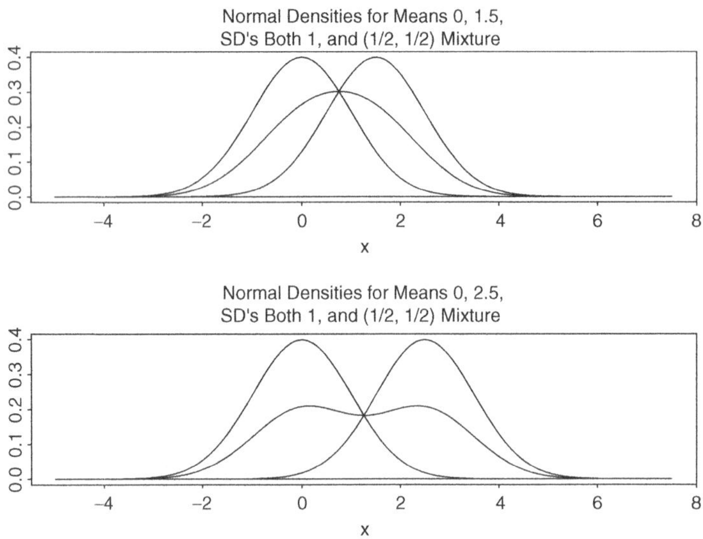


**FIGURE 4.3.14** Normal densities and their mixtures.

or 1 with probabilities $1/2$, $1/2$, $Z \sim \mathrm{N}(0, 1)$, and $B$ and $Z$ are independent. Show that $f$ has one mode (maximum point) at $\mu/2$ for $|\mu| \leq 2$, but that $f$ has two local maxima for $|\mu| > 2$. Thus, $f$ is unimodal for $|\mu| \leq 2$, bimodal for $|\mu| > 2$. If the male and female populations have the same standard deviations, heights among all adults have a unimodal (nonnormal) distribution if the difference in means is less than two standard deviations. For distributions like those in Problem 4.2.4, the standard deviations are close and the difference in means is approximately equal to twice this "common" standard deviation, so that the combined distribution is either uni- or bimodal.

CHAPTER FIVE

# Conditional Distributions

## 5.1 INTRODUCTION

We begin our discussion of conditional distributions with the discrete case.

***Example 5.1.1*** Consider the American box of Example 1.4.4, with nine balls, of which four are red, three are white, and two are blue. A simple random sample of three balls is chosen. That is, a subset of three balls is chosen in such a way that all 84 subsets are equally likely. Again let $R$ and $W$ be the numbers of red and white balls chosen. As determined in Section 1.4, the joint distribution of $R$ and $W$ is as shown in Table 5.1.1.

Suppose now that we learn that the event $[R = 1]$ has occurred. Then the revised probability of the event $[W = 2]$ is $P(W = 2 \mid R = 1) = P(R = 1, W = 2)/P(R = 1) = (12/84)/(40/84) = 12/40 = 0.3$. Similarly, $P(W = 0 \mid R = 1) = (4/84)/(40/84) = 0.1$ and $P(W = 1 \mid R = 1) = (24/84)/(40/84) = 0.6$. We say that given $R = 1$, the conditional probability function of $W$ is $f_W(w \mid R = 1) \equiv g(w)$, where $g(0) = 0.1$, $g(1) = 0.6$, $g(2) = 0.3$, and $g(w) = 0$ for other $w$ (see Figure 5.1.1). Obviously, $g$ is a probability mass function. We can compute the conditional expectation of $W$ as usual, writing $E(W \mid R = 1) = (0)g(0) + (1)g(1) + 2g(2) = 1.2$. For this example we could have thought instead of the conditional experiment of drawing $3 - 1 = 2$ balls at random from among the $9 - 4 = 5$ balls left after the red balls are removed, so that, for example, $P(W = 2 \mid R = 1) = \binom{3}{2} \Big/ \binom{5}{3} = 3/10$. □

With this introduction we are ready for a formal definition:

**Definition 5.1.1** Let $(X, Y)$ have joint probability mass function $f(x, y)$. The *conditional probability mass function* for $Y$, given $[X = x]$, where $P(X = x) \equiv f_X(x) > 0$, is $f_Y(y \mid X = x) \equiv f(x, y)/f_X(x)$, for each real number $y$. □

---

*Models for Probability and Statistical Inference: Theory and Applications,* By James H. Stapleton

**TABLE 5.1.1 The Joint Probability Mass Function $f(r, w)$**

| | $w$ | | | | |
|---|---|---|---|---|---|
| $r$ | 0 | 1 | 2 | 3 | $P(R = r) = f_R(r)$ |
| 0 | 0 | 3/84 | 6/84 | 1/84 | 10/84 |
| 1 | 4/84 | 24/84 | 12/84 | 0 | 40/84 |
| 2 | 12/84 | 18/84 | 0 | 0 | 30/84 |
| 3 | 4/84 | 0 | 0 | 0 | 4/84 |
| $P(W = w) = f_W(w)$ | 20/84 | 45/84 | 18/84 | 1/84 | 1 |

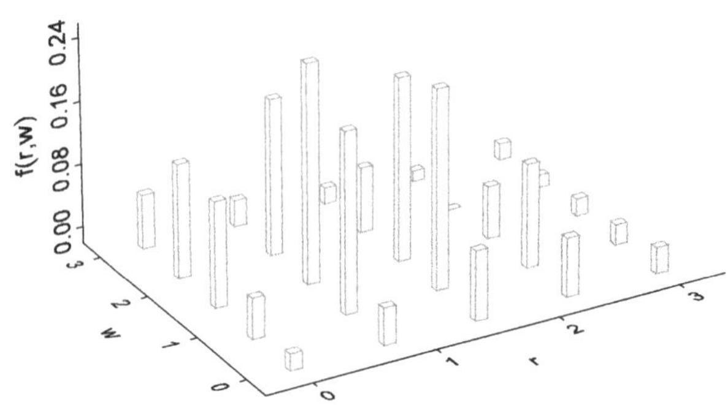


**FIGURE 5.1.1** Probability mass function for $(R, W)$.

COMMENTS: Since $(X, Y)$ is discrete $f_Y(y \mid X = x) > 0$ for a finite or countably infinite set $D(x)$ of values of $y$. $D(x)$ is the range of $Y$, given the event $[X = x]$. For Example 5.1.1, with $R$ and $W$ replacing $X$ and $Y$, $D(1) = \{0, 1, 2\}$ and $D(2) = \{0, 1\}$. Since $\sum_{y \in D(x)} f(x, y) = f_X(x)$, the function $f_Y(y \mid X = x)$ is a probability mass function for each $x$ for which $f_X(x) > 0$. As $x$ ranges over the set $B = \{x \mid f_X(x) > 0\}$, the conditional mass functions $f_Y(y \mid X = x)$ vary in general. However, if $X$ and $Y$ are independent, then since $f_{XY}(x, y) = f_X(x)f_Y(y)$, $f_Y(y \mid X = x) = f_Y(y)$, the same for all $x$ in $B$. We will sometimes omit either "mass" or "probability".

***Example 5.1.2*** Let $X_1$ and $X_2$ be independent, with Poisson distributions with parameters $\lambda_1$ and $\lambda_2$. As shown in Section 2.4, $T = X_1 + X_2$ has the Poisson distribution with parameter $\lambda_1 + \lambda_2$. Let us try to determine the conditional distribution of $X_1$, given $T = t$, defined for $t = 0, 1, 2, \ldots$. Let $p(x; \lambda)$ be the Poisson probability density function (pdf). Let $\theta = \lambda_1/(\lambda_1 + \lambda_2)$. We have, for $0 \le x_1 \le t$, with $x_1$ and $t$ nonnegative integers, $f_{X_1}(x_1 \mid T = t) = p(x_1; \lambda_1)p(t - x_1; \lambda_2)/p(t; \lambda_1 + \lambda_2) = e^{-\lambda_1}e^{-\lambda_2}\lambda_1^{x_1}\lambda_2^{x_2}/[x_1!(t - x_1)!]/[e^{-\lambda_1-\lambda_2}(\lambda_1 + \lambda_2)^2/t!] = (t!/[x_1!(t - x_1)!])\theta^{x_1}(1 - \theta)^{t-x_1} = b(x_1; t, \theta)$, the binomial pdf. Thus, conditionally on $T = X_1 + X_2 = t$, $X_1$ can be viewed as the sum of $t$ independent Bernoulli rv's, each with $\theta = \lambda_1/(\lambda_1 + \lambda_2)$. Using the fact that $p$ is a function of $R \equiv \lambda_1/\lambda_2$, in Chapter Twelve we show how this result may be exploited to estimate $R$ when we observe $X_1$ and $X_2$ but do not know $\lambda_1$ or $\lambda_2$. □

Suppose now that $T \sim$ Poisson($\lambda$) and that conditionally on $T = t, X_1 \sim$ Binomial$(t, \theta)$. For example, $T$ might be the number of automobile accidents that occur on a 5-mile stretch of freeway during a one-year period, and $X_1$ might be the number of these for which the driver was more than 60 years old. Then $(T, X_1)$ has joint probability function $g(t, x_1; \lambda, \theta) = p(t; \lambda)b(x_1; t, \theta)$ $= [e^{-\lambda}\lambda^t/t!]\binom{t}{x_1}\theta^{x_1}(1-\theta)^{t-x_1}$ for $x_1 = 0, \ldots, t$ and $t = 0, 1, 2, \ldots$. Summing across $t = x_1, x_1 + 1, \ldots$ we get the probability function for $X_1$, $f_{X_1}(x_1) = [e^{-\lambda}\lambda^{x_1}/x_1!]\theta^{x_1}\sum_{t=x_1}^{\infty}[\lambda(1-\theta)]^{t-x_1}/(t-x_1)! = [e^{-\lambda}\lambda^{x_1}/x_1!]\sum_{j=0}^{\infty}[\lambda(1-\theta)]^j/j!$ $= [e^{-\lambda}/x_1!]\lambda^{x_1}\theta^{x_1}e^{\lambda(1-\theta)} = e^{-\lambda\theta}(\lambda\theta)^{x_1}/x_1!$ for $x_1 = 0, 1, \ldots,$ the probability mass function for the Poisson($\lambda\theta$) distribution.

More generally, we can consider random vectors $\mathbf{X}$ and $\mathbf{Y}$.

**Definition 5.1.2** Let $\mathbf{X} = (X_1, \ldots, X_m)$ and $\mathbf{Y} = (Y_1, \ldots, Y_n)$, both discrete. Let $\mathbf{Z} = (\mathbf{X}, \mathbf{Y})$, an $(m+n)$-component random vector. The *conditional probability function* of $\mathbf{Y}$ given $\mathbf{X} = \mathbf{x} \equiv (x_1, \ldots, x_m)$ is

$$f_{\mathbf{Y}}(\mathbf{y} \mid \mathbf{X} = \mathbf{x}) = \frac{f_{\mathbf{Z}}(\mathbf{z})}{f_{\mathbf{X}}(\mathbf{x})}$$

for any $\mathbf{y} = (y_1, \ldots, y_n)$, any $\mathbf{x}$ for which $f_{\mathbf{X}}(\mathbf{x}) > 0$, and $\mathbf{z} = (\mathbf{x}, \mathbf{y})$. □

***Example 5.1.3*** Let $X_1, \ldots, X_n$ be the numbers occurring when $n$ six-sided fair dice are thrown. Let $U = X_{(1)}$ and $V = X_{(n)}$ be the minimum and maximum of these $X_i$'s. For $d = 1, 2, \ldots, 5$, let $H(d) = (d+1)^n - 2d^n + (d-1)^n$. Then $(U, V)$ has probability function $f(u, v) = H(v-u)/6^n$ for $1 \le u < v \le 6$ and $1/6^n$ for $1 \le u = v \le 6$. It follows that $\mathbf{X} = (X_1, \ldots, X_n)$ has, given $(U, V) = (u, v)$, conditional probability function $f(\mathbf{x} \mid (\mathbf{U}, \mathbf{V}) = (\mathbf{u}, \mathbf{v})) = (1/6^n)/p(u, v) = 1/H(v-u)$ for every $\mathbf{x} = (x_1, \ldots, x_n)$ for which $\min(\mathbf{x}) = u$, and $\max(\mathbf{x}) = v$.

Students should check this for the case $n = 3$ or $n = 4$. If, for example, $n = 4$, then $f(2, 5) = H(3)/6^3 = [256 - (2)(81) + 16]/1296 = 110/1296$. All 4-tuples $\mathbf{x} = (x_1, x_2, x_3, x_4)$ with minimum 2 and maximum 5 have probability 1/53. Notice that $H$ is a polynomial of degree $(n-2)$. For $n = 3$, $H(d) = 6d$. For $n = 4$, $H(d) = 12d^2 + 2$. □

### Discrete Markov Chains

Consider the following sequence of 200 0's and 1's:

```
1 1 0 1 1   1 1 0 1 0   1 0 1 0 1   0 1 1 0 0   1 0 1 1 0   1 0 1 1 0
1 0 1 0 1   0 1 1 0 0   1 1 0 1 1   0 1 1 0 1   0 1 0 0 1   0 1 1 1 0
1 0 1 1 1   0 0 1 0 0   1 0 1 0 1   1 1 0 0 0   0 1 0 1 1   0 1 0 0 0
1 1 0 0 1   0 0 0 1 0   1 0 0 1 0   1 0 1 0 1   0 1 0 1 0   1 0 1 0 1
0 1 0 0 1   0 1 0 1 0   0 1 0 0 1   0 1 0 0 1   0 0 0 0 1   0 1 0 1 0
1 0 0 1 0   1 0 1 1 0   1 1 0 1 0   0 1 0 1 0   1 1 0 0 0   1 1 0 0 0
0 1 0 1 0   1 0 0 0 1   0 1 0 0 1   0 1 0 1 0
```

There are 104 0's and 96 1's. A zero is followed by another zero 31 times, by a one 72 times. A 1 is followed by a zero 73 times, by a one 23 times. Quite obviously, both zeros and 1's tend to be followed by the other number, unlike coin tosses, which have no "memory." Actually, these "states" (0 and 1) were generated according to a *Markov chain*. Let $X(t)$ be the state at time $t$. $P(X(t) = 1 \mid X(t-1) = 0)$ was chosen to be 0.7, while $P(X(t) = 1 \mid X(t-1) = 1)$ was 0.2. This model might describe the two gender's as they stand in a theater line, with men and women tending to alternate. These probabilities may be summarized in the following table, called a $2 \times 2$ *Markov transition matrix* $P$, with rows summing to 1.

| | $X(t)$ | |
|---|---|---|
| $X(t-1)$ | 0 | 1 |
| 0 | 0.3 | 0.7 |
| 1 | 0.8 | 0.2 |

A coin toss was used to determine $X(1)$, so that $P(X(1) = 1) = 1/2$. To qualify as a Markov chain, conditional probabilities of the type $Q(t, x_t, x_{t-1}, \ldots, x_1) \equiv P(X(t) = x_t \mid X(t-1) = x_{t-1}, X(t-2) = x_{t-2}, \ldots, X(1) = x_1)$ must depend only on $x_t$ and $x_{t-1}$, that is, only on $x_t$ and the "immediate past." This sequence, an example of a "random process," is also stationary, in the sense that $Q$ does not depend on $t$. Use of a tree diagram should enable students to verify that two-step transition probabilities $P(X(t) = x_0 \mid X(t-2) = x_2)$ are given by the matrix $P^2$, whose first and second rows are (0.65, 0.35) and (0.40, 0.60). Similarly, the 16-step transition matrix $P^{16}$ has rows (0.5333405, 0.4666595) and (0.5333252, 0.4666748).

Under rather general conditions, $P^n$ converges to a matrix with equal rows $\mathbf{p}$, where $\mathbf{p}$ is the *stationary probability vector*, which satisfies the matrix equation $\mathbf{p}P = \mathbf{p}$. This equation can be solved to show that $\mathbf{p} = (8/15, 7/15)$, so that that the Markov chain will have sample proportions in states 0 and 1 which converge to 8/15 and 7/15. In a sequence of length 10,000, the proportion of zeros was 0.5375, slightly more than $8/15 = 0.5333\ldots$. The probabilities of each of the four outcomes (0, 0), (0, 1), (1, 0), (1, 1) at times $t$ and $t+1$ converges to those given by multiplying $\mathbf{p}$ coordinate-wise by each of the rows of $P$, so that, for example, $\lim_{t\to\infty} P(X(t) = 1, X(t+1) = 0) = (7/15)(0.8) = 56/150$. The counts of these pairs among 9999 consecutive pairs were (0, 0), 1572; (0, 1), 3731; (1, 0), 3731; (1, 1), 965. $3731/9999 = 0.3721$ is close to $56/150 = 0.3733$.

### Problems for Section 5.1

**5.1.1** In Example 5.1.1, let $T = R + W$. For each of $t = 0, 1, 2, 3$, give the conditional distribution of $R$ given $T = t$.

**5.1.2** Suppose that a die is tossed until two sixes have occurred. Let $X_1$ and $X_2$ be the numbers of the tosses on which the first and second sixes occur. Find the conditional probability function of $X_1$ given $X_2 = k$ for $k = 0, 1, 2, \ldots$. *Hint*: $X_1$ and $D = X_2 - X_1$ are independent.

**5.1.3** Three coins are tossed five times. For $k = 0, 1, 2, 3$, let $X_k$ be the number of times among these five trials in which $k$ heads are observed. Find the conditional probability function for $(X_0, X_1, X_2)$ given $X_3 = x_3$ for $x_3 = 0, 1, \ldots, 5$.

**5.1.4** Let $X_1$ and $X_2$ be independent binomial random variables with parameters $(n_1, p)$ and $(n_2, p)$ (the same $p$). Let $T = X_1 + X_2$. Show that conditionally on $T = t$, $X_1$ has the hypergeometric distribution.

**5.1.5** **(a)** Let $X \sim \text{B}(n, p_1)$, and conditionally on $X = k$, let $Y \sim B(k, p_2)$. Find the marginal distribution of $Y$. *Hint*: Let $k = n - y$ and consider the binomial expansion of $(1 + p_1 q_2/q_1)^{n-y}$.

**(b)** Consider an experiment with independent events $A$ and $B$ with $P(A) = p_1$ and $P(B) = p_2$. Suppose that this experiment is repeated independently $n$ times. Let $X$ be the number of occurrences of $A$, and let $Y$ be the number of occurrences of $A \cap B$. We might, for example, throw two six-sided fair dice, one green, one red, 10 times, with $A =$ [red die shows 6] and $B =$ [green die shows 6]. Argue that $X$ and $Y$ satisfy the conditions of part (a). This makes the conclusion of part (a) follow immediately by definition.

**5.1.6** Consider a Markov chain of three states, with transition matrix

$$B = \begin{pmatrix} 0.3 & 0.5 & 0.2 \\ 0.1 & 0.2 & 0.7 \\ 0.6 & 0.3 & 0.1 \end{pmatrix}.$$

Show that the stationary probability vector is $\mathbf{p} \equiv (51, 54, 59)/164$.

**5.1.7** A mole lives in four different holes in the ground 10 m apart in a straight line from west to east, numbered 1, 2, 3, 4. Being easily bored, the mole moves often, never more than one hole eastward or one hole westward. At hole 1 the next move is always to hole 2. Similarly, from hole 4 the next move must be to 3. However, from holes 2 and 3 the mole stays in the same hole with probability 0.5, moves westward with probability 0.3, and moves eastward with probability 0.2.

**(a)** Let $H(t)$ be the hole number after $t$ moves, beginning at hole 1. $H(t)$ is a Markov chain. What is its transition matrix $P$?

**(b)** What is its two-day transition matrix? Three-day matrix? If a computer is available, find the four-, eight-, 16-, and 32-day matrices. Note that all four rows of $P^n$ appear to be converging to the same four-component vector $\mathbf{p}$.

**(c)** Find the mole's stationary probability vector **p**. (Express the probabilities as multiples of $1/63$. In a simulation the frequencies for 10,000 days were $1415, 4851, 3106, 618$.) *Hint*: $p_1 = 4/63$.

## 5.2 CONDITIONAL EXPECTATIONS FOR DISCRETE RANDOM VARIABLES

Consider Example 5.1.1 The conditional distribution of $W$, given $R = 1$, has probability mass $4/40, 24/40$, and $12/40$ at $y = 0, 1, 2$. This distribution has mean $48/40 = 1.2$. We express this as $E(W \mid R = 1) = 1.2$. Similarly, $E(W \mid R = 0) = 1.8$, $E(W \mid R = 2) = 0.60$, and $E(W \mid R = 3) = 0$. Notice that if we write $g(r) = E(W \mid R = r)$, then $E(g(R)) = 1.8(10/84) + 1.2(40/84) + 0.6(30/84) + 0(4/84) = 1 = E(W)$. As we shall see, the fact that $E(g(R)) = E(W)$ is not a coincidence.

We need a definition.

**Definition 5.2.1** Let $(X, Y)$ be a pair of discrete random variables. For each $x$ for which $P(X = x) = f_X(x) > 0$, the conditional expectation of $Y$, given $X = x$, is the mean of the conditional distribution of $Y$ given $X = x$. This *conditional expectation* is denoted by $E(Y \mid X = x)$. The function $g(x) = E(Y \mid X = x)$, defined whenever $P(X = x) > 0$, is called the *regression function* for Y on X. □

Since $E(Y \mid X = x)$ is an ordinary expectation with respect to a probability distribution (the conditional distribution of $Y$, given $X = x$), all the same properties hold as they did for $E(Y)$ or $E(X)$. For example, $E(a + bY \mid X = x) = a + bE(Y \mid X = x)$, $E(h(Y) \mid X = x) = \sum_y h(y) f_Y(y \mid X = x)$. This enables us to define the conditional variance of $Y$ given $X = x$: Again, let $g(x) = E(Y \mid X = x)$. Then $\text{Var}(Y \mid X = x) = E[(Y - g(x)^2 \mid X = x]$. That is, $\text{Var}(Y \mid X = x)$, the conditional variance of $Y$ given $[X = x]$, is the variance of the conditional distribution of $Y$ given $X = x$.

For Example, let $v(r) = \text{Var}(W \mid R = r)$. Then $v(0) = v(1) = 0.36$, $v(2) = 0.24$, $v(3) = 0$. Notice that $E(v(R)) = 0.36(10/84 + 40/84) + 0.24(30/84) = 0.3 < 0.5 = \text{Var}(W)$. We will soon learn why the expectation of the conditional variance is never more than the unconditional variance, and in cases in which the regression function is not constant, is smaller. In fact, since $\text{Var}(g(R)) = 0.2$, we see that $\text{Var}(W) = E[v(R)] + \text{Var}(g(R))$.

***Example 5.2.1*** Each day a used car dealer sells $N$ cars, where $N$ takes the values $1, 2, 3$ with probabilities $0.3, 0.5, 0.2$. The profit from the sales of these $N$ cars are random, taking the values, in \$1000s, $X_1, X_2, X_3$. Each of the $X_i$ takes the values 1, 2, 3, 4 with probabilities 0.4, 0.3, 0.2, 0.1. Suppose also that $N$ and $(X_1, X_2, X_3)$ are independent. Let $T = X_1 + \cdots + X_N$. Thus, $Y$, the total profit in \$1000s, is random for two reasons: $N$ is a random variable, and the profits made on the $N$ cars sold are random. Intuitively, it might seem that $E(T) = E(N)E(X_1) = (1.9)(2.0) = 3.80$.

Here are $100(N, T)$ pairs:

| $(N,T)$ | $(N,T)$ | $(N,T)$ | $(N,T)$ | $(N,T)$ | $(N,T)$ | $(N,T)$ | $(N, T)$ | $(N,T)$ | $(N,T)$ |
|---|---|---|---|---|---|---|---|---|---|
| 2 5 | 2 6 | 3 7 | 2 2 | 2 3 | 2 4 | 2 4 | 2 | 1 1 | 3 6 |
| 2 4 | 2 2 | 2 5 | 2 6 | 2 5 | 2 4 | 1 2 | 2 3 | 3 6 | 2 5 |
| 1 2 | 3 4 | 2 3 | 1 3 | 1 3 | 2 3 | 2 2 | 3 5 | 2 5 | 2 5 |
| 2 5 | 3 9 | 3 4 | 1 1 | 2 3 | 1 3 | 3 4 | 3 6 | 2 6 | 3 3 |
| 3 9 | 2 6 | 1 1 | 2 3 | 2 2 | 1 1 | 2 3 | 3 3 | 2 5 | 2 5 |
| 2 3 | 3 5 | 3 5 | 3 6 | 1 2 | 3 6 | 2 4 | 2 3 | 1 2 | 1 3 |
| 1 2 | 3 6 | 3 7 | 1 1 | 1 1 | 1 2 | 3 3 | 1 2 | 1 2 | 2 3 |
| 1 1 | 3 4 | 2 6 | 2 7 | 2 6 | 3 5 | 2 2 | 2 4 | 3 6 | 2 3 |
| 2 3 | 2 5 | 3 5 | 2 7 | 2 3 | 1 3 | 2 5 | 3 6 | 2 4 | 2 3 |
| 1 1 | 2 5 | 2 2 | 3 6 | 3 4 | 1 2 | 3 8 | 2 5 | 1 2 | 2 2 |

**Frequency Table for 1000 pairs**

| | $t$ | | | | | | | | | | | |
|---|---|---|---|---|---|---|---|---|---|---|---|---|
| $n$ | 1 | 2 | 3 | 4 | 5 | 6 | 7 | 8 | 9 | 10 | 11 | Total |
| 1 | 123 | 81 | 56 | 25 | 0 | 0 | 0 | 0 | 0 | 0 | 0 | 285 |
| 2 | 0 | 92 | 126 | 130 | 99 | 48 | 22 | 5 | 0 | 0 | 0 | 525 |
| 3 | 0 | 0 | 11 | 29 | 42 | 45 | 34 | 20 | 8 | 3 | 1 | 193 |
| Total | 123 | 173 | 193 | 184 | 141 | 93 | 56 | 25 | 8 | 3 | 1 | 1000 |

*Sample statistics*:

$N$: mean $= 1.908$, variance $= 0.470$

$T$: mean $= 3.752$, variance $= 3.708$

Sample covariance $= 0.933$, sample correlation $= 0.707$

We can, with some patience, using a tree diagram, find the joint probability function for $(N, T)$. From $f(n, t) = P(N = n, T = t) = P(N = n)P(T = t \mid N = n)$, we get, for example, $f(2, 5) = (0.5)(0.20) = 0.10$, very close to the relative frequency observed. From this we can determine the marginal distribution of $T$, then $E(T)$, $\text{Var}(T)$. Using the joint distribution, we can also determine $\rho(N, T)$, the correlation coefficient of $N$ and $T$. However, as we shall see from the following theorem and its corollary, we can avoid the necessity of this heavy computation by determining $E(T)$, $\text{Var}(T)$, and $\rho(N, T)$ using the functions $g(n) = E(T \mid N = n)$ and $v(n) = \text{Var}(T \mid N = n)$. □

**Theorem 5.2.1** Let $(X, Y)$ be discrete random variables. Let $g(x) = E(Y \mid X = x)$ for each $x$ in the set $A$ of $x$ for which $P(X = x) > 0$. Let $v(x) = \text{Var}(Y \mid X = x)$ for $x \in A$. Let $h$ be any function on the range of $X$. Then:

**(a)** $E(h(X)Y) = E(h(X)g(X))$. In particular, if $h \equiv 1$, $E(Y) = E(g(X)) = E(E(Y \mid X))$.

**(b)** Let $D = Y - g(X)$. Then $\text{Cov}(h(X), D) = E(h(X)D) = 0$.
**(c)** $E(Y - h(X))^2 = E(v(X)) + E(g(X) - h(X))^2$. In particular, if $h$ is identically $E(Y)$, we get $\text{Var}(Y) = E(v(X)) + \text{Var}(g(X)) = E(v(X)) + \text{Var}(E(Y \mid X))$.

*Proof:*

**(a)** $E[h(X)g(X)] = \sum_x h(x)E(Y \mid X = x)f_X(x) = \sum_x f_X(x)h(x)\sum_y y[f_{XY}(x, y)/f_X(x)] = \sum_x \sum_y yh(x)f_{XY}(x, y) = E(h(X)Y)$. The third equality follows from Fubini's theorem, justifying the exchange of order of summation. Of course, that is trivial when either $X$ or $Y$ takes only a finite number of values.
**(b)** $E(D) = 0$ from part (a). Therefore, $\text{Cov}(h(X), D) = E(h(X)D) - E(h(X))E(D) = E(h(X)D) = E(h(X)Y) - E(h(X)g(X)) = 0$, again from (a).
**(c)** $Y - h(X) = D + (g(X) - h(X)) = (Y - g(X)) + (g(X) - h(X))$. The second term is a function of $X$. Therefore, from part (B), $E(D(g(X) - h(X))) = 0$. Hence, $E(Y - h(X))^2 = \text{Var}(D) + E(g(X) - h(X))^2$. But $\text{Var}(D) = E(v(X))$. □

COMMENTS: We have decomposed the mean squared error (MSE) in predicting $Y$ by $h(X)$ into two parts: the error $Y - g(X)$ made by predicting $Y$ by $g(X)$, which demands that $X$ be observed, and the error $g(X) - h(X)$ caused by the deviation of $h(X)$ from $g(X) = E(Y \mid X)$. From part (b), these two sources of error are uncorrelated.

**Corollary 5.2.1** Let $N, X_1, X_2, \ldots$ be independent random variables, where $E(X_j) = \mu_X$, $\text{Var}(X_j) = \sigma_X^2$, for $j = 1, 2, \ldots$. Let $\mu_N = E(N)$ and $\sigma_N^2 = \text{Var}(N)$. Let $T = \sum_{j=1}^{N} X_j$. Then:

**(a)** $E(T) = \mu_N \mu_X$.
**(b)** $\text{Var}(T) = \mu_N \sigma_X^2 + \sigma_N^2 \mu_X^2$.
**(c)** $\text{Cov}(N, T) = \mu_X \sigma_N^2$.
**(d)** $\rho(N, T) = 1/\sqrt{1 + \theta}$, where $\theta = \sigma_X^2/\left(\mu_X \sigma_N^2\right)$.

*Proof:* Use Theorem 5.2.1 with $N$ replacing $X$ and $T$ replacing $Y$. $g(n) = E(T \mid N = n) = n\mu_X$, so (a) follows from Theorem 5.2.1(a). To show (b), note that $v(n) = \text{Var}(T \mid N = n) = n\sigma_X^2$. Thus, (b) follows from Theorem 5.2.1(c). $E(TN) = E(NE(T \mid N)) = E(N(N\mu_X)) = \mu_X E(N^2)$. Thus, $\text{Cov}(N, T) = \mu_X E(N^2) - (\mu_N \mu_X)\mu_N = \mu_X \sigma_N^2$, proving (c). Part (d) follows from $\rho(N, T) = \text{Cov}(N, T)/\sqrt{\text{Var}(T)\sigma_N^2}$. □

***Example 5.2.1 Revisited*** $E(N) = 1.9$, $\text{Var}(N) = 0.49$, $\mu_X = 2.0$, $\sigma_X^2 = 1.0$. Therefore, $E(T) = (2.0)(1.9) = 3.8$, $\text{Var}(T) = (1.9)(1.0) + (0.49)(2^2) = 3.86$. The estimates of $E(T)$ and $\text{Var}(T)$ based on the 1000 trials were quite close.

$\mathrm{Cov}(N, T) = 0.98$, $\theta = 1/0.98 = 1.0205$, $\rho(N, T) = 1/\sqrt{1 + 1/0.98} = 0.7035$, close to the sample correlation 0.707. □

***Example 5.2.2*** Suppose that a population of size $n$ consists of $k$ *subpopulations* or *strata*. Let the $i$th stratum have $n_i$ elements with measurements $x_{ij}$. Let $\bar{x}_i$ be the mean of the measurements $x_{ij}$ in stratum $i$, and let $\bar{x}$, be the grand mean of all $n = \sum_{ij} n_{ij}$ elements. Suppose that one of the elements (*units* in the language of sample surveys) in the population of $n$ is chosen at random. Let $Y$ be the $x$-value and $I$ be the stratum number of the unit chosen. Then $E(Y \mid I = i) = \bar{x}_i$. $I$ takes the value $i$ with probability $n_i/n$. $E(Y) = \sum_{i=1}^{n}(n_i/n)\bar{x}_i = \bar{x}$ and $\mathrm{Var}(Y) = (1/n)\sum_{ij}(x_{ij} - \bar{x})^2$. The regression function for $Y$ on $I$ is $g(i) = \bar{x}_i$, and the conditional variance function is $v(i) = (1/n_i)\sum_j (x_{ij} - \bar{x}_i)^2$. It follows from Theorem 5.2.1 that $(1/n)\sum_{ij}(x_{ij} - \bar{x})^2 = \sum_{i=1}^{n_i} v(i)(n_i/n) + \sum_{i=1}^{k}(n_i/n)[g(i) - \bar{x}]^2$. Multiplying through by $n$, we get the identity

$$\sum_{ij}(x_{ij} - \bar{x})^2 = \sum_{ij}(x_{ij} - \bar{x}_i)^2 + \sum_{i=1}^{k} n_i(\bar{x}_i - \bar{x})^2. \qquad (5.2.1)$$

These three terms are usually called *total sum of squares*, *error sum of squares*, and *among means sum of squares*. Viewed from a vector-space point of view (see Chapter Eleven), this follows from the Pythagorean theorem. This is another example of a fundamental approach in statistics: the decomposition of a measure of variability into pieces, called the *analysis of variance* (AOV). Of course, (5.2.1) may be proved by a more direct approach involving the manipulation of summations. That the cross-product term $\sum_{ij} n_i(x_{ij} - \bar{x}_i)(\bar{x}_i - \bar{x}) = 0$ follows from Theorem 5.2.1(b) or simply by summing first on $j$. □

***Example 5.2.3*** Let $k = 3, n_1 = 2, n_2 = 3, n_3 = 2, x_{11} = 5, x_{12} = 9, x_{21} = 2, x_{22} = 3, x_{23} = 4, x_{31} = 8, x_{32} = 4$. Then $\bar{x} = 5, \bar{x}_1 = 7, \bar{x}_2 = 3, \bar{x}_3 = 6$, and the total, error, and among means sums of squares are 40, 18, and 22. (Students: Do the computations yourself, then tell your families and friends that you have performed your first AOV! It won't be your last.) □

### Prediction

If we observe $X$ and wish to predict $Y$ with $h(X)$, then $\mathrm{MSE}(h) \equiv E(Y - h(X))^2$ is the mean squared error of $h(X)$. To minimize $\mathrm{MSE}(h)$, it follows from Theorem 5.2.1(c) that we should choose $h(x) = g(x) \equiv E(Y \mid X = x)$ for every $x$. In that case, the error $D = Y - g(X)$ has mean zero, variance $E(v(X)) = \mathrm{MSE}(g)$. If $X$ is not observed and $h(x)$ must be identically a constant, then $\mathrm{MSE}(h)$ is minimized by choosing $h(x)$ to be identically $\mu_Y$.

### Linear Prediction

It often happens that the distribution of $(X, Y)$ is unknown, so that the function $g(x) = E(Y \mid X = x)$ is unknown. With enough observations on $(X, Y)$ we might be able to estimate $g$. However, it often happens that sufficient data are not available. When $g$ is roughly linear in such cases, we can sometimes make some progress by choosing $h$ to be linear, of the form $h(x) = a + bx$ over the range of $X$. Let's now see what the appropriate choices for $a$ and $b$ should be.

Again, let's use the MSE as a measure of the worth of the predictor $h(X)$. If $h(x) = a + bx$, then $\text{MSE}(h) \equiv G(a, b) = E(Y - (a + bX))^2$. Let $Z_X = (X - \mu_X)/\sigma_X$ and $Z_Y = (Y - \mu_Y)/\sigma_Y$ be the standardized versions of $X$ and $Y$. These random variables have zero means, zero variances, and $E(Z_X Z_Y) = \rho$, the correlation of $X$ and $Y$. Then $Y - (a + bX) = (Z_Y - cZ_X)\sigma_Y + (d - a)$, where $c = b\sigma_X/\sigma_Y$ and $d = \mu_Y - b\mu_X$. The first term has expectation zero. The second term is a constant. Therefore, $\text{MSE}(h) = E(Z_Y - cZ_X)^2\sigma_Y^2 + (d - a)^2 = [(1 - \rho^2) + (c - \rho)^2]\sigma_Y^2 + (d - a)^2$. This is minimized by taking $c = \rho$, hence $b = \rho\sigma_Y/\sigma_X = \sigma_{XY}/\sigma_X^2$, where $\sigma_{XY} = \text{Cov}(X, Y)$, and taking $a = d = \mu_Y - b\mu_X$. Thus, $h(x) = a + bx = \mu_Y + b(x - \mu_X)$. Notice that $h(x)$ is a straight line with slope $b$ passing through the *point of means* $(\mu_X, \mu_Y)$. For these choices of $a$ and $b$, the MSE becomes $(1 - \rho^2)\sigma_Y^2$. The proportion of variance in $Y$ unexplained by linear regression on $X$ is $\text{MSE}(h)/\text{Var}(Y) = 1 - \rho^2$. The proportion explained is $\text{Var}(h(X))/\text{Var}(Y) = \rho^2$. Slope $b$ is the *rescaled correlation coefficient* $b = \rho\sigma_Y/\sigma_X$. The slope of the regression line for $Z_Y$ on $Z_X$ is $\rho$. In fact, Francis Galton in about 1890 originally defined $\rho$ as the slope of the regression line of $Z_Y$ on $Z_X$.

***Example 5.2.4*** Suppose that $(X, Y)$ takes the values (0, 0), (0, 1), (1, 1), (2, 1), (2, 4), each with probability 1/5 (see Figure 5.2.1.) Then $\mu_X = 1, \mu_Y = 7/5, \sigma_X^2 = 4/5, \sigma_Y^2 = 46/25, \sigma_{XY} \equiv \text{Cov}(X, Y) = 1, \rho = 2\sqrt{5/46}$. Hence, $b = \sigma_{XY}/\sigma_X^2 = 1, a = \mu_Y - b\mu_X = 2/5$.

| | $X = 0$ | $x = 1$ | $x = 2$ |
|---|---|---|---|
| $P(X = x)$ | 2/5 | 1/5 | 2/5 |
| $g(x) = E(Y \mid X = x)$ | 1/2 | 1 | 5/2 |
| $v(x) = \text{Var}(Y \mid X = x)$ | 1/4 | 0 | 9/4 |
| $h(x) = a + bx$ | 2/5 | 7/5 | 12/5 |

Then $\text{Var}(Y) = E(v(X)) + \text{Var}(g(X)) = 1 + 21/25 = 46/25$, $\text{MSE}(h) = (1 - \rho^2)\sigma_Y^2 = (26/46)(46/25) = 26/25$, $\text{MSE}(g) = E(v(X)) = 1$, $E(h(X) - g(X))^2 = 1/25$. This verifies Theorem 5.2.1(c). □

### Problems for Section 5.2

**5.2.1** Let $(X, Y)$ have probability function $f(x, y) = C(x + y)$ for $x = 1, 2, 3$ and $y = 1, 2, 3$.

**(a)** Find $C$.

**(b)** Find the regression function $g(x) \equiv E(Y \mid X = x)$ and show that $E(g(X)) = E(Y)$.

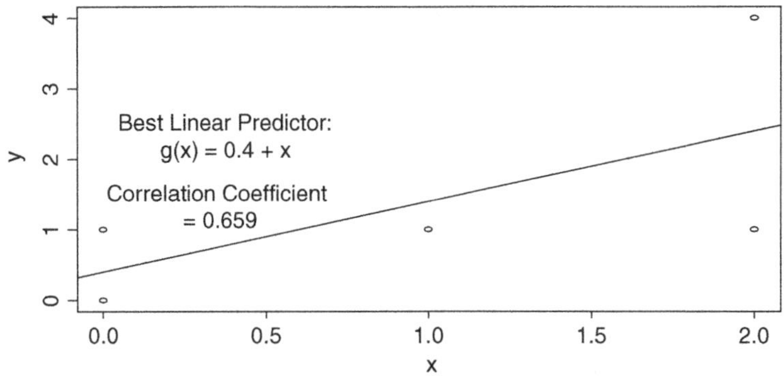


**FIGURE 5.2.1** Distribution on five points.

**(c)** Find $v(x) = \text{Var}(Y \mid X = x)$ and show that $\text{Var}(Y) = E[v(X)] + \text{Var}(g(X))$.

**5.2.2** Let $X_1, X_2, X_3$ be uncorrelated rv's with means $\mu_1, \mu_2, \mu_3$ and variances $\sigma_1^2, \sigma_2^2, \sigma_3^2$. Let $T_2 = X_1 + X_2$ and $T_3 = X_1 + X_2 + X_3 = T_2 + X_3$. Find $g(t_2) \equiv E(T_3 \mid T_2 = t_2)$ and $v(t_2) \equiv \text{Var}(T_3 \mid T_2 = t_2)$, and show that $\text{Var}(T_3) = E[v(T_2)] + \text{Var}(g(T_2))$.

**5.2.3** A box has five balls numbered 1, 2, 3, 4, 5. A ball with number $X$ is drawn at random. Balls with larger numbers are withdrawn from the box, and the ball with the number $X$ is replaced in the box. Then $X$ balls are drawn at random with replacement from the remaining $X$ balls. Let $Y$ be the total of the numbers on the $X$ balls drawn.

**(a)** Find the regression function $g(x) \equiv E(Y \mid X = x)$.

**(b)** Find $E(Y)$ and $\text{Var}(Y)$.

**5.2.4** A fair six-sided die is tossed until a 6 occurs. Let $X$ be the number of tosses necessary. Then the die is tossed until $X$ more 6's have occurred. Let the total number of tosses necessary, including the first $X$, be $Y$. Find $E(Y)$ and $\text{Var}(Y)$. *Hint*: Conditionally on $X = x$, $D = Y - X$ has the negative binomial distribution with parameters $1/6$ and $x$. The mean and variance of the negative binomial distribution with parameters $r$ and $p$ are $r/p$ and $r(1-p)/p^2$.

**5.2.5** Let $X$ take the values 0, 1, 2 with probabilities 0.3, 0.4, 0.3. Let $Y = 5 - X^2 + \varepsilon$, where $\varepsilon$ and $X$ are independent and $\varepsilon$ takes the values $-1, +1$ with probabilities $1/2, 1/2$.

**(a)** Find the functions $g(x) = E(Y \mid X = x)$ and $h(X)$, the least squares linear predictor of $Y$.

**(b)** Find the correlation coefficient $\rho = \rho(X, Y)$.

**(c)** Find $\text{MSE}(g)$ and $\text{MSE}(h)$.

**5.2.6** Use calculus to verify the formulas for $a$ and $b$ so that $h(X) = a + bX$ is the least squares predictor of $Y$.

**5.2.7** A used-car salesman makes 0, 1, 2, 3 sales each day with probabilities 0.4, 0.3, 0.2, 0.1. When he does make a sale, his profit in units of \$500 are 1, 2, 3 with probabilities 0.3, 0.5, 0.2. Let his profit (in units of \$500) for one day be $T$. Find $E(T)$ and $\text{Var}(T)$.

**5.2.8** Let $X \sim \text{Binomial}(n, p_1)$. Conditionally on $X = k$, let $Y \sim \text{Binomial}(k, p_2)$.

**(a)** What is the unconditional distribution of $Y$? Use your answer to give $E(Y)$ and $\text{Var}(Y)$.

**(b)** Find the functions $g(k) = E(Y \mid X = k)$ and $v(k) = \text{Var}(Y \mid X = k)$. Use these to find $E(Y)$ and $\text{Var}(Y)$.

**(c)** Find $\text{Cov}(X, Y)$ and $\rho(X, Y)$.

**5.2.9** Repeat Problem 5.2.8 for the case that $X \sim \text{Poisson}(\lambda)$ but everything else is the same.

**5.2.10** $A$ and $B$ play a game in which they take turns throwing two dice until one of them is successful. $A$ is successful if his two dice total 6. $B$ is successful if her two dice total 7. $A$ goes first. What is the probability that $A$ wins the game? This is the fourteenth of Huygens' "14 Propositions" (1657) (see Hald, 2003, p. 72).

**(a)** Find the probability that $A$ wins the game by (1) summing a geometric series, and by (2) conditioning on the outcome of the first two attempts, one for $A$, one for $B$.

**(b)** Show that for the general case that $A$ and $B$ have probabilities $p_a$ and $p_b$ of being successful when the two dice are thrown, $P(A \text{ wins}) = p_a/(1 - q_a q_b)$, where $q_a = 1 - p_a$ and $q_b = 1 - p_b$.

## 5.3 CONDITIONAL DENSITIES AND EXPECTATIONS FOR CONTINUOUS RANDOM VARIABLES

Let $(X, Y)$ have a continuous distribution, with density $f(x, y)$. We would like to define the conditional density of $Y$ given $X = x$. What we seek is a function $k(y; x)$ which has the property that:

$$P((X, Y) \in B) = \int_B k(y; x) f_X(x)\, dy\, dx \qquad \text{for every (Borel) subset } B \text{ of } R_2. \tag{5.3.1}$$

**Definition 5.3.1** Let $(X, Y)$ have joint density $f(x, y)$ and let $X$ have density $f_X(x)$. Then the *conditional density* of $Y$, given $X = x$, is $f_Y(y \mid X = x) \equiv f(x, y)/f_X(x)$, defined for every $x$ for which $f_X(x) > 0$, and for all $y$. □

*Notes:* This is a function $k(y; x)$ that satisfies (5.3.1). Recall that a density function is not unique, since it can be defined arbitrarily on a finite, or even countably infinite, set of points (more precisely, on a set of *Lebesgue measure zero*). Thus, there are an infinite number of versions of a conditional density which describe the same probability structure. Usually, we choose the one that is simplest in some sense. □

***Example 5.3.1*** Let $(X, Y)$ have density $f(x, y) = 3x$ for $0 < y < x < 1$. Then $f_X(x) = 3x^2$ for $0 < x < 1$, and $f_Y(y \mid X = x) = 1/x$ for $0 < y < x < 1$. This is the Unif(0, $x$) distribution. Without formally defining conditional expectation for the continuous case just yet, it should seem reasonable that the regression function of $Y$ on $X$ is $g(x) \equiv E(Y \mid X = x) = \int y f_Y(y \mid X = x)\, dy = x/2$ for $0 < x < 1$ and that $v(x) \equiv \text{Var}(Y \mid X = x) = \int [y - g(x)]^2 f_Y(y \mid X = x(y \mid X = x)\, dy = x^2/12$, defined for every $x$ for which $f_X(x) > 0$ and the integral exists.

Since $Y$ has density $f_Y(y) = \int_y^1 3x\, dx = (3/2)(1 - y^2)$ for $0 < y < 1$, the conditional density of $X$, given $Y = y$, is $f_X(x \mid Y = y) = 2x/(1 - y^2)$ for $0 < y < x < 1$. *Suggestion*: Sketch these densities (in $x$) for $y = 0.3$ and for $y = 0.8$. Verify that the regression function for $X$ on $Y$ is $k(y) \equiv E(X \mid Y = y) = \int_y^1 x f_X(x \mid Y = y)\, dx = (y + 2)/3$ and that $w(y) \equiv \text{Var}(X \mid Y = y) = (1 - y)^2/18$. To make the computation easier, notice that given $X = x$, $Y$ has the triangular density on the interval $(y, 1)$. Find the mean and variance for the triangular density on (0, 1) and "adjust"; that is, first consider $Z = (X - y)/(1 - y)$, conditionally on $Y = y$. □

***Example 5.3.2*** Let $X \sim \Gamma(\alpha, 1)$, $Y \sim \Gamma(\beta, 1)$, independent, where $\alpha > 0$, $\beta > 0$. As in Problem 4.2.5, define $U = X + Y$, $V = X/(X + Y)$. From Problem 4.3.5(a), $U$ and $V$ are independent and $V$ has the beta distribution with parameters $\alpha$ and $\beta$. It follows that the conditional distribution of $V$, given $U = u$, is the same for all $u$. Let $g(u) \equiv E(X \mid U = u)$ be the regression function of $X$ on $U$. Since, $E(V \mid U) = u = E(V) = \alpha/(\alpha + \beta)$, the same for all $u$, $g(u) = E(VU \mid U = u) = u\alpha/(\alpha + \beta)$. Since $g$ is linear, $g(U)$ is also the linear least squares predictor of $X$. When $\alpha = \beta = 1$, so that $X$ and $Y$ are both exponential random variables, the conditional distribution of $X$ given $U = u$ is Unif(0, $u$). □

### Discrete Mixtures

One way to build probability models is through *mixtures*. Suppose, for example, that the population of 10,000 students at Height University comprises 6000 women and 4000 men. The heights of the women in inches are normally distributed with mean $\mu_1 = 65$, standard deviation $\sigma_1 = 2.5$, whereas for men the mean is $\mu_2 = 70$, standard deviation $\sigma_2 = 2.7$. If a student is drawn at random, what is the distribution of her or his height $X$? That's easy if we think conditionally. Let $W$ be the event that the student chosen is a woman and $M = W^c$ be the event that the student is a man. Given that the student is a woman, $X$ has cdf $F_1(x) = \Phi((x - 65)/2.5)$. Given that the student is a man, $X$ has cdf $F_2(x) = \Phi((x - 70)/2.7)$.

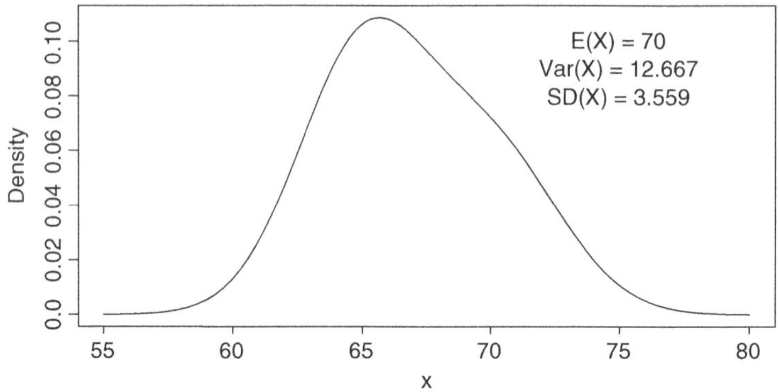


**FIGURE 5.3.1** Density of heights.

For any subset $A$ of the real line, $P(X \in A) = P([X \in A] \cap F) + P([X \in A] \cap M) = P(X \in A \mid F)P(F) + P(X \in A \mid M)P(M)$. Taking $A = (-\infty, x)$, we get $F_X(x) = (0.6)F_1(x) + (0.4)F_2(x)$, the (0.6, 0.4) mixture of $F_1$ and $F_2$. Thus, $X$ has the density $f_X(x) = f_1(x)(0.6) + f_2(x)(0.4)$, where $f_1$ and $f_2$ are the densities of the heights of women and men (see Figure 5.3.1).

We can represent $X$ succinctly as follows. Let $I = 1$ or 2, depending on whether the student is a woman or man. Let $X_1$ have cdf $F_1$, and let $X_2$ have cdf $F_2$. Then $X = X_I$.

We can compute expectations and variances using Theorem 5.2.1. The regression function of $X$ on $I$ is $g(i) = E(X_I \mid I = i) = \mu_i$. The conditional variance function is $v(i) \equiv \text{Var}(X_I \mid I = i) = \sigma_i^2$. Then $E(X) = E(g(I)) = \mu_1(0.6) + \mu_2(0.4) = 67$, $E(v(I)) = (0.6)(2.5)^2 + (0.4)(2.7)^2 = 6.667$, $\text{Var}(g(I)) = (0.6)(0.4)(\mu_1 - \mu_2)^2 = 6.0$, so that $\text{Var}(X) = 6.667 + 6.0 = 12.667$.

More generally, let $(I, X)$ be a pair of random variables such that $I$ takes the values $1, \ldots, k$ with probabilities $p_1, \ldots, p_k$ summing to 1, and conditionally, on $I = i$, $X$ has cdf $F_i$, mean $\mu_i$, and variance $\sigma_i^2$. Then $X$ has the cdf $F = p_1 F_1 + \cdots + p_k F_k$. If $f_i$ is a density corresponding to $F_i$ for each $i$, then $X$ has density $f = p_1 f_1 + \cdots + p_k f_k$. If $F_i$ corresponds to a probability mass function $f_i$, then, of course, the same formula for $f$ holds. Conditioning on $I$, and using Theorem 5.2.1, we get

$$\mu \equiv E(X) = \sum_{i=1}^{k} p_i \mu_i \quad \text{and} \quad \text{Var}(X) = \sum_{i=1}^{k} p_i \sigma_1^2 + \sum_{i=1}^{k} p_i (\mu_i - \mu)^2.$$

## Continuous Mixtures

Suppose that a certain type of worm has infested a 1-acre field (roughly, 60 meters $\times$ 60 meters) of tomato plants. To study the degree of the infestation, we want a model for the numbers of such worms on plots 3 feet $\times$ 3 feet. There are 4840 such subplots. A model for which the number of worms has a Poisson distribution would seem reasonable. However, there is reason to believe that the rate $\lambda$ varies from plot to plot because the moisture level varies across the field and because the source of the worms seems to an adjacent field in a northeasterly direction. So $\lambda$ itself seems to be random. What is the distribution of the number $X$ of worms infesting a plot chosen randomly?

Suppose that conditionally on $\lambda$, $X$ had the Poisson distribution with mean $\lambda$. Thus, conditionally on $\lambda$, $X$ has probability mass function $p(k;\lambda) = e^{-\lambda}\lambda^k/k!$ for $k = 0, 1, \ldots$. Suppose that $\lambda$ is random with the $\Gamma(m, \theta)$ distribution, $m > 0, \theta > 0$. That is, $\lambda$ has density $f(u; m, \theta) = [\Gamma(m)\theta^m]^{-1}u^{m-1}e^{-u/\theta}$ for $u > 0$. Then $(X, \lambda)$ has joint density $f(k, u; m, \theta) = p(k; u) f(u; m, \theta)$. The density is really a density with respect to the counting measure in the first argument, Lebesgue measure in the second in the sense that for a rectangular set $A = A_1 \times A_2$ in two-space, $P((X, \lambda) \in A) = \int_{A_2} \sum_{k \in A_1} f(k, u; m, \theta)\, du$. The marginal density probability function of $X$ is $f_X(k) = \int_0^\infty f(k, u; m, \theta)\, du = [\Gamma(k + m)/(k!\Gamma(m))]p^m q^k$, for $k = 0, 1, 2, \ldots$, where $p = 1/(1 + \theta)$ and $q = 1 - p = \theta/(1 + \theta)$. For the case that $m$ is an integer, the term in brackets is $\binom{m+k-1}{m-1}$, so that $X$ has the negative binomial distribution taking values $0, 1, 2, \ldots$ with parameters $m$ and $p$. We can extend the definition of the negative binomial distribution to the case that $m$ is any positive number by saying that any random variable with the mass function of $X$ has the negative binomial distribution with parameters $m$ and $p$.

Since the regression function for $X$ on $\lambda$ is $g(\lambda) \equiv E(X \mid \lambda) \equiv \lambda$, and $E(\lambda) = m\theta$, we find that $E(X) = m\theta = mq/p$. Since $v(\lambda) \equiv \text{Var}(X \mid \lambda) = \lambda$, we find, from the generalization of Theorem 5.2.1(c) to this discrete–continuous case, that $\text{Var}(X) = \text{Var}(g(\lambda)) + E(v(\lambda)) = \text{Var}(\lambda) + E(\lambda) = m\theta^2 + m\theta = m\theta(\theta + 1) = mq/p^2$, the same as was shown earlier when $m$ is a positive integer. Recall that this negative binomial random variable takes the values $0, 1, \ldots$, rather than $m, m + 1, \ldots$. Because both distributions are called a negative binomial, it is incumbent on the user to make the range clear. One is just a "shift" of the other, so the variance is the same for one as for the other.

### Estimating the Pair $(m, p)$

The following sample of 100 observations on $X$ was generated using the S-Plus function "rnbinom" with the parameters $m = 3$ and $p = 0.1$.

| | | | | | | | | | | | | | | | | | | |
|---|---|---|---|---|---|---|---|---|---|---|---|---|---|---|---|---|---|---|
| 8 | 51 | 14 | 38 | 27 | 19 | 21 | 30 | 61 | 11 | 6 | 32 | 15 | 22 | 35 | 12 | 23 | 25 | 14 |
| 41 | 16 | 48 | 51 | 66 | 6 | 53 | 25 | 11 | 48 | 45 | 9 | 15 | 40 | 16 | 40 | 29 | 58 | 11 |
| 25 | 11 | 65 | 7 | 21 | 23 | 91 | 69 | 53 | 49 | 77 | 17 | 33 | 43 | 5 | 11 | 13 | 22 | 11 |
| 33 | 24 | 9 | 11 | 12 | 37 | 30 | 24 | 11 | 3 | 25 | 32 | 44 | 54 | 20 | 16 | 30 | 42 | 37 |
| 15 | 38 | 38 | 23 | 9 | 31 | 26 | 81 | 11 | 11 | 33 | 19 | 16 | 11 | 15 | 30 | 49 | 12 | 30 |
| 46 | 29 | 24 | 5 | 7 | | | | | | | | | | | | | | |

The sample mean is 29.01. The sample variance is 336.55 ($n$ divisor). If $(m, p)$ were unknown, how could we estimate them? One method, the method of moments, is

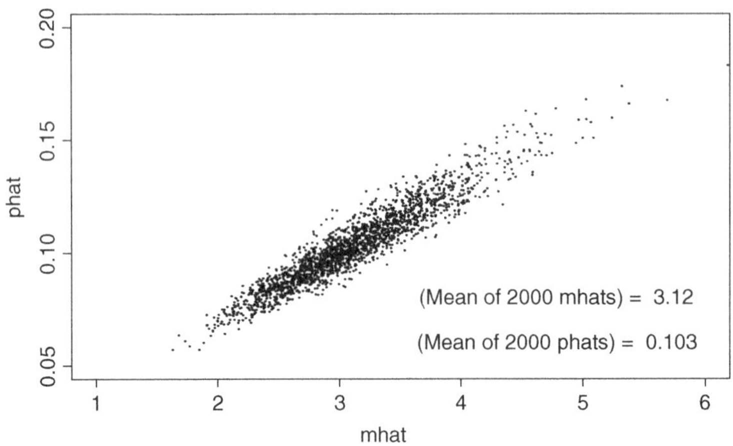


**FIGURE 5.3.2** 400 $(\hat{m}, \hat{p})$ pairs.

quite intuitive. It was discussed earlier for the gamma distribution. Since the mean $\mu = mq/p$ and the variance $\sigma^2 = mq/p^2$, we can solve uniquely for $m$ and $p$ in terms of $\mu$ and $\sigma^2$: $p = \mu/\sigma^2$ and $m = \mu^2/(\sigma^2 - \mu)$. Since the statistics $\bar{X}$ and $\hat{\sigma}^2$, the sample mean and variance, should be close to their corresponding parameters $\mu$ and $\sigma^2$ for large $n$ (more precisely, are consistent estimators of $\mu$ and $\sigma^2$), we should expect $\hat{p} \equiv \hat{X}/\hat{\sigma}^2$ and $\hat{m} \equiv \bar{X}^2/(\hat{\sigma}^2 - \bar{X})$ to be close to $p$ and $m$. For this sample of 100, we find that $\hat{p} = 0.086$ and $\hat{m} = 2.736$. This experiment was repeated 400 times, with results indicated by the scatter diagram in Figure 5.3.2.

Actually, we can do a bit better than this by using a method call maximum likelihood, but we'll postpone that to Chapter Seven.

### The Bivariate Normal Distribution

Consider a model for the heights $(Z_1, Z_2)$ of (father, son) pairs, with the son an adult. For simplicity, assume that $Z_1$ and $Z_2$ have been standardized to have means zero, variances 1. Furthermore, suppose that

$$Z_2 = a + bZ_1 + \varepsilon_2 \tag{5.3.2}$$

for some constants $a$ and $b$, where $E(\varepsilon_2) = 0$, $\text{Var}(\varepsilon_2) = d > 0$, and $\text{Cov}(Z_1, \varepsilon_2) = E(Z_1\varepsilon_2) = 0$. Taking expectations on both sides of (5.3.2), we find that $a = 0$. Multiplying both sides of (5.3.1) by $Z_1$ and taking expectations, it follows from Theorem 5.2.1(c), that $b = \rho(Z_1, Z_2) \equiv \rho$. In addition, $1 = \text{Var}(Z_1) = \rho^2 + d$, so that $d = 1 - \rho^2$. Thus, if the regression of $Z_2$ on $Z_1$ is linear and the "innovation" $\varepsilon_2$ is uncorrelated with $Z_1$, it follows that the regression function for $Z_2$ on $Z_1$ is $g(z_1) \equiv E(Z_2 \mid Z_1 = z_1) = \rho z_1$, and that $\text{Var}(\varepsilon_2) = 1 - \rho^2$. If, in addition, $Z_1$ and $\varepsilon_2$ have normal distributions with $Z_1$ and $\varepsilon_2$ independent, it follows that $Z_2$ also has

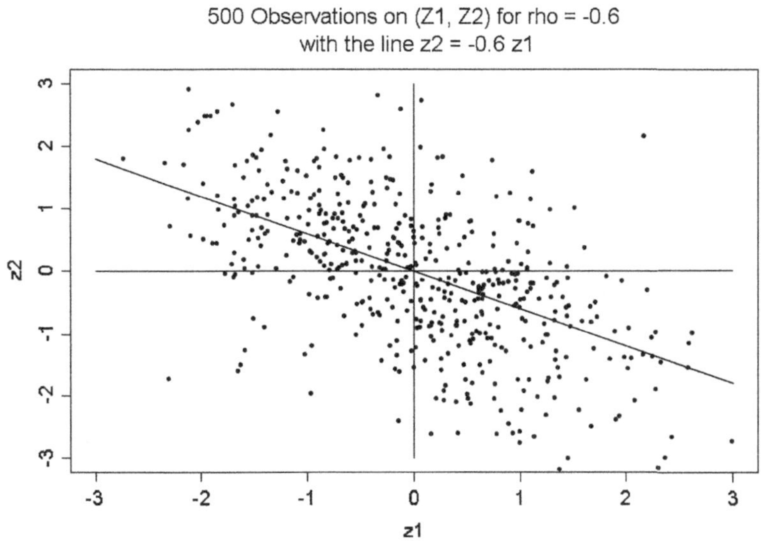


**FIGURE 5.3.3** Scatter plot of 500 observations from a bivariate normal distribution.

a normal distribution. The pair $(Z_1, Z_2)$ are said to have the *standardized bivariate normal distribution* with correlation $\rho$ (see Figure 5.3.3).

The conditional distribution of $Z_2$, given $Z_1 = z_1$, is $N(\rho z_1, d = 1 - \rho^2)$. $Z_1 \sim N(0, 1)$. It follows that the pair $(Z_1, Z_2)$ has joint density $f(z_1, z_2) = C \exp[-(1/2)Q(z_1, z_2)]$, where $Q(z_1, z_2) = (z_1^2 + (z_2 - \rho z_1)^2/d) = (z_1^2 - 2\rho z_1 z_2 + z_2^2)/2d$, and $C = 1/(2\pi d^{1/2})$. The quadratic form $Q$ can be written as $Q(z_1, z_2) = \mathbf{z}^T \Sigma^{-1} \mathbf{z}$, where $\mathbf{z}^T = (z_1, z_2)$, so that $\mathbf{z}$ is a column vector and $\Sigma$ is the $2 \times 2$ covariance matrix for $(Z_1, Z_2)$. Notice that $C = 1/[2\pi \det(\Sigma)^{1/2}]$. The case $\rho = 0$ corresponds to the independence of $Z_1$ and $Z_2$.

Now let's generalize a bit. Let $X_1 = \mu_1 + \sigma_1 Z_1$ and, conditionally on $X_1 = x_1$, equivalently, $Z_1 = z_1 \equiv (x_1 - \mu_1)/\sigma_1$, let $X_2 \sim \mathrm{N}(\mu_2 + \rho\sigma_2 z_1, d\sigma_2^2)$, equivalently, $Z_2 \equiv (X_2 - \mu_2)/\sigma_2 \sim \mathrm{N}(\rho z_1, d)$. The pair $(X_1, X_2)$ is said to have the bivariate normal distribution with parameters $\mu_1, \mu_2, \sigma_1^2, \sigma_2^2, \rho$. Equivalently, the pair of standardized random variables $Z_1 \equiv (X_1 - \mu_1)/\sigma_1$ and $Z_2 \equiv (X_2 - \mu_2)/\sigma_2$ has the standardized bivariate normal distribution with parameter $\rho$. $\rho = E(Z_1 Z_2)$ is the correlation coefficient of $X_1$ and $X_2$. The pair $(X_1, X_2)$ has the density $f(x_1, x_2) = (1/2\pi\sigma_1\sigma_2)\exp[-(1/2)Q(z_1, z_2)]$, where $Q(z_1, z_2) = (z_1^2 - 2\rho z_1 z_2 + z_2^2)/(1 - \rho^2)$ for $z_1 = (x_1 - \mu_1)/\sigma_1$, $z_2 = (x_2 - \mu_2)/\sigma_2$.

***Example 5.3.3*** Let the heights $(X_1, X_2)$ of adult father–son pairs in centimeters have a bivariate normal distribution with parameters $\mu_1 = 178$, $\mu_2 = 178$, $\sigma_1 = \sigma_2 = 7.0$, $\rho = 0.5$.

**(a)** Find the probability that the son is taller given that $X_1 = x_1$.

Conditionally on $X_1 = x_1$, $X_2 \sim \mathrm{N}(\mu_2 + \rho\sigma_2 z_1, \sigma_2^2 d = 49(1-\rho^2) = 36.75)$, where $z_1 = (x_1 - \mu_1)/\sigma_1$. $P(X_2 > X_1 \mid X_1 = x_1) = P(Z_2 > Z_1 \mid Z_1 = z_1)$, where $Z_1$ and $Z_2$ are the standardized versions of $X_1$ and $X_2$. Conditionally on $Z_1 = z_1$, $Z_2 \sim \mathrm{N}(\rho z_1, d = 1 - \rho^2)$. The fact that $E(Z_2 \mid Z_1 = z_1) = \rho z_1$ is closer than $z_1$ to zero is called *regression toward the mean*. Thus, tall fathers tend to have sons who are not as tall as their fathers, and short fathers tend to have sons who are taller than they are. In the same way, baseball players who have higher-than-average batting averages in their first year have conditional expected batting averages which are less than their first-year average. Unfortunately, this purely statistical phenomenon is often called the "sophomore slump" by those without understanding of probability and/or statistics. Your challenge, reader, is to explain all of this to your friends and family.

Thus, $P(X_2 > X_1 \mid X_1 = x_1) = 1 - \Phi((z_1 - \rho z_1)/\sqrt{d}) = \Phi(-z_1(1-\rho)/\sqrt{d}) = \Phi(-z_1\sqrt{(1-\rho)/(1+\rho)})$. If $z_1 > 0$ (father taller than average), $P$(son taller than father) $\equiv p < 1/2$. Replacing $z_1 > 0$ by $-z_1$, the probability becomes $1 - p > 1/2$. If, for example, $z_1 = 1.0$, $\rho = 1/2$, we get $\Phi(-1/\sqrt{3}) = 0.282$, and for $z_1 = -1.0$, $\rho = 1/2$, we get $\Phi(1/\sqrt{3}) = 1 - 0.282 = 0.718$.

**(b)** Find the probability that the son is taller unconditionally.

$D = X_2 - X_1 \sim \mathrm{N}(0, \sigma_1^2(1 + 1 - 1) = \sigma_1^2 = 49)$, so $P$(son is taller) $= P(D > 0) = 1 - \Phi(0) = 1/2$. See Problem 5.3.7 for extensions to the case in which the means $\mu_1$ and $\mu_2$ differ.

To display these data graphically, the 928 $(x, y)$ points were jittered; that is, Unif$(-1/2, 1/2)$ random variables were added independently to all $928(2) = 1856$ values (see Figure 5.3.4). Since the original values were rounded to the nearest inch, this would seem to provide a good approximation to the original heights, "mid-heights" for the parents. The statistics printed on the graph provide the equation of the least squares line $y = \mu_y + (\rho\sigma_y/\sigma_x)(x - \mu_x) = 45.98 + 0.328x$. □

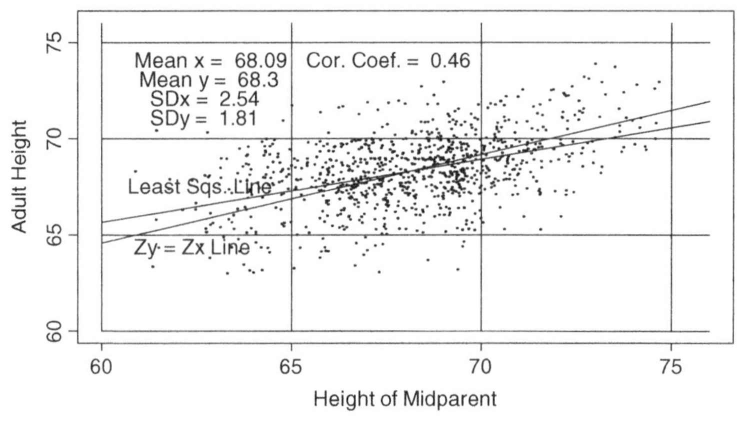


**FIGURE 5.3.4** Galton's height data for 918 (parent, adult child), pairs (heights in inches).

## Problems for Section 5.3

**5.3.1** Let $(X, Y)$ have joint density $f(x, y) = Cy$ for $0 \le y \le x \le 1$.

**(a)** Find the marginal density of $X$. Evaluate $C$.

**(b)** Find the conditional density of $Y$ given $x$, $0 < x \le 1$. Sketch the density for $x = 0.3$ and $x = 0.8$.

**(c)** Find the regression function $g(x) \equiv E(Y \mid X = x)$, and the conditional variance function $v(x) \equiv \text{Var}(Y \mid X = x)$.

**(d)** Show that $\text{Var}(Y) = \text{Var}(g(X)) + E(v(X))$.

**(e)** Suppose that you wished to generate an observation $(X, Y)$ on your computer. Present an algorithm for this. *Hint*: First generate $X$, then for $X = x$, generate $Y$ using the $F$-inverse method and your answer to part (b).

**5.3.2** Let $X$ have the exponential distribution with mean 1. Conditionally on $X = x$, let $Y$ have the exponential distribution with mean $1/x$.

**(a)** Find the marginal distribution of $Y$.

**(b)** Let $g(x) = E(Y \mid X = x)$, $v(x) = \text{Var}(Y \mid X = x)$. Show that $E(g(X)) = E(Y)$ and $\text{Var}(Y) = \text{Var}(g(X)) + E(v(X))$.

**5.3.3** Conditionally on $p$, $0 < p < 1$, let $X$ have the binomial distribution with parameters and $p$. Let $p$ have the uniform distribution on [0, 1]. Show that the marginal distribution of $X$ is the uniform distribution on the integers $0, 1, \ldots, n$.

**5.3.4** Let $A > 0$, and let $X$ have the uniform distribution on $[0, A]$. Conditionally on $X$, let $Y$ have the uniform distribution on $[0, X]$. Find the marginal density of $Y$.

**5.3.5** Let $X$ and $Y$ be independent, each with the exponential distribution with mean 1. Let $W = X + Y$. What is the conditional density of $W$ given $X = x$? Use this to determine the marginal density of $W$. *Be careful:* $X$ can never exceed $W$.

**5.3.6** Let $X_i$ have the uniform distribution on $[0, i]$ for $i = 1, 2$. A coin is tossed. If the coin shows heads, then $X = X_1$. Otherwise, $X = X_2$. Thus, $X$ has the distribution of a mixture of the uniform distributions on [0, 1] and on [0, 2]. What is the density $f$ of $X$?

**(a)** Let $I_H = I[\text{coin is heads}]$. Let $g(k) = E(X \mid I_H = k)$ and $v(k) = \text{Var}(X \mid I_H = k)$ for $k = 0, 1$. Find $g(k)$ and $v(k)$ for each $k$. Show that $E(g(I_H)) = E(X)$ and $\text{Var}(X) = \text{Var}(g(I_H)) + E(v(I_H))$.

**(b)** Suppose that you want to simulate an observation on $X$. Let $U_1$ and $U_2$ be independent, each Unif(0,1). Develop two algorithms that will produce observations on $X$. One should depend on $U_1$ only, the other on both $U_1$ and $U_2$.

**5.3.7** A fair coin is tossed twice. Let $X$ be the number of heads. Then the coin is tossed $X$ times. Let the number of heads among these be $Y$.

**(a)** Find the probability function of $Y$. Use a tree diagram. What is the name of this distribution?

**(b)** Find $g(k) = E(Y \mid X = k)$ and $v(k) = \text{Var}(Y \mid X = k)$ for $k = 0, 1, 2$. Show that $\text{Var}(Y) = \text{Var}(g(X)) + E(v(X))$.

**5.3.8** Let $(X, Y)$ be the heights in centimeters of randomly selected husband–wife pairs. Suppose that $(X, Y)$ has the bivariate normal distribution with parameters $\mu_X = 178$, $\mu_Y = 165$, $\sigma_X = 7.0$, $\sigma_Y = 6.0$, $\rho = 0.4$.

**(a)** Find the conditional probability that the wife is taller than 170 cm given that her husband has height 171 cm. Repeat for the husband's height of 185 cm.

**(b)** What are the equations of the regression functions $g(x) = E(Y \mid X = x)$ and $h(y) = E(X \mid Y = y)$? Sketch them on the same $(x, y)$ graph.

**(c)** Find $\text{Cov}(h(Y), X - h(Y))$, $\text{Var}(h(Y))$, and $\text{Var}(X - h(Y))$.

**(d)** What is the probability that the husband is at least 10 cm taller than his wife?

**5.3.9** Let $B$ take the values $-1$ and 1 with probabilities 1/2, 1/2. Let $(Z_1, Z_2)$ have the standardized bivariate normal distribution with correlation coefficient $\rho$, and suppose that $B$ and $(Z_1, Z_2)$ are independent. Define $Z_i^* = |Z_i| B$ for $i = 1, 2$. Show that $Z_i^* \sim \text{N}(0, 1)$ for $i = 1, 2$, but $(Z_1^*, Z_2^*)$ does not have the bivariate normal distribution.

CHAPTER SIX

# Moment Generating Functions and Limit Theory

## 6.1 INTRODUCTION

The primary purpose of this chapter is to discuss limit theory, especially limits in probability and in distribution. We have already shown, for example, using the Chebychev inequality, that the sequence $\{\bar{X}_n\}$ of sample means for random samples from a distribution with mean $\mu$ and variance $\sigma^2$ converges in probability to $\mu$. We showed that the sequence of binomial probabilities $b(k; n, p_n)$ with $\lim_{n\to\infty} np_n = \lambda$ converges to $f(k; \lambda) = e^{-\lambda}\lambda^k/k!$, the Poisson probability, for $k = 0, 1, 2, \ldots$. This last convergence is called *convergence in distribution*. We need to define such convergence carefully. We have hinted through histograms that in a sense to be discussed, the binomial distribution and more generally, the distributions of sums and of sample means, when properly standardized, converge to a normal distribution. This will be made clear, we hope, through the central limit theorem, which is certainly the most useful of the limit theorems to be discussed. The limit theorems become useful when they provide good approximations for large $n$. We provide empirical evidence, through computer simulation, that the approximations provided are good, often even when $n$ is of moderate size, only 10 or 20.

As a tool in providing proofs of some of these limit theorems, we need moment generating functions (mgf's), which are transforms of probability distributions. We have postponed presentation of material on mgf's until now because this will be the first time we really need this tool. Some instructors may have preferred to introduce them earlier because, as the name implies, mgf's can be used to find moments.

## 6.2 MOMENT GENERATING FUNCTIONS

**Definition 6.2.1** Let $X$ be a random variable. The *moment generating function* (mgf) for $X$ is $M(t) = E(e^{Xt})$, defined for all $t$ for which the expectation exists. □

*Models for Probability and Statistical Inference: Theory and Applications,* By James H. Stapleton

Notice that since $e^{Xt} = 1$ for $t = 0$ for all $X$, $M(0) = 1$. The mgf may not exist for any $t$ other than $t = 0$. For example, let $X$ have density $f(x) = 1/(2x^2)$ for $|x| > 1$. For that reason the characteristic function $C(t) = E(e^{iXt})$, where $i = \sqrt{-1}$, is a better tool, since $|e^{iXt}| = 1$ for all $X$ and $t$, so that $C(t)$ exists for all $t$. However, the use of characteristic functions requires the use of complex variables, probably not a familiar mathematical tool for most readers of this book.

***Example 6.2.1*** Let $X$ have the exponential distribution with mean 1. Then $M(t) = \int_0^\infty e^{xt} e^{-x} dx = \int_0^\infty e^{-x(1-t)} dx = 1/(1-t)$. This integral exists for $1 - t > 0$, equivalently, $t < 1$. We therefore say that the mgf exists for $t < 1$. □

***Example 6.2.2*** Let $X$ have the Poisson distribution with parameter $\lambda > 0$. Then $X$ has the mgf $M(t) = E(e^{Xt}) = \sum_{k=0}^\infty e^{-\lambda}\lambda^k e^{kt}/k! = e^{-\lambda}\sum_{k=0}^\infty (\lambda e^t)^k/k! = e^{\lambda(e^t-1)}$, existing for all real numbers $t$. □

Mgf's are useful for several reasons. The first and most obvious stems from their ability to produce moments.

**Property One** If an mgf $M_X(t)$ exists for values of $t$ in a neighborhood of $t = 0$, then all moments $\nu_k \equiv E(X^k)$ for $k = 1, 2, \ldots$ exist, and $M^{(k)}(0) \equiv \frac{d^k}{dt^k} M(t)|_{t=0} = \nu_k$.

*Proof:* Exchanging expectation and differentiation, we get $\frac{d^k}{dt^k} M(t) = E(X^k e^{Xt})$. Letting $t = 0$, we get Property One. □

If $M(t)$ can be expressed as a Taylor series $M(t) = \sum_{k=0}^\infty a_k t^k$ in a neighborhood of $t = 0$, it follows from calculus that $\nu_k = a_k$k!.

***Example 6.2.1 and 6.2.2 Continued*** For $M(t) = (1-t)^{-1}$, $t < 1$, $\frac{d^k}{dt^k} M(t) = k!(1-t)^{-k+1}$, so that $\nu_k = k!$. We can also see this from the Taylor series representation of $M(t)$. Since $M(t) = \sum_{k=0}^\infty t^k$ for $-1 < t < 1$, the coefficients $a_k$ of the $t^k$ are all 1, so that the moment $\nu_k = (1)k! = k!$.

For the Poisson distribution, $M(t) = e^{\lambda(e^t-1)}$, so that $\frac{d}{dt} M(t) = \lambda e^{\lambda t} M(t)$, which is $\nu_1 = E(X) = \lambda$ for $t = 0$. Taking another derivative and letting $t = 0$, we get $\nu_2 = E(X^2) = \lambda + \lambda^2$, so that $\text{Var}(X) = E(X^2) - [E(X)]^2 = \lambda$. Other moments may be determined by taking more derivatives. □

**Property Two** Let $Y = aX + b$ for constants $a, b$. The mgf for $Y$ is $M_Y(t) = E(e^{(aX+b)t}) = e^{bt} M_X(at)$.

**Property Three** Let $X_1, \ldots, X_n$ be independent random variables with mgf's $M_1(t), \ldots, M_n(t)$. Then $Y = X_1 + \cdots + X_n$ has mgf $M_Y(t) = E(e^{Yt}) = \prod_{i=1}^n E(e^{X_i t}) = \prod_{i=1}^n M_i(t)$.

We present the next property without proof.

**Property Four** (Uniqueness) Let $X_1$ and $X_2$ be random variables with mgf's $M_1(t)$ and $M_2(t)$, each existing and equal on some neighborhood $N$ (open set containing zero) of $t = 0$. Then $X_1$ and $X_2$ have the same probability distribution.

If $M(t)$ is the mgf of a random variable $X$, existing on a set $A$ of $t$-values, which includes a neighborhood $N$ of $t = 0$, it follows that $M(t)$ is completely determined on $A$ by its values on $N$. Thus, it is the behavior of $M(t)$ for $t$ near zero that determines its values on $A$. We demonstrate the value of this uniqueness property after considering Property Five.

**Property Five** Let $X_1, \ldots, X_n$ be independent, and suppose that $X_j$ has mgf $M_j(t)$, defined on a neighborhood $A_j$ of zero. Let $Y = X_1 + \cdots + X_n$. Then $Y$ has mgf $M_Y(t) = \prod_{j=1}^n M_j(t)$, defined for $t \in \bigcap_{j=1}^n A_j$.

*Proof:* $M_Y(t) = E(e^{Yt}) = E\left(\prod_{j=1}^n e^{tX_j}\right) = \prod_{j=1}^n E(e^{tX_j}) = \prod_{j=1}^n M_j(t)$. □

***Example 6.2.3*** Let $X_1, \ldots, X_n$ be independent, each Bernoulli with parameter $p$. Each $X_i$ has mgf $M(t) = (1-p)e^{(0)t} + pe^{(1)t} = q + pe^t$. Hence, $Y = X_1 + \cdots + X_n$ has mgf $M_Y(t) = (q + pe^t)^n$. Since $Y$ has the binomial distribution with parameters $n$ and $p$, it follows from Property Four that the mgf of this binomial distribution is $(q + pe^t)^n$. □

***Example 6.2.4*** Let $Z \sim \mathrm{N}(0, 1)$. $Z$ has the mgf $M_Z(t) = \int_{-\infty}^{\infty} e^{xt}\phi(x)\,dx$. The exponent of $e$ under the integral sign can be written as $(-1/2)(x^2 - 2xt + t^2) + t^2/2 = (-1/2)(x-t)^2 + t^2/2$. Therefore, $M_Z(t) = e^{t^2/2}\int_{-\infty}^{\infty}\phi(x-t)\,dx = e^{t^2/2}$ for all $t$. From Property Two if follows that $X = \mu + \sigma Z$ has mgf $M_x(t) = e^{\mu t + \sigma^2 t^2/2}$, the mgf of the $N(\mu, \sigma^2)$ distribution.

If $X_1, \ldots, X_n$ are independent with $X_i \sim N(\mu_i, \sigma_i^2)$, mgf $M_i(t)$, then, from Property Five, $Y = X_1 + \cdots + X_n$ has mgf $M_Y(t) = \prod_{j=1}^n M_j(t) = e^{\mu_Y t + \sigma_Y^2 t^2/2}$, where $\mu_Y = \mu_1 + \cdots + \mu_n$ and $\sigma_Y^2 = \sigma_1^2 + \cdots + \sigma_n^2$. By the uniqueness property, this proves Theorem 4.2.1 again. It was easier this time, although we needed to use Property Four (uniqueness), which was not proved.

One more very important property of mgf's will be discussed after we consider convergence in distribution in Section 6. 3. □

### Problems for Section 6.2

**6.2.1** Let $X \sim \mathrm{Unif}(0, 1)$.

**(a)** Find the mgf for $X$ and use it to find $E(X)$. *Hint*: Use the Taylor series expansion of $e^t$ about $t = 0$ to find $\frac{d}{dt}M(t)\,|_{t=0}$.

**(b)** Use the result of part (a) to find the mgf of the triangular distribution on $[0, 2]$ having density $f(x) = 1 - |x - 1|$ for $0 \le x \le 2$. *Hint*: The sum of two independent Unif(0,1) random variables has this density.

**(c)** Use Property Two to find the mgf of the Unif($-2$, 2) distribution.

**6.2.2** **(a)** Find the mgf for the $\Gamma(\alpha, 1)$ distribution.

**(b)** Let $X_1, X_2$ be independent, with $X_k \sim \Gamma(\alpha_k, 1)$. Use mgf's to show that $Y = X_1 + X_2 \sim \Gamma(\alpha_1 + \alpha_2, 1)$, as was shown in Section 4.3.

**(c)** Use $M_Y(t)$ to find $E(Y)$ and Var$(Y)$.

**6.2.3** Find expressions for the moments $\nu_k$ for each positive integer $k$ of the N$(0, \sigma^2)$ distribution by using mgf's.

**6.2.4** Let $Z$ have the standard normal distribution.

**(a)** Prove that $P(Z \geq x) \leq e^{-x^2/2}$ for all $x > 0$. For example, for $x = 2, 3, 4$ the upper bounds are 0.1353, 0.0111, 0.000335, while the actual probabilities are 0.0227, 0.0135, 0.000032. *Hint*: Let $Y = e^{Zt}$, use the Markov inequality for $Y$, and minimize with respect to $t$.

**(b)** Let $X_1, \ldots, X_n$ be a random sample from the N$(\mu, \sigma^2)$ distribution, and let $\bar{X}$ be their sample mean. Use the result of part (a) to give an upper bound for $P(|\bar{X} - \mu| > K\sigma)$, then find $n$ such that this probability is less than 0.001.

**6.2.5** Let $X$ have the negative binomial distribution with parameters $r$ and $p$.

**(a)** Find the mgf $M(t)$ for the case that $r = 1$ so that $X$ has the geometric distribution with parameter $p$. For which $t$ does $M(t)$ exist?

**(b)** Use part (a) to find the mgf $M(t)$ of $X$ for the case that $r$ is a positive integer.

**(c)** Find the mgf for $X$ for the case that $r$ is not necessarily an integer (see Section 2.4).

**(d)** Use the mgf to find $E(X)$ and Var$(X)$.

## 6.3 CONVERGENCE IN PROBABILITY AND IN DISTRIBUTION, AND THE WEAK LAW OF LARGE NUMBERS

The principal purpose of this section is to discuss types of convergence of sequences of random variables, in probability and in distribution. In addition, we briefly discuss almost sure convergence. The principal theorem of this section is the weak law of large numbers, discussed in Section 1.6 for discrete random variables.

To begin, we need to define precisely what we mean by almost sure convergence, convergence in probability, and convergence in distribution.

**Definition 6.3.1** Let $X$ be a random variable and let $\{X_n\}$ be a sequence of random variables.

**(a)** $\{X_n\}$ is said to *converge almost surely* to a random variable $X$ if

$$P(\{\omega | \lim_{n \to \infty} X_n(\omega) = X(\omega)\}) = 1.$$

**(b)** $\{X_n\}$ is said to *converge in probability* to $X$ if for every $\varepsilon > 0$,

$$\lim_{n\to\infty} P(|X_n - X| \leq \varepsilon) = 1.$$

**(c)** Let $X$ have cdf $F$, and for each $n$, let $X_n$ have cdf $F_n$. The sequence $\{X_n\}$ is said to *converge in distribution* (or *in law*) if for each $x$ at which $F$ is continuous,

$$\lim_{n\to\infty} F_n(x) = F(x). \qquad (6.3.1)$$

□

In many cases of interest the random variable $X$ is a constant $C$. For almost sure convergence and convergence in probability convergence of $\{X_n\}$ to $X$ is equivalent to convergence of $\{X_n - X\}$ to the constant 0. We will be most interested in convergence in probability and in distribution, since these are more useful than almost sure convergence in the study of statistics. For that reason we will say little about almost sure convergence. (a) implies (b) and (b) implies (c), but the converses do not necessarily hold.

To understand (a), almost sure convergence, fix $\omega$ and consider a sequence $\{X_n(\omega)\}$. Let $A = \{\omega | \lim_{n\to\infty} | X_n(\omega)| = X(\omega)\}$. Then, if (a) holds, $P(A) = 1$. For (b), convergence in probability, consider any $\varepsilon > 0$ and let $B_n(\varepsilon) = \{\omega | |X_n(\omega) - X(\omega)| \leq \varepsilon\}$. Then $\lim_{n\to\infty} P(B_n(\varepsilon)) = 1$. Item (c), convergence in distribution, concerns only the distributions $F_n$ of the sequence $\{X_n\}$, saying nothing about the closeness of $X_n$ to $X$.

***Example 6.3.1*** Suppose that a box contains $r$ red and $w$ while balls at stage $n = 0$. A ball is drawn randomly. That ball is then replaced by two balls of the same color. Then another ball is chosen randomly and the same rule is used to replace the ball chosen. Let $R_n$ be the number of red balls in the box after $n$ draws. The total number of balls in the box after $n$ draws is $T_n = r + w + n$. Let $X_n = R_n/T_n$, the proportion of red balls in the box after $n$ draws. It follows from the Martingale convergence theorem, not to be discussed here, that with probability 1 the sequence $\{X_n\}$ converges to a random variable $X$. By other means it can be shown that $X$ has the beta distribution with parameters $\alpha = r$ and $\beta = w$. Figure 6.3.1 shows three sample paths for the process $X_n$ [indicated as $X(n)$ in the figure] for the case that $r = 30$, $w = 20$. For each of the three, the paths seem to be converging, as they will with probability 1. Figure 6.3.2 is a histogram of $X_{2000} = X(2000)$ for 1000 repetitions. Although $X_{2000}$ is only an approximation of the limit $X$ for each path, $X_{2000}$ should be close enough to the limit to provide a good approximation of the distribution of $X$. □

If the limiting distribution has a continuous cdf $F$, then $\lim_{n\to\infty} F_n(x)$ must be $F(x)$ for all $x$. However, if $F$ is discontinuous at a point $x$, then (6.3.1) need not hold for that $x$. To see why the convergence need not hold for the $x$'s at which $F$ is discontinuous, consider the following example.

***Example 6.3.2*** Suppose that $X_n$ is uniformly distributed on the interval $[0, 1/n]$. Then for any $\varepsilon > 0$, $P(|X_n| < \varepsilon) = \varepsilon n$ for $1/n \geq \varepsilon$, equivalently, $n \leq 1/\varepsilon$

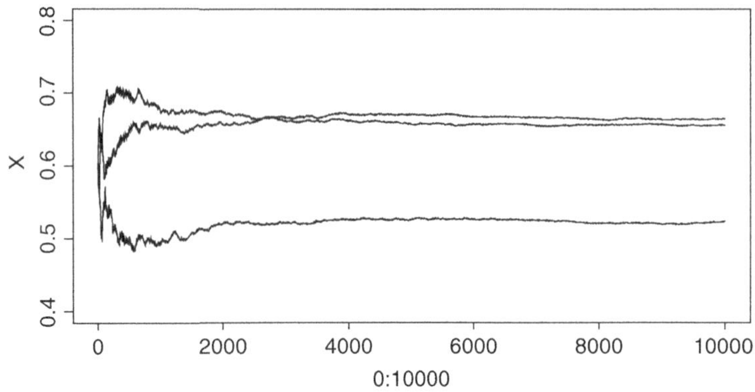


**FIGURE 6.3.1** Three martingale sequences.

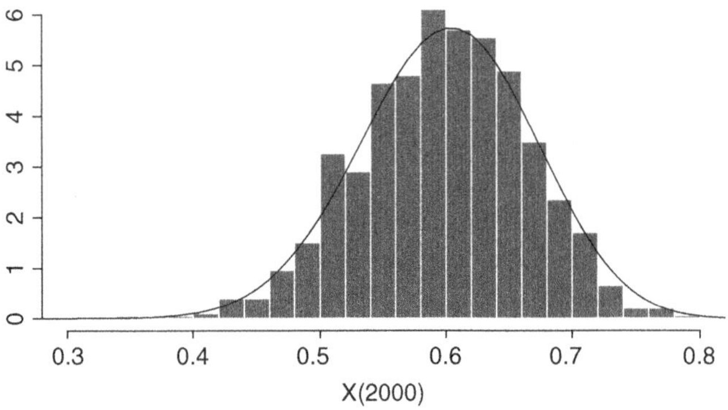


**FIGURE 6.3.2** Histogram of approximate limits.

and 1 for $n > 1/\varepsilon$. Thus, the sequence $\{X_n\}$ converges in probability to zero. That is, $\{X_n\}$ converges in probability to the random variable $X$ which takes the value zero with probability 1. As we will see, $\{X_n\}$ converges in distribution to $X$ as well. $X_n$ has cdf $F_n(x) = 0$ for $x < 0$, $= nx$ for $0 \le x < 1/n$, and 1 for $x \ge 1/n$. $X$ has cdf $F(x) = 0$ for $x < 0$ and 1 for $x \ge 0$. For each fixed $x \neq 0$, $\lim_{n\to\infty} F_n(x) = F(x)$. However, $\lim_{n\to\infty} F_n(0) = \lim_{n\to\infty} 0 = 0 \neq F(0)$. But $F$ has a discontinuity point at $x = 0$, so we are saved. $\{X_n\}$ does converge in distribution to $X$. □

***Example 6.3.3*** Let $X_1, \ldots, X_n$ be a random sample from the Unif(0, 1) distribution. Let $M_n = \max(X_1, \ldots, X_n)$. Then $M_n$ has cdf $F_n(u) = u^n$ for positive integers $n$, and for $0 < \varepsilon < 1$, $P(|M_n - 1| \le \varepsilon) = P(1 - \varepsilon \le M_n) = 1 - F_n(1 - \varepsilon) = 1 - (1 - \varepsilon)^n$, which converges to 1, as $n \to \infty$. Therefore, $\{M_n\}$ converges in probability to 1 and $\lim_{n\to\infty} F_n(u) = 0$ for $u < 1$, 1 for $u \ge 1$, the cdf of the point mass at $u = 1$. Now let $m_n = \min(X_1, \ldots, X_n)$. $m_n$ has cdf $G_n(u) = 1 - (1 - u)^n$ for $0 \le u \le 1$, so that $P(|m_n - 0| \leq \varepsilon) = P(m_n \leq \varepsilon) = G_n(\varepsilon) = 1 - (1 - \varepsilon)^n$ for $0 < \varepsilon < 1$, which converges to 1 as $n \to \infty$ for each $\varepsilon > 0$. Thus, $\{m_n\}$ converges in probability to zero. $\lim_{n\to\infty} G_n(u) = 0$ for $u \leq 0$, 1 for $u > 0$. The cdf $G$ for the point mass 1 at $x = 0$ is $G(u) = 0$ for $u < 0$, 1 for $u \geq 0$. Thus, $\lim_{n\to\infty} G_n(0) = 0 \neq G(0) = 1$. We still say that $\{m_n\}$ converges in distribution to the cdf $G$ because $\lim_{n\to\infty} G_n(u) = G(u)$ for every point $u$ at which $G$ is continuous. □

One useful way to demonstrate convergence in probability to a constant $C$ is to use first and second moments.

**Theorem 6.3.1** Let $\{X_n\}$ be a sequence of random variables with finite means and variances. A sufficient condition that $|X_n|$ converge in probability to a constant $C$ is that $\lim_{n\to\infty} E(X_n) = C$ and $\lim_{n\to\infty} \text{Var}(X_n) = 0$.

*Proof:* For any constants $C, \varepsilon > 0$ and random variable $Y$ with mean $\mu$, variance $\sigma^2$, it follows from the Markov inequality that $P(|Y - C| > \varepsilon) = P(|Y - C|^2 > \varepsilon^2) \leq E(Y - C)^2/\varepsilon^2 = [\text{Var}(Y) + (\mu - C)^2]/\varepsilon^2$. Replacing $Y$ by $X_n$ and taking limits, we get $\lim_{n\to\infty} P(|X_n - C| > \varepsilon) \leq [\lim_{n\to\infty} \text{Var}(X_n) + \lim_{n\to\infty}[E(X_n) - C]^2]\,\varepsilon^2 = 0$. □

If $g$ is continuous on a set $A$ of the real line such that $P(X \in A) = 1$, then convergence of $\{X_n\}$ to $X$ in any of the three ways almost sure, in probability, and in distribution implies the corresponding convergence of $\{g(X_n)\}$ to $g(X)$ (see Fabian and Hannan, 1985, p. 158).

In Section 1.5, for the discrete case, we stated a form of the weak law of large numbers (WLLN) due to Chebychev (1867). We restate and reprove the theorem here for completeness. We state another form of the WLLN due to Khintchine (1929), with just a hint of a proof. Both forms are concerned with the convergence of a sequence of sample means $\{\bar{X}_n\}$ to a mean $\mu$.

**Theorem 6.3.2 [Weak Law of Large Numbers (Chebychev)]** Let $X_1, X_2, \ldots, X_n$ be random variables, with equal means $\mu$, variances $\sigma^2$, and covariances all zero. Let $\bar{X}_n = (X_1 + \cdots + X_n)/n$. Then $\{\bar{X}_n\}$ converges in probability to $\mu$.

*Proof:* Let $\varepsilon > 0$. From the Chebychev inequality $P(|\bar{X}_n - \mu| > \varepsilon) \leq \text{Var}(\bar{X}_n)/\varepsilon^2 = \sigma^2/(n\varepsilon^2)$, which converges to zero as $n \to \infty$. □

Notice that Chebychev's form of the WLLN does not require independence, although it does require zero covariances. In the usual case we have zero covariances because the model states that the random variables are independent. In the Khintchine form the random variables are independent and identically distributed with finite mean $\mu$, but the existence of a common variance is not needed.

**Theorem 6.3.3 [Weak Law of Large Numbers (Khintchine)]** Let $X_1, X_2, \ldots, X_n$ be independent and identically distributed with mean $\mu$. Then $\{\bar{X}_n\}$ converges in probability to $\mu$.

*Hint of a Proof:* For each $k$ define $X^*_{nk} = X_k I[|X_k| < \delta n]$ for $\delta > 0$. $X^*_{nk}$ is the truncated version of $X_k$. Chebychev's version of the WLLN can be used to show that the sequence $\{\bar{X}^*_n - \mu^*_n\}$ converges in probability to zero. It can also be shown that $P(\bar{X}_n \neq \bar{X}^*_n) \leq \delta/n$ and that $E(\bar{X}^*_n) = E(X^*_{nk}) \equiv \mu^*_n$ converges to $\mu$ (see Feller, 1950). □

***Example 6.3.4*** Let $X_1, X_2, \ldots$ be iid, each with density $f(x) = 4x^{-5}$ for $x > 1$. Then each random variable has mean $\mu \equiv E(X_1) = 4/3$. In Figure 6.3.3 we present histograms that approximation the distributions of $\bar{X}_n$ for $n = 16, 64, 256$, and $1024$. For each $n$ the histogram was based on 10,000 values of $\bar{X}_n$. The numbers within $\varepsilon = 0.05$ of $\mu = 4/3$ are indicated. The scale is the same for the first three, changed for the last. The variance of the $X_i$ is $\sigma^2 = 2/9$, so that the standard deviations of the $\bar{X}$ for the sample sizes 16, 64, 256, 1024 are 0.1179, 0.0589, 0.0295, 0.0147. The sample standard deviations for the 10,000 values indicated in Figure 6.3.3 are very close to these. □

Although we use the language of random variables, convergence in distribution is a property of the sequence of marginal distributions of the random variables. Thus, if $X_n$ has the same distribution as $Y_n$ for each $n$ then $\{Y_n\}$ has the same limiting distribution as $\{X_n\}$ if $\{X_n\}$ has a limiting distribution. As suggested by Example 6.3.3, convergence in probability to a constant $C$ implies convergence in distribution

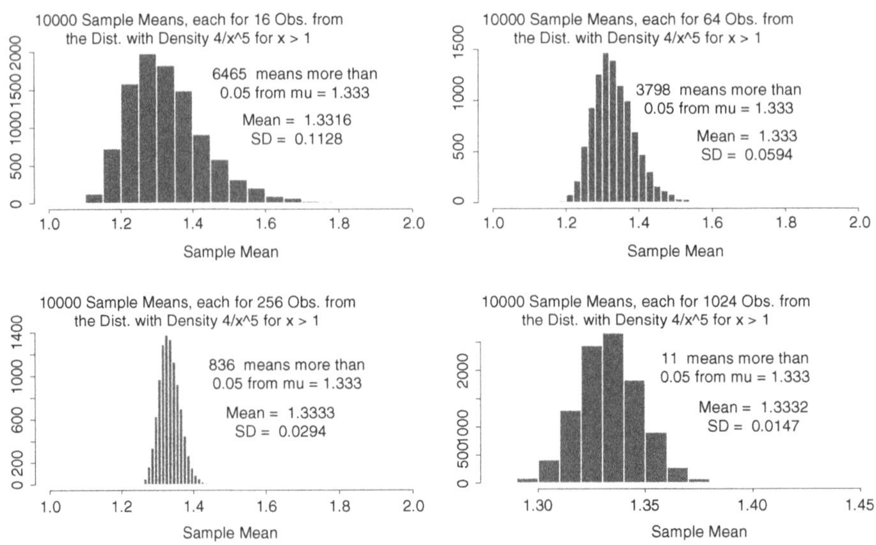


**FIGURE 6.3.3** Histogram of 10,000 sample means.

to the distribution with mass 1 at $C$. Students are asked to prove this in Problem 6.3.4. The following theorem establishes this in a more general case.

**Theorem 6.3.4** Let $\{X_n\}$ converge in probability to the random variable $X$. Then $\{X_n\}$ converges in distribution to $X$.

*Proof:* Let $X$ have cdf F and let $X_n$ have cdf $F_n$ for each $n$. Let $x$ be a point of continuity of $F$. For any $\varepsilon > 0$, $F_n(x) = P(X_n \le x) = P(X_n \le x, X > x + \varepsilon) + P(X_n \le x, X \le x + \varepsilon) \le P(|X_n - X| \ge \varepsilon) + P(X \le x + \varepsilon) = P(|X_n - X| \ge \varepsilon) + F(x + \varepsilon)$. By the convergence in probability of $X_n$ to $X$, the first term on the right converges to zero. It follows that $\overline{\lim}_{n\to\infty} F_n(x) \le F(x + \varepsilon)$. Since this is true for every $\varepsilon > 0$, it follows that $\overline{\lim}_{n\to\infty} F_n(x) \le F(x)$. (This inequality holds even if $F$ is discontinuous at $x$.)

Also, $F(x - \varepsilon) = P(X \le x - \varepsilon, X_n > x) + P(X \le x - \varepsilon, X_n \le x) \le P(|X_n - X| \ge \varepsilon) + F_n(x)$. Again, the first term converges to zero. It follows that $\underline{\lim}_{n\to\infty} F_n(x) \ge F(x - \varepsilon)$. By the arbitrariness of $\varepsilon > 0$ and continuity of $F$ at $x$ it follows that $\underline{\lim}_{n\to\infty} F_n(x) \ge F(x)$. By this and $\overline{\lim}_{n\to\infty} F_n(x) \le F(x)$, it follows that $\lim_{n\to\infty} F_n(x) = F(x)$. □

***Example 6.3.5*** Let $X_n$ have the uniform distribution on the integers 1, 2, ..., $n$. Then $X_n$ has cdf $F_n(x) = [x]/n$, for $x < n$, 1 for $x \ge n$, where $[x]$ is the greatest integer less than or equal to $x$. Therefore, $\lim_{n\to\infty} F_n(x) = 0$ for all $x$, so $\{X_n\}$ does not converge in distribution. The mass of $X_n$ "runs off to infinity." However, consider $Y_n = X_n/n$. $Y_n$ has cdf $G_n(y) = [ny]/n$, which converges to $y$ for $0 \le y < 1$, to 1 for $y \ge 1$. Thus, $\{Y_n\}$ converges in distribution to a random variable $U$ having the Unif(0, 1) distribution. □

***Example 6.3.6*** Let $U_1, \dots, U_n$ be independent, each Unif(0, 1), as in Example 6.3.3. Again, let $M_n = \max(U_1, \dots, U_n)$. Then $M_n$ has cdf $F_n(x) = x^n$ for $0 \le x \le 1$, so that $\lim_{n\to\infty} F_n(x) = 0$ for $x < 1$, 1 for $x \ge 1$, so that $\{M_n\}$ converges in distribution to a random variable $M$ taking the value 1 with probability 1. However, consider $Y_n = n(1 - M_n)$. Then $Y_n$ has cdf $G_n(y) = 1 - F_n((1 - y/n)) = 1 - (1 - y/n)^n$. Thus, $\lim_{n\to\infty} G_n(y) = 1 - e^{-y}$ for all $y > 0$, so that $\{Y_n\}$ converges in distribution to a random variable $Y$ having the exponential distribution with mean 1. Notice that $E(Y_n) = n(1 - n/(n+1)) = n/(n+1)$ converges to $E(Y) = 1$. Although this limiting distribution could be used to provide approximations of $P(a < M_n \le b) = F_n(b) - F_n(a)$ for $a < b$, in this case it is not needed because we have explicit expressions for the $F_n$. □

***Example 6.3.7*** Let $X_n$ take the values 0 and $n$ with probabilities $1 - 1/n$ and $1/n$. It is easy to show that $\{X_n\}$ converges in distribution to the trivial random variable $X$ taking the value zero with probability 1. However, $E(X_n) = 1$ for each $n$, and $E(X) = 0$.

Thus, although it may be tempting to think so, convergence in distribution does not imply convergence of the corresponding expected values $E(X_n)$ to $E(X)$ or $E(g(X_n))$

to $E(g(X))$ for some functions $g$. By definition, convergence in distribution of $\{X_n\}$ to $X$ does imply convergence of $E(g(X_n))$ to $E(g(X))$ when $g$ is the indicator of the interval $[-\infty, x]$ and $F$ is continuous at $x$. Theorem 6.3.5, given without proof, states that convergence in distribution of $\{X_n\}$ to $X$ is equivalent to convergence of $E(g(X_n))$ to $E(g(X))$ for functions $g$ that are "nice" in a certain sense. □

**Theorem 6.3.5** $\{X_n\}$ converges in distribution to $X$ if and only if $\lim_{n\to\infty} E(g(X_n)) = E(g(X))$ for all continuous bounded functions $g$ that are zero outside a bounded set.

We need mathematical tools that can determine limiting distributions in the case that the $F_n$ cannot be written in simple form. We have such a tool in the continuity theorem for mgf's. We present this without proof.

**Theorem 6.3.6 (The Continuity Theorem for MGFs)** Let $\{X_n\}$ be a sequence of random variables, with corresponding cdf's $F_n$ and mgf's $M_n$, where the $M_n$ are all defined on a neighborhood $N$ of zero. Let $X$ be a random variable with cdf $F$, mgf $M$, also defined on $N$. Then $\lim_{n\to\infty} M_n(t) = M(t)$ for all $t \in N$ implies that $\{X_n\}$ converges in distribution to $X$. That is, $\lim_{n\to\infty} F_n(x) = F(x)$ for all $x$ at which $F$ is continuous.

Notice that $\lim_{n\to\infty} M_n(t) = M(t)$ for all $t \in N$ is sufficient for convergence of $\{X_n\}$ to $X$ in distribution, but is not necessary. It is necessary if all $X_n$ and $X$ have ranges in a finite interval. For $X_n$ as in Example 6.3.5, $M_n(t)$ converges to $\infty$ for all $t > 0$, although $\{X_n\}$ does converge in distribution. Problem 6.3.6 asks students to show that for Example 6.3.5, $\lim_{n\to\infty} M_n(t) = (1 - e^t)/t$, the mgf for the Unif(0, 1) distribution.

## Problems for Section 6.3

**6.3.1** Let $X_1, \ldots, X_n$ be a random sample from the exponential distribution with mean 1. Let $m_n = \min(X_1, \ldots, X_n)$. Show that $\{m_n\}$ converges in probability to zero.

**6.3.2** Let $\{\eta_n\}$ converge in probability to 1. Let $X$ be a random variable and $Y_n = X\eta_n$. Show that $\{Y_n\}$ converges in probability to $X$. *Hint*: Define $D_n = |Y_n - X|$. For any $\delta > 0$, let $M$ satisfy $P(|X| > M) < \delta$. Let $\varepsilon > 0$. Then $[D_n > \varepsilon] = [D_n > \varepsilon, |X| > M] \cup [D_n > \varepsilon, X \leq M]$.

**6.3.3** Let $X_n \sim \text{Binomial}(n, p)$ and let $\hat{p}_n = X_n/n$ for $0 < p < 1$.

**(a)** Show that $\{\hat{p}_n\}$ converges in probability to $p$.

**(b)** Does $\{\hat{p}_n^2\}$ converge in probability to p$^2$?

**(c)** Let $g(u)$ be the indicator of $[0, 1/2]$. For which $p$ does $\{g(\hat{p}_n)\}$ converge in probability to $g(p)$?

**6.3.4** Let $\{X_n\}$ converge in distribution to the constant $C$. Prove that $\{X_n\}$ converges in probability to $C$.

**6.3.5** Suppose that $\{X_n\}$ converges in probability to $X$ and $\{Y_n\}$ converges in probability to $Y$. Let $W_n = X_n + Y_n$. Prove that $\{W_n\}$ converges in probability to $W = X + Y$.

**6.3.6** For each positive integer $n$, let $Y_n$ have the uniform distribution on $1/n, 2/n, \ldots, (n-1)/n, 1$.
**(a)** Find the mgf $M_n$ for $Y_n$.
**(b)** Show that $\lim_{n\to\infty} M_n(t) = (1 - e^t)/t$ for all $t$. Use this to show that $\{Y_n\}$ converges in distribution to Unif(0, 1) (see Example 6.3.5).

**6.3.7** Let $X_n$ have the uniform distribution on the interval $[-1/n, 1/n]$. Find the MGF for $X_n$, then use the continuity theorem for mgf's to prove that $\{X_n\}$ converges in probability to a random variable $X$ for which $P(X = 0) = 1$. (Use the result of Problem 6.3.4).

**6.3.8** Let $\lambda > 0$ and let $p_n = \lambda/n$ for $n = 1, 2, \ldots$. Let $X_n \sim$ Binimial $(n, p_n)$. Use the continuity theorem to prove that $\{X_n\}$ converges in distribution to the Poisson distribution with parameter $\lambda$.

**6.3.9** Let $X_1, \ldots, X_n$ be iid random variables with unique $\alpha$th quantile $x_\alpha$ for $0 < \alpha < 1$. Let $X_n^*$ be the $k_n$th order statistic, where $k_n = [n\alpha]$. Prove that $\{X_n^*\}$ converges in probability to $x_\alpha$. *Hint*: Let $U_n =$ (number of $X_i$ which are $\leq x_\alpha - \varepsilon$) and $V_n =$ (number of $X_i$ which are $\geq x_\alpha + \varepsilon$). Then $P(|X_n{}^* - x_\alpha| > \varepsilon) = P(U_n \geq k_n) + P(V_n < k_n)$. Now use the Chebychev inequality.

## 6.4 THE CENTRAL LIMIT THEOREM

We are now ready to discuss the central limit theorem, which will enable us to approximate the distribution of sums and means of independent random variables by the normal distribution. We have hinted at this informally several times. See the graphical displays of binomial mass functions and gamma densities which seem to have the normal distribution shape for larger $n$.

The central limit theorem has a history of approximately 270 years. It began with the work of DeMoivre on the approximation of the binomial distribution, was given by Laplace in 1810 in the form to be presented here, and was given great generality in the Lindeberg–Feller necessary and sufficient conditions of 1935.

We are now ready to consider limiting distributions for sums. Recall that every binomial random variable is the sum of independent Bernoulli random variables. If $X_n \sim$ Binomial$(n, p)$, with $0 < p < 1$, then $E(X_n) = np$, which converges to infinity. More generally, if $X_1, \ldots, X_n$ are independent, each with mean $\mu$, then for $S_n = X_1 + \cdots + X_n$, $E(S_n) = n\mu$, which "runs off to infinity" unless $\mu = 0$. We can remove

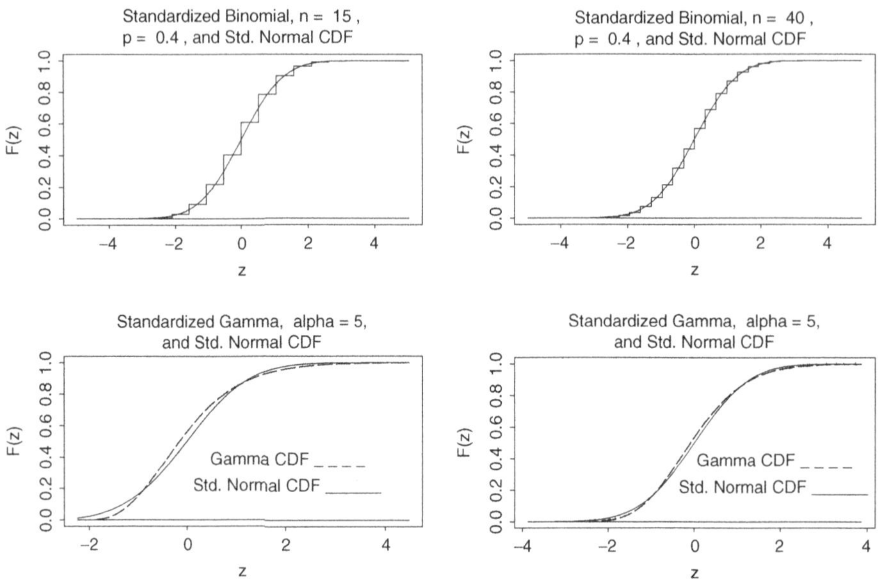


**FIGURE 6.4.1** Approximation of the CDF of standardized sums of rv's by the standard normal cdf.

this difficulty by subtracting $n\mu$, considering $D_n = S_n - n\mu$ instead. However, if all $X_n$ have variance $\sigma^2$, then $\text{Var}(S_n) = \text{Var}(D_n) = n\sigma^2$, so that the spread of the distribution of $D_n$ increases with $n$, so that the probability mass goes off to $-\infty$ and $+\infty$ as $n$ goes to infinity. We can get over this problem by standardizing, defining $Z_n = (S_n - n\mu)/(\sigma\sqrt{n})$. Dividing numerator and denominator by $n$ and letting $\bar{X}_n = S_n/n$, we get $Z_n = (\bar{X}_n - \mu)/(\sigma/\sqrt{n})$, so that $Z_n$ is both the standardized $S_n$ and the standardized $\bar{X}_n$. We will show that $\{Z_n\}$ converges in distribution to the standard normal random variable $Z$.

First, to convince the reader, consider two cases: (1) The $X_k$ are Bernoulli($p = 0.4$), and (2) the $X_k$ are each exponential, mean 1. For (1), $S_n$ has the Binomial($n, p$) distribution. For (2), $S_n$ has the $\Gamma(\alpha = n, 1)$ distribution (see Figure 6.4.1).

**The Central Limit Theorem** Let $X_1, \ldots, X_n$ be independent and identically distributed random variables, each with mean $\mu$, variance $\sigma^2$. Let $S_n = X_1 + \cdots + X_n$ and $Z_n = (S_n - n\mu)/(\sigma\sqrt{n})$. Then

$$\lim_{n\to\infty} P(Z_n \le u) = \Phi(u) \qquad \text{for every } u,\ -\infty < u < \infty.$$

**Comment:** Thus, if the $X_k$ are iid, $\{Z_n\}$ converges in distribution to $Z \sim \text{N}(0, 1)$.

To prove the central limit theorem (CLT), we need a lemma.

**Lemma 6.4.1** Let $\{b_n\}$ be a sequence of numbers converging to a number $b$. Then

$$\lim_{n\to\infty}\left(1+\frac{b_n}{n}\right)^n = e^b.$$

*Proof:* Taylor approximation gives $\log(1+x) = x + g(x)$, where $\lim_{x\to 0} g(x)/x = 0$. Therefore, $\log(1+b_n/n)^n = n(b_n/n + g(b_n/n)) = b_n + g(b_n/n)/(1/n)$ converges to $b$ as $n \to \infty$. Since $h(y) \equiv e^y = \exp(y)$ is a continuous function of $y$, it follows that $h\left(\log(1+b_n/n)^n\right) = (1+b_n/n)^n$ converges to $h(b) = e^b$. □

If $M(t)$ is the mgf of a random variable with mean $\mu$, second moment $\nu_2 \equiv E(X^2)$, defined on a neighborhood $N$ of zero, Taylor approximation gives $M(t) = 1 + \mu t + \nu_2 t^2/2 + k(t)$, for $t \in M$, where $\lim_{t\to 0} k(t)/t^2 = 0$. Therefore, if $\mu = 0$, in which case $\nu_2 = E(X)^2 = \text{Var}(X) \equiv \sigma^2$, we have $M(t) = 1 + \sigma^2 t^2/2 + k(t)$. We are ready now to prove the central limit theorem.

**Proof of the Central Limit Theorem:** Let $Y_j = (X_j - \mu)/\sigma$ for each $j$. Then $Z_n = (Y_1 + \cdots + Y_n)/\sqrt{n}$, and each $Y_j$ has mean zero, variance 1. Let $M_Y$ be the mgf for each $Y_i$. The Taylor approximation described above gives $M_Y(t) = 1 + t^2/2 + k(t)$, where $\lim_{t\to 0} k(t)/t^2 = 0$. From Properties Three and Two in Section 6.2 it follows that $Z_n$ has mgf $M_n(t) = M_Y(t/\sqrt{n})^n = \left(1 + t^2/2n + k\left(t/\sqrt{n}\right)\right)^n = \left(1 + [t^2 + (2n)k(t/\sqrt{n})]/2n\right)^n$. The second term in the numerator converges to 0 as $n \to \infty$. It follows from Lemma 6.4.1 that $\lim_{n\to\infty} M_n(t) = e^{t^2/2}$, the mgf of the standard normal distribution (see Example 6.2.4). It follows from the continuity theorem for mgf's that $\{Z_n\}$ converges in distribution to the standard normal distribution.

Let's see how well the approximation provided by the CLT works for various $n$.

***Example 6.4.1*** Let $X_1, \ldots, X_n$ be iid, each Unif(0, 1). Let $S_n$ be their sum. Since the Unif(0, 1) distribution has mean 1/2, variance 1/12, $Z_n = (S_n - n/2)/\sqrt{n/12}$. For the case $n = 12$, with the cdf $F_n$ for $Z_n$, we get the following:

| | $u$ | | | |
|---|---|---|---|---|
| | 0 | 1 | 2 | 3 |
| $F_n(u)$ | 0.5000 | 0.8393 | 0.9777 | 0.9990 |
| $\Phi(u)$ | 0.5000 | 0.8413 | 0.9772 | 0.9987 |

Because $\text{Var}(S_{12}) = 1$, $S_{12} - 6$, and the ease of adding Unif(0, 1) pseudorandom variables in a computer $S_{12} - 6$ is often used to simulate standard normal random variables. For more precision, $(S_{48} - 24)/2$ could be used instead. □

### The Binomial Approximation

Let $X_n \sim \text{Binomial}(n, p)$. Since $X_n$ has the distribution of the sum of $n$ independent Bernoulli($p$) random variables, $Z_n \equiv (X_n - np)/\sqrt{np(1-p)}$ converges in distribution to the standard normal. The approximation provided can be improved a bit by the 1/2-*correction*, as follows. For integer $k$, $0 \le k \le n$,

$$P(X_n \le k) = P(X_n \le k + 1/2) = P\left(Z_n \le \frac{k + 1/2 - np}{\sqrt{np(1-p)}}\right) \doteq \Phi\left(\frac{k + 1/2 - np}{\sqrt{np(1-p)}}\right),$$

where "$\doteq$" denotes approximation. Therefore,

$$P(a < X_n \le b) \doteq \Phi\left(\frac{b + 1/2 - np}{\sqrt{np(1-p)}}\right) - \Phi\left(\frac{a + 1/2 - np}{\sqrt{np(1-p)}}\right)$$

for integers $a$ and $b$ and

$$P(a \le X_n \le b) - P(a - 1 < X_n \le b) \doteq \Phi\left(\frac{b + 1/2 - np}{\sqrt{np(1-p)}}\right) - \Phi\left(\frac{a - 1/2 - np}{\sqrt{np(1-p)}}\right).$$

Let's see how well the approximation works for some pairs, $n$ and $p$. For $n = 10$, $p = 0.5$:

| | $k$ | | | | | | | |
|---|---|---|---|---|---|---|---|---|
| | 2 | 3 | 4 | 5 | 6 | 7 | 8 | 9 |
| $P(X_{10} \le k)$ | 0.0547 | 0.1719 | 0.3770 | 0.6230 | 0.8281 | 0.9453 | 0.9892 | 0.9990 |
| Approx. | 0.0569 | 0.1714 | 0.3756 | 0.6241 | 0.8286 | 0.9431 | 0.9866 | 0.9978 |

The approximation doesn't work as well if $p$ is close to 0 or 1. For $n = 20$, $p = 0.2$:

| | $k$ | | | | | | | |
|---|---|---|---|---|---|---|---|---|
| | 0 | 1 | 2 | 3 | 4 | 5 | 6 | 7 |
| $P(X_{20} \le k)$ | 0.0115 | 0.0692 | 0.2061 | 0.4114 | 0.6296 | 0.8042 | 0.9133 | 0.9678 |
| Approx. | 0.0551 | 0.1269 | 0.2468 | 0.4097 | 0.5903 | 0.7532 | 0.8731 | 0.9449 |

However, even for $p = 0.10$, $n = 400$, the approximation is reasonably good.

| | $k$ | | | | | | |
|---|---|---|---|---|---|---|---|
| | 28 | 32 | 36 | 40 | 44 | 48 | 52 |
| $P(X_{400} \le k)$ | 0.0235 | 0.1030 | 0.2849 | 0.5420 | 0.7763 | 0.9188 | 0.9783 |
| Approx. | 0.0276 | 0.1056 | 0.2798 | 0.5332 | 0.7734 | 0.9217 | 0.9814 |

Suppose that $X_1, \ldots, X_n$ are independent, each uniform on the population of integers $\{1, 2, \ldots, 9, 25\}$. Each $X_k$ has mean $\mu = 7$, variance $\sigma^2 = 42.0$. Let $S_n$ be their sum and let $\bar{X}_n = S_n/n$ be their sample mean. Since $\bar{X}_n$ takes values that are multiples of $1/n$, the 1/2-correction in the normal approximation for $\bar{X}_n$ is as follows. For $r$ a multiple of $1/n$, the approximation is

$$P(\bar{X}_n \le r) = P(\bar{X}_n \le r + 1/(2n)) = P\left(Z_n \le \frac{r + 1/(2n) - 7.0}{\sqrt{42/n}}\right)$$
$$\doteq \Phi\left(\frac{r + 1/(2n) - 7.0}{\sqrt{42/n}}\right).$$

For $n = 25$, we get

| | $r$ | | | | | | | | |
|---|---|---|---|---|---|---|---|---|---|
| | 4.5 | 5.0 | 5.5 | 6.0 | 6.5 | 7.0 | 7.5 | 8.0 | 8.5 |
| $P(\bar{X}_n \le r)$ | 0.013 | 0.049 | 0.119 | 0.236 | 0.374 | 0.531 | 0.669 | 0.791 | 0.874 |
| Approx. | 0.028 | 0.063 | 0.127 | 0.225 | 0.356 | 0.506 | 0.656 | 0.784 | 0.880 |

For $n = 100$,

| | $r$ | | | | | | | | |
|---|---|---|---|---|---|---|---|---|---|
| | 5.75 | 6.0 | 6.25 | 6.50 | 6.75 | 7.0 | 7.25 | 7.5 | 7.75 |
| $P(\bar{X}_n \le r)$ | 0.023 | 0.056 | 0.123 | 0.225 | 0.361 | 0.514 | 0.662 | 0.785 | 0.876 |
| Approx. | 0.027 | 0.062 | 0.125 | 0.222 | 0.353 | 0.503 | 0.653 | 0.782 | 0.878 |

The probabilities $P(\bar{X}_n \le r)$ were estimated by simulations, 100,000 times, so they may be wrong by 0.001 or 0.002. The approximation improves with $n$. If the population is "farther from normal," the convergence to normality is slower. For example, if the population becomes $\{1, 2, \ldots, 9, 100\}$, with mean $= 14.5$, variance 818.25, $n = 100$, then simulation (100,000 times) provides the estimate 0.3169 of $P(\bar{X}_n \le 13.00)$, while the normal approximation gives 0.3006. The approximation by the normal is worse because the value of $\bar{X}_n$ depends so much on the number of times the value 100 is in the sample. For example, for the third histogram of Figure 6.4.2, there are peaks (local maxima) at approximately $-1$ and $+1$, corresponding to samples that contain one and three 100's.

### Convergence of Binomial to Poisson

In Section 2.5 we showed that if $\{p_n\}$ is a sequence of probabilities satisfying $\lim_{n\to\infty} p_n n = \lambda > 0$, then $\lim_{n\to\infty} b(j; n, p_n) = e^{-\lambda}\lambda^j/j! \equiv f(j; \lambda)$ for all nonnegative integers $j$, the probability mass function for the Poisson distribution. If $F_n(x)$ is the cdf for the binomial distribution with parameters $n$ and $p_n$ and $F$ is the cdf for

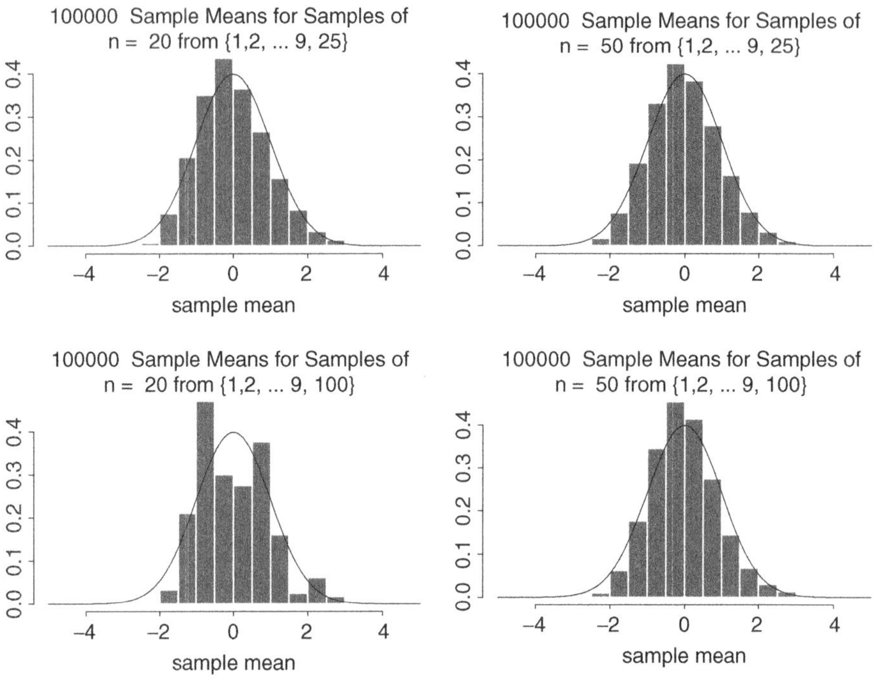


**FIGURE 6.4.2** Standardized sample means and the standard normal density.

the Poisson($\lambda$) distribution, then for $k = [x]$, the greatest integer less than or equal to $x$, $F_n(x) = F_n(k) = \sum_{j=0}^{k} b(j; n, p_n)$ and $F(x) = \sum_{j=0}^{k} f(j; \lambda)$. It follows that $\lim_{n\to\infty} F_n(x) = F(x)$ for every $x$, so the sequence of binomial distributions converges in distribution to the Poisson distribution.

### The Cauchy Distribution

Consider an infinitely long straight wall, running west to east. A soldier with a rifle is stationed distance 1 south of the wall. The wall is marked from $-\infty$ to $+\infty$, with zero being the point on the wall closest to the soldier. He is instructed to choose an angle from the perpendicular to the wall at random from $-\pi/2$ to $+\pi/2$, then shoot a bullet into the wall with his gun pointed at that angle from the perpendicular. What is the distribution of the position $X$ at which the bullet hits the wall?

Let $\theta$ be his chosen angle. $\theta$ has cdf $F(t) = (t + \pi/2)/\pi$ for $-\pi < t < \pi$ . $X = \tan\theta$. Therefore, $X$ has cdf $G(x) = P(X \le x) = P(\theta \le \arctan x) = F(\arctan x) = (\arctan x + \pi/2)/\pi$, so that $X$ has density $g(x) = (d/dx)G(x) = (1/\pi)[1/(1 + x^2)]$ for $-\infty < x < \infty$. This distribution is called the *Cauchy distribution,* although it seems that Poisson knew of it a few years earlier than Cauchy did (see Stigler, 1986).

The moments $\nu_k = E(X^k)$ of the Cauchy distribution and the mfg (for any $t \neq 0$) do not exist. In fact, as shown by Cauchy about 180 years ago, this distribution has the peculiar property that the sample mean for random samples has the same distribution as one observation! (We omit a proof.) . In fact, the Cauchy distribution is the only distribution with this property. It is obvious therefore that the sample mean for samples from the Cauchy distribution is not asymptotically normally distributed. In the case of iid samples the existence of the variance, and of course the mean, is necessary for asymptotic normality of the sample mean or sum. The random variable $Y = aX + b$, for constants $a \neq 0$ and $b$, having density $f_Y(y) = g((y - b)/a)/|a|$, is said to have the Cauchy distribution with location parameter $b$ and scale parameter $a$. The characteristic function for $X$ is $C(t) = E(e^{it}) = e^{-|t|}$ for all $t$, where $i = \sqrt{-1}$.

## Sampling Without Replacement

Given a finite population $\mathcal{P}_N$ of units $(1, \ldots, N)$ with corresponding measurements $(x_1, x_2, \ldots, x_N)$, suppose that a simple random sample of units $\mathbf{u} = (u_1, \ldots, u_n)$ is taken. That is, all $P(N, n) = N!/(N - n)!$ permutations of $n$ units from the units in $\mathcal{P}_N$ are equally likely. Let $Y_1, \ldots, Y_n$ be the corresponding measurements. Let $S_n = \Sigma Y_i$ and $\bar{Y}_n = S_n/n$, the sample mean. The $Y_i$ are not independent. In fact, it is relatively easy to show that $\rho(Y_i, Y_j) = -1/(N - 1)$ for all $i \neq j$. We cannot therefore conclude from the CLT that $S_n$ and $\bar{Y}_n$ are asymptotically normally distributed. In fact, if the population size $N$ remains fixed, the sample size $n$ cannot exceed $N$, and if $n = N$, $S_n$ and $\bar{Y}_n$ are fixed constants.

For many years prior to 1960, normal approximations to the distribution s of $S_n$ and $\bar{Y}_n$ were used, although no proof of asymptotic normally had been provided. Finally, in 1960, Jaroslav Hajek provided an ingenious proof. We state his theorem but do not provide a proof.

We must consider a sequence of populations of measurements $\mathcal{P}_N = (x_{N1}, \ldots, x_{NN})$. Let $\mathcal{P}_N$ have mean $\mu_N$ and variance $\sigma_N^2$ ($N$ divisor). Let $\tau_N = \max(|X_{Ni} - \mu_n| N\sigma_N^2)$, where the maximum is taken for all $i = 1, \ldots, N$. Let $\mathbf{Y} = (Y_{N1}, \ldots, Y_{Nn})$ be the measurements for a simple random sample of $n = n_N$. Although we consider a sequence of sample sizes $\{n_N\}$ as $N \to \infty$, we sometimes omit the subscript $N$ on $n$ for simplicity. Let $S_{Nn}$ and $\hat{Y}_{Nn}$ be the corresponding sample sum and sample means. Then $\text{Var}(S_{Nn}) = n\sigma_N^2\ (N - n_N)/(N - 1)$. Define $Z_{Nn} = (S_{Nn} - n\mu_N)/\sqrt{\text{Var}(S_{Nn})}$.

**Central Limit Theorem for Sampling Without Replacement** (Hajek, 1960) Let $\{n_N\}$ and $\{N - n_N\}$ converge to $\infty$ as $N \to \infty$. Let $\tau_N \to 0$ as $N \to \infty$. Then $\{Z_{Nn}\}$ converges in distribution to N(0, 1).

COMMENTS: In practice, the approximation of the distribution of $Z_{Nn}$ by the standard normal distribution depends very much on the three numbers $n_N, n - n_N$, and $\tau_N$. We illustrate this by approximating probabilities $P(Z_{Nn} \leq z)$ by (1) simulations and (2) $\Phi(z)$.

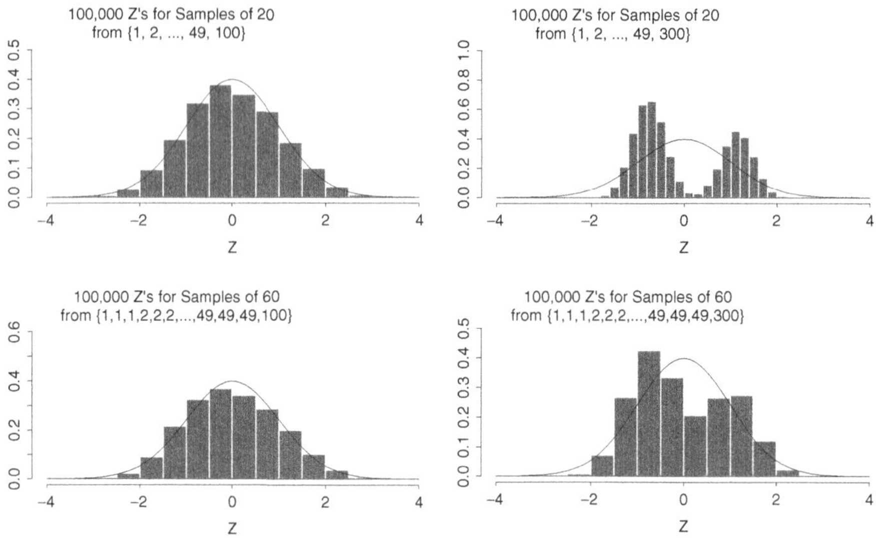


**FIGURE 6.4.3** Histograms of standardized sums of samples without replacements from finite populations.

Figure 6.4.3 presents histograms of $Z$ = standardized sum of simple random samples from populations with *outliers*, since one or more population values deviate from the population means by "large" amounts. For these two populations and two sample sizes (20 and 60), the values of $\tau_N$ are 0.353, 0.866, 0.082, 0.714. These suggest that the approximations by the standard normal distribution are better for small $\tau_N$. The bimodal distribution in the second graph reflects the fact that the sum of sample values tends to be large when 300 is in the sample, small otherwise. It is large with probability 20/50. Although $\tau_N$ is only slightly smaller for the fourth graph, and the distribution is still bimodal, the distribution is a bit closer to standard normal because of the larger sample size.

## Problems for Section 6.4

**6.4.1** Let $X$ have the binomial distribution with $n = 10$, $p = 1/2$.

- **(a)** Find the normal approximations for $P(X \le k)$ for $k = 3, 4, 6, 7$. (The actual values are 0.172, 0.377, 0.828, 0.945.)
- **(b)** Find the normal approximations for $P(X = 4)$ and $P(X = 7)$.

**6.4.2** Let $U_1, U_2, \ldots, U_{48}$ be independent, each with the uniform distribution on [0, 1]. Let $S_{48}$ be the sum of these $U_k$ and $\bar{U} = S_{48}/48$. Find the normal approximation for:

- **(a)** $P(20 \le S_{48} \le 28)$.
- **(b)** $P(0.4167 \le \bar{U} \le 0.5833)$.

**6.4.3** Having not studied all semester for his course on Greek literature for the years 356 through 983, but feeling lucky, a student uses coin flips and die throws to choose his answers for an exam. The exam has 40 multiple-choice questions, each with four possible answers, one of which is correct, and 50 true–false questions. The generous instructor has promised to pass students who get at least 45 correct. What is the approximate probability that the student passes? (Use the $1/2-$ correction. The central limit theorem as stated does not apply directly, but the normal approximation should provide a good approximation. Among 100,000 such student exams produced randomly by a computer, 1667 students passed.)

**6.4.4** Reprove the CLT for the special case that the $X_k$ each have the Bernoulli distribution with parameter $p$. Use the fact that the three-term Taylor approximation of $e^u$ around $u = 0$ is $1 + u + u^2/2$.

**6.4.5** Find an approximation for the probability that it will take more than 650 throws of a die to get at least 100 6's. In 100,000 computer-simulated observations, more than 650 throws were needed 17,814 times. Determine the approximation by:

**(a)** Using the normal approximation for the negative binomial distribution.

**(b)** Using the normal approximation for the binomial distribution.

**6.4.6** Let $Y \sim \Gamma(25, 1)$. Give the normal approximation for $P(Y \leq 20)$. The function "pgamma" in S-Plus gives 0.157.

**6.4.7** Vehicles pass a point on a freeway late at night in a Poisson process with rate three per minute. Use a normal approximation to find:

**(a)** The probability that at least 35 pass in a 10-minute period. [This can be done by the normal approximation of the Poisson (see Problem 6.4.9) or by the normal approximation of the gamma (as in Problem 6.4.6).]

**(b)** The probability that the time necessary to have 100 vehicles pass exceeds 36 minutes.

**6.4.8** For the population $\{1, 2, \ldots, 9, 100\}$ and random samples of sizes $n = 25$, with sample mean $\bar{X}$, find:

**(a)** $E(\bar{X})$ and $\text{Var}(\bar{X})$.

**(b)** The normal approximations for $P(\bar{X} \leq 12.0)$ and $P(12.0 \leq \bar{X} \leq 17.0)$. In 100,000 computer simulations the events occurred 30,529 and 43,965 times. (Use the 1/2-correction, remembering that $\bar{X}$ takes values that are multiples of 1/100. This is an example of a case in which the normal approximation is not very good.)

**(c)** Repeat part (b) for $n = 100$. In 100,000 computer simulations the events occurred 20,163 and 60,450 times. Normal approximations are much better this time.

**6.4.9** Let $X_n$ have the Poisson distribution with parameter $\lambda_n$, and suppose that $\lim_{n\to\infty} \lambda_n = \infty$. Let $Z_n = (X_n - \lambda_n)/\sqrt{\lambda_n}$ and let $M_n$ be the mgf for $Z_n$.

**(a)** Show that $\{Z_n\}$ converges in distribution to N(0, 1) by showing that $\lim_{n\to\infty} \log(M_n(t)) = t^2/2$, and therefore $\lim_{n\to\infty} M_n(t) = e^{t^2/2}$ for all $t$.

**(b)** Let **X** have the Poisson distribution with parameter $\lambda = 100$. Use the result of part (a) and the 1/2-correction to find an approximation for $P(85 \le X \le 115)$. Computation with the Poisson probabilities gave 0.8793.

**6.4.10** Let $X$ have the $\Gamma(\alpha, 1)$ distribution, and let $Z = (X - \alpha)/\sqrt{\alpha}$, the standardized version of $X$.

**(a)** Show that the mgf of the $\Gamma(\alpha, 1)$ distribution is $(1 - t)^{-\alpha}$, existing for $t < 1$.

**(b)** Show that the mgf for $Z$ is $M_Z(t) \equiv e^{-t\sqrt{\alpha}}(1 - t/\sqrt{\alpha})^{-\alpha}$ for $t < 1$.

**(c)** Let $X_n \sim \Gamma(\alpha_n, 1)$, and let $Z_n$ be the standardized version of $X_n$. Let $M_n(t)$ be the mgf for $Z_n$. Suppose that $\lim_{n\to\infty} \alpha_n = \infty$. Show that $\lim_{n\to\infty} M_n(t) = e^{t^2/2}$, so that $\{Z_n\}$ converges in distribution to N(0, 1) as $n \to \infty$.

**(d)** Use the result of part (c) to find approximations for $P(X > 30)$ for $\alpha = 25$ and $P(X > 60)$ for $\alpha = 50$. (More exact computation gives 0.1572 and 0.0844.)

**6.4.11** Let $U_1, \ldots, U_n$ be independent, each Unif(0, 1). Let $\bar{U}_n = (U_1 + \cdots + U_n)/n$. How large must $n$ be in order to have $P(|\bar{U}_n - 1/2| < 0.02) \ge 0.95$?

**6.4.12** An airline overbooks for flights on its 400-seat airplane, knowing that those with reservations are no-shows with probability 0.10, independently.

**(a)** Suppose that the airline accepts 434 reservations. Find an approximation for the probability that it will have enough seats.

**(b)** If the airline wants to have probability at least 0.98 that it will have enough seats, how many reservations can it accept?

**(c)** Repeat parts (a) and (b) for the case that passengers show or don't show in pairs, independently. That is, the members of each of $n$ pairs are both shows or both no-shows with probabilities 0.90, 0.10.

**6.4.13** Let $\{p_n\}$ be a sequence of positive probabilities converging to zero. For each $n$ let $X_n$ have the geometric distribution with parameter $p_n$, taking the values 1, 2, .... Define $Y_n = p_n(X_n - 1)$. Let $M_n$ be the mgf for $Y_n$ (see Figure 6.4.13).

**(a)** Show that $\lim_{n\to\infty} M_n(t) = (1 - t)^{-1}$ for $t$ in a neighborhood of zero. What is the limiting distribution $F$ for $\{Y_n\}$? [As a check, for

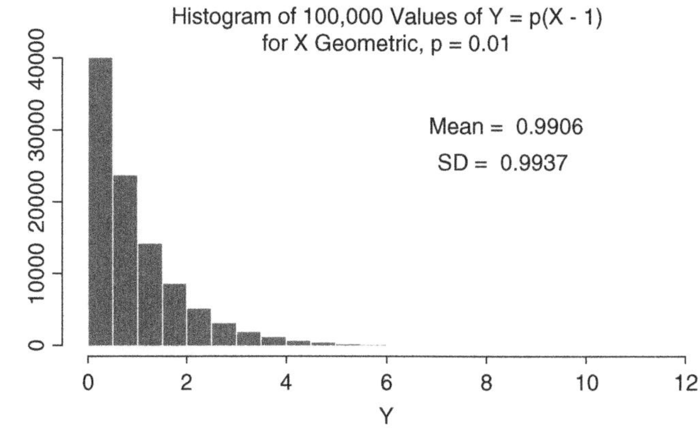


**FIGURE 6.4.13** Histogram of 10,000 values of $Y = p(X - 1)$ for $X \sim$ geometric.

$p_n = 1/n$, $P(X_{100} > 110) = (1 - 1/100)^{110} = 0.3310$. The probability for the limiting distribution is 0.3329.]

**(b)** Show that the cdf for $Y_n$ is $F_n(x) = 1 - (1 - p_n)^{k_n}$, where $k_n = [x/p_n]$, $x \geq 0$. Use the fact that $\lim_{m\to\infty}(1 - a_m/m)^m = e^a$ if $\lim_{m\to\infty} a_m = a$ to show directly, without using mgf's, that $\{Y_n\}$ converges in distribution to the $F$ of part (a).

**(c)** Let $W_n$ have the negative binomial distribution with parameters $r$ and $p_n$, with $\lim_{n\to\infty} p_n = 0$, $r$ a positive integer. Let $W_n^* = p_n(W_n - r)$. Does $\{W_n^*\}$ have a limiting distribution? If so, what is it?

**6.4.14** Consider the population $\mathcal{P} = \{1, 2, \ldots, 20, 50\}$.

**(a)** Find the mean $\mu$, variance, $\sigma^2$, and the parameter $\tau = \tau_N$.

**(b)** Let $\bar{X}$ be the sample mean for a simple random sample of $n = 12$ from $\mathcal{P}$. Find $\text{Var}(\bar{X})$.

**(c)** Find an approximation for $P(|\bar{X} - \mu| < 2.0)$. Use the $1/(2n)$-correction. In 10,000 simulations the event occurred 6362 times.

CHAPTER SEVEN

# Estimation

## 7.1 INTRODUCTION

Probability is concerned with the determination and approximation of the probability of events and with properties of random variables. We begin with a probability model, an assignment of probabilities to events, usually stated by assigning probabilities to a collection of simple events, then from this assignment, determining the probabilities of more complex events. For one of our first examples, the birthday problem, we supposed that for $n$ people and $D$ the collection of days of the year, all $n$-tuples birthdays, all points in the Cartesian product $S = D^{(n)}$ were equally likely. From this model we were then able to determine the probability that all birthdays would be different, and give an approximation.

In many situations there seems to be no natural way to select a model. Consider the very simple case of dropping a thumbtack on a hard surface 20 times from a fixed height, with the point landing "down" or "up." It should seem reasonable that there is some number $p$, $0 < p < 1$, the probability that the tack will have its point up, and that the outcomes of the 20 drops are independent. Therefore, there would seem to be a reasonable model corresponding to each $p$. Prior to the experiment, we might guess what the value of $p$ is, but we could not know a precise value of $p$. It would seem to be wishful thinking to take $p = 1/2$ arbitrarily, as we would for a coin. Once we have performed the experiment, we would at least have a clue as to the value of $p$. If, for example, $X$ of the drops resulted in "up" and we observed $X$ to be 15, it would seem doubtful that $p < 0.4$, although it is possible that for $p = 0.39$, for example, we could observe $X = 15$.

For another example, suppose that we are studying the lengths of life of automobile tires in 1000s of miles. It may be reasonable from past experience to suppose that these lengths have a gamma distribution with power parameter $\alpha$ and scale parameter $\lambda$. Unless we have studied this tire extensively, we cannot know the value of the pair $(\alpha, \lambda)$.

---

The subject of the remainder of the book is statistical inference. We consider probability models for which they are more than one possible probability measure assigned to the sample space. As a result of the experiment we will get some clues as to which of these probability measures is more reasonable, in a sense that we discuss. In this chapter we describe point and interval estimation.

## 7.2 POINT ESTIMATION

Suppose that $X_1, \ldots, X_n$ is a random sample (independent and identically distributed) from the uniform distribution on the interval $[0, \theta]$, for $\theta > 0$, with $\theta$ unknown. Suppose that for $n = 10$, these $X_i$'s are, to the nearest integer, 27, 11, 17, 23, 4, 11, 20, 31, 19, 9. What is your estimate of $\theta$? Could $\theta$ be 27.93? Do you think $\theta$ could be 56.3? In general, we should give a rule, a function $\hat{\theta} = \hat{\theta}(\mathbf{X})$, for $\mathbf{X} = (X_1, \ldots, X_n)$, called a point estimator of the parameter $\theta$. $\hat{\theta}$ determines an estimate of $\theta$ for each possible $\mathbf{X}$. To judge the performance of an estimator, we must consider the probability distribution of the estimator. This distribution depends on the unknown parameter and is therefore unknown, although in some cases the distribution of the error $\hat{\theta} - \theta$ or of the ratio $\hat{\theta}/\theta$ is the same for all $\theta$.

The parameter $\theta$ may be a vector. For the gamma distribution it may be the pair $(\alpha, \lambda)$. For the multinomial distribution with $k$ outcomes it could be $\mathbf{p} = (p_1, \ldots, p_k)$.We may be interested in estimating the entire parameter vector, some subset of components, or some other function of the parameter vector.

For the normal distribution with $\theta = (\mu, \sigma^2)$ we may be interested in estimating the 0.95 quantile (95th percentile) $\eta = \mu + 1.645\sigma$. For the multinomial we might be interested in estimating $p_1 + p_2$, or $p_1/(p_1 + p_2)$. In general, we wish to estimate a function $g(\theta)$, defined on the set $\Omega$ of all possible $\theta$. The set $\Omega$ (capital omega) is called the parameter space. The function $g$ on $\Omega$ takes its values on the real line or in some larger Euclidean space.

**Definition 7.2.1** Let $g(\theta)$ be a function on a parameter space $\Omega$. A point estimator of $g(\theta)$ is a function $T = T(\mathbf{X})$, defined for all possible $\mathbf{X}$, taking values in $g(\Omega) \equiv \{g(\theta) \mid \theta \in \Omega\}$. □

COMMENT: The term point estimator is used to distinguish it from an interval estimator, which we discuss later. We usually say "estimator" to mean point estimator.

For many examples the function $g$ will be the identity function and our estimator will be denoted by putting a "hat" on the parameter. Thus, $\hat{\theta}$ will denote an estimator of the parameter $\theta$.

**Definition 7.2.2** If $\mathbf{X} = \mathbf{x}$ is the value observed and $T(\mathbf{X})$ is an estimator of $g(\theta)$, then $T(\mathbf{x})$ is called the *estimate* of $g(\theta)$. □

COMMENTS: The estimator and estimate of $\theta$ are often written simply as $\hat{\theta}$ rather than as $\hat{\theta}(\mathbf{X})$ or $\hat{\theta}(\mathbf{x})$, not distinguishing between the estimator $\hat{\theta}(\mathbf{X})$ and the estimate $\hat{\theta}(\mathbf{x})$ obtained for the particular sample $\mathbf{x}$. Usually, the meaning can be determined from the context. The *estimator* is a random variable or random vector. The *estimate* is a number or vector of numbers.

The distribution of $\mathbf{X}$ and therefore of $T(\mathbf{X})$ depends on $\theta$, whose value is an unknown member of the parameter space $\Omega$. We are interested in choosing estimators $T$ whose distributions are concentrated about $g(\theta)$. That is, we want $T(\mathbf{X})$ to be close to $g(\theta)$ with high probability for each $\theta \in \Omega$. One measure of the performance of an estimator is its mean squared error.

**Definition 7.2.3** Let $T(\mathbf{X})$ be an estimator of a real-valued function $g(\theta)$ for $\theta \in \Omega$. The *mean squared error* for $T$ is

$$r_T(\theta) = E_\theta[(T(\mathrm{X}) - g(\theta))^2]. \qquad \square$$

COMMENT: The function $r_T(\theta)$ is called the *risk function for squared-error loss*. The subscript $\theta$ on the expectation operator $E$ has been used to remind us that the distribution of $\mathbf{X}$, and therefore of $T(\mathbf{X})$, depends on $\theta$. We do not always use this subscript, although its use often improves understanding.

As we learned earlier, for any random variable $Y$ and constant $c$, $E(Y - c)^2 = \mathrm{Var}(Y) + [c - E(Y)]^2$, so that $r_T(\theta) = \mathrm{Var}_\theta(T) + [E_\theta(T) - g(\theta)]^2$. It follows that $r_T(\theta)$ is small if both terms in the sum are small. The term in brackets on the right is the bias $b_T(\theta) = E_\theta(T) - g(\theta)$. An estimator is unbiased if $b_T(\theta) = 0$ for all $\theta \in \Omega$.

**Definition 7.2.4** Let $g(\theta)$ be a real-valued function on a parameter space $\Omega$, and let $T$ be a point estimator of $g(\theta)$. $T$ is said to be an *unbiased estimator* of $g(\theta)$ if

$$E_\theta(T) = g(\theta) \quad \text{for all } \theta \in \Omega. \qquad \square$$

***Example 7.2.1*** Let $X_1, \ldots, X_n$ be a random sample from the Unif(0, $\theta$) distribution for $\theta > 0$. The estimator $T_1 = 2\bar{X}$ is an unbiased estimator of $\theta$. Unfortunately, $T_1$ has the undesirable property that some values of $\mathbf{X}$ lead to ridiculous estimates. Consider, for example, for $n = 3$, $X_1 = 7$, $X_2 = 29$, $X_3 = 3$. Then $T_1 = 2\bar{X} = 26$, although it is obvious that $\theta \geq \max(X_1, \ldots, X_n) = 29$. For our estimator $T_1$, $r_{T_1}(\theta) = \mathrm{Var}_\theta(T_1) = 4\ \mathrm{Var}_\theta(\bar{X}) = (4\ \theta^2)/(12n) = \theta^2/3n$, since the variance of the uniform distribution on $[0, \theta]$ is $\theta^2/12$.

Note that $r_{T_1}(\theta)$ converges to zero as $n \to \infty$, another desirable property of an estimator. Intuitively at least, we like to feel that as $n \to \infty$, the probability that our estimator will deviate from its target $g(\theta)$ by any fixed amount will become closer and closer to 1 as $n \to \infty$.

Continuing with the Unif(0, θ) example, consider the estimator $T_2 = \max(X_1, \ldots, X_n)$. Note that $Y_i = X_i/\theta$ has the uniform distribution on [0, 1] and $W \equiv T_2/\theta = \max(Y_1, \ldots, Y_n)$. Since $Y_i$ and $W$ have distributions that do not depend on θ for all $\theta > 0$, we will determine the properties of $T_2$ by first studying those of $W$.

$W$ has cdf $F_W(w) = P(Y_1 \le w, \ldots, Y_n \le w) = \prod_{i=1}^n P(X_i \le w) = w^n$ for $0 \le w \le 1$, so that $W$ has density $f_W(w) = nw^{n-1}$ for $0 \le w \le 1$. Therefore, $E(W) = n/(n+1)$, $E(W^2) = n/(n+2)$, and $\text{Var}(W) = n/[(n+2)(n+1)^2]$. It follows that $E_\theta(T_2) = \theta(n/n+1)$ and $\text{Var}_\theta(T_2) = \theta^2(n/(n+2)(n+1)^2)$, $b_{T_2}(\theta) = -\theta/(n+1)$, and $r_{T_2}(\theta) = 2\theta^2/[(n+1)(n+2)]$.

Which estimator is better, $T_1$ or $T_2$? Consider the ratio $r_{T_1}(\theta)/r_{T_2}(\theta) = [(n+1)(n+2)]/(6n)$, which is 1 for $n = 1$ and 2 and is greater than 1 for $n > 2$. Thus, $T_2$ is uniformly better than $T_1$ in the sense of mean squared error. The word *uniformly* here refers to the fact that $r_{T_1}(\theta) > r_{T_2}(\theta)$ for all $\theta > 0$ when $n > 2$. For $n > 2$, $T_2$ is an example of an estimator that is biased but has smaller mean squared error than an unbiased estimator because it has smaller variance.

For a simple example of an estimator that has smaller mean squared error for some θ, larger for others, consider the estimator $T_3$, which is always 17 (a favorite number of the author). $T_3$ has bias $b_{T_3}(\theta) = \theta - 17$ for all $\theta > 0$, and $\text{Var}_\theta(T_3) = 0$. Thus, $r_{T_3}(\theta) = (\theta - 17)^2$ for all $\theta > 0$. Thus, $r_{T_3}(\theta)$ is less than $r_{T_2}(\theta)$ for θ close enough to 17, but is much larger for θ very far from 17. □

***Example 7.2.2*** Suppose that a distance θ is to be estimated. Two methods are available, with resulting measurements $X_1$ and $X_2$. A reasonable model may state that $X_1$ and $X_2$ are independent, each with mean θ, but possibly different known variances $\sigma_1^2$ and $\sigma_2^2$. Consider a linear estimator $T = aX_1 + bX_2$, where $a$ and $b$ are constants. Since $E(T) = a\theta + b\theta = (a+b)\theta$, $T$ is unbiased for θ if and only if $b = 1 - a$. In this case the mean square error of $T$ is $r_T(\theta) = \text{Var}(T) = a^2\sigma_1^2 + (1-a)^2\sigma_2^2$. Call this $g(a)$. We seek to minimize $g(a)$. Since $\frac{d}{da}g(a) = 2a\sigma_1^2 - 2(1-a)\sigma_2^2 = 2a[\sigma_1^2 + \sigma_2^2] - \sigma_2^2$ and $(d^2/da^2)g(a) > 0$, $g(a)$ is a minimum for $a = \sigma_2^2/(\sigma_1^2 + \sigma_2^2) = (1/\sigma_1^2)/(1/\sigma_1^2 + 1/\sigma_2^2)$. We gain additional insight into why $g$ is minimized by this choice for $a$ by *completing the square*. Let $\tau = \sigma_2^2/\sigma_1^2$. Then $g(a) = \sigma_1^2[a^2 + (1-a)^2\tau] = \sigma_1^2\{[a - \tau(1+\tau)]^2(1+\tau) + \tau\ (1+\tau)\}$. Thus, $g$ is minimized by $a = \tau/(1+\tau) = (1/\sigma_1^{2)}/(1/\sigma_1^2 + 1/\sigma_2^2)$. In this case the minimum value is $\sigma_1^2\tau/(1+\tau) = 1/(1/\sigma_1^2 + 1/\sigma_2^2)$. In practice, if τ is unknown, a guess at τ still produces a better estimator than the usual choice, $\tau = 1$, $a = 1/2$ (so the estimator is the mean of the two observations), *if* the guess is on the same side of 1 as τ is. □

**Problems for Section 7.2**

**7.2.1** **(a)** For the $U(0, \theta)$ example, let $T_4 = \max(X_1, \ldots, X_n)(n+1)/n = T_2(n+1)/n$. Show that $T_4$ is an unbiased estimator of θ and find its mean squared error $r_{T_4}(\theta)$.

**(b)** Let $T_5 = c_n T_2$. Find the constant $c_n$ for which $r_{T_5}(\theta)$ is minimum. Does $c_n$ depend on θ? If so, then $T_5$ is not an estimator, since it depends on an unknown parameter. Evaluate $r_{T_5}(\theta)$ for this minimizing value of $c_n$, and compare it to $r_{T_2}(\theta)$ and $r_{T_4}(\theta)$.

**7.2.2** Let $X_1, \ldots, X_n$ be a random sample from a distribution with mean $\mu$ and variance $\sigma^2$ Consider an estimator $T = \sum_{i=1}^{n} a_i X_i = \mathbf{a}\mathbf{X}$, for $\mathbf{a} = (a_1, \ldots, a_n)$ and $\mathbf{X} = (X_1, \ldots, X_n)^T$, so that $\mathbf{a}$ is a row vector and and $\mathbf{X}$ is a column vector.

**(a)** Find a condition on $\mathbf{a}$ so that $T$ is an unbiased estimator of $\mu$.

**(b)** Find $\mathbf{a} = \mathbf{a}_0$ so that the condition in part (a) is satisfied and $\text{Var}(T)$ is a minimum. $T_0 = \mathbf{a}_0^T\mathbf{X}$ is then the *best linear unbiased estimator* (the BLUE) of $\mu$. What is this minimum variance?

**7.2.3** Let $X$ have a binomial distribution with parameters $n$ and $p$, with $n$ known, $p$ unknown, $0 < p < 1$. Let $\hat{p} = X/n$. Find $r_{\hat{p}}(p)$.

**7.2.4** Let $X_1$, $X_2$ be a random sample from a discrete distribution with probability function $p(k; \theta)$ for $\theta \in \Omega = \{1, 2, 3\}$ as follows:

| | $k$ | | | |
|---|---|---|---|---|
| $\theta$ | 1 | 2 | 3 | 4 |
| 1 | 0.4 | 0.3 | 0.2 | 0.1 |
| 2 | 0.2 | 0.3 | 0.3 | 0.2 |
| 3 | 0.1 | 0.2 | 0.3 | 0.4 |

Suggest two reasonable estimators $T_1$ and $T_2$ of $\theta$, and determine the mean squared error for each. [First present the sample space for $(X_1, X_2)$ and for each possible pair of values give the value of $T_1$ and for $T_2$.]

**7.2.5** Let $\mathbf{X} = (X_1, \ldots, X_n)$ be a random sample from the exponential distribution with rate parameter $\lambda > 0$, mean $1/\lambda$. Consider the estimator $\hat{\lambda} = 1/\bar{X}$.

**(a)** Show that $\hat{\lambda}$ is a biased estimator of $\lambda$. What is its bias and its mean squared error?

**(b)** Find a constant $C_n$ (not depending on $\lambda$) so that $C_n/\bar{X}$ is an unbiased estimator of $\lambda$, and determine its mean squared error. *Hint*: $\sum X_i$ has a gamma distribution with parameters $n$ and $1/\lambda$.

**7.2.6** Let $\Omega = \{7, 8\}$ and for $\theta \in \Omega$ suppose that a box has five balls with numbers $\theta, \theta, \theta, \theta, \theta + 1$. One ball is chosen at random from the box. Let $X$ be the number on the ball chosen.

**(a)** Suggest a reasonable estimator $T(X)$ of $\theta$. What is its bias? Evaluate $r_T(\theta)$ for $\theta \in \Omega$.

**(b)** Suppose that two balls are chosen randomly without replacement. Let $X_1$ and $X_2$ be numbers on the two balls chosen. Suggest an estimator $T_2 = T_2(X_1, X_2)$ and find its mean squared error.

(c) Suppose that a random sample of three balls is taken without replacement. Suggest an estimator $W$ of $\theta$ so that the risk function $r_W(\theta) = 0$ for $\theta = 7, 8$.

(d) Suppose that a random sample of n balls is chosen with replacement. Choose an estimator $V_n$ in such a way that for each $\theta$ the risk function for $V_n$ converges to 0 as $n \longrightarrow \infty$.

**7.2.7** Let $X$ have the Poisson distribution with parameter $\lambda > 0$. Let $p(\lambda) = P(X = 0) = \exp(-\lambda)$. Let $T(X)$ be an estimator of $p(\lambda)$.

(a) Find the function $T(k)$ on the nonnegative integers so that $T(X)$ is an unbiased estimator of $p$. Show that it is the only such estimator. *Hint*: Express $E[T(X)]$ as a power series in $\lambda$. The condition that this must be $p(\lambda)$ for each $\lambda > 0$ determines the function $T$ uniquely.

(b) Consider an estimator $T_c \equiv \exp(-cX)$. Express $E(T_c(X))$ as a function of $c$ and $\lambda$. Is there any $c$ for which $T_c$ is unbiased for $p(\lambda)$? Suppose that $c = 10^9$ and we are sure that $\lambda < 10^6$. Is $T_c$ approximately unbiased?

**7.2.8** Let $X_1, \ldots, X_n$ be a random sample from a distribution with mean $\mu$ and variance $\sigma^2$. Let $\bar{X}$ be the sample mean and let $S^2 \equiv \sum_{i=1}^n (Xi - \bar{X})^2/(n-1)$, the sample variance. Show that $S^2$ is an unbiased estimator of $\sigma^2$. *Hint*: Recall that $= \sum_{i=1}^n (X_i - c)^2 - (\bar{X} - c)^2 n$ for any constant $c$.

**7.2.9** Let $X_1, \ldots, X_n$ be a random sample on the $U(\theta, \theta + 1)$ distribution for $\theta$ any real number. Let $U = \min(X_1, \ldots, X_n)$, $V = \max(X_1, \ldots, X_n)$, $T = V - n/(n+1)$, and $\hat{\theta} = (U + V)/2 - 1/2$. See Section 3.4, page 102.

(a) Show that $T$ is unbiased for $\theta$ and determine its variance. *Hint*: Define $Y_i = X_i - \theta$, and note that $V = \theta + \max(Y_1, \ldots, Y_n)$.

(b) Show that $\hat{\theta}$ is an unbiased estimator of $\theta$ and determine its variance.

(c) Compare $\text{Var}(T)$ and $\text{Var}(\hat{\theta})$.

## 7.3 THE METHOD OF MOMENTS

We need methods that will produce estimators. In this section we discuss the simplest of these, the *method of moments*, (MME). In Section 7.4 we discuss the maximum likelihood method. Consider, for example, the family of gamma distributions with shape parameter $\alpha$ and scale parameter $\theta$. As shown in Section 4.3, the mean is $\mu = \theta\alpha$ and the variance is $\mu_2 = \sigma^2 = \theta^2\alpha$. We can solve these for $\theta$ and $\alpha$ in terms of $\mu$ and $\sigma^2$. Thus,

$$\theta = \frac{\sigma^2}{\mu} \quad \text{and} \quad \alpha = \frac{\sigma^2}{\theta^2} = \frac{\mu^2}{\sigma^2}. \tag{7.3.1}$$

The moment estimators of $\mu$ and $\sigma^2$ are $\hat{\mu} = \bar{X}$ and $\hat{\sigma}^2 = (1/n) \sum_{i=1}^n (X_i - \bar{X})^2$. (Whether we use the denominator $n$ or $n - 1$ as in the definition of $S^2$ in Problem

7.2.8 is arbitrary.) The MME of the parameter vector $(\alpha, \theta)$ is given by replacing the parameters $\alpha$ and $\theta$ by their moment estimators in (7.3.1). Thus, the MME of $(\alpha, \theta)$ is $(\hat{\alpha}, \hat{\theta})$, where

$$\hat{\alpha} = \frac{\bar{X}^2}{\hat{\sigma}^2} \quad \text{and} \quad \hat{\theta} = \frac{\hat{\sigma}^2}{\bar{X}}. \tag{7.3.2}$$

As we discuss later, it seems reasonable to expect $\bar{X}$ to be close to $\mu$ and $\hat{\sigma}^2$ to be close to $\sigma^2$, so that $\hat{\alpha}$ will be close to $\alpha$ and $\hat{\theta}$ close to $\theta$.

The following sample of 25 was taken from the gamma distribution with $\theta = 10$, $\alpha = 4$:

26.3 47.7 48.4 23.9 23.0 27.7 34.1 30.4 25.3 43.9 63.0 18.2
17.1 62.1 25.6 44.5 76.9 68.1 42.0 53.8 21.4 28.3 27.0 79.0
84.9

For this sample, $\bar{X} = 41.71$ and $\hat{\sigma}^2 = 403.59$. (We are abusing notation a bit by using the same symbol for the estimates as for the estimators. The reader is supposed to determine the meaning from the context.) From (7.3.2) we obtain $\hat{\alpha} = 4.31$ and $\hat{\theta} = 9.68$.

This experiment was repeated 500 times, with results as given in Figure 7.3.1. These simulations took about 2 seconds using S-Plus. There seems to be a bias of about 0.55 for $\hat{\alpha}$ and $-0.37$ for $\hat{\theta}$. Both distributions are skewed to the right. From these simulations, can you estimate the mean squared errors for the two estimators

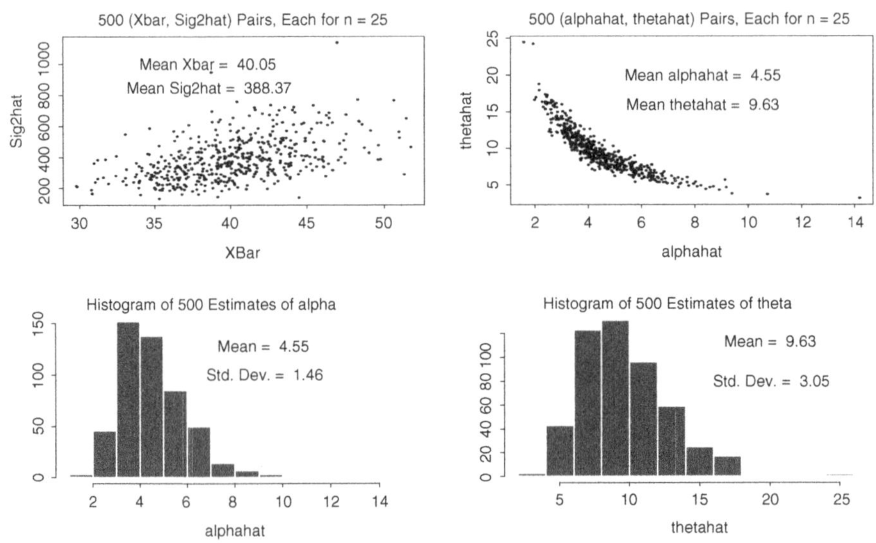


**FIGURE 7.3.1** 500 Estimates of $\hat{\alpha}$ and $\hat{\theta}$.

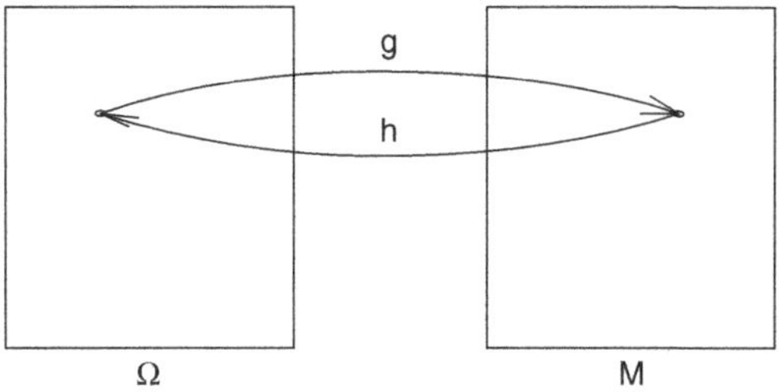


**FIGURE 7.3.2** The transformations $g$ and $h = g^{-1}$.

for these parameter values? What do you think would happen if the sample size were 100 rather than 25?

To consider the MME more formally, let the $k$th moment about zero of a random variable $X$ be denoted by $\nu_k$ for any positive integer $k$. Thus, $E(X) = \nu_1 = \mu$. Similarly, let the $k$th central moment for $X$ be $\mu_k = E[(X - \mu)^k]$. Thus, $\mu_2 = \text{Var}(X) = \sigma^2$. The sample moments for a sample $X_1, \ldots, X_n$ are defined by assigning probability $1/n$ to each $X_i$. Thus, $\hat{\nu}_k = (1/n)\sum_{i=1}^n X_i^k$ is the sample $k$th moment about zero and $\hat{\mu}_k = (1/n)\sum_{i=1}^n (X_i - \bar{X})^k$ is the sample $k$th central moment. We also use the notation $\bar{X} = \hat{\nu}_1$ and $\hat{\sigma}^2 = \hat{\mu}_2$.

By the weak law of large numbers, each $\hat{\nu}_k$ converges in probability to the corresponding $\nu_k$. It follows easily, as we show formally later, that each $\hat{\mu}_k$ converges in probability to the corresponding $\mu_k$. This implies that for large sample size $n$, $\hat{\nu}_k$ should be close to $\nu_k$ and $\hat{\mu}_k$ close to $\mu_k$, with probabilities close to 1.

Suppose now that $(X_1, \ldots, X_n) = \mathbf{X}$ is a random sample from a distribution with cdf $F(x; \theta)$ for $\theta \in \Omega \subset R_k$, where $\theta = (\theta_1, \ldots, \theta_k)$ is unknown. Suppose that the relationship between the moment vector $\nu = (\nu_1, \ldots, \nu_k)$ and $\theta$ is given by $\nu = g(\theta)$ and let $M = \{\nu | \nu = g(\theta), \theta \in \Omega\}$ be the range of $g$ (see Figure 7.3.2). We call $M$ the *moment space*. Suppose that $g$ is one-to-one on $\Omega$, and let $h$ be its inverse. Thus, $h(\nu) = \theta$ if and only if $g(\theta) = \nu$. Let $\hat{\nu} = (\bar{X}, \hat{\nu}, \ldots, \hat{\nu}_k)$ be the moment estimator of $\nu$. The *method of moments estimator* (MME) of $\theta$ is $h(\hat{\nu})$.

The MME may instead be determined from the central moments. Let $\boldsymbol{\mu} = (\mu, \sigma^2, \mu_3, \ldots, \mu_k)$ be the vector of central moments (the first is the first moment about zero). Let $g^*(\theta) = \boldsymbol{\mu}$ and let $M^* = \{\mu | \mu = g^*(\theta), \theta \in \Omega\}$. We call $M^*$ the *central moment space*. Suppose that $g^*$ is one-to-one on $\Omega$, and let $h^*$ be the inverse of $g^*$. Let $\hat{\boldsymbol{\mu}} = (\bar{X}, \hat{\sigma}^2, \hat{\nu}_3, \ldots, \hat{\nu}_k)$ be the sample central moment vector. Then the method of moments estimator of $\theta$ is $h^*(\hat{\boldsymbol{\mu}})$.

The MME is the same whether moments about zero or central moments are used. The next paragraph proves this. The paragraph may be skipped with impunity (always remembering that your instructor is the boss).

It is easy to see that there is a one-to-one correspondence between the moment vectors $\boldsymbol{\nu}$ and the central moment vectors $\boldsymbol{\mu}$. Suppose that $k(\boldsymbol{\nu}) = \boldsymbol{\mu}$ for each $\boldsymbol{\nu} \in M$. Thus, $\theta = h^*(\mu) = h^*(k(\boldsymbol{\nu}))$. But also, $\boldsymbol{\theta} = h(\boldsymbol{\nu})$ for $\boldsymbol{\nu} \in M$, so that $h^*(k(\boldsymbol{\nu})) = h(\boldsymbol{\nu})$ for all $\boldsymbol{\nu} \in$ M. It follows that $h^*(\hat{\boldsymbol{\mu}}) = h^*(k(\hat{\boldsymbol{\nu}})) = h(\hat{\boldsymbol{\nu}})$ for all observation vectors $\mathbf{X}$.

Note that MME estimators may take values that are not in the parameter space and in some cases may not even make sense (see Problems 7.3.1 and 7.3.4).

***Example 7.3.1*** Let $\mathbf{X} = (X_1, \ldots, X_n)$ be a random sample from the exponential distribution, with density $f(x; \lambda) = \lambda e^{-\lambda x}$, for $x > 0$, $\lambda > 0$. The parameter space is $\Omega = \{\lambda \in R_1 | \lambda > 0\} = R_1^+$ is a subset of $R_1$, so we need to consider only one moment, $\mu = E(X_1) = 1/\lambda$, and $h(\mu) = 1/\mu$ is its inverse. Thus, $\mu = g(\lambda) = 1/\lambda$ and $\hat{\lambda} = 1/\bar{X}$ is the MME estimator for $\lambda$. □

***Example 7.3.2*** Let $(X_1, \ldots, X_n)$ be a random sample from the $N(\mu, \sigma^2)$ distribution. Take $\Omega = \{(\mu, \sigma^2) \mid \mu \in R_1, \sigma^2 \in R_1^+\} = R_1 \times R_1^+$. Then $g^*(\mu, \sigma^2) = (\mu, \sigma^2)$ and its inverse $h$ are both the identity function. It follows that the MME estimator of $(\mu, \sigma^2)$ is the vector of sample moments $(\bar{X}, \hat{\sigma})$. □

## Problems for Section 7.3

**7.3.1** Let $(X_1, \ldots, X_n)$ be a random sample from the Unif$(\theta_1, \theta_2)$ distribution, with $\theta_1 < \theta_2$.

**(a)** Find the MME estimator of the pair $(\theta_1, \theta_2)$. Find its value for the sample $X_1 = 1$, $X_2 = 0$, $X_3 = 0$. (Consider these observations as values obtained after rounding.)

**(b)** Does your estimate of $(\theta_1, \theta_2)$ make sense?

**7.3.2** Each day a saleswoman calls on customers until she makes a sale, then quits. Suppose that the probability of a sale is $\theta > 0$ for each customer and that the events of sales for different customers are independent. Let $X_1, \ldots, X_n$ be the numbers of failures the saleswoman has on $n$ consecutive days. What is the MME estimator of $\theta$?

**7.3.3** The negative binomial distribution is often used as a model for count data. For example, we may suppose that the number of tomato worms on a tomato plant in a square meter may have this distribution. Samples of 1-square meter plots planted in tomatoes may then be taken and the number of worms counted for each such plot. The probability function is $f(k; r, p) = \binom{k+r-1}{k} p^r q^k = \binom{-r}{k} p^r (-q)^k$, for $k = 0, 1, 2, \ldots, q = 1 - p, 0 < p < 1$, since

$$\binom{-r}{k} \equiv \frac{(-r)(-r-1)\cdots(-r-k+1)}{k!} = \frac{(r+k-1)\cdots(r+1)r}{k!}(-1)^k.$$

When $r$ is a positive integer, this is $\binom{r+k-1}{k}(-1)^k$. In this case $f(k; r, p)$ is the probability function of the number of failures that occur before the $r$th success in independent Bernoulli trials, each with probability $p$ of success. Students who have previously studied the negative binomial should note that we are discussing the distribution of the number $X$ of failures necessary to get $r$ successes, not the total number of trials, which is $X + r$. Both $X$ and $X + r$ are said to have a *negative binomial distribution*. We can extend the definition of the negative binomial distribution to the case $r$ is any positive number by defining $f(k; r, p) = \binom{-r}{k} p^r (-q)^k$ for $k = 0, 1, 2, \ldots$. As determined in Chapter Two, $X$ has mean $rq/p$ and variance $rq/p^2$. For the parameter space $\Omega = R_1^+ \times (0, 1)$, with $\theta = (r, p)$, and a random sample $(X_1, \ldots, X_n)$, find the MME estimator $\hat{\theta}$ of $\theta$. Evaluate $\hat{\theta}$ for the following sample:

| $k$ | 0 | 1 | 2 | 3 | 4 | 5 | 6 | 7 | 8 | 9 |
|---|---|---|---|---|---|---|---|---|---|---|
| Freq. | 80 | 103 | 83 | 49 | 38 | 27 | 10 | 4 | 4 | 2 |

**7.3.4** Let $\mathbf{X} = (X_1, \ldots, X_n)$ be a random sample from the shifted exponential distribution, with density $f(x; \lambda, \eta) = \lambda e^{-\lambda(x-\eta)}$ for $x > \eta$. Use the sample central moments to determine an estimator of the pair $(\lambda, \eta)$.

**7.3.5** Consider Problem 7.2.6. Find the MME estimator $\hat{\theta}$ of $\theta$ based on a sample $(X_1, X_2)$. Evaluate $r_{\hat{\theta}}(\theta)$ for random sampling with and without replacement.

**7.3.6** Let $\mathbf{X} = (X_1, \ldots, X_n)$ be a random sample from the binomial distribution with parameters $m$ and $p$, both unknown. Let $\Omega = I^+ \times [0, 1]$, where $I^+$ is the set of positive integers. Find the method of moments estimator of the pair $\boldsymbol{\theta} = (m, p)$. Be careful to distinguish between $n$ and $m$.

**7.3.7** Let $\mathbf{X} = (X_1, \ldots, X_n)$ be a random sample from the Poisson distribution with parameter $\lambda > 0$. Let $\theta = g(\lambda) = P(X_1 = 0) = e^{-\lambda}$. See Problem 7.2.7 for the case that $n = 1$.

**(a)** Find the MME $\hat{\theta}$ of $g(\lambda)$. Show that the bias $\hat{\theta} - \theta$ converges to zero as $n \to \infty$. *Hint*: The mgf for the Poisson distribution with parameter $\lambda$ is $e^{\lambda(e^t - 1)}$.

**(b)** Show that $\hat{\theta}_2 = (1/n) \sum_{i=1}^n \mathrm{I}\,[X_i = 0]$ is unbiased for $\theta$.

**(c)** Which estimator is better for large $n$? Show that $E(\hat{\theta}) \to \theta$ as $n \to \infty$ and that $\mathrm{Var}(\hat{\theta}_2)/\mathrm{Var}(\hat{\theta}) \to e^{\lambda} - 1$ as $n \to \infty$. *Hint*: Find expressions for the variances of $\hat{\theta}$ and $\hat{\theta}_2$, then use a Taylor approximation $e^{-y} \sim 1 - y + y^2/2$ for small $y$.

## 7.4 MAXIMUM LIKELIHOOD

The maximum likelihood method provides another way to generate estimators. Often, these MLEs have better properties than those provided by the method of moments. The first solid evidence of the use of the method of maximum likelihood seems to have been a paper of Daniel Bernoulli (Bernoulli, 1778). However, the modern theory was developed by A. W. Edgeworth (1908–1909), R. A. Fisher (1912, 1922, 1934),

Wald (1949), Le Cam (1953), and many others. See the discussion in Chapter 16 of the book by Stephen M. Stigler, *Statistics on the Table* (1999).

Suppose that a box contains 10 balls, that $\theta$ of them are white, and that the remaining are black. We want to estimate $\theta$. To do so we take a simple random sample (without replacement) of two balls from the box and note the number $X$ that are white. First consider the special case that $\Omega = \{2, 8\}$. That is, we know that $\theta = 2$ or 8. Suppose that our experiment yields $X = 0$. A reasonable estimate of $\theta$ would then be 2. A justification for this is that $X = 0$ is more likely for $\theta = 2$ than for $\theta = 8$. The method (or principle) of maximum likelihood generalizes this idea. Broadly speaking, following this method, if $X = x$ is observed, one determines the likelihood (the probability or the density at $x$) under all possible values of $\theta$ and chooses as an estimate the $\theta$ that maximizes this likelihood. It turns out that this method works very well in many situations.

**Definition 7.4.1**

*Discrete case:* Let $\mathbf{X} = (X_1, \ldots, X_n)$ be a discrete random vector with joint probability function $f_\theta$ with $\theta \in \Omega$. For each $\mathbf{x}$ in the range space $R(\theta)$ of $\mathbf{X}$ for some $\theta$ the function $L_\mathbf{x}$ defined on $\Omega$ by

$$L_\mathbf{x}(\theta) = f_\theta(\mathbf{x}) = P_\theta(\mathbf{X} = \mathbf{x})$$

is called the *likelihood function* for $\mathbf{X}$.

*Continuous case:* Let $\mathbf{X} = (X_1, \ldots, X_n)$ be a continuous random vector with joint density function $f_\theta(\mathbf{x})$, with $\theta \in \Omega$. For each $\mathbf{x}$ in the union over $\theta \in \Omega$ of the sets $R(\theta)$ for which $f_\theta(\mathbf{x}) > 0$, the function $L_\mathbf{x}(\theta)$ is called the *likelihood function* for $\mathbf{X}$. □

Comments: The functions $L_\mathbf{x}(\theta)$ and $f_\theta(\mathbf{x})$, considered as functions of both variables, have identical values. However, the notation $f_\theta(\mathbf{x})$ emphasizes the function of $\mathbf{x}$ for each $\theta$, with domains $R(\theta)$, while the notation $L_\mathbf{x}(\theta)$ emphasizes the function of $\theta$, with domain $\Omega$, for each $\mathbf{x}$. Therefore, when considering the likelihood function, think of $\mathbf{x}$ as fixed, defined as a function of $\theta$, $\theta \in \Omega$. We sometimes write the probability function as $f(\mathbf{x}; \theta)$ or $f(x|\theta)$ rather than $f_\theta(\mathbf{x})$, and the likelihood function as $L(\theta\,;\,\mathbf{x})$.

**Definition 7.4.2** The maximum likelihood estimate for observed $\mathbf{X} = \mathbf{x}$ (the MLE) $\hat\theta = \hat\theta(\mathbf{x})$ of $\theta \in \Omega$ is a value of $\boldsymbol\theta$ in $\Omega$ that maximizes the likelihood function. That is, $L_\mathbf{x}(\hat\theta) = L_\mathbf{x}(\hat\theta(\mathbf{x})) = \sup_{\theta\in\Omega} L_\mathbf{x}(\theta)$. The maximum likelihood estimator of $\boldsymbol\theta$ is $\hat\theta(\mathbf{X})$. If $g$ is a function on $\Omega$, the ML estimator of $g(\boldsymbol\theta)$ is $g(\hat\theta(\mathbf{X}))$. □

Comments: We have used "sup" rather than "max" because there may be some cases for which the maximum is not achieved. Notice also that $\hat\theta$ is called "a value" rather than "the value" because there may be more than one $\theta$ that maximizes the likelihood function.

Let us return to the box example with a simple random sample of two balls drawn from a box with θ white, 10 − θ black, with parameter space $\Omega = \{0, 1, 2, .., 10\}$. Since $\binom{10}{2} = 45$, we have and $f(x;\theta) = \binom{\theta}{x}\binom{10-\theta}{2-x}\Big/\binom{10}{2} = L_x(\theta)$ for $x = 0, 1, 2$ and $\theta \in \Omega$.

| | θ | | | | | | | | | | |
|---|---|---|---|---|---|---|---|---|---|---|---|
| x | 0 | 1 | 2 | 3 | 4 | 5 | 6 | 7 | 8 | 9 | 10 |
| 0 | 45 | 36 | 28 | 21 | 15 | 10 | 6 | 3 | 1 | 0 | 0 |
| 1 | 0 | 9 | 16 | 21 | 24 | 25 | 24 | 21 | 16 | 9 | 0 |
| 2 | 0 | 0 | 1 | 3 | 6 | 10 | 15 | 21 | 28 | 36 | 45 |

For example, $L_2(5) = f(2; 5) = 25/45$. Therefore, $\hat{\theta}(0) = 0$, $\hat{\theta}(1) = 5$, and $\hat{\theta}(2) = 10$. Thus, since Ω is finite, the value of the likelihood could be determined for every pair $(x, \theta)$, and the value of θ for which $L_x(\theta)$ was maximum could be determined by inspection. In most of the examples that we consider, **X** will consist of $n$ iid random variables. In such cases the likelihood function is the product of the individual densities or probability functions. The MLE can often be found using calculus. Since the function $\log(y)$ is an increasing function of $y$, the MLE can be found by maximizing $\log(L_{\mathbf{x}}(\theta))$ rather than $L_{\mathbf{x}}(\theta)$. This has the computational advantage of replacing a product by a sum.

Formally, $\log(L_{\mathbf{x}}(\theta))$ is called the *log-likelihood function*. When it is differentiable for all θ, the equation $\frac{d}{d\theta}\log(L_{\mathbf{x}}(\theta)) = 0$ is called the *log-likelihood equation*. The function $\frac{d}{d\theta}\log(L_{\mathbf{x}}(\theta)) \equiv \ell_{\mathbf{x}}(\theta)$ is called the *score function*. If $\hat{\theta}$ satisfies the likelihood equation, is unique, and $\frac{d^2}{d\theta^2}\log(L_{\mathbf{x}}(\theta))|_{\theta=\hat{\theta}} < 0$, then $\hat{\theta}$ is the MLE. Notice that in the case that the components of $\mathrm{X} = (X_1, \ldots, X_n)$ are independent, the score function is a sum, and when the $X_i$ are identically distributed, we should expect the score function to be approximately normally distributed for any fixed θ and true parameter value $\theta_0$.

Before we look at some examples, a few points should be made. A MLE may not always exist, and when it does, it may not be unique. The likelihood function may not be differentiable, and even then, the solution of the log-likelihood equation may not be a MLE.

***Example 7.4.1*** Let $\mathbf{X} = (X_1, \ldots, X_n)$, where the $X_i$ are iid Bernoulli(θ) and $\Omega = [0, 1]$ (the closed unit interval). Then $f_\theta(x_1, \ldots, x_n) = \theta^t\,(1-\theta)^{n-t}$, where $t = \sum_{i=1}^n x_i$ for $x_i = 0$ or $1$, $i = 1, \ldots, n$. If $t = 0$, $L_{\mathbf{x}}(\theta) = (1-\theta)^n$, which is maximum for $\theta = \hat{\theta} = 0$. If $t = n$, $L_{\mathbf{x}}(\theta) = \theta^n$, which is maximum for $\theta = \hat{\theta} = 1$ for $0 < \theta < 1$. Suppose then that $0 < t < n$. Then $0 < \theta < 1$ and $\log(L_{\mathbf{x}}(\theta)) = t\log(\theta) + (n-t)\log(1-\theta)$. Differentiating with respect to θ, we get the log-likelihood equation $\frac{d}{d\theta}\log(L_{\mathbf{x}}(\theta)) = \ell_{\mathbf{x}}(\theta) = t/\theta - [(n-t)/(1-\theta)] = 0$ for $0 < \theta < 1$. Differentiating again, we get $\frac{d^2}{d\theta^2}\log(L_{\mathbf{x}}(\theta)) = -t/\theta^2 - (n-t)/(1-\theta)^2$, which is negative for all $\theta$, $0 < \theta < 1$ and $t = 0, \ldots, n$. Solving for $\theta$, we get $\theta = \hat{\theta} = \hat{\theta}(\mathbf{x}) = t/n = (1/n)\sum_{i=1}^{n} X_i$, the sample proportion of successes. This same formula holds for $t = 0$ and $t = n$. □

***Example 7.4.2*** Let $\mathbf{X} = (X_1, \ldots, X_n)$ be a random sample from the N($\mu_0$, $\sigma^2$) distribution with $\mu_0$ known, $\sigma^2$ unknown. Take $\Omega = (0, \infty)$. Then $\log(L_\mathbf{x}(\sigma^2)) = -(n/2)\log(2\pi) - (n/2)\log(\sigma^2) - (1/2)\,\Sigma(x_i - \mu_0)^2$. Differentiating with respect to $\sigma^2$, we get the likelihood equation $\ell_\mathbf{x}(\sigma^2) = n/(2\sigma^2) - \Sigma(x_i - \mu_0)^2/(2\sigma^4) = 0$. Solving for $\sigma^2$, we get $\sigma^2 = \hat{\sigma}^2 = (1/n)\,\Sigma\,(x_i - \mu_0)^2$. Note that $\hat{\sigma}^2$ is a function of the vector $\mathbf{x}$ of sample values, so that to be more rigorous we should write $\hat{\sigma}^2(\mathbf{x})$. As stated in Section 7.2, this dependence on the sample is often not displayed. Since $E(X_i - \mu_0)^2 = \sigma^2$, $E(\hat{\sigma}^2\ \mathbf{X})) = E[(1/n)\Sigma(X_i - \mu_0)^2] = \sigma^2$, so $\hat{\sigma}^2\ (\mathbf{X})$ is an unbiased estimator of $\sigma^2$. □

***Example 7.4.3*** Let $\mathbf{X} = (X_1, \ldots, X_n)$ be a random sample from the Unif(0, $\theta$) distribution for $\theta > 0$. Then the likelihood function is

$$L_\mathbf{x}(\theta) = f(x_1, \ldots, x_n; \theta) = \begin{cases} 1/\theta^n & \text{for } 0 \le x_i \le \theta, \quad i = 1, \ldots, n \\ 0 & \text{otherwise} \end{cases}$$

Let $M = M(\mathbf{x}) = \max(x_1, \ldots, x_n)$. Thus, $L_\mathbf{x}(\theta) = 1/\theta^n$ for all $x_i \ge 0$, and $M(\mathbf{x}) \le \theta$.

As a function of $\theta$ for any fixed $\mathbf{x}$, $L_\mathbf{x}(\theta)$ jumps from 0 for $\theta < M(\mathbf{x})$ to $1/\mathrm{M}(\mathbf{x})^n$ at $\theta = M(\mathbf{x})$. For $\theta \ge M(\mathbf{x})$, $L_\mathbf{x}(\theta) = 1/\theta^n$ is a decreasing function of $\theta$. Therefore, the maximum likelihood *estimate* of $\theta$ for observed $\mathbf{X} = \mathbf{x}$ is $M(\mathbf{x})$. The maximum likelihood *estimator* is $M(\mathbf{X})$.

As shown in Example 7.2.1, $M(\mathbf{X})$ has bias $-\theta/(n + 1)$ as an estimator of $\theta$. In general, MLEs need not be unbiased, although under "smoothness conditions" the bias approaches zero as $n \to \infty$. □

***Example 7.4.4*** Let $\mathbf{X} = (X_1, \ldots, X_n)$ be a random sample from the Poisson distribution with parameter $\lambda$. Let $\Omega = (0, \infty)$. Then $\log(L_\mathbf{x}(\lambda)) = -n\lambda + \Sigma\, x_i \log(\lambda) - \log(x_1!x_2! \cdots x_n!)$. Differentiating with respect to $\lambda$ we get the score function $\ell_\mathbf{x}(\lambda) = -n + \Sigma\, x_i/\lambda$ and the MLE $\hat{\lambda} = \bar{X}$. □

***Example 7.4.5*** Let $\mathbf{X} = (X_1, \ldots, X_n)$ be a random sample from the Cauchy distribution with median $\theta$. Each $X_i$ has density $f(x;\theta) = (1/\pi)[1/1 + (x - \theta)^2]$. It is not possible to give an explicit mathematical expression for the MLE $\hat{\theta}(\mathbf{X})$ for $\theta$. Nevertheless, the MLE can be found by numerical means (see Figure 7.4.1). For $\theta = 3$ and sample size $n = 100$, the likelihood function is graphed for $0 \le \theta \le 6$ in (part (a).) In parts (b) to (d) results are presented for sample medians and for the MLE. □

**Interesting Features of Samples from a Cauchy Distribution**

1. The sample mean has the same distribution as a single observation $X_1$, so it is a poor estimator of $\theta$.

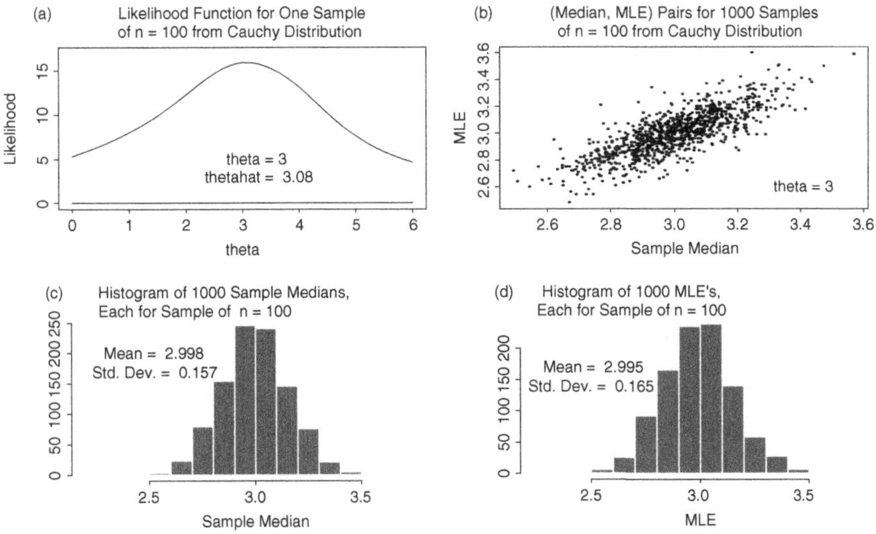


**FIGURE 7.4.1** Estimating θ for samples from the Cauchy distribution.

2. The likelihood function seems to be approximately quadratic in a neighborhood of the true parameter value $\theta = 3$, so the score function would seem to be approximately linear in θ around $\theta = 3$.
3. The joint distribution of Median(**X**) and the MLE $\hat{\theta}(\mathbf{X})$ seems to be bivariate normal. Figure 7.4.1(c) and (d) back this up with histograms indicating normality.
4. Both Median(**X**) and $\hat{\theta}(\mathbf{X})$ seem to have expectations that are close to the true parameter value $\theta = 3$.
5. The standard deviations for Median(**X**) and $\hat{\theta}(\mathbf{X})$ seem to be close.
6. Another possible estimator is the *trimmed mean*. The α-trimmed mean for a sample of $n$ is the mean of the reduced (trimmed) sample on $X_i$'s obtained by discarding the smallest and largest $k = [\alpha n]$ among the components of **X**. For α close to ½, the resulting statistic is close to Median(**X**). For α close to zero, the trimmed mean behaves too much like the sample mean. For 1000 samples of 100, the 0.1, 0.2, 0.3, 0.4 trimmed means and the median had standard deviations of 0.230, 0.170, 0.15, 0.150, 0.155, so it seems to make little difference whether a trimmed mean for $\alpha \geq 0.2$ or the MLE is used.

***Example 7.4.6*** Let $\mathbf{X} = (X_1, \ldots, X_n)$ be a random sample from the $N(\mu, \sigma^2)$ distribution. Let $\Omega = R_1 \times R_1^+$. The log-likelihood function is $\log(L_{\mathbf{x}}(\mu, \sigma^2) = -(n/2)$ $\log(2\pi) - (n/2)\log(\sigma^2) - [1/(2\sigma^2)]\ \Sigma\ (x_i - \mu)^2$. This is maximum for any $\sigma^2$ for $\mu = \bar{X}$. For this choice for μ, we can use the result of Example 7.4.2 to determine the MLE$(\bar{X}, \hat{\sigma}^2)$ of $(\mu, \sigma^2)$, where $\hat{\sigma}^2 = (1/n)\Sigma(X_i - \bar{X})^2$. In Chapter nine we show that $\bar{X}$ and $\hat{\sigma}^2$ are independent and that $(n/\sigma^2)\hat{\sigma}^2 = \sum_{i=1}^n (X_i - \bar{X})^2/\sigma^2$ has a chi-square distribution with $n - 1$ degrees of freedom. As shown in Problem 7.2.8 $E(\hat{\sigma}^2) = [(n-1)/n]\sigma^2$. □

***Example 7.4.7*** Let $\mathbf{X} = (X_1, \ldots, X_n)$ have the gamma distribution with power parameter $\alpha > 0$, scale parameter $\theta > 0$, density $f(\mathbf{x}; \theta) = [1/(\Gamma(\alpha)\theta^\alpha)]x^{\alpha-1}e^{-x/\theta}$ for $x > 0$. The log-likelihood function is therefore $\log(L_{\mathbf{x}}(\alpha, \theta)) = -n\log(\Gamma(\alpha)) - \alpha n \log(\theta) + (\alpha - 1)\Sigma \log(x_i) - (1/\theta)\Sigma x_i$.

Taking partial derivatives with respect to $\alpha$ and $\theta$, we get the simultaneous likelihood equations (1) $-n\Gamma'(\alpha)/\Gamma(\alpha) - n\log(\theta) + \Sigma\log(x_i) = 0$, and (2) $-\alpha n/\theta + (1/\theta^2)\Sigma x_i = 0$. Solving the second, we get $\theta = \Sigma x_i/n\alpha = \bar{x}/\alpha$. Replacing $\theta$ in the first equation by $\bar{x}/\alpha$, we get a single equation in $\alpha$. Unfortunately, there is no simple expression for $\Gamma'(\alpha)$, and therefore no simple expression for the MLE $(\hat{\alpha}, \hat{\theta})$ of $(\alpha, \theta)$. Numerical methods can be used. This was done for the case of $\alpha = 3$, $\theta = 5$. It was repeated 1000 times. Each time, the MLE pair and the MME pair were determined (see Figure 7.4.2). □

As in Example 7.4.5, each pair of estimators $(\hat{\alpha}, \hat{\theta})$, for both the ML and MME estimators, seems to have a bivariate normal distribution. In fact, there is theoretical support for that for large $n$. The last two histograms for 1000 values of $\hat{\alpha}$ indicate that the MLE has smaller variance than does the MME estimator. Both estimators have a slight positive bias, since both estimators have expectations that exceed $\alpha$ when $\alpha = 3.0$.

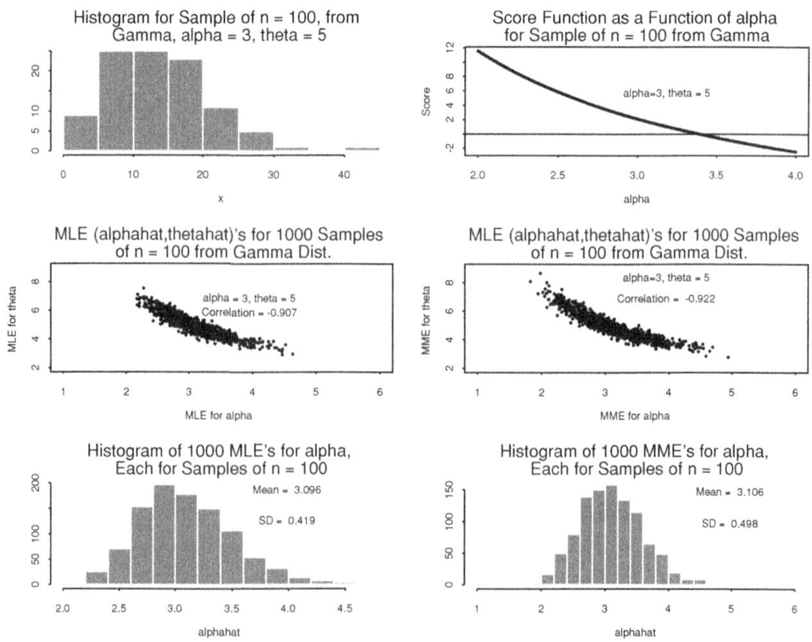


**FIGURE 7.4.2** Estimation of $(\alpha, \theta)$.

## Problems for Section 7.4

**7.4.1** A box contains eight eggs, of which an unknown number $R$ are rotten. You take a simple random sample of three eggs. Let $X$ be the number of rotten eggs in the sample.

**(a)** Find the MLE for $R$. (*Warning*: Don't use calculus.)

**(b)** What is the MLE for $E(X) = 3(R/8)$?

**7.4.2** Let $X_1, \ldots, X_n$ be a random sample from the geometric distribution with parameter $\theta$, $0 < \theta < 1$. Show that the MLE and the MME of $\theta$ are the same.

**7.4.3** Let $X_1$ and $X_2$ be independent, with $X_1 \sim$ Binomial$(n_1, p)$, $X_2 \sim$ Binomial$(n_2, p)$ for $n_1$ and $n_2$ known, but $p$ unknown, $0 \le p \le 1$. Find the MLE for $p$.

**7.4.4** Let $(X_1, \ldots, X_n)$ be a random sample from the Laplace distribution (also called the *double exponential distribution*), with density $f(x;\, \theta) = (1/2)e^{-|x-\theta|}$ for all $x$, $\Omega = R_1$. Show that $\hat{\theta} = \text{median}(X_1, \ldots, X_n)$ is the MLE for $\theta$.

**7.4.5** Let $X_1, \ldots, X_n$ be a random sample from the uniform distribution on $[\theta_1, \theta_2]$ for $-\infty < \theta_1 < \theta_2 < \infty$. Let $\hat{\theta}_{1n} = \min(X_1, \ldots, X_n)$ and $\hat{\theta}_{2n} = \max(X_1, \ldots, X_n)$. Show that $(\hat{\theta}_{1n}, \hat{\theta}_{2n})$ is the MLE for $(\theta_1, \theta_2)$.

**7.4.6** Consider the exponential distribution of Problem 7.3.4 with parameters $\eta$ and $\lambda$. Find the MLE for the pair $(\eta, \lambda)$ for a random sample $X_1, \ldots, X_n$.

**7.4.7** Let $\mathbf{X} = (X_1, \ldots, X_n)$ be a random sample from the Pareto distribution with parameter $\alpha$. That is, each $X_i$ has cdf $F(x;\alpha) = 1 - x^{\alpha}$ for $x \ge 1$, $\alpha > 0$.

**(a)** Find the MME and MLE of $\alpha$.

**(b)** Do they exist for all $\alpha > 0$?

**7.4.8** A genetic model states that under random mating the three genotypes *aa*, *Aa*, and *AA* should have probabilities $p_0 = (1-\theta)^2$, $p_1 = 2\theta(1-\theta)$, and $p_2 = \theta^2$ for $0 \le \theta \le 1$. The parameter $\theta$ is the proportion of allele $A$ in the population. A random sample of $n$ is taken from the population. Let $X_0, X_1, X_2$ be the frequencies of these three genotypes in the sample.

**(a)** Find the MLE $\hat{\theta}$ for $\theta$. Is it unbiased for $\theta$? Can you define $\hat{\theta}$ in simple word form?

**(b)** Give the MLE $\hat{\mathbf{p}}$ for the vector $\mathbf{p} = (p_0, p_1, p_2)$. Is $\hat{\mathbf{p}}$ unbiased for $\mathbf{p}$?

**(c)** Show that $\hat{\mathbf{p}}^* \equiv (X_0/n, X_1/n, X_2/n)$ is an unbiased estimator of $\mathbf{p}$.

**7.4.9** Let $Y_i = \beta x_i + \varepsilon_i$ for $i = 1, \ldots, n$, where the $x_i$ are known constants and the $\varepsilon_i$ are iid $N(0, \sigma^2)$.

**(a)** Find the MLE $(\hat{\beta}, \hat{\sigma}^2)$ for the pair $(\beta, \sigma^2)$. $\hat{\beta}$ is also called the *least squares estimator of* $\beta$. Why? Notice that $\hat{\beta}$ is a linear function of the $Y_i$.

**(b)** Determine $(\hat{\beta}, \hat{\sigma}^2)$ for $n = 3$ and the $(x_i, Y_i)$ pairs (1, 2), (2, 5), (3, 10).

**(c)** What is $(\hat{\beta}, \hat{\sigma}^2)$ for the case that each $x_i$ is 1?

**(d)** Show that $\hat{\beta}$ is unbiased for $\beta$.

**(e)** Determine $\text{Var}(\hat{\beta})$.

**(f)** Consider any other linear estimator $T = \Sigma a_i Y_i$. What condition must the $a_i$ satisfy in order that $T$ be unbiased for $\beta$?

**(g)** Suppose that $T$ as in (f) is unbiased for $\beta$. Show that $\text{Var}(T) > \text{Var}(\hat{\beta})$ unless $T$ is identically equal to $\hat{\beta}$ [the same for all $\mathbf{Y} = (Y_1, \ldots, Y_n)$].

**7.4.10** Suppose that $\theta \in \Omega = \{1, 2, 3\}$. Let $p(k; \theta)$ be as follows:

| | $k$ | | |
|---|---|---|---|
| | 0 | 1 | 2 |
| $p(k; 1)$ | 0.1 | 0.3 | 0.6 |
| $p(k; 2)$ | 0.1 | 0.8 | 0.1 |
| $p(k; 3)$ | 0.6 | 0.3 | 0.1 |

**(a)** Let $X_1, X_2$ be a random sample from one of these distributions. Find the MLE $\hat{\theta}$ for $\theta$. For each of $\theta = 1, 2, 3$, find $C(\theta) \equiv P_\theta(\hat{\theta} = \theta)$.

**(b)** Repeat part (a) for the case of a random sample $X_1, X_2, X_3$.

**(c)** Suppose that $X_1, \ldots, X_{10}$ is a random sample and that four of these are 0's, five are 1's, and one is 2. What is the ML estimate for $\theta$?

**(d)** What is the MLE for the case of a random sample of $n$? Express it in terms of the frequencies $f_0, f_1, f_2$.

## 7.5 CONSISTENCY

Now that we have two methods for the determination of estimators, let us turn to a property of estimators. It is desirable that as the sample size becomes larger and larger, an estimator of a parameter $\theta$ or $g(\theta)$ be more and more likely to be close to the parameter. This can be made precise by making use of *convergence in probability,* defined in Chapter Six. Recall that a sequence of random variables $\{Y_n\}$ is said to converge in probability to a constant $c$ if for every $\varepsilon > 0$, $\lim_{n\to\infty} P(|Y_n - c| < \varepsilon) = 1$. If we replace $Y_n$ by an estimator and the constant $c$ by the parameter it estimates, we get the definition of *consistency*.

**Definition 7.5.1** Let $\{T_n\}$ be a sequence of estimators of a parametric function $g(\theta)$ for $\theta \in \Omega$. $\{T_n\}$ is said to be a *consistent sequence of estimators of* $g(\theta)$ if for every

$\theta \in \Omega$ and every $\varepsilon > 0$

$$\lim_{n\to\infty} P_\theta(|T_n - g(\theta)| < \varepsilon) = 1. \tag{7.5.1}$$

□

COMMENTS: Note that consistency is a property of a sequence of estimators, not simply of an estimator for one fixed value of $n$. In that sense the fact that in practice $n$ is fixed should make us wonder whether the definition has any real relevance. However, it is often possible to establish that certain estimators, which may be rather difficult to deal with analytically, are consistent when considered as a sequence and have limiting distributions (after standardizing) which we know, even though we do not know their distributions for any $n > 1$. The fact that a sequence of estimators is consistent does not establish just how large $n$ must be before the probability above is as close to 1 as we wish without further (usually more difficult) analytic work or computer simulations. Consistency at least provides the hope that if we take a large enough sample, our estimator will have satisfactory properties.

We may, from time to time, slip, as is common, and say that "$T_n$ is consistent for $\theta$." Always remember that consistency is a property of a sequence of estimators. Note also that the sequence of probabilities in (7.5.1) must converge to 1 for *every* $\theta$. Thus, a sequence $\{T_n\}$ might converge in probability to $g(\theta)$ for some $\theta \in \Omega$, but fail for others. Such a sequence would not be consistent relative to the entire set $\Omega$.

Consistency may often be shown by using expected mean squares and the Markov inequality: For any random variable $Y$ for which $P(Y \geq 0) = 1$, $E(Y)$ exists, and any $\eta > 0$, $P(Y \geq \eta) \leq E(Y)/\eta$. Replacing $Y$ by the random variable $Y_n = |T_n - g(\theta)|^2$ and $\eta$ by $\varepsilon^2$, we get $P\theta(|T_n - g(\theta)| \geq \varepsilon) = P_\theta(|T_n - g(\theta)|^2 \geq \varepsilon^2) \leq E_\theta(Y_n)/\varepsilon^2 = r_{T_n}(\theta)/\varepsilon^2$. It follows that $\{T_n\}$ is a consistent sequence of estimators of $g(\theta)$ if the corresponding sequence of expected mean squares converges to zero. Since, for any estimator $T$, $r_T(\theta) = \text{Var}_\theta(T) + [E_\theta(T) - g(\theta)]^2$, it follows that we can demonstrate consistency by showing that the bias and the variance both converge to zero as $n \to \infty$, although this is not the only way to demonstrate consistency.

***Example 7.5.1*** Let $X_1, \ldots, X_n$ be a random sample from the Unif(0, $\theta$) distribution with $\theta > 0$, and let $T_n = \max(X_1, \ldots, X_n)$. As shown at the beginning of Section 7.2, $T_n$ has bias $-\theta/(n+1)$. In addition, $\text{Var}_\theta(T_n) = [\theta^2 n]/[(n+1)^2(n+2)]$. It follows that the bias and the MSE (mean squared error) converge to zero as $n \to \infty$, so that $\{T_n\}$ is a consistent sequence of estimators of $\theta$. For this example we could have made a more direct approach. Since $T_n$ has cdf $F_{T_n}(x) = (x/\theta)^n$ for $0 \leq x \leq \theta$, $P_\theta(|T_n - \theta| \geq \varepsilon) = P_\theta(T_n \leq \theta - \varepsilon) = [(\theta - \varepsilon)/\theta]^n \to 0$ as $n \to \infty$. □

***Example 7.5.2*** Let $X_1, \ldots, X_n$ be a random sample from the Cauchy distribution with density $f(x;\theta) = (1/\pi)/[1 + (x - \theta)^2]$ for any real $x$ and $\theta$. As stated in Example 7.4.5, the mean does not exist, and $\bar{X}_n$ has the same distribution as any single observation $X_i$. Thus, $\{\bar{X}_n\}$ is not consistent for $\theta$. However, $M_n \equiv \text{Median}(X_1, \ldots, X_n)$ *is* consistent for $\theta$. This is indicated by Figure 7.5.1, in which histograms of

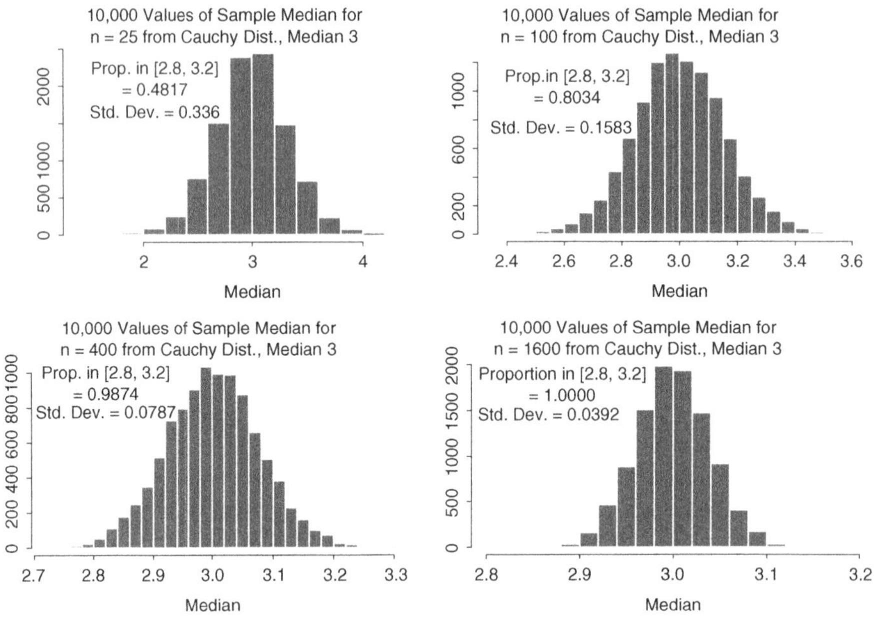


**FIGURE 7.5.1** Estimating θ for samples from the Cauchy distribution.

10,000 samples are taken from this distribution with $\theta = 3$ for the cases $n = 25$, $n = 100$, $n = 400$, and $n = 1600$. Notice that as $n$ increases, the proportions within the fixed interval [2.8, 3.2] approach 1. Notice also that as the sample sizes are multiplied by 4, the standard deviations are halved. Theory not to be presented here states that an approximation for the variance of a sample median θ is given by $1/[4nf(\theta)^2]$, where $f$ is the density sampled. In our case this is $\pi^2/[4n]$, which are 0.0987, 0.0247, 0.0062, 0.0015 for $n = 25$, 100, 400, 1600 with corresponding standard deviations 0.314, 0.157, 0.078, 0.039, close to those given in the histograms in Figure 7.5.1.

If $\{T_n\}$ is a consistent sequence of estimators of a parameter $\theta \in \Omega$ and $g(\theta)$ is a continuous function on $\Omega$, it follows that $\{g(T_n)\}$ is a consistent sequence of estimators of $g(\theta)$. To see this, note that continuity of $g$ at $\theta$ implies that for any $\varepsilon > 0$ there exists a constant $\delta = \delta(\theta, \varepsilon)$ such that $|g(\eta) - g(\theta)| < \varepsilon$ for $|\eta - \theta| < \delta$. Therefore, $P_\theta(|g(T_n) - g(\theta)| < \varepsilon) \geq P_\theta(|T_n - \theta| < \delta)$. Since the right side converges to 1 as $n \to \infty$, it follows that the left side does also.

If $\boldsymbol{\theta} = (\theta_1, \ldots, \theta_k)$ is a vector of parameters in the parameter space $\Omega$ and $\{T_{nj}\}$ is a sequence of consistent estimators of $\theta_j$ for $j = 1, \ldots, k$, we say that the sequence of vectors $\{\hat{\mathbf{T}}_n = (\hat{T}_{n1}, \ldots, \hat{T}_{nk})\}$ of estimators is consistent for θ. If $g$ is a continuous function on $\Omega$, it follows by an argument similar to that given above that $\{g(\mathbf{T}_n)\}$ is consistent for $g(\boldsymbol{\theta})$. In particular, since moment estimators are consistent for the corresponding population moments by the weak law of large numbers, when a function $h$ on the moment space to the parameter space is continuous, the sequences of estimators produced by the MME are consistent for the parameters they estimate. □

***Example 7.5.3*** Let $\mathbf{X}_n = (X_{n1}, \ldots, X_{nk})$ have the multinomial distribution with parameters $n$ and $\mathbf{p} = (p_1, \ldots, p_k)$. For example, if two coins are tossed $n = 10$ times and $X_{nj}$ is the number of occurrences of $j$ heads, then $X_n = (X_{n0}, X_{n1}, X_{n2})$ has the multinomial distribution with parameters $n = 10$ and $\mathbf{p} = (1/4, 1/2, 1/4)$. The marginal distribution of $X_{nj}$ is Binomial$(n, p_j)$. $\mathbf{X}_n$ has the probability function

$$f(x_1, \ldots, x; p) = \binom{n}{n_1 \cdots n_k} p_1^{x_1} p_2^{x_2} \cdots p_k^{x_k} = \frac{n!}{x_1! \cdots x_k!} \prod_{j=1}^{n} p_j^{x_j}.$$

Let $\hat{p}_{nj} = X_{nj}/n$. Then each $\{\hat{p}_{nj}\}$ is consistent for $p_j$. It follows that $\{\hat{\mathbf{p}}_n \equiv (\hat{p}_{n1}, \ldots, \hat{p}_{nk})\}$ is a consistent sequence of estimators of $\mathbf{p}$. □

## Problems for Section 7.5

**7.5.1** Let $X_1, \ldots, X_n$ be a random sample from the shifted exponential distribution having density $f(x; \alpha, \lambda) = e^{-(x-\alpha)/\lambda}$ for $x > \alpha$.

**(a)** For $\lambda$ known, show that the MME $\hat{\alpha}_n$ determines a consistent sequence of estimators of $\alpha$.

**(b)** For both $\alpha$ and $\lambda$ unknown, let $\{(\alpha_n, \lambda_n)\}$ be the sequence of MMEs for $(\alpha, \lambda)$. Show that this sequence is a consistent sequence of estimators of $(\alpha, \lambda)$.

**7.5.2** Let $X_1, \ldots, X_n$ be a random sample from the uniform distribution on $[\theta_1, \theta_2]$ for $-\infty < \theta_1 < \theta_2 < \infty$ . Let $\hat{\theta}_{1n} = \min(X_1, \ldots, X_n)$ and $(\hat{\theta}_{2n} = \max(X_1, \ldots, X_n)$. Show that the MLE $\hat{\theta}_n = (\hat{\theta}_{1n}, \hat{\theta}_{2n})$ for $\boldsymbol{\theta} = (\theta_1, \theta_2)$ determines a consistent sequence of estimators of $\theta$. (See "Maxima and Minima," Section 3.5.)

**7.5.3** Let $X_1, \ldots, X_n$ be a random sample from a distribution with mean $\mu$ and variance $\sigma^2$. Let $\hat{\sigma}_n^2 \equiv (1/n)\sum_{i=1}^n (X_i - \bar{X})^2$ and $S_n^2 = [1/(n-1)]\sum_{i=1}^n (X_i - \bar{X})^2$.

**(a)** Show that $\{\hat{\sigma}_n^2\}$ is consistent for $\sigma^2$.

**(b)** Show that $\{S_n^2\}$ is consistent for $\sigma^2$.

**(c)** Show that $\hat{T}_n = \bar{X} + 3S_n$ defines a consistent sequence of estimators of $\mu + 3\sigma$.

**7.5.4** Let $X_1, \ldots, X_n$ be a random sample from the uniform distribution on $(0, \theta)$ for $\theta > 0$. Let $T_n = \max(X_1, \ldots, X_n)$. Let $g(\theta) = 0$ for $0 < \theta \leq 1$ and $g(\theta) = 1$ for $\theta > 1$.

**(a)** Is $\{g(T_n)\}$ consistent for $g(\theta)$?

**(b)** Would the answer be different if $g(1) = 1$? More generally, if $g$ is the indicator of a closed set $[\theta_0, \infty]$, is $\{g(T_n)\}$ consistent for $\theta$?

**(c)** Is $\{g(T_n)\}$ consistent for $g(\theta)$, an indicator of an open set $(\theta_0, \infty)$?

**7.5.5** Is the sequence of MMEs of the parameter pair $(\alpha, \lambda)$ of the gamma distribution consistent?

**7.5.6** Let $f(x; \alpha) = \alpha x^{-\alpha-1}$ for $x \geq 1$, $\alpha > 2$ (the Pareto density). Show that both the MME and the MLE for $\alpha$ are consistent for $\alpha$. Would this be true if $\alpha$ were allowed to be 1?

**7.5.7** Let $X_1, \ldots, X_n$ be a random sample from a distribution with cdf $F$. Let the parameter space $\Omega$ be the collection of all cdf's. For each real number $x$, let $F_n(x) = (1/n)\sum_{i=1}^{n} I[X_i \leq x]$, the sample cdf. Show that for any $x_0$, $\{F_n(x_0)\}$ is consistent for the parameter $F(x_0)$. Although we will not prove it here, it is also true (by the Glivenko–Cantelli theorem) that if the function $F$ is the parameter, and the distance between $F_n$ and $F$ is defined by $D(F_n, F) = \sup_{-\infty<x<\infty}|F_n(x) - F(x)|$, then $P(D(F_n, F) > \varepsilon)$ converges to zero for each $\varepsilon > 0$, so that $\{F_n\}$ is a consistent sequence of estimators of $F$. Note that $\{F_n\}$ is a sequence of functions, while $\{F_n(x_0)\}$ is a sequence of numbers for each $x_0$.

**7.5.8** $X_1, \ldots, X_n$ be a random sample from the cdf $F$. Let $0 < \alpha < 1$ and let $F(x_\alpha) = \alpha$. $x_\alpha$ is the $\alpha$th-quantile of $F$. Suppose that $F$ has density $f$, that $f(x_\alpha) > 0$, and that $f$ is continuous at $x_\alpha$. Let $k_n = [n\alpha]$, the largest integer less than or equal to $n\alpha$. Let $X_{nk_n}$ be the $k_n$th order statistic. That is, one $X_j$ is exactly equal to $X_{nk_n}$ and $k_n - 1$ of the $X_j$'s are less than $X_{nk_n}$. Prove that $\{X_{nk_n}\}$ is a consistent sequence of estimators of $x_\alpha$. *Hint*: For $\varepsilon > 0$, show that $P(X_{nk_n} < x_\alpha - \varepsilon) \to 0$ and $P(X_{nk_n} > x_\alpha + \varepsilon) \to 0$. These probabilities can be expressed in terms of binomial random variables. (See Problem 6.3.9.)

## 7.6 THE δ-METHOD

Suppose that a surveyor must determine the height of a tree. He measures a distance $\rho$ from the base of the tree to a point and then measures the angle $\theta$ from 1.7 meters above ground level to the top of the tree (see Figure 7.6.1). The height of the tree is then $h = 1.7 + \rho\tan(\theta)$. However, measurements of this sort are never perfect, and at least some consideration should be given to the errors in estimation which may be caused by errors in measurements of $\rho$ and of $\theta$, or of both.

Suppose that $\rho$ is measured relatively precisely but the measurement $X$ of $\theta$ is a random variable with mean $\mu$ and variance $\sigma^2$. What can be said about the mean and

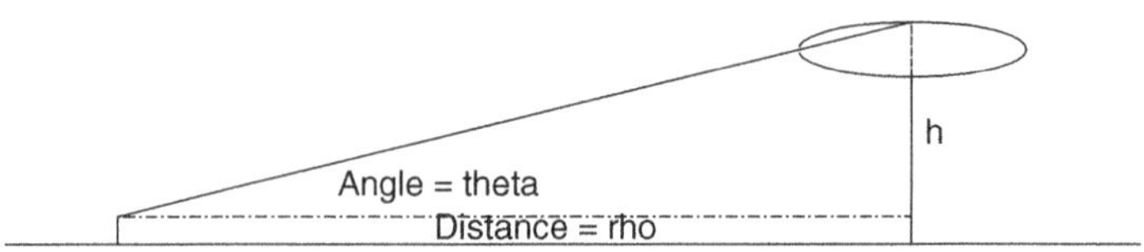


**FIGURE 7.6.1** Measuring the height of a tree.

variance of the estimator $\hat{h}(X) = 1.7 + \rho \tan X$ of $h$? Since the tangent function is not linear in its argument, we cannot simply replace $X$ by $\mu$ to determine $E(\hat{h})$. The δ-method provides a way to compute an approximate mean and variance of a nonlinear function of a random variable or random variables.

Consider the Taylor approximation of a function $h(x)$ about a point $x_0$. Suppose that $h$ has a continuous first derivative in a neighborhood $A$ of $x_0$. Let $h_1(x) = h(x_0) + h'(x_0)(x - x_0)$. Then for any $x \in A$ there exists a point $\eta$, $x_0 < \eta < x$ or $x < \eta < x_0$ such that $h(x) = h_1(x) + h''(\eta)(x - x_0)^2$. If $x$ is close to $x_0$, then $h(x) - h_1(x)$ is small relative to $h_1(x)$. If $h$ has a continuous third derivative in a neighborhood $A$ of $x_0$, a better Taylor approximation of $h(x)$ is given by $h_2(x) = h_1(x) + h''(x_0)(x - x_0)^2/2 = h(x_0) + h'(x_0)(x - x_0) + h''(x_0)(x - x_0)^2/2$. Then there exists a point $\eta^*$, $x_0 < \eta^* < x$ or $x < \eta^* < x_0$, such that $h(x) = h_2(x) + h'''(\eta^*(x - x_0)^3/3!$. We call $h_1$ and $h_2$ the *linear* and *quadratic approximations* of $h$ at $x_0$.

Replace the number $x$ by the random variable $X$, having mean $\mu$ and variance $\sigma^2$. It should seem reasonable to approximate the expected value of $h(X)$ by the expected value of $h_1(X)$ or, if we can determine it, by the expected value of $h_2(X)$. We have $E(h_1(X)) = h(x_0) + h'(x_0)(\mu - x_0)$ and $E(h_2(X)) = h(x_0) + h'(x_0)(\mu - x_0) + [h''(x_0)/2]E(X - x_0)^2$. Of course, $E(X - x_0)^2 = \sigma^2 + (\mu - x_0)^2$. The variance of $h(X)$ may be approximated by $\text{Var}(h_1(X)) = [h'(x_0)]^2\sigma^2$. Of course, these will be reasonable approximations only if $X$ will be close enough to $x_0$ to make $h_1$ or $h_2$ a good approximation of $h$ with high probability. This is true if both $|\mu - x_0|$ and $\sigma^2$ are small relative to changes in $h$ near $x_0$. It is often difficult to compute $\text{Var}(h_2(X))$, so we will not try to derive an explicit formula. Usually, we choose $x_0 = \mu$ so that $E(h_1(X)) = h(\mu)$ and $E(h_2(X)) = h(\mu) + h''(\mu)\sigma^2$.

Return now to the surveyor example. We consider the estimator $\hat{h}(X) = 1.7 + \rho \tan X$ of $h(\theta) = 1.7 + \rho \tan\theta$. We use the Taylor approximations about $\theta$. Since $h'(\theta) = \rho \sec^2\theta$, and $h''(\theta) = -2\rho \tan\theta \sec^2\theta$, it follows that when $|\theta - \mu|$ and $\sigma^2$ are reasonably small, $E(\hat{h}(X))$ is approximately $E(h_2(X)) = h(\theta) + \rho \sec^2\theta(\mu - \theta) + (-\rho \tan\theta \sec^2\theta)[\sigma^2 + (\mu - \theta)^2]$. If $\mu = \theta$, this simplifies to $h(\theta) - (\rho \tan\theta \sec^2\theta)\sigma^2 = h(\theta)(1 - \sigma^2 \sec^2\theta)$.

Thus, in approximation, the bias of $\hat{h}(X)$ is $-\rho \tan\theta (\sec^2\theta)\sigma^2$. To approximate $\text{Var}(h(X))$ we use $\text{Var}(h_1(X)) = (\rho \sec^2\theta)^2\sigma^2$.

For simulations we chose $\theta = 15$ degrees $= (15\pi/180 = 26.796)$ radians, $\rho = 100$, so that $h = 1.7 + 26.795$. For $\sigma = 2$ degrees $= (2\pi/180 = 0.0349)$ radians, the approximate bias is $-\rho \tan\theta(\sec^2\theta)\sigma^2 = -0.03497$. The approximate variance is $\text{Var}(h_1(X)) = 13.06$. With $X \sim N(h, \sigma^2)$ 10,000 observations were made on $X$. For each $X$, $\hat{h}(X)$ was determined. The mean of these 10,000 values was $1.7 + 26.760$, 0.036 less than $h$, very close to the bias of $h_2(X)$. The sample variance of these 10,000 values of $\hat{h}(X)$ was 13.82, fairly close to $\text{Var}(h_1(X)) = 13.06$. The distribution of $\hat{h}(X)$ is close to normal because $h$ is nearly linear in a neighborhood of $\theta$.

Suppose now that *both* the measurements $R$ on $\rho$ and $X$ on $\theta$ are random variables, with means $\mu_R$ and $\mu_X$, with variances $\sigma_R^2$ and $\sigma_X^2$. We can use the Taylor approximation for functions of two variables to approximate $h(X, R) = 1.7 + R \tan X$: $h_{11}(x, r) = h(\theta, \rho) + h'_x(\theta, \rho)(x - \theta) + h'_r(\theta, \rho)(r - \rho)$.

The two 1's indicate that the approximation uses only linear terms in $x$ and $r$. $h'_x(\theta, r)$ and $h_r(\theta, \rho)$ are partial derivatives of $h$ with respect to the variables indicated in the subscripts. It is convenient to write $h_{11}$ in vector form: Let $\mathbf{V} = \begin{pmatrix} X \\ R \end{pmatrix}$, $\mu = \begin{pmatrix} \mu_X \\ \mu_R \end{pmatrix}$, $\eta = \begin{pmatrix} \theta \\ \rho \end{pmatrix}$, and let $\delta$ be the two-component column vector of partial derivatives, evaluated at $\eta$. Then $h_{11}(\mathbf{V}) = h(\eta) + \delta^T(\mathbf{V} - \eta)$. Then $E(h_{11}(\mathbf{V})) = h(\eta) + \delta^{\mathrm{T}}(\mu - \eta)$ and $\mathrm{Var}(h_{11}(V)) = \delta^{\mathrm{T}}\Sigma\delta$, where $\Sigma$ is the covariance matrix of the random vector V. That is, $\Sigma$ is the $2 \times 2$ symmetric matrix with terms $c_{11} = \mathrm{Var}(X)$, $c_{12} = c_{21} = \mathrm{Cov}(X, R)$, and $c_{22} = \mathrm{Var}(R)$.

***Example 7.6.1*** Suppose that $(X, Y)$ are uniformly distributed on the square $[3, 4] \times [4, 5]$. Let $h(x, y) = xy$. Then the Taylor approximation of $h$ about $\eta = \begin{pmatrix} 3.4 \\ 4.6 \end{pmatrix}$ is $h_{11}(\mathbf{V}) = h(\eta) + \delta^{\mathrm{T}}(\mu - \eta)$, where $\mathbf{V} = \begin{pmatrix} X \\ Y \end{pmatrix}$, $\delta = \begin{pmatrix} 4.6 \\ 3.4 \end{pmatrix}$, $\mu = \begin{pmatrix} 3.5 \\ 4.5 \end{pmatrix}$. Thus, $E(h_{11}(\mathrm{V})) = (3.4)(4.6) + \delta^{\mathrm{T}}(\mu - \eta) = 15.64 + 0.1 = 15.74$. The covariance matrix for $\mathbf{V}$ is $(1/12)I_2$. It follows that $\mathrm{Var}(h_{11}(\mathbf{V})) = (4.5^2 + 3.5^2)/12 = 2.708$. 100,000 simulations of $Z = h(X, Y) = XY$, produced mean 15.753, variance 2.703. Mathematical computations produced 15.75 and 2.715. □

The following theorem is useful in that it provides a way to make probability statements about the behavior of estimators when it is not possible to determine the distribution by analytic means.

**Theorem 7.6.1** Let $\{T_n\}$ be a sequence of random variables, and let $\theta$ and $\sigma > 0$ be constants. Suppose that $Z_n \equiv n^{1/2}(T_n - \theta)/\sigma$ converges in distribution to N(0, 1). Let $h(t)$ have a continuous first derivative at $t = \theta$. Then $W_n \equiv n^{1/2}[h(T_n) - h(\theta)]/(\sigma|h'(\theta)|)$ also converges in distribution to N(0, 1).

The essential idea of the theorem is that $h(T_n)$ and $h_1(T_n)$, its linear Taylor approximation, are, after standardization, distributed the same asymptotically. Since $h_1(T_n)$ is linear in $T_n$, the theorem follows. A proof is omitted.

Since $Z_n$ converges in distribution and $|T_n - \theta| = \sigma|Z_n|/n^{1/2}$, it follows that $\{T_n\}$ converges in probability to $\theta$. If this is true for every $\theta$ in a parameter space $\Omega$, then $\{T_n\}$ is consistent for $\theta$. Stated another way, if $T_n$ is approximately distributed normally with mean $\theta$, variance $\sigma^2/n$, then $h(T_n)$ is approximately distributed as $\mathrm{N}(h(\theta), \sigma^2 h'(\theta)^2/n)$. The term $h'(\theta)^2\sigma^2$ is the variance of the asymptotic distribution of $W_n$. It is not necessarily the limit of $\mathrm{Var}(W_n)$.

Consider a random sample $X_1, \ldots, X_n$ from a distribution with mean $\mu$ and variance $\sigma^2$. We know by the central limit theorem that $Z_n \equiv (\bar{X}_n - \mu)/(\sigma/\sqrt{n})$ converges in distribution to the standard normal distribution. From this we know, for example, that

$$P(-1.96 \leq Z_n \leq 1.96) = P(|\bar{X}_n - \mu| \leq 1.96\sigma/\sqrt{n}) = P(\mu \in [\bar{X}_n \pm 1.96\,\sigma/\sqrt{n}]) \tag{7.6.1}$$

will be close to 0.95 for large $n$. In fact, the probability is 0.950004 to six decimal places for *every* $n$ if the distribution sampled is normal. For this case the random interval $I = [\bar{X}_n \pm 1.96\,\sigma/\sqrt{n}]$ is called a 95% *confidence interval* on $\mu$. A problem with this is that $\sigma$ is usually unknown. In this case it is tempting to replace $\sigma$ by an estimator $S_n$ in $Z_n$, hoping that the probabilities in (7.6.1) remain approximately the same. That is, if $Z_n$ is replaced by $\hat{Z}_n \equiv (\bar{X}_n - \mu)/(S_n/\sqrt{n})$, and therefore the interval $I$ by the interval $\hat{I} = [\bar{X}_n \pm 1.96S_n/\sqrt{n}]$, the hope is that the probability will remain approximately the same.

In 1908, William Gosset determined the distribution of $\hat{Z}_n$ for the case that sampling is from a normal distribution. The distribution, usually called Student's $t$-distribution, is discussed in Chapter Nine. However, it is true that as $n \to \infty$ $\hat{Z}_n$ does converge in distribution to the standard normal whenever the variance $\sigma^2$ exists. This follows by Slutsky's theorem, which we do not prove.

**Theorem 7.6.2 (Slutsky's Theorem)** Let $h(x, y)$ be a function defined on subset $A$ of $R_1 \times R_1 = R_2$. Let $\{T_n\}$ and $\{W_n\}$ be sequences of random variables. Suppose that $\{T_n\}$ converges in distribution and that for some constant $c$, $\{h(T_n, c)\}$ converges in distribution to a distribution $F$. Suppose also that $\{W_n\}$ converges in probability to $c$, a constant, and that $h$ is continuous on $\{(x, c)\colon (x, c) \in A\}$. Then $\{h(T_n, W_n)\}$ converges in distribution to $F$.

In the example above, take $h(x, y) = x/y$ for $(x, y) \in R_1 \times R_1^+ = A$, $T_n = n^{1/2}(\bar{X}_n - \mu)/\sigma$, $c = 1$, and $W_n = S_n/\sigma$. Then $\{W_n\}$ converges in probability to $c = 1$, and by the CLT, $\{T_n\}$ converges to N(0, 1). We conclude that $h(T_n, W_n) = \hat{Z}_n$ converges in distribution to N(0, 1).

Consider the special case that $X_1, \ldots, X_n$ are independent Bernoulli random variables with parameter $p$, $0 < p < 1$. Let $V_n$ be their sum, $\hat{p}_n = V_n/n$, the sample proportion. Then $Z_n = (\bar{X}_n - \mu)/(\sigma/\sqrt{n}) = (\hat{p}_n - p)/(\sigma/\sqrt{n}) = (V_n - np)/\sqrt{np(1-p)}$, since $\sigma^2 = p(1-p)$. By the argument above we may replace $np(1-p)$ by $S_n^2 = [n/(n-1)]\hat{p}_n(1-\hat{p}_n)$. Since the factor $n/(n-1)$ converges to 1, we conclude that $\hat{Z}_n \equiv (\hat{p}_n - p)/\sqrt{\hat{p}_n(1-\hat{p})/n}$ converges in distribution to N(0, 1). This enables us to say that for large $n$, $p$ not too close to zero or 1, $P(p \in [\hat{p}_n \pm 1.96\sqrt{\hat{p}_n(1-\hat{p}_n)/n}])$ is approximately 0.95. For the case $p = 0.2$ in 10,000 simulations, the interval contained $p$ 8965 times when $n = 100$, but 9523 times for $n = 400$. Thus, the approximation is poor for $n = 100$, good for $n = 400$. For $p = 0.5$ these frequencies were 9173 for $n = 100$, 9599 for $n = 400$.

When we think that $Z_n$ is approximately distributed as standard normal, we can write $P(|\bar{X}_n - \mu| \le d) \doteq 2\Phi(d/(\sigma/n^{1/2})) - 1$. If $\sigma$ is unknown and $S_n$ is close to $\sigma$ with high probability, we can replace $\sigma$ by $S_n$ in $\Phi$ to get an approximation of the probability. We must be a bit cautious, however, because two approximations are being used. For example, for a sample of $n = 100$ from the exponential distribution with mean 1 for $n = 100$, $d = 0.1$, the true probability is 0.6935, while the approximation has an expected value of approximately 0.692 and a standard deviation of approximately 0.065.

By the CLT the $k$th sample moment $\hat{\nu}_k$ is asymptotically normally distributed if the $2k$th moment $\nu_{2k}$ exists. It follows that if the function $h$ from the moment space to the parameter space has continuous first derivatives, the estimators obtained by the MME are asymptotically normally distributed. Thus, if $T_n$ is an estimator of a parameter $\theta$

obtained by the MME, then $\sqrt{n}(T_n - \theta)/(|h'(\theta)|\sigma)$ will be asymptotically standard normal. This would enable us the make probability statements about the error $T_n - \theta$, except that $\theta$ and $\sigma$ are unknown. However, as noted above, we can replace $\theta$ by its consistent estimator $T_n$ and $\sigma$ by a consistent estimator $\hat{\sigma}_n$ or $S_n$, to obtain a consistent estimator of $|h'(\theta)|\sigma$. Thus, $P_\theta(n^{1/2}|T_n - \theta| \leq d)$ may be approximated by $2\,\Phi(d/(|h'(T_n)\hat{\sigma}_n|) - 1$. If $T_n$ is the MLE, the function $|h'(\theta)|^2\sigma^2$ may be replaced by a function $I(\theta)$, the "information" (described later) if certain smoothness conditions on the distributions sampled as a function of $\theta$ (described later). In general, estimators provided by the method of ML have asymptotic distributions with smaller variance than those provided by the MME.

### Problems for Section 7.6

**7.6.1** Let $X$ have the uniform distribution on the interval [8, 10] and let $Y = 1/X$. Use the $\delta$-method to find approximations for the mean and variance of $Y$. Use $h_2$ for the mean, $h_1$ for the variance. Find exact values for $E(Y)$ and Var$(Y)$. Compare these with their approximations.

**7.6.2** A surveyor wishes to estimate the number of square meters in a rectangular field of dimensions $\theta_1$ and $\theta_2$. She makes estimates $X_1$ of $\theta_1$ and $X_2$ of $\theta_2$, so that $E(X_1) = \theta_1, E(X_2) = \theta_2, \text{Cov}(X_1, X_2) = \sigma_{12}, \text{Var}(X_1) = \sigma_1^2, \text{Var}(X_2) = \sigma_2^2$. Give approximations of the mean (using $h_2$) and variance (using $h_1$) of $\hat{A} = X_1X_2$ as an estimator of $A = \theta_1\theta_2$.

**7.6.3** In the analysis of frequency data it is common to suppose that the frequency $W$ of some event has a Poisson distribution with parameter $\lambda$. Let $Y = \sqrt{W}$. To see why this transformation is made, find approximations for the mean (using the quadratic approximation $h_2$) and variance (using the linear approximation $h_1$) of $Y$ for *large* values of $\lambda$. Show that $E(Y) \doteq \sqrt{\lambda}[1 - 1/(8\lambda)]$ and Var$(Y) \doteq 1/4$ *Hint*: Let $X = W/\lambda$ and $g(x) = x^{1/2}$. Since $E(X) = 1$ and Var$(X) = 1/\lambda$, the $\delta$-method produces a good approximation for Var$(Y)$ for large $\lambda$. In fact, Var$(Y)$ for $\lambda$ = 1, 2, 3, 4, 5, 10, 20, 40, 80, 160, 320, 640 are 0.40217, 0.39000 0.33999, 0.30562, 0.28609, 0.26130, 0.25507, 0.25243, 0.25119, 0.25059, 0.25029, 0.25015, so that 1/4 is a good approximation even for moderate $\lambda$.

**7.6.4** A model for the analysis of frequency data supposes that $X$ has a binomial distribution with parameters $n$ and $p$, with $n$ known. Let $\hat{p} = X/n$. Let $Y = \arcsin(\hat{p}^{1/2}) = \sin^{-1}(\hat{p}^{1/2})$ (this is the *arcsin transformation* of $\hat{p}$). Find approximations for the mean and variance of $Y$. *Hint*: $\dfrac{d}{dx}\arcsin x = (1/2)/\sqrt{x(1-x)}$.

**7.6.5** Suppose that the surveyor takes 25 independent readings $X_i$ on the angle, each $X_i$ having mean $\theta$, standard deviation 0.5 degree. If the true angle is $\theta = 30$ degrees and $\rho = 100$ is known, give an approximation for the probability that her estimator $\hat{h}$ differs from $h$ by less than 1.0 meter. Use $\bar{X}$and the CLT. Be careful to express angles in radians. In 1000 simulations the event occurred 604 times.

**7.6.6** Let $X_1, \ldots, X_n$ be a random sample from the exponential distribution with parameter $\lambda$, mean $1/\lambda$. Let $T_n = T_n(X) = 1/\bar{X}_n$. Show that $T_n$ is asymptotically normally distributed with mean $1/\lambda$, variance $1/n\lambda^3$. That is, $Z_n \equiv (T_n - 1/\lambda)/\sqrt{1/(n\lambda^3)}$ is distributed asymptotically as N(0, 1).

**7.6.7** If $p$ is a probability, $0 < p < 1$, of an event $B$, the odds for $B$ is $p/(1-p)$, and the log odds is $h(p) \equiv \log((p/(1-p))$. Since $h$ takes values on the entire real line it is sometimes a useful transformation, taking the interval (0, 1) onto the entire real line with inverse $h^{-1}(u) = e^u/(1+e^u)$. Let $X$ have the binomial distribution with parameters $n$ and $p$, and let $\hat{p} = X/n$.

**(a)** For $n = 200$ give a lower bound on the probability that $\hat{h} = \log(\hat{p}/(1-\hat{p}))$ differs from $h(p)$ by 0.2 or more. You may assume that $0.3 \le p \le 0.7$. For 5000 observation on $X$, each for $n = 200$, $p = 0.4$, the event occurred 4191 times. First use the $\delta$-method to find approximations for $E(\hat{h})$ and Var($\hat{h}$).

**(b)** For the data $X = 83$, $n = 200$, find a 95% confidence interval $[\hat{h} \pm 1.96\,\hat{\sigma}_{\hat{h}}]$ on $h(p)$, then use $h^{-1}$ to get a confidence interval on $p$.

**7.6.8** Let $X_1, \ldots, X_n$ be a random sample from the distribution with density $f(x;\theta) = \theta x^{\theta-1}$ for $0 \le x \le 1$, for $\theta > 0$. Let $\hat{\theta}_1$ and $\hat{\theta}_2$ be the MME and MLE of $\theta$.

**(a)** Find approximations for $E(\hat{\theta}_1)$ (using a quadratic approximation) and Var($\hat{\theta}_1$) (using a linear approximation). For $n = 100$, $\theta = 3$, simulations produced mean 3.025, variance 0.999.

**(b)** Find exact and approximate values for $E(\hat{\theta}_2)$ and Var($\hat{\theta}_2$). For $n = 100$, $\theta = 3$, simulations produced mean 3.031, variance 0.093.

**(c)** For $\theta = 2.0$ and $n = 100$, find approximations for $P_\theta(|\hat{\theta}_1 - \theta| < 0.1)$ and $P_\theta(|\hat{\theta}_2 - \theta| < 0.1)$. In 10,000 simulations these events happened 3617 and 3838 times.

## 7.7 CONFIDENCE INTERVALS

To this point we have discussed methods for finding and evaluating point estimators. Point estimators have the undesirable property that they are almost never exactly equal to parameters they estimate and by themselves do not carry any information about their precision. It is good practice to present an estimate of the *standard deviation* (often called the *standard error* or the *estimate of the standard error*) along with a point estimate. At least when the estimator is approximately normally distributed, we can then make approximate probability statements about the size of the error we have made.

A more satisfactory method was developed by Egon Pearson and Jerzy Neyman (Neyman and Pearson, 1933), that of confidence intervals. Consider a simple example: The Michigan Department of Agriculture wishes to estimate the total number of acres planted in corn on July 1 among its 50,000 farms. It has a list of farms, so it takes a simple random sample (without replacement) of 400 farms, contacts all 400 farms, and determines the number of acres planted in corn for each. The mean turns out to be 22.48 acres, with sample standard deviation $s = 35.36$ acres. What can be said about the total number of acres planted in corn in Michigan?

Let $\mu$ be the mean number of acres in corn among the $N = 50,000$ farms in Michigan. Let $\sigma^2$ be the corresponding variance. Then the total planted in corn is $\tau = N\mu = 50{,}000\mu$. Let us first try to make a confidence interval statement about $\mu$. By the CLT (or its equivalent for without-replacement sampling) and Slutsky's theorem we know that $\hat{Z} \equiv (\bar{X} - \mu)/(S/\sqrt{n})$ is approximately distributed as N(0, 1). This should be true in approximation even though the population sampled is far from normally distributed, consisting of a large proportion of zeros and some values exceeding 500. Therefore, $P(-1.96 \leq \bar{Z} \leq 1.96) \doteq 0.95$. [We have ignored the finite correction factor $(N - n)/(N - 1)$, because $n$ is small relative to $N$.] By manipulating the inequalities for the event $A$ in parentheses, we see that $A$ is equivalent to $\bar{X} - 1.96S/\sqrt{n} \leq \mu \leq \bar{X} + 1.96S/\sqrt{n}$.

The interval $I = [\bar{X} \pm 1.96S/\sqrt{n}]$ is said to be a 95% *confidence interval* on $\mu$. In this case the confidence coefficient 0.95 is only approximate. The interval $I$ is a random interval having the property that $P(\mu \in I) \doteq 0.95$ for any $\mu$. For these data we get $I = [22.48 \pm 1.96\ (35.36/20)] = [22.48 \pm 3.47]$, and the resulting 95% confidence interval on $\tau = N\mu$ is 50,000 $[22.48 \pm 3.47] = [1{,}124{,}000 \pm 73{,}500]$. We are *not* justified in saying that $P(\mu \in [22.48 \pm 3.48]) \doteq 0.95$. The interval $[22.48 \pm 3.48]$ is fixed, although before the sample is taken, the interval $I$ is random. The event $A$ that $\mu \in I$ has probability approximately 0.95. $\mu$ is contained in the interval $[22.48 \pm 3.48]$ or it is not, so its probability is zero or 1. We do not know which. The procedure used to take the sample and determine the sample has the property that about 95% of samples will provide intervals $I$ that contain $\mu$. We rarely learn whether the sample we obtained provided one of the "good intervals" (one that contains $\mu$), but are 95% confident that we have such an interval because we know that 95% of all possible samples will produce good intervals.

Figure 7.7.1 includes a histogram of the 30,000 nonzero values in a fictitious population of 50,000 values. Since few values were above 300, they were omitted from the histogram plot. A second histogram shows the 1000 samples of 400 that were taken, and for each the $\hat{Z}$ that was determined, indicating that $\hat{Z}$ is approximately distributed as standard normal. 946 of these 1000 values of $\hat{Z}$ were between $-1.96$ and $+1.96$. These 946 samples would have provided confidence intervals containing $\mu = 24.27$.

Let us now carefully define the term *confidence interval*.

**Definition 7.7.1** Let $\mathbf{X}$ be a random vector taking values in $R_n$ whose distribution depends on a parameter $\theta \in \Omega$. Let $L(\mathbf{x})$ and $U(\mathbf{x})$ be real-valued functions of $\mathbf{x}$ defined on the range $R_{\mathbf{X}}$ of $\mathbf{X}$ such that $L(\mathbf{x}) \leq U(\mathbf{x})$ for every such $\mathbf{x}$. Let $g$ be a real-valued function defined on $\Omega$. The interval $[L(\mathbf{X}), U(\mathbf{X})]$ is said to be a $100\gamma\%$

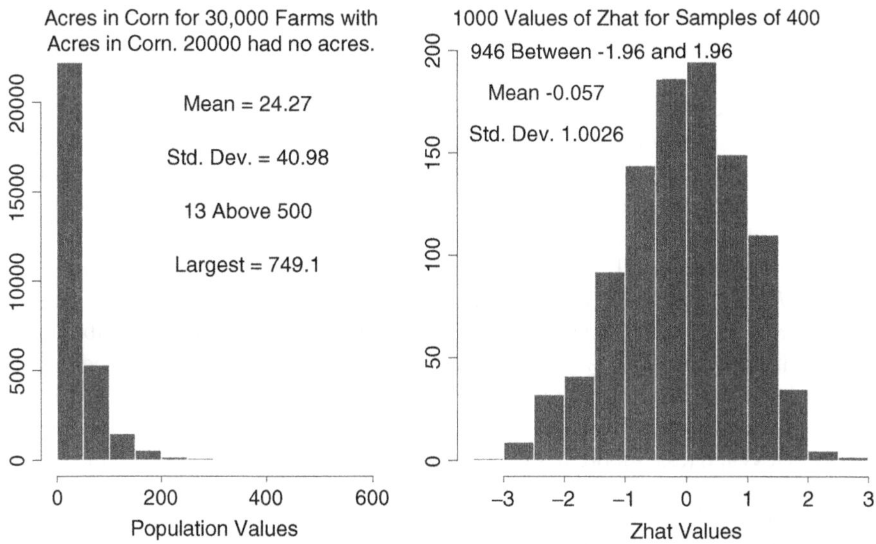


**FIGURE 7.7.1** Histograms for the population of corn acreages and 1000 values of $\hat{Z}$.

*confidence interval* on $g(\theta)$ if

$$P_\theta(L(\mathbf{X}) \le g(\theta) \le U(\mathbf{X})) = \gamma \qquad \text{for all } \theta \in \Omega. \qquad \square$$

COMMENTS: The definition is somewhat demanding in that the probability must be $\gamma$ for every $\theta$. In practice, $100\gamma\%$ is also used when the equality is replaced by "$\ge$" or by "$\doteq$". We can allow $L$ to take the value $-\infty$ for all $\mathbf{x}$. Then the interval $(-\infty, U(\mathbf{X}))$ is said to be *one-sided with upper limit U*. Similarly, we can take $U(\mathbf{x}) = +\infty$ for all $\mathbf{x}$ and $[L, +\infty)$ is a *one-sided interval with lower limit L*. Of course, the interval should always be a subset of the parameter space. The number $\gamma$ is called the *confidence coefficient*. The principal method used to determine confidence intervals is the pivotal method. A function $h(\mathbf{x}, \theta)$ is found, defined on $R_X \times \Omega$, such that the distribution of $h(\mathbf{X}, \theta)$ is the same for all $\theta \in \Omega$. Then constants $c_1 < c_2$ are found such that $P_\theta(c_1 \le h(\mathbf{X}, \theta) \le c_2) = \gamma$ for all $\theta \in \Omega$. The set $A(\mathbf{X}) \equiv \{\theta | c_1 \le h(\mathbf{X}, \theta) \le c_2\}$ is then a $100\,\gamma\%$ *confidence region on* $\theta$. Similarly, a set $B(\mathbf{X}) = \{g(\theta) | c_1 \le h(\mathbf{X}, \theta) \le c_2\}$ is a $100\,\gamma\%$ confidence region on $g(\theta)$. If $A(\mathbf{X})$ or $B(\mathbf{X})$ is an interval, it is then a $100\,\gamma\%$ confidence interval on $\theta$ or $g(\theta)$.

For the corn example above, $h(\mathbf{X}, \theta)$ was $\hat{Z} \equiv (\bar{X} - \mu)/(S/\sqrt{n})$, $\theta = \mu$, and the distribution of $\hat{Z}$ was only approximately N(0, 1) for the case that $n$ was large. Student's $t$-statistic is $t_n = (\bar{X} - \mu)/(S/\sqrt{n})$. We have been discussing this statistic as an approximation of the random variable $Z$ obtained when $\sigma$ appears in the denominator. We called it $\hat{Z}$ for that reason. Technically, neither $Z$ nor $\hat{Z} = t_n$ are statistics if either depends on an unknown parameter. although $t_n$ was known to be approximately distributed as standard normal during the nineteenth century, its distribution for small $n$ was unknown. In 1908, William Gosset published a paper in the *Journal of the Royal Statistical Society* in which he gave the density function for any positive $n > 1$ for the case that the sample was taken from a normal distribution. He published the paper (Gosset, 1908) under the name "A Student," since he was an employee on leave from the Guiness Brewery, which wanted its employees to remain anonymous. Since that time $t_n$ has been called *Student's t-statistic*. In fact, any time a random variable is standardized by subtracting its expected value and is then divided by an estimator of its standard deviation, it is said to have been *Studentized*. Student's $t$-distribution (really distributions) depends on a parameter $\nu$, called the *degrees of freedom*. For $t_n$ as defined above, $\nu = n - 1$. Important quantiles (corresponding to, for example, 0.90, 0.95, 0.975, 0.99) are given in almost all statistics books for at least $\nu = 1, 2, \ldots, 30$.

If $X_1, \ldots, X_n$ is a random sample from $N(\mu, \sigma^2)$, then since $t_n$ has a $t$-distribution, it follows that $P_{(\mu,\sigma)}(-t_{(1+\gamma)/2} \leq t_n \leq t_{(1+\gamma)/2}) = \gamma$, and manipulating the inequalities we get the $100\,\gamma\%$ confidence interval $[\bar{X} \pm t_{(1+\gamma)/2} S/n^{1/2}]$ on $\mu$. If $n$ is large, say $n \geq 25$, it makes little difference if $t_{(1+\gamma)/2}$ or $z_{(1+\gamma)/2}$ is used in this formula, and it becomes less and less important that the population sampled is normal as $n$ increases.

***Example 7.7.1*** The following random sample of $n = 10$ observations was taken from a normal distribution. (The author tells you secretly that S-Plus was used with $\mu = 50$, and $\sigma = 8$. In real life you wouldn't know these parameter values.) The sample values were 54.24, 64.06, 61.25, 41.39, 45.88, 43.18, 48.12, 53.08, 49.86. The sample mean and standard deviation were $\bar{X} = 49.81$ and $s = 8.57$. For $\gamma = 0.95$, the $t$-table for $\nu = n - 1 = 9$ degrees of freedom provided $t_{0.975} = 2.26$, so that $t_{(1+\gamma)/2} S/n^{1/2} = 6.13$, and $[\bar{X} \pm t_{(1+\gamma)/2} S/n^{1/2}] = [43.67, 55.94]$. This time the interval contained $\mu$. □

***Example 7.7.2*** Let $\mathbf{X} = (X_1, \ldots, X_n)$ be a random sample from the uniform distribution on $[0, \theta]$ for $\theta > 0$. As shown earlier, $h(\mathbf{X}, \theta) = M/\theta$ for $M = \max(X_1, \ldots, X_n)$ has the cdf $F(u) = u^n$ for $0 \leq u \leq 1$. We will use the pivotal quantity $h(\mathbf{X}, \theta)$ to find a 90% confidence interval on $\theta$. We have for $0 < c < 1$, $P_\theta(c \leq M/\theta \leq 1) = 1 - F(c) = 1 - c^n$. We want this to be $\gamma = 0.90$. Solving for $c$, we get $c = (1 - 0.90)^{1/n}$.

The event in parentheses is equivalent to $M \leq \theta \leq M/c$. It follows that $[M, U(\mathbf{X})]$ is a 90% confidence interval on $\theta$ for $U(\mathbf{X}) = M/(1 - 0.90)^{1/n}$. If, for example, $n = 100$, then $c = 0.97724$, and if we observe $M = 53.73$, then the upper 90% confidence limit on $\theta$ is $U = 54.98$. The 90% confidence interval is [53.73, 54.98]. □

### Estimation of $p$

Consider a population of $N$ elements, of which a number $N_1$ have a certain characteristic. The population might be the collection of registered voters in Michigan, and the characteristic might be that the voter favors a new gun law. Or the population might be the collection of people in the United States who develop lung cancer in 2007, and the characteristic might be that the person lives at least another five years when a new drug is used. We are often interested in estimating the proportion $p = N_1/N$. Suppose that a random sample of $n$ members of the population is taken. Usually, sampling is without replacement, so that all $\binom{N}{n}$ subsets of the population are equally likely. However, the probability theory is simpler for sampling with replacement and is a good approximation if $n \ll N$. We consider the case of with-replacement sampling first.

Let $X$ be the number in the sample with the characteristic. Under sampling with replacement, $X$ has the binomial distribution with parameters $n$ and $p$. It is important to note that for this model the population size $N$ plays no role beyond that determined by $n$ and $p$, in determining the distribution of $X$. The MM and ML estimators of $p$ are both $\hat{p} = X/n$. Since $X \sim \text{Binomial}(n, p)$, $E(X) = np$ and $\text{Var}(X) = np(1-p)$. It follows that $E(\hat{p}) = p$ and $\text{Var}(\hat{p}) = pq/n$ for $q = 1 - p$. This implies that $\{\hat{p}_n\}$ is consistent for $p$. (We have added the subscript $n$ to emphasize the dependence of $\hat{p}$ on $n$.)

By the DeMoivre–Laplace theorem (the CLT for Bernoulli random variables) $Z_n \equiv (X - np)/(npq)^{1/2} = (\hat{p}_n - p)/(pq/n)^{1/2}$ converges in distribution to the standard normal. Thus, for $n$ moderately large and $p$ not too close to 0 or 1,

$$P(|\hat{p}_n - p| < \varepsilon) \doteq \Phi(h) - \Phi(-h) = 2\Phi(h) - 1 \quad \text{for}$$
$$h = \frac{\varepsilon}{(pq/n)^{1/2}} = \varepsilon\left(\frac{n}{pq}\right)^{1/2}. \tag{7.7.1}$$

It follows that we can make this probability approximately equal to some prescribed number $\gamma$ (0.95, for example), by letting $h = z_{(1+\gamma)/2}$ satisfy $\Phi(z_{(1+\gamma)/2}) = (1+\gamma)/2$, so that $2\Phi(h) - 1 = \gamma$. Solving for $n$, we get

$$n = n_w = (pq)\left[\frac{z_{(1+\gamma)/2}}{\varepsilon}\right]^2. \tag{7.7.2}$$

This formula for the sample size depends on the unknown parameter $p$, a rather unsatisfactory condition, since we are trying to decide what the sample size should be. Fortunately, the function $g(p) \equiv pq = 1/4 - (p - 1/2)^2$ is maximum for $p = 1/2$. Therefore, if we simply take $p = 1/2$ in the formula for $n$, we will have assumed the worst and will have an unnecessarily larger sample size when $p \neq 1/2$. In fact, if it is quite obvious that $0 < p \le p_0 \le 1/2$ for some $p_0$, we can use $p_0(1 - p_0)$ rather than $pq$ in the formula for $n$. The same is true if $p_0$ and $p$ satisfy $1/2 \le p_0 \le p < 1$.

In the usual case that sampling is without replacement, $X$ has instead a hypergeometric distribution. We still have $E(X) = np$ and $E(\hat{p}) = p$, but the variance is smaller than it is for the binomial distribution: $\text{Var}(X) = np(1-p)[(N-n)/(N-1)]$. The factor in brackets is called the *finite correction factor*. It is still true that $X$ is approximately normally distributed for large $n$, $p$ not too close to zero or 1, but we need the additional condition that $N - n$ is also large. If we replace $pq/n$ in (7.7.1) by $(pq/n)[(N-n)/(N-1)]$ and determine $n = n_{wo} = n_w/[N/(N-1) + n_w/N]$, which in good approximation is

$$n_{wo} = \frac{n_w}{1 + n_w/N}. \tag{7.7.3}$$

Notice that as $N$ becomes large relative to $n$, $n_{wo}$ approaches $n_w$.

### Generalization to μ

The sample-size formulas (7.7.2) and (7.7.3) can be generalized to populations with numerical characteristics other than zero and 1 as they were for the estimation of $\mu = p$. We need only that in the case of sampling with replacement, $Z_n \equiv (\bar{X} - \mu)/\sqrt{\sigma^2/n}$ be approximately distributed as N(0, 1) to get the formula

$$n = n_w = \sigma^2 \left[ \frac{z_{(1+\gamma)/2}}{\varepsilon} \right]^2. \tag{7.7.4}$$

The normal approximation holds reasonably well if there are not extremes in the population, values that are far from the mean $\mu$ relative to $\sigma$ (i.e., if the range is not too large relative to $\sigma$).

The problem here is that $\sigma^2$ is seldom known. Similar data may have been collected in the past and an estimate $\hat{\sigma}^2$ may replace $\sigma^2$ in (7.7.4). Or a preliminary sample of (say) 100 make be taken to estimate $\sigma^2$. Equation (7.7.3), relating the sample sizes needed for sampling with and without replacement, still holds.

### Confidence Intervals on μ and *p*

As stated earlier, the interval $[\bar{X} \pm z_{(1+\gamma)/2} S/n^{1/2}]$ is at least in approximation a 100 $\gamma$% confidence interval on $\mu$ if $n$ is large and there are no extreme observations in the case of sampling with replacement. If sampling is without replacement, we need only multiply $S/n^{1/2}$ by $\sqrt{(N-n)/(N-1)}$. If the population is 0–1, so that $\mu = p$, the proportion of 1's in the population, $S^2$ may be replaced by $\hat{p}(1-\hat{p})$. Figure 7.7.2 shows histograms of the values of the $t$-statistic for 10,000 samples of sizes 10 and 20, together with the densities of the $t$-distribution for 9 and 19 degrees of freedom. In the first row the population has an extreme value of 160, so that the corresponding $t$-confidence intervals, with 2.263 and 2.093 as multipliers rather than 1.96, produced by the samples would contain $\mu$ 9332 and 9170 times. For the second row the extreme value is only 80, so that the histograms are closer to standard normal and the resulting confidence intervals would be correct 9422 and 9446 times. closer to the nominal 95%.

***Example 7.7.3*** A few years ago an article on autodeaths among young children appeared in a local newspaper. Four sentences from that article follow.

> On the positive side, fewer young children died in car crashes. The number of those under age 4 who were killed fell 5.7 percent (from 2001 to 2002), while there was an 8 percent decline for youngsters 4 to 7. For the first time since the federal agency began keeping records in the 1970s, fewer than 500 children in each age category died in a year. There were 484 fatalities in the youngest age group and 496 for those age 4 to 7.

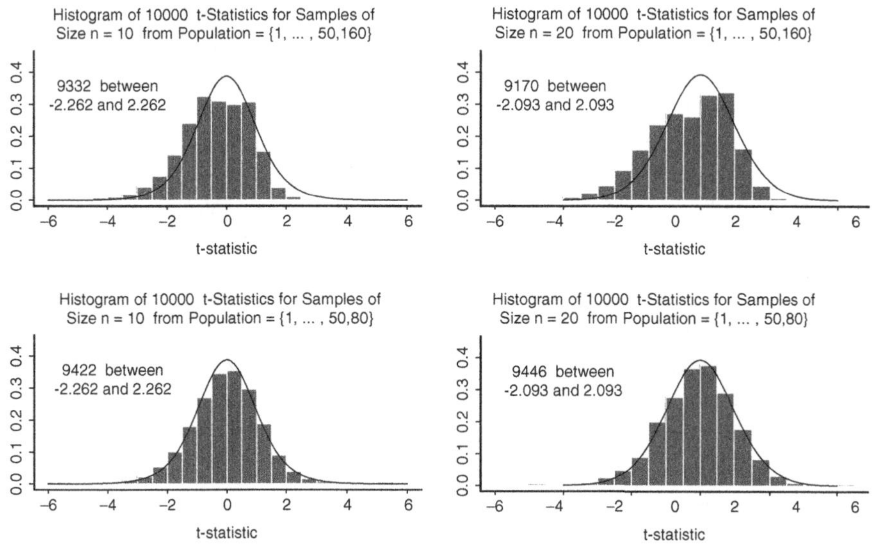


**FIGURE 7.7.2** Histograms for the $t$-statistic for samples from populations with outliers.

There is a tendency among newspaper writers to overinterpret changes that are often due to chance. To see this, let the numbers of deaths among children under 4 in 2001 and 2002 be $X_1$ and $X_2$. It is reasonable to suppose that $X_1$ and $X_2$ are independent and that they have Poisson distributions with parameters $\lambda_1$ and $\lambda_2$. In fact, since multiple deaths sometimes occur in accidents, the Poisson model may be only a rough approximation, but it should be close enough for our purposes. We would like to estimate $\lambda_2/\lambda_1 \equiv R$. The author of the article seemed to be convinced that $R < 1$.

From Chapter Five we know that conditionally on $T = t = X_1 + X_2$, $X_2$ has the binomial distribution with parameters $n = t$ and $p = \lambda_2/(\lambda_1 + \lambda_2) = R/(1 + R)$. A point estimate of $p$ is given by $\hat{p} = X_2/(X_1 + X_2) = \hat{R}/(1 + \hat{R})$, where $\hat{R} = X_2/X_1$. In this case we observe that $X_2 = 484$, $X_1 = X_2/0.943 \doteq 513$, $\hat{R} = 0.943$, $\hat{p} = 0.485$. An approximate 95% confidence interval on $p$ is given by $[\hat{p} \pm 1.96\sqrt{\hat{p}(1 - \hat{p})/n}] = [0.485 \pm 0.031] = [0.454, 0.516]$. Since the inverse function $R = h(p) = p/(1 - p)$ is monotone in $p$, the interval $[h(0.454), h(0.516)] = [0.832, 1.066]$ is a 95% confidence interval on $R$. The conclusion is that the real death rates may not have changed at all. As is common with such data, the author jumped to conclusions not warranted by the data. The 8% drop in deaths among children aged 4 to 7 may be enough to convince us that the change is real, and combining the data for both age groups may be even more convincing (see Problem 7.6.3).

We might wish to find a confidence interval on $\lambda_2$ alone or on $\delta = \lambda_2 - \lambda_1$. As mentioned in Chapter Six, since $\lambda$ is large and $X \sim \text{Poisson}(\lambda)$, $Z \equiv (X - \lambda)/\sqrt{\lambda}$ is approximately distributed as N(0, 1), so that $P(-1.96 \le Z \le 1.96) \doteq 0.95$. The event in parentheses is $[Z^2 \le 1.96^2] = [G(\lambda) \le 0]$, where $G(\lambda) = \lambda^2 + B\lambda + C$, where $B = 2X + 1.96^2$ and $C = X^2$. Solving the quadratic we find that a confidence interval on $\lambda$ is given by $[L(X), U(X)] = [(X + 1.96^2/2) \pm 1.96\sqrt{X + 1.96^2/4}]$. For $X$ large as for $X_2$ in our example, it makes little difference whether we use the pivotal quantity $Z$ as we have here or $\hat{Z} \equiv (X - \lambda)/\sqrt{X}$, which produces the 95% interval $[X \pm 1.96\sqrt{X}]$. For $X_2 = 484$ as in the child-death example, we get the intervals [442.8, 529.1] and [440.9, 527.1] for this last rougher approximation. In 10,000 simulations with $\lambda = 500$, the intervals produced confidence intervals containing $\lambda$ 9447 times for the more complex method, 9432 for the simpler method.

To get a confidence intervals on $\delta = \lambda_2 - \lambda_1$, we can use the pivotal quantity $Z \equiv [(X_2 - X_1) - (\lambda_2 - \lambda_1)]/(X_2 + X_1)^{1/2}$ to produce the interval $[(X_2 - X_1) \pm 1.96(X_2 + X_1)^{1/2}]$. For these data we get the 95% interval $[-90.9, 32.9]$. Since the interval includes zero, we have again indicated that the author of the article overinterpreted the change. □

## Problems for Section 7.7

**7.7.1** Let $f(x, \theta) = (1/\theta)e^{-x/\theta}$ for $x > 0$, $\theta > 0$.

**(a)** For one observation $X$ from this density, find an upper 90% confidence limit on $\theta$.

**(b)** Let $X_1, \ldots, X_{10}$ be a random sample from $f$. Give a formula for a 90% confidence interval on $\theta$. Hint: Consider their sum $T$. A $\gamma$-quantile $x_\gamma$ for the gamma distribution with shape parameter $\alpha$, scale parameter 1 may be found using chi-square tables, given in almost all statistics books. If $Y$ has the $\Gamma(\alpha, 1)$ distribution, $W = 2Y \sim \chi^2_{2\nu}$ and $x_\gamma = w_\gamma/2$, where $w_\gamma$ is the $\gamma$-quantile for $W$. Denote the 0.95 and 0.05 quantiles of the $\chi^2_{2\nu}$ distribution by $\chi^2_{2\nu}(0.95)$ and $\chi^2_{2\nu}(0.05)$.

**(c)** Evaluate the confidence interval for the sample 7.9, 2.6, 8.7, 30.7, 1.5, 8.9, 0.5, 6.5, 5.7, 21.3. Their sum is 94.3. (The true value of $\theta$ used to generate these data was 5.0. )

**7.7.2** Let $X_1, \ldots, X_n$ have the shifted exponential distribution, having density $f(x; \eta) = e^{-(x-\eta)}$ for $x > \eta$, $\eta$ any real number. Find a pivotal quantity that depends on the MLE and use this to give a lower 95% confidence limit $L$ on $\eta$. Evaluate it for the sample of 9: 16.56, 14.30, 13.29, 14.22, 13.46, 16.38, 16.30, 13.48.

**7.7.3** A village has 2000 adult residents. A political scientist wishes to estimate the proportion of these who would be willing to pay \$100 more each year for three years in taxes in order to build a park. He wants the sample size $n$ to be large enough to make the probability at least 0.90 that the sample proportion will be within 0.04 of the population proportion. How large should $n$ be if:

**(a)** Sampling is with replacement?

**(b)** Sampling is without replacement?

**7.7.4** The state of Michigan suspects that a medical clinic is overcharging on bills sent to the state for laboratory work. To estimate the proportion $p$ of bills sent to the state for which there were overcharges, the state decides to take a simple random sample of the 1763 bills received over a six-month period. It is too expensive to examine all 1763 bills (a "census"), since medical people must determine whether there have been overcharges.

(**a**) How large must the sample size be to estimate $p$ within 0.03 with probability 0.95? Assume that $p \leq 0.20$.

(**b**) If the sample size determined in part (a) were used but $p$ were actually 0.40, what would the probability be that $\hat{p}$ differed from $p$ by more than 0.03?

(**c**) Suppose that the sample size $n_{wo} = 500$ was used and the number of bills observed for which there were overcharges was 47. Give a 95% upper confidence limit on $p$. [This means that the upper confidence limit $U = U(X)$ is determined by a method which has the property that $P(U \geq p) = 0.95$.]

**7.7.5** A newspaper wishes to take a simple random sample of $n$ from the 2 million voters for governor of the state in order to estimate the difference $\Delta = p_A - p_B$ in the proportions who favor the two candidates $A$ and $B$. There are only two candidates. How large must the sample size be to have probability at least 0.90 that the estimator $\hat{\Delta}$ differs from $\Delta$ by less than 0.04?

**7.7.6** Let $X_1, X_2, X_3, X_4$ be a random sample from $\mathrm{N}(\mu, \sigma^2)$, where $\mu$ and $\sigma^2$ are both unknown. Suppose that the values observed were 7, 7, 3, 7. Find a 90% confidence interval on $\mu$. This example was invented before the days of calculators, so that square roots could easily be obtained.

**7.7.7** The number of cases of lung cancer reported over a three-year period among 8791 men of ages 50 to 59 living in a county in which a nuclear reactor was located was 81. During the same time period in other counties in that same state there were 62,547 men of ages 50 to 59, of which 483 were reported to have lung cancer over the same period. Let $\theta_1$ and $\theta_2$ be the underlying (long-term) rates per 1000 men of this age group in the two regions.

(**a**) State a model for the numbers $X_1$ and $X_2$ with cancer in the two regions. Then find a 90% confidence interval on $R = \theta_1/\theta_2$. Use the conditional binomial method.

(**b**) Find 90% confidence intervals on $\theta_1$ and on $\theta_2$.

**7.7.8** To measure the effect of a new drug on blood pressure, 20 people with blood pressure over 170 were asked to take part in an experiment. The 20 people were paired according to age and gender, those in the same pair having the same gender and approximately the same age. Then one member of each pair was chosen randomly to receive the drug (the treatment). The other member received a placebo (the control), a pill that tasted and looked like the drug, but which had no real effect on people. Blood pressure (bp) was measured both before and two weeks after the 20 began taking the pills. The bp decreases for the 20 people were as follows:

| Pair | 1 | 2 | 3 | 4 | 5 | 6 | 7 | 8 | 9 | 10 |
|---|---|---|---|---|---|---|---|---|---|---|
| Treatment | 8 | 2 | −4 | 12 | 1 | −2 | 6 | −4 | 10 | 3 |
| Control | 1 | −5 | 3 | 5 | −6 | 0 | 7 | −8 | 2 | 4 |

Let $Y_i$ and $X_i$ be the treatment and drug decreases for the $i$th pair. Thus, $Y_3 = -4$, $X_3 = 3$, indicates that in the third pair, for the person receiving the drug the bp decreased −4, equivalently increased 4, whereas for the person receiving the placebo the bp decreased 3. Let $D_i = Y_i - X_i$, and suppose that $D_1, \ldots, D_{10}$ is a random sample from $N(\mu_D, \sigma_D^2)$, with both parameters unknown. Use the $t$-method to find a 95% confidence interval on $\mu_D$. The method used here is called the *paired sample t-method.*

**7.7.9** Let $I_i$ be a $100(1 - \alpha_i)\%$ confidence interval on a parameter $\theta_i$ for $i = 1, \ldots, k$.

**(a)** Prove that $P(\theta_i \in I_i \text{ for } i = 1, \ldots, k) \geq 1 - \sum_{i=1}^{k} \alpha_i$. *Hint:* Let $A_i$ be the event $[\theta_i \in I_i]$. Then $P(A_i) = 1 - \alpha_i$ for each $i$.

**(b)** What is the probability if the $I_i$ are independent?

**7.7.10** **(a)** A box has 10 balls, of which $R$ are red and $10 - R$ are white. A simple random sample of four balls is chosen and $X = 3$ are red. Find a lower 85% (approximately) confidence limit on $R$. Although we won't prove it here, such a lower limit is $L(X)$, where $L(x)$ for each $x$ is the smallest $R$ such that $P(X \geq x | R) \leq 0.15$.

**(b)** Another box has 100 balls, of which $R$ are red. A simple random sample of $n = 36$ balls is chosen, of which $X = 27$ are red. Use a normal approximation to give an approximate 85% lower confidence limit on $R$.

**7.7.11** To investigate the effects of a special program to keep students in high school in Metropolis, 800 students were chosen randomly from among the 30,000 tenth graders to take part in the study. Four hundred of these were randomly chosen to receive special counseling. The other 400 were to receive no extra counseling. At the end of the school year the numbers among the two groups who had dropped out of school were determined. Some of those in the 800 moved out of the city, so that the sample sizes were reduced somewhat. Among the 359 in group 1 (the treatment group, receiving extra counseling), $X_1 = 53$ dropped out. Among the 367 in group 2 (the control group), $X_2 = 74$ dropped out.

Suppose that $X_1$ has a binomial distribution with parameters $n_1 = 359$ and $p_1$, and that $X_2$ has the binomial distribution with parameters $n_2 = 367$ and $p_2$. Suppose also that $X_1$ and $X_2$ are independent. We want a confidence interval on $\Delta \equiv p_1 - p_2$.

**(a)** Interpret $p_1$ and $p_2$ relative to the 30,000 tenth graders.

**(b)** Suggest an unbiased estimator $\hat{\Delta}$ of $\Delta$.

**(c)** Express $\text{Var}(\hat{\Delta}) = \sigma^2_{\hat{\Delta}}$ in terms of $n_1, p_1, n_2, p_2$.

**(d)** Suggest an estimator of $\text{Var}(\hat{\Delta})$. Call this $\hat{\sigma}^2_{\hat{\Delta}}$.

**(e)** Define a pivotal quantity which for large $n_1$ and $n_2$ has an approximate standard normal distribution. Use this to give an approximate 95% confidence interval on $\Delta$. Interpret this interval for the benefit of the school board, consisting of six people who have never studied statistics and one who took Statistics 201 48 years ago.

**7.7.12** You observe $Y_i = U_i^{\theta}$ for $i = 1, \ldots, n$, where $\theta > 0$ is unknown and the $U_i$ are independent, each Unif(0, 1). Give a formula for a $100\gamma\%$ confidence interval on $\theta$.

**7.7.13** Independent Bernoulli trials, each with probability $\theta$ of success, are performed until 1000 successes have occurred. Let $T$ be the total number of trials necessary.

**(a)** What is the distribution of $T$?

**(b)** Give formula for a $100\gamma\%$ confidence interval on $\theta$. Evaluate it for $\gamma = 0.90$, $T = 5937$.

## 7.8 FISHER INFORMATION, CRAMÉR–RAO BOUND, AND ASYMPTOTIC NORMALITY OF MLEs

Let $\mathbf{X}$ be a random variable, or random vector, with density $f(\mathbf{x}; \theta)$, defined for every $\theta \in \Omega$, where $\Omega$ is an open subset of the real line. As defined in Section 7.4, the likelihood function for $\mathbf{X}$ is $L(\theta; \mathbf{x})$. Suppose that $\frac{\delta}{\delta\theta} f(\mathbf{x}; \theta) = f'(\mathbf{x}; \theta)$ exists for each $\mathbf{x}$ in the range of $\mathbf{X}$. The derivative of $\log(L(\theta;\mathbf{x}))$ is the *score function*:

$$\ell(\theta; \mathbf{x}) = \frac{\delta}{\delta\theta}\log(\ell\theta; \mathbf{x})) = \frac{\delta}{\delta\theta}\log(L(\theta; \mathbf{x})) = \frac{f'(\mathbf{x}; \theta)}{f(\mathbf{x}; \theta)}.$$

$\ell(\theta; \mathbf{x})$ is a measure of the relative effect of a small change of $\theta$ on the likelihood function when $\mathbf{x}$ is observed. Consider the random variable $\ell(\theta; \mathbf{X})$. Then $E_\theta(\ell(\theta; \mathbf{X})) = \int_{-\infty}^{\infty} \ell(\theta; \mathbf{x})\, f(\mathbf{x}; \theta)\, d\mathbf{x} = \int_{-\infty}^{\infty} f'(\mathbf{x}; \theta)\, d\mathbf{x}$. (If $\mathbf{x}$ is a vector of length $n$, the integral is an $n$-fold integral.) If we can interchange integration and differentiation, then this last term is $\frac{\delta}{\delta\theta}(1)=0$. Now consider the second partial derivative:

$$\dot{\ell}(\theta;\mathbf{x}) = \frac{\delta^2}{\delta\theta^2}\log(L(\theta;\mathbf{x})) = \frac{\delta}{\delta\theta}\ell(\theta;\mathbf{x}) = \frac{f''(\mathbf{x};\theta)}{f(\mathbf{x};\theta)} - \left[\frac{f'(\mathbf{x};\theta)}{f(\mathbf{x};\theta)}\right]^2$$
$$= \frac{f''(\mathbf{x};\theta)}{f(\mathbf{x};\theta)} - [\ell(\mathbf{x};\theta)]^2.$$

Again, $f(\mathbf{x};\,\theta)$ in the numerator and denominator cancel, so that if integration and differentiation may be exchanged, $E_\theta(\dot{\ell}(\theta;\,\mathbf{X})) = 0 - E_\theta([\ell(\theta;\,\mathbf{X}))]^2) = -E_\theta(\ell(\theta;\,\mathbf{X})^2)$. The function

$$I(\theta) = E_\theta(\ell(\theta;\mathbf{X})^2) = \mathrm{Var}_\theta(\ell(\theta;\mathbf{X})) = -E_\theta(\dot{\ell}(\theta;\mathbf{X}))$$

is called the *information function*, as defined by R. A. Fisher (Fisher, 1922). It is the mean curvature of the log-likelihood function at $\theta$. As will be shown, this function has some interesting properties relating to MLEs and to estimators in general.

Notice that if $\mathbf{X} = (X_1, \ldots, X_n)$ is a random sample from such a density $f(x;\theta)$, the log-likelihood function is the sum of the log-likelihood functions for the $X_i$. It follows that the information function for $\mathbf{X}$ is $I_n(\theta) = I_1(\theta)n$, where $I_1(\theta) = I_1(\theta)$ is the information function for one observation.

If $\mathbf{X}$ has a discrete mass function $f(\mathbf{x};\theta)$, we can repeat the definitions above for $\ell(\theta;\,\mathbf{x})$, for $\dot{\ell}(\theta;\,\mathbf{X})$ and $I(\theta)$ just by replacing integrals by sums. The examples will make this clear.

***Example 7.8.1*** Let $X \sim \mathrm{N}(\mu, 1)$. The log-likelihood function is $\log(f(x;\,\mu)) = C - (1/2)(x-\mu)^2$, where $C$ is a constant. Then $\ell(\mu;x) = (x-\mu)$, the score function, and $\dot{\ell}(\theta;\mathbf{X}) = -1$. Notice that $E_\mu(\ell(\mu;\mathbf{X})) = 0$. Since $\dot{\ell}(\theta;\,\mathbf{X})$ is a constant, its expectation is $I(\mu) = 1 = \mathrm{Var}_\mu(\ell(\mathbf{X};\mu))$ . Thus, one observation $X$ carries information 1. It's easy to show that if $\mathrm{Var}(X) = \sigma^2$ rather than 1, $I(\mu) = 1/\sigma^2$. Less information on $\mu$ is given by an observation with larger variance. □

***Example 7.8.2*** Let $X \sim$ Poisson($\lambda$). Then $\log(L(\lambda;x)) = -\lambda + x\,\log(\lambda) + C$, where $C$ does not depend on $\lambda$. Thus, $\ell(\lambda;x) = -1 + x/\lambda$, and $\dot{\ell}(\theta;\mathbf{X}) = -x/\lambda^2$. Notice that $E_\lambda(\ell(\lambda;X)) = 0$ and $I(\lambda) = -E_\lambda(\dot{\ell}(\theta;\mathbf{X})) = \lambda/\lambda^2 = 1/\lambda = \mathrm{Var}(\ell(X;\lambda))$. □

***Example 7.8.3*** Let $X \sim$ Unif(0, $\theta$) for $\theta > 0$. The likelihood function is $L(\theta;x) = 1/\theta$ for $0 \le x \le \theta$. Its score function is $\ell(\theta;x) = -1/\theta$ for $0 < x < \theta$ and is undefined otherwise. But $E_\theta(\ell(\theta;\mathbf{X})) = -1 \neq 0$. The range of the integral determining $E_\theta(\ell(\theta;\,\mathbf{X}))$ depends on $\theta$, so the order of differentiation and expectation cannot be exchanged. $\dot{\ell}(\theta;\mathbf{X}) = 1/\theta^2$ for $0 < x < \theta$, but $\mathrm{Var}_\theta(\ell(\theta;X)) = 0 \neq -E_\theta(\dot{\ell}(\theta;\mathbf{X})) = 1/\theta^2$. □

**The Cramér–Rao Inequality** Let **X** be a random variable or vector with density or probability function $f(\mathbf{x};\theta)$ for $\theta \in \Omega$. Let $T(\mathbf{X})$ be an estimator of $\theta$. Let $g(\theta) \equiv E_\theta(T(\mathbf{X}))$, and let $g'(\theta) = \frac{\delta}{\delta\theta} g(\theta)$ for all $\theta \in \Omega$. Suppose that as in the definition of the information function $I(\theta)$, the order of differentiation with respect to $\theta$ and integration with respect to **x** may be exchanged. Also suppose that

$$\int_{-\infty}^{\infty} T(\mathbf{x}) \frac{\delta}{\delta\theta} f(\mathbf{x};\theta) d\mathbf{x} = \frac{\delta}{\delta\theta} \int_{-\infty}^{\infty} T(\mathbf{x}) f(\mathbf{x};\theta)\, d\mathbf{x} \qquad \text{for} \quad \theta \in \Omega.$$

Then we have the inequality

$$\text{Var}_\theta(T(\mathbf{X})) \geq \frac{[g'(\theta)]^2}{I(\theta)}$$

with equality if and only if there exist constants $a(\theta)$ and $b(\theta)$ such that

$$T(\mathbf{X}) = a(\theta) + b(\theta)\ell(\theta;\mathbf{X}) \tag{7.8.1}$$

with probability 1 for all $\theta \in \Omega$. $[g'(\theta)]^2/I(\theta)$ is the *Cramér–Rao lower bound* on $\text{Var}_\theta(T(X))$. The inequality is called the *Cramér–Rao inequality* or the *information inequality*.

COMMENTS:

1. If $E_\theta(T(\mathbf{X})) = \theta$ for $\theta \in \Omega$ then $g(\theta) = \theta$, and $g'(\theta) = 1$, so that $\text{Var}_\theta(T(\mathbf{X})) \geq 1/I(\theta)$. This means that every unbiased estimator of $\theta$ that satisfies the exchange conditions has variance at least $1/I(\theta)$ and that if $T(\mathbf{X})$ attains this lower bound, $T(\mathbf{x})$ is a linear function of $\ell(\theta;\mathbf{x})$, where the coefficients may depend on $\theta$.
2. The function $T(\mathbf{x})$ in comment 1 is an estimator only if it is the same for all $\theta \in \Omega$.
3. The inequality holds in the case that **X** has probability mass function $f(\mathbf{x};\theta)$. For the proof, replace integration by summation.
4. If $\mathbf{X} = (X_1, \ldots, X_n)$ is a random sample from $f(x;\theta)$ and $I(\theta)$ is the information for one observation, the information function for **X** is $I_n(\theta) = nI(\theta)$, so the Cramér–Rao lower bound on $\text{Var}_\theta(T(\mathbf{X}))$ is $[g'(\theta)]^2/[nI(\theta)]$.

***Example 7.8.1 Revisited*** For the case that $\mathbf{X} = (X_1, \ldots, X_n)$ is a random sample from $N(\mu, \sigma^2)$ with $\mu$ unknown, $\sigma^2$ known, the information function is $I_n(\mu) = n/\sigma^2$, the same for all $\mu$. It follows that every unbiased estimator $T(\mathbf{X})$ of $\mu$ has variance at least $\sigma^2/n$.

The estimator $\bar{X}$ attains this lower bound. We know, for example, that Median(**X**) must have a larger variance for every $\mu$. It can be shown that $n$ Var(Median(**X**)) converges to $\pi/2$ for samples from the normal distribution. The median can do better than the mean for samples from other distributions, especially for those with "heavy tails." □

***Example 7.8.2*** Let $X \sim$ Binomial$(n, p), 0 < p < 1$. The log-likelihood function is $\log(L(p; x)) = C + x \log(p) + (n - x) \log(1 - p)$, so that $\ell(p; X) = x/p - (n - x)/(1 - p)$ and $\dot{\ell}(p; \mathbf{X}) = -x/p^2 + (n - x)/(1 - p)^2$. $C$ does not depend on $p$. Thus, $I(p) = -E_p(\dot{\ell}(p; \mathbf{X})) = n/p + n/(1 - p) = n/[p(1 - p)]$. This could also have been found by determining $\text{Var}_p(\ell(p; X)) = \text{Var}_p(X/[p(1 - p)])$ directly. Thus, every unbiased estimator has variance at least $1/I(p) = p(1 - p)/n$, the variance of $\hat{p} = X/n$. □

*Proof of the Cramér–Rao Inequality:* The inequality follows directly from the fact that the square of the correlation coefficient $\rho = \rho(T(X), \ell(\theta; X))$ satisfies $\rho^2 \leq 1$. Equivalently, $\text{Cov}(T(X), \ell(X; \theta))/\ \text{Var}_\theta(\ell(X; \theta)) \leq \text{Var}_\theta(T(X))$, a consequence of the Cauchy–Schwartz inequality. Since $E_\theta(\ell(\theta; X)) = 0$, the left side is $E_\theta(T(X)\ell(X; \theta))$, which, after exchange of the order of differentiation and integration, is $\frac{\delta}{\delta\theta} E_\theta(T(X)) = \frac{\delta}{\delta\theta} g(\theta)$. Since $\text{Var}_\theta(\ell(\theta; \text{X})) = I(\theta)$ the inequality is proved. Since $\rho(U, V)^2 = 1$ for any two random variables if and only if there exist constants $a, b$ such that $V = a + bU$ with probability 1, statement (7.8.1) holds. □

### Asymptotic Normality of MLEs

In a variety of cases, MLEs are asymptotically normally distributed. We state a theorem largely due to Cramér indicating this. The result, taken from Thomas Ferguson's *A Course in Large Sample Theory* (1996), is rather technical and will not be proved here. It does not specifically concern MLEs, but instead, refers to roots of the likelihood equation $\ell(\theta; \mathbf{x}) = 0$ when $\mathbf{X} = (X_1, \ldots, X_n)$ consists of iid random variables.

**Theorem 7.8.3** Let $X_1, \ldots, X_n$ be iid with density $f(x; \theta), \theta \in \Omega$, and let $\theta_0$ denote the true parameter. If the following hold:

- **(a)** $\Omega$ is an open subset of $R_1$.
- **(b)** Second partial derivatives of $f(x; \theta)$ with respect to $\theta$ exist and are continuous for all $x$ and may be passed under the integral sign in $\int f(x; \theta) dx$.
- **(c)** There exists a function $K(x)$ such that $E_{\theta_0}[K(X)] < \infty$ and $\dot{\ell}(\theta; x)$ is bounded in absolute value by $K(x)$ uniformly in some neighborhood of $\theta_0$ for each $x$.
- **(d)** $I(\theta_0) = -E_{\theta_0}[\dot{\ell}(\theta;\mathbf{X})] > 0$.
- **(e)** $P_\theta(f(X; \theta_1) = f(X; \theta_0))$ for all $\theta \in \Omega$ implies that $\theta_1 = \theta_0$.

Then there exists a consistent sequence $\{\hat{\theta}_n\}$ of roots of the likelihood equation such that

$$Z_n \equiv \sqrt{n}\ (\hat{\theta}_n - \theta_0) \tag{7.8.2}$$

converges in distribution to $N(\theta, 1/I(\theta))$.

COMMENTS:

1. If a root $\hat{\theta}_n$ of the likelihood equation exists and is unique for $n > \mathrm{N}(\theta_0)$, then $\hat{\theta}_n$ is the MLE and (7.8.2) holds. It is possible that there may be several roots, and choosing one of these for each $n$ may not produce a sequence for which (7.8.2) holds.
2. The sequence $\{\hat{\theta}_n\}$ may be chosen to be *strongly consistent* in the sense that $P_{\theta_0}(\lim_{n\to\infty}\hat{\theta}_n = \theta_0) = 1$.
3. $f(x;\theta)$, as before, may be a probability mass function.

Theorem 7.8.1 may be generalized as follows. Suppose that $\boldsymbol{\theta} = (\theta_1, \ldots, \theta_k)$ takes its values in $\Omega$, an open subset of $R_k$. Let the score function be $\ell(\boldsymbol{\theta};\mathbf{x})$, the $k$-component vector of partial derivatives with respect to $\theta_1, \ldots, \theta_k$. Let $\dot{\ell}(\boldsymbol{\theta};\mathbf{X})$ be the $k \times k$ matrix of second partial derivatives. Define $I(\boldsymbol{\theta}_0) = -E_{\boldsymbol{\theta}_0}[\dot{\ell}(\boldsymbol{\theta};\mathbf{X})]$ as before. $I(\boldsymbol{\theta}_0)$ is the information matrix. Then $Z_n$ is a $k$-component vector and $1/I(\theta_0)$ in (7.8.2) becomes $I(\boldsymbol{\theta}_0)^{-1}$, the inverse of the information matrix.

***Example 7.8.3*** Let $X_1, \ldots, X_n$ be a random sample from $\mathrm{N}(\mu, \sigma^2)$ with $\sigma^2$ known. Then for one observation $X$, $\ell(\mu;x) = (x-\mu)/\sigma$, and $\dot{\ell}(\theta;\mathbf{X}) = -1/\sigma^2$, so $I(\mu) = 1/\sigma^2$. Since the MLE $\bar{X}_n$ is the unique solution to the likelihood equation, $Z_n = \sqrt{n}(\bar{X}_n - \mu)$ converges in distribution to $\mathrm{N}(0, \sigma^2)$.

Now suppose that both $\mu$ and $\sigma^2$ are unknown, so that $\Omega = R_1 \times R_1^+$. Letting $\eta = \sigma^2$, we get $\frac{\delta}{\delta\mu}\log(L(\mu,\eta;x)) = -(x-\mu)/\eta$ and $\frac{\delta}{\delta\eta}\log(L(\mu,\eta;x)) = -1/(2\eta) + (x-\mu)^2/(2\eta^2)$ and the second partial derivatives are $\frac{\delta^2}{\delta\mu^2}\log(L(\mu,\eta;x)) = -1/\eta$, $\frac{\delta^2}{\delta\mu\,\delta\eta}\log(L(\mu,\sigma^2;x)) = \frac{\delta^2}{\delta\eta\,\delta\mu}\log(L(\mu,\eta;x)) = (x-\mu)/\eta$, $\frac{\delta^2}{\delta\eta}\log(L(\mu,\eta;x)) = -1/(2\eta^2) - (x-\mu)^2/\eta^3$. Taking expectations of the $2\times 2$ matrix of partial derivatives, we get $I(\mu,\sigma^2) = \begin{pmatrix} 1/\sigma^2 & 0 \\ 0 & 1/(2\sigma^4)\end{pmatrix}$. Thus, the MLE pair $(\bar{X}, \hat{\sigma}^2)$ is asymptotically normally distributed in the sense that $\sqrt{n}[(\bar{X}, \hat{\sigma}^2) - (\mu, \sigma^2)]$ converges in distribution to the bivariate normal with mean vector (0, 0), covariance matrix $I(\mu,\sigma^2)^{-1} = \begin{pmatrix} \sigma^2 & 0 \\ 0 & 2\sigma^4\end{pmatrix}$. Actually, $\mathrm{Var}(\sqrt{n}\hat{\sigma}^2) = 2\sigma^4[(n-1)/n]$, which converges to $2\sigma^4$. □

### Problems for Section 7.8

**7.8.1** Let $X_1, \ldots, X_n$ be a random sample from the exponential distribution with mean $1/\theta > 0$.

**(a)** Find the Cramér–Rao lower bound on the variance of unbiased estimators of $\theta$.

**(b)** Find the MLE $\hat{\theta}$ of $\theta$.

**(c)** Find a constant $C(n)$ so that $\hat{\theta}_2 = C(n)\hat{\theta}$ is an unbiased estimator of $\theta$. Find $\text{Var}_\theta(\hat{\theta}_2)$ and show that it satisfies the inequality in part (a) (see Problem 7.2.5).

**(d)** Find a constant $d(\theta)$ such that $Z_n = (\hat{\theta}_2 - \theta))/[n^{1/2}d(\theta)]$ converges in distribution to N(0, 1).

**(e)** Use $\hat{Z}_n = (\hat{\theta} - \theta))/[n^{1/2}b(\hat{\theta})]$ as a pivotal quantity to find a formula for a 95% confidence interval on $\theta$ for the case that $n$ is "large." That $\hat{Z}_n$ and $Z_n$ have the same limiting distribution follows from Slutsky's theorem.

**7.8.2** Repeat problem 7.8.1 for the case that $X_1, \ldots, X_n$ is a random sample from the Poisson distribution with mean $\theta > 0$.

**7.8.3** Let $\mathbf{X} = (X_1, \ldots, X_n)$ be a random sample from the geometric distribution with parameter $\theta, 0 < \theta < 1$.

**(a)** Find the information function $I_n(\theta)$ and use this to give an upper bound on the variance of unbiased estimators of $\theta$.

**(b)** Find the Cramér–Rao bound on the variance of unbiased estimators of $\theta$.

**(c)** Show that the MLE and MME for $\theta$ are both $1/\bar{X}$.

**(d)** Use the $\delta$-method to find an approximation for $\text{Var}(\hat{\theta})$ for $n$ large.

**(e)** Give a formula for an approximate 95% confidence interval on $\theta$ for $n$ large.

**7.8.4** Let $\mathbf{X} = (X_1, \ldots, X_n)$ be a random sample from the Pareto distribution with cdf $F(x; \alpha) = 1 - x^{-\alpha}$ for $x > 1, \alpha > 0$.

**(a)** Find the Cramér–Rao bound on the variance of unbiased estimators of $\alpha$.

**(b)** Show that the MLE for $\alpha$ is $\hat{\alpha}_n = 1/\bar{Y}$, where $Y_i = \log(X_i)$ for each $i$, and $Y_i$ has the exponential distribution with mean $1/\alpha$. Use the fact that $\Sigma Y_i \sim \Gamma(n, 1/\alpha)$ to show that $E(\hat{\alpha}_n) = \alpha n/(n-1)$ and $\hat{\alpha}_n^* \equiv \hat{\alpha}_n(n-1)/n$ is an unbiased estimator of $\alpha$. Give an upper bound on $\text{Var}(\hat{\alpha}_n)$.

**(c)** Use the $\delta$-method to find a constant $C(\alpha)$ and to prove that $\sqrt{n}C(\alpha)(\hat{\alpha}_n - \alpha)$ converges in distribution to standard normal. Use this and an estimator of $C(\alpha)$ to give a formula for an approximate 95% confidence interval on $\alpha$.

**7.8.5** Let $X_1, \ldots, X_n$ be a random sample from the normal distribution with known mean $\mu$, unknown variance $\eta = \sigma^2 > 0$.

**(a)** Find the Cramér–Rao lower bound on the variance of unbiased estimators of $\eta$. *Hint*: Determine the score function and $I_1(\eta)$.

**(b)** Find the MLE $\hat{\eta}$ of $\theta$.

**(c)** Find a constant $C(n)$ so that $\hat{\eta}_2 = C(n)\hat{\eta}$ is an unbiased estimator of $\eta$. Find $\text{Var}_\eta(\hat{\eta}_2)$ and show that it satisfies the inequality in part (a).

**(d)** Find a constant $d(\eta)$ such that $Z_n = (\hat{\eta} - \eta))/[n^{1/2}d(\eta)]$ converges in distribution to N(0, 1).

**(e)** Give a formula for a 95% confidence interval on $\eta$, assuming that $n$ is large.

**7.8.6** A genetic theorem states that for $n$ twin pairs the number having property $A$ (statistical brilliance, for example) has a binomial distribution for some $\theta, 0 < \theta < 1$. To test this theory, $n = 1000$ pairs are observed. Let $X_k$ be the number of such pairs having $k$ with property $A$. Let $\mathbf{X} = (X_0, X_1, X_2)$.

**(a)** Give the likelihood function for an observed vector $\mathbf{X}$. *Hint*: Each $X_k$ has a binomial distribution with $n = 1000$. What is the distribution of $\mathbf{X}$?

**(b)** Find the ML estimator $\hat{\theta}$ for $\theta$, and use this to estimate $\theta$ for $X_0 = 94$, $X_1 = 412$, $X_2 = 494$. Estimate $p(k;\theta) =$ probability that $k$ members of a pair have property $A$ for $k = 0, 1, 2$ and compare these to $X_k/1000$ for each $k$. Does the fit seem to be good?

**(c)** Show that the MLE for $\hat{\theta}$ is unbiased for $\theta$.

**(d)** Give the Cramér–Rao lower bound $B(\theta, n)$ on $\text{Var}_\theta(\hat{\theta})$ and show that $\text{Var}_\theta(\hat{\theta}) = B(\theta, n)$.

**(e)** Give a formula for a 95% CI on $\theta$. Find it for the data of part (b).

**7.8.7** Let $X_{ij}$ for $i = 0, 1$ and $j = 0, 1$ be the frequency of occurrences of $(i, j)$ in a $2 \times 2$ table with rows labeled 0, 1 and columns labeled 0, 1. Suppose that $\mathbf{X} = (X_{00}, X_{01}, X_{10}, X_{11})$ has the multinomial distribution with parameters $n$ (known) and $\mathbf{p} = \mathbf{p}(\theta_1, \theta_2) = (p_{00}, p_{01}, p_{10}, p_{11})$, where $p_{ij} = p_{ij}(\theta_1, \theta_2) = \theta_1^i(1-\theta_1)^{1-i}\,\theta_2^j(1-\theta_2)^{1-j}$ for $i = 0, 1$ and $j = 0, 1$. Thus, row and columns are independent, with probabilities $(1-\theta_1)$ and $\theta_1$ for rows, $(1-\theta_2)$ and $\theta_2$ for columns. Suppose that $\mathbf{X}$ is observed, $0 < \theta_1 < 1, 0 < \theta_2 < 1$, with both $\theta_1$ and $\theta_2$ unknown.

**(a)** Find the MLE $\hat{\theta}$ for $\theta = (\theta_1, \theta_2)$. Evaluate it for $n = 1000, \mathbf{X} = (115, 185, 285, 415)$. For its intuitive value, write $\mathbf{X}$, $\mathbf{p}$, and $\hat{\mathbf{p}} = \mathbf{p}(\hat{\theta})$ as $2 \times 2$ matrices.

**(b)** Use the asymptotic properties of MLEs to determine 95% CIs on $\theta_1$ and on $\theta_2$ for the data of part (a).

## 7.9 SUFFICIENCY

Let $X_1, X_2$ be iid Poisson($\lambda$) for $\lambda > 0$, and let $T = T(X_1, X_2) = X_1 + X_2$. As was shown in Section 5.1, the conditional distribution of $X_1$, given $T = t$, is binomial with parameters $n = t$ and $p = \lambda/(\lambda + \lambda) = 1/2$. Knowledge that $X_1 = x_1$, given that we know that $T = t$, provides no additional information about $\lambda$. Thus, in recording the results we need only record $T$, as long as we believe that $X_1$ and $X_2$ satisfy the model. Put another way, on the contours $C(t) = \{(x_1, x_2)|x_1 + x_2 = t\}$ the distribution of

$(X_1, X_2)$ is the same for all $\lambda$. We will show that it is enough (sufficient), when we consider estimators of $\lambda$, to consider estimators that are constant on these contours [i.e., depend on $(X_1, X_2)$ only through $T = T(X_1, X_2)$. To do otherwise will increase the variance of the estimator.

Consider another example. Let $X$ have probability mass function $f(k; \theta)$ for $\theta = 1, 2, 3$.

| | $k$ | | | | |
|---|---|---|---|---|---|
| | −2 | −1 | 0 | 1 | 2 |
| $f(k;1)$ | 0.1 | 0.2 | 0.1 | 0.4 | 0.2 |
| $f(k;2)$ | 0.0 | 0.3 | 0.1 | 0.6 | 0.0 |
| $f(k;3)$ | 0.2 | 0.0 | 0.4 | 0.0 | 0.4 |

The conditional distribution of $X$, given $T = |X| = t$, puts mass 1/3, 2/3 on $-t$ and $+t$, whenever defined, for each of $\theta = 1, 2, 3$. $T = |X|$ is therefore sufficient for $\theta$. Note that the conditional distribution of $X$, given $|X| = 1$, is not defined for $\theta = 3$.

We're now ready for a formal definition.

**Definition 7.9.1** Let $\mathbf{X} = (X_1, \ldots, X_n)$ have density or probability mass function $f(x; \theta)$ for $\theta \in \Omega$, and let $T = T(\mathbf{X})$ be a statistic. Let $f(\mathbf{x} \mid T(\mathbf{X}) = t; \theta)$ be the conditional density or mass function of $\mathbf{X}$, given $T(\mathbf{X}) = t$. If for all $\mathbf{x}$ and $t$, $f(\mathbf{x} \mid T(\mathbf{X}) = t; \theta)$ is the same, whenever defined, for all $\theta \in \Omega$, then $T(\mathbf{X})$ is said to be *sufficient* for $\theta \in \Omega$. □

***Example 7.9.1*** Let $\mathbf{X} = (X_1, \ldots, X_n)$ be a random sample from the Bernoulli distribution with parameter $\theta, 0 \le \theta \le 1$. Let $T = T(\mathbf{X}) = \Sigma X_i$. Then $T \sim$ Binomial$(n, \theta)$ and the conditional mass function for $\mathbf{X}$, given $T = t$, is $f(\mathbf{x} \mid T = t; \theta) = f(\mathbf{x}; \theta)/b(t; n, \theta) = \prod_{i=1}^{n} \theta_i^{x}(1-\theta)^{1-x_i} / \binom{n}{t}\theta^t(1-\theta)^{n-t} = 1/\binom{n}{t}$, where $\mathbf{x} = (x_1, \ldots, x_n)$ is a vector of 0's and 1's with sum $t$. Thus, given $T(X) = t$, $\mathbf{X}$ has, for all $\theta$, the uniform distribution on the collection of $\mathbf{x}$ with sum $t$, a *simplex* in $n$-space. Thus, $T(\mathbf{X})$ is sufficient for $\theta$. □

COMMENTS:

1. $T(\mathbf{X})$ can be a vector. In particular, we could take $T(\mathbf{X}) = \mathbf{X}$. Since, in this case, $f(\mathbf{x} \mid T(\mathbf{X}) = \mathbf{t}; \theta) = 1$ for all $\mathbf{t}$ and $\mathbf{x}$ and $\theta \in \Omega$, $\mathbf{X}$ itself is sufficient of $\theta \in \Omega$. If $T(\mathbf{X})$ instead is defined to be the vector of order statistics $(X_{(1)}, X_{(2)}, \ldots, X_{(n)})$ and $f(\mathbf{x}; \theta)$ is symmetric in the components of $\mathbf{x}$ for every $\theta \in \Omega$, then for all $\theta \in \Omega$, $f(\mathbf{x} \mid T(\mathbf{X}) = \mathbf{t}; \theta) = 1/n!$ for all $\mathbf{x}$ whose vector of order statistics is $\mathbf{t}$, so that the vector of order statistics is sufficient for $\theta \in \Omega$.
2. If $\mathbf{X}$ is discrete and $T(\mathbf{X})$ is sufficient for $\theta \in \Omega$ and $A$ any event, then $P(A \mid T(\mathbf{X}) = t; \theta) = \Sigma_{\mathbf{x} \in A} f(\mathbf{x} \mid T(\mathbf{X}) = t; \theta)$ is the same for every $\theta \in \Omega$. If $\mathbf{X}$ is continuous, the argument is the same with an $n$-fold integral replacing summation.

3. If $T(\mathbf{X})$ is sufficient for $\theta \in \Omega$ and there are functions $U(\mathbf{x})$ and $g(u)$ such that $T(\mathbf{x}) = g(U(\mathbf{x}))$ for all $\mathbf{x}$, then $U(\mathbf{X})$ is also sufficient for $\theta \in \Omega$. For example, if $T(\mathbf{X})$ is sufficient for $\theta$, so is $cT(\mathbf{X})$ for any constant $c \neq 0$ and so is any pair $(T(\mathbf{X}), h(\mathbf{X}))$. It is the partition of the sample space that the statistic $T(\mathrm{X})$ provides that is important, not the specific values taken. The following proof for the discrete case should make this clear.

   Let $\mathbf{x}$ be in the range of $\mathbf{X}$, let $t = T(\mathbf{x})$, and let $u = U(\mathbf{x})$. Then, since $T(\mathbf{x}) = g(U(\mathbf{x})), t = g(u)$. Let $A = [\mathbf{X} = \mathbf{x}]$, $B = [U(\mathbf{X}) = u]$, and $C = [T(\mathbf{X}) = t]$. Then $A \subset B \subset C$, so that $P(A|B;\theta) = P(A;\theta)/P(B;\theta) = P(A|C;\theta)/P(B|C;\theta)$. But from the definition of sufficiency and from comment 2, both the numerator and denominator are the same for all $\theta \in \Omega$. Thus, $U(\mathbf{X})$ is sufficient for $\theta \in \Omega$. The argument for the continuous case requires ideas from measure theory.
4. The sufficiency property for a statistic $T$ depends crucially on the parameter space. If, we increase the size of $\Omega$, then $T$ may no longer be sufficient. If, for the example described in the second paragraph of the opening text in this section, we add the parameter value $\theta = 4$, with the corresponding probabilities 0.1, 0.5, 0.1, 0.2, 0.1 for $x = -2, -1, 0, 1, 2$, then $|X|$ (or $X^2$) is no longer sufficient. On the other hand, $|X|$ **is** sufficient if these probabilities are 0, 0.2, 0.4, 0.4, 0.

***Example 7.9.2*** Let $\mathbf{X} = (X_1, \ldots, X_n)$ be a random sample from the Unif$(\theta_1, \theta_2)$ distribution for $\theta_1 < \theta_2$. Let $T_1 = \min(X_1, \ldots, X_n)$ and $T_2 = \max(X_1, \ldots, X_n)$. It was shown in Section 3.5 for the case $\theta_1 = 0, \theta_2 = 1$ that $(T_1, T_2)$ have joint density $n(n-1)(t_2 - t_1)^{n-2}$ for $0 \le t_1 < t_2 \le 1$, so that in the general case $(T_1, T_2)$ has density $n(n-1)(t_2 - t_1)^{n-2}/\Delta^n$ for $\theta_1 \le t_1 < t_2 \le \theta_2$, where $\Delta = \theta_2 - \theta_1$. Since $\mathbf{X}$ has density $1/\Delta^n$ for all $\mathbf{x}$ in the $n$-cube $[\theta_1, \theta_2]^{(n)}$, and the ratio of the density of $\mathbf{X}$ to that of $(T_1, T_2)$ is uniformly $n(n-1)(t_2 - t_1)^{n-2}$ for those $\mathbf{x}$ for which $\min(\mathbf{x}) = t_1$, and $\max(\mathbf{x}) = t_2$ , the pair $(T_1(\mathbf{X}), T_2(\mathbf{X}))$ is sufficient for $\theta = (\theta_1, \theta_2)$. □

***Example 7.9.3*** Let $\mathbf{X} = (X_1, \ldots, X_n)$ be a random sample from the exponential distribution with mean $\theta > 0$. Let $T = T(\mathbf{X}) = \sum_{i=1}^n X_i$. Then $T \sim \Gamma(n, \theta)$, with density $f_T(t;\theta) = \theta^{-n}\Gamma(n)^{-1}t^{n-1}e^{-t/\theta}$, so that the conditional density of $\mathbf{X}$, given $T(\mathbf{X}) = t$, is $f(\mathbf{x}\,|T(\mathbf{X}) = t;\theta) = \Gamma(n)^{-1}$ on the simplex $\{\mathbf{x}|\sum_{i=1}^n x_i = t, \text{all } x_i > 0\}$. Since this is the same for all $\theta > 0$, $T(\mathbf{X})$ is sufficient for $\theta$. By comment 3,$1/T(\mathbf{X})$ is also sufficient. $\sum_{i=1}^n X_i^2$ is not. □

For our examples it was necessary to select a candidate for sufficiency, find its distribution for each parameter value $\theta$, then show that the conditional distribution of $\mathbf{X}$ was the same for all $\theta$ in the parameter space. The following theorem, due to Jerzy Neyman (1935), reduces the difficulty.

**Theorem 7.9.1 (Neyman Factorization)** Let $f(\mathbf{x};\theta)$ be the density or probability function for $\mathbf{X} = (X_1, \ldots, X_n)$ for $\theta \in \Omega$, and let $T(\mathbf{X})$ be a statistic. $T(\mathbf{X})$ is sufficient for $\theta \in \Omega$ if and only if there exist functions $g(t; \theta)$ and $h(\mathbf{x})$ such that

$$f(\mathbf{x}; \theta) = g(T(\mathbf{x}); \theta)h(\mathbf{x}) \tag{7.9.1}$$

for all $\mathbf{x}$ and $\theta \in \Omega$.

We postpone a proof for the discrete case until we consider two examples.

***Example 7.9.3 Revisited*** $\mathbf{X}$ has density $f(\mathbf{x};\theta) = \theta^{-n}e^{-(\Sigma x_i)/\theta}$ for all $x_i > 0$. Let $g(t, \theta) = \theta^{-n}e^{-t/\theta}$ for $t > 0$ and $h(\mathbf{x}) = 1$ for all $x_i > 0$, 0 otherwise. Let $T(\mathbf{x}) = \Sigma x_i$. Then $f(\mathbf{x};\theta) = g(T(\mathbf{x});\theta)h(\mathbf{x})$ for all $\mathbf{x}$ and $\theta > 0$. Thus, by Neyman factorization, $T(\mathbf{X})$ is sufficient for $\theta \in \Omega$. □

***Example 7.9.2 Revisited*** $\mathbf{X} = (X_1, \dots, X_n)$ has joint density $f(\mathbf{x}; \theta_1, \theta_2) = \Delta^{-n}$ for $\theta_1 \le x_i \le \theta_2$, for each $i$, where $\Delta = \theta_2 - \theta_1$. Let $T(\mathbf{x})$ be the pair ($T_1(\mathbf{x}) = \min(\mathbf{x})$, $T_2(\mathbf{x}) = \max(\mathbf{x})$). Let $g((t_1, t_2); (\theta_1, \theta_2)) = \Delta^{-n}$ for $\theta_1 \le t_1 \le t_2 \le \theta_2$, zero otherwise, and let $h(\mathbf{x})$ be identically 1. Then $f(\mathbf{x}; \theta_1, \theta_2) = g(T(\mathbf{x}); (\theta_1, \theta_2))h(\mathbf{x})$ for all $\mathbf{x}$ and all $\theta_1 < \theta_2$, so that $T(\mathbf{X})$ is sufficient for $\theta = (\theta_1, \theta_2)$. □

## Proof of Neyman Factorization for the Discrete Case

*Necessity:* Suppose that $T(\mathbf{X})$ is sufficient for $\theta \in \Omega$. Then for each $t$ and $\mathbf{x}$, $f(\mathbf{x} \mid T(\mathbf{X}) = t; \theta)$ is the same for all $\theta \in \Omega$ and is positive only if $T(\mathbf{x}) = t$. Define $h(\mathbf{x}) = f(\mathbf{x} \mid T(\mathbf{X}) = T(\mathbf{x}); \theta)$. Define $g(T(\mathbf{x});\theta) = P(T(\mathbf{X}) = T(\mathbf{x});\theta)$. Then (7.9.1) follows by the definition of conditional probability. □

*Sufficiency:* Suppose that (7.9.1) holds. $P(T(\mathbf{X}) = t; \theta) = \sum_{T(\mathbf{x})=t} g(T(\mathbf{x}), \theta) h(\mathbf{x}) = g(t; \theta) \sum_{T(\mathbf{x})=t} h(\mathbf{x})$. Define this last sum to be $k(t)$. Then, for each $\theta \in \Omega$, $f(\mathbf{x} \mid T(\mathbf{X}) = t; \theta) = [g(T(\mathbf{x}); \theta)h(\mathbf{x})]/P(T(\mathbf{X}) = t; \theta) = h(\mathbf{x})/k(t) = h(\mathbf{x})/k(T(\mathbf{x}))$ when $T(\mathbf{x}) = t$ and is zero otherwise. Since this is the same for all $\theta \in \Omega$, $T(\mathbf{X})$ is sufficient for $\theta \in \Omega$. □

***Example 7.9.4*** Let $\mathbf{X} = (X_1, \dots, X_n)$ be a random sample from the $\Gamma(\alpha, \theta)$ distribution for $\alpha > 0, \theta > 0$. $X$ has density $f(x; \alpha, \theta) = [\Gamma(\alpha)\theta]^{-n} T_1(\mathbf{x})^{\alpha-1} e^{-T_2(\mathbf{x})}$ for $T_1(\mathbf{x}) = \prod x_i$, and $T_2(\mathbf{x}) = \Sigma x_i$ for $\mathbf{x} \in R_n^+ \equiv \{\mathbf{x} \mid \text{all } x_i > 0\}$. For $t = (t_1, t_2)$, let $g(t; \alpha, \theta) = [\Gamma(\alpha)\theta]^{-n} t_1^{\alpha-1} e^{-t_2}$ for $t_1 > 0, t_2 > 0$. Let $h(\mathbf{x})$ be 1 whenever $\mathbf{x} \in R_n^+$, zero otherwise. Then (7.9.1) holds for $T(\mathbf{x}) = (T_1(\mathbf{x}), T_2(\mathbf{x}))$, so that $T(\mathbf{X})$ is sufficient for $(\alpha, \theta)$.

When $T(\mathbf{X})$ is sufficient for $\theta \in \Omega$, and therefore (7.9.1) holds for some $g$ and $h$, $f(\mathbf{x}; \theta)$ is maximum as a function of $\theta$ if and only $g(T(\mathbf{x}); \theta)$ is maximum. It follows that a maximum likelihood estimator of $\theta$ is a function of each sufficient statistic. In fact, if $T(\mathbf{X})$ is sufficient for $\theta \in \Omega$, any estimator $\hat{\theta}(\mathbf{X})$ of $\theta$ that is not a function of $T(\mathbf{X})$ is "wasteful" in the sense that it "too random." That is, $\hat{\theta}(\mathbf{X})$ should be constant on the contours of $T(\mathbf{X})$ and should be constant when $T(\mathbf{X})$ is constant.

To make this precise, recall two identities established in Theorem 5.2.1. For a pair $(X, Y)$ of random variables for which $\text{Var}(Y)$ exists, with $g(x) \equiv E(Y|X = x)$ and $v(x) = \text{Var}(Y|X = x)$,

$$E(g(X)) = E(Y) \quad \text{and} \quad \text{Var}(Y) = E(v(X)) + \text{Var}(g(X)). \tag{7.9.2}$$

Also from Theorem 5.2.1, $\text{Cov}(Y - g(X), g(X)) = 0$.

Let $T(\mathbf{X})$ be sufficient for $\theta \in \Omega$ and let $\hat{\theta}(\mathbf{X})$ be an estimator of $\theta$. Define $\hat{\theta}^*(\mathbf{X}) = E(\hat{\theta}(\mathbf{X})|T(\mathbf{X}); \theta)$. Notice that although this conditional expectation appears to depend on $\theta$, it follows from the sufficiency of $T(\mathbf{X})$ for $\theta$ that $\hat{\theta}^*(\mathbf{X})$ is the same for all $\theta$, so that $\hat{\theta}^*(\mathbf{X})$ **is** an estimator of $\theta$. Replacing $\mathbf{X}$ by $T(\mathbf{X})$ and $Y$ by $\hat{\theta}(\mathbf{X})$, we have $g(T(\mathbf{x})) = E(\hat{\theta}(\mathbf{X})|T(\mathbf{X}) = T(\mathbf{x})) = \hat{\theta}^*(\mathbf{x})$ and $v(T(\mathbf{x})) = \text{Var}(\hat{\theta}(\mathbf{X})|T(\mathbf{X}) = T(\mathbf{x}))$, $E(\hat{\theta}^*(\mathbf{X})) = E(\hat{\theta}(\mathbf{X}); \theta)$, and $\text{Var}(\hat{\theta}(\mathbf{X}); \theta) = E(v(T(\mathbf{X})); \theta) + \text{Var}(\hat{\theta}^*(\mathbf{X}); \theta)$. Thus, $\text{Var}(\hat{\theta}(\mathbf{X}); \theta) \geq \text{Var}(\hat{\theta}^*(\mathbf{X}); \theta)$ with equality if and only if $\hat{\theta}(\mathbf{X})$ is a constant on each contour $T(\mathbf{x}) = t$ with probability 1. Since $\hat{\theta}(\mathbf{X})$ and $\hat{\theta}^*(\mathbf{X})$ have the same expectations for each $\theta$, $\hat{\theta}^*(\mathbf{X})$ is unbiased for $\theta$ if $\hat{\theta}(\mathbf{X})$ is. $\hat{\theta}^*(\mathbf{X})$ is a "smoothed" version of $\hat{\theta}(\mathbf{X})$. We have proved the famous Rao–Blackwell theorem (see Lehmann and Casella, 1998, pp. 47–48). □

**Theorem 7.9.2** (C. R. Rao and David Blackwell) Let $\mathbf{X}$ have probability mass function or density $f(\mathbf{x}; \theta)$ for $\theta \in \Omega$. Let $T(\mathbf{X})$ be sufficient for $\theta \in \Omega$ and let $\hat{\theta}(\mathbf{X})$ be an estimator of $\theta$. Define $\hat{\theta}^*(\mathbf{X}) = E(\hat{\theta}(\mathbf{X})|T(\mathbf{X}))$. Then:

**(a)** $E(\hat{\theta}^*(\mathbf{X});\theta) = E(\hat{\theta}(\mathbf{X});\theta)$ for all $\theta \in \Omega$.
**(b)** $\text{Var}(\hat{\theta}(\mathbf{X}); \theta) = E(\text{Var}(\hat{\theta}(\mathbf{X})|T(\mathbf{X}))) + \text{Var}(\hat{\theta}^*(\mathbf{X}); \theta)$ for all $\theta \in \Omega$.

***Example 7.9.5*** Let $\mathbf{X} = (X_1, X_2)$ be a random sample from the Poisson distribution with parameter $\lambda > 0$. Then $T(\mathbf{X}) = X_1 + X_2$ is sufficient for $\lambda$. Let $\hat{\lambda} = X_1$. Then $\hat{\lambda}$ is an unbiased estimator of $\lambda$ but is not a function of $T(\mathbf{X})$. Conditionally on $T(\mathbf{X}) = t$, $\hat{\lambda}$ has the binomial distribution with parameters $n = t$, and $p = 1/2$. It follows that for $\hat{\lambda}^*(\mathbf{x}) \equiv E(\hat{\lambda}(\mathbf{X})\,|T(\mathbf{X}) = T(\mathbf{x})) = T(\mathbf{x})/2 = (x_1 + x_2)/2 = \bar{x}$. Of course, $E(\hat{\lambda}^*(\mathbf{X})) = E(\hat{\lambda}) = \lambda$. In addition, $\text{Var}(\hat{\lambda}(\mathbf{X}); \lambda) = E\,(\text{Var}(\hat{\lambda}(\mathbf{X})\mid T(\mathbf{X}))) + \text{Var}(\hat{\lambda}^*(\mathbf{X});\ \theta) = E((1/2)(1 - 1/2)T(\mathbf{X}); \lambda) + \text{Var}(\bar{X}; \lambda) = \lambda/2 + \lambda/2 = \lambda$. On the contour $\{(x_1, x_2)|x_1 + x_2 = t\}$ $\hat{\lambda}$ has conditional expectation $t/2$, conditional variance $t(1/2)(1 - 1/2)$, while $\hat{\lambda}^*$ is constant on each such contour. □

***Example 7.9.6*** Let $\mathbf{X} = (X_1, \ldots, X_n)$ be a random sample from the Bernoulli distribution with parameter $\theta$, $0 \leq \theta \leq 1$. For $1 \leq n_0 < n$, define $S_{n_0} = \Sigma_{i=1}^{n_0} X_i$, $S_n = \Sigma_{i=1}^{n} X_i$, and $D = S_n - S_{n_0} = X_{n_0+1} + \cdots + X_n$, $\hat{\theta} = S_{n_0}/n_0$. Then $\hat{\theta}$ is unbiased for $\theta$, and $\text{Var}(\hat{\theta}; \theta) = \theta(1 - \theta)/n_0$. $S_n$ is sufficient for $\theta$. Let $\hat{\theta}^*(S_n) = E(\hat{\theta}|S_n)$. Conditionally on $S_n = s$, $S_{n_0}$ has the hypergeometric distribution with parameters $n, s$, and $n_0$. It follows that $\hat{\theta}^*(S_n) = S_n/n$ is also unbiased for $\theta$ (see Figure 7.9.1). The conditional expectation of $\hat{\theta}$, given $S_n = s$, is $g(s) = s/n$. The conditional variance of $\hat{\theta}$, given $S_n = s$, is $v(s) = (s/n)(1 - s/n)(n - n_0)/(n - 1)$. $\text{Var}(\hat{\theta}) = \theta(1 - \theta)/n_0 = \text{Var}(\hat{\theta}^*; \theta) + E(v(S_n); \theta)$, so that $E(v(S_n); \theta)) = \text{Var}(\hat{\theta}; \theta) - \text{Var}(\hat{\theta}^*;$

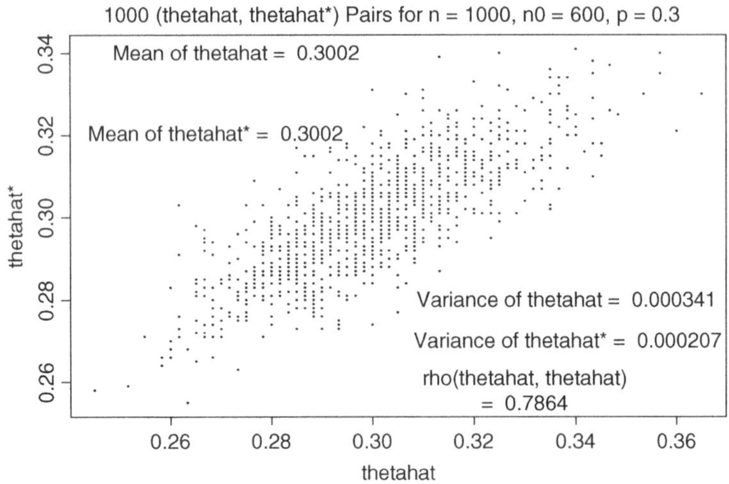


**FIGURE 7.9.1** Estimation of $\theta$ for $n_0$ and $n$ observations.

$\theta) = \theta(1-\theta)(1/n_0 - 1/_n)$. From the remark under (7.9.2) $\text{Cov}(\hat{\theta}^*, \hat{\theta} - \hat{\theta}^*) = 0$. Therefore, $\text{Cov}(\hat{\theta}^*, \hat{\theta}) = \text{Var}(\hat{\theta}^*) = \theta(1-\theta)/n$ and $\rho(\hat{\theta}^*, \hat{\theta}) = [\text{Var}(\hat{\theta}^*)/\text{Var}(\hat{\theta})]^{1/2} = [n_0/n]^{1/2}$. □

## Problems for Section 7.9

**7.9.1** For the example at the beginning of this section, suppose that $f(k; \theta = 3)$ for $k = -2, -1, 0, 1, 2$, are:

**(a)** 0.1, 0.1, 0.4, 0.2, 0.2. Is $|X|$ sufficient for $\theta \in \{1, 2, 3\}$?

**(b)** 0.2, 0.2, 0.4, 0.1, 0.1. Is $|X|$ sufficient for $\theta \in \{1, 2, 3\}$?

**7.9.2** Let $\mathbf{X} = (X_1, \ldots, X_n)$ be a random sample from the geometric distribution with parameter $\theta$, $0 < \theta < 1$.

**(a)** Show that $T(\mathbf{X}) = \Sigma X_i$ is sufficient for $\theta$. Is the pair $(X_1, \sum_{i=2}^n X_i)$ sufficient for $\theta$?

**(b)** Define the functions $g$ and $h$ in the Neyman factorization of the mass function $f(\mathbf{x}; \theta)$ of $\mathbf{X}$.

**7.9.3** Let $\mathbf{X} = (X_1, \ldots, X_n)$ be a random sample from the $N(\mu, \sigma^2)$ distribution with $-\infty < \mu < \infty$, and $\sigma^2 > 0$. Let $\bar{X}$ and $S^2$ be the sample mean and variance. Show that the pair $(\bar{X}, S^2)$ is sufficient for $(\mu, \sigma^2)$.

**7.9.4** Let $\mathbf{X} = (X_1, \ldots, X_n)$ be a random sample from the Poisson distribution with parameter $\lambda > 0$. Let $\theta = e^{-\lambda}$ and $\hat{\theta} = I[X_1 = 0]$ and $T(\mathbf{X}) = \Sigma X_i$.

**(a)** Show that $T(\mathbf{X})$ is sufficient for $\lambda$.

**(b)** Use the Rao–Blackwell theorem to find an unbiased estimator $\hat{\theta}^* = \hat{\theta}^*$ (**X**), which is a function of $T(\mathbf{X})$. *Hint:* If $\mathbf{Y} = (Y_1, \ldots, Y_n)$ has the multinomial distribution with parameters $t$ and $\mathbf{p} = (p_1, \ldots, p_n)$, then $X_1 \sim \text{Binomial}(t, p_1)$ (see Problem 7.2.7).

**(c)** Find the variances of $\hat{\theta}$ and $\hat{\theta}^*$.

**(d)** Are the sequences $\{\hat{\theta} = \hat{\theta}_n\}$ and $\{\hat{\theta}^* = \hat{\theta}^*_n\}$ consistent for $\theta$?

**(e)** What is the MLE for $\theta$?

**7.9.5** An urn contains $B$ black and $R$ red balls, with $N = B + R$ known but $B$ and $R$ unknown. Two balls are drawn randomly and consecutively without replacement. Let $X_i$ be the indicator of the event [black on $i$th draw] for $i = 1, 2$. Show that $T = X_1 + X_2$ is sufficient for $B$.

**7.9.6** Let $x_1, \ldots, x_k$ be $n$-known constants, let $\mu(x) = \beta_0 + \beta_1 x$, where $\beta_0$ and $\beta_1$ are unknown parameters, and let $p(x) = e^{\mu(x)}/(1 + e^{\mu(x)})$ . Suppose that $Y_i \sim \text{Binomial}(n_i, p(x_i))$ is observed for each $i$, independently. $Y_i$ might, for example, be the number of rats that die when $n_i$ are fed dosage $x_i$ of a poison. This is a logistic-regression model.

**(a)** Show that $\mu(x) = \log(p(x)/[1 - p(x)])$, the log odds corresponding to $x$.

**(b)** Let $\mathbf{x_0}$ be the vector of all 1's, let $\mathbf{Y} = (Y_1, \ldots, Y_k)$, and let $\mathbf{x_1} = (x_1, \ldots, x_k)$. Show that the pair of inner products $(\mathbf{Y}, \mathbf{x}_0) = \Sigma\, Y_i$ and $(\mathbf{Y}, \mathbf{x}_1) = \Sigma Y_i x_i$ are sufficient for $(\beta_0, \beta_1)$.

**(c)** Now suppose that $Y_i \sim \text{N}(\mu(x_i), \sigma^2)$, again independent for $i = 1, 2, \ldots, n$. Suppose that $\sigma^2$ is known. Again show that the two inner products in part (b) are sufficient for $(\beta_0, \beta_1)$.

**7.9.7** Let $\mathbf{Y} = (Y_0, Y_1, Y_2)$ be the frequencies of genotypes $aa$, $aA$, and $AA$ of $n$ offspring under random mating in a population with frequency $\theta$, $0 < \theta < 1$ for allele $A$. Thus, the proportions of genotypes $aa$, $aA$, and $AA$ should be $(1 - \theta)^2$, $2\theta(1 - \theta)$, and $\theta^2$. Find a sufficient statistic $T$ for $\theta$, and express the MLE for $\theta$ as a function of $T$.

**7.9.8** Give an example for which a method of moments estimator $\hat{\theta}$ of a parameter $\theta$ and a sufficient statistic $T(\mathbf{X})$ for $\theta$ such that $\hat{\theta}$ is not a function of $T(\mathbf{X})$.

**7.9.9** Let $\mathbf{X} = (X_1, \ldots, X_n)$ be a random sample from the density $f(x; \alpha, \theta) = (1/\theta)e^{-(x-\alpha)/\theta}$ for $x > \alpha$, $\alpha \in R_1$, $\theta \in R_1^+$.

**(a)** Find a sufficient statistic $T(\mathbf{X}) = (T_1(\mathbf{X}), T_2(\mathbf{X}))$ for $(\alpha, \theta)$.

**(b)** Find the MLE for $(\alpha, \theta)$.

**7.9.10** Let $f(w)$ be a probability mass function taking positive values at $w = w_1, \ldots, w_m$, where the $w_j$ are distinct (all different). Let $p_j = f(w_j)$ for each $j$. Suppose that $\mathbf{X} = (X_1, \ldots, X_n)$ is a random sample from $f$. Let $Y_j =$ (no. of $X_i$ equal to $w_j$) $= \sum_{i=1}^n \Delta_{ij}$, where $\Delta_{ij} = I[X_i = w_j]$, the indicator of the event $[X_i = w_j]$.

**(a)** Show that $\mathbf{Y} = (Y_1, \ldots, Y_m)$ is sufficient for $\mathbf{p} = (p_1, \ldots, p_m)$. *Hint*: For $\mathbf{x} = (x_1, \ldots, x_n)$, $P(X_1 = x_j) = \sum_{k=1}^m p_k^{\delta_{jk}}$, where $\delta_{jk} = 1$ if $k = j$ and $\delta_{jk} = 0$ otherwise.

**(b)** Suppose that $f$ takes positive values on a countably infinite set as it does for the Poisson distribution. If $M$ is the maximum $j$ for which any $Y_j$ is positive, is the vector $\mathbf{Y} = (Y_1, \ldots, Y_M)$ (having random length) sufficient for $\mathbf{p} = (p_1, p_2, \ldots)$? Suppose that Neyman factorization also holds for this case.

CHAPTER EIGHT

# Testing of Hypotheses

## 8.1 INTRODUCTION

Suppose that a cancer drug has had a "cure" (having no further symptoms after five years) rate of $p = 0.3$ and that we have developed a new drug which we hope will increase this rate. We decide to choose six new patients at random and try to determine whether the new drug increases the cure rate. In practice we would need to assign patients randomly to old and new drugs, but let's first consider this one-sample experiment. We plan to observe the number $X$ among the six who have been cured after five years. On the basis of this observation $X$, we would like to choose between the null hypothesis $H_0$: $p \leq 0.3$ and the alternative hypothesis $H_a$:$p > 0.3$. For which values of $X$ should we reject the null hypothesis $H_0$? We would like the probability that we reject $H_0$ to be small for $p \leq 0.3$, but at the same time have large probability for $p > 0.3$.

Suppose that we arbitrarily decide to reject $H_0$ for $X = 5$ or 6. For any value of $p$ the probability that we will reject $H_0$ is $P(X \geq 5; p) = 6p^5(1-p) + p^6$. The function $\gamma(p) = P(\text{rejecting } H_0; p)$ is called the *power function*. This function, defined on the parameter space, facilitates the study of the performance of the decision rule (to reject $H_0$ if $X \geq 5$). The power function $\gamma(p)$ takes its maximum value for $p$ satisfying $H_0$ for $p = 0.3$. Its value at $p = 0.3$ is $\gamma(0.3) = 0.01094$. This maximum value for values of $p$ that satisfy the null hypothesis is called the *level of significance* or *the size* of the decision rule (or test) and is usually denoted by $\alpha$. Thus, $\alpha = 0.01094$. Since $\alpha$ is quite small, this seems to be satisfactory. Notice, however, that although we would like $\gamma(p)$ to be large for $p$ satisfying $H_a$:$p > 0.3$, $\gamma(p)$ is quite small for $p$ near 0.3. In fact, $\gamma(0.5) = 7/64 = 0.109$ and $\gamma(0.6) = 0.233$. These small probabilities would be quite unsatisfactory for the manufacturer of the drug that increases the cure rate, but would be very likely be labeled as a drug that does *not* improve the rate.

The test that rejects $H_0$ for $X \geq 4$ has power function $\gamma_4(p) = P(X = 4) + \gamma(p) = 15p^4(1-p)^2 + \gamma(p)$. Its graph is given in Figure 8.1.1. We find

*Models for Probability and Statistical Inference: Theory and Applications*, By James H. Stapleton

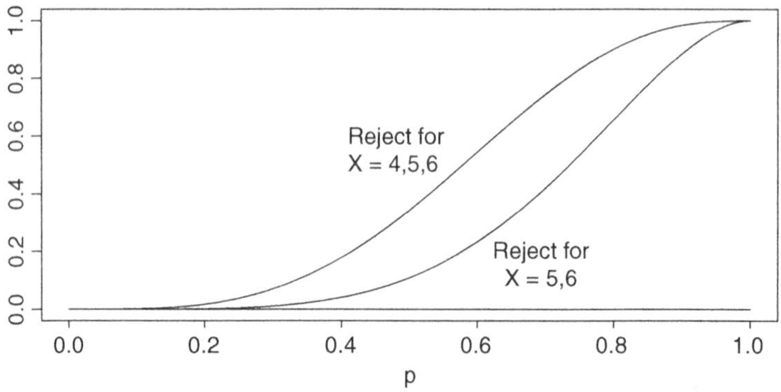


**FIGURE 8.1.1** Power functions for two tests of hypotheses.

$\gamma_4(0.3) = 0.070$, $\gamma_4(0.5) = 0.344$, $\gamma_4(0.6) = 0.544$. Although the level of significance of this test is $\alpha = 0.070$, so that we are more likely to reject a true null hypothesis, the test does perform better for $p > 0.3$. On balance we may prefer the test that rejects for $X \geq 4$.

In general, a *statistical hypothesis* is a statement about the value of a parameter $\theta$, which may be one-, many-, or even infinite-dimensional. For example, we may consider a model for which the distribution of our observations (random variables) is $F$, with $F$ continuous. We might like to test the hypothesis that $F$ is a normal distribution. In this case the parameter $\theta$ is a distribution function. A *null hypothesis* is a statement $H_0$ that $\theta \in \Omega_0$, where $\Omega_0$ is a subset of the parameter space $\Omega$. The *alternative hypothesis* $H_a$ is the statement that $\theta \in \Omega_1 \equiv \Omega_0^c$, the complement of $\Omega_0$. We then choose a decision rule by dividing the sample space into two sets $C$, the critical region, and $C^c$, its complement. If $\mathbf{X} \in C$, we reject $H_0$. If $\mathbf{X} \in C^c$, we say that "we fail to reject $H_0$." It is better to avoid using "accept $H_0$." The philosophy should be that the evidence is insufficient to reject $H_0$, not that we have established the truth of $H_0$.

If $\theta \in \Omega_0$ but our decision is to reject $H_0$, we say that we have made a *type I error* or an *error of the first kind*. If $\theta \notin \Omega_0$ (so $\theta \in \Omega_a$) but we decide not to reject $H_0$, we say that we have made a *type II error* or an *error of the second kind*. The power function corresponding to a decision rule (or *test*) is the function $\gamma(\theta) = P(\mathbf{X} \in C; \theta)$, defined on the parameter space $\Omega$.

***Example 8.1.1*** Let $\mathbf{X} = (X_1, \ldots, X_{25})$ be a random sample from the $N(\mu, \sigma^2)$ distribution with $\sigma^2 = 100$ known but $\mu$ unknown. Suppose that we would like to test the null hypothesis $H_0$: $\mu \geq 80$ versus the alternative hypothesis $H_a$: $\mu < 80$. Suppose also that we would like the level of significance $\alpha$ to be 0.05. For which values of $\mathbf{X}$ should $H_0$ be rejected?

Since $\bar{X}$ is sufficient for $\mu$ it should seem reasonable to base the decision rule on its value. It should also be intuitively clear that we should reject $H_0$ for small $\bar{X}$, say $\bar{X} < c$. Later we show that our intuition is correct in that no other rule can do better, in a certain sense. How should we choose $c$, the *critical value*? The power function is $\gamma(\mu) = P(\bar{\mathbf{X}} < c; \mu) = \Phi((c - \mu)/\sigma/\sqrt{n})$. We want this to be less than or equal

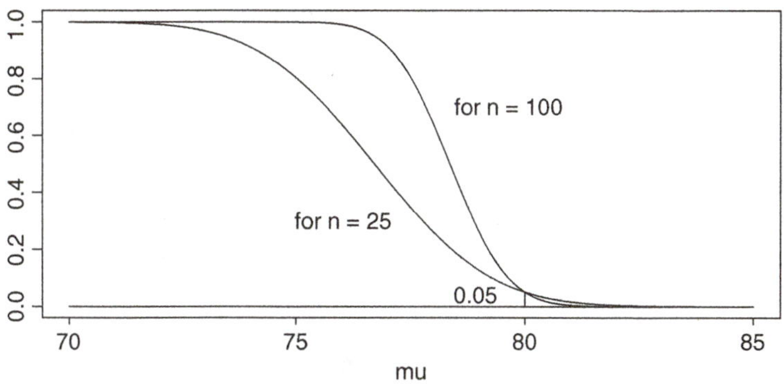


**FIGURE 8.1.2** Power functions for tests of hypotheses on $\mu$.

to $\alpha = 0.05$ whenever $H_0$: $\mu \geq 80$ is true. Since $\gamma(\mu)$ is monotone decreasing in $\mu$, it takes its maximum at $\mu = 80$ for $\mu \geq 80$. Therefore, we need $(c - 80)/(\sigma/\sqrt{n}) = -1.645$, and $c = 80 - 1.645\sigma/\sqrt{n} = 80 - 3.29 = 76.71$. If we change $n$ to 100, for example, the critical value becomes $c^* = 80 - (1.645)(10/\sqrt{100}) = 78.355$. The corresponding power functions for $n = 25$ and $n = 100$ are plotted in Figure 8.1.2. We find, for example, that for $n = 25$, $\gamma(76) = 0.639$ and $\gamma(78) = 0.259$, and for $n = 100$, $\gamma(76) = 0.991$ and $\gamma(78) = 0.639$. □

***Example 8.1.2*** A large manufacturing company has found that the accident rate has been relatively constant at 2.4 per day. The distribution of the numbers of accidents per day may be approximated by the Poisson distribution. After a safety program was begun, the numbers of accidents for the next 30 workdays were observed. Let $X_1, \ldots, X_{30}$ denote the numbers of accidents on these days. The company would like to test $H_0$: $\lambda \geq 2.4$ versus $H_a$: $\lambda < 2.4$, where $\lambda$ is the daily rate under the safety program.

$T = \sum_{i=1}^{n} X_i$ is a sufficient statistic for $\lambda$, so we should choose a decision rule (a test) that is a function of $T$, which has a Poisson distribution with parameter $30\lambda$. We should reject $H_0$ for $T \leq c$, where the critical value $c$ is chosen so that $\gamma(\lambda) \leq \alpha \doteq 0.10$ (an arbitrary choice) for $\lambda \geq 2.4$. Since $\gamma(\lambda)$ is a decreasing function of $\lambda$, we may choose $c$ so that $\gamma(2.4) \doteq 0.10$. For $\lambda = 2.4$, $T \sim \text{Poisson}(30(2.4) = 72)$. Since $30\lambda$ is large, we can use the normal approximation. For $c$ an integer we have $\gamma(\lambda) \doteq \Phi((c + 0.5 - 30\lambda)/\sqrt{30\lambda})$. Thus, we should take $c = 30(2.4) - 0.5 - (1.282)\sqrt{30(2.4)} = 60.62$. In reasonable approximation we should therefore reject $H_0$ for $T \leq 61$, equivalently for $\bar{X} \leq 61/30 = 2.033$. Then $\gamma(2.0) = 0.577$, $\gamma(2.2) = 0.290$, and $\gamma(2.4) = 0.108$. Using S-Plus, we get the more precise values 0.585, 0.295, 0.106. □

## Sample Size Necessary to Achieve a Specific Power

A statistician is often asked to provide advice concerning the sample size to be used in an experiment. He or she is asked, "Is my sample size large enough?" Consider the binomial experiment involving a cancer drug that was discussed at the beginning of the chapter. The experimenter wanted to test $H_0$: $p \leq 0.3$ versus $H_a$: $p > 0.3$. In answer to the question the statistician should ask the experimenter: "What would you like to accomplish? If, for example, $p = 0.50$, what would you like the probability of rejection of $H_0$ (the power) to be?" The experimenter might answer, "100%." After some further discussion about the impossibility of that, she might back off to 99%. Let us determine the sample size necessary to accomplish that. The experimenter may find that the sample size needed to be too large because of time and expense and that she should be less demanding.

Let $n$ be the sample size. (In practice we almost certainly would have to do a study in which we randomly assign available patients to new and old drugs. It is almost impossible to choose a random sample from a population that could be viewed as the same, as was used to determine the original 0.3 cure rate.) Let $X$ be the number among the $n$ patients who are cured. Let us choose $\alpha = 0.05$. We should reject if $X \geq k_n$, where $k_n$ is chosen so that the level of significance is $\alpha = 0.05$. The normal approximation is justified because the sample size needed is large and $p$ is thought not to be very close to zero or 1. This leads to $k_n = 0.3n + 1/2 + 1.645[n(0.3)(0.7)]^{1/2}$ or its nearest integer approximation. In good approximation the power function is then $\gamma(p) = 1 - \Phi(a_n)$, where $a_n = (k_n - 1/2 - np)/[np(1-p)]^{1/2} = n^{1/2}[(0.3 - p)(pq)^{1/2}] - 1.645[0.21/(pq)]^{1/2}$, where $q = 1 - p$. We have used the expression above for $k_n$ rather than its nearest integer approximation.

In order to have $\gamma(0.5) = 0.99$, we need $a_n = z_{0.99} = -2.326$. Solving for $n$, we get $n = [1.645(0.21)^{1/2} + 2.326(pq)^{1/2}]^2/(0.3 - p)^2$. For $p = 0.5$ we get $n = 91.86$, so that a sample size of $n = 92$ seems reasonable. We then get $k_n = 35.33$. The test that rejects for $X \geq 35$ has level of significance 0.0604. For this test $\gamma(0.5) = 0.992$. This was found using an exact binomial computation. In order to have $\alpha < 0.05$, we would need to reject for $X \geq 36$, making $\alpha = 0.0384$. The power for $p = 0.5$ is then 0.986.

The experimenter may decide that she cannot afford such a sample size and ask what the sample size must be to have power 0.9 for $p = 0.5$. Substituting $\Phi^{-1}(0.90) = z_{0.9} = 1.282$ for 2.326, we find that $n = 48.64$, so that a sample size of 49 seems to be appropriate. The test that rejects for $X \geq 21$ gives $\alpha = \gamma(0.3) = 0.0302$, with $\gamma(0.5) = 0.8736$, while the test that rejects for $X \geq 20$ has $\alpha = 0.057$, power $\gamma(0.5) = 0.924$. The choice between these two tests is somewhat arbitrary unless the medical journal in which the experimenter wishes to publish or the U.S. Food and Drug Administration demands $\alpha = 0.05$ level tests. The power functions for the test that rejects $H_0$ for $X \geq 36$ for $n = 92$ and the test that rejects $H_0$ for $X \geq 21$ for $n = 49$ are graphed in Figure 8.1.3.

### Two-Sided Alternatives and Tests

Return to the cancer example, with $n = 20$ patients, and suppose that we aren't sure whether the new drug will decrease or increase the probability that a patient will survive. We might then test $H_0$: $p = 0.3$ versus $H_a$: $p \neq 0.3$. Although it is traditional to set this up as a two-decision problem, it really is, or should be, a three-decision problem. That is, we really wish to decide among three possibilities: $H_0$, $H_1$: $p < 0.3$

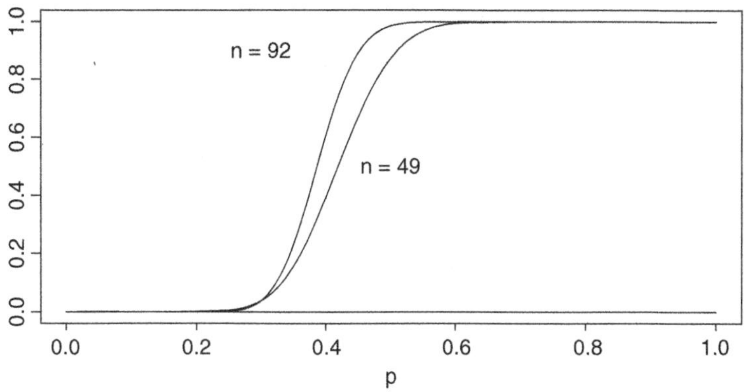


**FIGURE 8.1.3** Power functions for sample sizes 49 and 82.

(the new drug is worse than the old), and $H_2$: $p > 0.3$ (the new drug is better than the old). We use the language of the two-decision problem and show how it relates to the three-decision problem.

Since $P(X \geq 10; p = 0.3) = 0.048$ and $P(X \leq 2; p = 0.3) = 0.035$, the test of $H_0$ versus $H_a$ which rejects for $X \geq 10$ and for $X \leq 2$ has level of significance (size) 0.083. Of course, if, for example, we observed that $X = 11$, it would be rather silly merely to state that we think that $H_a$ is true. A statistician who submitted a report to the vice-president for research of his drug company which stated that "$p$ is not 0.3" is likely to be asked: "Is $p$ larger than 0.3 or smaller?" *if* the VP managed to keep her temper. The statistician should say that in the case that $X \leq 2$ that $H_1$ is true, and if $X \geq 10$, say that $H_2$ is true, and go on from there to discuss error probabilities, if the VP will listen.

Let's study the power function for the two-decision problem and see how that relates to the three-decision problem. In Figure 8.1.4 three graphs are presented: (1) $\gamma_1(p) = P(X \leq 2; p)$, (2) $\gamma_2(p) = P(X \geq 10; p)$, and (3) $\gamma(p) = \gamma_1(p) + \gamma_2(p)$, which is the power function for the test of $H_0$ versus $H_a$.

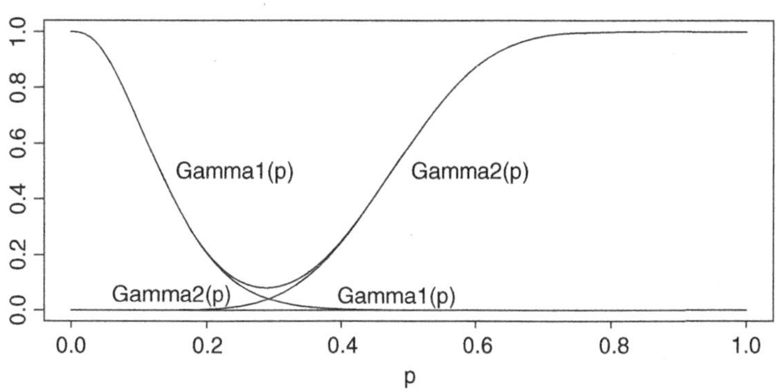


**FIGURE 8.1.4** Power functions for one- and two-sided tests.

Notice that $\gamma_1(p) < 0.035$ for $p > 0.3$ and approaches zero rapidly as $p$ increases. Similarly, $\gamma_2(p) < 0.048$ for $p < 0.3$ and approaches zero rapidly as $p$ decreases. Therefore, $\gamma(p)$ is close to $\gamma_1(p)$ for $p$ a bit larger than 0.3, and $\gamma(p)$ is close to $\gamma_2(p)$ for $p$ a bit smaller than 0.3. For example, $\gamma_1(0.32) = 0.0235$, $\gamma_2(0.32) = 0.0790$, $\gamma_1(0.34) = 0.0152$, $\gamma_2(0.34) = 0.1032$. Thus, the graphs for $\gamma_1$ and $\gamma$ almost coincide for $p > 0.3$, and the graphs for $\gamma_2$ and for $\gamma$ almost coincide for $p < 0.3$. Therefore, not much harm is done by considering what really should be a three-decision problem as a two-decision problem, $H_0$ versus $H_a$.

### Problems for Section 8.1

**8.1.1** Let $X_1, X_2$ be a random sample from $f(k;\theta)$ for $\theta = 1, 2, 3$, where $f$ is as follows:

| | $k$ | | |
|---|---|---|---|
| | 0 | 1 | 2 |
| $f(k;\theta=1)$ | 0.1 | 0.3 | 0.6 |
| $f(k;\theta=2)$ | 0.3 | 0.4 | 0.3 |
| $f(k;\theta=3)$ | 0.5 | 0.4 | 0.1 |

Let $H_0$: $\theta = 1$ or 2 and $H_a$: $\theta = 3$. Consider the test that rejects for $T \equiv X_1 + X_2 \leq 1$.

**(a)** Give the power function $\gamma(\theta)$.

**(b)** Give the level of significance of this test.

**(c)** Consider the test that rejects for $\max(X_1, X_2) \leq 1$. Give its power function. Is this a better test?

**(d)** Can you improve on the test that rejects for $T \leq 1$ by taking one more observation $X_3$ by using the statistic $T_3 = X_1 + X_2 + X_3$?

**8.1.2** Let $X$ have the binomial distribution with $n = 100$, unknown $p$. Suppose that we wish to test $H_0$: $p \geq 0.8$ versus $H_a$: $p < 0.8$.

**(a)** For $\alpha =$ level of significance $\doteq 0.10$, for which values of $X$ should $H_0$ be rejected? Use the normal approximation.

**(b)** Express the power function $\gamma(p)$ in terms of the standard normal cdf $\Phi$ and $p$.

**(c)** Evaluate the power for $p = 0.65, 0.70, 0.75, 0.80, 0.85$ and sketch $\gamma(p)$.

**8.1.3** See Example 8.1.2.

**(a)** If the numbers of accidents for the 30 days were: 4, 1, 2, 2, 5, 3, 1, 1, 0, 1, 0, 3, 0, 3, 0, 1, 1, 1, 0, 2, 2, 1, 1, 0, 3, 1, 1, 4 For $\alpha = 0.10$, should $H_0$: $\lambda \geq 2.4$ be rejected? Use a normal approximation.

**(b)** Actually, these data were generated using S-Plus for $\lambda = 1.8$. What was the probability of rejecting $H_0$?

**8.1.4** Boxes A, B, and C each contain six numbered balls as follows:

| | Number | | |
|---|---|---|---|
| Box | 1 | 2 | 3 |
| *A* | 3 | 2 | 1 |
| *B* | 2 | 2 | 2 |
| *C* | 1 | 2 | 3 |

You are presented with one of the three boxes, none of which has any indication of which it is. To help you decide which box it is, you are allowed to choose two balls randomly with replacement from the box. Let $X_1$ and $X_2$ be the numbers on the balls chosen.

**(a)** What is the parameter space? What is the sample space?

**(b)** Suppose that you wish to test $H_0$: (you have box $A$) versus $H_a$: (you have box $B$ or $C$). For $\alpha = 0.25$, which sample points $(x_1, x_2)$ would you put in the critical region $C$? For this $C$, what is the power function?

**(c)** Repeat part *b*) for the case that sampling is without replacement.

**(d)** Suppose that you were allowed to take a simple random sample of three balls. What critical region would you choose? What is its power function?

**8.1.5** An inspector suspects that the "2-pound" cans of coffee made by the Cooper Coffee Company (CCC, named for the founder Cornelius Cornwallis Cooper) really had mean weight $\mu$ less than 2 pounds. The inspector wishes to have probability no more than 0.05 that he will accuse CCC of cheating when really $\mu \geq 2.0$, but wants to have probability at least 0.90 that he will accuse CCC of cheating if $\mu = 1.98$. From past experience he believes that the contents of coffee cans made by CCC are normally distributed with $\sigma = 0.05$. He plans to take a random sample of $n$ cans.

**(a)** State $H_0$ and $H_a$.

**(b)** Determine $n$ and a decision rule that depends on the sample mean $\bar{X}_n$ and a critical value $c_n$. What is the resulting value of $c_n$?

**(c)** Express the power function $\gamma(\mu)$ in terms of $\Phi$ and $\mu$. Find $\Phi$ (1.99).

**8.1.6** Let $X_1, \ldots, X_{10}$ be a random sample from the uniform distribution on $[0, \theta]$ for $\theta > 0$.

**(a)** Suppose that we wish to test $H_0$: $\theta \geq 20$ versus $H_a$: $\theta < 20$ for $\alpha = 0.05$. Suggest a test and give the power function $\gamma(\theta)$. Sketch $\gamma(\theta)$.

**(b)** Repeat part (a) for $H_0$: $\theta = 20$ versus $H_a$: $\theta \neq 20$.

**8.1.7** Consider the cancer drug example at the beginning of Section 8.1. Suppose that $n = 25$ patients are studied and that $X$ is the number who survive for five years. Consider $H_0$: $p = 0.3$ and the two-sided alternative $H_a$: $p \neq 0.3$.

**(a)** Suppose that we will reject $H_0$ for $X \leq k_1$ and for $X \geq k_2$, where $k_1$ and $k_2$ are chosen integers, and that we want $\alpha \doteq 0.078$. Use the binomial table to choose $k_1$ and $k_2$. (There is some arbitrariness in this, but try to make the tail probabilities approximately equal.)

**(b)** Find the power function $\gamma(p)$ for $p = 0.1, 0.2, 0.3, 0.4, 0.5$ and sketch it.

**(c)** Suppose that $n = 100$. Use the normal approximation to find $k_1$ and $k_2$ so that $\alpha \doteq 0.078$.

**(d)** Give the power function $\gamma(p)$ for the test in part (c) in terms of the standard normal cdf $\Phi$ and $p$. Sketch it for $0.15 \leq p \leq 0.45$.

**(e)** Suppose that we consider this as a three-decision problem as discussed. What is the probability of deciding that $p < 0.3$ if $p = 0.25$? How does that differ from $\gamma(0.25)$?

## 8.2 THE NEYMAN–PEARSON LEMMA

In Section 8.1 we discussed a few tests that were chosen on somewhat intuitive grounds. In this section and the next we introduce a more systematic approach with the Neyman–Pearson lemma (Neyman and Pearson, 1933) in this section, followed by the likelihood ratio test in Section 8.3. The Neyman–Pearson lemma is simpler in that it pertains directly only to the case that $H_0$ and $H_a$ are *simple hypotheses*, hypotheses that completely specify the distribution of the random variables observed. The hypotheses considered in Section 8.1 were composite. The null hypothesis of Problem 8.1.4 is simple, while the alternative hypothesis $H_a$ is composite. In our cancer example we might have instead let $H_0$: $p = 0.3$ and $H_a$: $p \neq 0.3$. Then $H_0$ is simple whereas $H_a$ is composite. As is implied by the word *simple*, it is somewhat easier to find an optimal test when both $H_0$ and $H_a$ are simple, although that situation is rather rare in practical examples. The Neyman–Pearson lemma provides a critical region in the simple versus simple situation, which is optimal in a certain sense.

Before we state the Neyman–Pearson lemma in its more general form, consider an example. Let $f(k; \theta)$ be a probability mass function for $\theta = 0$ or 1, where $f$ is as follows.

| | $k$ | | | |
|---|---|---|---|---|
| | 1 | 2 | 3 | 4 |
| $f(k; 0)$ | 0.4 | 0.3 | 0.2 | 0.1 |
| $f(k; 1)$ | 0.1 | 0.2 | 0.3 | 0.4 |

Suppose that a random sample $X_1, X_2$ is taken from one of these two distributions and that we wish to test $H_0$: $\theta = 0$ versus $H_a$: $\theta = 1$. The two joint distributions of $\mathbf{X} = (X_1, X_2)$ are: for $\theta = 0$,

| | $x_2$ | | | |
|---|---|---|---|---|
| $x_1$ | 1 | 2 | 3 | 4 |
| 1 | 0.16 | 0.12 | 0.08 | 0.04 |
| 2 | 0.12 | 0.09 | 0.06 | 0.03 |
| 3 | 0.08 | 0.06 | 0.04 | 0.02 |
| 4 | 0.04 | 0.03 | 0.02 | 0.01 |

and for $\theta = 1$,

| | $x_2$ | | | |
|---|---|---|---|---|
| $x_1$ | 1 | 2 | 3 | 4 |
| 1 | 0.01 | 0.02 | 0.03 | 0.04 |
| 2 | 0.02 | 0.04 | 0.06 | 0.08 |
| 3 | 0.03 | 0.06 | 0.09 | 0.12 |
| 4 | 0.04 | 0.08 | 0.12 | 0.16 |

Which sample points should be included in the critical region $C$ if we wish $\alpha \leq 0.15$? The sum of the probabilities under $\theta = 0$ must not exceed 0.15. We should try to choose points so that although the sum under $H_0$ is 0.15 or less, the sum under $H_a$ is as large as possible. That is, we have 0.15 probability to spend and wish to buy as much power as possible. The principle is the same as that of a frugal shopper. It seems reasonable to put points in the critical region if they provide a large ratio of probability under $H_1$ to that under $H_0$. For our 16 sample points these ratios $r(x_1, x_2) = f(x_1; \theta = 1)/f(x_1, x_2; \theta = 0)$ are:

| | $x_2$ | | | |
|---|---|---|---|---|
| $x_1$ | 1 | 2 | 3 | 4 |
| 1 | 1/16 | 1/6 | 3/8 | 1 |
| 2 | 1/6 | 4/9 | 1 | 8/3 |
| 3 | 3/8 | 1 | 9/4 | 6 |
| 4 | 1 | 8/3 | 6 | 16 |

If we let $C = \{(x_1, x_2) \,|\, r(x_1, x_2) \geq 9/4\} = \{(4,4), (3,4), (4,3), (2,4), (3,3), (4,2)\}$, then $P((X_1, X_2) \in C; \theta = 0) = 0.15$, while $P((X_1, X_2) \in C; \theta = 1) = 0.69$. That is, the power function is $\gamma(0) = 0.15$, $\gamma(1) = 0.64$.

The Neyman–Pearson lemma, first stated in the 1933 paper of Jerzy Neyman and Egon Pearson, shows that critical regions that place points in the critical region for which the ratio $r(x_1, \ldots, x_n) = r(\mathbf{x}) = f(\mathbf{x}; \theta_1)/f(\mathbf{x}; \theta_0)$ is large are optimal in that they maximize the power at $\theta_1$ for given power at $\theta_0$ for the test of $H_0\colon \theta = \theta_0$ versus $H_a; \theta = \theta_1$.

**The Neyman–Pearson Lemma** Let **X** have density or probability mass function $f(\mathbf{x}; \theta)$ for $\theta = \theta_0$ or $\theta = \theta_1$. Let $C^*$ be a critical region for $H_0\colon \theta = \theta_0$ versus $H_a\colon \theta = \theta_1$ having the property that for some positive constant $k$, $f(\mathbf{x}; \theta_1) \geq kf(\mathbf{x}; \theta_0)$ for $\mathbf{x} \in C^*$ and $f(\mathbf{x}; \theta_1) \leq kf(\mathbf{x}; \theta_0)$ for $\mathbf{x} \notin C^*$. Let $C$ be any other critical region. Let $\gamma(\theta)$ and $\gamma^*(\theta)$ be the power functions for $C$ and $C^*$. Let $\alpha = \gamma(\theta_0)$ and $\alpha^* = \gamma^*(\theta_0)$. Then

$$\gamma^*(\theta_1) - \gamma(\theta_1) \geq k[\alpha^* - \alpha],$$

so that $\alpha^* \geq \alpha$ implies that $\gamma^*(\theta_1) \geq \gamma(\theta_1)$.

Comments:

1. Define $r(\mathbf{x}) = f(\mathbf{x}; \theta_1)/f(\mathbf{x}; \theta_0)$ whenever the denominator is positive. Otherwise, define $r(\mathbf{x})$ to be $\infty$. Then $r(\mathbf{x}) \geq k$ for $x \in C^*$ and $r(\mathbf{x}) \leq k$ for $\mathbf{x} \in \bar{C}^*$, the complement of $C^*$. We will exploit these inequalities to determine a simpler form of $C^*$. $r(\mathbf{x})$ is the *likelihood ratio*.
2. The Neyman–Pearson (NP) lemma implies that among all tests of $H_0$ versus $H_a$ of level $\alpha^*$ the critical region $C^*$ is *most powerful*.

*proof* A proof will be given for the continuous case. The discrete case follows by substitution of summations for integrals. For ease of notation we use $f_1$ to denote $f(\mathbf{x}; \theta_1)$ and $f_0$ to denote $f(\mathbf{x}; \theta_0)$. We also omit the **dx** from the integrals. Then $\gamma^*(\theta_1) - \gamma(\theta_1) = \int_{C^*} f_1 - \int_C f_1 = \int_{C^* \cap \bar{C}} f_1 - \int_{C \cap \bar{C}^*} f_1 \geq \int_{C^* \cap \bar{C}} kf_0 - \int_{C \cap \bar{C}^*} kf_0 = k[\int_{C^*} f_0 - \int_C f_0] = k[\alpha^* - \alpha]$. □

Comments: The first inequality follows from $f_1(\mathbf{x}) \geq kf_0(\mathbf{x})$ for $\mathbf{x} \in C^*$ and $f_1(\mathbf{x}) \leq kf_0(\mathbf{x})$ for $\mathbf{x} \in \bar{C}^*$. The second and third equalities follow by adding and subtracting integrals over $C^* \cap C$ (see Figure 8.2.1).

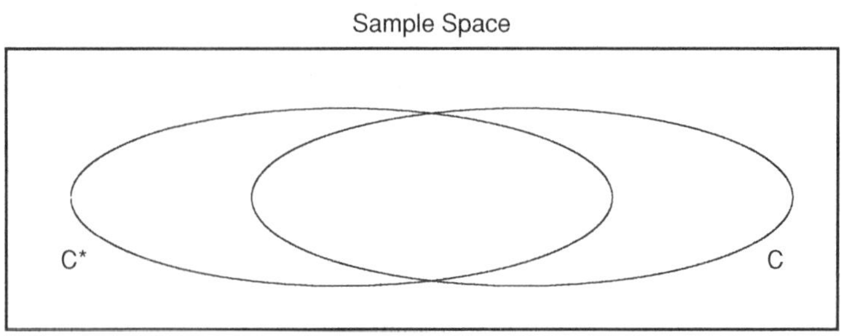


**FIGURE 8.2.1** Venn diagram illustrating the proof of the Neyman–Pearson lemma.

***Example 8.2.1*** Let $X \sim \text{Binomial}(n, p)$ and suppose that we wish to test $H_0$: $p = p_0$ versus $H_a$: $p = p_1$, where $p_1 > p_0$. The likelihood ratio is

$$r(x) = \frac{\binom{n}{x} p_1^x (1-p_1)^{n-x}}{\binom{n}{x} p_0^x (1-p_0)^{n-x}} = \left[\frac{p_1(1-p_0)}{p_0(1-p_1)}\right]^x \left(\frac{1-p_1}{1-p_0}\right)^n.$$

Since $p_1 > p_0$, $r(x)$ is an increasing function of $x$, so that $r(x) \geq k$ if and only if $x \geq k'$ for some $k'$ which we can determine as follows. Suppose that we want $\alpha$ to be approximately 0.10, $p_0 = 0.2$ and $n = 20$. Since $P(X \geq 6; p = 0.2) = 0.196$ and $P(X \geq 7; p = 0.2) = 0.087$, we choose the test that rejects $H_0$ for $X \geq 7$. The power function is $\gamma(p) = P(X \geq 7; p)$. We find, for example, using computer software or a binomial table that $\gamma(0.3) = 0.392$, $\gamma(0.4) = 0.750$, $\gamma(0.5) = 0.942$. The test we chose depended on $p_1$ only because $p_1 > p_0$. Therefore, the test that rejects for $X \geq 7$ is *uniformly most powerful* against all tests of $H_0$ versus $H_a$ with $\alpha \leq 0.087$. □

***Example 8.2.2*** Let $\mathbf{X} = (X_1, \ldots, X_n)$ be a random sample from the exponential distribution with mean $\theta > 0$. Suppose that we would like to test $H_0$: $\theta = 5$ versus $H_a$: $\theta = 8$ for $\alpha = 0.05$. The Neyman–Pearson lemma indicates that we should place points $\mathbf{x} = (x_1, \ldots, x_n)$ in the critical region $C$ if $f(\mathbf{x}; 8) \geq kf(\mathbf{x}; 5)$. We will try to find $k$ later. The inequality holds if and only if $r(\mathbf{x}) = f(\mathbf{x}; 8)/f(\mathbf{x}; 5) = \prod_{i=1}^n [(1/8)e^{-x_i/8}]/\prod_{i=1}^n [(1/5)e^{-x_1/5}] = (8/5)^n e^{-(1/8-1/5)\Sigma x_i} \geq k$. Equivalently, $\mathbf{x} \in C$ if and only if $\Sigma x_i \geq [\log k + n\log(5/8)](1/5 - 1/8)$. Call this last constant $k'$. We have concluded that the test should reject $H_0$ for large values of $T \equiv \Sigma X_i$. We have yet to determine the constant $k'$. To determine $k'$ we need to know the distribution of $T$ for $\theta = 5$. We should choose $k'$ so that $\alpha = 0.05 = P(T \leq k'; \theta = 5)$. We know that $T$ has a gamma distribution with parameters $n$ and $\theta$. From the relationship between the chi-square and gamma distributions we conclude that $2T/\theta$ has a chi-square distribution with $2n$ degrees of freedom. From chi-square tables we can find the 0.95-quantile for $2T/\theta$. Call this $c_{0.95}$. Thus, $P(2T/5 \geq c_{0.95}; \theta = 5) = 0.05 = P(T \geq 5c_{0.95}/2; \theta = 5)$.

The power function for this test is $\gamma(\theta) = P(T \geq 5c_{0.95}/2; \theta) = P(2T/\theta \geq 5c_{0.95}/\theta) = P(\chi^2_{2n} \geq 5c_{0.95}/\theta)$, where $\chi^2_{2n}$ is a chi-square random variable with $2n$ degrees of freedom. The power function can be evaluated for specific choices of $n$ and $\theta$ by using a chi-square table or a statistical package such as S-Plus.

For $n = 5$, we find $c_{0.95} = \chi^2_{10.95} = 18.307$ (the 0.95-quantile of the chi-square distribution with 10 df), so that we reject $H_0$ for $T \geq 45.768$, equivalently for $\bar{X} \geq 9.152$. Then $\gamma(8) = 0.3241$. For $n = 25$, $c_{0.95} = 67.505$, we reject for $\bar{X} \geq 6.7504$, and $\gamma(8) = 0.7758$ (see Figure 8.2.2). For larger $n$, the normal distribution should provide good approximations. $E(T) = n\theta$, and $\text{Var}(T) = n\theta^2$, so $P(T \geq k'; \theta) \doteq 1 - \Phi(z)$ for $z = (k - n\theta)/(\theta n^{1/2})$. The power for $\theta = 5$ should be 0.05 so that $z = 1.645$, $k' = 5n + 1.645(5)n^{1/2}$. Thus, we reject for $\bar{X} \geq 5 + 1.645(5)/n^{1/2}$. Then $\gamma(\theta) = 1 - \Phi((k' - \theta n)/[\theta n^{1/2}])$. For $n = 25$ we get the test: Reject for $\bar{X} \geq 6.645$ and $\gamma(8) = 0.801$, not far from the exact values given above.

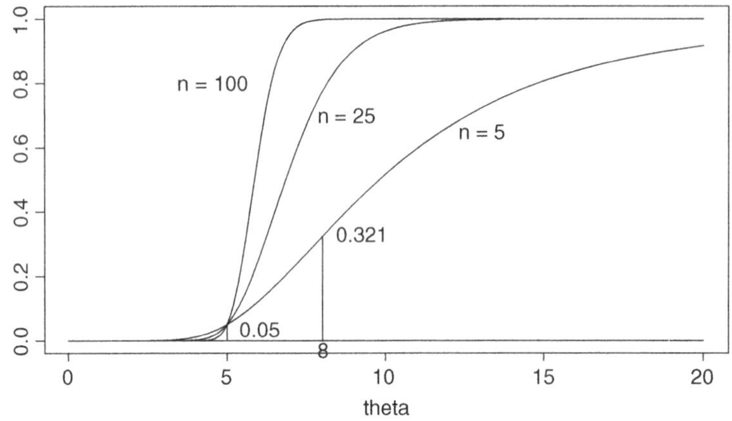


**FIGURE 8.2.2** Power functions for $n = 5, 25$, and 100.

Notice that the test depends on the alternative hypothesis only through the fact that $8 > 5$. That is, if $H_a$ had been $\theta = \theta_1$ for any $\theta_1 > 5$, the test would have been the same. Since the Neyman–Pearson (NP) lemma (the likelihood ratio test) is most powerful among tests with the same or smaller $\alpha$ -level, this implies that the test we found is *uniformly most powerful* against the alternative hypothesis $H_a$: $\theta > 5$. If the alternative hypothesis had instead been $H_a$: $\theta < 5$, we would reject for small $T = \Sigma X_i$, and the power functions would be large for $\theta < 5$. However, for any $\theta_0$ there is no uniformly most powerful test of $H$: $\theta = \theta_0$ versus $H_a$: $\theta \neq \theta_0$. □

***Example 8.2.3*** Consider the example at the beginning of Section 8.2, with two discrete distributions on 1, 2, 3, 4. Suppose that we have 10 observations $X_1, \ldots, X_{10}$ from one of these and wish to test $H_0$: $\theta = 0$ versus $H_a$: $\theta = 1$. One way to do this would be to make the decision based solely on the number $Y$ of 1's observed. Then the NP lemma suggests that we reject $H_0$ for small $Y$. For $\theta = 0$, $Y \sim$ Binomial(10, 0.4), so that $P(Y \leq 1; \theta = 0) = 0.0464$. Then $P(Y \leq 1; \theta = 1) = 0.7361$. Not too bad? But can we do better? The NP lemma based on the $X_i$'s suggests that we can.

Let $\delta_{ik}$ be the indicator of $[x_i = k]$. For example, $\delta_{43} = 1$ if $x_4 = 3$ and is zero otherwise. Then $P(X_i = k; \theta) = \prod_{k=1}^{4} f(k;\theta)^{\delta_{ik}}$. Let $g(k) = f(k;\theta_1)/f(k;\theta_0)$. Then the likelihood ratio is $r(\mathbf{x}) = \prod_{i=1}^{10} \prod_{k=1}^{4} g(k;\theta)^{\delta_{ik}} = \prod_{k=1}^{4} g(k)^{m_k}$, where $m_k = \sum_{i=1}^{10} \delta_{ik}$ is the frequency of $k$ among the $10x_i$'s. For example, if $m_1 = 3, m_2 = 4, m_3 = 2$, and $m_4 = 1$, then $r(\mathbf{x}) = (1/4)^3(2/3)^4(3/2)^2(4)$, a small value which would seem to indicate that $\mathbf{x}$ should not be in the critical region.

Consider $w(\mathbf{x}) \equiv \log(r(\mathbf{x})) = \sum_{k=1}^{4} m_k \log(g(k))$. Let $b_k = \log(g(k))$. We wish to include in the critical region $C$ those $\mathbf{x}$ for which $w(\mathbf{x}) = \sum_{k=1}^{4} b_k m_k$ is large. The vector $\mathbf{m} = (m_1, m_2, m_3, m_4)$ has the multinomial distribution with $n = 10$, $\mathbf{p} = \mathbf{p}_0 \equiv (0.4, 0.3, 0.2, 0.1)$ when $\theta = 0$, and $\mathbf{p} = \mathbf{p}_1 \equiv (0.1, 0.2, 0.3, 0.4)$ when $\theta = 1$. From Section 2.7, $\mathbf{m}$ has expectation $E(m;\theta) = 10\mathbf{p}_\theta$ and covariance matrix

$\text{Cov}(\mathbf{m}; \theta) = n[\text{diag}(\mathbf{p}_\theta) - \mathbf{p}_\theta\ \mathbf{p}_\theta^T]$ for $\theta = 0$ or 1. Let $\mathbf{b} = (b_1, \ldots, b_4)$. It follows that $E(w(\mathbf{X}); \theta) = \mathbf{b}\mathbf{p}_\theta^T$ and $\text{Var}(w(X); \theta) = \mathbf{b}\ \text{Cov}(\mathbf{m}; \theta)\mathbf{b}^T$. Since $w(\mathbf{X})$ is a sum of independent random variables, it follows that for $\theta = 0$ or 1, $w(\mathbf{X})$ is approximately normally distributed. We will use this to decide which points $\mathbf{x}$ to include in $C$.

Computations using S-Plus yield $\mathbf{b} = (1.386, 0.405, -0.405, -1.386)$, $E(w(\mathbf{X}); \theta = 0) = 4.5643$, $E(w(\mathbf{X}); \theta = 1) = -4.5643$, $\text{Var}(w(\mathbf{X}); \theta = 0) = \text{Var}(w(\mathbf{X}); \theta = 1) = 8.3477$. For $\alpha = 0.05$ we should reject for $w(\mathbf{X}) \leq k = 4.5643 - 1.645 \times 8.3477^{1/2} = -0.1885$. Then the power for $\theta = 1$ is $\Phi((-0.1885 + 4.5643)/8.3477^{1/2}) = 0.935$.

Since the distribution of $w(\mathbf{X})$ is discrete, the experiment was simulated in the computer 50,000 times for both $\theta = 0$ and $\theta = 1$. For $\theta = 0$, $H_0$ was rejected 2296 times, indicating that $\alpha$ is approximately 0.0460. For $\theta = 1$, $H_0$ was rejected 45,786 times, indicating that the power for $\theta = 1$ is approximately 0.916. □

## Problems for Section 8.2

**8.2.1** Let $X_1, \ldots, X_n$ be a random sample from the $N(\mu, \sigma^2)$ distribution with $\mu$ unknown, $\sigma^2$ known. Suppose that you wish to test $H_0$: $\mu = \mu_0$ versus $H_a$: $\mu = \mu_1$, where $\mu_1 > \mu_0$.

- **(a)** Use the NP lemma to give the most powerful $\alpha = 0.05$-level test for $H_0$ versus $H_a$.
- **(b)** Express the power function $\gamma(\mu)$ in terms of $\Phi$, $\mu - \mu_0$, $\sigma$, and $n$.
- **(c)** Is the test given in uniformly most powerful? Why?
- **(d)** Show that the test given in part (a) is not a uniformly most powerful level $\alpha = 0.05$ test of $H_0$: $\mu = \mu_0$ versus $H_a$: $\mu \neq \mu_0$.
- **(e)** Consider the case that $\mu$ is known but $\sigma^2$ is unknown, $n = 25$. Use the NP lemma and the fact that $\Sigma(X_i - \mu)^2/\sigma^2$ has the chi-square distribution with $n$ df to give a $\alpha = 0.05$-level test of $H_0$: $\sigma^2 = 100$ versus $H_a$: $\sigma^2 = 140$. Evaluate its power for $\sigma^2 = 140$.

**8.2.2** For the $\theta = 0, 1$ example at the beginning of this section, suppose that $X_1, X_2, X_3$ is a random sample from one of two discrete distributions: having masses 0.5, 0.4, 0.1 on 1, 2, 3 for $\theta = 0$, and masses 0.1, 0.3, 0.6 on 1, 2, 3 for $\theta = 1$. Use the NP lemma to determine the most powerful $\alpha = 0.076$-level test. What are $\gamma(0)$ and $\gamma(1)$?

**8.2.3** Let $X_1, X_2, X_3$ be a random sample from the Poisson distribution with parameter $\lambda > 0$. Suppose that we wish to test $H_0$: $\lambda = 2$ versus $H_a$: $\lambda = 3$.

- **(a)** Use the Neyman–Pearson lemma to find the most powerful $\alpha = 0.083$-level test.
- **(b)** Find $\gamma(\lambda)$ for $\lambda = 2.5, 3.5$, and 4.5. Use these to sketch a graph of the function $\gamma$.

**(c)** Is the test you found in part (a) uniformly most powerful for $H_a$: $\lambda > 2$? Why?

**(d)** Suppose that we have 100 rather than three observations. For which observation vectors $\mathbf{X} = (X_1, \ldots, X_{100})$ should $H_0$ be rejected for $\alpha \doteq 0.05$? Evaluate the power function $\gamma(\lambda)$ for $\lambda = 2.4, 2.6, 2.8$.

**8.2.4** Let $\mathbf{X} = (X_1, \ldots, X_n)$ be a random sample from $f(\mathbf{x}; \theta)$ for $\theta \in \Omega$. Let $H_0$: $\theta = \theta_0$ and $H_a$: $\theta = \theta_0$. Let $T(\mathbf{X})$ be a sufficient statistic for $\theta \in \Omega$. Show that the likelihood ratio $r(\mathbf{x})$ is a function of $T(\mathbf{x})$ and corresponding to any critical region of the form $C = C_k = \{\mathbf{x} \mid r(\mathbf{x}) \geq k\}$ there exists a set $B$ such that $C = \{\mathbf{x} \mid T(\mathbf{x}) \in B\}$.

**8.2.5** Let $\mathbf{X} = (X_1, \ldots, X_5)$ be a random sample from the Unif$(0, \theta)$, $\theta > 0$.

**(a)** Find the most powerful $\alpha = 0.10$-level test of $H_0$: $\theta = 20$ versus $H_a$: $\theta = 24$.

**(b)** Find the power of this test for $\theta = 24$ and for $\theta = 22$.

**(c)** Is the test uniformly most powerful for $H_a$: $\theta > 20$?

**(d)** What is the most powerful $\alpha = 0.05$-level test of $H_0 : \theta = 20$ versus $H_a$: $\theta = 15$? What is the power for $\theta = 15$?

**8.2.6** Let $f(k; \theta)$ be defined as follows:

| | $k$ | | |
|---|---|---|---|
| | 1 | 2 | 3 |
| $f(k; \theta = 0)$ | 0.5 | 0.3 | 0.2 |
| $f(k; \theta = 1)$ | 0.6 | 0.3 | 0.1 |

Suppose that we wish to test $H_0$: $\theta = 0$ versus $H_a$: $\theta = 1$. Let $X_1, \ldots, X_{100}$ be a random sample from one of these distributions.

**(a)** Give an approximate $\alpha = 0.05$ level test, and give an approximation of the power for $\theta = 1$. In 10,000 simulations $H_0$ was rejected 479 times when $\theta = 0$ and 8973 times when $\theta = 1$.

**(b)** Would you reject $H_0$ for a sample for which there were 58 1's, 27 2's, and 15 3's?

**(c)** How large would $n$ have to be to have power 0.95 or more for $\alpha = 0.05$?

## 8.3 THE LIKELIHOOD RATIO TEST

Although the NP lemma will generate tests in the case that $H_0$ and $H_a$ are both simple and in some cases for which $H_a$ is one-sided, there are many situations in which it is not helpful at all. See Problem 8.1.7 for the case that $H_a$: $p \neq 0.3$. Consider the case that we have independent random samples of $X$'s and $Y$'s from normal distributions with mean $\mu_1$ and $\mu_2$, equal variances $\sigma^2$, all unknown. How can we test $H_0$: $\mu_1 = \mu_2$ versus $H_a$: $\mu_1 \neq \mu_2$?

Let $\mathbf{X} = (X_1, \ldots, X_n)$ be a random sample from a density or probability mass function $f(\mathbf{x}; \theta)$, $\theta \in \Omega$ and suppose that we wish to test $H_0$: $\theta \in \Omega_0$ versus $H_a$: $\theta \in \Omega_1 \equiv \Omega - \Omega_0 = \Omega_0^c$. Let $\hat{\theta}_0 = \hat{\theta}_0(\mathbf{x})$ and $\hat{\theta} = \hat{\theta}(\mathbf{x})$ maximize the likelihood function $L(\theta; \mathbf{x}) \equiv \prod_{i=1}^n f(x_i; \theta)$ for $\theta$ in the parameter spaces $\Omega_0$ and $\Omega$. Thus, $\hat{\theta}_0(\mathbf{X})$ and $\hat{\theta}(\mathbf{X})$ are the MLEs for $\theta$ corresponding to $\Omega_0$ and $\Omega$. Define $\Lambda_n = \Lambda_n(\mathbf{x}) = L(\hat{\theta}_0(\mathbf{x}); \mathbf{x})/L(\hat{\theta}(\mathbf{x}); \mathbf{x})$. $\Lambda_n(\mathbf{X})$ is called the *likelihood ratio statistic*. We should reject $H_0$ for *small* observed $\Lambda_n(\mathbf{x})$. Notice that this ratio is unlike the ratio of the NP lemma in that the numerator refers to the null hypothesis and $\Lambda_n(x) \leq 1$ for all $\mathbf{x}$.

Let $C_n \equiv -2\log(\Lambda_n(\mathbf{X}))$. Suppose that $\Omega$ is a subset of $R_k$ and that $\Omega_0$ is a subset of $\Omega$ of "dimension" $k$. Suppose that $f(\mathbf{x}; \theta)$ satisfies the smoothness conditions given in Section 7.8, which were sufficient for the asymptotic normality of MLEs. Then, assuming that $H_0$: $\theta \in \Omega$ is true, $C_n$ is asymptotically distributed as chi-square with $k - r$ degrees of freedom (df). We should therefore reject $H_0$ for $C_n \geq \chi^2_{(k-r)(1-\alpha)}$, the $(1 - \alpha)$-quantile of the chi-square distribution with $k - r$ df. [see Ferguson (1996) for a proof]. $k - r$ is usually the number of "restrictions" on $\theta$ under $H_0$. This was proved in 1938 by S. S. Wilks. We have been deliberately vague about the conditions and do not provide a proof.

***Example 8.3.1*** Let $X_1, \ldots, X_n$ be a random sample from the $N(\mu, \sigma^2)$ distribution with both $\mu$ and $\sigma^2$ unknown. Thus, the parameter is $\theta = (\mu, \sigma^2)$ and $\Omega = R_1 \times R_1^+$, the upper half of the plane. Suppose that we wish to test $H_0$: $\mu = \mu_0$, where $\mu_0$ is a known constant. Then $\hat{\theta}(\mathbf{X}) = (\bar{X}, \hat{\sigma}^2)$, where $\hat{\sigma}^2 = (1/n)\Sigma(X_i - \bar{X})^2$ and $\hat{\theta}_0(\mathbf{X}) = (\mu_0, \hat{\sigma}_0^2)$, where $\hat{\sigma}_0^2 = (1/n)\Sigma(X_i - \mu_0)^2$. We get $\log(\Lambda_n(X)) = (n/2)\log(\hat{\sigma}^2/\hat{\sigma}_0^2) - n/2 + n/2 = (n/2)\log(1 - n(\bar{X} - \mu_0)^2/\hat{\sigma}^2)$, so that $C_n = n\log(1 + (\bar{X} - \mu_0)^2/\hat{\sigma}^2) = n\log(1 + Z_n^2/n)$, where $Z_n \equiv (\bar{X} - \mu_0)/\sqrt{\hat{\sigma}^2/n}$. For $Z_n^2/n$ small, $C_n$ is approximately $Z_n^2$. Since $k = 2, r = 1$, $C_n$ is asymptotically distributed as chi-square with 1 df. We can do a little better than this. Since $C_n$ is a monotone increasing function of $|Z_n|$, we should reject for large $|Z_n|$. By Slutsky's theorem, $Z_n$ is asymptotically distributed as $N(0, 1)$. In fact, that is true whether or not the distribution sampled is normal. It follows that $Z_n^2$ is asymptotically distributed as chi-square with 1 df. Rejection for large $Z_n^2$ is equivalent to rejection for $|Z_n|$ large, so for a test of level $\alpha$, we should reject for $|Z_n| > z_{1-\alpha/2}$. □

In Chapter Nine we show that $Z_n$ (except for a constant multiple) has Student's $t$-distribution with $n - 1$ df.

***Example 8.3.2*** Suppose that a random sample of 400 adults was taken from the population of Frequency City, which has 100,000 adult residents. The 400 were

**TABLE 8.3.1 Frequency Data**

| | Age Group | | | | |
|---|---|---|---|---|---|
| | 1 | 2 | 3 | 4 | Total |
| Agree | 27 | 39 | 47 | 58 | 171 |
| Neutral | 44 | 25 | 20 | 10 | 99 |
| Against | 56 | 48 | 18 | 8 | 130 |
| Total | 127 | 112 | 85 | 76 | 400 |

cross-classified according to their view on a bill before Congress that would increase Social Security benefits and taxes with the categories "Agree," "Neutral," and "Against" and according to their age: (1) 18–35; (2) 36–50; (3) 51–65, and (4) 66–99. The frequencies are listed in Table 8.3.1.

The following would seem to be a suitable model for the $3 \times 4$ table (often called a *contingency table*) $\mathbf{X}$ of frequencies. Let $\mathbf{X}$ be a $3 \times 4$ table $(X_{ij})$. Suppose that $\mathbf{X} \sim \text{Multinomial}(n, \mathbf{p})$, where $\mathbf{p} = (p_{ij})$ is a $3 \times 4$ table of probabilities, summing to 1. Suppose that we wish to test $H_0$: (rows and columns are independent) $\Leftrightarrow$ ($p_{ij} = p_{i+}p_{+j}$ for all $i$ and $j$), where $p_{i+}$ and $p_{+j}$ are row and column sums. Let $a$ and $b$ be the numbers of rows and columns, 3 and 4 for this example. Then $\Omega$ is a $k = [a(b) - 1]$-dimensional subset of $R_{12}$ (a simplex in 12-space). Thus, for our example, $k = 11$. $\Omega_0$ has dimension $r = (a-1) + (b-1) = 5$. Under the full model that $\mathbf{p} \in \Omega$, the MLE of $\mathbf{p}$ is $\mathbf{X}/n \equiv \hat{\mathbf{p}} = (\hat{p}_{ij})$, the table of relative frequencies. Under $H_0$ the MLE of $\mathbf{p}$ is the table $\hat{\mathbf{p}}_0 = (\hat{p}_{0ij} = (\hat{p}_{i+}\hat{p}_{+j}))$, where $\hat{p}_{i+} = X_{i+}/n$ and $\hat{p}_{+j} = X_{+j}/n$. Then $\Lambda_n = \prod_{ij}(\hat{p}_{0ij}/\hat{p}_{ij})^{X_{ij}}$, so that $C_n = -2\sum_{ij} X_{ij}\log((\hat{p}_{i+}\hat{p}_{+j}/\hat{p}_{ij})) = 2\sum_{ij} X_{ij}\log(X_{ij}/\hat{m}_{ij})$, where $\hat{m}_{ij} = n\hat{p}_{0ij} = X_{i+}X_{+J}/n$, an unbiased estimator of $m_{ij} = np_{ij}$ under $H_0$. With large probability, $C_n$ will be close to Pearson's goodness-of-fit statistic, $\chi^2 = \sum_{ij}(X_{ij} - \hat{m}_{ij})^2/\hat{m}_{ij}$. Under $H_0$ both $C_n$ and $\chi^2$ are asymptotically distributed as chi-square with $k - r = (ab-1) - [a - 1 + b - 1] = (a-1)(b-1) = 6$ degrees of freedom.

For the Social Security data, we get $\hat{\mathbf{p}} = \begin{pmatrix} 0.068 & 0.098 & 0.118 & 0.145 \\ 0.110 & 0.062 & 0.050 & 0.025 \\ 0.140 & 0.120 & 0.045 & 0.020 \end{pmatrix}$, $\hat{\mathbf{p}}_0 = \begin{pmatrix} 0.136 & 0.120 & 0.091 & 0.081 \\ 0.079 & 0.069 & 0.052 & 0.047 \\ 0.103 & 0.091 & 0.069 & 0.062 \end{pmatrix}$, $\hat{\mathbf{m}}_0 = (\hat{m}_{ij}) = \begin{pmatrix} 54.3 & 47.9 & 36.3 & 32.5 \\ 31.4 & 27.7 & 21.0 & 18.8 \\ 41.3 & 36.4 & 27.6 & 24.7 \end{pmatrix}$. We get $C_{400} = 2\sum_{ij} X_{ij}\log(X_{ij})/\hat{m}_{ij} = 74.7$ and $\chi^2 = 71.3$. These are far out in the tail of the chi-square distribution for 6 df, with probability to the right of about $4.4 \times 10^{-14}$, so that it is very clear that $H_0$ is not true.

Further analysis shows that as people become older they tend to agree in larger proportions, as might be expected. We return to the analysis of such data in Chapter twelve.

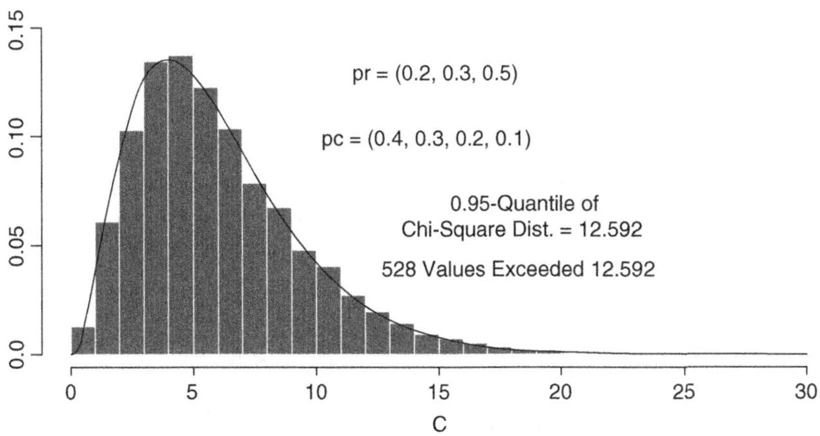


**FIGURE 8.3.1** Histogram of 10,000 log-chi-square statistics.

For Figure 8.3.1 the row and column probabilities were taken as given in the figure. The cell probabilities were then the corresponding products, so $H_0$ was true. For $n = 400$, $\mathbf{X}$ was observed 10,000 times. Each time the log-chi-square statistic $C$ was determined. The chi-square density with 6 df is superimposed, indicating a good fit. □

## Problems for Section 8.3

**8.3.1** Consider four probability mass functions $f(k;\theta)$ on the integers 0, 1, 2, 3 corresponding to $\theta = 1, 2, 3, 4$ listed in Table 8.3.1.
Suppose that we wish to test $H_0$: $\theta = 1$ or 2 versus $H_a$: $\theta = 3$ or 4, and that $X_1, X_2$ is a random sample from one of these distributions.

**(a)** Determine the likelihood ratio $\Lambda$ for each possible pair $(x_1, x_2)$ of values for $(X_1, X_2)$.

**(b)** For which values $\Lambda$ should $H_0$ be rejected for $\alpha = 0.09$?

**(c)** Determine the power function $\gamma(\theta)$ for $\theta = 1, 2, 3, 4$.

**TABLE 8.3.1**

| | $k$ | | | |
|---|---|---|---|---|
| | 0 | 1 | 2 | 3 |
| $f(k;1)$ | 0.1 | 0.2 | 0.4 | 0.3 |
| $f(k;2)$ | 0.2 | 0.1 | 0.3 | 0.4 |
| $f(k;3)$ | 0.4 | 0.3 | 0.1 | 0.2 |
| $f(k;4)$ | 0.3 | 0.4 | 0.2 | 0.1 |

**8.3.2** Let $\mathbf{X} = (X_1, \ldots, X_n)$ be a random sample from the Unif(0, $\theta$) distribution for $\theta > 0$.

**(a)** Let $H_0$: $\theta = \theta_0$, $H_a$: $\theta \neq \theta_0$, where $\theta_0$ is a fixed known constant. Express the likelihood ratio $\Lambda$ statistic as a function of $\max(\mathbf{x}) = \max(x_1, \ldots, x_n)$.

**(b)** Let $M = \max(\mathbf{X})$. Find a constant $c$ so that the test which rejects for $M < c$ and $M > \theta_0$ has level $\alpha$. Give its power function $\gamma(\theta)$. Apply the test for the case that $\theta_0 = 20$, $n = 7$, $M = 13.1$, $\alpha = 0.05$.

**(c)** Give the $\alpha$-level likelihood ratio test of $H_0$: $\theta \leq \theta_0$ versus $H_a$: $\theta > \theta_0$. Give the power function $\gamma(\theta)$.

**(d)** Repeat part (c) for $H_0$: $\theta \geq \theta_0$ versus $H_a$: $\theta < \theta_0$.

**8.3.3** Let $X_1, \ldots, X_n$ be a random sample from $N(\mu, \sigma^2)$ with $\sigma^2$ known, $\mu$ unknown. Let $H_0$: $\mu \leq \mu_0$ and $H_a$: $\mu > \mu_0$, where $\mu_0$ is a known constant.

**(a)** Find the likelihood ratio statistic $C_n \equiv -2\log(\Lambda_n(\mathbf{X}))$. *Hint:* $\Lambda_n(\mathbf{X})$ is 1 with probability 1/2 when $H_0$ is true.

**(b)** Show that the test which rejects for $C_n \geq \chi_1^2(1-\alpha)$ is equivalent to the test that rejects for $(\bar{X} - \mu_0)/\sqrt{\sigma^2/n} \geq \Phi^{-1}(1-\alpha)$. Use the fact that the square of a standard normal random variable has the chi-square distribution with 1 df.

**8.3.4** Let $X_1, \ldots, X_n$ be a random sample from $N(\mu, \sigma^2)$ with both $\mu$ and $\sigma^2$ unknown. Let $H_0$: $\sigma^2 = \sigma_0^2$, $H_a$: $\sigma^2 \neq \sigma_0^2$, where $\sigma_0^2$ is a known constant.

**(a)** Show that $\log(\Lambda_n) = [\log(W) - W + 1](n/2)$, where $W = \hat{\sigma}^2/\sigma_0^2$ with $\hat{\sigma}^2$ as defined in Example 8.3.1.

**(b)** Show that $\log(\Lambda_n)$ is maximum for $W = 1$.

**(c)** For $\alpha = 0.05$, $n = 100$, $\sigma_0^2 = 25$, $\hat{\sigma}^2 = 20.1$, make the decision.

**8.3.5** Let $\mathbf{X} = (X_0, X_1, X_2)$ be the vector of frequencies of genotypes $aa$, $aA$, and $AA$ in a random sample of $n$ from a population of rats. Suppose that $X \sim$ Multinomial($n$, $\mathbf{p} = (p_0, p_1, p_2)$). Under random mating, $\mathbf{p}$ should be of the form $p = ((1-\theta)^2, 2\theta(1-\theta), \theta^2)$ where $\theta$ is the $A$-allele proportion in the population sampled.

**(a)** Show that $C_n = 2\sum_{i=0}^{2} X_i \log(X_i/\hat{E}_i)$, where $\hat{\theta} = (2X_2 + X_1)/2n$, $\hat{E}_0 = (1-\hat{\theta})n$, $\hat{E}_1 = 2\hat{\theta}(1-\hat{\theta})$, $\hat{E}_2 = \hat{\theta}^2$.

**(b)** Apply the test for $\alpha = 0.05$, $\mathbf{X} = (43, 152, 205)$.

**8.3.6** Let $X_i \sim$ Poisson($\lambda_i$) for $i = 1, 2, 3$ with $X_1, X_2, X_3$ independent. Suppose that we wish to test $H_0$: $\lambda_1 = \lambda_2 = \lambda_3$.

**(a)** Let $\Lambda_n$ be the likelihood ratio statistic. Show that $C_n = -2\log(\Lambda_n) = 2\Sigma X_i \log(X_i/\bar{X})$.

**(b)** Test $H_0$ for $\alpha = 0.05$, $X_1 = 46$, $X_2 = 57$, $X_3 = 38$.

**8.3.7** Let $X_1, \ldots, X_n$ and $Y_1, \ldots, Y_n$ be independent random samples from uniform distributions on $[0, \theta_1]$ and $[0, \theta_2]$ for $\theta_1 > 0$ and $\theta_2 > 0$. Let $M_1 = \max(X_1, \ldots, X_n)$ and $M_2 = \max(Y_1, \ldots, Y_n)$.

(**a**) Show that the level-$\alpha$ likelihood ratio test of $H_0$: $\theta_1 = \theta_2$ versus $H_a$: $\theta_1 \neq \theta_2$ rejects $H_0$ for $(M_1/M_2)^n < \alpha$ and for $(M_2/M_1)^n < \alpha$. *Hint:* $M_1/\theta_1$ and $M_2/\theta_2$ have the same joint distribution as $U_1^{1/n}$ and $U_2^{1/n}$, where $U_1$ and $U_2$ are independent, each Unif(0, 1).

(**b**) Show that the power function is $\gamma(\theta_1, \theta_2) = (\alpha/2)[(\theta_2/\theta_1)^n + (\theta_1/\theta_2)^n]$ for this test for the case that $\eta_1 = \alpha(\theta_2/\theta_1)^n < 1$ and $\eta_2 = \alpha(\theta_1/\theta_2)^n < 1$.

**8.3.8** Suppose that $T(\mathbf{X})$ is a sufficient statistic for a parameter $\theta \in \Omega$, and we wish to test $H_0$: $\theta \in \Omega_0$ versus $H_a$: $\theta \in \Omega - \Omega_0$. Show that the likelihood ratio test is a function of $T(\mathbf{X})$.

## 8.4 THE *P*-VALUE AND THE RELATIONSHIP BETWEEN TESTS OF HYPOTHESES AND CONFIDENCE INTERVALS

Our discussion of testing hypotheses has focused on decision making, establishing rules that lead to one of two decisions, the choice of $H_0$ or $H_a$. Often, we don't really want to make a firm decision as to the truth of $H_0$ or $H_a$, but instead, want a measure of the "believability" of $H_0$. Consider the example with which we began Chapter Eight, with $H_0$: $p \leq 0.3$ and $H_a$: $p > 0.3$. Suppose that we observe $X$, the number of patients who survive among the six, to be five. Since we will reject for large $X$, the observed $p$-value, or simply the $p$-value, is the maximum of $P(X \geq 5; p)$ for all $p$ that satisfy $H_0$. In this case it is $P(X \geq 5; p = 0.3) = 0.0109$. Since this is small, it leads us to suspect strongly that $H_0$ is not true. This $p$-value indicates that if we tested $H_0$ versus $H_a$ at any $\alpha$-level greater than or equal to 0.0109, at level 0.015, for example, we would reject $H_0$. If we tested at any level less than 0.0109, we would not reject $H_0$. Thus, the $p$-value makes it possible for the reader of a report with a stated $p$-value to test $H_0$ versus $H_a$ at any $\alpha$-level the reader might choose.

**Definition 8.4.1**

(**a**) Consider a test of $H_0$: $\theta \in \Omega_0$ versus $H_a$: $\Omega - \Omega_0$, and suppose that $\mathbf{X} = (X_1, \ldots, X_n)$ is observed. Let $T(\mathbf{X})$ be a test statistic. Consider the collection of critical regions of the form $C_\alpha = \{\mathbf{x} \mid T(\mathbf{x}) \geq t_\alpha\}$ such that $C_\alpha$ has level $\alpha$. Suppose that $\mathbf{X} = \mathbf{x}$ is observed. The *observed p-value* corresponding to $\mathbf{x}$ is $\sup_{\theta \in \Omega 0} P(T(\mathbf{X}) \geq T(\mathbf{x}); \theta) = \hat{\alpha} = \hat{\alpha}(\mathbf{X})$.

(**b**) For critical regions $C_\alpha = (x \mid T(x) \leq t_\alpha)$ and observed $X = x$, the *observed p-value* is $\hat{\alpha} = \hat{\alpha}(\mathbf{X}) = \sup_{\theta \in \Omega 0} P(T(\mathbf{X}) \leq t; \theta)$.

The observed $p$-value is often simply called the *p-value* or *observed level of significance*. For students new to the subject, especially when the term *p-value* is applied to tests on a parameter $p$, it can be rather confusing. **Warning**: It is *not* the probability that $H_0$ is true, although unfortunately, it is often interpreted that way. □

***Example 8.4.1*** Let $X_1, \ldots, X_{25}$ be a random sample from a normal distribution with unknown mean $\mu$, unknown variance $\sigma^2$. Suppose that we wish to test $H_0$: $\mu \geq 50$ versus $H_a$: $\mu < 50$. Let $T(\mathbf{X}) = (\bar{X} - 50)/\sqrt{S^2/25}$, the one-sample $t$-statistic. For given $\alpha$ we would reject for $T(\mathbf{X}) \leq t_\alpha$, where $t_\alpha$ is the $\alpha$-quantile of the $t$-distribution with 24 degrees of freedom. For observed $\mathbf{X} = \mathbf{x}$ and $t = T(\mathbf{x})$, the observed $p$-value $\hat{\alpha}$ is therefore $P(T(\mathbf{X}) \leq t; \mu = 50)$. Since $T(\mathbf{X})$ has Student's $t$-distribution with 24 df, this is the area to the left of $t$ under the $t_{24}$ density. To determine this we need a more complete $t$-table than is provided in most textbooks, or a statistical computing package that provides such tail probabilities. If, for example, for $n = 25$ we observe that $\bar{X} = 47.30$, $S = 6.63$, then $t = (47.30 - 50)/\sqrt{6.63/5} = -2.036$, so that $\hat{\alpha} = 0.026$, found by using "pt(−2.035, 24)" in S-Plus. This indicates that anyone testing $H_0$ versus $H_a$ at the level $\alpha \geq 0.026$ would reject $H_0$, whereas anyone testing with $\alpha < 0.026$ would not reject. For the two-sided problem, with $H_0$: $\mu = 50$ versus $H_a$: $\mu \neq 50$ with test statistic $|\,T(\mathbf{X})\,|$, $\hat{\alpha} = 2(0.026) = 0.052$. It is often tempting to determine the value of the test statistic first, then choose the null and alternative hypotheses so that $\hat{\alpha}$ is as small as possible. This is intellectually dishonest, and good statisticians do their best to avoid it, both in their own analyses and in those of their clients. □

***Example 8.4.2*** Consider the cancer example at the beginning of Chapter Eight, with $H_0$: $p \leq 0.3$ and $H_a$: $p > 0.3$. Suppose that $n = 25$ and we observe $X = 12$. Then $\hat{\alpha} = \hat{\alpha}(12) = P(X \geq 12; p = 0.3) = 0.0442$. For $\alpha$ chosen to be 0.05, we would reject, while for $\alpha = 0.04$, we would not. What is $\hat{\alpha}$ for $X = 4$? A common mistake among beginning students is to take $\hat{\alpha} = P(X \leq 4; p = 0.3)$ or even to take $\hat{\alpha} = P(X = 4; p = 0.3)$. Since $p > 0.3$ under $H_a$, reasonable critical regions are of the form $X \geq k$ for some $k$, so that $\hat{\alpha} = P(X \geq 4; p = 0.3) = 0.967$, indicating that $H_0$ is quite believable. □

The author once performed a taste test in a class. Ten small cups were filled with Coke or Pepsi independently, with probabilities 1/2 and 1/2. A student volunteer, C.L., claimed that she could distinguish between the two soft drinks. We let $\theta$ be the probability that she could guess correctly and supposed that the events of correctness for the 10 cups were independent. We let $H_0$: $\theta \leq 1/2$ and $H_a$: $\theta > 1/2$. We counted the number $X$ for which C.L. was correct. We observed that $X = 1$! Since $P(X \geq 1; \theta = 1/2) = 1013/1024$, the null hypothesis seems to be quite believable. On the other hand, there is evidence that she could distinguish between Coke and Pepsi but had the wrong names. Our (actually, my) mistake was in making $H_0$ versus $H_a$ one-sided. It had not occurred to me prior to the experiment that she might be able to distinguish the soft drinks but not know which was which. I should have gotten a clue when she turned to another student during the experiment and asked, "Which is sweeter, Coke or Pepsi?" As the experimenter, I demanded immediately that no one answer.

### Relationship Between Confidence Intervals and Test of Hypotheses

Suppose that $I(\mathbf{X})$ is a random interval having the property $P(\theta \in I(\mathbf{X}); \theta) = \gamma$ for each $\theta \in \Omega$. $I(\mathbf{X})$ is a $100\gamma\%$ confidence interval on $\theta$. Now suppose that we wish to test $H_0$: $\theta = \theta_0$ versus $H_a$: $\theta \neq \theta_0$, where $\theta_0$ is a known constant. In testing the null hypothesis that a coin is fair, we might set $P(\text{head}) = \theta$ and $\theta_0 = 1/2$. In a paired sample $t$-test, we might choose $\theta_0 = \mu_0 = 0$. Consider the test of $H_0$ versus $H_a$ which rejects $H_0$ whenever $\theta_0 \notin I(\mathbf{X})$. This test has the critical region $C(\theta_0) = \{\mathbf{x} \mid \theta_0 \notin I(\mathbf{x})\}$, acceptance region $A(\theta_0) = \{\mathbf{x} \mid \theta_0 \in I(\mathbf{x})\}$. Then $P(\mathbf{X} \in C(\theta_0); \theta_0\} = P(\theta_0 \notin I(\mathbf{X}); \theta_0) = 1 - P(\theta_0 \in I(\mathbf{X}); \theta_0) = 1 - \gamma$. Thus, a 95% confidence interval $[0.32, 0.41]$ on a binomial parameter $p$ indicates that for any $\alpha = 0.05$ level test of $H_0$: $p = p_0$ versus $H_a$; $p \neq p_0$ the test would reject $H_0$ for $p_0$ not in the interval, accept (fail to reject) for $p_0$ in the interval. Vice versa, if $C(\theta_0)$ is an $\alpha$-level critical region for a test of $H_0$: $\theta = \theta_0$ versus $H_a$: $\theta \neq \theta_0$ define $I(\mathbf{x}) = \{\theta_0 \mid \mathbf{x} \notin C(\theta_0)\}$. Then $P(\theta_0 \in I(\mathbf{X}); \theta_0) = 1 - P(\theta_0 \notin I(\mathbf{X}); \theta_0) = 1 - P(\mathbf{X} \in C(\theta_0); \theta_0) = 1 - \alpha$, so that $I(\mathbf{X})$ is a $100(1 - \alpha)\%$ confidence interval on $\theta$.

The power function $\gamma(\theta)$ for an $\alpha$-level test of $H_0$: $\theta = \theta_0$ versus $H_a$: $\theta \neq \theta_0$ can be useful in studying the performance of the corresponding confidence interval $I(\mathbf{X})$, since $\gamma(\theta) = P(\mathbf{X} \in C(\theta_0); \theta) = P(\theta_0 \notin I(\mathbf{X}); \theta)$. If $|\theta - \theta_0|$ is large, we want this probability to be large.

Now consider $H_0$: $\theta \leq \theta_0$ versus $H_a$: $\theta > \theta_0$ and let $L(\mathbf{X})$ be a $100\gamma\%$ lower confidence limit on $\theta$. This mean that for each $\theta \in \Omega$, $P(L(\mathbf{X}) \leq \theta; \theta) = \gamma$. Consider the test that rejects $H_0$ whenever $L(\mathbf{X}) > \theta_0$. This test has level $P(L(\mathbf{X}) > \theta_0; \theta_0) = 1 - P(L(\mathbf{X}) \leq \theta_0; \theta_0) = 1 - \gamma$. Similarly, the upper $100\gamma\%$ confidence limit $U(\mathbf{X})$ corresponds to the test of $H_0$: $\theta \geq \theta_0$ versus $H_a$: $\theta < \theta_0$ which rejects at level $(1 - \gamma)$ for $U(\mathbf{X}) < \theta_0$.

***Example 8.4.3*** For the sample of $n = 25$ discussed in Example 8.4.1, suppose that we wish to test $H_0$: $\mu \geq 50$ versus $H_a$: $\mu < 50$. An upper 95% confidence limit on $\mu$ is given by $U(\mathbf{X}) = \bar{X} + t_{24.95}S/\sqrt{25} = \bar{X} + 1.711S/5$. For $\bar{X} = 47.3$, $S = 6.63$, we get $U(\mathbf{X}) = 49.57$. Since $49.57 < 50$, we reject $H_0$ at level $\alpha = 0.05$. We would not reject at level 0.05 if 50 were replaced by $\theta_0 < 49.57$. □

### Problems for Section 8.4

**8.4.1** Let $X \sim \text{Binomial}(n, p)$ and suppose that we wish to test $H_0$: $p \geq 0.4$ versus $H_a$: $p < 0.4$.

**(a)** For $n = 10$, give the $p$-value for $X = 0, 1, 2$.

**(b)** For $n = 100$, give the approximate $p$-value for $X = 25, 35, 40$ and 45. Use the 1/2-correction.

**8.4.2** Let $X_1, \ldots, X_n$ be a random sample from the Poisson distribution with parameter $\lambda$. Suppose that we wish to test $H_0$: $\lambda \le 3$ versus $H_a$: $\lambda > 3$. Let $T_n = X_1 + \cdots + X_n$.

**(a)** For $n = 2$ observations and $\alpha$ approximately 0.08 for which observation vectors $\mathbf{X} = (X_1, X_2)$, should $H_0$ be rejected? What is $\alpha$ for this test?

**(b)** Suppose that $T_2 = 12$. What is the $p$-value $\hat{\alpha}$?

**(c)** What is the power of the test in part (a) for $\lambda = 4.0$?

**(d)** For $n = 100$ use the normal approximation to find an $\alpha = 0.08$ level test based on $T_{100}$.

**(e)** For observed $T_{100} = 329$, what is the observed $p$-value?

**(f)** What is the power for the test in part (d) for $\lambda = 3.3$?

**8.4.3** Let $(X_1, \ldots, X_n) = \mathbf{X}$ be a random sample from the $N(\mu, \sigma^2)$ distribution with $\sigma^2$ known to be 64. Suppose that we wish to test $H_0$: $\mu = 80$ versus $H_a$: $\mu \ne 80$.

**(a)** For $n = 25$, $\bar{X} = 84.0$, what is the $p$-value $\hat{\alpha}$?

**(b)** Give an $\alpha = 0.05$ level test of $H_0$ versus $H_a$.

**(c)** Find a 95% confidence interval on $\mu$. Use the interval to decide whether you should reject $H_0$ at level $\alpha = 0.05$ when $\bar{X} = 84.0$.

**(d)** Let $I(\mathbf{X})$ be the interval found in part (a) for observation vector $\mathbf{X}$. For $\theta = 84$ find the power $\gamma(83)$ and use it to determine $P(80 \in I(\mathbf{X}); \theta = 83)$.

**8.4.4** Let $\mathbf{X} = (X_1, \ldots, X_n)$ be a random sample from the Unif$(0, \theta)$ distribution for $\theta > 0$.
Let $H_0$: $\theta = \theta_0$ and $H_a$: $\theta \ne \theta_0$ for a known constant $\theta_0$. Let $M = \max(X_1, \ldots, X_n)$.

**(a)** For $0 < \alpha < 1$, give a size-$\alpha$ critical region $C$ for a test of $H_0$ versus $H_a$. The test should have high power for $\theta$ small and also for $\theta$ large.

**(b)** Use the result of part (a) to give a formula for a $100(1 - \alpha)$% confidence interval $I(\mathbf{X})$ on $\theta$.

**(c)** For $n = 10$, $\theta_0 = 46$, $\alpha = 0.05$ find $P(\theta_0 \in I(\mathbf{X}); \theta = 50)$.

**(d)** Suppose that $n = 10$, $\theta_0 = 46$, and $M = 39.2$. Give the $p$-value.

**8.4.5** Consider the four probability mass functions of Problem 8.3.1 and $H_0$: $\theta = 1$ or 2 versus $H_a$: $\theta = 3$ or 4.

**(a)** Suppose that $(X_1, X_2)$ is a random sample from one of these distributions. Suppose also that $H_0$ is to be rejected for small values of $T = X_1 + X_2$. Give the $p$-value corresponding to each possible value of $T$.

**(b)** Consider a test based on a random sample $X_1, \ldots, X_9$ which rejects for large $T = \sum_{i=1}^{9} X_i$. Find an approximate $p$-value for $T = 21$, $\bar{X} = T/9 = 2.333$. In 10,000 simulations for $\theta = 1$,559 of the $\bar{X}$ were 2.333

or less. For $\theta = 2,921$ were 1.333 or less. Use the 1/2-correction. If the random variable is $\bar{X}$, the correction is actually $1/(2n)$.

**8.4.6** Let $\mathbf{X} = (X_1, \ldots, X_{25})$ be a random sample from the density $f(x;\eta) = (1/5)e^{-(x-\eta)/5}$ for $x > \eta$, $\eta$ any real number. Let $m = \min(X_1, \ldots, X_n)$. Suppose that you wish to test $H_0$: $\eta \le 10$ versus $H_a$: $\eta > 10$.

**(a)** Give an $\alpha = 0.10$ level test depending on $m$.

**(b)** Give the $p$-value for $m = 10.53$.

**8.4.7** Let $\mathbf{X}$ be a random variable or vector with a distribution depending on $\theta$. Let $H_0$: $\theta \ge \theta_0$ and $H_a$: $\theta < \theta_0$, let $T(\mathbf{X})$ be a test statistic whose distribution is continuous for $\theta = \theta_0$, and define $\hat{\alpha}(\mathbf{x}) = P(T(\mathbf{X}) \le T(\mathbf{x}); \theta_0)$ for observed $\mathbf{X} = \mathbf{x}$. Show that $\hat{\alpha}(\mathbf{X})$ has the Unif(0, 1) distribution. *Hint:* $\hat{\alpha}(\mathbf{x}) = F_T(T(\mathbf{x}); \theta_0)$, where $F_T$ is the cdf of $T(\mathbf{X})$ when $\theta = \theta_0$.

CHAPTER NINE

# The Multivariate Normal, Chi-Square, *t*, and *F* Distributions

## 9.1 INTRODUCTION

In this chapter we study four distributions that are closely related to the normal. Each is defined in terms of normally distributed random variables. These distributions are useful is making inferences about the parameters of one or more normal distributions. Consider the case of one sample from the $N(\mu, \sigma^2)$ distribution with both $\mu$ and $\sigma^2$ unknown. In Chapters Seven and Eight it was claimed, without proof, that $T = (\bar{X} - \mu)/\sqrt{S^2/n}$ has the $t$-distribution with $n - 1$ degrees of freedom. That is proved in Section 9.4 by (1) defining what it means to state that a random variable has a $t$-distribution, (2) showing that $\bar{X}$ and $S^2$ are independent, and (3) showing that $(n - 1)S^2/\sigma^2$ has the chi-square distribution with $n - 1$ degrees of freedom. We do not prove every statement made in this chapter. Readers looking for a detailed discussion should consult the book by the author (Stapleton, 1995), or much more famous books by, for example, Rao (1952), Bickel and Doksum (1977), and Casella and Burger (2002). We begin with a "once over lightly" discussion of the multivariate normal distribution.

## 9.2 THE MULTIVARIATE NORMAL DISTRIBUTION

**Definition 9.2.1** A random vector $\mathbf{X} = (X_1, \ldots, X_k)$ is said to have the *multivariate normal distribution* if every linear combination $\sum_{i=1}^{k} a_i X_i$ has the univariate normal distribution. □

*Models for Probability and Statistical Inference: Theory and Applications,* By James H. Stapleton

COMMENTS:

1. Like the univariate normal distribution, the MV-normal distribution is characterized by two parameters, its mean vector $\boldsymbol{\mu} = E(\mathbf{X})$ and its $k \times k$ covariance matrix $\boldsymbol{\Sigma} \equiv \text{Cov}(\mathbf{X})$. We write $\mathbf{X} \sim \text{N}_k(\boldsymbol{\mu}, \boldsymbol{\Sigma})$.
2. If $\mathbf{X} \sim$ MV-normal, any subset of components also has an MV-normal distribution. In particular, any single component has a univariate normal distribution and any pair has a bivariate normal distribution.
3. If $\mathbf{X} \sim N_k(\boldsymbol{\mu}, \boldsymbol{\Sigma})$, then $\mathbf{X}$ takes all of its values in the set Range $(\mathbf{X}) = \{\mathbf{x}|\mathbf{x} = \boldsymbol{\mu} + \mathbf{u}, \mathbf{u} \in \text{Col}(\boldsymbol{\Sigma})\}$. Col($\boldsymbol{\Sigma}$) is the column space of $\boldsymbol{\Sigma}$. If $\boldsymbol{\Sigma}$ has rank $r < k$, Range($\mathbf{X}$) is a hyperplane in $R_k$ and $\mathbf{X}$ does not have a density in $R_k$. For example, let $Y_1, \dots, Y_k$ be independent, each $\text{N}(\mu, \sigma^2)$. Define $X_j = Y_j - \bar{Y}$ for $j = 1, \dots, k$. Then $\mathbf{X} = (X_1, \dots, X_n)$ has covariance matrix $\text{Cov}(\mathbf{X}) = \sigma^2[\mathbf{I}_k - (1/n)\mathbf{J}_k]$, where $\mathbf{J}_k$ is the $k \times k$ matrix of all 1's. Cov($\mathbf{X}$) has rank $(k-1)$. Range($\mathbf{X}$) $= \{\mathbf{x}|\mathbf{x}\perp\mathbf{1}_k\}$, where $\mathbf{1}_k$ is the length-$k$ vector of all 1's, and $\mathbf{x} \perp \mathbf{1}_k$ means that $\mathbf{x}$ is orthogonal to $\mathbf{1}_k$ (has inner product zero). That is, the sum of the components of $\mathbf{x}$ is zero.
4. Let $\mathbf{X} \sim \text{N}_k(\boldsymbol{\mu}, \boldsymbol{\Sigma})$. Let $\mathbf{Y} = \mathbf{A}\mathbf{X}$, where $\mathbf{A}$ is an $m \times k$ matrix of constants. Then $\mathbf{Y} \sim N_m(\mathbf{A}\boldsymbol{\mu}, \mathbf{A}\boldsymbol{\Sigma}\mathbf{A}^{\text{T}})$.
5. If $\mathbf{X} \sim \text{N}_k(\boldsymbol{\mu}, \boldsymbol{\Sigma})$ and $\boldsymbol{\Sigma}$ is nonsingular, $\mathbf{X}$ has the density $f(\mathbf{x}; \boldsymbol{\mu}, \boldsymbol{\Sigma}) = [2\pi]^{-n/2} \det(\Sigma)^{-1/2} e^{-(1/2)Q(\mathbf{x})}$, where $Q(\mathbf{x}) = (\mathbf{x} - \boldsymbol{\mu})^{\text{T}}\boldsymbol{\Sigma}^{-1}(\mathbf{x} - \boldsymbol{\mu})$ is a quadratic form in $(\mathbf{x} - \boldsymbol{\mu})$, where $\mathbf{x}$ and $\boldsymbol{\mu}$ are written as column vectors.
6. $\mathbf{X}$ has $k$-dimensional moment generating function (mgf) $m(\mathbf{t}) \equiv E(e^{t_1X_1+\cdots+t_kX_k}) = e^{t^{\text{T}}\boldsymbol{\mu}+(1/2)\mathbf{t}^{\text{T}}\Sigma t}$ for $\mathbf{t}$ the column vector with components $t_1, \dots, t_k$.
7. If $\mathbf{Y}_1, \dots, \mathbf{Y}_n$ are independent $k$-component MV-normal random vectors, it follows that $\sum_{i=1}^n \mathbf{Y}_i \sim$ MV-normal.
8. Let $\mathbf{X} = (\mathbf{X}_1, \mathbf{X}_2)$, where $\mathbf{X}_1$ has $k_1$ and $\mathbf{X}_2$ has $k_2 = k - k_1$ components. Then $\mathbf{X}_1$ and $\mathbf{X}_2$ are independent if and only if $m(\mathbf{t})$ factors into the product of the mgf's of $\mathbf{X}_1$ and of $\mathbf{X}_2$. But this is true if and only Cov($\mathbf{X}_1, \mathbf{X}_2$) is the $k_1 \times k_2$ matrix of all zeros. It is this feature of the MV-normal distribution that makes it so mathematically tractable and used so often as a model for multivariate data. Consider the example in comment 3. Let $\mathbf{W} = (\bar{Y}, \mathbf{X})$, a vector of $k+1$ components. $\mathbf{W}$ has the MV-normal distribution because every linear combination of its components has the univariate normal distribution. $\text{Cov}(\bar{Y}, \mathbf{X}_j) = \text{Cov}(\bar{Y}, \mathbf{Y}_i) - \text{Cov}(\bar{Y}, \bar{Y}) = (\sigma^2/k) - \sigma^2/k = 0$. Thus, $\bar{Y}$ and the vector $\mathbf{X}$ of deviations of the $Y_i$ from $\bar{Y}$ are uncorrelated and therefore independent. Without the assumption that $\mathbf{Y}$ has the MV-normal distribution $\text{Cov}(Y_i - \bar{Y}, \bar{Y}) = 0$ for each $i$, but $Y_i - \bar{Y}$ and $\bar{Y}$ will not necessarily be independent.
9. For any covariance matrix $\boldsymbol{\Sigma}$, let $\mathbf{B}$ be any matrix such that $\mathbf{B}\mathbf{B}^{\text{T}} = \Sigma$. There are an infinity of such $B$ since $\mathbf{B}\mathbf{B}^{\text{T}} = \boldsymbol{\Sigma}$ implies that for any orthogonal matrix $\mathbf{U}$, $\mathbf{B}\mathbf{U}\mathbf{U}^{\text{T}}\mathbf{B}^{\text{T}} = \boldsymbol{\Sigma}$. $\mathbf{B}$ can be chosen, for example, to be lower-triangular or to be symmetric. If $\mathbf{B}$ is $k \times r$, let $\mathbf{Z}$ be a column vector of $r$ independent standard

normal random variables. Then $\mathbf{Y} = \mathbf{BZ} + \boldsymbol{\mu} \sim \mathrm{N}_k(\boldsymbol{\mu}, \boldsymbol{\Sigma})$. Most statistical computer packages have functions that generate pseudo-N(0, 1) independent random variables. Thus, this procedure can be used to generate pseudorandom vectors $\mathbf{Y}$ such that $\mathbf{Y} \sim N_k(\boldsymbol{\mu}, \boldsymbol{\Sigma})$.

**10.** Suppose that $\mathbf{X} \sim \mathrm{N}_k(\boldsymbol{\mu}, \boldsymbol{\Sigma})$, with $\boldsymbol{\Sigma}$ nonsingular. Suppose that $\mathbf{A}^{\mathrm{T}}\mathbf{A} = \boldsymbol{\Sigma}^{-1}$ so that $\boldsymbol{\Sigma} = A^{-1}(\mathbf{A}^{\mathrm{T}})^{-1}$. Then $\mathbf{Y} = \mathbf{AX} \sim \mathrm{N}_k(\mathbf{A}\boldsymbol{\mu}, \mathbf{A}\boldsymbol{\Sigma}\mathbf{A}^{\mathrm{T}} = \mathbf{I}_k)$, so the components of $\mathbf{Y}$ are independent N(0, 1) random variables.

## Problems for Section 9.2

**9.2.1** Let $(X, Y)$ have the bivariate normal distribution with parameters $\mu_x, \mu_y, \sigma_x, \sigma_y$, correlation $\rho$. Use comment 5 above to determine the density of $(X, Y)$ in nonmatrix form.

**9.2.2** The heights of married couples [$X$ = wife's height (cm), $Y$= husband's height (cm)] in a large population are approximately distributed as bivariate normal with $\mu_x = 165$, $\mu_y = 178$, $\sigma_x = 6.0$, $\sigma_y = 7.7$, $\rho = 0.5$. Find an approximation for

**(a)** $P(Y > X)$. *Hint*: Consider $Y - X$.

**(b)** $P(|Y - X| > 5)$.

**9.2.3** Let $\mathbf{X} = (X_1, X_2, X_3) \sim N_3(\boldsymbol{\mu}, \boldsymbol{\Sigma})$, where $\boldsymbol{\mu} = (8, 10, 12)$,

$$\boldsymbol{\Sigma} = \begin{pmatrix} 4 & 3 & -1 \\ 3 & 9 & -5 \\ -1 & -5 & 16 \end{pmatrix}.$$

**(a)** Give the correlation matrix for X.

**(b)** Let $Y_1 = X_1 + X_2 - X_3$ and $Y_2 = 4X_1 + 2X_2 + 3X_3$. What is the distribution of $\mathbf{Y} = (Y_1, Y_2)$?

**(c)** Find $\rho(Y_1, Y_2)$.

**(d)** Find $P(Y_1 > 9, Y_2 > 100)$.

**9.2.4** Let $Z_1, Z_2$ be independent standard normal random variables. Let $\mu_x, \mu_y, \sigma_x > 0, \sigma_y > 0$, and $\rho, -1 < \rho < 1)$ be constants. Let $X = \mu_x + \sigma_x Z_1$ and $Y = \mu_y + \sigma_y(\rho Z_1 + \sqrt{1-\rho^2}Z_2)$. Show that $(X, Y)$ has the bivariate normal distribution with parameters $\mu_x, \mu_y, \sigma_x, \sigma_y, \rho$.

**9.2.5** Let $\mathbf{X} = (X_1, X_2, X_3) \sim N_3(\boldsymbol{\mu}, \boldsymbol{\Sigma})$, where $\boldsymbol{\mu} = (10, 15, 20)$ and

$$\boldsymbol{\Sigma} = \begin{pmatrix} 4 & 3 & 1 \\ 3 & 9 & -6 \\ 1 & -6 & 7 \end{pmatrix}.$$

**(a)** Show that $\mathbf{a} \equiv (1, -1, 1)^{\mathrm{T}}$ (so that $\mathbf{a}$ is a column vector) is in the nullspace of $\boldsymbol{\Sigma}$.

**(b)** Find $\text{Var}(\mathbf{a}^T\mathbf{X})$.

**(c)** Is $(12, 25, 12)^T$ in Range($\mathbf{X}$)? $(12, 25, 13)^T$?

**9.2.6** Let $X_1, X_2$ be independent, each N(50, 100). Find $P(X_1 + X_2 > 90, X_1 - X_2 > 20)$.

**9.2.7** Let $\boldsymbol{\Sigma} = \begin{pmatrix} 9 & 3 \\ 3 & 5 \end{pmatrix}$. Find matrices $\mathbf{B}$ such that $\mathbf{B}\mathbf{B}^T = \Sigma$, where:

**(a)** $\mathbf{B}$ is lower triangular.

**(b)** $\mathbf{B}$ is symmetric.

**(c)** How could $\mathbf{B}$ be used to generate pseudo bivariate normal rv's with mean vector (10, 20) and covariance matrix $\boldsymbol{\Sigma}$?

## 9.3 THE CENTRAL AND NONCENTRAL CHI-SQUARE DISTRIBUTIONS

The chi-square distribution plays an important role in statistics. It is the exact or approximate distribution of some very useful statistics. In Chapter Eight we said that the asymptotic distribution of $Q_n \equiv -2\log(\Lambda_n)$, where $\Lambda_n$ was the likelihood ratio statistic, is chi-square under certain conditions. As we will show, the distribution of $S^2(n-1)/\sigma^2$, when $S^2$ is the sample variance for a random sample from a normal distribution, is distributed as chi-square with $n - 1$ degrees of freedom. Its density is a special case of the gamma distribution, so students should already be familiar with some of its properties. As will be obvious from its definition, the sum of independent chi-square random variables also has a chi-square distribution. To consider the power of tests, we need to consider the noncentral chi-square distribution.

**Definition 9.3.1** Let $\delta > 0$ and let $Z_1, \ldots, Z_n$ be independent, each with the N(0, 1) distribution. Let $Q = (Z_1 + \delta^{1/2})^2 + Z_2^2 + \cdots + Z_n^2$. Q is said to have the *noncentral chi-square distribution* with noncentrality parameter $\delta$ and $n$ degrees of freedom. We write $Q \sim \chi_n^2(\delta)$. If $\delta = 0$, $Q$ has the *central chi-square distribution* with $n$ degrees of freedom. For that case we write $Q \sim \chi_n^2$. □

COMMENTS:

1. $E(Q) = (1 + \delta) + (n - 1)(1) = \delta + n$. Let $V$ be the first term of $Q. E(V^2) = E(Z_1 + \delta^{1/2})^4 = 3 + 6(\delta) + \delta^2$, so that $\text{Var}(V) = 2 + 4\delta$. Therefore, $\text{Var}(Q) = 2 + 4\delta + (n-1)(2) = 2n + 4\delta$.
2. The central chi-square distribution with $n$ df has density $f(y; n) = (y^{n/2-1}e^{-y/2}/\,\Gamma(n/2)\cdot 2^{n/2}$ for $y > 0$. This can be shown using mgf's or by a convolution argument (see Problem 9.3.1). This is the density of the $\Gamma(n/2, 2)$ distribution. Appendix Table 5 presents quantiles $x_p$ for selected values values of $p$ and $n$. The noncentral chi-square density is

$f(y; n, \delta) = \sum_{k=0}^{\infty} p(k; \delta/2) f(y; n + 2k)$, where $p(k; \delta/2)$ is the Poisson probability mass function for mean $\lambda = \delta/2$.

## Linear Algebra and Random Vectors

1. Let $\mathbf{a}_1, \ldots, \mathbf{a}_n$ be an orthonormal basis for $R_n$, where these $\mathbf{a}_i$ are column vectors. That is, the $\mathbf{a}_i$ are orthogonal ($\mathbf{a}_i^T \mathbf{a}_j = 0$ for $i \neq j$), and each $\mathbf{a}_i$ has length 1 ($||\mathbf{a}_i||^2 = \mathbf{a}_i^T \mathbf{a}_i = 1$ for each $i$). Equivalently, if $\mathbf{A}$ is the $n \times n$ matrix with columns $\mathbf{a}_1, \ldots, \mathbf{a}_n$, then $\mathbf{A}^T\mathbf{A} = \mathbf{I}_n$. Then for each $\mathbf{x} \in R_n$, $\mathbf{x} = \sum_{i=1}^n b_i \mathbf{a}_i$, where $b_i = \mathbf{a}_i^T \mathbf{x}_j$. To prove this, multiply by $\mathbf{a}_i^T$ on the left sides of both $\mathbf{x}$ and $\sum_{i=1}^n b_i \mathbf{a}_i$. By the orthogonality of the $\mathbf{a}_i$, $||\mathbf{x}||^2 = \mathbf{x}^T\mathbf{x} = \sum_{i=1}^n b_i{}^2$.
2. Consider the special case that $\mathbf{a}_1 = \mathbf{1}_n/\sqrt{n}$. Then $b_1 = \sqrt{n}\ \bar{\mathbf{x}}$ and $b_1^2 = n\bar{x}^2$. It follows that $||\mathbf{x}||^2 = b_1^2 + \sum_{i=2}^n b_i^2$ and $||\mathbf{x}||^2 - n\bar{x}^2 = \sum_{i=1}^n (x_i - \bar{x})^2 = \sum_{i=2}^n b_i^2$.
3. Let $\mathbf{X}$ be a random vector with covariance matrix $\text{Cov}(\mathbf{X}) = \sigma^2 \mathbf{I}_n$. That is, the $X_i$ have common variance $\sigma^2$ and zero covariances. Let $\mathbf{a}$ and $\mathbf{b}$ be vectors of constants. Let $Y_a = \mathbf{a}^T\mathbf{X}$ and $Y_b = \mathbf{b}^T\mathbf{X}$, two linear combinations of the components of $\mathbf{X}$. Then $\text{Cov}(Y_a, Y_b) = \mathbf{a}^T(\sigma^2\mathbf{I}_n)\ \mathbf{b} = \sigma^2\mathbf{a}^T\mathbf{b}$. If $\mathbf{a} = \mathbf{b}$, then $\text{Var}(Y_a) = \sigma^2||\mathbf{a}||^2$. If the inner product $\mathbf{a}^T\mathbf{b} = 0$, then $\text{Cov}(Y_a, Y_b) = 0$. If, in addition, $\mathbf{X}$ has a multivariate distribution, then $Y_a$ and $Y_b$ are independent.
4. Let $\mathbf{X} = (X_1, \ldots, X_n)$ be a random sample from a distribution with mean $\mu$, variance $\sigma^2$. Let $\mathbf{a}_1, \ldots, \mathbf{a}_n$ be as in case 2, and let $B_1 = \mathbf{a}_1{}^T\mathbf{X}$. Then $E(B_1) = \mathbf{a}_1^T \mu \mathbf{1}_n = \sqrt{n}\,\mu$, and $E(B_i) = 0$ for $i > 1$. $B_i$ has a variance $\sigma^2||\mathbf{a}_i||^2 = \sigma^2$ for each $i$. Another argument: The matrix $\mathbf{B}$ with column vectors $B_1$ is $\mathbf{A}^T\mathbf{X}$, so that $E(\mathbf{B}) = A^T\mu \cdot \mathbf{1}_n = (\sqrt{n}\,\mu, 0, \ldots, 0)^T$ and $\text{Cov}(\mathbf{B}) = A^T(\sigma^2 n)A = \sigma^2 I_n$, so that $E(\mathbf{B}) = \mathbf{A}^T\mu \cdot \mathbf{1}_n = (\sqrt{n}\,\mu, 0, \ldots, 0)^T$ and $\text{Cov}(\mathbf{B}) = A^T(\sigma^2\, n)A = \sigma^2\mathbf{I}_n$.
5. From case 2, $\sum_{i=1}^n (X_i - \bar{X})^2 = \sum_{i=2}^n B_i^2$, so that $E\left(\sum_{i=1}^n (X_i - \bar{X})^2\right) = E\left(\sum_{i=2}^n B_i^2\right) = (n-1)\sigma^2$, and $E(S^2) = \sigma^2$. Thus, $S^2$ is an unbiased estimator of $\sigma^2$. This explains the use of $n - 1$ as the divisor rather than the more intuitive $n$.

We can extend the usefulness of the definition of noncentral chi-square distribution as follows:

**Theorem 9.3.1** Let $\mathbf{X} = (X_1, \ldots, X_n)$, where $X_i \sim \text{N}(\mu_i, \sigma^2)$, and the $X_i$ are independent. Then $Q = \sum_{i=1}^n X_i^2/\sigma^2 \sim \chi_n^2(\delta)$ for $\delta = (\sum_{i=1}^n \mu_i^2)/\sigma^2$.

*Proof:* Let $\boldsymbol{\mu}$ be the vector of $\mu_i$'s, let $\mathbf{a}_1 = \boldsymbol{\mu}/||\boldsymbol{\mu}||$, and let $\mathbf{a}_1, \ldots, \mathbf{a}_n$ be an orthonormal basis for $R_n$. Define $\mathbf{B} = (B_1, \ldots, B_n)^T$ as in case 4. Then $\text{Cov}(\mathbf{B}) = \sigma^2\mathbf{I}_n$, $E(\mathbf{B}) = \mathbf{A}^T\boldsymbol{\mu} = (||\boldsymbol{\mu}||, 0, \ldots, 0)^T$, and $Q = \sum_{i=1}^n (B_i/\sigma)^2$. Therefore, $Q \sim \chi_n^2(\delta)$ for $\delta = (||\boldsymbol{\mu}||^2 + 0^2 + \cdots + 0^2)/\sigma^2 = \left(\sum_{i=1}^n \mu_i^2\right)/\sigma^2$. □

Now we are ready to consider the sample variance $S^2$ when a sample is taken from a normal distribution.

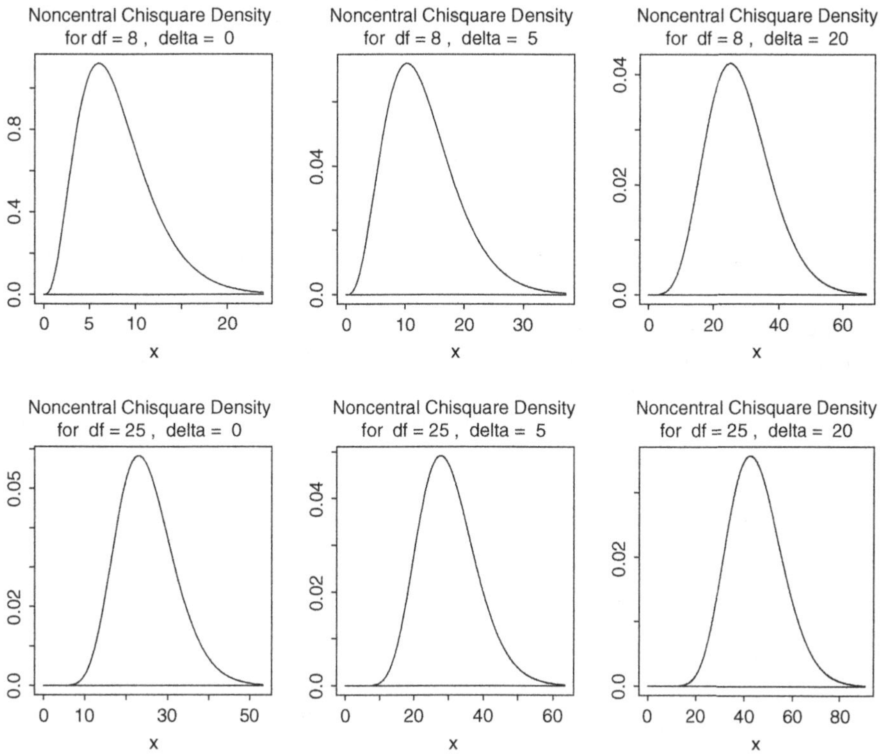


**FIGURE 9.3.1** Central and noncentral chi-square densities.

**Theorem 9.3.2** Let $\mathbf{X} = (X_1, \ldots, X_n)$ be a random sample from the N($\mu, \sigma^2$) distribution. Let $\bar{X}$ and $S^2 = \sum_{i=1}^{n}(X_i - \bar{X})^2/(n-1)$ be the sample mean and sample variance. Then:

**(a)** $S^2$ and $\bar{X}$ are independent.
**(b)** $S^2(n-1)/\sigma^2 \sim \chi^2_{n-1}$, the central chi-square distribution with $n-1$ df (see Figure 9.3.1).

*Proof:* By comment 8 in Section 9.2, $\bar{X}$ and the vector of deviations $X_i - \bar{X}$ are uncorrelated and therefore, by normality, independent. It follows that $\bar{X}$ and any function of the vector of deviations are independent. $S^2$ is one such function. This proves case 1.

To prove case 2, let $\mathbf{X} = (X_1, \ldots, X_n)$, and define $\mathbf{B}$ as in case 4 on the previous page. Then, from case 2, $(n-1)S^2/\sigma^2 = (1/\sigma^2)\sum_{i=1}^{n}(X_i - \bar{X})^2 = \sum_{i=2}^{n}(B_i/\sigma)^2$. By the definition of the chi-square distribution, this last random variable has the central chi-square distribution with $n-1$ df. $\square$

### Confidence Intervals on $\sigma^2$

Suppose that we observe $\mathbf{X} = (X_1, \ldots, X_n)$, a random sample from a $N(\mu, \sigma^2)$ distribution with $\mu$ and $\sigma^2$ unknown. Since $W \equiv S^2(n-1)/\sigma^2 \sim \chi^2_{n-1}$, we can use $W$ as a pivotal quantity to produce a confidence interval on $\sigma^2$. Let $C_{\alpha/2}$ and $C_{1-\alpha/2}$ be the $\alpha/2$ and $(1-\alpha/2)$ quantiles of the central chi-square distribution with $n-1$ df. We provide these values for many combinations of $n$ and $\alpha$ in Appendix Table 5. Then $1-\alpha = P(C_{\alpha/2} \le W \le C_{1-\alpha/2}; \sigma^2) = P(S^2(n-1)/C_{1-\alpha/2} \le \sigma^2 \le S^2(n-1)/C_{\alpha/2;}\sigma^2)$. Thus, $[S^2(n-1)/C_{1-\alpha/2}, S^2(n-1)/C_{\alpha/2}]$ is a $100(1-\alpha)\%$ CI on $\sigma^2$.

WARNING: The correctness of the probability statement for this CI depends strongly on the distribution of the observations $X_i$. For example, for samples of size 25 from a double-exponential distribution, the "95%" CI contained $\sigma^2$ for only 824 of 1000 trials. For samples of 100 the CIs were even worse: 784 of 1000 intervals contained $\sigma^2$. Similar sampling from a normal distribution produced 958 and 955 successful coverages of $\sigma^2$. In his consulting the author has very seldom (never?) found such intervals useful, partially because distributions sampled are often not close to normal.

Tables of the chi-square distribution do not usually include quantiles for df > 60. One rough approximation merely uses the fact that a chi-square random variable with $\nu$ df is the sum of the squares of $\nu$ independent standard normal random variables, so that $\chi^2_{\nu\gamma} = \nu + z_\gamma\sqrt{2\nu}$. Another approximation is useful. Let $U_n$ have the central chi-square distribution with n df. Let $Z_n = [(U_n/n)^{1/3} - a_n]/b_n$, where $a_n = 1 - 2/(9n)$ and $b_n = [2/(9n)]^{1/2}$. Then $Z_n$ is approximately N(0, 1) for $n$ as small as 5 or 10. The $\gamma$-quantile $x_\gamma$ for $U_n$ is therefore given in good approximation by $d_n = n(a_n + z_\gamma b_n)^3$, where $z_\gamma$ is the $\gamma$-quantile of the standard normal distribution. For example, for $n = 10$, $\gamma = 0.95$, $a_{10} = 0.9778$, $b_{10} = 0.1491$, $z_{0.95} = 1.645$, $d_{10} = 18.293$, while a more exact method using S-Plus gave 18.307. The simpler approximation $\nu + z_\gamma\sqrt{2\nu}$ gives 17.36.

### Problems for Section 9.3

**9.3.1** Let $Z \sim N(0, 1)$. Show that $Z^2$ has the density function given in comment 2 in Section 9.2. for $n = 1$.

**9.3.2** Let $Z_1, Z_2$ be independent, each N(0, 1).

**(a)** Show that $U \equiv Z_1^2 + Z_2^2$ has the exponential distribution with mean 2.

**(b)** Find the 0.95-quantile for $U$.

**(c)** Find $P(U \ge 7 | U \ge 3)$.

**9.3.3** Let $X_1, X_2, X_3$ be independent, with $X_k \sim N(k, 3)$ for $k = 1, 2, 3$.

**(a)** Find a constant $C$ such that $C(Y_1^2 + Y_2^2 + Y_3^2)$ has the noncentral chi-square distribution. What are the df and noncentrality parameters?

**(b)** Let $Y_1 = X_1 + X_2 + X_3$ and $Y_2 = X_1 + X_2 - 2X_3$. Let $V = Y_1^2 + Y_2^2$. Find constants $C_1$ and $C_2$ such that $V = C_1Y_1^2 + C_2Y_2^2$ has a noncentral chi-square distribution. What are the degrees of freedom and noncentrality parameter?

**9.3.4** Let $\mathbf{a}_1 = (1/2)(1, 1, 1, 1)^T$ and $\mathbf{a}_2 = (1, -1, 0, 0)^T/\sqrt{2}$.

**(a)** Find vectors $\mathbf{a}_3, \mathbf{a}_4$ such that $\mathbf{a}_1, \mathbf{a}_2, \mathbf{a}_3, \mathbf{a}_4$ is an orthonormal basis for $R_4$. Check your computations by defining $\mathbf{A} = (\mathbf{a}_1, \mathbf{a}_2, \mathbf{a}_3, \mathbf{a}_4)$ and verifying that $\mathbf{A}^T\mathbf{A} = \mathbf{I}_4$. *Hint*: To make computations simpler, first find such $\mathbf{a}_i$ so they are orthogonal, then divide by constants so they each have length 1.

**(b)** Let $\mathbf{x} = (1, 3, 7, 9)^T$. Find constants $b_j$ such that $\mathbf{x} = \sum_{i=1}^4 b_j\mathbf{a}_j$.

**(c)** Verify that $||\mathbf{x}||^2 = \sum_{i=1}^4 b_j^2$ and that $\sum_{i=1}^4 (x_i - \bar{x})^2 = \sum_{i=2}^4 b_j^2$.

**(d)** If $\mathbf{X} \sim N_4(\boldsymbol{\mu} = (10, 20, 30, 40)^T, 25\mathbf{I}_4)$, what is the distribution of $\mathbf{U} = \mathbf{A}^T\mathbf{X}$? Find a constant $K$ so that $K||\mathbf{U}||^2$ has a noncentral chi-square distribution. Give the df value and noncentrality parameter.

**9.3.5** Using the S-Plus function command "pchisq(40, 20, 10)," we find that the 0.95-quantile of the $\chi^2_{20}$ distribution is 31.41. Find the values of the two approximations discussed in the final paragraph of this section.

**9.3.6** Let $X_1, \ldots, X_{25}$ be a random sample from the N(50, 100) distribution. Find a good approximation for $P(\bar{X} > 53, S^2 > 120)$.

**9.3.7** Let $X_1, \ldots, X_n$ be a random sample from a normal distribution with mean $\mu$, variance $\sigma^2$, both parameters unknown.

**(a)** For $n = 9$, $S^2 = 20.0$, find a 95% confidence interval on $\sigma^2$.

**(b)** Suppose that a second sample $Y_1, \ldots, Y_m$ is taken from a normal distribution with mean $\nu$ known, variance $\sigma^2$ (same as for the $X_i$'s). For $m = 6$, sample variance $S_Y^2 = 14.0$ combine the data of part (a) with the $Y_i$ data to give a point estimate and a 95% CI on $\sigma^2$.

## 9.4 STUDENT'S $t$-DISTRIBUTION

In 1907, William Sealey Gosset, an employee of the Guiness Brewery in Dublin, Ireland, began a year of study with Karl Pearson at the University of London. Gosset had studied in the New College of Oxford University, earning first-class degrees in mathematics and chemistry. Under Karl Pearson, Gosset set to work to study the probability distribution of the standardized mean, when the scale parameter $\sigma$ had to be estimated. Specifically, he studied the distribution of $T = (\bar{X} - \mu)/(S/\sqrt{n})$, for samples of $n$ from the N$(\mu, \sigma^2)$ distribution. It had long been known that $Z = (\bar{X} - \mu)/(\sigma/\sqrt{n}) \sim$ N(0, 1), that this was true in approximation for large $n$ even when sampling from (most) nonnormal distribution, and that this was true for $T$ as well for $n$ "large enough." But in his work analyzing data for Guiness, Gosset often had small samples. He attacked the problem in two ways: (1) mathematically, and (2) by simulation.

For the first approach he concluded that $S^2$ (after multiplication by a constant) is approximately distributed as chi-square by considering moments, showed that $\bar{X}$ and $S^2$ have correlation zero, concluding that they were therefore independent (in those days correlation zero and independence were often confused), then determined the distribution of the ratio $T$. For the second approach he used a method that has become known as the *Monte Carlo method*. He obtained 3000 measurements on criminal heights that seemed to be at least approximately normally distributed. He wrote the numbers on 3000 pieces of cardboard. He divided these 3000 randomly into 750 piles of four each, and for each he determined $Z \equiv (\bar{X} - \mu)/S = T/\sqrt{n}$. That was not easy in those days of hand computation. The histogram of 750 values of $Z$ was closely approximated by the density that he had obtained using the first approach. Reading the paper today is difficult, and from the modern point of view not rigorous, but at that time it was a remarkable paper and is justifiably famous.

Gosset published his results under the name "A Student," since Guiness asked its employees to remain anonymous. His paper "On the Probable Error of the Mean," appeared in *Biometrika*, Vol. 5, in 1908. It took some time before the value of the paper was fully appreciated, but by the middle of the century, every statistics book contained Student's $t$-table. The same volume of *Biometrika* contains another paper of Student's, "The Probable Error of a Correlation Coefficient." That paper was based completely on simulation, using the pairs (criminal heights, middle-finger lengths.) Six years later, R. A. Fisher was able to determine the distribution of the sample correlation coefficient for samples from bivariate normal distributions by a geometric argument.

We extend the definition of the $T$-distribution to the noncentral case.

**Definition 9.4.1** Let $Z \sim \mathrm{N}(0, 1)$, and $V \sim \chi^2_\nu$, with $Z$ and $V$ independent. Let $\theta$ be a constant. Let $T \equiv (Z + \theta)/\sqrt{V/\nu}$. Then $T$ is said to have *Student's t-distribution* with noncentrality parameter $\theta$ and $\nu$ degrees of freedom. We write $T \sim t_{n-1}(\theta)$. When $\theta = 0$, we say that $T$ has the *central t- distribution* with $\nu$ degrees of freedom and write $T \sim t_\nu$. □

Since $Z$ and $V$ are independent and the density of each is known, the density of $T$ for the case $\theta = 0$ can be expressed by first determining the density of the denominator $W \equiv \sqrt{V/\nu} : f_W(w) = 2wm^2 f_V(m^2w^2)$. Then the density of $T$ is $f_T(t; \nu) = \int_{-\infty}^{\infty} f_Z(z) f_W(tz)\, dz$. After some manipulation (see Problem 9.4.1) we get

$$f_T(t;\nu) = \left[\sqrt{\nu\pi}\,\Gamma\left(\frac{\nu}{2}\right)\right]^{-1} \Gamma\left(\frac{\nu}{2} + \frac{1}{2}\right)\left(1 + \frac{t^2}{\nu}\right)^{-(\nu+1)/2} \quad \text{for } -\infty < t < +\infty. \tag{9.4.1}$$

Making use of Stirling's formula, $\Gamma(\alpha + 1)/[e^{-\alpha}\alpha^\alpha\sqrt{2\pi\alpha}] \to 1$ as $\alpha \to \infty$, we get $\lim_{\nu\to\infty} f_T(t;\nu) = \phi(t)$, the standard normal density, for each real number $t$, as illustrated in Figure 9.4.1. Thus, if $\nu$ is large, the central $t$-distribution and the standard

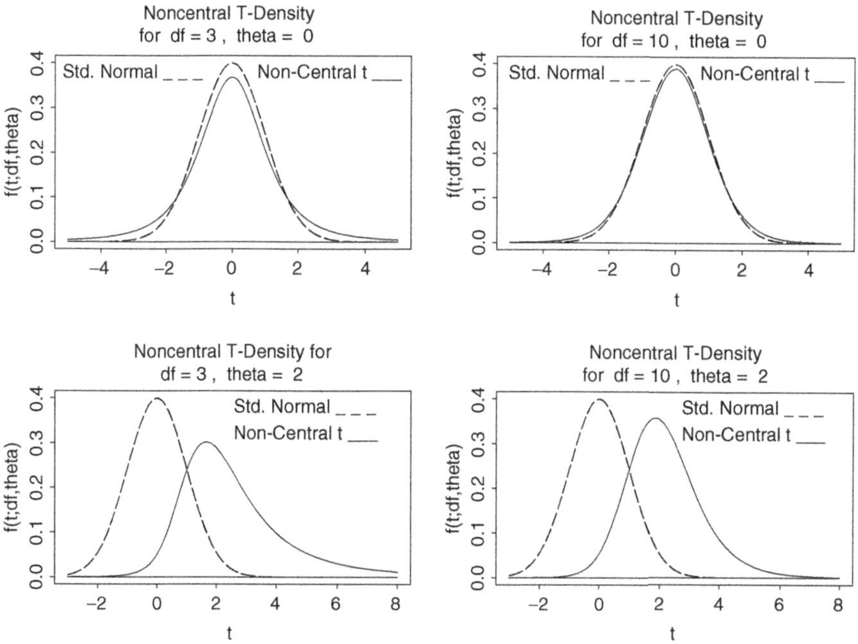


**FIGURE 9.4.1** Central and noncentral $t$-densities with the standard normal density.

normal are close. The noncentral $T$-density may be expressed as an infinite series in the noncentrality parameter $\theta$. We omit the rather complex expression (see Kennedy and Gentle, 1980).

***Example 9.4.1*** Let $X_1, \ldots, X_n$ be a random sample from $N(\mu, \sigma^2)$. Let $C$ be a constant. Then $Z \equiv (\bar{X} - \mu)/\sqrt{\sigma^2/n} \sim N(0, 1)$, $V \equiv S^2(n-1)/\sigma^2 \sim \chi^2_{n-1}$ and $Z$ and $V$ are independent. Therefore, $(\bar{X} - C)/(S/\sqrt{n}) = (z + \theta)/\sqrt{V/(n-1)n} = t_{n-1}(\theta)$ for $\theta = (\mu - C)/\sigma/\sqrt{n}$. Taking $C = \mu$, we get the result obtained by Student, that $T = (\bar{X} - \mu)/(S/\sqrt{n}) \sim t_{n-1}$. Letting $t_{0.975}$ be the 0.975-quantile of this distribution, so that, by symmetry, the 0.025-quantile is $-t_{0.975}$, we get $0.950 = P(-t_{0.975} \leq T \leq t_{0.975}) = P(\mu \in [\bar{X} \pm t_{0.975}S/\sqrt{n}])$, so that $[\bar{X} \pm t_{0.975}S/\sqrt{n}]$ is a 95% confidence interval on $\mu$ (see Figure 9.4.2). □

In his 1908 paper, Student discussed an experiment comparing two drugs, labeled LHH and DHH, as reported in the *Journal of Physiology* in 1904, comparing the numbers of "additional" hours of sleep produced by each drug for 10 patients:

| | Patient | | | | | | | | | |
|---|---|---|---|---|---|---|---|---|---|---|
| | 1 | 2 | 3 | 4 | 5 | 6 | 7 | 8 | 9 | 10 |
| DHH | 0.7 | −1.6 | −0.2 | −1.2 | −1.0 | 3.4 | 3.7 | 0.8 | 0.0 | 2.0 |
| LHH | 1.9 | 0.8 | 1.1 | 0.1 | −0.1 | 4.4 | 5.5 | 1.6 | 4.6 | 3.4 |
| Diff. | 1.2 | 2.4 | 1.3 | 1.3 | 0.0 | 1.0 | 1.8 | 0.8 | 4.6 | 1.4 |

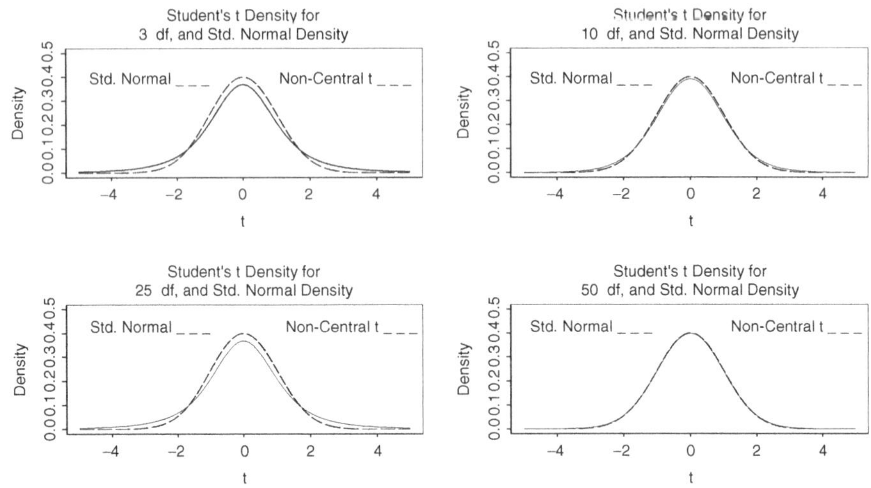


**FIGURE 9.4.2** Student's t densities and the standard normal density.

Let $D_i$ = (LHH value) – (DHH value) for patient $i$. A reasonable model is that these $D_i$ constitute a random sample from a $N(\mu_D, \sigma_D^2)$ distribution. We find that $\bar{D} = 1.58$, $S = 1.23$ (this differed slightly from Student's value because he used the denominator $n$ in determining $S^2$). Since $t_{0.975} = 2.262$ for 9 df, obtained from Student's $t$-table, we get the 95% CI $[1.58 \pm 0.880] = [0.70, 2.46]$. The concept of CIs was not developed until 25 years later, so instead, Student presented the one-tailed $p$-value corresponding to the observed value $T = (1.58 - 0)/(1.23/\sqrt{10}) = 4.06$. He obtained 0.0015. Using S-Plus, the author obtained 0.00142, so Student's work was remarkably precise. Student stated that "the odds are about 666 to one that LHH is the better soporific." These days we would test $H_0{:}\mu_D = 0$ versus $H_a$: $\mu_D \neq 0$ and obtain the $p$-value 2(0.00142) = 0.00284. We would then say "Given that $H_0$ is true, the odds against obtaining a $t$-value as large in absolute value as the one obtained are about $(1 - 0.00284)/0.00284 = 351$ to 1." The practical conclusion is the same. We conclude that "on average, LHH results in more sleep."

Student suggested that the distribution of $T$ should be close to the $t$-distribution he found if the distribution sampled is close to normal. To show that his intuition was correct, consider Figures 9.4.3 and 9.4.4. In Figure 9.4.3 four distributions are considered, each symmetric around zero. For each, 10,000 samples were taken and $T = \bar{X}/(S/\sqrt{n})$ determined for each. The numbers of values of $T$ less than the 0.025-quantile and greater than the 0.975-quantile of the $t$-distribution with $(n - 1)$ df is given for each case. The first shows, for example, that for samples of 10 from the double-exponential distribution with mean $\mu$, $100(1 - 0.0220 - 0.0211)\% = 95.69\%$ of nominal 95% CIs would have contained $\mu$.

In Figure 9.4.4 samples are taken from the gamma distribution with shape parameter $\alpha = 5$ and $T = (\bar{X} - 5)/S/\sqrt{n}$. Even for $n = 5$, the coverage probability of a nominal 95% CI is about 0.938, still approximately 95%. Since this distribution is

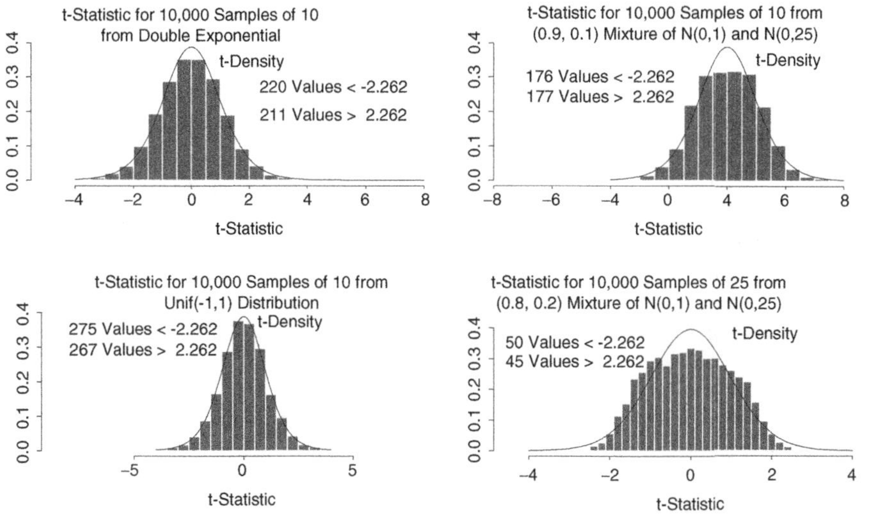


**FIGURE 9.4.3** Histograms of 10,000 values of $t$-statistics for samples from nonnormal distributions.

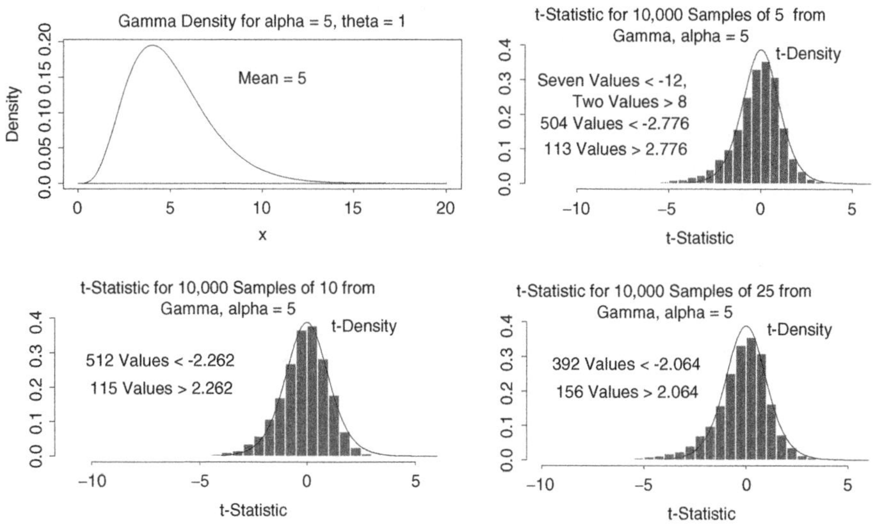


**FIGURE 9.4.4** Histograms of the $t$-statistic for samples from gamma distributions.

a bit skewed to the right, not symmetric as are the distributions in Figure 9.4.3, the distribution of $T$ is also skewed to the right, although not as much, with the amount of skewness decreasing as $n$ increases.

***Example 9.4.2*** To determine the effect of a gasoline additive on the miles per gallon of an automobile, the following experiment was conducted. A high-powered

**TABLE 9.4.1 Gasoline Consumption Data**

| Run | Additive | Number of Gallons | Run | Additive | Number of Gallons |
|---|---|---|---|---|---|
| 1 | × | 7.71 | 11 | × | 8.54 |
| 2 | × | 8.65 | 12 | × | 8.72 |
| 3 | | 9.04 | 13 | | 8.76 |
| 4 | | 8.39 | 14 | | 9.28 |
| 5 | × | 8.94 | 15 | × | 8.09 |
| 6 | | 9.67 | 16 | × | 8.48 |
| 7 | | 8.73 | 17 | × | 8.79 |
| 8 | | 9.06 | 18 | × | 9.09 |
| 9 | | 8.89 | 19 | | 9.09 |
| 10 | × | 8.37 | 20 | | 7.84 |

automobile was driven 100 miles 20 times (*runs*) over a 2-mile track. For each run the auto was gradually accelerated to 60 miles per hour, then kept at that speed for the entire 100 miles. The same driver was used for all 20 runs. The additive was added to the gasoline for 10 randomly chosen runs. At the end of each run the number of gallons consumed was measured. Neither the driver nor the person who measured the gasoline consumption knew for which runs the additive was used. Thus, the experiment was *double blind*. The run numbers with the additive were: 17, 18, 15, 11, 1, 2, 12, 5, 16, 10. These were chosen using the S-Plus function "sample." The measurements are listed in Table 9.4.1.

It seems reasonable to consider the following model. Let the 10 measurements for the nonadditive runs be $X_1, \ldots, X_{10.}$ Let the measurements for the additive runs be $Y_1, \ldots, Y_{10}$. Suppose that the 20 measurements are independent and that $X_i \sim \mathrm{N}(\mu_1, \sigma^2)$, $Y_i \sim \mathrm{N}(\mu_2, \sigma^2)$. Thus, the means may be different, but the variances are equal.

First we find a 95% CI on $\Delta \equiv \mu_2 - \mu_1$, the additive effect. Negative values of $\Delta$ correspond to a decrease in gasoline consumption when the additve is used. Let $\bar{X}$ and $\bar{Y}$ be the sample means, and let $S_1^2$ and $S_2^2$ be the sample variances for the $X$'s and $Y$'s. Let the sample sizes be $n_1 = 10$ and $n_2 = 10$. Then (1) $\bar{X} \sim \mathrm{N}(\mu_1, \sigma^2)$, (2)$\bar{Y} \sim \mathrm{N}(\mu_2, \sigma^2)$, (3)$S_1^2(n_1 - 1)/\sigma^2 \sim \chi^2_{n_1-1}$, (4) $S_2^2(n_2 - 1)/\sigma^2 \sim \chi^2_{n_2-1}$, and (5) these four random variables are independent. It follows from (1), (2) and (5) that $Z \equiv (\bar{X} - \bar{Y} - \Delta)/\sqrt{\sigma^2(1/n_1 + 1/n_2)} \sim \mathrm{N}(0, 1)$. Let $S_p^2 = [S_1^2(n_1 - 1) + S_2^2(n_2 - 1)]/(n_1 - 1 + n_2 - 1)$, the *pooled estimator* of the common variance $\sigma^2$. From (3), (4), and (5) it follows that $V \equiv S_p^2(n_1 + n_2 - 2)/\sigma^2 \sim \chi^2_{n_1+n_2-2}$. Thus, from Definition 9.3.1, $T \equiv Z/\sqrt{V/(n_1 + n_2 - 2)} \sim t_{n_1+n_2-2}$. $\sigma^2$ in the numerator and denominator of $T$ cancel so that $T = (\bar{Y} - \bar{X} - \Delta)/\sqrt{S_p^2(1/n_1 + 1/n_2)}$. Letting $t_{0.975}$ be the 0.975-quantile of the $t$-distribution with $n_1 + n_2 - 2$ df, we get $0.95 = P(-t_{0.975} \le T \le t_{0.975}) = P(\Delta \in I)$, where $I = [\bar{Y} - \bar{X} \pm t_{0.975}\sqrt{S_p^2(1/n_1 + 1/n_2)}]$, so that I is a 95% CI on $\Delta$.

For these data we observe that $\bar{X} = 8.875$, $\bar{Y} = 8.538$, $S_1^2 = 0.249$, $S_2^2 = 0.166$, $S_p^2 = 0.208$, $t_{0.975} = 2.1009$, $\sqrt{S_p^2(1/n_1 + 1/n_2)} = 0.204$, and the 95% CI $[-0.337 \pm 0.428] = [-0.765, 0.091]$. Since the 95% CI includes zero, we do not have enough evidence to conclude that $\Delta \neq 0$. On the other hand, if we are quite sure that the additive cannot increase the amount of gasoline needed, it may be more appropriate to determine an upper 95% bound on $\Delta$. We find that $U = [\bar{Y} - \bar{X} + t_{0.95}\sqrt{S_p^2(1/n_1 + 1/n_2)}] = 0.017$, so there is not enough evidence to conclude that $\Delta < 0$ at the $\alpha = 0.05$ level.

More formally, we might wish to test $H_0$: $\Delta \geq 0$ versus $H_a$: $\Delta < 0$, since then $H_a$ is equivalent to the statement that the additive decreases the amount of gasoline used. It can be shown (Problem 9.4.2) that the likelihood ratio test is equivalent to the rejection of $H_0$ for small $T^* = (\bar{Y} - \bar{X} - 0)/\sqrt{S_p^2(1/n_1 + 1/n_2)}$. For $\Delta = 0$, $T^* \sim t_{n_1+n_2-2}$, so the 0.05-level test rejects for $T < -t_{0.95} = -1.734$. We observe $T^* = (-0.337/0.204) = -1.654$, so again we cannot reject $H_0$. The observed $p$-value is $P(T \leq -1.654) = 0.058$. These data were actually generated using S-Plus, with $\mu_1 = 8.9$, $\mu_2 = 8.3$, $\sigma = 0.5$. The power function for the test of $H_0$ versus $H_a$ is a function of the noncentrality parameter

$$\theta = E\left(\frac{\bar{Y} - \bar{X} - 0}{\sqrt{\sigma^2(1/n_1 + 1/n_2)}}\right) = \frac{\Delta}{\sqrt{\sigma^2(1/n_1 + 1/n_2)}} = \frac{-0.6}{\sqrt{0.05}} = -2.683.$$

Using the function "pnoncentt" in S-Plus, we find power 0.824. □

Although the probability theory of Example 9.4.2 is exact, it does rest on the equal-variance model as well as on the normality. These $t$-methods are quite robust against nonnormality even for moderate sample sizes by the central limit theorem. They are a bit more sensitive to unequal variances, so it is often wise to use the estimator $\hat{\sigma}^2(\bar{Y} - \bar{X}) = S_1^2/n_1 + S_2^2/n_2$ of $\text{Var}(\bar{X} - \bar{X})$. The methods resulting, although not exact, provide better approximations than those given by the methods that assume equal variances for cases for which that is not true. In 1947, Welch proposed the use of the statistic $T_w \equiv (\bar{Y} - \bar{X})/\hat{\sigma}(\bar{Y} - \bar{X})$, with approximation by the $t$-distribution with $\nu$ df, where $\nu$ is the closest integer to $(V_1 + V_2)^2/[V_1^2/(n_1 + 1) + V_1^2/(n_2 + 1)] - 2$, where $V_1 = S_1^2/n_1$ and $V_2 = S_2^2/n_2$, so that $\hat{\sigma}^2(\bar{Y} - \bar{X}) = V_1 + V_2$. For the data of Example 9.4.3, we get $\hat{\sigma}^2(\bar{Y} - \bar{X}) = 0.200$, $T_w = -1.853$, $\nu = 19$, almost the same statistics as were obtained when common variances were assumed. The resulting 95% CI on $\Delta$ is $[-0.367 \pm (2.093)(0.200)] = [-0.486, 0.052]$. In practice it makes little difference which method is used when $S_1^2$ and $S_2^2$ are relatively close.

In general, if $\sigma_1^2$ and $\sigma_2^2$ are close, it matters little whether the method used is based on the pooled sample variance $S_p^2$ or on Welch's approximation. The same is true if the sample sizes are equal. However, if the sample sizes differ *and* $\sigma_1^2$ and $\sigma_2^2$ differ very much, the Welch approximation is better. This is illustrated in Table 9.4.2, giving the numbers $N_p$ and $N_w$ of CIs, for 10,000 simulations, which covered $\Delta$ for nominal 95% intervals for both methods. The mean lengths $L_p$ and $L_w$ of the

**TABLE 9.4.2 Performances of Two-Sided, Two-Sample $t$-CIs for Differing Standard Deviations and Sample Sizes**

| | $n_1 = 10, n_2 = 10$ | $n_1 = 14, n_2 = 6$ | $n_1 = 6, n_2 = 14$ |
|---|---|---|---|
| $\sigma_1 = 1$, $\sigma_2 = 2$ | 9876, 9518, 3.72, 2.97, 2.77 | 9691, 9381, 9.30, 8.99, 3.37 | 9880, 9514, 3.52, 2.79, 2.64 |
| $\sigma_1 = 1$, $\sigma_2 = 5$ | 9549, 9467, 6.47, 6.97, 6.32 | 8243, 9390, 5.57, 9.56, 8.07 | 9904, 9470, 7.26, 5.83, 5.48 |
| $\sigma_1 = 1$, $\sigma_2 = 10$ | 9402, 9456, 11.89, 6.97, 12.46 | 7437, 9394, 9.30, 18.99, 16.04 | 9911, 9489, 14.02, 11.37, 10.60 |

CIs are also given. As should be expected, when $\Delta$ is covered more often, the mean length is larger. For purposes of comparison, fixed-length CIs, for the case that $\sigma_1^2$ and $\sigma_2^2$ are known, have lengths $L_k = 2(1.96)\sqrt{\sigma_1^2/n_1 + \sigma_1^2/n_2}$. Each of the nine cells in Table 9.4.2 presents $N_p$, $N_w$, $L_p$, $L_w$, and $L_k$ in this order. Had the standard deviations both been multiplied by $C$, the corresponding lengths would also have been multiplied by $C$.

Notice that the pooled method works poorly when $n_2 < n_1$ and $\sigma_2 > \sigma_1$. Whenever possible, it is better to use the larger sample size for the population with the larger standard deviation. When knowledge of the comparative sizes of the standard deviations is lacking, use the Welch approximation with equal sample sizes.

A MISTAKE: Rarely (we hope) users of two-sample methods fail to note that observations were taken in pairs, so that the paired difference analysis of Example 9.4.2 is appropriate. Suppose, for example, that a company, the Prep Company, offers courses that prepare students for standardized tests, the SAT tests given to high school students, for example. It has designed a new one-week course and wishes to evaluate it. The company might perform an experiment as follows. Thirty volunteers are found from among high school juniors, each paid \$100 for taking part. First they take a one-hour older version of the test, and the results are graded. Then the 30 students are paired, pair 1 having the two highest scores, pair 2 having the next two highest, . . . , pair 15 having the two lowest. Then a coin is flipped 15 times to decide which in each pair is assigned to the treatment group, which to the control group. Those in the treatment group take the one-week course. Those in the control group do not. Then after the week all 30 students take the new test, and their scores $X_i$ for the control student in pair $i$ and $Y_i$ for the treatment student in pair $i$, $i = 1, \ldots, n = 15$. The following model seems appropriate. Suppose that $X_i = A_i + \varepsilon_i$ and $Y_i = \Delta + A_i + \eta_i$, where the $A_i$ are constants and the $\varepsilon_i$ and $\eta_i$ are independent $N(0, \sigma^2)$ random variables. $\Delta$ is the *additive treatment effect*. Let $D_i = Y_i - X_i = \Delta + \eta_i - \varepsilon_i$. Then $D_i \sim N(\Delta, 2\sigma^2 \equiv \sigma_D^2)$. Thus, a $100(1-\alpha)\%$ CI on $\Delta$ is given by $[\bar{D} \pm t_{1-\alpha/2}\sqrt{S_D^2/n}]$. The paired difference $t$-test of $H_0$: $\Delta = 0$ versus $H_a$: $\Delta \neq 0$ rejects when this interval does not include zero.

The two-sample $t$-test is sometimes used mistakenly in this situation. For simplicity, suppose that the pooled estimator $S_p^2(1/n + 1/n) = 2S_p^2/n$ of Var $(\bar{Y} - \bar{X})$ is used.

Let us determine $E(S_p^2)$. Since $(X_i - \bar{X})^2 = [(A_i - \bar{A}) + (\varepsilon_i - \bar{\varepsilon})]^2 = (A_i - \bar{A})^2 + 2(A_i - \bar{A})(\varepsilon_i - \bar{\varepsilon}) + (\varepsilon_i - \bar{\varepsilon})^2$ and the $A_i$'s are constants, $E\left[\sum_{i=1}^n (X_i - \bar{X})^2\right] = \sum_{i=1}^n (A_i - \bar{A})^2 + (n-1)\sigma^2$. Similarly, $\sum_{i=1}^n (Y_i - \bar{Y})^2$ has the same expectation. Therefore, $E(S_p^2) = [1/(n-1)] \sum_{i=1}^n (A_i - \bar{A})^2 + \sigma^2$. Call the first term $\sigma_A^2$. Letting $t_{1-\alpha/2}^*$ be the $(1-\alpha/2)$-quantile of the $t$-distribution with $2(n-1)$ df, we find that the mistaken two-sample CI on $\Delta$ will have length that is approximately $K \equiv (t_{1-\alpha/2}^*/t_{1-\alpha/2})\sqrt{(\sigma_A^2 + \sigma^2)/(\sigma^2)}$ times as long as the paired different CI. Unless $\sigma_A^2$ is quite small, which is unlikely to be the case, the two-sample CI will be much longer than is needed. This is only approximate because $E(S_p) < \sigma_p$ and $E(S_D) < \sigma_D$. The coverage probability will in general be larger than claimed.

Suppose, for example, that scores range from 0 to 100 and that $\sigma_A = 5$ and $\sigma = 5$. Then, for $\alpha = 0.05, n = 15, t_{1-\alpha/2}^* = 2.048, t_{1-\alpha/2} = 2.145$, and $K = 1.35$. That is, the ratio of the mean lengths for the mistaken two-sample CIs to the mean length for the paired difference CIs should be approximately 1.35. For 10,000 simulations the ratio of the mean lengths was 1.36. See Problem 9.4.5 to see what the cost is if the paired sample $t$-test is used even though the two-sample $t$-test is justified.

## Problems for Section 9.4

**9.4.1** Let $X_1, X_2, X_3$ be a random sample from the $N(\mu, \sigma^2)$ distribution. Let $Y_1 = X_1 + X_2 + X_3$. Let $Y_2 = X_1 - X_2$ and $Y_3 = X_1 + X_2 - 2X_3$. Find constants $k_1$ and $k_2$ such that $W \equiv k_2 Y_1/(Y_2^2 + k_1 Y_3^2)^{1/2}$ has a noncentral $t$-distribution. Give the degrees of freedom and the noncentrality parameter.

**9.4.2** Let $f(t; \nu)$ be the t-density for $\nu$ df. Use Stirling's approximation to show that $\lim_{\nu\to\infty} f(t; \nu) = \phi(t)$, the standard normal density, for every real $t$. *Hint:* To make computations easier, let $K = \nu/2$.

**9.4.3** Let $Z \sim N(0, 1)$, and $V \sim \chi_\nu^2$, and suppose that $Z$ and $V$ are independent. Show that $T = Z/\sqrt{V/\nu}$ has the density $f(t; \nu)$ given by (9.4.1).

**9.4.4** **(a)** Show that for $\nu = 1$ df, the $t$-density $f(t; \nu)$ is a Cauchy density, so that its mean does not exist.

**(b)** For $X_1, X_2$ as in Problem 9.4.1 with $\mu = 0$, let $W = (X_1 + X_2)/|X_1 - X_2|$. Find $P(W > 1)$. *Hint*: First find a simple description for $\{(x_1, x_2) | (x_1 + x_2)/|x_1 - x_2| > 1\}$.

**(c)** Use part (a) to find $P(W > 1)$. *Hint*: The standard Cauchy distribution has cdf $(\tan^{-1} x)/\pi + 1/2$.

**9.4.5** Let $X_1, \ldots, X_n$ be a random sample from the $N(\mu, \sigma^2)$ distribution. Let $H_0$: $\mu = \mu_0$ and $H_a$: $\mu \neq \mu_0$, where $\mu_0$ is a known constant. Show that the log-likelihood ratio statistic $-2\log(\Lambda(X))$ is a monotone function of $T^2$, where $T$ is the one-sample $t$-statistic used to test $H_0$ versus $H_a$.

**9.4.6** Nine type A flashlight batteries were chosen randomly from among 2000 in a shipment and tested until failure, with their lifetimes in hours $X_i$ recorded. The $X_i$ were: 27.8, 26.2, 29.8, 24.2, 24.7, 29.2, 22.8, 33.2, 23.7.

**(a)** Assuming that these $X_i$ constitute a random sample from a normal distribution with mean $\mu_A$, find a 90% CI on $\mu_A$.

**(b)** Give an approximation of the $p$-value corresponding to these data for $H_0$: $\mu_A = 24$ versus $H_a$: $\mu_A \neq 24$.

**(c)** A random sample of eight batteries was chosen from a shipment of 5000 type B batteries and tested until failure. Let their lifetimes be $Y_1, \ldots, Y_n$. Suppose that these $Y_i$ are independent, each $\text{N}(\mu_B, \sigma^2)$, where $\sigma^2 = \text{Var}(X_i) = \text{Var}(Y_i)$. Let $\Delta = \mu_B - \mu_A$. Give a 90% CI on $\Delta$ for $X_i$'s as above and $Y_i$'s: 29.7, 25.5, 21.7, 28.1, 33.8, 29.0, 26.1, 31.2.

**(d)** Give an approximate $p$-value for the two-sample $t$-test of $H_0$: $\Delta = 0$ versus $H_a$: $\Delta \neq 0$.

**(e)** A lazy experimenter decided to throw away the last observation, $X_9$, then use the paired-sample $t$-test to test $H_0$ as in part (c). Give the ratio $\theta_y/\theta_x$ of the noncentrality parameters for the two-sample $t$-statistics as used in part (c) and for the "lazy man's" $t$-test.

**(f)** Carry out the lazy method to find a 90% CI and to find an approximate $p$-value for $H_0$: $\Delta = 0$.

**9.4.7** Let $\mathbf{X} = (X_1, \ldots, X_n)$ be a random sample from $\text{N}(\mu_1, \sigma^2)$ and let $\mathbf{Y} = (Y_1, \ldots, Y_m)$ be a random sample from $\text{N}(\mu_2, \sigma^2)$. Suppose that we wish to test $H_0$: $\mu_1 = \mu_2$ versus $H_a$: $\mu_1 \neq \mu_2$.

**(a)** Show that the likelihood function $\Lambda(\mathbf{X}, \mathbf{Y})$ is a monotone function of $T^2$, where $T$ is the two-sample $t$-statistic that uses the pooled estimator $S_p^2$ of the common variance $\sigma^2$.

**(b)** Let $S_1^2$ be the sample variance for the $X_i$'s. Let $W = (\bar{Y} - \bar{X} - \Delta_0)/S_1$, where $\Delta_0$ is a constant. Find a constant $K$ such that $KW$ has the noncentral $t$-distribution. Give the degrees of freedom and noncentrality parameter $\theta$.

## 9.5 THE $F$-DISTRIBUTION

To compare the unknown variances $\sigma_1^2$ and $\sigma_2^2$ of two populations it should seem reasonable to compare their sample variance $S_1^2$ and $S_2^2$. If we have samples of sizes $n_1$ and $n_2$ and the distributions sampled are normal, we know that for $i = 1, 2$, $V_i \equiv S_i^2(n_i - 1)/\sigma_i^2 \sim \chi^2_{n_i-1}$. From this and the independence of $V_1, V_2$ we should be able to determine the distribution of the ratio $V_1/V_2$. That in turn should produce methods that allow us to compare $S_1^2$ and $S_2^2$. We extend this idea a bit by considering a noncentral chi-square random variable in the numerator.

**Definition 9.5.1** Let $V_1 \sim \chi^2_{\nu 1}(\delta)$, $V_2 \sim \chi^2_{\nu 2}$ with $V_1, V_2$ independent. Let $F = (V_1/\nu_1)/(V_2/\nu_2)$. Then $F$ is said to have the *noncentral F-distribution* with $\nu_1$ and $\nu_2$ degrees of freedom and noncentrality parameter $\delta$. □

COMMENTS:

1. The symbol $F$ was adopted to remind us of the great statistician and geneticist Sir Ronald A. Fisher, who found the density for the central $F$-distribution, for which $\delta = 0$.
2. We write $F \sim F(\nu_1, \nu_2; \delta)$, and for the central $F$-distribution, $F \sim F(\nu_1, \nu_2)$. We denote the $\gamma$-quantile of the central $F$-distribution by $F_\gamma(\nu_1, \nu_2)$, or simply as $F_\gamma$ when the values of $\nu_1$ and $\nu_2$ are clear.
3. Let $T \sim t_\nu(\theta)$. Then, by definition, $T$ has the distribution of $(Z + \theta)/\sqrt{V/\nu}$ with $Z, V$ independent, $Z \sim \text{N}(0, 1)$, and $V \sim \chi^2_\nu$. Therefore, $T^2$ has the distribution of $(Z + \theta)^2/(V/\nu)$. It follows from the definition of the $F$-distribution that $T^2 \sim F_{1,\nu}(\theta^2)$.
4. $E(F) = [\nu_2/(\nu_2 - 2)](1 + \delta/\nu_1)$ for $\nu_2 > 2$ and $\text{Var}(F) = 2\nu_2^2/[\nu_1^2(\nu_1 - 2)^2(\nu_2 - 4)][(\nu_2 + 2\delta)(\nu_2 - 2) + (\nu_1 + \delta)^2]$ for $\nu_2 > 4$. (These are two of those formulas that are rarely useful and should *never* be memorized, since there would be considerable danger that the brain will become "too full.")
5. Most statistics books contain 0.90-, 0.95-, and 0.99-quantiles for the central $F$-distribution for various choices of $\nu_1$ and $\nu_2$.
6. Define

$$h(v; \nu_1, \nu_2) = C(\nu_1, \nu_2) \frac{v^{\nu_1/2-1}}{[1 + v(\nu_1/\nu_2)]^{(v_1+v_2)/2}}$$

for

$$C(\nu_1, \nu_2) = \frac{\Gamma((\nu_1 + \nu_2)/2)}{\Gamma(\nu_1/2)\Gamma(\nu_2/2)} \left(\frac{\nu_1}{\nu_2}\right)^{n_1/2},$$

where $h$ is the central $F$-density. See Problem 9.5.1 for an outline of a proof. See Figure 9.5.1 for the graphs of $h$ for a few pairs $(\nu_1, \nu_2)$.

7. As $\nu_2 \to \infty$, the denominator $V_2/\nu_2$ of $F$ converges in probability to 1, so that for large $\nu_2$, $F$ is distributed approximately as $\chi^2_{\nu_1}(\delta)/\nu_1$.
8. If $F \sim F(\nu_1, \nu_2)$, then $1/F \sim F(\nu_2, \nu_1)$, so that lower quantiles of $F$ can be found using the upper quantiles. Thus, $F_\gamma(\nu_1, \nu_2) = 1/F_{1-\gamma}(\nu_2, \nu_1)$. For example, from Figure 9.5.2, $F_{0.05}(5, 2) = 1/F_{0.95}(2, 5) = 1/5.05 = 0.198$.
9. The noncentral $F$-density can be expressed as a Poisson-weighted sum of central-$F$ densities: $h(v; \delta, \nu_1, \nu_2) = \sum_{k=0}^{\infty} p(k; \delta/2)\ h(v; \nu_1, \nu_2)$, where $p(k; \delta/2)$ is the probability mass function for the Poisson distribution with mean $\delta/2$.
10. From Problem 4.3.5, $X \sim \Gamma(\alpha, 1)$, $Y \sim \Gamma(\beta, 1)$, with $X, Y$ independent, implies that $U = X + Y \sim \Gamma(\alpha + \beta, 1)$, $V = X/(X + Y) \sim \text{Beta } (\alpha, \beta)$, and $U, V$ are independent. But when $\alpha$ and $\beta$ are positive integers $2X \sim \chi^2_{2\alpha}$,

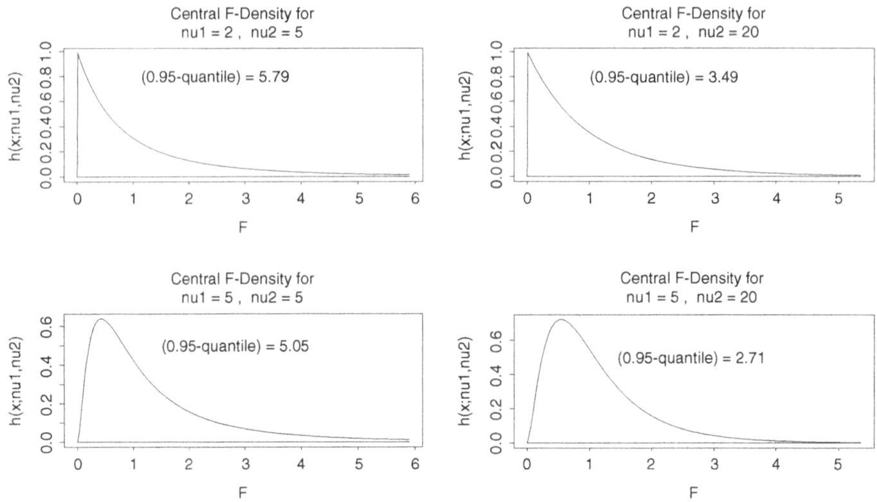


**FIGURE 9.5.1** Central $f$ densities.

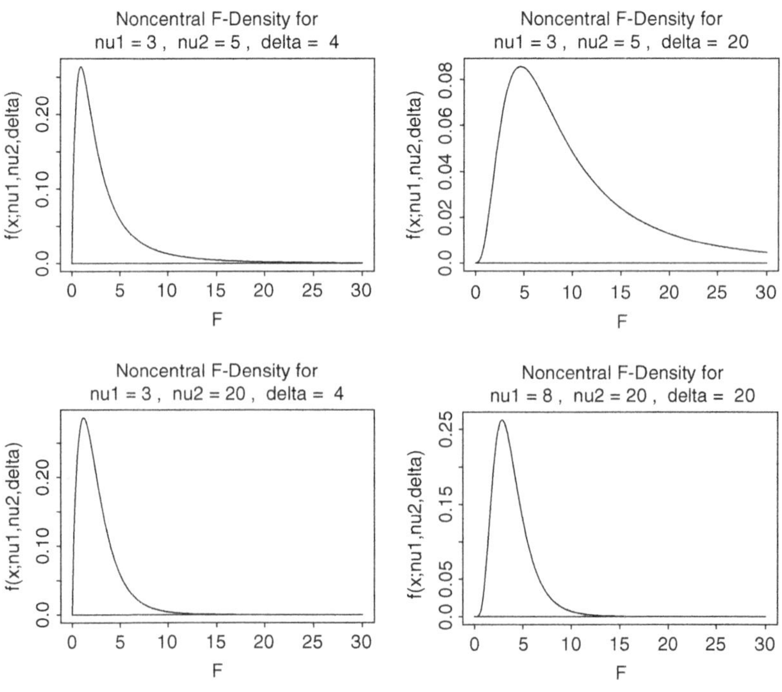


**FIGURE 9.5.2** Noncentral $F$-densities.

$2Y \sim \chi^2_{2\beta}$, so that $F \equiv (2X/2\alpha)/(2Y/2\beta) = (\beta/\alpha)(X/Y) = (\beta/\alpha)[V/(1-V)] \sim F(2\alpha, 2\beta)$. In reverse, $V = (\alpha/\beta)F/[1 + (\alpha/\beta)F] \sim$ Beta $(\alpha, \beta)$.

### Confidence Intervals on $R \equiv \sigma_2^2/\sigma_1^2$

Let $\mathbf{X} = (X_1, \ldots, X_m)$ be a random sample from $N(\mu_1, \sigma_1^2)$ and let $\mathbf{Y} = (Y_1, \ldots, Y_n)$ be a random sample from $N(\mu_2, \sigma_2^2)$ with $\mathbf{X}$ and $\mathbf{Y}$ independent Let $S_1^2$ and $S_2^2$ be the corresponding sample variances. From Section 9.3, $V_1 \equiv S_1^2(m-1)/\sigma_1^2 \sim \chi^2_{m-1}$ and $V_2 \equiv S_2^2(n-1)/\sigma_2^2 \sim \chi^2_{n-1}$, with $V_1, V_2$ independent. It follows from Definition 9.5.2 that $R \equiv [V_1/(m-1)]/[V_2/(n-1)] = (S_1^2/\sigma_1^2)/(S_2^2/\sigma_2^2) \sim F(m-1, n-1)$. Letting $F_{\alpha/2}$ and $F_{1-\alpha/2}$ be the $\alpha/2$- and $(1-\alpha/2)$-quantiles of this distribution, we get $1 - \alpha = P(F_{\alpha/2} \le R \le F_{1-\alpha/2}) = P((S_2^2/S_1^2)/(F_{1-\alpha/2}) \le \sigma_2^2/\sigma_1^2 \le (S_2^2/S_1^2)/(F_{\alpha/2}))$. Therefore, $[(S_2^2/S_1^2)/F_{1-\alpha/2}, (S_2^2/S_1^2)/F_{\alpha/2}]$ is a $100(1-\alpha)\%$ CI on $\sigma_2^2/\sigma_1^2$.

***Example 9.5.1*** Consider the data of Problem 9.4.6. We get $S_1^2 = 11.73$, $S_2^2 = 13.88$, $S_2^2/S_1^2 = 1.18$, $F_{0.025}(8, 7) = 1/F_{0.975}(7, 8) = 4.53 = 0.221$, $F_{0.975}(8, 7) = 4.90$, and the 95% CI [0.242, 5.360]. The interval is rather wide. In general, it is more difficult to estimate variances and ratios of variances than it is to estimate means or differences of means. □

WARNING: This method is quite sensitive to the distributional shape, especially to the value of $\beta = \mu_4/\sigma^4$. For example, for samples from normal distributions, $\beta = 3$, and of course, the actual coverage probabilities are the nominal value $(1-\alpha)$. However, for the exponential distribution, $\beta = 9$. For 10,000 simulations with $m = 5, n = 6$, the "95%" CI actually covered the ratio $\sigma_2^2/\sigma_1^2$ only 8175 times. Larger sample sizes don't help—for samples of $m = 200, n = 300$, the interval contained the ratio only 6758 times. For $\beta$ smaller than 3, the true coverage exceeds the nominal value. For samples from the uniform, for which $\beta = 9/5$, the proportions of coverages were 9758/10,000 for $m = 5, n = 6$, and 9982/10,000 for $m = 200$, $n = 300$.

A two-stage procedure to test $H_0$: $\mu_1 = \mu_2$ is sometimes used, by which a test of the null hypothesis that $\sigma_1^2 = \sigma_2^2$ is first performed, then the decision to pool or not pool in a test of $H_0$ depends on this preliminary test, pooling when this preliminary test does not reject. The author recommends against this. If equality of the population variances $\sigma_1^2$ and $\sigma_2^2$ is at all doubtful, do not pool the sample variances.

The behavior of $R$ is in contrast to that of the $t$-statistic in that for large $m$ and $n$ the $t$-statistic is approximately distributed as normal, even though the populations sampled are not. In general, $R$ is not approximately distributed as $F$, unless the population distributions are themselves close to normal. More precisely, if $\beta$ is not close to 3, the distribution of $R$ is not close to the central $F$-distribution. We will see in Chapter Eleven that the $F$-distribution is quite useful in drawing inferences about the parameters (the $\beta_j$'s) of a linear model.

## Problems for Section 9.5

**9.5.1** Let $X \sim \Gamma(\alpha, 1)$ and $Y \sim \Gamma(\beta, 1)$ for $\alpha > 0$, $\beta > 0$, with $X$, $Y$ independent.

**(a)** Use the result of Problem 4.3.5 to determine the density of $W \equiv Y/X$, then use this to determine the density of $U \equiv (Y/\beta)/(X/\alpha)$.

**(b)** Use part (a) to determine the density $h(v; \nu_1, \nu_2)$ of the central $F$-distribution.

**(c)** Let $\mathbf{X} = (X_1, \ldots, X_m)$ and $\mathbf{Y} = (Y_1, \ldots, Y_n)$ be random samples from $\mathrm{N}(\mu_1, \sigma^2)$ and $\mathrm{N}(\mu_2, \sigma^2)$ distributions. Let $S_p^2$ be the pooled estimator of $\sigma^2$ and let $R = S_2^2/S_1^2$. Use the independence of $X + Y$ and $Y/X$ as proved in Problem 4.3.5 to prove that $S_p^2$ and R are independent.

**(d)** Let $m = n = 9$, $\sigma^2 = 25$. Find $P(S_p^2 > 36.784, R < 3.438)$.

**9.5.2** Let $\mathbf{X}$ and $\mathbf{Y}$ be distributed as in Problem 9.4.1(c) with $\mu_1 = \mu_2$ and $m = n$. Let $D_i = Y_i - X_i$ for each $i$ and let $S_D^2$ be the sample variance for the $D_i$'s. Let $W \equiv [(\bar{Y} - \bar{X})/S_D]K$, where $K$ is a constant.

**(a)** Find $K$ so that $W \sim t_\nu(\theta)$ for some $\nu$ and $\theta$. Give the values of $\nu$ and $\theta$.

**(b)** What is the distribution of $W^2$?

**9.5.3** Let $X_1, \ldots, X_m$ be a random sample from $\mathrm{N}(\mu_1, \sigma_1^2)$ and let $Y_1, \ldots, Y_n$ be a random sample from $\mathrm{N}(\mu_2, \sigma_2^2)$.

**(a)** (Not easy) Show that the likelihood ratio test of $H_0: \sigma_1^2 \le \sigma_2^2$ versus $H_a: \sigma_1^2 > \sigma_2^2$ rejects for large values of $F \equiv S_1^2/S_2^2$. Note that the MLE for $(\sigma_1^2, \sigma_2^2)$ under $H_0$ must satisfy $\sigma_1^2 \le \sigma_2^2$. In the case that $\Sigma(X_i - \bar{X})^2/m \equiv \hat{\sigma}_1^2 > \hat{\sigma}_2^2 \equiv \Sigma(Y_i - \bar{Y})^2/n$, first show that the MLE under $H_0$ is $(S_p^2, S_p^2)$. This is the more difficult part.

**(b)** For $\alpha = 0.05$, $m = 10$, $n = 16$, for which values of $F$ should $H_0$ be rejected?

**(c)** Let $\tau = \sigma_1^2/\sigma_2^2$. Express the power function for the test in part (b) in terms of the cdf $F(u; \nu_1, \nu_2)$ of the central $F$-distribution with these parameters.

**(d)** Suppose that $m = 5$, $n = 4$, and we observe $X_1 = 23$, $X_2 = 27$, $X_3 = 20$, $X_4 = 28$, $X_5 = 22$, $Y_1 = 31$, $Y_2 = 21$, $Y_3 = 40$, $Y_4 = 36$. Find a 95% CI on $\tau = \sigma_1^2/\sigma_2^2$.

**9.5.4** Let $W \sim F(\nu_1, \nu_2)$. Let $U = (\nu_2/\nu_1)W$. Show that $U/(1 + U) \equiv Q$ has a beta distribution.

**(a)** What are the parameters $\alpha$ and $\beta$?

**(b)** Use $F$-tables to find the 0.95-quantile of the Beta(5, 6) distribution.

**9.5.5** Suppose that you wished to estimate the 0.60-quantile $x_{0.60}$ of the $F(10, 15)$ distribution but had no computer program to compute them directly.

**(a)** Assuming that you have a computer program that will produce independent N(0, 1) pseudo random variables, how could you estimate $x_{0.60}$?

**(b)** Let $F \sim F(10, 15)$. Suppose that you wished to estimate $P(F \leq 2.0)$ within 0.005 with probability at least 0.95. How many pseudo standard normal random variables would you need?

**9.5.6** Verify the formula under comment 4: $E(F) = [\nu_2/(\nu_2 - 2)](1 + \delta/\nu_1)$.

**9.5.7** Suppose that $x_1, \ldots, x_n$ are known constants and that $Y_i = \beta x_i + \varepsilon_i$, where the $\varepsilon_i$ are independent N(0, $\sigma^2$) random variables and $\beta$ is an unknown parameter. (This is an example of a linear model.)

**(a)** Show that $Q(\beta) \equiv \sum_{i=1}^{n}(Y_i - \beta x_i)^2$ is minimized by the least squares estimator $\hat{\beta} \equiv \sum_{i=1}^{n} Y_i x_i / \sum_{i=1}^{n} x_i^2$.

**(b)** Show that $\hat{\beta} \sim N(\beta, \sigma^2 / \sum_{i=1}^{n} x_i^2)$.

**(c)** Let $\hat{Y}_i = \hat{\beta} x_i$ and $\text{SSE} = \sum_{i=1}^{n}(Y_i - \hat{Y}_i)^2$. Show that $\text{SSE}/\sigma^2 \sim \chi_{n-1}^2$ and that $\hat{\beta}$ and SSE are independent.

**9.5.8** Use comment 10 and $F$-tables to find the 0.95-quantile of the Beta(4, 10) distribution.

CHAPTER TEN

# Nonparametric Statistics

## 10.1 INTRODUCTION

The statistical models considered in Chapters Seven, Eight, and Nine were parametric, in the sense that the members of the families of distributions considered were distinguished by one or more real parameters. Chapter Nine, for example, was concerned almost entirely with one or two normal distributions, so there were two, three, or four unknown parameters. In general, the procedures discussed should work well if the models employed are correct at least to some degree of approximation. It is often the case, however, that the distributions sampled are unknown, not only in their parameter values but also in their shapes. In other cases the sample values are only random in the sense that a treatment is randomly assigned to some of the subjects. Consider an example.

A small study is designed to determine whether tutoring in an introductory statistics class is effective. Nine student volunteers are chosen. After exam 1 five of these are chosen randomly to be members of the treatment group. The other four students are members of the control group. Those in the treatment group receive special tutoring for two hours each week for the four weeks between exams 1 and 2. Those in the control group receive no additional help. After exam 2 the improvement (exam 2 score) – (exam 1 score) is recorded for each of the nine students. The scores $Y$ for the treatment group, $X$ for the control group, were:

| Student Number | 1 | 2 | 3 | 4 | 5 | 6 | 7 | 8 | 9 |
|---|---|---|---|---|---|---|---|---|---|
| Treatment Group | | 11 | | 7 | 16 | | −2 | 9 | |
| Control Group | 4 | | −9 | | | 10 | | | −13 |

Consider the hypotheses $H_0$: tutoring has no effect on performance and $H_a$: tutoring tends to improve performance. Consider also a test based on the test statistic

*Models for Probability and Statistical Inference: Theory and Applications,* By James H. Stapleton

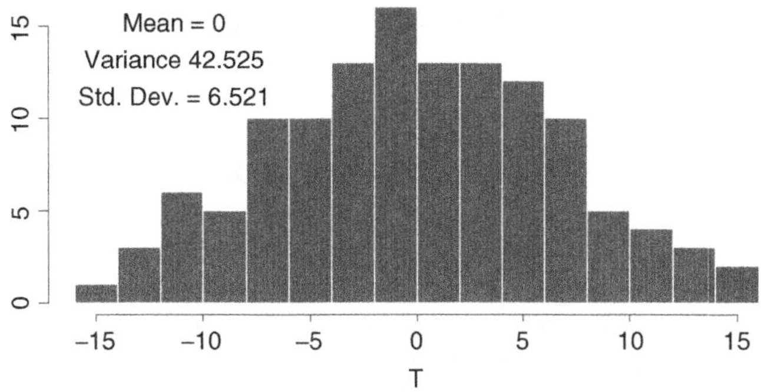


**FIGURE 10.1.1** Permutation distribution of $T = \bar{Y} - \bar{X}$.

$T = \bar{Y} - \bar{X}$. Under $H_0$ all nine of these values were unaffected by the treatment. Therefore, each subset of five of the nine values was equally likely to be those of the treatment group under $H_0$. Under $H_0$, conditionally on the nine improvement scores, $T$ has the *permutation distribution*, the distribution determined by assigning probability $1\Big/\binom{9}{5} = 126$ to each subset of five from $1, 2 \ldots, 9$ (see Figure 10.1.1). For example, if the treatment group students had been 1, 2, 3, 8, 9, then $T = 0.4 - 7.75 = -7.35$. For the scores actually obtained, $T = 8.2 - (-2) = 10.2$. The largest possible value for $T$ is $(16 + 11 + 10 + 9 + 7)/5 - (4 - 2 - 9 - 13)/4 = 10.6 - (-5) = 15.6$.

Notice that the null distribution of $T$ is roughly normal. We can determine $E(T)$ and $\text{Var}(T)$ under $H_0$ as follows. Let $\mu$ and $\sigma^2$ be the mean and variance of the nine improvement scores, considered here to be the population. We find $\mu = 3.67$ and $\sigma^2 = 84$. Then $E(T) = E(\bar{Y}) - E(\bar{X}) = \mu - \mu = 0$. Since $T = \bar{Y} - \bar{X} = \bar{Y} - (9\mu - 5\bar{Y})/4 = \bar{Y}(1 + 5/4) - (9/4)\mu = (9/4)(\bar{Y} - \mu)$, $\text{Var}(T) = (9/4)^2(\sigma^2/5)$ $[(9 - 5)/(9 - 1)] = 9^2\sigma^2/(4)(5)(9 - 1) = 42.525$. The factor $(9 - 5)/(9 - 1)$ is the finite correction factor for sampling without replacement. More generally, for samples of size $n_t$ and $n_c$ for treatment and control groups, and $N = n_t + n_c$, $T = \bar{Y} - \bar{X} = (N/n_c)(\bar{Y} - \mu)$, so that $\text{Var}(T) = N^2\sigma^2/n_c n_t(N - 1) \doteq \sigma^2(1/n_c + 1/n_t)$. The $p$-value is the proportion $9/126 = 0.071$ of the 126 possible samples for which $T$ is at least the value observed. The normal approximation provides $1 - \Phi((10.2 - 0)/6.52) = 1 - \Phi(1.564) = 0.059$. The corresponding pooled two-sample $t$-statistic is 1.756, giving a $p$-value of 0.061. That $T$ is approximately normally distributed under the hypothesis of no treatment effect follows from Hajek's theorem on without-replacement sampling (see Section 6.3) if $\tau = \max(x_i - \mu)^2/N\sigma^2$ is small, with better approximations for smaller $\tau$. For example, if the population values are 1, 2, $\ldots$, $N$, then $\tau = [1 - (N + 1)/2]^2/[N(N^2 - 1)/12] = 3(N - 1)/[N(N + 1)]$, which converges to 0 as $N \to \infty$. For this case the approximation is quite good.

Although in this case the interpretations provided by the permutation and two-sample $t$-tests are approximately the same, the permutation test rests on more solid ground, depending as it does only on the randomness provided by the experimenter.

In the next section we study a particular two-sample permutation test for which the scores are replaced by ranks.

## 10.2 THE WILCOXON TEST AND ESTIMATOR

If we replace the scores in the example above by "ranks," we obtain the well-known *rank-sum* or *Wilcoxon statistic*. Assume for simplicity for a moment that $v_1, \ldots, v_N$ are distinct numbers. The *rank* $r_i$ of $v_i$ among $\{v_1, \ldots, v_N\}$ is the number of the $v_j$ that are less than or equal to $v_i$. The smallest has rank 1, the next smallest rank 2, and so on. Thus, $r_i = \sum_{j=1}^{N} I[v_j \leq v_i]$. For the improvement scores the nine students had the following ranks:

| Student $i$ | 1 | 2 | 3 | 4 | 5 | 6 | 7 | 8 | 9 |
|---|---|---|---|---|---|---|---|---|---|
| Rank $= r_i$ | 4 | 8 | 2 | 5 | 9 | 7 | 3 | 6 | 1 |

The Wilcoxon statistic $W$ is the sum of the ranks for the treatment group, which in this case is $W = 8 + 5 + 9 + 3 + 6 = 31$. The Wilcoxon test rejects $H_0$ for this one-sided alternative for large $W$. The observed $p$-value is then the probability under the permutation distribution that $W$ is equal to or greater than 31. That is, in this case, an easy computation. The following subsets of five ranks from $\{1, 2, \ldots, 9\}$ are the only ones that provide $W \geq 31$. The last is the one observed (see Table 10.2.1). The observed $p$-value is therefore 10/126 = 0.079, approximately the same as was obtained for the permutation test based on the improvement scores themselves, and as obtained for the two-sample $t$-statistic.

Figure 10.2.1 presents the null distribution of $W$ and the approximating normal distribution. The mean and variance are determined easily as follows. Let $\mu$ and $\sigma^2$ be the mean and variance of the population of ranks $1, \ldots, N$. Then $\mu = (N + 1)/2$ and $\sigma^2(N^2 - 1)/12$. If $n$ $Y$'s and $m$ $X$'s are ranked, and if $W$ is the sum of the ranks of the $Y$'s among the $N = m + n$ observations, then when the $Y$ ranks are a random subset of the $N$ ranks, $E(W) = n\mu = n(m + n + 1)/12$ and $\text{Var}(W) = n\sigma^2(N - n)/(N - 1) = mn(m + n + 1)/12$. That $W$ is asymptotically normally distributed as $m \to \infty$ and $n \to \infty$ follows by Hajek's theorem on sampling without replacement, as discussed in Section 6.4 and earlier in this section. Using the

**TABLE 10.2.1 Wilcoxon Data**

| Sample | $W$ | Sample | $W$ |
|---|---|---|---|
| 9, 8, 7, 6, 5 | 35 | 9, 8, 7, 5, 3 | 32 |
| 9, 8, 7, 6, 4 | 34 | 9, 8, 6, 5, 4 | 32 |
| 9, 8, 7, 6, 3 | 33 | 9, 8, 7, 5, 2 | 31 |
| 9, 8, 7, 5, 4 | 33 | 9, 7, 6, 5, 4 | 31 |
| 9, 8, 7, 6, 2 | 32 | 9, 8, 6, 5, 3 | 31 |

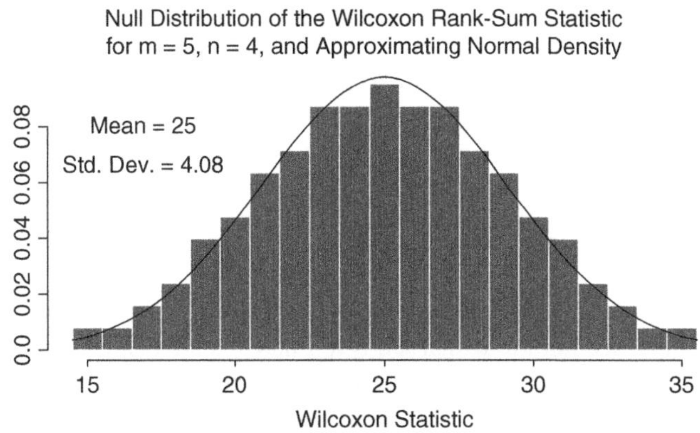


**FIGURE 10.2.1** Null distribution of $W$ for $m = 5, n = 4$.

normal approximation, we find that $P(W \geq 31) \doteq 1 - \Phi((30.5 - 25)/4.0825) = \Phi(-1.347) = 0.089$.

The rank-sum statistic $W$ was invented by Frank Wilcoxon, a chemist, in a 1945 *Biometrics* paper, as an alternative to the two-sample $t$-statistic. That paper gave birth to the development of a large theory of nonparametric statistics showing that such methods were often robust, having good asymptotic properties not only for testing but for estimation as well. In this chapter we present only some of that theory, often without proof.

Independently, in 1947, Mann and Whitney, professor and student, respectively, at Ohio State, suggested another rank statistic, which, it turned out, is closely related to $W$. As before, let $X_1, \ldots, X_m$ and $Y_1, \ldots, Y_n$ be the samples of $X$'s and $Y$'s. Let $W_{XY} = \sum_{i=1}^m \sum_{j=1}^n I(X_i \leq Y_j)$. This is the Mann–Whitney statistic. For the improvement data the five $Y_j$'s exceed four, three, four, two, and three $X_i$'s, so that $W_{XY} = 16$. $Y_{(1)}$, the smallest $Y_j$, is greater than or equal to (GTET) two $X_i$'s and one $Y_j$. For that reason, the rank corresponding to $Y_{(1)}$ is 3. Similarly, $Y_{(2)}$ GTET three $X_i$'s and two $Y_j$'s, so it has rank 5. $Y_{(3)}$ has rank 6 because it is GTET three $X_i$'s and three $Y_j$'s. For similar reasons $Y_{(4)}$ has rank $4 + 4 = 8$, and $Y_{(5)}$ has rank $4 + 5 = 9$.

More generally, $W_{XY} = \sum_{j=1}^n S_j$, where $S_j = \sum_{i=1}^m I[X_i \leq Y_{(j)}]$. The rank $R_j$ of $Y_{(j)}$ among all $N = (m + n)X$'s and $Y$'s is $S_j + j$. Thus, $W = \sum_{j=1}^n R_j = \sum_{j=1}^n (S_j + j) = W_{XY} + n(n + 1)/2$. For example, the smallest possible value of $W_{XY}$ is zero, occurring when the $Y$'s are the $n$ smallest among the $N$, so the smallest possible value of $W$ is $n(n + 1)/2$. When the ranks of the $Y$'s are a simple random sample from $\{1, \ldots, N\}$, $W_{XY}$ is distributed symmetrically about $mn/2$, so that $W$ is distributed symmetrically about $mn/2 + n(n + 1)/2 = n(N + 1)/2$. Since $W$ and $W_{XY}$ differ by a constant, $\text{Var}(W_{XY}) = \text{Var}(W) = mn(m + n + 1)/12$.

See Appendix Table 7 for values of $P(W_{XY} \leq k)$ for most combinations of $n$ $Y$'s and $m$ $X$'s from 3 to 10, $n \leq m$, and all possible $k$. If the number $m$ of $Y$'s is greater than the number $n$ of $X$'s, the symmetry of $W_{XY}$ about $mn/2$ can be exploited.

That is, $P(W_{XY} \le k \mid nY\text{'s and } mX\text{'s}) = P(W_{XY} \le k \mid mY\text{'s and } nX\text{'s})$ for each $k$. For example, from the table, with five $X$'s and seven $Y$'s, $P(W_{XY} \le 13)$ is found by letting $n = 5$, $m = 7$, so we get 0.2652. The normal approximation works well, especially with the 1/2-correction, for larger $n$ and $m$. For $n = 7$, $m = 5$, we get

$$P(W_{XY} \le 13) \doteq \Phi\left(\frac{13.5 - 17.5}{\sqrt{(7)(5)(13/12)}}\right) = \Phi(-0.6496) = 0.2580.$$

Software is readily available for the computation of such probabilities.

We have supposed until now that the $N$ $X$'s and $Y$'s are distinct, so that there are no ties. In the case that $k$ $X$'s and/or $Y$'s are tied at some value $u$, define the midranks of all these observations to be (no. $X$'s and $Y$'s $< u$) + $(k + 1)/2$. For example, if the observations are 2, 4, 4, 4, 6, 7, 7, 9, their corresponding midranks are 1, 3, 3, 3, 5, 6.5, 6.5, 8.

Define $W^*$ to be the sum of the midranks of the $Y$'s, and let $W_{XY}^* = \sum_{ij} I_{ij}$, where $I_{ij} = 1$ if $Y_j > X_i$, $I_{ij} = 1/2$ if $X_i = Y_j$ and $I_{ij} = 0$ for $X_i > Y_j$. Some effort is needed to show that $W^* = W_{XY}^* + n(n+1)/2$ (see Problem 10.2.3). The null distribution of $W^*$ may be determined by enumeration. Even in the case of a large number of ties, when $n$ and $m$ are large, the null distributions of $W^*$ and of $W_{XY}^*$ may be approximated well by the normal. The expected values remain the same as for the no-ties case, but a correction is needed for $\text{Var}(W^*) = \text{Var}(W_{XY}^*)$. Let there be $e$ ties, with $d_i$ tied for the $i$th tie. For the numbers 2, 4, 4, 4, 6, 7, 7, 9, $e = 2$, $d_1 = 3$, $d_2 = 2$. Let $C = 1 - \sum_{i=1}^{e} d_i(d_i^2 - 1)/N(N^2 - 1)$. Then

$$\text{Var}(W^*) = \text{Var}(W_{XY}^*) = \text{Var}(W)C = \frac{mn(m+n+1)}{12}C. \tag{10.2.1}$$

For these eight numbers, with $m = 3, n = 5$, $\text{Var}(W^*) = \text{Var}(W_{XY}^*) = (5)(3)(9/12)[1 - 30/504] = 10.58$. The correction factor $C$ is very close to 1, so that it could probably be ignored in any normal approximation.

***Example 10.2.1*** One hundred volunteers who had previously had problems with headaches were assigned randomly into treatment and control groups of 50 each. Those in the control group were asked to take an aspirin every four waking hours for one month. Those in the treatment group took an identical-looking aspirin that also contained a special additive ZQP, also for one month. None of the volunteers were told which group they were in. One of the members of the control group dropped out of the study after one week due to a broken leg. After the month all 99 remaining subjects were asked to rate themselves with respect to headaches, with 1 = much better, 2 = somewhat better, 3 = about the same, 4 = somewhat worse, 5 = much worse. The frequencies were as given in Table 10.2.1.

The midranks for the 99 ratings were determined as follows. The 11 subjects with rating 1 were each assigned the mean of the integers 1, 2, ..., 11. The 24 subjects with rating 2 were each assigned the mean of the integers 12, ..., 35. Those with

**TABLE 10.2.1 Frequency Data**

| | Rating | | | | |
|---|---|---|---|---|---|
| | 1 | 2 | 3 | 4 | 5 |
| Treatment group | 7 | 15 | 21 | 5 | 2 |
| Control group | 4 | 9 | 25 | 7 | 4 |
| Total | 11 | 24 | 46 | 12 | 6 |
| Midrank | 6 | 23.5 | 58.5 | 87.5 | 96.5 |

rating 3 were assigned the mean of 36, ..., 81; and so on. The Wilcoxon statistic is then $W^* = 7(6) + 15(23.5) + 21(58.5) + 5(87.5) + 2(96.5) = 2253.5$. Under the null hypothesis that the additive ZQP had no effect, $W^*$ is approximately normally distributed with mean $n(N + 1)/2 = 50(100)/2 = 2500$ and variance given by (10.2.1). The number of ties is $e = 5$, of sizes 11, 24, 46, 12, 6. Thus, $C = 0.88215$, $\text{Var}(W^*) = 18{,}011$, $\text{SD}(W^*) = 134.2$. We reject for small $W^*$, so that the observed $p$-value is $\Phi(z)$, where $z = (2253.75 - 2500)/134.2 = -1.8335$. We have used the 1/2-correction, remembering that $W^*$ takes values which are multiples of 1/2. We find the $p$-value 0.0333. We find that there is reason to believe that ZQP is effective, although much further testing and improvements or additional strength may be needed. All this suggests the extensive work that must be done in the development of new drugs and fortunately for those who have studied statistics, requires the employment of statisticians at very high salaries (often more than those of their professors). □

### Power of the Wilcoxon Test

Use of the Wilcoxon test does not require any model at all for the distribution of the $X$'s and $Y$'s, only that under the null hypothesis the ranks of the $Y$'s are a simple random sample of all $N = m + n$ ranks. However, suppose that the $X$'s are a random sample from a distribution $F$ and the $Y$'s are a random sample from a distribution $G$. If $F$ and $G$ are normal distributions, for example, do we lose anything by using the Wilcoxon rather than the two-sample $t$-test?

Consider the special case that $G(x) = F(x - \Delta)$ for a constant $\Delta$ and all $x$. That is, if $X \sim F$ and $Y \sim G$, then $P(Y \le y) = P(X \le y - \Delta) = P(X + \Delta \le y)$, so that $Y$ and $X + \Delta$ have the same distribution $G$, equivalently $X$ and $Y - \Delta$ have the same distribution $F$. Let $Y_j^* = Y_j - \Delta$. So that the number of ties will be zero with probability 1, suppose that $F$, and therefore $G$, are continuous. If $F$ has density $f$, $G$ has density $g(y) = f(y - \Delta)$. Then $p_1(\Delta) \equiv P(X_1 < Y_1; \Delta) = P(X_1 - Y_1 < 0; \Delta)$ $= \int_{-\infty}^{\infty} F(y)g(y)\,dy = 1 - \int_{-\infty}^{\infty} f(y)G(y)\,dy = \int_{\infty}^{-\infty} F(y)f(y - \delta)\,dy = 1 - \int_{-\infty}^{\infty} f(y)F(y - \Delta)\,dy$. This is an increasing function of $\Delta$, so that $E(W_{XY}; \Delta)$ is an increasing function of $\Delta$. $\text{Var}(W_{XY}; \Delta)$ is also a function of $\Delta$ and $F$, although for $\Delta$ near zero it is close to that given by (10.2.1), so we will use that in determining approximations of the power (see Lehmann, 1975, pp. 69–71).

Suppose that $F(x) = \Phi((x - \mu)/\sigma)$. Then $p_1(\Delta) = P(X_1 - Y_1 < 0; \Delta) = \Phi(\Delta/\sqrt{2\sigma^2})$. For large $m, n$, $W_{XY}$ is approximately distributed as $N(nm\, p_1(\Delta), \sigma_W^2)$, where $\sigma_W^2$ is given by (10.2.1). Therefore, an $\alpha$-level test of $H_0$: $\Delta \le 0$ versus $H_a$: $\Delta > 0$ rejects for $W_{XY} > k = (mn/2) + z_{1-\alpha}\sigma_W$. The power is therefore $\gamma(\Delta) = 1 - \Phi((k - mp_1(\Delta))/\sigma_W) = \Phi(mn(p_1(\Delta) - 1/2)/\sigma_W - z_{1-\alpha}) = \Phi\left(\sqrt{12mn}/(m+n+1)(p_1(\Delta) - 1/2) - Z_{1-\alpha}\right)$. The argument $z_{mn}$ of $\Phi$ converges to $+\infty$ for $\Delta > 0$ as $m \to \infty$ and $n \to \infty$ so that $\gamma(\Delta)$ converges to 1. For example, for $m = n = 100, \sigma = 10, \alpha = 0.05, \Delta = 3$, we find $z_{1-\alpha} = z_{0.95} = 1.645, k = 5673.2, p_1(3) = 0.584, z_{mn} = 0.40745, \gamma(3) = 0.658$. Similarly, we find that $p_1(4) = 0.6114, p_1(2) = 0.5562, p_1(1) = 0.5282, \gamma(4) = 0.859, \gamma(2) = 0.393, \gamma(1) = 0.170$.

The power of the two-sample $t$-test is in good approximation for large $m$ and $n$ the same as the power of the two-sample $z$-test, based on known $\sigma^2$. Thus, the power is approximately $\Phi(\Delta/\sqrt{\sigma^2/(1/n + 1/m)} - 1.645)$. We find that $\gamma(1) = 0.174, \gamma(2) = 0.408, \gamma(3) = 0.683, \gamma(4) = 0.881$. The power for the two-sample $t$-test is just a bit larger than for the two-sample Wilcoxon test, uniformly in $\Delta$, *when the distributions sampled are normal.* To verify these approximations, using S-Plus, 10,000 simulations were performed. The 0.05-level $t$-test rejected 6883 times, while the Wilcoxon test rejected 6687 times, in each case a few more than predicted by the asymptotic theory.

If $F$ is the double-exponential distribution with location parameter $\theta$ and standard deviation $\sigma$, some patient work shows that $p_1(\Delta) = e^{-\eta}(1 + \eta/2)/2$, where $\eta = \sqrt{2}\Delta/\sigma$. Again for $\sigma^2 = 100$, we get $p_1(1) = 0.5352, p_1(2) = 0.5699, p_1(3) = 0.6035, p_1(4) = 0.6357$. Therefore, $\gamma(1) = 0.217, \gamma(2) = 0.525, \gamma(3) = 0.812, \gamma(4) = 0.953$. These are uniformly higher than for the two-sample $t$-test, for which the powers remain as they were for samples from normal distributions.

We can judge the relative efficiency of the Wilcoxon versus the $t$-test by determining the sample sizes necessary to achieve the same power. Suppose that we want to have power $\gamma$ (0.90, for example) for the case that $\Delta = \Delta_0$ (3, for example) for the case that $m = n$. We need $z_{mn} = z_\gamma$, the $\gamma$-quantile of the standard normal distribution. We find that in good approximation we need $n = m = n_w = [(z_\gamma + z_{1-\alpha})/(p_1(\Delta_0) - 1/2)]^2/6$. For samples from the normal distribution we get sample sizes 1770, 452, 202, and 115 for $\Delta_0 = 1, 2, 3, 4$. Similarly, for the two-sample $t$-test we need $n = m = n_t = 2[\sigma(z_\gamma + z_{1-\alpha})/\Delta_0]^2$. For these $\Delta_0$ we need sample sizes 1713, 428, 190, 107. Notice that the ratio of the sample sizes needed $n_t/n_w = \sigma^2(p_1(\Delta_0) - 1/2)^2$, the same for all $\alpha$ and $\gamma$. This simplifies still further for $\Delta_0$ close to zero, for then $p_1(\Delta_0) - 1/2 \sim \Delta_0 p_1'(0) = \Delta_0 \int_{-\infty}^{\infty} f(y)^2\, dy$, so that in approximation the *relative efficiency* of the Wilcoxon test to the $t$-test is $\text{eff}(f) \equiv n_t/n_w = 3\sigma^2 C(f)^2$, where $C(f)$ is this last integral. For the case that $f$ is a normal density we get $C(f) = \sigma/\sqrt{\pi}$, so that $\text{eff}(f) = 3/\pi \doteq 0.9549$. That is, if $\Delta_0$ is close to zero and the distributions sampled are normal, we need about 95.5% as many observations using the $t$-test as we need for the Wilcoxon test to achieve the same power and the same $\alpha$-level. $\text{eff}(f)$ is called *the asymptotic relative efficiency* or *Pitman efficiency*.

For the case that $f$ is a double-exponential density, we find $C(f) = 1/(\sqrt{2}\sigma)$, so that $\text{eff}(f) = 3/2$. For samples from the Cauchy distribution the $t$-test is relatively worthless, so that eff$(f) = \infty$. On the other hand, it has been proved that eff$(f) \geq 108/125 = 0.864$ for every density $f$ (see Lehmann, 1975, pp. 377–378).

### Estimation

Consider the shift model again. That is, each $X_i \sim F$ and each $Y_i \sim G$, where $G(y) = F(y - \Delta)$ for all $y$. Suppose that we wish to estimate $\Delta$. Since $V_{ij} = X_i - (Y_j - \Delta)$ and $-V_{ij}$ have the same distribution, $D_{ij} = X_i - Y_j$ is symmetrically distributed about $\Delta$. It follows that med(**X**)−med(**Y**) and $\bar{X} - \bar{Y}$ are each distributed symmetrically about $\Delta$, and each is an unbiased estimator of $\Delta$. Still another estimator can be defined by "inverting" the statistic $W_{XY}$. Define $W_{X,Y-\Delta}$ to be the $W_{XY}$ statistic for the case that each $Y_j$ is replaced by $Y_j - \Delta$. Since $X_i$ and $Y_j - \Delta$ have the same cdf $F$, $E(W_{X,Y-\Delta}) = mn/2$. Let $\hat{\Delta}$ be a value of $\Delta$ for which $W_{X,Y-\Delta}$ is as close as possible to $mn/2$. That is, we want to find $\hat{\Delta}$ such that $\sum_i \sum_j I[X_i < Y_j - \hat{\Delta}] = \sum_i \sum_j I[Y_j - X_i > \hat{\Delta}]$ is approximately $mn/2$. For $D_{ij} = Y_j - X_i$, $\hat{\Delta} = \text{median}(D_{ij})$, where the median is taken for all $mn$ differences $D_{ij}$. If $m$ and $n$ are both odd, there is a unique such value. Otherwise, take $\hat{\Delta}$ to be the mean of the two middle $D_{ij}$.

***Example 10.2.2*** Consider the improvement scores of Example 10.2.1: The $X_i$ are −13, −9, 4, 10 and the $Y_j$ are −2, 7, 9, 11, 16. The $X_i$ and $Y_j$ have been ordered for computational convenience. The $D_{ij}$ are then

| | | | | |
|---|---|---|---|---|
| 11 | 20 | 22 | 24 | 29 |
| 7 | 16 | 18 | 20 | 25 |
| −6 | 3 | 5 | 7 | 12 |
| −12 | −3 | −1 | 1 | 6 |

The ninth- and tenth-order statistics, $D_{(9)}$ and $D_{(10)}$ among these 20 $D_{ij}$ are each seven. Since the eleventh order statistic is $D_{(11)} = 11$, take $\hat{\Delta} = (7 + 11)/2 = 9.0$. The two other estimates are $\bar{Y} - \bar{X} = 8.2 - (-2.0) = 10.2$ and median(**Y**) − median(**X**) $= 9 - (-2.5) = 11.5$. We state without proof that $Z^*_{mn} \equiv (\hat{\Delta} - \Delta)/\sqrt{(m + n + 1)/12mn}$ converges in distribution to N(0, 1) as $m \to \infty$ and $n \to \infty$. It can be shown that as $n \to \infty$ and $m \to \infty$, $\lim_{n\,m}[\text{Var}(\bar{Y} - \bar{X})/\text{Var}(\hat{\Delta})] = 3\sigma^2 C(f)^2 = \text{eff}(f)$, the asymptotic relative efficiency defined above. □

Confidence intervals on $\Delta$ can be determined from the ordered $D_{ij}$ as follows. We first state Theorem 4 from Lehmann (1975, p. 87). Suppose that the differences $D_{ij}$ are ordered, with their values being $D_{(1)} < D_{(2)} < \cdots < D_{(mn)}$. Then for any integer $k$, $1 \leq k \leq mn$, and any real number $a$,

$$D_{(k)} \leq a \text{ if and only if } W_{X,Y-a} \leq mn - k \tag{10.2.2}$$

and therefore,

$$D_{(k)} > a \text{ if and only if } W_{X,Y-a} \geq mn - k + 1. \tag{10.2.3}$$

To prove this, note that (10.2.2) holds if and only if at least $k$ of the differences $D_{ij} - a = (Y_j - a) - X_i$ are less than or equal to zero, and that this is true if and only if at most $mn - k$ of the $D_{ij}$ are greater than zero (i.e., if $W_{X,Y-a}$ does not exceed $mn - k$).

From (10.2.2) and (10.2.3) we can determine CIs on $\Delta$. From these it follows that for each $k, 0 \le k \le mn$, $D_{(k)} \le \Delta \le D_{(k+1)}$ if and only if $W_{X,Y-\Delta} = mn - k$, where $D_{(0)} = -\infty$ and $D_{(mn+1)} = +\infty$. Therefore, $P(D_{(k)} \le \Delta \le D_{(k+1)}; \Delta) = P(W_{X,Y-\Delta} = mn - k; \Delta)$. The distribution of $(Y_1 - \Delta, Y_2 - \Delta, \ldots, Y_n - \Delta)$ is the same for every $\Delta$. Therefore, we can, for example, take $\Delta = 0$ in this last probability statement. Summing over $k = s, \ldots, t-1, s < t$, we get $P(D_{(s)} \le \Delta \le D_{(t)}, \Delta) = P(mn - t + 1 \le W_{XY} \le mn - s; \Delta = 0) = P(s \le W_{XY} \le t - 1; \Delta)$. We can either use tables of the Wilcoxon distribution for small $m$ and $n$, or the normal approximation, to choose $s$ and $t$ so that this probability is suitably large. We can take advantage of the symmetry of the distribution of $W_{XY}$ about $mn/2$ when $\Delta = 0$. From Appendix Table 6, $P(3 \le W_{XY} \le 18; \Delta = 0) = 0.9851 - 0.0317 = 0.9534$, so that since $mn - s = 18$ and $mn - t + 1 = 3$, $[D_{(2)}, D_{(18)}]$ is a 95.34% CI on $\Delta$. For large samples take $t$ to be the closest integer to $mn/2 + z_{1-\alpha/2}\sigma_W$ and $s$ the closest integer to $mn/2 + 1/2 - z_{1-\alpha/2}\sigma_W$, where $\sigma_W^2 = mn(m + n + 1)/12$. For the improvement data $D_{(3)} = -6$ and $D_{(18)} = 24$, so that $[-6, 24]$ is a 95.34% CI on $\Delta$. We might be more interested in a lower bound on $\Delta$. Since $P(D_{(3)} \le \Delta; \Delta) = P(W_{XY} \le mn - 3; \Delta = 0) = 0.0556$, $D_{(3)}$ is a 95.56% lower confidence bound on $\Delta$. For these data we get $D_{(3)} = -1$.

### Problems for Section 10.2

**10.2.1** Find the null distributions of $W$ and of $W_{XY}$ for the case that $n = 3$, $m = 2$.

**10.2.2** The Zebbo Drug Company has developed a new drug, Zebto, designed to lower cholesterol for people with high levels. Fifteen volunteers, all with levels above 250, are chosen to take part in a preliminary study. Eight of these are chosen at random to be part of the treatment group. The other seven were placed in the control group. Those in the treatment group took Zebto every day. Those in the control group took a placebo, a pill that appeared to be the Zebto pill, one having the same look and taste but without any physical effect on the people taking them. The drops in cholesterol level were as follows:

| Subject | 1 | 2 | 3 | 4 | 5 | 6 | 7 | 8 | 9 | 10 | 11 | 12 | 13 | 14 | 15 |
|---|---|---|---|---|---|---|---|---|---|---|---|---|---|---|---|
| Control | | 33 | | | | 19 | −1 | | 13 | | 51 | −4 | | 43 | |
| Treatment | 41 | | 60 | 58 | 34 | | | 68 | | 21 | | | 47 | | 66 |

**(a)** Determine the $p$-value for the Wilcoxon test of $H_0$: no effect of Zebto versus $H_a$: Zebto causes a larger drop. Use both the Wilcoxon table and the normal approximation.

**(b)** Assuming the shift model, find the Wilcoxon estimate $\hat{\Delta}$ of the shift parameter $\Delta$.

**(c)** Find $s$ and $t$, $s < t$, so that $[D_{(s)}, D_{(t)}]$ is an approximate 95% CI on $\Delta$. Evaluate the interval for these data.

**(d)** Find a 95% CI on $\Delta$ using the pooled-variance $t$-method.

**(e)** The data were generated using S-Plus, with $X_i \sim N(25, 400)$ and $Y_j \sim$ N(50, 400), then rounding to the nearest integer. Find the approximate power of the Wilcoxon 0.05-level test. Use the fact that in this case the variance of $W$ is about 56% of what it is under the null hypothesis. In 10,000 simulations $H_0$ was rejected 7026 times.

**(f)** Assuming the model in part (e) and common variances $\sigma^2 = 400$ were known, find the power of the $Z$-test based on $\bar{Y} - \bar{X}$. In 10,000 simulations $H_0$ was rejected 7771 times.

**(g)** Repeat part (a) for the case that the treatment observations are changed as follows: 41 to 43, 34 to 43, 60 to 61, 68 to 66, 47 to 43. Use the normal approximation.

**10.2.3** Suppose that $X_i \sim F$ for $i = 1, 2, 3$, and $Y_j \sim G$ for $j = 1, 2$, with $X_1, X_2, X_3, Y_1, Y_2$ independent. Suppose that we wish to test $H_0$: $F =$ Unif(0, 1), $G =$ Unif(0, 1) versus $H_a$: $F =$ Unif(0, 1), $G =$ Unif(0, 2).

**(a)** What is the distribution of $W$ under $H_0$? What is the distribution under $H_a$?

**(b)** Determine $E(W)$ and Var$(W)$ under both $H_0$ and $H_a$.

**(c)** For $\alpha = 0.2$, for which $W$ should $H_0$ be rejected? What is the power of this test under $H_a$?

**(d)** Let $M = \max(Y_1, Y_2)$. Find $c$ so that the test which rejects for $M > c$ has level $\alpha$, where $0 < \alpha < 1$. Express the power under $H_a$ as a function of $\alpha$. For $\alpha = 0.20$, is this test more powerful than the test in past (c)? Is this a test given by the NP lemma?

**(e)** Is the test in part (d) uniformly most powerful for $H_a$: $F$ the cdf of Unif(0, 1), $G$ the cdf of Unif(0, $\theta$), $\theta > 1$?

**10.2.4** Two methods, $A$ and $B$, have been developed to determine the content of the rare element Klubyl in specimens of ore. The methods have been shown to be accurate in the sense that $X = \theta + \varepsilon_A$ is a measurement using method $A$, and $Y = \theta + \varepsilon_B$ using $B$, where $\theta$ is the "true value," $\varepsilon_A \sim F_A$, $\varepsilon_B \sim F_B$, and both $F_A$ and $F_B$ are continuous and symmetric about zero. That is, $F_A(u) = 1 - F_A(-u)$, and $F_B(v) = 1 - F_B(-v)$ for all $u$ and $v$. Also suppose that $F_B(v) = F_A(v/\eta)$ for some $\eta > 0$. That is, $\varepsilon_B$ has the same distribution as $\eta\varepsilon_A$. Let $X_1, \ldots, X_m$ and $Y_1, \ldots, Y_n$ be random samples of $X$ and $Y$. That is, $X_i = \theta + \varepsilon_{Ai}$ and $Y_j = \theta + \varepsilon_{Bj}$, where the $\varepsilon_{Ai}$ have the cdf $F_A$ and the $\varepsilon_{Bj}$ have cdf $F_B$ and the $(n + m)\varepsilon$'s are independent. Suppose that we wish to test $H_0$: $\eta \leq 1$ versus $H_a$: $\eta > 1$. If $F_A$ and therefore $F_B$

are normal distributions, we could use the $F$-statistic $S_A^2/S_B^2$. However, the distribution of this ratio depends strongly on normality, so that a rank test may be appropriate. The Siegel–Tukey test is designed for this. Rather than assigning ranks in the usual way, let the smallest among the combined sample have rank 1, the largest have rank 2, the next-to-largest rank 3, the second-to-smallest rank 4, and so on, switching from small to large after assigning two ranks. Thus, for $m + n = 9$, the ordered values have ranks 1, 4, 5, 8, 9, 7, 6, 3, 2. The Siegel–Tukey statistic is then the sum $W_{ST}$ of the ranks of the $Y$'s. The null distribution of $W_{\text{ST}}$ is the same as that of $W$. For $X_i$:23, 31, 27, 25 and $Y_j$: 18, 28, 36, 19, 33, find the $p$-value for the Siegel–Tukey test of $H_0$ versus $H_a$.

**10.2.5** To determine whether a proposed TV political ad for Senator Snider would have a favorable effect, 75 volunteers were chosen. Forty were assigned randomly to the treatment group, the other 35 to the control group. The 75 subjects were not told to which group they had been assigned. All 75 were asked to watch two hours of TV. Those in the treatment group were shown the political ad for Senator Snider twice during the two hours, along with many other political ads and commercials. Those in the control group were shown the same ads and commericals with the exception of the ads for Senator Snider. After the two hours all 75 were asked to state their viewpoint toward the senator, with 1 = strongly favorable, 2 = moderately favorable, 3 = neutral, 4 = moderately opposed, 5 = strongly opposed. The results were:

| Viewpoint | 1 | 2 | 3 | 4 | 5 |
|---|---|---|---|---|---|
| Treatment Group | 13 | 10 | 7 | 6 | 4 |
| Control Group | 8 | 7 | 8 | 5 | 7 |

Use the Wilcoxon statistic with the normal approximation to test $H_0$: Snider ad had no effect versus $H_a$: Snider ad had a negative or a positive effect. Determine the $p$-value and state your conclusions.

**10.2.6** **(a)** For $m = 4$, $n = 2$, with combined sample 3, 4, 4, 7, 7, 9, find the null distribution of the Wilcoxon statistic $W^*$ based on midranks.

**(b)** Determine $E(W^*)$ and $\text{Var}(W^*)$ using the distribution obtained in part (a) and verify that these are as given by the formulas of this section.

**10.2.7** **(a)** Let $m = 5$, $n = 7$. Determine $p \equiv P(W \geq 46)$ under the assumption that the ranks of the $Y$'s are a simple random sample from $\{1, 2, \ldots, 12\}$.

**(b)** Use the normal approximation with 1/2-correction to estimate $p$.

## 10.3 ONE-SAMPLE METHODS

We begin with what would seem to be a relatively simple problem, the estimation of a cdf by a sample cdf. Let $X_1, \ldots, X_n$ be a random sample from a distribution $F$. Define $F_n(x) = (1/n)\sum_{i=1}^{n} I[X_i \le x] = (1/n)$ (no. $X_i$'s $\le x$) for each $x$. $F_n$ is called the *sample cumulative distribution function* or *empirical cumulative distribution function*. Since $Y_i = I[X_i \le x]$ is a Bernoulli random variable with parameter $p = F(x)$ and $T \equiv \sum_{i=1}^{n} Y_i \sim \text{Binomial}(n, p = F(x))$, $E(F_n(x)) = p = F(x)$ and $\text{Var}(F_n(x)) = p(1-p)/n = F(x)(1-F(x))/n$ for each real number $x$. The random interval $I_n(x) \equiv [F_n(x) \pm z_{1-\alpha/2}\sqrt{F_n(x)(1-F_n(x))/n}]$ is an approximate $100(1-\alpha)\%$ CI on $F(x)$ for each fixed $x$. That is,

$$P(F(x) \in I_n(x)) \doteq 1 - \alpha \tag{10.3.1}$$

for each $x$, with the approximation being better for $F(x)$ nearer $1/2$ and $n$ large. A graphical display of $F_n(x)$ may be helpful. The vertical lines are there only to make $F_n$ easier to graph and to see. Notice that the probability statement does not say anything about

$$P(F(x) \in I_n(x) \quad \text{for all } x). \tag{10.3.2}$$

Notice that the phrase "for all $x$" is inside the parentheses in this case, whereas in (10.3.1) it was not. That is, the event in question in (10.3.1) depends on $x$, whereas in (10.3.2) it does not. In Figure 10.3.1 the 95% interval for $x = 58$ in the fourth graph

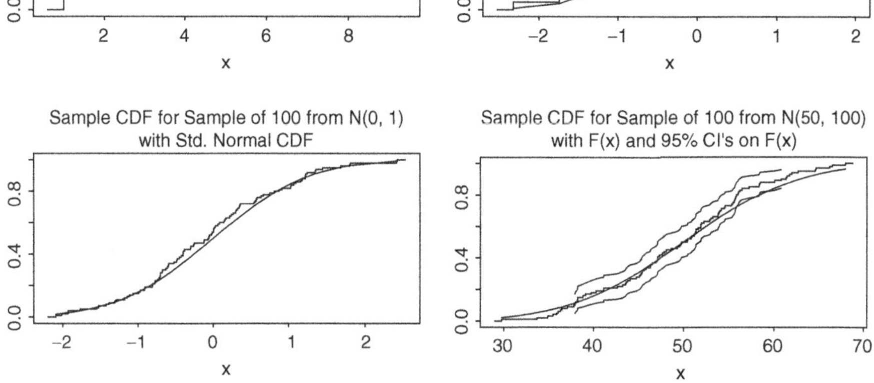


**FIGURE 10.3.1** Sample CDFs compared to standard normal CDFs.

seems to fail to contain $F(58)$. We will say more about this in Section 10.4 when we discuss the Kolmogorov statistic.

### Confidence Intervals on Quantiles

Let $F$ be a continuous cdf, let $0 < \alpha < 1$, and let $x_\alpha$ be a $\alpha$-quantile. That is, $F(x_\alpha) = \alpha$. Suppose that $X_1, \ldots, X_n$ is a random sample from $F$, with corresponding order statistics $X_{(1)} < \cdots < X_{(n)}$. We will try to determine integers $s < t$ so that $P(x_\alpha \in [X_{(s)}, X_{(t)}]) \doteq \gamma$, where $\gamma$ is a chosen probability (say, 0.95 or 0.99). Let $Y_\alpha =$ (no. of $X_i \le x_\alpha$). Then $A_{st} \equiv [x_\alpha \in [X_{(s)}, X_{(t)}]] = [X_{(s)} \le x_\alpha \le X_{(t)}] = [s \le Y_\alpha < t]$. Since $Y_\alpha \sim \text{Binomial}(n, F(x_\alpha) = \alpha)$, $P(A_{st}) = \sum_{k=s}^{t-1} b(k; n, \alpha)$. For smaller $n$ or $\alpha$ close to 0 or 1, binomial tables or the Poisson approximation can be used to determine a pair $(s, t)$ so that $P(A_{st})$ is close to $\gamma$. For example, from the binomial table, for $\alpha = 0.2$, $n = 20$, $P(A_{1,8}) = P(1 \le Y_{0.2} < 8) = 0.968 - 0.012 = 0.956$, so that $[X_{(1)}, X_{(8)}]$ is a 95.6% CI on $x_{0.2}$. Similarly, $X_{(8)}$ is a 96.8% upper confidence bound on $x_{0.2}$.

For large $n$ we can use the normal approximation of the distribution of $Y_{0.2}$. Since $P(A_{st}) \doteq \Phi((t - 1/2 - n\alpha)/(\sqrt{\alpha(1-\alpha)n}) - \Phi((s - 1/2 - n\alpha)/\sqrt{\alpha(1-\alpha)n})$, we can take $(s - 1/2 - n\alpha) = -(t - 1/2 - n\alpha) \equiv k$, so that $P(A_{st}) \doteq \Phi(k/(\sqrt{\alpha(1-\alpha)n}) - \Phi(-k/\sqrt{\alpha(1-\alpha)n}) = 2\Phi(k/\sqrt{\alpha(1-\alpha)n}) - 1$. This will be approximately $\gamma$ for $k \doteq z_{(1+\gamma)/2}\sqrt{\alpha(1-\alpha)n}$, $t = 1/2 + n\alpha + k$, $s = 1/2 + n\alpha - k$. For $\alpha = 0.2$, $\gamma = 0.95$, $n = 100$, we get $k = 7.84$, $s = 12.66$, $t = 28.34$. Computations using the binomial distribution show that $P(A_{13,29}) = P(13 \le Y_{0.2} < 29) = 0.9780 - 0.0253 = 0.9537$, so that $[X_{(12)}, X_{(29)}]$ is a 95.37% CI on $x_{0.2}$.

### Nonparametric Tolerance Intervals

The *coverage* of a random interval $I_{st} \equiv [X_{(s)}, X_{(t)}]$ is $C(X_{(s)}, X_{(t)}) = F(X_{(t)}) - F(X_{(s)}) = P(X \in I_{st} \mid X_{(s)}, X_{(t)})$, where $X$ is a future observation from $F$, independent of $X_1, \ldots, X_n$. $C(X_{(s)}, X_{(t)})$ is a random variable because the interval $I_{st}$ is random. If $F$ is continuous, then $U_i \equiv F(X_i)$ is distributed uniformly on [0, 1] for each $i$, and, because $F$ is monotone nondecreasing, the order statistics for the $U_i$'s and the $X_i$'s satisfy $U_{(i)} = F(X_{(i)})$. That is, the $U_i$'s have the same ordering as do the $X_i$'s. Since $U_i \sim \text{Unif}(0, 1)$, $U_{(1)}, \ldots, U_{(n)}$ are the order statistics of a random sample from the Unif(0, 1) distribution. Let $Y_i = U_{(i)} - U_{(i-1)}$ for $i = 1, \ldots, n$, the $i$th "gap" size, where $U_{(0)} = 0$. Then $\mathbf{Y} = (Y_1, \ldots, Y_n)$ has joint density $f(\mathbf{y}) = n!$ for all $\mathbf{y} \in \{\mathbf{y} \mid \text{for all } i, \text{and } \Sigma y_i = 1\}$. Since this is symmetric in the arguments of $y$, $C(s, t) = Y_{s+1} + \cdots + Y_t$ has the same distribution as $Y_1 + \cdots + Y_{t-s} = U_{(t-s)}$. From Section 3.6, $U_{(t-s)} \sim \text{Beta}(t - s, n + 1 - t + s)$. Quantiles of the distribution of $U_{(t-s)}$ can be determined from the relationship between the beta and $F$-distributions given in comment 10 of Section 9.5. $F \sim F(2\alpha, 2\beta)$ implies that $V = (\alpha/\beta)F/(1 + (\alpha/\beta)F) \sim \text{Beta}(\alpha, \beta)$. We can also exploit the relationship between $U_{(t-s)}$ and $W_u$, the number of $U_i$ that are less than or equal to $u$. Since $W_u \sim$ Binomial$(n, u)$, we have $P(C(s, t) > u) = P(W_u < t - s) = \sum_{k=0}^{t-s-1} b(k; n, u)$.

Suppose, for example, that $n = 400$ and we want a 90% tolerance interval $I_{st}$ with confidence coefficient 0.95. That is, for $C_{st} = F(X_{(t)}) - F(X_{(s)})$ we want $P(C_{st} \geq 0.90) = 0.95$. By the normal approximation of the binomial for $n = 400$, $u = p = 0.90$, we find that $t - s$ must be approximately $un + z_{1-\alpha}\sqrt{nu(1-u)} + 1/2 = 370.4$. We can take $t$ and $s$ arbitrarily with $t - s = 370$, say $t = 385$, $s = 15$. More exact computations show that $P(C_{15,385} \geq 0.90) = 0.948$. Using the relationship between $V$ and $F$, we have $\alpha = t - s = 370$, $\beta = n - t + s - 1 = 29$, $F_{0.95}(2\alpha, 2\beta) = F_{0.95}(740, 58)$, $V_{0.95} = 0.947$. One-sided tolerance intervals, of the form $[-\infty, X_{(t)}]$ or $[X_{(s)}, \infty]$ with coverage probability $\gamma$ are equivalent to one-sided CI's on quantiles $x_\gamma$. so that the previous discussion is appropriate.

### The One-Sample Sign and Wilcoxon Tests

A drug in the form of a pill, XL7, was developed to lower weight in adult males. Thirty volunteers, all of whom were at least 50 pounds overweight, were chosen. These 30 were paired, with those with approximately the same overweight scores being placed in the same pair. One member of each pair was randomly assigned to the treatment group, the other to the control group. Weights were taken before the experiment began. Those in the treatment group received the XL7 pill each morning for six months. Those in the control group received a pill, a placebo, which in appearance was the same as XL7 but which had no real physiological effect. None of the 30 knew whether they were receiving XL7 or the placebo. All were instructed to keep the same exercise and eating habits. At the end of the six months, weights were determined for all 30 again. The losses in weight for all 30 are given in Table 10.3.1.

Let $D_i$ be XL7 loss minus placebo loss for the $i$th pair. We would like to test $H_0$: XL7 and placebo effects are identical versus $H_a$: XL7 tends to increase weight loss. Let $Y$ be the number of pairs for which $D_i > 0$. Under $H_0$: $Y \sim$ Binomial(15, 1/2), since we observe $Y = 11$, the $p$-value is 0.0592.

The beauty of this approach is that it depends only on the randomness of the choices of which members of the pairs enter the treatment group. The weight losses of the

**TABLE 10.3.1 Weight Loss Data**

| Pair | Placebo | XL7 | Difference | Pair | Placebo | XL7 | Difference |
|---|---|---|---|---|---|---|---|
| 1 | −7 | 7 | 14 | 9 | −12 | 6 | 18 |
| 2 | −4 | 6 | 10 | 10 | −11 | −3 | 8 |
| 3 | 7 | 0 | −7 | 11 | −2 | 8 | 10 |
| 4 | 15 | 12 | −3 | 12 | 2 | 7 | 5 |
| 5 | −13 | −2 | 11 | 13 | 5 | 7 | 2 |
| 6 | 11 | 2 | −9 | 14 | 10 | 13 | 3 |
| 7 | 3 | −1 | −4 | 15 | −1 | 8 | 9 |
| 8 | 2 | 8 | 6 | | | | |

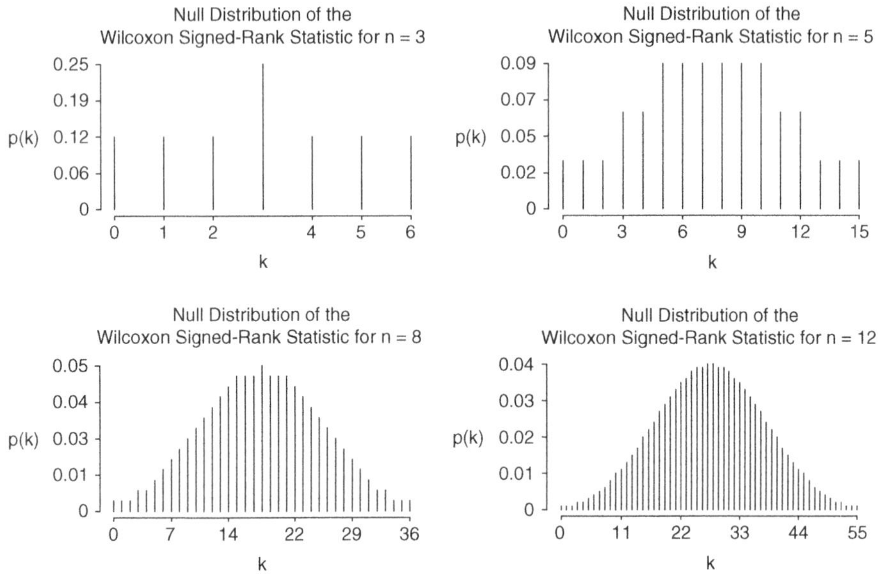


**FIGURE 10.3.2** Null distributions for the Wilcoxon signed-rank statistic.

30 men are considered as fixed, so the only randomness occurs because of the random choice of the treatment group. Another such test, the Wilcoxon signed-rank test, has this same property and has the additional property that it is often more powerful in cases in which the numerical values (the $D_i$ in this case) can be considered to be a random sample from some distribution (see Figure 10.3.2).

Let the absolute values of the $D_i$'s be ranked from smallest to larger, with rank $R_i$ for the $i$th pair. Thus, we have

| Pair = $i$ | 1 | 2 | 3 | 4 | 5 | 6 | 7 | 8 | 9 | 10 | 11 | 12 | 13 | 14 | 15 |
|---|---|---|---|---|---|---|---|---|---|---|---|---|---|---|---|
| Midrank $R_i$ | 13 | 11.5 | 7 | 2.5 | 14 | 9.5 | 4 | 6 | 15 | 8 | 11.5 | 5 | 1 | 2.5 | 9.5 |

Only pairs 3, 4, 6, 7 correspond to negative $D_i$. We will take the Wilcoxon signed-rank statistic to be $W^+$ = sum of midranks corresponding to positive $D_i$. It is slightly easier to sum the ranks corresponding to negative $D_i$. We get 23, so that $W^+ = (15)(16)/2 - 23 = 97$.

Under $H_0$ the 15 midranks are equally likely to correspond to correspond to positive or negative $D_i$'s, independently. Thus, $W^+ = \Sigma R_i I_i$, where $I_i = I[D_i > 0]$ is a Bernoulli(1/2) random variable and the $I_i$ are independent. Thus, $E(W^+ \mid H_0) = (1/2)\Sigma R_i = (1/2)N(N+1)/2 = (1/4)N(N+1) = 60$, and $\text{Var}(W^+ \mid H_0) = \Sigma_i R_i^2(1/2)(1-1/2) = 1238.5/4 = 309.6$. We do not prove it here, but in general $E(W^+ \mid H_0) = [N(N+1) - d_0(d_0+1)]/4$, where $d_0$ is the number of $D_i$'s which are zero, and $\text{Var}(W^+ \mid H_0) = [N(N+1)(2N+1) - d_0(d_0+1)(2d_0+1)]/24 - (1/48)\sum_{i=1}^{e} d_i(d_i-1)(d_i+1)$, where $e$ and $d_i$ are the number of ties and sizes of ties as they were for the Wilcoxon two-sample statistic. In addition, in good approximation, even for relatively small $N$, when the 1/2-correction is used, $W^+$ is approximately normally distributed. For these data, the $p$-value is therefore approximately $1 - \Phi((96.5 - 6)/\sqrt{309.6}) = 1 - \Phi(2.075) = 0.019$. For the one-sample $t$-test of $H_0\colon \mu_D \leq 0$ versus $H_a\colon \mu_D > 0$, we have $\bar{D} = 4.87$, $S_D = 7.84$, $t = \bar{D}/(S/\sqrt{n}) = 2.40$, with corresponding $p$-value 0.0152.

We will not discuss the relative power of the sign, Wilcoxon, and $t$-tests for the one-sample case except to say that they are the same asymptotically as for the two-sample case. If for example, the distribution of the $D_i$'s is normal, the *asymptotic relative efficiency* (ARE) of the Wilcoxon test to the $t$-test is $3/\pi$, and the ARE for the sign test to the Wilcoxon test is 2/3. If the $D_i$'s are samples from the double-exponential distribution, the AREs of the Wilcoxon to the $t$-test and sign test to the $t$-test are 1.5 and 2. Lehmann (1975, p. 129) shows that $W^+ =$ (no. pairs $ij, i \leq j$, such that $D_i + D_j > 0$). From this it follows that $E(W^+) = N(N-1)p^*/2 + Np$, where $p^* = P(D_i + D_j > 0)$ and $p = P(D_i > 0)$.

As for the two-sample Wilcoxon statistic, the signed-rank Wilcoxon statistic can be used to estimate $\Delta$ in the model for which the distribution of the $D_i$'s is symmetric about $\Delta$. We will not go into detail here, except to say that the Wilcoxon signed-rank estimator is the median of all the pairwise averages $T_{ij} = (D_i + D_j)/2$, and the statistical properties relative to the sample mean are similar to the properties of tests based on $W^+$ relative to the $t$-test. A corresponding CI method is based on the order statistics among the $T_{ij}$ (see Lehmann, 1975, Chap. 3).

### Problems for Section 10.3

**10.3.1** For the random sample $X_1 = 13, X_2 = 5, X_3 = 9, X_4 = 3, X_5 = 9$, sketch the sample cdf.

**10.3.2** Let $F_n$ be the sample cdf for a random sample of $n$ from cdf F. Let $k$ be an integer $1 \leq k \leq n$ and $x_1 < x_2 < \cdots < x_k$, with $0 < F(x_j) < 1$ for $j = 1, \ldots, k$. Let $D_j = n(F_n(x_j) - F_n(x_{j-1}))$ for $j = 2, \ldots, n$, and $D_1 = nF_n(x_1)$.

**(a)** What is the joint distribution of $(D_1, \ldots, D_n)$?

**(b)** For $n = 10$ and $F$ the Unif(0, 3) distribution, find $P(F_n(2) - F_n(1) > 0.5)$.

**(c)** For $n$ and $F$ as in part (b), find $P(F_n(1) = 0.4, F_n(2) = 0.9)$.

**10.3.3** Let $X_1, \ldots, X_n$ be a random sample from cdf $F$.

**(a)** Prove that for each $x$, $\{F_n(x)\}$ converges in probability to $F(x)$.

**(b)** (Much harder) Prove that $D_n \equiv \{\sup_x \mid F_n(x) - F(x)|\}$ converges in probability to zero, that is, $\{F_n(x)\}$ converges uniformly in probability to $F(x)$. The famous *Glivenko–Cantelli theorem* states that $\{D_n\}$ converges almost certainly to 0; that is, $P(\{D_n\}$ converges to $0) = 1$.

**10.3.4** Let $X_1, \ldots, X_{1600}$ be a random sample from a continuous cdf $F$. Find $s$ and t so that $[X_{(s)}, X_{(t)}]$

**(a)** Is an approximate 95% CI on the 10th percentile $x_{0.1}$ of $F$.

**(b)** Is an approximate 90% tolerance interval with confidence coefficient 0.80.

**10.3.5** Find the null distribution of the Wilcoxon signed-rank distribution for $n = 4$. Verify that $W^+$ is symmetrically distributed about $n(n+1)/4$ and that $\text{Var}(W^+) = n(n+1)(2n+1)/24$.

**10.3.6** A random sample $Y_1, \ldots, Y_8$ was taken from a distribution that is continuous and symmetric about $\Delta$. The values to the nearest integer were 21, 15, 7, 18, 28, 16, 11.

**(a)** Let $H_0$: $\Delta = 12$ versus $H_a$: $\Delta \neq 12$. Find the $p$-value corresponding to $W^+$, based on the differences $D_i = Y_i - 12$.

**(b)** Use the normal approximation to find the same $p$-value.

**10.3.7** Let $X_1, \ldots, X_n$ be a random sample from the $N(\mu, \sigma^2)$ distribution. Suppose that $\sigma^2$ is known to be 100. Suppose that we wish to test $H_0$: $\mu \leq 50$ versus $H_a$: $\mu > 50$. Define $D_i = X_i - 50$, let $Y$ be the number of positive $D_i$'s, let $W^+$ be the Wilcoxon signed-rank statistic for the $D_i$'s, and let $\bar{D}$ be the means of the $D_i$'s.

**(a)** For $n = 400$, state $\alpha = 0.05$ level tests based on $Y$, $W^+$, and $\bar{D}$.

**(b)** Find approximations of the power of these tests for $\mu = 51.0$. For this suppose that $\text{Var}(W^+)$ is the same as it is under $H_0$. You will need to determine $P(D_i > 0)$ and $P(D_i + D_j > 0)$ for $\mu = 51$.

**10.3.8** For $D_i$'s 7, 3, −4, 6, −1, show that $W^+$ is the number of pairs $i \leq j$ for which $D_i + D_j > 0$.

**10.3.9** To compare two weight-loss methods, $A$ and $B$, 40 overweight people were paired by gender and age. One member of each pair was chosen randomly to use $A$, the other to use $B$. The differences in weight losses in pounds after six months ($A$ loss minus $B$ loss) were 3, 5, 7, −3, 13, −6, 5, 7, 11, −5, 6, 0, 2, 15, −4, 5, 6, 17, −3, 14. Find the $p$-value for both the sign and Wilcoxon signed-rank tests of the null hypothesis of no difference in effects against the alternative of some effect. Notice that the effective sample size is 19, since one value of $D_i$ is zero.

**10.3.10** Suppose that $D_1, D_2, D_3$ are uniformly distributed on $[-1, 2]$.

**(a)** Find the distribution of $W^+$. *Hint:* Think conditionally on the intervals $[-1, 0]$, $(0, 1]$, and $(1, 2]$.

**(b)** Find $E(W^+)$ and $\text{Var}(W^+)$.

**10.3.11** Show that the observed $p$-value for a one-sided test of the null hypothesis that the $D_i = \text{LHH}_i - \text{DHH}_i$ following Example 9.4.1 are symmetrically distributed about 1.0, using the test based on the $W^+$ statistic, is 8/512. (Omit the zero for any $D_i = 1$.) Use $H_a$: median of distribution of $d_i$'s exceeds 1.0.

## 10.4 THE KOLMOGOROV–SMIRNOV TESTS

Suppose that we have a random sample $X_1, \ldots, X_n$ from a continuous cdf $F$ and wish to decide whether $F$ is some specified cdf $F_0$. It should be intuitively clear that $D_n = \sup_x | F_n(x) - F(x) |$ should tend to be small if $F = F_0$, larger if $F \neq F_0$. In fact, as mentioned in Problem 10.3.3, the Glivenko–Cantelli theorem states that $\{D_n\}$ converges to zero with probability 1 (see Figure 10.3.1). Since $F_n(x)$ is a jump function, taking all its jumps at the order statistics $X_{(i)}$, $D_n$ is the maximum of $D_n^+ \equiv \max_i(F_n(X_{(i)}) - F_0(X_{(i)}-)) = \max_i((i/n) - F_0(X_{(i)}))$ and $D_n^- \equiv \max_i(F_0(X_{(i)}) - F_n(X_{(i)})) = \max_i(F_0(X_{(i)}) - (i-1)/n)$. [By $F_0(x-)$ for any $x$, we mean the limit of $F(x-d)$ as $d \to 0$, for $d > 0$.] For example, for $n = 3$, $F(x) = x$ for $0 < x < 1$, $X_{(1)} = 0.4$, $X_{(2)} = 0.5$, $X_{(3)} = 0.8$, $D_3^+ = (2/3 - 0.5) = 1/6$, $D_3^- = (0.4 - 0) = 0.4$, and $D_3 = 0.4$. Kolmogorov showed that $\lim_{n\to\infty} P(D_n\sqrt{n} \leq w) = 1 - 2\sum_{k=1}^{\infty}(-1)^{k+1}e^{-2k^2w^2}$. For larger values of $w$, this is close to $1 - 2e^{-2w^2}$. Thus, if we want this probability to be $1 - \alpha$ for large $n$, we need $w = w_\alpha = \sqrt{-\log(\alpha/2)/2}$. The $\alpha$-level Kolmogorov one-sample goodness-of-fit test of $H_0$: $F(x) = F_0(x)$ for all $x$ versus $H_a$: $F(x) \neq F_0(x)$ for some $x$, rejects for large $n$ for $D_n^- \geq w_\alpha/\sqrt{n}$. Similarly, $\lim_{n\to\infty} P(\sqrt{n}D_n^+ \leq w) = \lim_{n\to\infty} P(\sqrt{n}D^-{}_n \leq w) = 1 - e^{-2w^2}$. Suppose that $n$ is large. For a test of $H_0$: $F(x) \leq F_0(x)$ for all $x$ versus $H_a$: $F(x) > F_0(x)$ for some $x$, we reject for large $D_n^+$. For a $\alpha$-level test, reject for $D_n^+\sqrt{n} \geq w_\alpha^+ = \sqrt{-\log(\alpha)/2}$. Similarly, for $H_0$: $F(x) \geq F_0(x)$ for all $x$ versus $H_a$: $F(x) < F_0(x)$, reject $H_0$ for $D_n\sqrt{n} \geq w_\alpha^+$.

These Kolmogorov–Smirnov (KS) tests are "consistent" against all alternatives. That is, if $F$ satisfies $H_a$, the power converges to 1 as $n \to \infty$. That is not true for the sign and Wilcoxon tests. However, these KS tests have lower power against shift alternatives than do the Wilcoxon and sign tests. For example, for $n = 100$, for $F_0$ the standard normal distribution, $\alpha = 0.05$, the Wilcoxon two-sided test rejected 4914 in 10,000 simulations, while the KS test rejected 3842 times when $F$ was the N(0.2, 1) distribution. However, for $F$ double exponential (DblExp), mean zero, variance 4, the Wilcoxon test rejected 513 among 10,000 simulations, while the KS test rejected 7028 of 10,000. When the underlying distribution was DblExp, mean zero, variance 1, but when $F_0$ was the N(0, 1) distribution, the Wilcoxon and KS tests rejected 554 and 2604 times. Thus, the KS test is sensitive to deviations from normality, although $n$ may have to be quite large to achieve satisfactory power.

The Lilliefors statistic $D_n^*$ replaces the observations $X_i$ by the corresponding standardized observations $Z_i = (X_i - \bar{X})/S$, where $\bar{X}$ and $S$ are the sample mean and standard deviation. Then $D_n^*$ is the KS statistic when $F_n$ is the sample cdf of the $Z_i$'s and $F_0$ is the standard normal cdf. The null distribution of $D_n^*$ is, of course, different. Lilliefors (1967) provided tables for small $n$. He showed that for large $n$, $P(D_n^* \geq d_n^*)$ is approximately 0.01, 0.05, and 0.10 for $d_n^* = 1.031/\sqrt{n}, 0.886/\sqrt{n}$, and $0.805/\sqrt{n}$. For the same 10,000 simulations the Lilliefors test rejected at level $\alpha = 0.05$ when $F$ was the N(0.2, 1) distribution 524 times, close to 0.05, as it should be. For $F$ the DblExp, mean zero, variance 4, it rejected 7119 times, and for $F$ DblExp, mean zero, variance 1, it rejected 7124 times, so that the Lilliefors test is considerably more powerful than the KS test for deviations of shapes (functional forms) from the normal.

The Lilliefors test is sometimes suggested as a preliminary test for normality of a distribution, with acceptance of the null hypothesis of normality to be followed by $t$-tests or other procedures that require normality for exact probability statements. The author does *not* recommend this. First, the acceptance (or failure to reject) of a null hypothesis does not prove that the null hypothesis is true, especially when the sample is small. In fact, $t$-tests and $t$-CIs are quite robust for large sample sizes, and it is for large sample sizes that the Lilliefors or other tests for normality are more likely to reject. In his consulting experience the author has never found a case for which he thought a preliminary test for normality was appropriate.

### Simultaneous Confidence Bands on $F$

Since $D_n = \sup_x \mid F_n(x) - F(x) \mid$ has a distribution that is the same for all continuous $F$, it can be used as a pivotal quantity in order to establish CIs on $F(x)$, holding simultaneously for all $x$. For example, the events $[D_n \leq d]$ and $[F(x) \in [F_n(x) \pm d]$ for all $x]$ are equal, so that if $w$ is chosen to make $P(D_n \leq d) = \gamma$, then $[F_n(x) \pm d]$ defines an infinite collection of CIs, holding simultaneously for all $x$ with probability $\gamma$. We call these it simultaneous *confidence bands* (see Figure 10.4.1). For $n$ large we can take $d = w_\alpha/\sqrt{n}$, where $w_\alpha = \sqrt{-\log(\alpha/2)/2}$, which is 1.36 for $\alpha = 0.05$. The KS test of $H_0$: $F = F_0$ versus $H_a$: $F \neq F_0$ rejects $H_0$ at the $\alpha$ level if $F_0$ does not lie

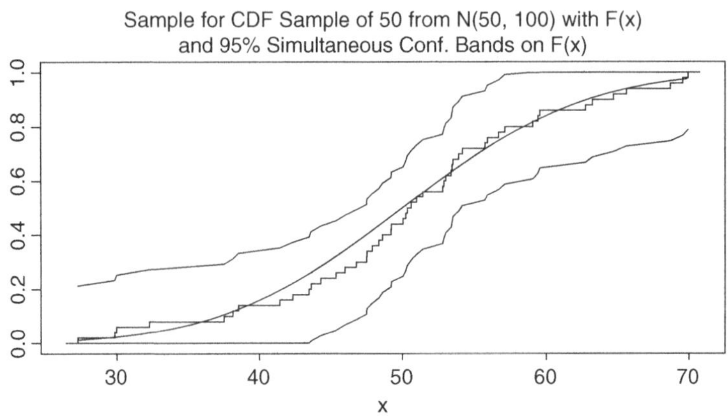


**FIGURE 10.4.1** True CDF, sample CDF, and 95% confidence bands.

within the bands for all $x$. The bands are so wide for small $n$ that they hardly seem worthwhile.

### The Two-Sample Kolmogorov–Smirnov Test

Let $X_1, \ldots, X_m$ be a random sample from $F$, and let $Y_1, \ldots, Y_n$ be a random sample from $G$, where $F$ and $G$ are continuous and the $X_i$'s and $Y_j$'s are independent. Let $F_m$ and $G_n$ be the corresponding sample cdf's. Define $S_{mn}^+ = \sup_x[F_m(x) - G_n(x)]$, $S_{mn}^- = \sup_x[G_n(x) - F_m(x)]$, and $S_{mn} = \max(S_{mn}^+, S_{mn}^-) = \sup_x | F_n(x) - G_m(x) |$. For example, if $m = 3, n = 2$, and the $X$'s and $Y$'s have the ordering $XXYYX$, then $[F_m(x) - G_n(x)]$ takes its largest value at the second $X$, with $S_{mn}^+ = 2/3 - 0 = 2/3$, and $G_n(x) - F_m(x)$ takes its largest value at the second $Y$, with $S_{32}^- = 1 - 2/3 = 1/3$. Therefore, $S_{32} = 2/3$.

More generally, $S_{mn}^+ = \max_i(i/m - (R_i - i)/m)$, where $R_i$ is the rank of $X_{(i)}$, the $i$th-order statistic, among all $X_i$'s and $Y_j$'s. Similarly, $S_{mn}^- = \max_j(j/n - (R_j^* - j)/n$, where where $R_j^*$ is the rank of $Y_{(j)}$, the $j$th-order statistic, among all $X_i$'s and $Y_j$'s. Thus, $S_{mn}^+$ and $S_{mn}^-$ and therefore $S_{mn}$ are rank statistics. Define $\tau_{nm} = (1/n + 1/m)^{-1/2}$. Then for limits as $n \to \infty$ and $m \to \infty$, $\lim_{nm} P(\tau_{nm} S_{mn}^+ \le u) = \lim_{nm} P(\tau_{nm} S_{mn}^- \le u) = 1 - e^{-2u^2}$ and $\lim_{nm} P(\tau_{nm} S_{nm} \le u) = \sum_{k=-\infty}^{\infty}(-1)^k e^{-2k^2u^2}$. For $u > 1$, as we usually need, this last sum is $1 - 2e^{-2u^2}$ in good approximation, so that the $(1 - \alpha)$-quantile is approximately $s_\alpha = w_\alpha = \sqrt{(-\log(\alpha/2)/2)}$, which is 1.36 for $\alpha = 0.05$. To show how good the approximation is, consider the case $n = m = 10$. Tables of the distribution of $S_{nn}$ indicate that $P(S_{nn} \ge 0.6) = 0.0524$, while the approximation indicates that the 0.95-quantile is $s_{0.05}\tau_{10,10} = 0.608$. Because of the discreteness of $S_{mn}$, $P(S_{10,10} > 0.608) = P(S_{10,10} \ge 0.7) = 0.0123$. Simulations show that this test is conservative, in the sense that it rejects with probability roughly 0.03 or 0.04 for nominal $\alpha = 0.05$ for smaller $n$ and $m$, but with probabilities close to 0.05 for larger $n(>100$, for example). In practice, the KS test has low power and is therefore somewhat useless unless $n$ and $m$ exceed 25 or 50.

### Problems for Section 10.4

**10.4.1** Let $X_1, X_2, \ldots, X_5$ be a random sample from a distribution with cdf $F$.

**(a)** For observed $X_1 = 7.8, X_2 = 11.5, X_3 = 0.7, X_4 = 13.3, X_5 = 8.3$, $F_0(x) = 1 - e^{-x/10}$ for $x > 0$, sketch graphs of the sample cdf and $F_0$ on the same axes.

**(b)** Find the statistics $D^+$, $D^-$, and $D$ for $H_0$: $F = F_0$.

**10.4.2** Fifty randomly chosen 60-watt light bulbs were tested until failure from among the 40,000 manufactured by GB Company in November. Suppose that the lengths of life $X_i$ constitute a random sample from a continuous cdf F. The order statistics $X_{(1)}, \ldots, X_{(50)}$ of their lengths of life were:

| | | | | | | | | | |
|---|---|---|---|---|---|---|---|---|---|
| 43 | 102 | 123 | 137 | 183 | 189 | 201 | 202 | 228 | 230 |
| 236 | 237 | 241 | 256 | 258 | 259 | 259 | 263 | 291 | 291 |
| 311 | 313 | 318 | 320 | 324 | 339 | 341 | 345 | 353 | 360 |
| 369 | 373 | 378 | 411 | 424 | 465 | 468 | 475 | 485 | 511 |
| 532 | 533 | 589 | 593 | 628 | 659 | 665 | 683 | 690 | 720 |

**(a)** Sketch the sample cdf $F_{50}$. Give 95% individual (not simultaneous) confidence intervals on $F(x)$ for $x = 200, 400, 600$.

**(b)** Give a 95% confidence interval on $x_{0.6}$ (see Section 10.3).

**(c)** Find the approximate $p$-value for the sign test of $H_0$: $F(500) \leq 0.6 \Leftrightarrow$: $x_{0.6} \geq 500$ versus $H_a$: $F(500) > 0.6 \Leftrightarrow x_{0.6} < 500$.

**(d)** Studies for a light bulb manufactured by an older process showed that in good approximation $F = F_0$ was the cdf of the $N(350, 100^2)$ distribution. Thus, $F_0(x) = \Phi((x - 350)/100)$. The values of $F_0(x)$ for the order statistics $X_{(i)}$ observed were:

| | | | | | | | | | |
|---|---|---|---|---|---|---|---|---|---|
| 0.0011 | 0.0066 | 0.0116 | 0.0116 | 0.0475 | 0.0537 | 0.0681 | 0.0694 | 0.1112 | 0.1151 |
| 0.1271 | 0.1292 | 0.1379 | 0.1736 | 0.1788 | 0.1814 | 0.1814 | 0.1922 | 0.2776 | 0.2776 |
| 0.3483 | 0.3557 | 0.3745 | 0.3821 | 0.3974 | 0.4562 | 0.4641 | 0.4801 | 0.5120 | 0.5398 |
| 0.5753 | 0.5910 | 0.6103 | 0.7291 | 0.7704 | 0.8749 | 0.8810 | 0.8944 | 0.9115 | 0.9463 |
| 0.9656 | 0.9664 | 0.9916 | 0.9925 | 0.9973 | 0.9990 | 0.9992 | 0.9996 | 0.9997 | 0.9999 |

Determine the value of the statistic $D_{50}$ and use this to find an approximate $p$-value for the Kolmogorov test of $H_0$: $F = F_0$.

**10.4.3** Random samples of sizes 8 $X$'s and 10 $Y$'s were taken from cdf's $F$ and $G$.

**(a)** Find the KS statistics, $S^+_{8,10}$ and $S^-_{8,10}$, for the following data:

| | | | | | | | | | | |
|---|---|---|---|---|---|---|---|---|---|---|
| Ordered $X$'s | 13.6 | 16.5 | 24.4 | 28.4 | 29.4 | 30.6 | 34.2 | 36.8 | | |
| Ordered $Y$'s | 28.8 | 31.1 | 34.1 | 35.3 | 36.4 | 40.0 | 43.4 | 44.7 | 56.2 | 62.7 |

**(b)** Find an approximate $p$-value for the KS test of $H_0$: $F = G$ versus $H_a$: $F \neq G$.

**(c)** For $H_0$ and $H_a$ as in part (b), find an approximate $p$-value for the Wilcoxon two-sample test.

CHAPTER ELEVEN

# Linear Statistical Models

## 11.1 INTRODUCTION

Over a 12-year period the author collected data in his class on linear models. Students were asked to record $y$ (their height in centimeters), $x_1$ (the indicator for males), $x_2$ (the height in centimeters of their father), and $x_3$ (the height of their mothers in centimeters). The observation vectors $(y, x_1, x_2, x_3)$ were recorded for 129 males and 78 females. Figure 11.1.1 shows four plots, corresponding to the four combinations of father and mother versus student heights for males and females. We notice that males tend to be taller. Although there seems to be a tendency for taller mothers and fathers to have taller children, the relationship is not strong, being especially weak for female students.

The straight lines were fit using the principle of least squares. In Section 11.2 we discuss this principle, show how the equations were determined, and consider multiple regression, by which we can use both mothers' and fathers' heights to predict sons' or daughters' heights.

Figure 11.1.2 indicates that there is a slight tendency for taller people to marry taller people.

## 11.2 THE PRINCIPLE OF LEAST SQUARES

Consider $n$ pairs of points $(x_i, y_i)$ for $i = 1, \ldots, n$ as in any of the five plots of Figures 11.1.1 and 11.1.2. Suppose that we would like to approximate the $y_i$ by $\hat{y}_i = b_0 + b_1 x_i$, a linear function of $x_i$. How should $b_0$ and $b_1$ be chosen? In most cases if $n > 2$, no matter how these constants are chosen there must be nonzero values of the residuals $e_i = y_i - \hat{y}_i$. That is, no matter which line is fit to the data, not all points will usually fall on the line. Some compromise would seem to be in order. We could, for example, fit a line through any pair of points, so

---

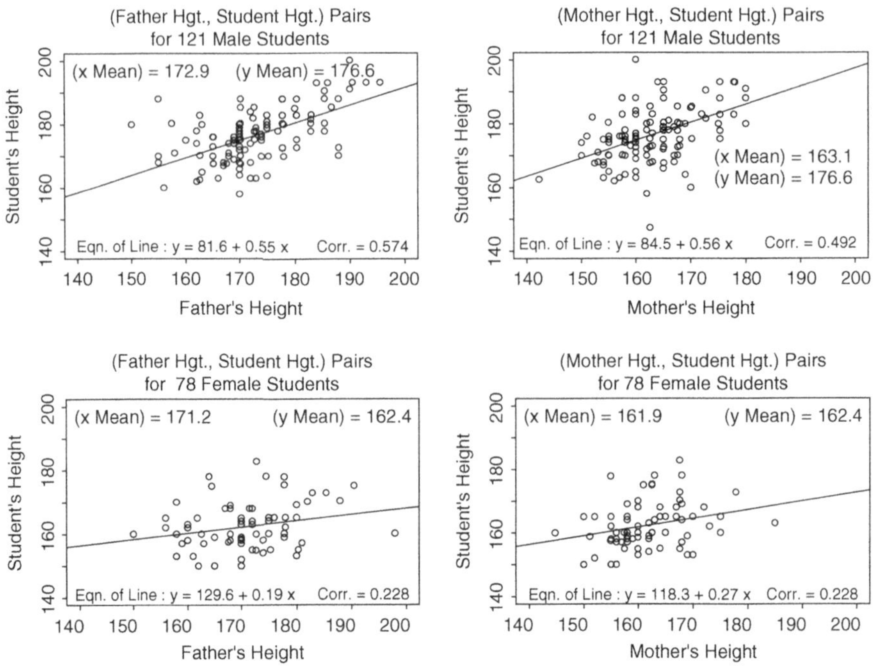


**FIGURE 11.1.1** Heights of students and their parents.

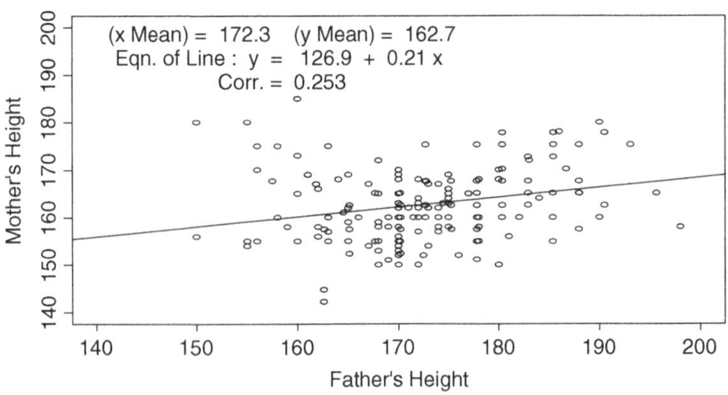


**FIGURE 11.1.2** 209 (father height, mother height) pairs.

that if we use this criterion there are $n(n-1)/2$ possible lines. We could find the line that minimizes $G(b_0, b_1) \equiv \sum_{i=1}^{n} |y_i - \hat{y}_i|$. This is the sum of the absolute values of the distances of the points from the line in a vertical direction. Linear programming techniques can be employed to find the minimizers $(b_0, b_1)$. For the $x =$ (father's height), $y =$ (male student's height) data in the first plot of Figure

11.1.1, the function "l1fit" in S-Plus found $(b_0, b_1) = (57.59, 0.693)$. For these values, $G(b_0, b_1) = 680.95$. Another criterion employs the *principle of least squares*, by which $H(b_0, b_1) = \sum_{i=1}^{n}(y_i - \hat{y}_i)^2$ is minimized. Use of the S-Plus function "lm" produced, for the same data, $(b_0, b_1) = (81.57, 0.550)$. Values of $G$ and $H$ for these two choices of $(b_0, b_1)$ were $G(57.59, 0.693) = 680.95$, $G(81.57, 0.550) = 691.91$, $H(57.59, 0.693) = 6818.12$, $H(81.57, 0.550) = 6526.77$. Thus, $H$ is smaller for the least squares solution, and $G$ is smaller for the $L_1$ solution, as should be expected.

As we will show, the method of least squares leads to relatively simple formulas for the minimizing pair $(b_0, b_1)$, and the extension to multiple regression with several *explanatory variables* (both mother's and father's height, for example) is relatively straightforward, at least for those who have studied linear algebra. We refer to the case of just one explanatory $x$-variable as *simple linear regression*, and the case of more than one is called *multiple regression*. The method of least squares was first used by Adrien-Marie Legendre (1752–1833) in his 80-page book *Nouvelles méthodes pour le détermination des orbites des comètes* (New methods for determination of the orbits of the planets) with a nine-page supplement, *Sur la méthode des moindres carrés* (On the method of least squares) (see Stigler, 1986).

## The One-Predictor Variable

We begin by considering a simple problem: For $n$-component vectors $\mathbf{y} = (y_1, \ldots, y_n)$ and $\mathbf{x} = (x_1, \ldots, x_n)$, not all zero, find the constant $b$ for which $Q(b) = \sum_{i=1}^{n}(y_i - bx_i)^2$ is minimum. Let the *inner product* of any two $n$-component vectors $\mathbf{u}$ and $\mathbf{v}$ be $(\mathbf{u}, \mathbf{v}) = \sum u_i v_i$. Let the squared length of any vector $\mathbf{u}$ be $\|\mathbf{u}\|^2 = (\mathbf{u}, \mathbf{u})$. We use these ideas to find $b$ so that $Q(b)$ is minimum.

**Definition 11.2.1** $n$-component vectors $\mathbf{u}$ and $\mathbf{v}$ are *orthogonal* if $(\mathbf{u}, \mathbf{v}) = 0$. We write $\mathbf{u} \perp \mathbf{v}$. Thus, $\mathbf{u} = (1, 1, 2)$ and $\mathbf{v} = (-3, 1, 1)$ are orthogonal. □

**The Pythagorean Theorem** Let $\mathbf{u}$ and $\mathbf{v}$ be orthogonal. Then $\|\mathbf{u} + \mathbf{v}\|^2 = \|\mathbf{u}\|^2 + \|\mathbf{v}\|^2$.

*Proof:* $\|\mathbf{u} + \mathbf{v}\|^2 = (\mathbf{u} + \mathbf{v}, \mathbf{u} + \mathbf{v}) = (\mathbf{u}, \mathbf{u}) + (\mathbf{v}, \mathbf{v}) + 2(\mathbf{u}, \mathbf{v}) = \|\mathbf{u}\|^2 + \|\mathbf{v}\|^2 + 0$. □

In the example above, $\|\mathbf{u} + \mathbf{v}\|^2 = \|(-2, 2, 3)\|^2 = 17 = \|\mathbf{u}\|^2 + \|\mathbf{v}\|^2 = 6 + 11$.

**Definition 11.2.2** The projection of a vector $\mathbf{y}$ on a vector $\mathbf{x}$ is a multiple $\hat{\mathbf{y}} = b\mathbf{x}$ such that $(\mathbf{y} - \hat{\mathbf{y}}) \perp \mathbf{x}$ (see Figure 11.2.1). □

We denote the projection of $\mathbf{y}$ on $\mathbf{x}$ by $p(\mathbf{y} \mid \mathbf{x})$. There should be no danger of confusion with conditional probability. We can determine $b$ as follows. $0 = (\mathbf{y} - \hat{\mathbf{y}}, \mathbf{x}) = (\mathbf{y} - b\mathbf{x}, \mathbf{x}) = (\mathbf{y}, \mathbf{x}) - b(\mathbf{y}, \mathbf{x})$, $b = (\mathbf{y}, \mathbf{x})/\|\mathbf{x}\|^2$ if $\mathbf{x} \neq \mathbf{0}$. If $\mathbf{x} = \mathbf{0}$, then $\hat{\mathbf{y}} = \mathbf{0}$. Then $\|\hat{\mathbf{y}}\|^2 = (\hat{\mathbf{y}}, \hat{\mathbf{y}}) = (\mathbf{y}, \hat{\mathbf{y}}) = b(\mathbf{y}, \mathbf{x}) = (\mathbf{y}, \mathbf{x})^2/\|\mathbf{x}\|^2$.

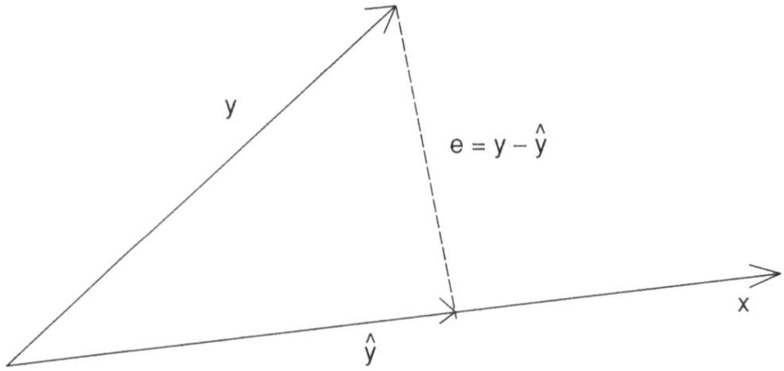


**FIGURE 11.2.1**

Note that it is easy to verify that $p(a\mathbf{y} \mid \mathbf{x}) = ap(\mathbf{y} \mid \mathbf{x})$ for any scalar $a$ and $p(\mathbf{y}_1 + \mathbf{y}_2 \mid \mathbf{x}) = p(\mathbf{y}_1 \mid \mathbf{x}_1) + p(\mathbf{y}_2 \mid \mathbf{x}_2)$ so that the projection function $p(\mathbf{y} \mid \mathbf{x})$ is a linear function of **y**, taking vectors in $R_n$ into vectors in $R_n$.

***Example 11.2.1*** Let $\mathbf{x} = (1, 2, 3, 5)$, $\mathbf{y} = (3, 2, 10, 16)$. Then $b = 117/39 = 3$, $\hat{\mathbf{y}} = 3\ \mathbf{x} = (3, 6, 9, 15)$, and the residual vector is $\mathbf{y} - \hat{\mathbf{y}} = (0, -4, 1, 1) \perp \mathbf{x}$. By the Pythagorean theorem, $\|\mathbf{y}\|^2 = \|\mathbf{y} - \hat{\mathbf{y}} + \hat{\mathbf{y}}\|^2 = \|\mathbf{y} - \hat{\mathbf{y}}\|^2 + \|\hat{\mathbf{y}}\|^2$. Here $\|\mathbf{y}\|^2 = 379 = 351 + 18$, and $\|\hat{\mathbf{y}}\|^2 = 117^2/39 = (117)(3) = 351$. □

Among all scalar multiples, $c\mathbf{x}$ of $\mathbf{x}$, $\hat{\mathbf{y}} = p(\mathbf{y} \mid \mathbf{x})$ is the closest. That is, $Q(c) = \sum_{i=1}^{n}(y_i - cx_i)^2 = \|\mathbf{y} - c\mathbf{x}\|^2$ is minimum for $c = b = (\mathbf{y}, \mathbf{x})/\|\mathbf{x}\|^2$, so that $c\mathbf{x} = b\mathbf{x} = \hat{\mathbf{y}}$.

*Proof:* $\|\mathbf{y} - c\mathbf{x}\|^2 = \|\mathbf{y} - b\mathbf{x} + (b - c)\mathbf{x}\|^2 = \|\mathbf{y} - b\mathbf{x}\|^2 + (b - c)^2\|\mathbf{x}\|^2$, since $(\mathbf{y} - b\mathbf{x}) \perp \mathbf{x}$. This is minimum for $c = b$. □

### The $k$-Predictor Problem

Let $\tilde{x}_i = (x_{i1}, x_{i2}, \ldots, x_{ik})$ be a row vector of real numbers, and let $y_i$ be any real number for $i = 1, \ldots, n$. These might take the values, for example, $y_i = i$th student's height, $x_{i1} \equiv 1$, $x_{i2} =$ indicator of males, $x_{i3} =$ father's height, $x_{i4} =$ mother's height for the $i$th student. Let **b** be a $k$-component column vector with $j$th component $b_j$. Consider the function $Q(\mathbf{b}) = \sum_{i=1}^{n}[y_i - \tilde{x}_i\mathbf{b}]^2$. The method of least squares chooses that **b** which minimizes $Q$. To translate the case of simple linear regression to this more general problem, we can take $x_{i1} = 1$, $x_{i2} = x_i$, for all $i$, and replace $b_0$ by $b_1$, $b_1$ by $b_2$. Let $g(\tilde{x}_i, \mathbf{b}) = \tilde{x}_i\mathbf{b}$.

The minimation problem may be translated into a problem in linear algebra as follows. Let $\mathbf{x}_j = (x_{1j}, x_{2j}, \ldots, x_{nj})^T$. Let $\mathbf{y} = (y_1, \ldots, y_n)^T$ Thus, **y** and all $\mathbf{x}_j$ are vectors in $R_n$. Then $Q(\mathbf{b}) = \|\mathbf{y} - (b_1\mathbf{x}_1 + \cdots + b_k\mathbf{x}_k)\|^2$ . We seek coefficients $b_1, \ldots, b_k$ so that $\hat{\mathbf{y}} = b_1\mathbf{x}_1 + \cdots + b_k\mathbf{x}_k$ is the closest vector in $V \equiv \mathcal{L}(\mathbf{x}_1, \ldots, \mathbf{x}_k)$ to **y**, where $V$ is the subspace spanned by $\mathbf{x}_1, \ldots, \mathbf{x}_k$. As it did for the case $k = 1$, this closest vector turns out to be the unique projection of **y** onto $V$. We need two definitions.

**Definition 11.2.3** Let $V$ be a subspace of $R_n$. A vector $\mathbf{y}$ is orthogonal to $V$ if $\mathbf{y} \perp \mathbf{v}$ for all $\mathbf{v} \in V$. □

**Definition 11.2.4** Let $\mathbf{V}$ be a subspace of $R_n$. The projection of a vector $\mathbf{y}$ onto $V$ is the unique vector $\hat{\mathbf{y}} \in V$ satisfying $(\mathbf{y} - \hat{\mathbf{y}}) \perp \mathbf{V}$. □

To justify the definition we need to prove (1) that such a vector exists, and (2) that it is unique. Let $\mathbf{x}_1, \ldots, \mathbf{x}_k$ be a basis for $V$. That is, these vectors span $V$ and are linearly independent. We seek coefficients $b_1, \ldots, b_k$ such that $\mathbf{y} - \hat{\mathbf{y}} = (\mathbf{y} - \Sigma_j b_j \mathbf{x}_j) \perp V$. It is enough to find such $b_j$ so that

$$(\mathbf{y}, \mathbf{x}_i) = \left(\sum_j b_j \mathbf{x}_j, \mathbf{x}_i\right) = \sum_j b_j(\mathbf{x}_j, \mathbf{x}_i) \qquad \text{for } i = 1, \ldots, k.$$

These $k$ linear equations in $k$ unknowns are the *normal equations*. ("Normal" here is concerned with orthogonality, *not* with the normal distribution.) The equations may be written in a simpler form. Let $\mathbf{X}$ be the $n \times k$ matrix formed by letting the $j$th column vector be $\mathbf{x}_j$ for each $j$. $\mathbf{X}$ is often called *the design matrix,* because in some applications the elements of $\mathbf{X}$ may be chosen by the user. The normal equations are $\mathbf{X}^T\mathbf{y} = \mathbf{X}^T\mathbf{X}\mathbf{b}$, where $b$ is the column vector $(b_1, \ldots, b_k)^T$ . The inner-product matrix $\mathbf{M} \equiv \mathbf{X}^T\mathbf{X}$ is nonsingular if and only if the $\mathbf{x}_j$ are linearly independent (see Problem 11.2.3). We will assume that to be the case. Otherwise, we can drop one or more of the $\mathbf{x}_j$ from the analysis. In that case $\mathbf{M}$ has an inverse $\mathbf{M}^{-1}$, so that $\mathbf{b} = \mathbf{M}^{-1}\mathbf{U}$, where $\mathbf{U} = \mathbf{X}^T\mathbf{y}$ is the inner product vector of $\mathbf{y}$ with the $\mathbf{x}_j$. Note that when the $\mathbf{x}_j$ are mutually orthogonal, $p(\mathbf{y} \mid V) = \Sigma p(\mathbf{y} \mid x_j) = \Sigma((\mathbf{y}, \mathbf{x}_i)/\|\mathbf{x}_j\|^2)\mathbf{x_j}$. In fact, when $p(\mathbf{y} \mid V) = \Sigma p(\mathbf{y} \mid \mathbf{x}_j)$ for all $\mathbf{y}$, the $\mathbf{x}_j$ must be mutually orthogonal. (Proofs are left to the student.)

Since the residual vector $\mathbf{e} = \mathbf{y} - \hat{\mathbf{y}}$ is orthogonal to the columns of $\mathbf{X}$ (equivalently to the subspace $V$ spanned by the columns),

$$\begin{aligned}\text{Total Sum of Squares (TSS)} &= \|\mathbf{y}\|^2 = \|\mathbf{y} - \hat{\mathbf{y}} + \hat{\mathbf{y}}\|^2 = \|\mathbf{e}\|^2 + \|\hat{\mathbf{y}}\|^2 \\ &= \text{Residual SSqs} + \text{Regression SSqs}\end{aligned}$$

(Later we define the total sum of squares as the sum of squares of the $\mathbf{y}_j$'s after their mean is subtracted. For now, we use this definition.) Let $\mathbf{P}_V = \mathbf{X}(\mathbf{X}^T\mathbf{X})^{-1}\mathbf{X}^T$. $\mathbf{P}_V$ is the projection matrix onto the subspace $V$ spanned by the columns of $\mathbf{X}$. Regression SSqs can be written in a simpler form: $\|\hat{\mathbf{y}}\|^2 = (\mathbf{P_V y}, \mathbf{P_V y}) = \mathbf{y}^T\mathbf{P}_\mathbf{V}^T\mathbf{P_V y} = \mathbf{y}^T\mathbf{P_V y} = \mathbf{y}^T\mathbf{X}\mathbf{b} = \mathbf{U}^T\mathbf{b}$, where $\mathbf{U} = \mathbf{X}^T\mathbf{y}$, the column vector of inner products of the columns of $\mathbf{X}$ with $\mathbf{y}$. The third equality follows from the fact that the projection matrix is symmetric and idempotent. That is, $\mathbf{P}_V^\mathbf{T} = \mathbf{P}_V$ and $\mathbf{P}_V\mathbf{P}_V = \mathbf{P}_V$. Since $\mathbf{U} = \mathbf{X}^T\mathbf{y}$ has been determined in order to determine $\mathbf{b}$, the additional computation necessary to determine $\|\hat{\mathbf{y}}\|^2$ can therefore be done with a hand calculator, although modern software packages such as S-Plus, SPSS, and SAS may provide more precise values by determining $\hat{\mathbf{y}}$ itself, then its squared length.

That the vector $\hat{\mathbf{y}} = \mathbf{Xb}$ is unique follows from the nonsingularity of **M**, but also by the following argument. If $\hat{\mathbf{y}}_1$ and $\hat{\mathbf{y}}_2$ were both projections of **y** on $V$, then $\|\hat{\mathbf{y}}_1 - \hat{\mathbf{y}}_2\|^2 = (\mathbf{y} - \hat{\mathbf{y}}_2, \hat{\mathbf{y}}_1 - \hat{\mathbf{y}}_2) - (\mathbf{y} - \hat{\mathbf{y}}_1, -\hat{\mathbf{y}}_2)$. Each term is zero because $(\hat{\mathbf{y}}_1 - \hat{\mathbf{y}}_2) \in \mathbf{V}$. Thus, $\hat{\mathbf{y}}_1 - \hat{\mathbf{y}}_2 = \mathbf{0}$. That $G(\mathbf{v}) \equiv \|\mathbf{y} - \mathbf{v}\|^2$ is minimized for $\mathbf{v} \in V$ by $\mathbf{v} = \hat{\mathbf{y}}$ follows from $\|\mathbf{y} - \mathbf{v}\|^2 = \|\mathbf{y} - \hat{\mathbf{y}}\|^2 + \|\hat{\mathbf{y}} - \mathbf{v}\|^2$, the last equality following from $(\hat{\mathbf{y}} - \mathbf{v}) \in V$ and $(\mathbf{y} - \hat{\mathbf{y}} \perp \mathbf{V}$. (See Figure 11.4.1 to gain some intuition.)

***Example 11.2.2*** **(Simple Linear Regression)** Suppose that we observe the following $(x, y)$ pairs: (1, 9), (2, 10), (3, 6), (4, 2), (5, 3). These points are shown in Figure 11.2.2.

Then $\mathbf{y} = (9, 10, 6, 2, 3)^T$, $\mathbf{X} = \begin{pmatrix} 1 & 1 \\ 1 & 2 \\ 1 & 3 \\ 1 & 4 \\ 1 & 5 \end{pmatrix}$, $\mathbf{M} = \mathbf{X}^T\mathbf{X} = \begin{pmatrix} 5 & 15 \\ 15 & 55 \end{pmatrix}$, $\mathbf{M}^{-1} =$ $\begin{pmatrix} 55 & -15 \\ -15 & 5 \end{pmatrix}(1/50) = \begin{pmatrix} 11 & -3 \\ -3 & 1 \end{pmatrix}(1/10)$, $\mathbf{U} = \mathbf{X}^T\mathbf{y} = \begin{pmatrix} 30 \\ 70 \end{pmatrix}$, $\mathbf{b} = \mathbf{M}^{-1}\mathbf{U} =$ $\begin{pmatrix} 12 \\ -2 \end{pmatrix}$, $\mathbf{Xb} = \begin{pmatrix} 10 \\ 8 \\ 6 \\ 6 \\ 2 \end{pmatrix}$, $\mathbf{e} = \mathbf{y} - \hat{\mathbf{y}} = \begin{pmatrix} -1 \\ 2 \\ 0 \\ -2 \\ 1 \end{pmatrix}$, $\|\hat{\mathbf{y}}\|^2 = 220, \|\mathbf{e}\|^2 = 10, \|\mathbf{y}\|^2 = 230$.

Notice that $\mathbf{X}^T\mathbf{e} = \begin{pmatrix} 0 \\ 0 \end{pmatrix}$. That is, **e** is orthogonal to the columns of **X**. □

Although it is possible to give nonmatrix formulas for **b** for simple linear regression, by subsituting symbols $x_1, \ldots, x_n$ for the components of the vector **x**, using the formula $\mathbf{b} = \mathbf{M}^{-1}\mathbf{X}^T\mathbf{y}$, the derivation is awkward. Instead, replace the second column of **X** by a vector $\mathbf{x}^*$, also in $V$, which is orthogonal to the

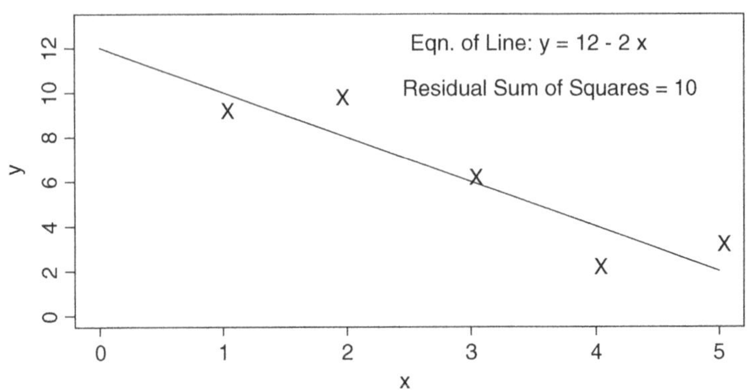


**FIGURE 11.2.2** Five $(x, y)$ points and their least squares line.

$n$-component column vector $\mathbf{J}_n$ of all 1's. Let $\mathbf{x}^* = \mathbf{x} - p(\mathbf{x} \mid \mathbf{J}_n) = \mathbf{x} - \overline{x}\mathbf{J}_n$. $\mathbf{x}^*$ is the vector of deviations of the components of $\mathbf{x}$ from $\overline{x}$. Then, since $\mathbf{J}_n$ and $\mathbf{x}^*$ are orthogonal, we can express $\hat{\mathbf{y}}$ easily in terms of $\mathbf{J}_n$ and $\mathbf{x}^*$. Let $\hat{\mathbf{y}} = a_0\mathbf{J}_n + a_1\mathbf{x}^*, \mathbf{a} = \begin{pmatrix} a_0 \\ a_1 \end{pmatrix}$, $\mathbf{X}^*$ an $n \times 2$ matrix with columns $\mathbf{J}_n$ and $\mathbf{x}^*$. The inner-product matrix $\mathbf{M}^* = \mathbf{X}^{*\mathrm{T}}\mathbf{X}^* = \begin{pmatrix} n & 0 \\ 0 & \|\mathbf{x}^*\|^2 \end{pmatrix}$, where $\|\mathbf{x}^*\|^2 \equiv S_{xx} = \sum(x_i - \overline{x})^2$. $\mathbf{U}^* \equiv \mathbf{X}^{*\mathrm{T}}\mathbf{y} = \begin{pmatrix} (\mathbf{J}_n, \mathbf{y}) \\ (\mathbf{x}^*, \mathbf{y}) \end{pmatrix} = \begin{pmatrix} \Sigma y_i \\ S_{xy} \end{pmatrix}$, where $S_{xy} = (x^*, y) = \sum(x_i - \overline{x})y_i$. Therefore, $\mathbf{a} = \mathbf{M}^{*-1}\mathbf{U}^* = \begin{pmatrix} \overline{y} \\ a_1 \end{pmatrix}$, where $a_1 = S_{xy}/S_{xx}$. The $i$th predicted value for $y$ is therefore $\hat{y}_i = b_0 + b_1x_i = a_0 + a_1(x_i - \overline{x})$, so that $b_1 = a_1 = S_{xy}/S_{xx}$ and $b_0 = a_0 - b_1\overline{x} = \overline{y} - b_1\overline{x}$. The regression line has the equation $y = \overline{y} + b_1(x - \overline{x})$. It passes through the "point of means" $(\overline{x}, \overline{y})$ with the slope $b_1 = S_{xy}/S_{xx}$. The vector of fitted values is $\hat{\mathbf{y}} = \overline{y}\mathbf{J}_n + b_1\mathbf{x}^*$. By the Pythagorean theorem, $\|\hat{\mathbf{y}}\|^2 = \overline{y}^2\|\mathbf{J}_n\|^2 + b_1^2\|\mathbf{x}^*\|^2 = \overline{y}^2 n + S_{xy}^2/S_{xx}$. Then the residual sum of squares is $\|\mathbf{y}\|^2 - \|\hat{\mathbf{y}}\|^2 = \Sigma y_i^2 - \overline{y}^2 n + S_{xy}^2/S_{xx} = S_{yy} - S_{xy}^2/S_{xx}$, where $S_{yy} = \Sigma(y_i - \overline{y})^2$.

***Example 11.2.2 Continued*** From Example 11.2.2, $\overline{y} = 6$, $S_{xy} = -20$, $S_{xx} = 10$, so that $\mathbf{a} = \begin{pmatrix} 6 \\ -2 \end{pmatrix}$, and the regression line has the equation $y = 6 - 2(x - 3) = 12 - 2x$. Also, $S_{yy} = 50$, so that $\|\mathbf{e}\|^2 = S_{yy} - S_{xy}^2/S_{xx} = 50 - (-20)^2/10 = 10$, as before. The *correlation coefficient* of $\mathbf{x}$ and $\mathbf{y}$ is $r = S_{xy}/\sqrt{S_{xx}S_{yy}} = -20/\sqrt{10(50)} = -2/\sqrt{5} \doteq -0.894$.

***Example 11.2.3 (Multiple Regression)*** Let us now try to find a linear predictor of male student heights using fathers' and mothers' heights. All heights are measured in centimeters. Again let $y_i$ be the height in centimeters of the $i$th student for $i = 1, 2, \ldots, n = 129$. Let $x_{1i}$ and $x_{2i}$ be the heights of the father and mother of the $i$th student. We seek a column vector $\mathbf{b} = (b_0, b_1, b_2)^{\mathrm{T}}$ such that $\|\mathbf{y} - \hat{\mathbf{y}}\|^2 = \sum(y_i - \hat{y})^2$ is minimized, where $\hat{y}_i = b_0 + b_1x_{1i} + b_2x_{2i}$. Let $\mathbf{J}_n$ be the $n = 129$ component column vector of 1's, and let $\mathbf{x}_1$ and $\mathbf{x}_2$ be the $n = 129$ component column vectors of $x_{1i}$'s and $x_{2i}$'s. We seek the vector $\hat{\mathbf{y}}$ in the subspace spanned by $\mathbf{J}_n, \mathbf{x}_1, \mathbf{x}_2$ which is closest to $\mathbf{y}$. Let $\mathbf{X} = (\boldsymbol{J}_n, \mathbf{x}_1, \mathbf{x}_2)$ be the $n \times 3$ matrix with columns $\boldsymbol{J}_n, \mathbf{x}_1, \mathbf{x}_2$. Then $\mathbf{b} = \mathbf{M}^{-1}\mathbf{U}$, where $\mathbf{M} = \mathbf{X}^{\mathrm{T}}\mathbf{X}$, $\mathbf{U} = \mathbf{X}^{\mathrm{T}}\mathbf{y}$.

The first four rows of $\mathbf{X}$, the first four components of $\mathbf{y}$, and $\mathbf{M}, \mathbf{U}, \mathbf{b}$ are:

| $x_0$ | $x_1$ | $x_2$ | $y$ | $M$ | | | $U$ | $b$ |
|---|---|---|---|---|---|---|---|---|
| 1 | 170.0 | 162.0 | 158.0 | 129.0 | 22,307.0 | 21,045.1 | 22,779.94 | 43.049 |
| 1 | 175.0 | 166.0 | 181.0 | 22,307.0 | 3,867,996.2 | 3,642,657.6 | 3,944,995.82 | 0.431 |
| 1 | 183.0 | 172.0 | 183.0 | 21,045.1 | 3,642,657.6 | 3,440,690.1 | 3,720,496.58 | 0.362 |
| 1 | 177.8 | 157.5 | 188.0 | | | | | |

The first four components of $\hat{\mathbf{y}}$ are 174.9, 178.5, 175.6, 184.1, so that the first student was considerably shorter than predicted, while the others were taller than predicted: Error SSqs $= \|\mathbf{y} - \hat{\mathbf{y}}\|^2 = 5711.0$, Total SSqs $= \|\mathbf{y} - \overline{y}\mathbf{J}_n\|^2 = 9726.7$, Regression SSqs $= \|\hat{\mathbf{y}} - \overline{y}\mathbf{J}_n\|^2 = 4015.7$. The *coefficient of determination is* $R^2 =$ Regression SSqs/Total SSqs $= 0.4128$. $R$ itself is 0.643. $R$ is called the *multiple correlation coefficient.* $R^2$ is often called the *proportion of variation* in $y$ *explained by linear regression on* $x_1$ *and* $x_2$. □

**Definition 11.2.5** Let $V$ be a subspace of $R_n$ which includes the vector $\mathbf{J}_n$ of all 1's and is spanned by $\mathbf{x}_1, \ldots, \mathbf{x}_k$. The multiple correlation coefficient of $\mathbf{y}$ with respect to $\mathbf{x}_1, \ldots, \mathbf{x}_k$ is $R = \|\hat{\mathbf{y}} - \overline{\mathbf{y}}\mathbf{J}_n\|/\|\mathbf{y} - \overline{\mathbf{y}}\mathbf{J}_n\| = [(\Sigma(\hat{y}_i - \overline{y})^2/\Sigma(y_i - \overline{y})^2]^{1/2}$. □

***Example 11.2.4 (Analysis for Female Students)*** The least squares fit of the height data for the 78 female students gave $y = 82.11 + 0.201x_1 + 0.283x_2$. The coefficients for $\mathbf{x}_1$ and $\mathbf{x}_2$ seem to be quite different than those for the male students. It might seem to those of us who know little biology that the coefficients $b_1$ and $b_2$ for $x_1$ and $x_2$ for males and females should be approximately equal, differing only in $b_0$. We look at this again more formally when we consider probabilistic properties of these sample coefficients. Other interesting statistics for the female students: $\overline{y} = 162.37$, total SSqs $= 3916.7$, residual SSqs $= 3432.3$, regression SSqs $= 484.4$, $R^2 = 484.4/3916.7 = 0.124$, $R = 0.352$. A smaller proportion of the variation in heights among female students than among male students is explained by the heights of their parents. □

## Problems for Section 11.2

**11.2.1** Let $\mathbf{x} = (1, 2, 4, 5)$, $\mathbf{y} = (2, 7, 13, 14)$. Find:

**(a)** The projection $\hat{\mathbf{y}} = p(\hat{\mathbf{y}} \mid \mathbf{x}) = b\mathbf{x}$.

**(b)** Sketch a graph of the points $(x_i, y_i)$ and the straight line $y = bx$.

**(c)** Show that $\|\hat{\mathbf{y}}\|^2 = (\mathbf{x}, \mathbf{y})^2/\|\mathbf{x}\|^2$.

**(d)** Show that $\mathbf{x}$ and $\mathbf{e} = \mathbf{y} - \hat{\mathbf{y}}$ are orthogonal.

**(e)** Show that the Pythagorean theorem holds: $\|\mathbf{y}\|^2 = \|\hat{\mathbf{y}}\|^2 + \|\mathbf{e}\|^2$.

**(f)** Use the formula $\mathbf{b} = (\mathbf{X}^T\mathbf{X})^{-1}\mathbf{X}^T\mathbf{y}$ (with $\mathbf{y}$ written as a column vector) to find $\mathbf{b}$.

**11.2.2** Use calculus to show that $Q(b) = \sum_{i=1}^n (y_i - bx_i)^2$ is minimized by $b = (\mathbf{x}, \mathbf{y})/\|\mathbf{x}\|^2$.

**11.2.3** Show that the inner product matrix $\mathbf{M} = \mathbf{X}^T\mathbf{X}$ is nonsingular if and only if the columns of $\mathbf{X}$ are linearly independent.

**11.2.4** Let $\mathbf{y} = (6, 5, 7, 3, 4)^T$, $\mathbf{x}_1 = (1, 1, 1, 0, 0)^T$, and $\mathbf{x}_2 = (0, 0, 1, 1, 1)^T$ (so that these are column vectors).

**(a)** Find $b_1, b_2$ so that $\hat{\mathbf{y}} = b_1\mathbf{x}_1 + b_2\mathbf{x}_2$ is the least squares approximation to $\mathbf{y}$.

**(b)** Show that $\mathbf{e} = \mathbf{y} - \hat{\mathbf{y}}$ is orthogonal to $\mathbf{x}_1$ and $\mathbf{x}_2$, and therefore to all vectors in the subspace $V$ spanned by $\mathbf{x}_1$ and $\mathbf{x}_2$.

**(c)** Show that $\|\mathbf{y}\|^2 = \|\hat{\mathbf{y}}\|^2 + \|\mathbf{e}\|^2$.

**(d)** Show that $\hat{\mathbf{y}} \neq p(\mathbf{y}\,|\,\mathbf{x}_1) + p(\mathbf{y}\,|\,\mathbf{x}_2)$.

**11.2.5** Let $\mathbf{y}$ be a column vector with $n = n_1 + n_2$ components, where $n_1$ and $n_2$ are positive integers. Let $\mathbf{x}_1$ be the indicator of the first $n_1$ components. Let $\mathbf{x}_2$ be the indicator of the last $n_2$ components.

**(a)** Find simple formulas for the projection of $\mathbf{y}$ onto the subspace $V$ spanned by $\mathbf{x}_1$ and $\mathbf{x}_2$.

**(b)** Find $\hat{\mathbf{y}} = p(\mathbf{y}\,|\,V)$ for $n_1 = 3$, $n_2 = 2$, $\mathbf{y} = (2, 7, 3, 6, 12)^T$.

**11.2.6** Consider (weight, price) pairs for 105 brands of 1978 automobiles. The 105 $(x_i, y_i)$ pairs are plotted in Figure P. 11.2.6, together with the least squares line. The summary statistics are $n = 105, \overline{x} = 2949.48, \overline{y} = 15,805.2$, $S_{xx} = 32112196$, $S_{yy} = 7118267246$, $S_{xy} = 312050392$.

**(a)** Find the equation of the least squares line.

**(b)** Find total SSqs, regression SSqs, residual SSqs, $R^2$, and $R$.

**(c)** The curved line was fit as follows. The least squares straight line was fit to the $(w_i = \log(x_i), z_i = \log(y_i))$ pairs (see Figure 11.2.6). The equation of the curved line obtained was then obtained by converting from the logscale to the x–y scale. What is the equation of the curved line?

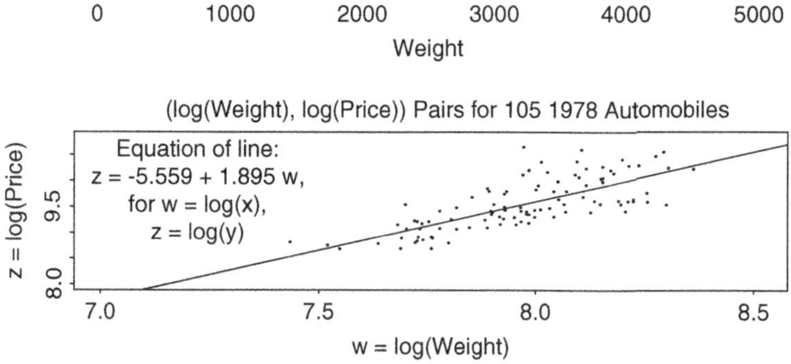


**FIGURE 11.2.6** Fitting a curve by using log transformations.

**11.2.7** Let $\mathbf{J}_4 = (1, 1, 1, 1)^{\mathrm{T}}$, $\mathbf{x}_1 = (1, 1, 0, 0)^{\mathrm{T}}$, $\mathbf{x}_2 = (1, 0, 1, 0)^{\mathrm{T}}$, $\mathbf{y} = (7, 3, 2, 2)^{\mathrm{T}}$. Find the $4 \times 4$ projection matrix $\mathbf{P}$ and the vectors $\hat{\mathbf{y}}$ and $\mathbf{e} = \mathbf{y} - \hat{\mathbf{y}}$ for the subspaces spanned by the following vectors. In each case write the vector or matrix as a multiple of $1/4$.
Verify that $\mathbf{e}$ is orthogonal to each of the vectors given.

**(a)** $\mathbf{J}_4$ only.

**(b)** $\mathbf{x}_1$ only.

**(c)** $\mathbf{J}_4$ and $\mathbf{x}_1$.

**(d)** $\mathbf{J}$, $\mathbf{x}_1$, and $\mathbf{x}_2$.

**(e)** Find the multiple correlation coefficient of $\mathbf{y}$ with respect to the vectors in part (d).

**11.2.8** A balance scale has two pans. If a weight $W_1$ is put in scale 1 and a weight $W_2$ is put in scale 2, the measurement shown on the scale is $Y = W_1 - W_2 + \varepsilon$, where $\varepsilon$ is a random error in the measurement. Suppose that four independent measurements $Y_1, Y_2, Y_3, Y_4$ were taken as follows: (1) Both weights were put in pan 1, (2) $W_1$ in pan 1, (3) $W_2$ in pan 2, (4) Weight $W_1$ in pan 1, $W_2$ in pan 2. For example, $Y_4 = W_1 - W_2 + \varepsilon_4$.

**(a)** Express the least squares estimates $\hat{W}_1$ of $W_1$ and $\hat{W}_2$ of $W_2$ as linear functions $(\mathbf{a}_1, \mathbf{Y})$ and $(\mathbf{a}_2, \mathbf{y})$, where $\mathbf{a}_1$ and $\mathbf{a}_2$ are four-component vectors of constants and $\mathbf{Y} = (Y_1, Y_2, Y_3, Y_4)^{\mathrm{T}}$.

**(b)** Find the values of the estimates for $\mathbf{Y} = \mathbf{y} = (8, 4, -1, 3)^{\mathrm{T}}$.

## 11.3 LINEAR MODELS

To determine whether the relationship between a variable $y$ and one or more predictors $x_1, \ldots, x_k$ is "real" and how much is due to chance, we need to impose a probabilistic model. Let's begin with a very simple model. Suppose that we observe $(x_i, Y_i)$ pairs for $i = 1, \ldots, n$, where the $x_i$ are positive, $Y_i = \beta x_i + \varepsilon_i$, where $\beta$ is an unknown parameter and the $\varepsilon_i$ are random variables, with $E(\varepsilon_i) = 0$ and $\mathrm{Var}(\varepsilon_i) = \sigma^2$, also an unknown parameter. Also suppose that the $\varepsilon_i$ have zero covariances. If we let $\mathbf{Y}$, $\mathbf{x}$, and $\boldsymbol{\varepsilon}$ be the $n$-component vectors of $Y_i$, $x_i$, and $\boldsymbol{\varepsilon}_i$ values, we can express the model as $\mathbf{Y} = \beta\mathbf{x} + \boldsymbol{\varepsilon}$, and the least squares estimator of $\beta$ as $\hat{\beta} = (\mathbf{x}, \mathbf{Y})/\|\mathbf{x}\|^2 = \beta(\mathbf{x}, \mathbf{x})/\|\mathbf{x}\|^2 + (\boldsymbol{\varepsilon}, \mathbf{x})/\|\mathbf{x}\|^2 = \beta + (\boldsymbol{\varepsilon}, \mathbf{x})/\|\mathbf{x}\|^2$.

Thus, $\hat{\boldsymbol{\beta}}$ is $\beta$ plus a random variable that is a linear combination of the $\boldsymbol{\varepsilon}_i$, which have expectations zero. It follows that under this model $E(\hat{\beta}) = \beta$, so that $\hat{\beta}$ is an unbiased estimator of $\beta$. Also, $\mathrm{Var}(\hat{\beta}) = \mathrm{Var}((\boldsymbol{\varepsilon}, \mathbf{x})/\|\mathbf{x}\|^2) = \mathrm{Var}((\boldsymbol{\varepsilon}, \mathbf{x}))/\|\mathbf{x}\|^4 = \sigma^2\|\mathbf{x}\|^2/\|\mathbf{x}\|^4 = \sigma^2/\|\mathbf{x}\|^2$.

### Expectations and Covariance Matrices for Random Vectors

To determine the properties of $\hat{\boldsymbol{\beta}}$ and of other statistics for models for the case of more than one predictor, we need to review the calculus of expectations and of covariance matrices as discussed at the end of Section 1.7. For ready reference, some of those properties are summarized here, and others are presented. Let $\mathbf{X}$ and $\mathbf{Y}$ be random vectors, written as column vectors with $m$ and $n$ components. Let $\mu = E(\mathbf{X})$ and $\nu = E(\mathbf{Y})$. Let $\mathbf{A}$ and $\mathbf{B}$ be $r \times m$ and $s \times n$ matrices of constants. Then:

1. $E(\mathbf{A}\mu) = \mathbf{A}\mu$ and $E(\mathbf{BY}) = \mathbf{B}\nu$.
2. $\text{Cov}(\mathbf{X}, \mathbf{Y})$ is the $m \times n$ matrix $(\sigma_{ij} = \text{cov}(X_i, Y_j))$. $\text{Cov}(\mathbf{X}, \mathbf{Y}) = E[(\mathbf{X} - \mu)(\mathbf{Y} - \nu)^{\text{T}}]$.
   We define $\text{Cov}(\mathbf{X}) = \text{Cov}(\mathbf{X}, \mathbf{X})$.
3. $\text{Cov}(\mathbf{AX}, \mathbf{BY}) = E[\mathbf{A}(\mathbf{X} - \mu)(\mathbf{B}(\mathbf{Y} - \nu))^{\text{T}}] = \mathbf{A}\ \text{Cov}(\mathbf{X}, \mathbf{Y})\mathbf{B}^T$. Therefore, $\text{Cov}(\mathbf{AX}) = \mathbf{A}\ \text{Cov}(\mathbf{X})\mathbf{A}^{\text{T}}$.
4. If $\mathbf{X}_1, \ldots, \mathbf{X}_k$ are $m$-component random vectors with $E(\mathbf{X}_i) = \mu_i$ and $\mathbf{Y}_1, \ldots, \mathbf{Y}_t$ are $n$-component random vectors with $E(\mathbf{Y}_j) = \boldsymbol{\nu}_j$, $\mathbf{S} = \sum_{i=1}^k \mathbf{X_i}$, $\mathbf{T} = \sum_{i=1}^t \mathbf{Y_i}$, then $E(\mathbf{S}) = \sum_{i=1}^k \mu_{\mathbf{i}}$, $E(\mathbf{T}) = \sum_{j=1}^t \nu_j$, $\text{Cov}(\mathbf{S}, \mathbf{T}) = \sum_{i=1}^k \sum_{j=1}^t \text{Cov}(\mathbf{X}_i, \mathbf{Y}_j)$, and $\text{Cov}(\mathbf{S}) = \sum_{i=1}^k \sum_{j=1}^t \text{Cov}(\mathbf{X}_i, \mathbf{X}_j)$. If $\text{Cov}(\mathbf{X}_i, \mathbf{X}_j) = 0$ for $i \neq j$, this reduces to $\sum_{i=1}^k \text{Cov}(\mathbf{X}_i)$, a familiar formula for the case that $m = 1$, so that $\text{Cov}(\mathbf{X}_i) = \text{Var}(\mathbf{X}_i)$.
5. Let $Q(\mathbf{X}, \mathbf{Y}) = \mathbf{X}^{\text{T}}\mathbf{AY} = \sum a_{ij} X_i Y_j$, where $\mathbf{A}$ is an $m \times n$ matrix of constants. $Q$ is a bilinear form, since it is linear in the $X_i$ and the $Y_j$. If $\mathbf{Y} = \mathbf{X}$, then $Q(\mathbf{X}) \equiv Q(\mathbf{X}, \mathbf{X})$ is a quadratic form in $\mathbf{X}$. Since $E(X_i Y_j) = \sigma_{ij} + \mu_i \nu_j$, where $\sigma_{ij} = \text{Cov}(X_i, Y_j)$, $E(Q(\mathbf{X}, \mathbf{Y})) = \sum a_{ij}\sigma_{ij} + \sum a_{ij}\mu_i\nu_j = \sum a_{ij}\sigma_{ij} + Q(\mu, \nu)$. The first term is an inner product of the matrices $\text{Cov}(\mathbf{X}, \mathbf{Y})$ and $\mathbf{A}$. The second term is $Q$ evaluated for the mean vectors. In the case that $\mathbf{Y} = \mathbf{X}$, with $Q(\mathbf{X}) = Q(\mathbf{X}, \mathbf{X})$, $E(Q(\mathbf{X})) = \sum a_{ij}\sigma_{ij} + Q(\mu) = \text{trace}\ (\mathbf{A}\ \text{Cov}(\mathbf{X})) + Q(\mu)$. The trace of a square matrix is the sum of the diagonal elements.

***Example 11.3.1*** Let $\mathbf{X} = (X_1, \ldots, X_n)$ be a vector of uncorrelated random variables. Suppose that $E(X_i) = \mu$, the same for all $i$, so that $E(\mathbf{X}) = \mu\mathbf{J}_n$. Let $\sigma_i^2 = \text{Var}(X_i)$ for each $i$. Then $\text{Cov}(\mathbf{X}) = \text{diag}(\sigma_1^2, \ldots, \sigma_n^2)$. Let $Q(\mathbf{X}) = \sum_{i=1}^n (X_i - \overline{X}) = \Sigma_{i=1}^n X_i^2 - n\overline{X}^2 = \mathbf{X}^{\text{T}}\mathbf{AX}$, where $\mathbf{A} = \mathbf{I}_n - (1/n)\mathbf{J}_n\mathbf{J}_n^{\text{T}}$ and $\mathbf{J}_n$ is the $n \times 1$ matrix of all 1's. $\mathbf{A}$ is the projection matrix onto the subspace of all vectors orthogonal to $\mathbf{J}_n$. $Q(\mathbf{X}) = \|\mathbf{AX}\|^2$, where $\mathbf{AX}$ is the vector of deviations $X_i^* = X_i - \overline{X}$. Thus, from property 5, $E(Q(\mathbf{X})) = \sum a_{ij}\sigma_{ij} + Q(\mu\mathbf{J}_n) = [(n-1)/n]\sum \sigma_i^2 - 0$. Let $S^2 = Q(\mathbf{X})/(n-1)$, usually called the *sample variance*. Then $E(S^2) = (\sum \sigma_i^2)/n$. Of course, if $\sigma_i^2 = \sigma^2$ for all $i$, then $E(S^2) = \sigma^2$. This explains why the denominator $n - 1$ is used in the definition of $S^2$. $\text{Var}(\overline{X}) = \text{Var}((1/n)\mathbf{J}_n^{\text{T}}\mathbf{X}) = (1/n^2)\mathbf{J}_n^{\text{T}}\ \text{Cov}(\mathbf{X})\mathbf{J}_n = \Sigma\sigma_i^2/n^2$, so that $S^2/n$ is an unbiased estimator of $\text{Var}(\bar{X})$, whether or not the $\sigma_i^2$ are equal. □

### Multiple Regression Models

As for the student height example, the following model relating an $n$-component vector $\mathbf{Y}$ to $n$-component vectors $\mathbf{x}_1, \ldots, \mathbf{x}_k$ of constants is often reasonable:

$$\mathbf{Y} = \beta_1\mathbf{x}_1 + \cdots + \beta_k\mathbf{x}_k + \boldsymbol{\varepsilon} = \boldsymbol{\theta} + \varepsilon.$$

This is called a *linear model* because it is linear in the β's. Often, $\mathbf{x}_1$ will be the vector of all 1's. In fact, most software multiple regression procedures and functions automatically include the vector of all 1's unless the programmer demands otherwise.

We begin by supposing that $E(\boldsymbol{\varepsilon}) = \mathbf{0}$ and that $\text{Cov}(\boldsymbol{\varepsilon}) = \sigma^2\mathbf{I}_n$. That is, we suppose that the components $\boldsymbol{\varepsilon}_i$ of $\boldsymbol{\varepsilon}$ are uncorrelated with expected values all zero, with equal variances $\sigma^2$. Let $\mathbf{X} = (\mathbf{x}_1, \ldots, \mathbf{x}_k)$, so that $\mathbf{X}$ is an $n \times k$ matrix of constants. We suppose that the $\mathbf{x}_j$ are linearly independent, so that $\mathbf{X}$ has full column rank $k$. Thus, $\mathbf{Y} = \mathbf{X}\boldsymbol{\beta} + \boldsymbol{\varepsilon}$, and the least squares estimator of $\boldsymbol{\beta}$ is $\hat{\boldsymbol{\beta}} = (\mathbf{X}^{\mathrm{T}}\mathbf{X})^{-1}\mathbf{X}^{\mathrm{T}}\mathbf{Y} = (\mathbf{X}^{\mathrm{T}}\mathbf{X})^{-1}\mathbf{X}^{\mathrm{T}}(\mathbf{X}\boldsymbol{\beta} + \boldsymbol{\varepsilon}) = \boldsymbol{\beta} + A\boldsymbol{\varepsilon}$, where $\mathbf{A} = (\mathbf{X}'\mathbf{X})^{-1}\mathbf{X}^{\mathrm{T}}$. $\mathbf{A}$ is called the *pseudo inverse* of $\mathbf{X}$ because $\mathbf{AX} = \mathbf{I}_k$ and $\mathbf{XA}$ is the projection matrix onto the column space of $\mathbf{X}$. $\mathbf{A}$ is not the inverse unless $k = n$. If $k < n$, $\mathbf{X}$ is singular. $\mathbf{A}$ is often called the *generalized inverse* or *Penrose inverse* of $X$.

Since $\hat{\boldsymbol{\beta}} = \boldsymbol{\beta} + \mathbf{A}\boldsymbol{\varepsilon}$ and $\mathbf{E}(\boldsymbol{\varepsilon}) = \mathbf{0}$, it follows that $E(\hat{\boldsymbol{\beta}}) = \boldsymbol{\beta}$, and $\text{Cov}(\hat{\boldsymbol{\beta}}) = \text{Cov}(\mathbf{A}\boldsymbol{\varepsilon}) = \mathbf{A}\mathbf{A}^{\mathrm{T}}\sigma^2 = (\mathbf{X}^{\mathrm{T}}\mathbf{X})^{-1}\sigma^2$.

If the vectors $\mathbf{x}_1, \ldots, \mathbf{x}_k$ are mutually orthogonal, it follows that the components $\hat{\beta}_j$ of $\hat{\boldsymbol{\beta}}$ are uncorrelated. From the properties of the multivariate normal distribution, as discussed in Section 9.2, it follows that when $\boldsymbol{\varepsilon}$ has the multivariate normal distribution, so does $\hat{\boldsymbol{\beta}}$, and that when the $\mathbf{x}_j$ are orthogonal, the $\hat{\beta}_j$ are independent random variables. Let $\mathbf{P}_V = \mathbf{X}(\mathbf{X}^{\mathrm{T}}\mathbf{X})^{-1}\mathbf{X}^{\mathrm{T}}$ be a projection onto the subspace spanned by the columns of $\mathbf{X}$. The matrix $\mathbf{I}_n - \mathbf{P}_V$ is the projection matrix onto $V^{\perp} = \{\mathbf{x} \mid \mathbf{x} \perp V\}$. Thus, $\hat{\mathbf{Y}} = \mathbf{P}_V\mathbf{Y}$ and the vector of residuals is $\mathbf{e} = \mathbf{Y} - \hat{\mathbf{Y}} = (\mathbf{I} - \mathbf{P}_V)\mathbf{Y}$. $\text{Cov}(\hat{\mathbf{Y}}, \mathbf{e}) = \text{Cov}(\mathbf{P}_V\mathbf{Y}, (\mathbf{I} - \mathbf{P}_V)\mathbf{Y}) = \mathbf{P}_V\ \text{Cov}(\mathbf{Y})(\mathbf{I} - \mathbf{P}_V)^{\mathrm{T}} = \mathbf{P}_V(\sigma^2 I_n))(\mathbf{I} - \mathbf{P}_V)^{\mathrm{T}} = \mathbf{0}$. Thus, the components of $\hat{\mathbf{Y}}$ and those of $\mathbf{e}$ are uncorrelated. It follows that when $\boldsymbol{\varepsilon}$, and therefore $\mathbf{Y}$, has the multvariate normal distribution, $\hat{\mathbf{Y}}$ and $\mathbf{e}$ are independent random vectors. Since $\hat{\boldsymbol{\beta}} = \mathbf{X}^{\mathrm{T}}\hat{\mathbf{Y}}$, this implies that $\text{Cov}(\hat{\boldsymbol{\beta}}, \mathbf{e}) = 0$, and under the MV normality of $\boldsymbol{\varepsilon}$, $\hat{\boldsymbol{\beta}}$ and $\mathbf{e}$ are independent. The error sum of squares is $\|\mathbf{e}\|^2 = \sum e_i^2 = \mathbf{Y}^{\mathrm{T}}(\mathbf{I}_n - \mathbf{P}_V)\mathbf{Y} = \mathbf{Q}(\mathbf{Y})$, a quadratic form in $\mathbf{Y}$. By property 5, $E(Q(\mathbf{Y})) = \text{trace}((\mathbf{I}_n - \mathbf{P}_V)(\sigma^2\mathbf{I}_n) + Q(\mathbf{X}\beta) = \sigma^2(n-k) + 0$. This follows from the fact that, in general, $\text{trace}(\mathbf{AB}) = \text{trace}(\mathbf{BA})$ if both $\mathbf{AB}$ and $\mathbf{BA}$ are defined, and therefore $\text{trace}(\mathbf{P}_V) = \text{trace}(\mathbf{X}(\mathbf{X}^{\mathrm{T}}\mathbf{X})^{-1}\mathbf{X}^{\mathrm{T}}) = \text{trace}(\mathbf{X}^{\mathrm{T}}\mathbf{X})^{-1}\mathbf{X}^{\mathrm{T}}\mathbf{X}) = k$. Thus, if we let $S^2 = \|\mathbf{e}\|^2/(n-k) = (\text{error sum of squares})/(n-k)$, then $E(S^2) = \sigma^2$. This is a generalization of the case that $\mathbf{X} = \mathbf{J}_n$, so that $S^2$ is the sample variance and $k = 1$. We will use this definition of $S^2$ throughout our discussion of multiple linear models. In the case that the components of $\boldsymbol{\varepsilon}$ are normally distributed, so that $\boldsymbol{\varepsilon} \sim \text{N}_n(\mathbf{0}, \sigma^2\mathbf{I}_n)$, it follows that $S^2$ and $\hat{\boldsymbol{\beta}}$ are independent.

### The formula $\hat{\beta}_k = (\mathbf{Y}, \mathbf{X}_k^{\perp})/\|\mathbf{X}_k^{\perp}\|^2$

Let $V = \mathcal{L}(\mathbf{x}_1, \ldots, \mathbf{x}_k)$ and $V_{k-1} = \mathcal{L}(\mathbf{x}_1, \ldots, \mathbf{x}_{k-1})$ have dimensions $k$ and $k-1$. Define $\hat{\mathbf{x}}_k = p(\mathbf{x}_k \mid V_{k-1})$, and $\mathbf{x}_k^{\perp} = \mathbf{x}_k - \hat{\mathbf{x}}_k$ so that $\mathbf{x}_k^{\perp}$ is orthogonal to $V_{k-1}$ and therefore to each of $\mathbf{x}_1, \ldots, \mathbf{x}_{k-1}$. Let $\hat{\mathbf{Y}} = \sum_{j=1}^k \hat{\beta}_j \mathbf{x}_j$. Then $(\hat{\mathbf{Y}}, \mathbf{x}_k^\perp) = \sum_{j=1}^k \hat{\beta}_j (x_j, x_k) = \hat{\beta}_k (\mathbf{x}_k, \mathbf{x}_k^\perp) = \hat{\beta}_k \|\mathbf{x}_k^\perp\|^2$. Therefore, $\hat{\beta}_k = (\mathbf{Y}, \mathbf{x}_k^\perp)/\|\mathbf{x}_k^\perp\|^2$. By replacing $\mathbf{Y}$ by $\boldsymbol{\theta} = \sum_{j=1}^k \beta_j x_j$, we get $\beta_j = (\boldsymbol{\theta}, \mathbf{x}_k^\perp)/\|\mathbf{x}_k^\perp\|^2$. If, in addition, $\mathbf{Y} = \boldsymbol{\theta} + \boldsymbol{\varepsilon}$, with $\text{Cov}(\boldsymbol{\varepsilon}) = \sigma^2 \mathbf{I}_n$, then $\text{Var}(\hat{\beta}_k) = \sigma^2/\|\mathbf{x}_k^\perp\|^2$.

By defining $V_j$ = subspace spanned by all $x_i$ except $\mathbf{x}_j$), $\mathbf{x}_j^\perp = \mathbf{x}_j - p(\mathbf{x}_j \mid V_j)$, we get $\hat{\beta}_j = (\mathbf{Y}, \mathbf{x}_j^\perp)\|\mathbf{x}_j^\perp\|^2$, $\beta_j = (\boldsymbol{\theta}, \mathbf{x}_j^\perp)/\|\mathbf{x}_j^\perp\|^2$ and $\text{Var}(\hat{\beta}_j) = \sigma^2/\|\mathbf{x}_j^\perp\|^2$. Thus, $\hat{\beta}_j$ and $\beta_j$ depend on the relationship of $\mathbf{Y}$ and $\boldsymbol{\theta}$ to $\mathbf{x}_j^\perp$, the part of $\mathbf{x}_j$ that measures a quantity which is "new" in a linear sense relative to the other $\mathbf{x}_i$. It also follows that $\text{Cov}(\hat{\beta}_i, \hat{\beta}_j) = \sigma^2 C_{ij}$, where $C_{ij} = (\mathbf{x}_i^\perp, \mathbf{x}_j^\perp)/[\|\mathbf{x}_i^\perp\|^2 \|\mathbf{x}_j^\perp\|^2]$. Thus, $C_{ij}$ is the *(ij)*th term of $(X^T X)^{-1}$, a surprising formula to the author when he first learned of it. Notice also that $\rho(\hat{\beta}_i, \hat{\beta}_j) = (\mathbf{x}_i^\perp, \mathbf{x}_j^\perp)/[\|\mathbf{x}_i^\perp\| \|\mathbf{x}_j^\perp\|]$.

***Example 11.3.2*** Let $\mathbf{x}_1 = (1, 1, 1, 1, 1)^T$, $\mathbf{x}_2 = (1, 1, 1, 0, 0)^T$, $\mathbf{x}_3 = (1, 0, 1, 0, 1)^T$, $V = \mathcal{L}(\mathbf{x}_1, \mathbf{x}_2, \mathbf{x}_3)$, $V_2 = \mathcal{L}(\mathbf{x}_1, \mathbf{x}_2)$. $V$ is spanned by $\mathbf{x}_2$ and $\mathbf{w} = \mathbf{x}_1 - \mathbf{x}_2 = (0, 0, 0, 1, 1)^T$ and $\mathbf{w} \perp \mathbf{x}_2$. Thus, $p(\mathbf{x}_3 \mid V_2) = (2/3)\mathbf{x}_2 + (1/2)\mathbf{w} = (1/6)(4, 4, 4, 3, 3)^T$ and $\mathbf{x}_3^\perp = \mathbf{x}_3 - p(\mathbf{x}_3 \mid V_2) = (1/6)(2, -4, 2, -3, 3)^T$. Thus, $\hat{\beta}_3 = (1/6)(2Y_1 - 4Y_2 + 2Y_3 - 3Y_4 + 3Y_5)/(7/6)$ and $\text{Var}(\hat{\beta}_3) = \sigma^2/\|\mathbf{x}_3^\perp\|^2 = \sigma^2/(7/6) = (6/7)\sigma^2$. □

**Distributions of $\mathbf{P}_V\mathbf{Y}$ and $\|\mathbf{P}_V\mathbf{Y}\|^2$ when $\mathbf{Y} \sim \mathbf{N}_n\theta, \sigma^2\mathbf{I}_n)$**

Suppose for this discussion that $\boldsymbol{\theta}$ is any $n$-component vector of constants and that $V$ is any subspace of $R_n$ of dimension $k$. Suppose that $\mathbf{Y} \sim N_n(\boldsymbol{\theta}, \sigma^2 \mathbf{I}_n)$. There exists an orthogonal basis of length-1 vectors $\mathbf{a}_1, \ldots, \mathbf{a}_k$ spanning $V$. Let $\mathbf{A}$ be the $n \times k$ matrix $(\mathbf{a}_1, \ldots, \mathbf{a}_k)$. Then $\mathbf{A}^T\mathbf{A} = \mathbf{I}_k$. Unless $k = n$, $\mathbf{A}$ is singular and does not have an inverse. $\mathbf{P}_V = \mathbf{A}(\mathbf{A}^T\mathbf{A})^{-1}\mathbf{A}^T = \mathbf{A}\mathbf{A}^T$. Thus, $\hat{\mathbf{Y}} \equiv P_V \mathbf{Y} = \mathbf{A}\mathbf{A}^T\mathbf{Y} = \sum_{j=1}^k \mathbf{a}_j b_j$, where $b_j = (\mathbf{a}_j, \mathbf{Y})$, the inner product of $\mathbf{a}_j$ and $\mathbf{Y}$. $E(\hat{\mathbf{Y}}) = \mathbf{P}_V\boldsymbol{\theta}$, and $\text{Cov}(\hat{\mathbf{Y}}) = \mathbf{P}_V(\sigma^2\mathbf{I}_n)\mathbf{P}_V^T = \sigma^2\mathbf{P}_V$. If $\mathbf{Y}$ has a multivariate normal distribution, so does $\hat{\mathbf{Y}}$. Thus, $\hat{\mathbf{Y}} \sim \text{N}_n(\mathbf{P}_V\boldsymbol{\theta}, \sigma^2\mathbf{P}_V)$.

$\|\hat{\mathbf{Y}}\|^2 = \sum_{j=1}^k b_j^2 \|\mathbf{a}_j\|^2 = \sum_{j=1}^k b_j^2$. We have made use of the orthogonality of the $\mathbf{a}_j$ and the Pythagorean theorem. The column vector $\mathbf{b} = (b_1, \ldots, b_k)^T = \mathbf{A}^T\mathbf{Y} \sim \text{N}_k(\mathbf{A}^T\boldsymbol{\theta}, \mathbf{A}\mathbf{A}\sigma^2 = \mathbf{I}_k\sigma^2)$. That is, the $b_j$ are uncorrelated and therefore independent, each with variance $\sigma^2$. It follows by Definition 9.3.1 that $\|\hat{\mathbf{Y}}\|^2/\sigma^2 \sim \chi_k^2(\delta)$, with $\delta = \|\mathbf{A}\mathbf{A}^T\boldsymbol{\theta}\|^2/\sigma^2 = \|\mathbf{P}_V\boldsymbol{\theta}\|^2/\sigma^2$. If $\theta \perp V$, $\delta = 0$, so that $\hat{\mathbf{Y}}\|^2/\sigma^2$ has the central chi-square distribution with $k$ degrees of freedom. Let $V_1$ and $V_2$ be orthogonal subspaces (every pair of vectors $\mathbf{v}_1 \in V_1$ and $\mathbf{v}_2 \in V_2$, is orthogonal) of dimensions $k_1$ and $k_2$. Let $\hat{\mathbf{Y}}_1$ and $\hat{\mathbf{Y}}_2$ be the projections of $\mathbf{Y}$ onto $V_1$ and $V_2$. Letting $\mathbf{A}_1$ and $\mathbf{A}_2$ be the corresponding $n \times k_1$ and $n \times k_2$ matrices defined for $V_1$ and $V_2$ as $\mathbf{A}$ was defined for $V$. Then $\text{Cov}(\hat{\mathbf{Y}}_1, \hat{\mathbf{Y}}_2) = \text{Cov}(\mathbf{A}_1\mathbf{A}_1^T\mathbf{Y}, \mathbf{A}_2\mathbf{A}_2^T\mathbf{Y}) = \mathbf{A}_1\ \mathbf{A}_1^T\ \mathbf{A}_2\ \mathbf{A}_2^T\ \sigma^2 = \mathbf{0}$, where $\mathbf{0}$ is the $n \times n$ matrix of all zeros. Thus, under the MV normal model, $\hat{\mathbf{Y}}_1$ and $\hat{\mathbf{Y}}_2$ are independent random vectors, and the squared lengths $\|\hat{\mathbf{Y}}_1\|^2$ and $\|\hat{\mathbf{Y}}_2\|^2$ are independent random variables. We exploit this when we consider $F$-tests on the vector $\boldsymbol{\beta}$.

***Example 11.3.3*** Let $Y_{11}, \ldots, Y_{1n_1}$ be a random sample from $N(\mu_1, \sigma^2)$ and let $Y_{21}, \ldots, Y_{2n_1}$ be a random sample from $N(\mu_2, \sigma^2)$. Suppose that the $Y_{1i}$ and $Y_{2j}$ are independent. Let $\mathbf{Y}$ be the $n = n_1 + n_2$ component vector of $Y_{ij}$, with the $Y_{1i}$ written first. Let $\mathbf{x}_1$ and $\mathbf{x}_2$ be the indicators of the first $n_1$ and last $n_2$ components. Then $E(\mathbf{Y}) \equiv \boldsymbol{\theta} = \mu_1\ \mathbf{x}_1 + \mu_2\ \mathbf{x}_2$, and by letting $\varepsilon = \mathbf{Y} - \boldsymbol{\theta}$, we have the linear model $\mathbf{Y} = \boldsymbol{\theta} + \varepsilon$, with $\varepsilon \sim N_n(\mathbf{0}, \sigma^2\ \mathbf{I}_n)$. The least squares estimators of $\mu_1$ and $\mu_2$ are $\overline{Y}_{1\cdot}$, the mean of the $Y_{1i}$, and $\overline{Y}_{2\cdot}$, the mean of the $Y_{2j}$. Then $\hat{\mathbf{Y}} = \overline{Y}_{1\cdot}\ \mathbf{x}_1 + \overline{Y}_{2\cdot}\ \mathbf{x}_2$ and Error SSqs $= \sum_{i=1}^{n_1}(Y_{1j} - \overline{Y}_{1\cdot})^2 + \sum_{j=1}^{n_2}(Y_{2j} - \overline{Y}_{2\cdot})^2$, which is independent of the pair $(\overline{Y}_{1\cdot}, \overline{Y}_{2\cdot})$. $S^2 =$ (Error SSqs)$/(n-2)$ is called the *pooled estimator of* $\sigma^2$. □

**Confidence Intervals on** $\eta = c_1\beta_1 + \cdots + c_k\beta_k$ Let $c_1, \ldots, c_k$ be constants, specified by the person who wishes to analyze the data. For example, for Example 11.3.1 we might be interested in $\eta = \mu_1 - \mu_2$, so that $c_1 = 1, c_2 = -1$. For simple linear regression we might be particularly interested in the value of the regression function $g(x) = \beta_0 + \beta_1 x$ for $x = x_0$, a specified point. In this case, changing the subscript notation slightly, $c_0 = 1, c_1 = x_0$, so that $\eta = c_0\beta_0 + c_1\ \beta_1 = g(x_0)$.

Letting $\mathbf{c} = (c_1, \ldots, c_k)^T, \eta = \mathbf{c}^T\boldsymbol{\beta}$. The least squares estimator of $\eta$ is then $\hat{\eta} = \mathbf{c}^T\hat{\boldsymbol{\beta}}$. Since this is a linear function of $\hat{\boldsymbol{\beta}}, \hat{\eta}$ is an unbiased estimator of $\eta$, with $\text{Var}(\hat{\eta}) = \mathbf{c}^T\ \text{Cov}\ (\hat{\boldsymbol{\beta}})\mathbf{c} = h(\mathbf{c})\sigma^2$, where $h(\mathbf{c}) = \mathbf{c}^T(\mathbf{X}^T\mathbf{X})^{-1}\mathbf{c}$. We can determine CIs on $\eta$ as follows. $Z = (\hat{\eta} - \eta)/\sqrt{\sigma^2 h(c)} \sim N(0, 1)$. $U \equiv \nu S^2/\sigma^2 \sim \chi^2_\nu$, where $\nu = n - k$. In addition, $\hat{\boldsymbol{\beta}}$ and $S^2$ are independent, so that $Z$ and $U$ are independent. It follows that $T = Z/\sqrt{U/\nu} = (\hat{\eta} - \eta)/\sqrt{S^2 h(c)} \sim t_\nu$, Student's $t$-distribution for $\nu$ degrees of freedom. From this it follows that $[\hat{\eta} \pm t_{(1+\gamma)/2}\sqrt{S^2 h(\mathbf{c})}]$ is a $100\gamma\%$ CI on $\eta$.

***Example 11.3.4*** Consider the simple linear regression model $Y_i = \beta_0 + \beta_1 x_i + \varepsilon_i$, where the $\varepsilon_i$ are independent, $N(0, \sigma^2)$ (see Figure 11.3.1). Letting $x_i^* = x_i - \bar{x}$ and $\gamma_0 = \beta_0 - \beta_1\bar{x}$, $Y_i = \gamma_0 + \beta_1 x_i^* + \varepsilon_i$. The least squares estimator of the pair $(\gamma_0, \beta_1)$ is $(\overline{Y}, \hat{\boldsymbol{\beta}}_1)$, where $\hat{\boldsymbol{\beta}}_1 = S_{xy}/S_{xx}$. Since the vectors $\mathbf{J}_n$ (all 1's) and $\mathbf{x}^*$ of $x_i^*$'s are orthogonal, $\overline{Y}$ and $\hat{\beta}_1$ are independent. Suppose that we wish to estimate the value of the regression function $g(x) = \beta_0 + \beta_1 x = \gamma_0 + \beta_1(x - \bar{x})$ at a point $x = x_0$, chosen by the data analyst. Then $g(x_0) = \gamma_0 + \beta_1(x_0 - \bar{x}) = c_0\gamma_0 + c_1\beta_1$ for $c_0 = 1, c_1 = x_0 - x$. The least squares estimator of $\eta = g(x_0)$ is $\hat{\eta} = \mathbf{c}^T\hat{\boldsymbol{\beta}} = \hat{g}(x_0) = \overline{Y} + \hat{\beta}_1(x_0 - x)$. $\text{Var}(\hat{\eta}) = \sigma^2\ \text{Var}(\overline{Y}) + (x_0 - \bar{x})^2\ \text{Var}(\hat{\beta}_1) = \sigma^2[1/n + (x_0 - \bar{x})^2/S_{xx}] = \sigma^2 h(\mathbf{c})$. Call $h(\mathbf{c})$ (the term in brackets) $h^*(x_0)$. Thus, $[\hat{g}(x_0) \pm t_{(1+\gamma)/2}\sqrt{S^2 h^*(x_0)}]$ is a $100\gamma\%$ CI on $g(x_0) = \beta_0 + \beta_1 x_0$.

Notice that the widths of the CIs, being a multiple of $\sqrt{h^*(x_0)}$, become larger as $(x_0 - x)^2$ increases (i.e., the distance of $x_0$ to the center of the data increases). These intervals are not simultaneous. That is, if $I(x_0)$ is the interval for $g(x_0)$, then $P(I(x_0) \in g(x_0)) = \gamma$. We are not saying that $P(I(x_0) \in g(x_0)$ for all $x_0) = \gamma$. The probability that the event $A(x_0) = [I(x_0) \in g(x_0)]$ is $\gamma$, but the probability of the intersection of all these $A(x_0)$ is smaller. □

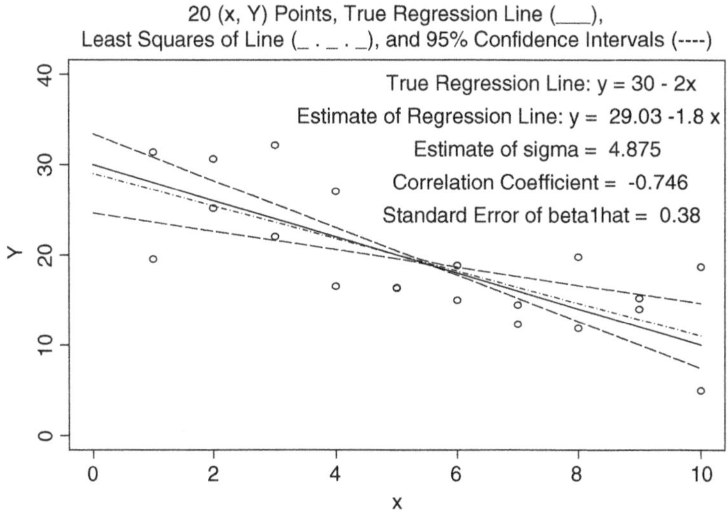


**FIGURE 11.3.1** True regression line, data points, least squares line, and 95% confidence intervals.

***Example 11.3.5*** Consider the height data of Example 11.2.2 for the 129 male students. Let $Y_i$ be the height for the ith student, and let $x_{1i}$ and $x_{2i}$ be the heights of his father and mother. Suppose that $Y_i = \beta_0 + \beta_1 x_{1i} + \beta_2 x_{2i} + \varepsilon_i$, where the $\varepsilon_i$ are independent, N(0, $\sigma^2$). S-Plus was used to fit the model, using the command "hgtanalysis = lm($y \sim x1 + x2$)$''$ (S-Plus doesn't like subscripts). The author chose the name "hgtanalysis." Here **y**, **x1**, and **x2** are vectors of length 129. The function "lm" adds the vector **J** of all 1's automatically unless told not to. The result "hgtanalysis" in S-Plus language is a "list," a collection of results of the least squares fit, including the ($\hat{\beta}$'s , the vector of fitted values $\hat{\mathbf{y}}$, and the residual vector $\mathbf{e} = \mathbf{y} - \hat{\mathbf{y}}$. The matrix **M** of inner products among the vectors $\mathbf{J}$, $\mathbf{x}_1$, $\mathbf{x}_2$ is given in Example 11.2.2. Its inverse is

$$\begin{pmatrix} 46,481 & -107.2 & -170.8 \\ -107.2 & 1.116 & 0.525 \\ -170.8 & 0.525 & 1.604 \end{pmatrix} \times 10^{-4}.$$

"hgtanalysis$coef" is the least squares estimate of beta: (43.05869, 0.4310, 0.3617)$^T$. The error SSqs is $||\mathbf{e}||^2 = 5711.029$, $S^2 = 5711.029/(129 - 3) = 45.320$. The estimate of Cov($\hat{\boldsymbol{\beta}}$) is $S^2\mathbf{M}^{-1}$. We might be interested in comparing $\beta_1$ and $\beta_2$. Let $\eta = \beta_1 - \beta_2 = \mathbf{c}^T\beta$, where $\mathbf{c}^T = (0, 1, -1)$. Then ($\hat{\eta} = \mathbf{c}^T\beta = \hat{\beta}_1 - \hat{\beta}_2 = 0.0593$. The estimate of Var($\hat{\eta}$) is $S^2(\hat{\eta}) = \mathbf{c}^T\mathbf{M}^{-1}\mathbf{c}S^2 = [1.116 + 1.604 + 2(0.525)](10^{-4})(45.320) = 0.01708$, so that the 95% CI of $\eta$ is $[0.0593 \pm (1.979)(0.1307)] = [0.0593 \pm 0.2586]$. Since this interval includes zero, we cannot conclude that $\beta_1$ and $\beta_2$ differ.

We might ask whether the coefficients of the father's height $\beta_{1m}$ for the males and $\beta_{1f}$ for the females differ. The estimate of $\text{Var}(\hat{\beta}_{1m})$ is $S^2(\hat{\beta}_{1m}) = S_m^2(1.116)(10^{-4}) = 0.00508$. Similar computation for the female data gives $S^2(\hat{\beta}_{1f}) = 0.00825$. Since the estimators for males and females are independent, we get $S^2(\hat{\beta}_{1m} - \hat{\beta}_{1f}) = 0.00506 + 0.00825 = 0.01331$, so that the 95% CI on $\beta_{1m} - \beta_{1f}$ is $[\hat{\beta}_{1m} - \hat{\beta}_{1f} \pm 1.96S(\hat{\beta}_{1m} - \hat{\beta}_{1f})] = [(0.4310 - 0.2012) \pm 0.2261] = [0.2298 \pm 0.2261]$. Since the interval does not include zero, a two-sided test of $\beta_{1m} = \beta_{1f}$ rejects at level 0.05, with a p-value just a bit less than 0.05. Later we test the null hypothesis that both $\beta_{1m} - \beta_{1f} = 0$ and $\beta_{2m} - \beta_{2f} = 0$. We need more theory. □

### The Gauss–Markov Theorem

Suppose that $\mathbf{Y} = \mathbf{X}\boldsymbol{\beta} + \boldsymbol{\varepsilon}$, with $E(\boldsymbol{\varepsilon}) = \mathbf{0}$, $\text{Cov}(\boldsymbol{\varepsilon}) = \sigma^2\mathbf{I}_n$. The least squares estimator of $\eta = c_1\beta_1 + \cdots + c_k\beta_k = \mathbf{c}^T\boldsymbol{\beta}$ is $\hat{\eta} = \mathbf{c}^T\hat{\boldsymbol{\beta}} = \mathbf{c}^T\mathbf{M}^{-1}\mathbf{X}^T\ \mathbf{Y} = \mathbf{a}^T\mathbf{Y}$, where $\mathbf{a} = \mathbf{X}\mathbf{M}^{-1}\mathbf{c}$. $\hat{\eta}$ is unbiased, linear in $\mathbf{Y}$, with variance $||\mathbf{a}||^2\sigma^2 = \mathbf{c}^T\ \mathbf{M}^{-1}\mathbf{c}\sigma^2$. Are there other linear unbiased esimators with smaller variance? For example, if $\mathbf{d}$ is a vector orthogonal to the subspace $V$ spanned by the columns of $\mathbf{X}$, then $E((\mathbf{a}+\mathbf{d})^T\mathbf{Y}) = (\mathbf{a}+\mathbf{d})^T\mathbf{X}\boldsymbol{\beta} = \mathbf{a}^T\mathbf{X}\boldsymbol{\beta} = \eta$, so that $\hat{\eta} \equiv (\mathbf{a}+\mathbf{d})^T\mathbf{Y}$ is another linear unbiased estimator of $\eta$. Does it have smaller variance than $\hat{\eta}$? The answer is no, as shown by the famous Gauss–Markov theorem.

**The Gauss–Markov Theorem** Suppose that $\mathbf{Y} = \mathbf{X}\boldsymbol{\beta} + \boldsymbol{\varepsilon}$, with $E(\boldsymbol{\varepsilon}) = \mathbf{0}$, $\text{Cov}(\boldsymbol{\varepsilon}) = \sigma^2\mathbf{I}_n$, where $\mathbf{X}$ has rank $k$. Let $\eta = \mathbf{c}^T\boldsymbol{\beta}$. Let $\hat{\eta} = \mathbf{c}^T\ \hat{\boldsymbol{\beta}} = \mathbf{c}^T\mathbf{M}^{-1}\mathbf{X}^T\ \mathbf{Y} = \mathbf{a}^T\mathbf{Y}$ be the least squares estimator of $\eta$ , where $\mathbf{a} = \mathbf{X}\mathbf{M}^{-1}\mathbf{c}$. Let $\hat{\eta}^* = \mathbf{h}^T\ \mathbf{Y}$ be any linear unbiased estimator of $\eta$. Then $\text{Var}(\hat{\eta}^*) \geq \text{Var}(\hat{\eta})$, with equality only if $\mathbf{h} = \mathbf{a}$, so that $\hat{\eta}^* = \hat{\eta}$ for all $\mathbf{Y}$.

COMMENTS: Since $\hat{\eta}$ is the best linear unbiased estimator of $\eta$, it is called the BLUE for $\eta$.

*Proof:* Since $E(\hat{\eta}^*) = \mathbf{h}^T\ \mathbf{X}\boldsymbol{\beta} = \mathbf{c}^T\boldsymbol{\beta}$ for all $\boldsymbol{\beta}$, it follows that $\mathbf{h}^T\mathbf{X} = \mathbf{c}^T$ and $\mathbf{X}^T\mathbf{h} = \mathbf{c}$. That is, linear unbiased estimators $\mathbf{h}^T\mathbf{Y}$ of $\eta$ have inner products $\mathbf{c}$ with the columns of $\mathbf{X}$; equivalently, $\mathbf{a}$ and $\mathbf{h}$ have the same inner products with every vector in the column space $\mathbf{V}$ of $\mathbf{X}$. The vector $\mathbf{a}$ lies in $V$. Since $\mathbf{d} = \mathbf{h} - \mathbf{a}$ has inner products zero with all vectors in $V$, and $\mathbf{a}$ lies in $V$, $\mathbf{a}$ is the orthogonal projection of $\mathbf{h}$ onto $\mathbf{V}$. Thus, $\mathbf{h} = \mathbf{a} + \mathbf{d}$, with $\mathbf{a} \perp \mathbf{d}$. Therefore, $\text{Var}(\hat{\eta}^*) = ||\mathbf{h}||^2\ \ \sigma^2 = (||\mathbf{a}||^2 + ||\mathbf{d}||^2)\sigma^2 = \text{Var}(\hat{\eta}) + \text{Var}(\mathbf{d}^T\mathbf{Y})$, so that $\text{Var}(\hat{\eta}^*)\ \ \geq \text{Var}(\hat{\eta})$, with equality only if $\mathbf{d} = \mathbf{h} - \mathbf{a} = \mathbf{0}$ (i.e., $\hat{\eta}^* = \hat{\eta}$ for all $\mathbf{Y}$). □

COMMENTS: As shown by the proof, every linear unbiased estimator $\hat{\eta}^*$ of $\eta$ is the sum of $\hat{\eta}$ and an unbiased estimator $\mathbf{d}^T\mathbf{Y}$ of zero which has covariance zero with $\hat{\eta}$. This "part" of $\hat{\eta}^*$ contributes nothing toward the estimation of $\eta$ except variability: "noise."

## Problems for Section 11.3

**11.3.1** Let $Y_i = \beta x_i + \varepsilon_i$ for $i = 1, \ldots, n$, for constants, $\varepsilon_i \sim \mathrm{N}(0, \sigma^2)$, the $\varepsilon_i$ independent.

**(a)** Give a formula for a $100\gamma\%$ confidence interval on $\beta$. Apply it for $n = 3$, with $(x_i, Y_i)$ pairs: $(1, 1), (2, 3), (3, 7)$.

**(b)** Let $g(x) = \beta x$. Determine two functions $U(x_0, \mathbf{Y}, \mathbf{x})$ and $L(x_0, \mathbf{Y}, \mathbf{x})$, where $\mathbf{Y}$ and $\mathbf{x}$ are three-component vectors, such that $P(L(x_0, \mathbf{Y}, \mathbf{x}) \le g(x_0) \le U(x_0, \mathbf{Y}, \mathbf{x})) = 0.95$ for each $x_0$. Graph $L$ and $U$ as functions of $x_0$ for $\mathbf{Y}$ and $\mathbf{x}$ as in part (a).

**11.3.2** The $t$-statistic for a test of $H_0 : \beta_1 = 0$ for the simple linear regression model is $T = \hat{\beta}_1 / S(\hat{\beta}_1)$. Express the correlation coefficient $r = S_{xy}/\sqrt{S_{xx} S_{yy}}$ in terms of $T$ and $n$. Then express $T$ as a function of $n$ and $r$.

**11.3.3** Let $\mathbf{x}_1 = (1, 1, 0, 1, 1)^{\mathrm{T}}, \mathbf{x}_2 = (1, 2, 3, 4, 5)^{\mathrm{T}}, \mathbf{y} = (6, 6, 6, 10, 14)^{\mathrm{T}}$. For the model $\mathbf{Y} = \boldsymbol{\theta} + \varepsilon$, with $\boldsymbol{\theta} = \beta_1 \mathbf{x}_1 + \beta_2 \mathbf{x}_2$, $\boldsymbol{\varepsilon} \sim \mathrm{N}_5(0, \sigma^2 I_5)$, $\mathbf{Y} = \mathbf{y}$:

**(a)** Find the least squares estimate $\hat{\beta}$ of $\beta$ and the estimates $\hat{\mathbf{y}} = \hat{\boldsymbol{\theta}}$ of $\boldsymbol{\theta}$, $S^2$ of $\sigma^2$.

**(b)** Find a 95% confidence interval on $\eta \equiv 2\beta_1 - \beta_2$.

**(c)** Determine the vector $\mathbf{a}$ such that $\mathbf{a}^{\mathrm{T}}\mathbf{Y} = \hat{\eta}$. Find a vector $\mathbf{h}$ such that $\hat{\eta}^* = \mathbf{h}^{\mathrm{T}}\mathbf{Y}$ is an unbiased estimator of $\eta$. Show that $(\mathbf{h} - \mathbf{a}) \perp V = \mathcal{L}(\mathbf{x}_1, \mathbf{x}_2)$, the subspace spanned by $\mathbf{x}_1$ and $\mathbf{x}_2$, and that $\mathrm{Var}(\hat{\eta}^*) = \mathrm{Var}(\hat{\eta}) + ||\mathbf{h} - \mathbf{a}||^2\sigma^2$.

**11.3.4** Let $Y_{ij} \sim \mathrm{N}(\mu_i, \sigma^2)$ for $j = 1, \ldots, n_i$ for $i = 1, 2, 3$, with $n_1 = 2, n_2 = 3, n_3 = 2$, with all seven $Y_{ij}$ independent. This is called the model for a *one-way layout* or for *one-way analysis of variance*.

**(a)** Define seven component vectors $\mathbf{Y}, \mathbf{x}_1, \mathbf{x}_2, \mathbf{x}_3, \boldsymbol{\varepsilon}$ so that $\mathbf{Y} = \mu_1 \mathbf{x}_1 + \mu_2 \mathbf{x}_2 + \mu_3 \mathbf{x}_3 + \boldsymbol{\varepsilon}$, with $\boldsymbol{\varepsilon} \sim \mathrm{N}_7(\mathbf{0}, \sigma^2 \mathbf{I}_7)$.

**(b)** Give nonmatrix formulas for the least squares estimators of the $\mu_i$. Apply the formulas to the following data: $Y_{11} = 5, Y_{12} = 9, Y_{21} = 8, Y_{22} = 12, Y_{23} = 10, Y_{31} = 12, Y_{32} = 14$.

**(c)** Give a nonmatrix formula for $S^2$. Determine the value for the data in part (b).

**(d)** Let $\eta_{23} = \mu_2 - \mu_3$. For the data in part (b) find a 95% confidence interval $I_{23}$ on $\eta_{23}$.

Also determine a 95% CI $I_{13}$ on $\eta_{13} = \mu_1 - \mu_3$. What can you say about $P(\eta_{13} \in I_{13}, \eta_{23} \in I_{23})$? (Here $I_{13}$ and $I_{23}$ are the random intervals given by the method, so that they depend on the random vector $\mathbf{Y}$. As is common in statistical inference, we have used the same notation for both the random intervals and for the intervals obtained for the observed $\mathbf{Y}$.)

**(e)** For the data of part (b) find the multiple correlation coefficient $R$ of $\mathbf{Y}$ with $\mathbf{x}_1, \mathbf{x}_2, \mathbf{x}_3$. Give a nonmatrix formula for $R$ that would apply for any values of $n_1, n_2, n_3$. What properties would the $Y_{ij}$ have to satisfy in order to have $R = 1? R = 0?$

**11.3.5** For the data of Problem 11.2.7 (b), with the usual linear model with normally distributed errors, find a 95% CI on $\eta = \beta_1 - \beta_2$.

**11.3.6** Consider two hybrids of corn, 1 and 2. $N = m + n$ plots of land are available for planting. $m$ of these are randomly chosen to be planted with hybrid 1 seed. The other $n$ receive hybrid 2 seed. A fertility measurement $x_{ij}$ is made on the $j$th plot receiving hybrid $i$ seed. The yield in kilograms of corn is then $Y_{ij}$. Let $\mathbf{Y}$ be the $N$-component vector $(Y_{11}, \ldots, Y_{1m}, Y_{21}, \ldots, Y_{2n})$, and let $\mathbf{x}$ be the corresponding vector of $x_{ij}$. Let $\mathbf{w}_1$ and $\mathbf{w}_2$ be indicator vectors for the first $m$ and last $n$ components, those corresponding to hybrids 1 and 2. Then a reasonable model is $Y_{ij} = \beta_i + \beta_3 x_{ij} + \varepsilon_{ij}$, where the $\varepsilon_{ij}$ are independent, each $N(0, \sigma^2)$ . In vector form this is $\mathbf{Y} = \beta_1 \mathbf{w}_1 + \beta_2 \mathbf{w}_2 + \beta_3 \mathbf{x} + \boldsymbol{\varepsilon}$. Find $\mathbf{x}^\perp = \mathbf{x} - p(\mathbf{x} \mid \mathcal{L}(\mathbf{w}_1, \mathbf{w}_2))$, then use $\mathbf{x}^\perp$ to give a nonmatrix formula for $\hat{\beta}_3$ and for $\text{Var}(\hat{\beta}_3)$. Also give formulas for $\hat{\beta}_1$ and $\hat{\beta}_2$ and their variances.

**11.3.7** Let $\mathbf{x}_1 = (1, 0, 0)$, $\mathbf{u} = (0, 0, 1)$, $\mathbf{x}_2 = \mathbf{x}_1 + \alpha \mathbf{u}$, $\mathbf{v} = \mathcal{L}(\mathbf{x}_1, \mathbf{x}_2)$. Suppose that $\mathbf{Y} = \beta_1 \mathbf{x}_1 + \beta_2 \mathbf{x}_2 + \boldsymbol{\varepsilon}$, where $E(\boldsymbol{\varepsilon}) = \mathbf{0}$, and $\text{Cov}(\boldsymbol{\varepsilon}) = \sigma^2 \mathbf{I}_3$. Find $\mathbf{x}_1^\perp, \mathbf{x}_2^\perp, ||\mathbf{x}_1^\perp||^2, ||\mathbf{x}_2^\perp||^2, (\mathbf{x}_1^\perp, \mathbf{x}_2^\perp), \text{Var}(\hat{\boldsymbol{\beta}}_1), \text{Var}(\hat{\boldsymbol{\beta}}_2), \text{Cov}(\hat{\boldsymbol{\beta}}_1, \hat{\boldsymbol{\beta}}_2)$, and $\rho(\hat{\boldsymbol{\beta}}_1, \hat{\boldsymbol{\beta}}_2)$. What happens to the variances and the correlation coefficient as $\alpha \to 0$ or $\alpha \to \infty$?

**11.3.8** During the 2005–2006 season the 30 National Hockey League (NHL) teams each played 82 games. Their goals scored per game ($x_1$), goals scored against ($x_2$), wins, losses, ties, and percentage of points ($y$) are given in Table 11.3.8. The NHL uses a rather strange system for deciding whether a game is a win or a tie. Games between teams $A$ and $B$ that end with the score tied in regulation time, with team $A$ scoring an extra goal in overtime, become wins for $A$, ties for $B$. Two points are awarded for a win, one point for a tie. The percentage is 100 (no. points)/[2 (no. games played)].

**(a)** Fit a multiple regression model for $y$ versus $x_1$ and $x_2$. Use a constant term.

**(b)** Find the multiple correlation coefficient $R$.

**(c)** Let $\beta_1$ and $\beta_2$ be the coefficients of $x_1$ and $x_2$. Find a 95% confidence interval on $\eta = \beta_1 + \beta_2$. Show that $\eta = 0$ (goals made are equally important with goals given up) corresponds to a simple linear regression model.

**TABLE 11.3.8 NHL Data**

| Team | $x_1$ | $x_2$ | Wins | Losses | Ties | $y$ |
|---|---|---|---|---|---|---|
| Detroit | 3.67 | 2.51 | 58 | 16 | 8 | 75.6 |
| Ottowa | 3.80 | 2.50 | 52 | 21 | 9 | 68.9 |
| Carolina | 3.49 | 3.15 | 52 | 22 | 8 | 68.3 |
| Dallas | 3.08 | 2.65 | 53 | 23 | 6 | 68.3 |
| Buffalo | 3.37 | 2.85 | 52 | 24 | 6 | 67.1 |
| Nashville | 3.08 | 2.73 | 49 | 25 | 8 | 64.6 |
| Calgary | 2.63 | 2.35 | 46 | 25 | 11 | 62.8 |
| New Jersey | 2.84 | 2.74 | 46 | 27 | 9 | 61.6 |
| Philadelphia | 3.21 | 3.08 | 45 | 26 | 11 | 61.6 |
| NY Rangers | 3.05 | 2.57 | 44 | 26 | 12 | 61.0 |
| San Jose | 3.23 | 2.87 | 44 | 27 | 11 | 60.4 |
| Anaheim | 3.06 | 2.71 | 43 | 27 | 12 | 59.8 |
| Colorado | 3.42 | 3.06 | 43 | 30 | 9 | 57.9 |
| Edmonton | 3.04 | 2.95 | 41 | 28 | 13 | 57.9 |
| Montreal | 2.94 | 2.98 | 42 | 31 | 9 | 56.7 |
| Tampa Bay | 3.00 | 3.12 | 43 | 33 | 6 | 56.1 |
| Vancouver | 3.07 | 3.06 | 42 | 32 | 8 | 56.1 |
| Toronto | 3.10 | 3.21 | 41 | 33 | 8 | 54.9 |
| Atlanta | 3.37 | 3.29 | 41 | 33 | 8 | 54.9 |
| Los Angeles | 2.96 | 3.28 | 42 | 35 | 5 | 54.3 |
| Florida | 2.88 | 3.07 | 37 | 34 | 11 | 51.8 |
| Minnesota | 2.76 | 2.58 | 38 | 36 | 8 | 51.2 |
| Phoenix | 2.95 | 3.27 | 38 | 39 | 5 | 49.4 |
| NY Islanders | 2.70 | 3.35 | 36 | 40 | 6 | 47.6 |
| Boston | 2.78 | 3.15 | 29 | 37 | 16 | 45.1 |
| Columbus | 2.62 | 3.37 | 35 | 43 | 4 | 45.1 |
| Washington | 2.80 | 3.66 | 29 | 41 | 12 | 42.7 |
| Chicago | 2.55 | 3.40 | 26 | 43 | 13 | 39.6 |
| Pittsburgh | 2.96 | 3.78 | 22 | 46 | 14 | 35.4 |
| St. Louis | 2.35 | 3.46 | 21 | 46 | 15 | 34.8 |

## 11.4 *F*-TESTS FOR $H_0 : \theta = \beta_1 X_1 + \cdots + \beta_K \mathbf{X}_K \in \mathbf{V}_0$, A SUBSPACE OF $V$

Let $V_0$ be a subspace of $V = \mathcal{L}(\mathbf{x}_1, \ldots, \mathbf{x}_k)$, the subspace spanned by $\mathbf{x}_1, \ldots, \mathbf{x}_k$ of dimension $k_0 < k$. Often, $V_0$ is defined by $V_0 = \{\mathbf{v} \mid \mathbf{v} = \mathbf{X}\boldsymbol{\beta}, \mathbf{C}\boldsymbol{\beta} = \mathbf{0}\}$, where $\mathbf{C}$ is a $(k - k_0) \times k$ matrix of constants of rank $r = k - k_0$ chosen by the statistician or analyst. For example, for the one-way layout of Problem 11.3.4, we might wish to test $H_0 : \mu_1 = \mu_2 = \mu_3$ versus $H_a$:$H_0$ not true. $H_0$ is equivalent to $\boldsymbol{\theta} = E(\mathbf{Y}) = \mu_1 \mathbf{J}_n$ for some parameter $\mu_1$, that $\boldsymbol{\theta} \in \mathcal{L}(\mathbf{J}_n)$, the one-dimensional subspace of all vectors with all components the same. For another example, suppose that we observe $n$ $(x_i, Y_i)$ pairs and wish to consider the cubic model with $E(Y_i) = \beta_0 + \beta_1 x_i + \beta_2 x_i^2 + \beta_3 x_i^3$, equivalently, $\mathbf{Y} = \sum_{j=0}^{3} \beta_j \mathbf{x}_j + \boldsymbol{\varepsilon}$, where $\mathbf{x}_j$ has ith component $x_i^j$. We might wish

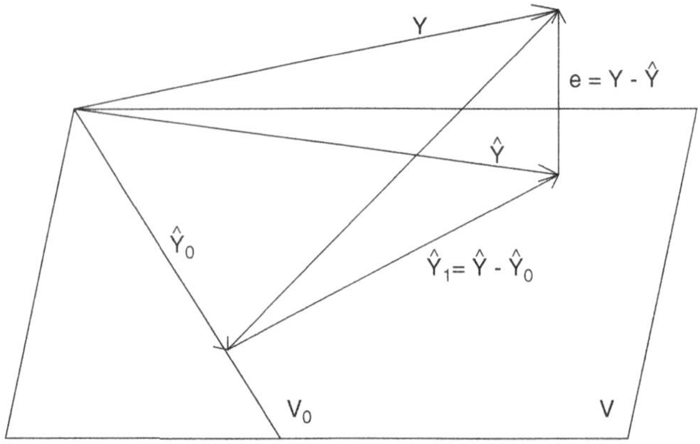


**FIGURE 11.4.1** Subspaces $V$, $V_0$, $V_1 = V \cap V_0^{\perp}$ and corresponding projections.

to test $H_0 : \beta_2 = \beta_3 = 0$, equivalently, that $E(\mathbf{Y}) = \boldsymbol{\theta} \in V_0$, the subspace spanned by $\mathbf{x}_0$ (the vector of all 1's) and $\mathbf{x}_1$.

Let $\hat{\mathbf{Y}} = p(\mathbf{Y} \mid V)$ and $\hat{\mathbf{Y}}_0 = p(\mathbf{Y} \mid V_0)$ (see Figure 11.4.1). Since $(\mathbf{Y}, \mathbf{v}) = (\hat{\mathbf{Y}}, \mathbf{v})$ for all $\mathbf{v} \in V$, and $(\hat{\mathbf{Y}}, \mathbf{v}) = (\hat{\mathbf{Y}}_0, \mathbf{v})$ for all $\mathbf{v} \in V_0$, it follows that $p(\hat{\mathbf{Y}} \mid \mathbf{V}_0) = \hat{\mathbf{Y}}_0$ for all possible $\mathbf{Y}$. It should seem intuitively clear that whenever $H_0$ is not true, $||\hat{\mathbf{Y}} - \hat{\mathbf{Y}}_0||^2$ should tend to be large. $\hat{\mathbf{Y}}_1 \equiv \hat{\mathbf{Y}} - \hat{\mathbf{Y}}_0$ is the projection of $\mathbf{Y}$ on the subspace $V_1 \equiv V \cap V_0^{\perp}$, the collection of vectors in $V$ that are orthogonal to $V_0$. The dimension of $V_1$ is $k - k_0 \equiv k_1$. Let $\boldsymbol{\theta}_0 = p(\boldsymbol{\theta} \mid V_0)$ and $\boldsymbol{\theta}_1 \equiv p(\boldsymbol{\theta} \mid V_1) = \boldsymbol{\theta} - \boldsymbol{\theta}_0$. From Section 9.3 we know that $Q_1 \equiv ||\hat{\mathbf{Y}}_1||^2/\sigma^2 \sim \chi^2_{k_1}(\delta)$, where $\delta = ||\boldsymbol{\theta} - \boldsymbol{\theta}_0||^2/\sigma^2 = ||\boldsymbol{\theta}_1||^2/\sigma^2$. Also from Section 9.3 we know that $\mathbf{e} \equiv \mathbf{Y} - \hat{\mathbf{Y}} = p(\mathbf{Y} \mid V^{\perp})$ and $\hat{\mathbf{Y}}$, being projections on orthogonal subspaces, are independent random vectors. It follows that $Q_1$ and $Q_2 \equiv ||\mathbf{e}||^2/\sigma^2$ are independent. Since $Q_2 \sim \chi^2_{n-k}$, and $S^2 = ||\mathbf{e}||^2/(n-k)$, it follows that $F \equiv [Q_1/(k-k_0)]/[Q_2/(n-k)] = [||\hat{\mathbf{Y}}_1||^2/(k-k_0)]/S^2 \sim F(k - k_0, n-k; \delta)$, where $\delta = ||\boldsymbol{\theta}_1||^2/\sigma^2$.

Under the null hypothesis $H_0 : \boldsymbol{\theta} \in V_0$, equivalently, $\delta = 0$, $F$ has the central $F$-distribution with $k - k_0$ and $n - k$ degrees of freedom. Thus, the test that rejects for $F > F_{1-\alpha}(k - k_0, n - k) \equiv F_{1-\alpha}$ is an $\alpha$-level test. The test that rejects $H_0$ for large $F$ is a likelihood ratio test (see Problem 11.4.4).

Figure 11.4.1 tells the story. The $F$-statistic is the ratio of the squared lengths of the vectors $\hat{\mathbf{Y}}_1$ and $\mathbf{e} = \mathbf{Y} - \hat{\mathbf{Y}}$, each divided by the dimension of their corresponding subspace.

***Example 11.4.1 (One-Way Analysis of Variance)*** Suppose that a crop scientist wished to compare the yields among three hybrids of corn: $C_1$, $C_2$, and $C_3$.$C_1$ was the standard hybrid, used on most of the farms in that county. The scientist had a 12-acre field available for growing the corn. He decided to plant 5 acres with $C_1$, 4 with $C_2$, and 3 with $C_3$. These were assigned randomly to the 12 acres so that the experimental design was completely randomized. The yields in bushels for the 12 one-acre plots were:

| | | |
|---|---|---|
| $C_1$: | 131, 158, 125, 141, 141 | sample mean = 139.2 |
| $C_2$: | 144, 142, 199, 203 | sample mean = 172.0 |
| $C_3$: | 225, 230, 170 | sample mean = 208.3 |

It seems that yields are highest for $C_3$, lowest for $C_1$. But this may be due to chance. How can we determine whether the differences are due to chance fluctuation? This is often the statistician's job—to separate real effects from "noise." We need a model. □

### Model for One-Way Analysis of Variance

Suppose that for $i = 1, 2, \ldots, k$, $Y_{i1}, \ldots, Y_{in_i}$ is a random sample from $\mathrm{N}(\mu_i, \sigma^2)$ and that all $n = n_1 + \cdots + n_k$ $Y_{ij}$ are independent. For the example above, $k = 3, n_1 = 5, n_2 = 4, n_3 = 3$. Define $\mathbf{Y}$ to be $n$-component vector of $Y_{ij}$'s, ordered in dictionary order with the $Y_{1j}$ first, $Y_{kj}$ last. Define $\mathbf{x}_i$ to be an $n$-component vector that indicates the components of $Y_{ij}$. Thus, $\mathbf{x}_i$ has $n_i$ ones, $n - n_i$ zeros. For the example above, $\mathbf{x}_2$ has 12 components. The first five and last three are zero. The sixth through ninth components are 1. In general, the $k$ $\mathbf{x}_i$ are mutually orthogonal, and $||\mathbf{x}_i||^2 = n_i$.

Let $\varepsilon_{ij} = Y_{ij} - \mu_i$. Then the $\varepsilon_{ij}$ are independent, and $\varepsilon_{ij} \sim \mathrm{N}(0, \sigma^2)$. By defining $\boldsymbol{\varepsilon}$ as the corresponding $n$-component vector, using the same ordering as for $\mathbf{Y}$, we can write the model in standard linear form:

$$\mathbf{Y} = \boldsymbol{\theta} + \boldsymbol{\varepsilon}, \text{ where } \boldsymbol{\theta} = \mu_1\mathbf{x}_1 + \cdots + \mu_k\mathbf{x}_k \in V = \mathcal{L}(\mathbf{x}_1, \ldots, \mathbf{x}_k).$$

Since the $\mathbf{x}_i$ are mutually orthogonal, the least squares esimator of $\mu_i$ is $\hat{\mu}_i = \mathbf{x}_i^T\mathbf{Y}/||\mathbf{x}_i||^2 = \sum_{j=1}^{n_i} Y_{ij}/n_i \equiv \overline{Y}_i$. Then $\hat{\mathbf{Y}} = \sum_{i=1}^{k} \bar{Y}_i.\mathbf{x}_i$. The residual vector $\mathbf{e} = \mathbf{Y} - \hat{\mathbf{Y}}$ has components $Y_{ij} - \overline{Y}_i.$. Its squared length is: Error sum of squares $=$ SSE $= ||\mathbf{e}||^2 = \sum_{ij} e_{ij}^2 = \sum_{ij}(Y_{ij} - \overline{Y}_i.)^2$. By the theory of Section 11.4, $S^2 \equiv ||\mathbf{e}||^2/(n-k)$ is an unbiased estimator of $\sigma^2$. This follows from the model $E(\mathbf{Y}) \in V$ and $\mathrm{Cov}(\mathbf{Y}) = \sigma^2\mathbf{I}_n$, and does not depend on normality. If $\mathbf{Y}$ has the MV-normal distribution, then $||\mathbf{e}||^2/\sigma^2 \sim \chi^2_{n-k}$, equivalently $S^2(n-k)/\sigma^2 \sim \chi^2_{n-k}$. In addition, the vector of sample means and $S^2$ are independent. Suppose now that we wish to test $H_0 : \mu_1 = \cdots = \mu_k$. $H_0$ is equivalent to $E(\mathbf{Y}) = \boldsymbol{\theta} = \sum_{i=1}^{k} \mu_1\mathbf{x}_i = \mu_1\mathbf{x}_0$, where $\mathbf{x}_0 = \sum_{i=1}^{k} \mathbf{x}_i$ is the vector of all 1's. Thus, under $H_0$, $\boldsymbol{\theta}$ lies in the one-dimensional subspace $V_0$ spanned by $\mathbf{x}_0$. The projection of $\mathbf{Y}$ onto $V_0$ is $\hat{\mathbf{Y}}_0 = \overline{Y}_{..}\mathbf{x}_0$, where $\overline{Y}_{..} \equiv \sum_{ij} Y_{ij}/n$, the grand mean. It follows that $||\mathbf{Y} - \hat{\mathbf{Y}}_0||^2$, usually called the *Corrected Total sum of squares*, or simply the *Total sum of squares*, is $\sum_{ij}(Y_{ij} - \overline{Y})^2 \equiv$ SST. Finally, the *Among means sum of squares* $\equiv$ SSA $= ||\hat{\mathbf{Y}} - \hat{\mathbf{Y}}_0||^2 = \sum_i n_i(\overline{Y}_i. - \overline{Y}_{..})^2$.

The projection of the mean vector $\boldsymbol{\theta}$ may be determined by substituting $\boldsymbol{\theta}$ for $\mathbf{Y}$ in the formula for $\hat{\mathbf{Y}}_0$. Thus, $\boldsymbol{\theta}_0 \equiv p(\boldsymbol{\theta} \mid V_0) = \overline{\mu}\mathbf{x}_0$, so that $E(\mathrm{SSA}) = (k-1)\sigma^2 + \sum_i n_i(\mu - \overline{\mu})^2$. Finally, $\mathrm{SSA}/\sigma^2 \sim \chi^2_{k-1}(\delta)$ with $\delta = \sum_i n_i(\mu_i - \overline{\mu})^2/\sigma^2$. It follows from the definition of the $F$-distribution that $F \equiv [\mathrm{SSA}/(k-1)]/S^2 \sim F(k-1, n-k)$ when $H_0$ holds. We should reject $H_0$ at level $\alpha$ when $F > F_{1-\alpha}(k-1, n-k)$.

**TABLE 11.4.1 Analysis of Variance**

| Source | Subspace | Degrees of Freedom | Sum of Squares | Mean Square | $F$ | Expected Mean Square |
|---|---|---|---|---|---|---|
| Among means | $V_1 = V \cup V_0^{\perp}$ | $k - 1 = 9$ | 90, 987.45 | 4, 543.72 | 6.575 | $\sigma^2(1 + \delta)/9$ |
| Error | $V^{\perp}$ | $n - k = 9$ | 6,219.47 | 691.05 | | $\sigma^2$ |
| Corr. total | $V_0^{\perp}$ | $n - 1$ | 15,306.92 | | | |

***Example 11.4.1 Continued*** For the data of Example 11.4.1, computations using S-Plus provided sample means $\overline{Y}_{i\cdot}$ : 139.20, 172.00, 208.33, $\overline{Y}_{\cdot\cdot} = 167.42$. The sums of squares were: SST $= 15, 306.92$, SSA $= 90, 987.45$, SSE $= 6219.47$ (Table 11.4.1). The computations were done by defining $\mathbf{X}$ as the $12 \times 3$ matrix consisting of columns $\mathbf{x}_1$, $\mathbf{x}_2$, and $\mathbf{x}_3$ (Table 11.4.2).

The analysis-of-variance table could have been produced easily using S-Plus:

```
> C = as.factor(c( rep(" C1", 5), rep(" C2", 4), rep(" C3", 3)))
# The quotation marks around C1, C2, C3 tell S-Plus that the values of the
#components of C should be treated as "factors" (names), not numbers.
> C
[1] C1 C1 C1 C1 C1 C2 C2 C2 C2 C3 C3 C3
```

C is then a vector of factor names. The numbers 1, 1, 1, 1, 1, 2, 2, 2, 2, 3, 3, 3 could have been used as components of C if they had been defined as factors. For example, with C = as.factor(rep(1:3,c(4,4,4))

```
> a = aov( y ~ C)
```

**TABLE 11.4.2**

| $\mathbf{x}_1$ | $\mathbf{x}_2$ | $\mathbf{x}_3$ | $\mathbf{Y}$ | $\hat{\mathbf{Y}}$ | $\hat{\mathbf{Y}}_0$ | $\mathbf{e} = \mathbf{Y} - \hat{\mathbf{Y}}$ |
|---|---|---|---|---|---|---|
| 1 | 0 | 0 | 131 | 139.20 | 167.42 | −8.20 |
| 1 | 0 | 0 | 158 | 139.20 | 167.42 | 18.80 |
| 1 | 0 | 0 | 125 | 139.20 | 167.42 | −14.20 |
| 1 | 0 | 0 | 141 | 139.20 | 167.42 | 1.80 |
| 1 | 0 | 0 | 141 | 139.20 | 167.42 | 1.80 |
| 0 | 1 | 0 | 144 | 172.00 | 167.42 | −28.00 |
| 0 | 1 | 0 | 142 | 172.00 | 167.42 | −30.00 |
| 0 | 1 | 0 | 199 | 172.00 | 167.42 | 27.00 |
| 0 | 1 | 0 | 203 | 172.00 | 167.42 | 31.00 |
| 0 | 0 | 1 | 225 | 208.33 | 167.42 | 16.67 |
| 0 | 0 | 1 | 230 | 208.33 | 167.42 | 21.67 |
| 0 | 0 | 1 | 170 | 208.33 | 167.42 | −38.33 |

We used the name "a." Any other name (other than the names of S-Plus functions) could have been used.

> summary(a)

| | Df | Sum of Sq | Mean Sq | F Value | Pr(F) |
|---|---|---|---|---|---|
| C | 2 | 9087.450 | 4543.725 | 6.575085 | 0.01737378 |
| Residuals | 9 | 6219.467 | 691.052 | | |

The observed $p$-value for $F$ is 0.0174, so that for any chosen $\alpha > 0.0174$, $H_0$ is rejected.

### A Short SAS Program

```
DATA CornExample;
INPUT Hybrid $ Yield @ @;
DATALINES; C1 131 C1 158 C1 125 C1 141 C1 141 C2 144 C2 142
C2 199 C2 203 C3 225 C3 230 C3 170 ;
PROC GLM;
Class Hybrid;
MODEL Yield = Hybrid;
run;
```

### SAS Output

The SAS System 16:19 Monday, February 23, 2004 10

The GLM Procedure

Dependent Variable: Yield

| Source | DF | Sum of Squares | Mean Square | F Value | Pr > F |
|---|---|---|---|---|---|
| Model | 2 | 9087.45000 | 4543.72500 | 6.58 | 0.0174 |
| Error | 9 | 6219.46667 | 691.05185 | | |
| Corrected Total | 11 | 15306.91667 | | | |

| R-Square | Coeff Var | Root MSE | Yield Mean |
|---|---|---|---|
| 0.593683 | 15.70206 | 26.28787 | 167.4167 |

R-Square is SSA/SST $= 9087.45/15306.92 = 0.5936 =$ proportion of variation in $Y$ explained by the hybrid variable.

□

We might be interested in comparing the three sample means by finding confidence intervals on $\eta_{ij} = \mu_i - \mu_j$ for pairs $i < j$. Since $\hat{\eta}_{ij} = \overline{Y}_{i\cdot} - \overline{Y}_{j\cdot} \sim \mathrm{N}(\eta_{ij}, \sigma^2(1/n_i + 1/n_j))$, 95% confidence intervals are given by $I_{ij} = [\overline{Y}_{i\cdot} - \overline{Y}_{j\cdot} \pm t_{9,0.975} S\sqrt{1/n_i + 1/n_j}]$ . That is, $P(\eta_{ij} \in I_{ij}) = 0.95$. However, $P(\eta_{ij} \in I_{ij}$ for all $i \neq j) < 0.95$. We can say that this last probability is at least $1 - 3(0.05) = 0.85$. If we wish the three confidence intervals to be true simultaneously, $t_{9,0.975}$ may be replaced by $\sqrt{(k-1)F_{0.95}(k-1, n-k)} = \sqrt{2F_{0.95}(2, 9)}$. The resulting intervals, due

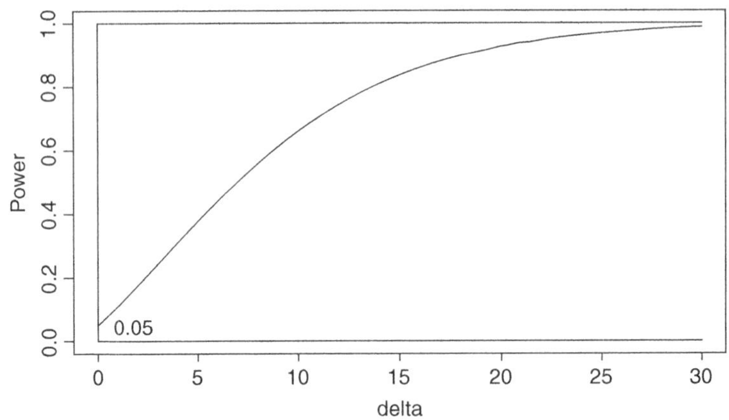


**FIGURE 11.4.2** Power function for the $F$-test for equality of means.

to Henry Scheffé, are called *Scheffé simultaneous confidence intervals*. A proof is not given here. A simultaneous method that provides shorter intervals on the $\eta_{ij}$ is due to John Tukey. See the text by Stapleton (1995) on linear models.

The corn data of Example 11.4.1 were produced in a computer using $\mu_1 = 150$, $\mu_2 = 170$, $\mu_3 = 190$, $\sigma = 20$. The power function for the $\alpha = 0.05$ level test is $\gamma(\delta) = P(F > F_{0.95}(2, 9) \mid \delta)$. Since $\bar{\mu} = [5(150) + 4(170) + 3(190)]/12 = 2000/12 = 166.67$, $\delta = [(150 - 166.67)^2(5) + (170 - 166.67)^2(4) + (190 - 166.67)^2]/400 = 7.67$ (see Figure 11.4.2). Since $F_{0.95}(2, 9) = 4.256$ and using the S-Plus command "$1 - pf(4.256, 2, 9, 7.667)$," we get power 0.541. We were lucky this time in that we made the correct decision for any $\alpha > 0.0174$. If we multiply the sample sizes by 2, so that $\delta$ becomes 15.33, the resulting power is 0.844. Multiplying by 3 increases power to 0.956. Such considerations can help in the design of an experiment.

For those without access to such functions as "qf" and "pf" in S-plus, which determine quantiles and cumulative probability values for the noncentral chi-square distribution, Pearson–Hartley charts may be used (see the appendix tables). Power is given for specified numerator and denominator degrees of freedom $\nu_1$ and $\nu_2$ for $\alpha = 0.05$ and 0.01 for each value of $\phi = \sqrt{\delta/(v_1 + 1)}$. For the corn example with $\alpha = 0.05$, $\nu_1 = 2$, $\nu_2 = 9$, $\delta = 7.667$, $\phi = 1.599$, we read power approximately 0.54. Trial and error can be used to determine the sample sizes necessary to achieve a given power for specified values of the $\mu_i$ and $\sigma$ .

***Example 11.4.2 (Testing for Linearity of a Regression Function)*** Suppose that for $0 \le x \le 10$, $E(Y \mid x) = g(x) = 15 + x^2$. Suppose that we observe $Y_{ij} = g(x_{ij}) + \varepsilon_{ij}$ for $i = 1, 2, \ldots, 20$, where $x_{ij} = i$, $\varepsilon_{ij} \sim \mathrm{N}(g(x_{ij}), \sigma^2 = 9)$ for $j = 1, 2$, $i = 1, 2, \ldots, 10$, with the $\varepsilon_{ij}$ independent. That is, we have two observations on $Y$ at each point $x = 1, 2, \ldots, 10$, and the distribution of $Y$, given $x$ is $N(g(x), 9)$. In practice we would not know the form of the regression function and might assume, for example, that the regression function is linear, of the form $h(x) = \beta_0 + \beta_1 x$. Assuming a full model for which $g(x)$ may be any function of $x$, let us (1) test

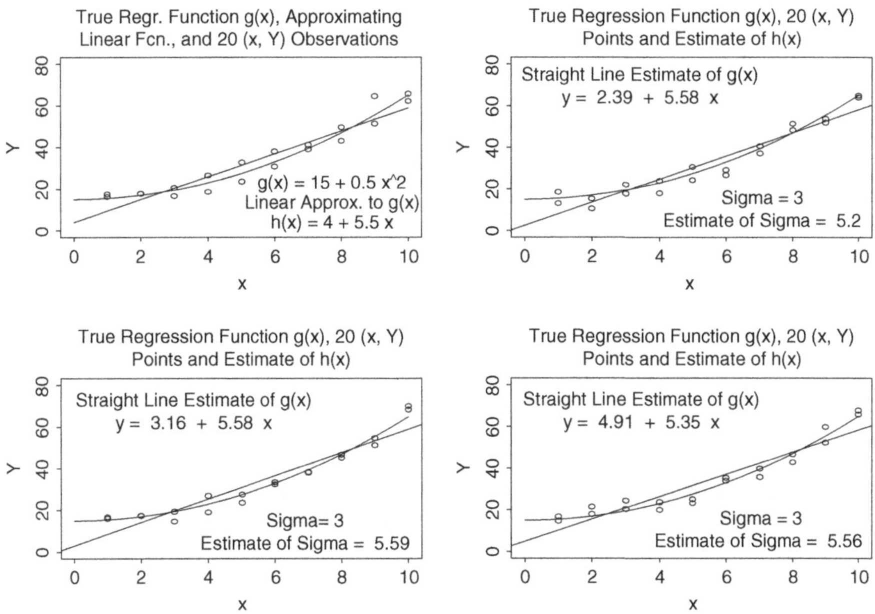


**FIGURE 11.4.3** Straight-line estimates of a quadratic regression function.

$H_0$:$g(x)$ is linear, and (2) measure the harm, or good, done by assuming that $g(x)$ is linear.

Figure 11.4.3 contains four plots of $g(x)$ together with 20 $(x, Y)$ pairs. The plot on the upper left also shows $h(x) = 4 + 5.5x$, the linear approximation of $g(x)$. The plot on the upper right again shows $g(x)$, together with the same $(x, Y)$ points and the least squares linear estimate of $h(x)$ corresponding to these $(x, Y)$. The last two plots repeat the second plot for new randomly generated $Y_i$, the same $x_i$..

The full model, which allows $g(x)$ to be any function of $x$, is equivalent to the model for one-way analysis of variance. That is, if we define $\mu_i = g(x_{i1}) = g(x_{i2})$, then $Y_{ij} \sim N(\mu_i, \sigma^2)$. Then error SSqs under this full model is given by the formula developed for one-way analysis of variance. In this case of two observations for each of the 10 different $x$-values $\text{SSE}_{FM} = \sum_{i=1}^{10} (Y_{i1} - Y_{i2})^2/2$. Under $H_0 : g(x)$ is linear, Error SSqs is $\text{SSE}_0 = S_{yy} - S_{xy}^2/S_{xx}$, as defined in Section 11.2. The $F$-statistic is therefore $F = [(\text{SSE}_0 - \text{SSE}_{FM})/(k-2)]/S^2$, where $k = 10$. Under $H_0$, $F \sim F(k-2 = 8, n-k = 20-10 = 10)$. The noncentrality parameter $\delta$ may be obtained by determining the numerator SSqs for the case that $\theta = E(Y)$ replaces $\mathbf{Y}$. For $\boldsymbol{\theta}$, $\text{SSE}_{FM} = 0$ and $\text{SSE}_0 = S_{\theta\theta} - S_{x\theta}^2/S_{xx}$. For $\mathbf{x}$ and $\boldsymbol{\theta}$ as given here, we get $S_{\theta\theta} = 5525.25$, $S_{x\theta} = 907.5$, $S_{xx} = 165$, $\text{SSE}_0 = 264$, $\delta = 264/9 = 29.33$. The $F$-test rejects for $F > F_{0.95}(8, 10) = 3.072$. The power is therefore $P(F > 3.072 \mid \delta = 29.333) = 0.798$.

Figure 11.4.4 indicates the results of 2000 repetitions of the experiment described. For the given parameters $\text{SSE}/9 \sim \chi_{10}^2$, $(\text{SSE}_0 - \text{SSE})/9 \sim \chi_8^2(\delta = 29.33)$, the pairs

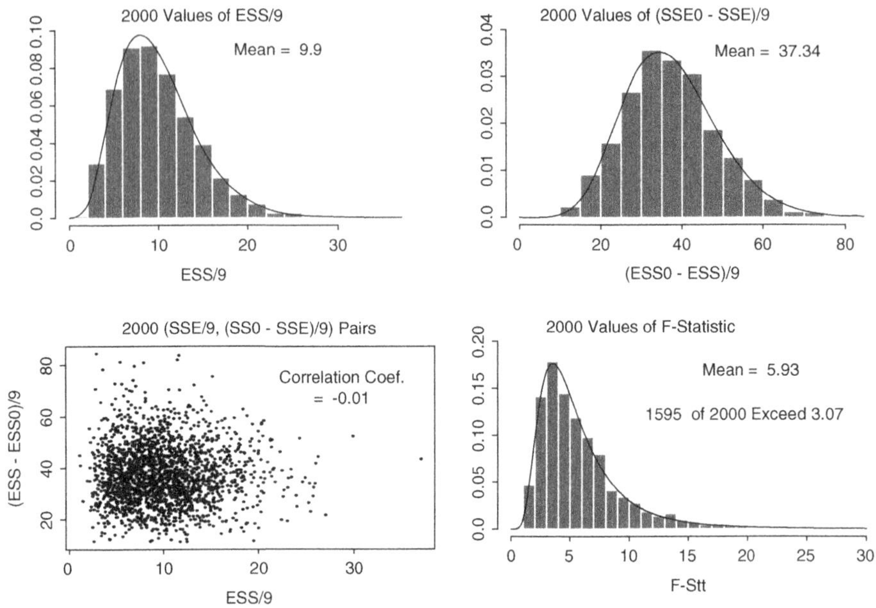


**FIGURE 11.4.4** Results of 2000 analyses of variance.

(SSE, $\text{SSE}_0 - \text{SSE}$) are independent and $F \sim F(8, 10, \delta)$. The appropriate, chi-square, noncentral chi-square, and noncentral $F$-densities are plotted with the histograms, giving good approximations in each case. Of the 2000 $F$-statistics, 1595 exceeded $F_{0.95}(8, 10)$, so the empirical evidence produced by the simulations is consistent with the power 0.798. In fact, the agreement seems almost too good. The reader will have to trust the author that these simulations were "honest." The histogram of $F$-statistics was improved a bit by omitting five values exceeding 30. □

## Problems for Section 11.4

**11.4.1** Consider the following vectors, each with five components: $\mathbf{x}_1 = (1, 1, 1, 1, 1)^T$, $\mathbf{x}_2 = (1, 1, 1, 0, 0)^T$, $\mathbf{x}_3 = (3, 1, 2, 1, 3)^T$, $\mathbf{Y} = (18, 8, 13, 8, 14)^T$.

**(a)** For the model $\mathbf{Y} = \boldsymbol{\theta} + \varepsilon$, for $\boldsymbol{\theta} = \beta_1\mathbf{x}_1 + \beta_2\mathbf{x}_2 + \beta_3\mathbf{x}_3$, find the least squares estimates of $\boldsymbol{\beta} = (\beta_1, \beta_2, \beta_3)^T$ and of $\boldsymbol{\theta}$. *Hint:* To simplify computations, first express vectors in $V = \mathcal{L}(\mathbf{x}_1, \mathbf{x}_2, \mathbf{x}_3)$ in terms of convenient multiples of $\mathbf{x}_i^* = \mathbf{x}_i - p(\mathbf{x}_i|\mathbf{x}_1)$ for $i = 2, 3$. The estimates of $\boldsymbol{\beta}$ and $\boldsymbol{\theta}$ consist entirely of integers.

**(b)** Find the residual vector $\mathbf{e}$ and $S^2$.

**(c)** Find a 95% CI on $\eta = \beta_1 + 2\beta_2 - \beta_3$.

**(d)** Find the $F$-statistic for $H_0 : \beta_1 = \beta_2 = \beta_3$. Test $H_0$ at level $\alpha = 0.10$.

(e) Find the noncentrality parameter for the $F$-statistic of part (d) for the case that $\beta_1 = 2.4$, $\beta_2 = 3.5$, $\beta_3 = 4.5$, $\sigma = 2$.

**11.4.2** (a) Determine the analysis-of-variance table for the following data.

Treatment 1: 3, 5, 7
Treatment 2: 10, 12, 16, 14
Treatment 3: 12, 16
Treatment 4: 6, 8, 14, 10, 9

(b) Test the null hypothesis $H_0$ that the means $\mu_i$ are equal for $i = 1, \ldots, 4$, assuming the usual model with normally distributed independent errors, equal variances.

(c) Find a 95% CI on $\eta = \mu_2 - (\mu_1 + \mu_3 + \mu_4)/3$. Since the coefficients of the $\mu_i$ sum to zero, $\eta$ is called a *contrast* on the $\mu_i$.

**11.4.3** Let $\mathbf{Y} = \mathbf{X}\boldsymbol{\beta} + \boldsymbol{\varepsilon}$, where $\mathbf{X}$ is an $n \times k$ matrix of constants of rank $k$. $\mathbf{X}$ is called a design matrix. Suppose that $\boldsymbol{\varepsilon} \sim N_n(\mathbf{0}, \sigma^2 \mathbf{I}_n)$. For $1 \leq k_0 < k$, let $\mathbf{C}$ be a $r \times k$ matrix of rank $r$. Suppose that we wish to test $H_0 : \mathbf{C}\boldsymbol{\beta} = 0$.

(a) Define a subspace $V_0$ of $V =$ (column space of $\mathbf{X}$) so that $H_0$ is equivalent to $\boldsymbol{\theta} = \mathbf{X}\beta \in V_0$. *Hint:* Express $\boldsymbol{\beta}$ in terms of $\mathbf{X}$ and $\boldsymbol{\theta}$.

(b) Let $V_1 = V \cap V_0^{\perp}$. Express $\hat{\mathbf{Y}}_1 = p(\mathbf{Y} \mid V_1)$ and $\text{SSE}_0 - \text{SSE} = ||\hat{\mathbf{Y}}_1||^2$ in terms of $\mathbf{X}$, $\mathbf{C}$, and $\hat{\boldsymbol{\beta}}$, where $SSE_0 = \|\mathbf{Y} - p(\mathbf{Y}|V_0)|)^2$.

(c) Determine a matrix $\mathbf{C}$ (there are an infinity of choices) for Problem 11.4.1 (d) and use the formula you obtained in part (b) to determine $\text{SSE}_0 - \text{SSE}$. Verify that it has the same value that you obtained in Problem 11.4.1 (b).

**11.4.4** Show that the likelihood ratio test of $H_0 : \theta \in V_0$ for the model of Problem 11.4.3 rejects for large $F$.

**11.4.5** Suppose that the model of Problem 11.4.3 holds with $k \geq 2$. Let the first column of $\mathbf{X}$ be the vector of all 1's.

(a) Let $F$ be the $F$-statistic for the test of $H_0 : \beta_2 = \cdots = \beta_k = 0$. Let $R$ be the multiple correlation coefficient for $\mathbf{Y}$ with respect to $\mathbf{x}_2, \ldots, \mathbf{x}_k$. Show that $F = [(n-k)/(k-1)][R^2/(1-R^2)]$.

(b) Use the relationship between $R^2$ and $F$ of part (a) to find $P(R^2 \geq 0.2525)$ for $n = 25$, $k = 4$, again assuming that $H_0$ is true.

**11.4.6** Consider a one-way analysis of variance with $k = 4$ treatment levels (Four $\mu_i$'s), with $n_0$ observations per treatment level. If two of the $\mu_i$'s differ by 20, the other two are half way between the first two, and $\sigma = 15$, about how large must $n_0$ be in order for the $\alpha = 0.05$-level test to have power at least 0.90? (Use a computer package or the Pearson–Hartley charts. Don't forget that the denominator df for the $F$-statistic depends on $n_0$.)

**11.4.7** Suppose that we observe $n_i$ pairs $(x_{ij}, Y_{ij})$ for $j = 1, 2, \ldots, n_i$ for $i = 1, 2$, where $Y_{ij} = g_i(x_{ij}) + \varepsilon_{ij}$, $g_i(x) = \beta_{i0} + \beta_{i1}x$ for all $x$, and $\varepsilon_{ij} \sim \text{N}(0, \sigma^2)$, independent for all $j$ and $i$. That is, the $Y_{ij}$ satisfy simple linear regression models, with different β's for the $Y_{1j}$ and $Y_{2j}$, with the same σ of all $Y_{ij}$. Let $S_{ixx} = \sum_{j-1}^{n_i}(x_{ij} - \bar{x}_i)^2$ and define $S_{ixy}$ and $S_{iyy}$ similarly.

**(a)** Express the $F$-statistic for the test of $H_0 : (\beta_{10} = \beta_{20}$ and $\beta_{11} = \beta_{21})$ in terms of the six sums of squares and cross-products.

**(b)** Perform the test with $\alpha = 0.05$ for the following data: $x_{1j} = j$ for $j = 1, 2, 3$, $x_{21} = 2$, $x_{22} = 3$, $x_{23} = 5$, $x_{24} = 6$, $Y_{11} = 7$, $Y_{12} = 2$, $Y_{23} = 3$, $Y_{21} = 9$, $Y_{22} = 6$, $Y_{23} = 4$, $Y_{24} = 5$.

**(c)** Plot the seven points, the two straight lines, and the fitted line under $H_0$.

**11.4.8** The amount $Y$ of gloxil produced by a chemical process depends on the temperature $t$ in Celsius at which it is formed. A model states that $E(Y \mid t) = g(t)$ is linear for $50 \le t \le 150$, with the slope possibly changing at $t = 100$, although $g$ is continuous for all $t$ in this interval. Such a function $g$ is called a *spline function*, with a *knot* at $t = 100$.

**(a)** Express the model as a linear model with three $\beta_j$. Do not forget that $g$ must be a continuous function of $t$.

**(b)** For the model $Y_i = g(t_i) + \varepsilon_i$ with the $\varepsilon_i$ independent, $\text{N}(0, \sigma^2)$, find the least squares estimate of $g$ for observation pairs $(t_i, Y_i)$: (50, 144), (80, 190), (110, 270), (130, 345), (150, 381). Plot the points and the function $\hat{g}$.

**(c)** For $\alpha = 0.05$, test the null hypothesis that the slopes are equal for $t \le 100$ as for $t > 100$.

**11.4.9** For the hockey data of Problem 11.3.6, test the null hypothesis that $E(Y \mid x_1, x_2)$ is a linear function of $d = x_1 - x_2$, using an $F$-test. Use $\alpha = 0.05$ or, better, find the $p$-value for the $F$-test.

## 11.5 TWO-WAY ANALYSIS OF VARIANCE

Consider an experiment designed to investigate the effects of three hybrids $(H_1, H_2, H_3)$ of corn and four types of fertilizer $(F_1, F_2, F_3, F_4)$ on the yields. A field was divided into 24 1/2-acre plots. (An acre is 43,560 square feet, which is approximately 4047 square meters.) Each of the $(H_i, F_j)$ combinations was assigned at random to two plots. All of the possible $(24!)/(2!)^{12} = 1.51 \times 10^{20}$ partitions of the set of plots were equally likely. Such a design is called *completely randomized*. If the set of plots had first been divided into "north" and "south" plots of 12 each, the 12 combinations assigned randomly with the North plots, and independently assigned randomly to the south plots, the design would have been called a *randomized block design*. That design would have been more desirable if, for example, the south

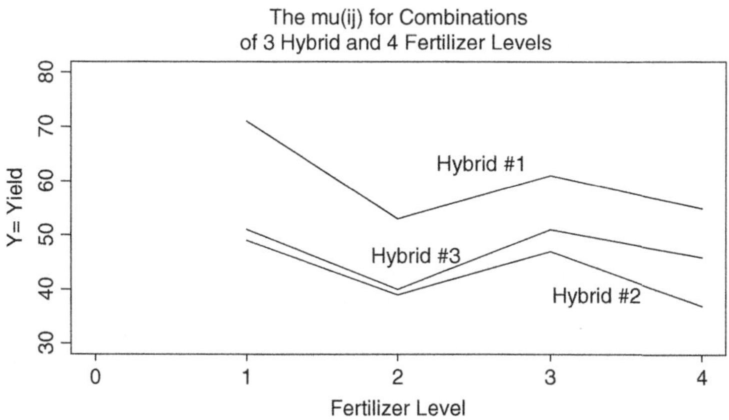


**FIGURE 11.5.1** Plot of the $\mu_{ij}$.

plots had been more fertile and therefore would tend to produce higher yields. Either design would also have been called a *complete factorial*, since there would be at least one observation for each treatment combination. For simplicity we discuss the more simple completely randomized design.

Let $Y_{ijk}$ be the $k$th observation, yield in bushels of corn, for treatment combination $(H_i, F_j)$, $i = 1, 2, 3$, $j = 1, 2, 3, 4$. Suppose that $Y_{ijk} \sim \mathrm{N}(\mu_{ij}, \sigma^2)$, independent for all $24 Y_{ijk}$. Our purpose is to estimate the $\mu_{ij}$, to describe how they may vary, and to compare them. Suppose, for example, that the $\mu_{ij}$ are as follows:

| | $F_1$ | $F_2$ | $F_3$ | $F_4$ |
|---|---|---|---|---|
| $H_1$ | 71 | 53 | 61 | 55 |
| $H_2$ | 49 | 39 | 47 | 37 |
| $H_3$ | 51 | 40 | 51 | 46 |

We can get some understanding of the effects of the two factors hybrids and fetilizers graphically, as in Figure 11.5.1.

Define $\mu = (1/12) \sum_{ij} \mu_{ij} = 50$, $\alpha_i = \overline{\mu}_{i+} - \mu$, $\beta_j = \overline{\mu}_{+j} - \mu$, $(\alpha\beta)_{ij} = \mu_{ij} - (\mu + \alpha_i + \beta_j)$. Then

$$\mu_{ij} = \mu + \alpha_i + \beta_j + (\alpha\beta)_{ij}, \tag{11.5.1}$$

where $\mu$ is called the *overall mean* or *grand mean*. The $\alpha_i$ are the main effects for the hybrids. The $\beta_j$ are the main effects for the fertilizers. The $(\alpha\beta)_{ij}$ are the interaction effects. We find $\alpha_1 = 60 - 50 = 10$, $\alpha_2 = 43 - 50 = -7$, $\alpha_3 = 47 - 50 = -3$. Similarly, $\beta_1 = 7$, $\beta_2 = -6$, $\beta_3 = 3$, $\beta_4 = -4$. Notice that the $\alpha_i$ sum to zero, as do the $\beta_j$, and that this is true in general. Finally, the interaction terms, the

$(\alpha\beta)_{ij}$, are

$$\begin{array}{rrrr} 4 & -1 & -2 & -1 \\ -1 & 2 & 1 & -2 \\ -3 & -1 & 1 & 3 \end{array}$$

The row and column sums must be zero. The fact that the $(\alpha\beta)_{ij}$ are small in absolute value is consistent with the "almost parallel" property of the graphs in Figure 11.5.1. Much of the variation among the $\mu_{ij}$ is caused by the magnitude of the $\alpha_i$. We can represent the full model in vector-space form as follows: Let $\mathbf{Y}$ be the 24-component column vector with the $Y_{ijk}$ represented in dictionary order $Y_{111}, Y_{112}, Y_{121}, Y_{122}, Y_{131}, \ldots, Y_{341}, Y_{342}$. Let $\mathbf{x}_0$ be the 24-component column vector of all 1's. Let $\mathbf{R}_1, \mathbf{R}_2, \mathbf{R}_3, \mathbf{C}_1, \mathbf{C}_2, \mathbf{C}_3, \mathbf{C}_4$ be the row and column indicators. For example, $\mathbf{C}_2 = (0, 0, 0, 0, 0, 0, 1, 1, 1, 1, 1, 1, 0, 0, 0, 0, 0, 0, 0, 0, 0, 0, 0, 0)^T$. Let $\mathbf{C}_{ij}$ be the indicator for cell $ij$. To understand this more clearly, write the vectors in table form rather than as column vectors. From (11.5.1)

$$E(\mathbf{Y}) \equiv \boldsymbol{\theta} = \sum_{ij} \mu_{ij}\mathbf{C}_{ij} = \mu\mathbf{x}_0 + \sum_i \alpha_i \mathbf{R}_i + \sum_i \beta_j \mathbf{C}_j + \sum_{ij} (\alpha\beta)_{ij}\mathbf{C}_{ij}. \tag{11.5.2}$$

Our full model states that $\boldsymbol{\theta}$ lies in the subspace $V$ spanned by the 12 $\mathbf{C}_{ij}$.

Let $\varepsilon_{ijk} = Y_{ijk} - \mu_{ij}$. Then $Y_{ijk} = \mu_{ij} + \varepsilon_{ijk}$. Let $\boldsymbol{\varepsilon}$ be the vector of $\varepsilon_{ijk}$, using the same ordering as before. Then $\mathbf{Y} = \boldsymbol{\theta} + \boldsymbol{\varepsilon} = \boldsymbol{\theta}_0 + \boldsymbol{\theta}_R + \boldsymbol{\theta}_C + \boldsymbol{\theta}_{RC} + \boldsymbol{\varepsilon}$, where $\boldsymbol{\theta}_0, \boldsymbol{\theta}_R, \boldsymbol{\theta}_C, \boldsymbol{\theta}_{RC}$ are the vectors in (11.5.2). The MLE for $\boldsymbol{\theta}$ under the full model is the least squares estimator $\hat{\mathbf{Y}} = \sum_{ij} \overline{Y}_{ij+}\mathbf{C}_{ij}$, the vector obtained by replacing each $Y_{ijk}$ by the mean $\overline{Y}_{ij+}$ for each $i$ and $j$. The subspace $V$ can be decomposed into four orthogonal subspaces. Let $V_0 = \mathcal{L}(\mathbf{x}_0)$, the subspace consisting of all multiples of $\mathbf{x}_0$. Let $V_R = \mathcal{L}(\mathbf{R}_1, \mathbf{R}_2, \mathbf{R}_3) \cap V_0^\perp = \{v \mid v = \sum_i a_i \mathbf{R}_i, \Sigma a_i = 0\}$, $V_C = \mathcal{L}(\mathbf{C}_1, \mathbf{C}_2, \mathbf{C}_3) \cap V_0^\perp = \{\mathbf{v} \mid \mathbf{v} = \sum_j b_j \mathbf{C}_j, \Sigma b_j = 0\}$, $V_{RC} = \mathcal{L}(\{\mathbf{C}_{ij}\}) \cap (V_0 + V_R + V_C)^\perp = \mathcal{L}(\{\mathbf{C}_{ij}\}) \cap V_0^\perp \cap V_R^\perp \cap V_C^\perp$, the collection of all vectors in $V$ that are orthogonal to $V_0$, $V_R$, and $V_C$. Thus, each vector in $V_{RC}$ is of the form $\sum_{ij}(ab)_{ij}\mathbf{C}_{ij}$, where $\sum_i (ab)_{ij} = 0$ for each $j$ and $\sum_j (ab)_{ij} = 0$ for each $i$. Again, we advise the reader to write these vectors as tables. In this form, row and column sums of vectors in $V_{RC}$ are all zero. The subspaces $V_0$, $V_R$, $V_C$, and $V_{RC}$ are mutually orthogonal. This follows directly from their definitions for all but the pair $V_R$ and $V_C$. It is easy to check that each $\mathbf{R}_i$ is orthogonal to every vector in $V_C$ and therefore $V_R \perp V_C$. This orthogonality depends crutially on the fact that the inner product of $\mathbf{R}_i$ and $\mathbf{C}_j$ is the number $K_{ij} = K$ of observations in cell $ij$ is the same for all cells. The vectors $\boldsymbol{\theta}_0, \boldsymbol{\theta}_R, \boldsymbol{\theta}_C, \boldsymbol{\theta}_{RC}$ are the projections of $\boldsymbol{\theta}$ on these four subspaces.

The projections of $\mathbf{Y}$ on these four subspaces are: $\hat{\mathbf{Y}}_0 = \overline{Y}_{+++}\mathbf{x}_0$, $\hat{\mathbf{Y}}_R = \sum_i \hat{\alpha}_i \mathbf{R}_i$, where $\hat{\alpha}_i = (\overline{Y}_{i++} - \overline{Y}_{+++})$, $\hat{\mathbf{Y}}_C = \sum_j \hat{\beta}_j \mathbf{C}_j$, where $\hat{\beta}_j = \overline{Y}_{+j+} - \overline{Y}_{+++}$, and $\hat{\mathbf{Y}}_{RC} = \sum_{ij} (\alpha\beta)_{ij}\mathbf{C}_{ij}$, where $\widehat{(\alpha\beta)}_{ij} = \overline{Y}_{ij+} - (\overline{Y}_{+++} + \hat{\alpha}_i + \hat{\beta}_j) = \overline{Y}_{ij+} - \overline{Y}_{i++} - \overline{Y}_{+j+} + \overline{Y}_{+++}$. An analysis of variance is a decomposition of the squared length

**TABLE 11.5.1** $\mu_{ij}$ **Data**

| | $F_1$ | $F_2$ | $F_3$ | $F_4$ |
|---|---|---|---|---|
| $H_1$ | 80.8, 73.1 | 48.6, 51.1 | 63.4, 53.3 | 58.0, 51.1 |
| | (76.95) | (49.85) | (58.35) | (54.55) |
| $H_2$ | 40.8, 49.4 | 34.9, 36.7 | 53.7, 44.3 | 33.4, 37.9 |
| | (45.10) | (35.80) | (49.00) | (35.65) |
| $H_3$ | 53.2, 54.2 | 40.5, 36.0 | 47.7, 46.8 | 35.8, 43.4 |
| | (53.70) | (38.75) | (47.25) | (39.60) |

into the sum of the squared lengths of several orthogonal vectors, using the Pythagorean theorem. In this case we have $\mathbf{Y} - \hat{\mathbf{Y}}_0 = \hat{\mathbf{Y}}_R + \hat{\mathbf{Y}}_C + \hat{\mathbf{Y}}_{RC}$. It follows by the Pythagorean theorem that

$$\begin{aligned}&\text{“Among cell means sum of squares”}\\ &\equiv ||\mathbf{Y} - \hat{\mathbf{Y}}_0||^2 = ||\hat{\mathbf{Y}}_R||^2 + ||\hat{\mathbf{Y}}_C||^2 + ||\hat{\mathbf{Y}}_{RC}||^2\\ &= (\text{SQS for rows}) + (\text{SQS for columns}) + (\text{SQS for interaction}). \quad (11.5.3)\end{aligned}$$

Since $\mathbf{Y} - \hat{\mathbf{Y}}_0 = \sum_{ij} \overline{Y}_{ij+}\mathbf{C}_{ij} - \overline{Y}_{+++}\mathbf{x}_0 = \sum_{ij}(\overline{Y}_{ij+} - \overline{Y}_{+++}), \mathbf{C}_{ij}$ and the $\mathbf{C}_{ij}$ are orthogonal, Among cell means SQS $= \sum_{ij}(\overline{Y}_{ij+} - \overline{Y}_{+++})^2||\mathbf{C}_{ij}||^2 = K\sum_{ij}(\overline{Y}_{ij+} - \overline{Y}_{+++})^2$, where $K$ is the number of observations in each cell, two for our example.

Row (or hybrid) SQS $= ||\hat{\mathbf{Y}}_R||^2 = \sum_i(\overline{Y}_{i++} - \overline{Y}_{+++})^2||R_i||^2 = 8\sum_i(\overline{Y}_{i++} - \overline{Y}_{+++})^2$, Column (or fertilizer) SQS $= ||\hat{\mathbf{Y}}_C||^2 = \sum_j(\overline{Y}_{+j+} - \overline{Y}_{+++})^2||\mathbf{C}_j||^2 = 6\sum_j(\overline{Y}_{+j+} - \overline{Y}_{+++})^2$.

Interaction SQS $= ||\hat{\mathbf{Y}}_{RC}||^2$ may be computed using the identity (11.5.3).

***Example 11.5.1*** A table of $Y_{ijk}$ was generated for $\mu_{ij}$ as given in Table 11.5.1. The $\varepsilon_{ijk}$ were a random sample from N$(0, \sigma^2 = 25)$. The cell sample means are given in parentheses and are plotted in Figure 11.5.2. $\sigma = 5$ was chosen to be small enough so that the graph of the cell means $\bar{Y}_{ij+}$ is somewhat like the corresponding graph of the $\mu_{ij}$ in Figure 11.5.1. Of course, in practice we must accept what nature hands us.

The hybrid and fertilizer sample means were:

| | |
|---|---|
| Hybrids: | 59.92 41.39 44.70 |
| Fertilizers: | 58.58 41.30 51.53 43.27 |
| Grand Mean: | 48.67 |

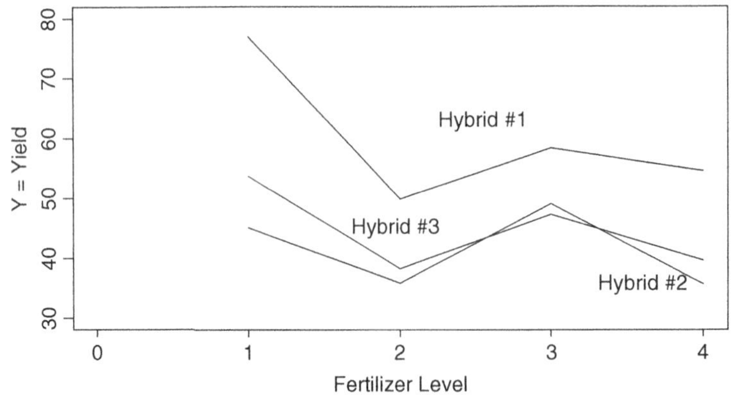


**FIGURE 11.5.2** Sample means for the 12 hybrid × fertilizer combinations.

$$\text{Among cells SQS} = \|\hat{Y} - \hat{Y}_0\|^2 = 2[(76.95 - 48.67)^2 + \cdots + (39.60 - 48.67)^2] = 2991.20$$

$$\text{Hybrid SQS} = \|\hat{\mathbf{Y}}_R\|^2 = 8[(59.92 - 48.67)^2 + \cdots + (44.70 - 48.67)^2] = 1563.77$$

$$\text{Fertilizer SQS} = \|\hat{\mathbf{Y}}_C\|^2 = 6[(58.58 - 48.67)^2 + \cdots + (43.27 - 48.67)^2] = 1139.92$$

$$\text{Corrected total SQS} = \|Y - \hat{Y}_0\|^2 = (80.8 - 48.67)^2 + \cdots + (43.4 - 48.67)^2 = 3231.60$$

Therefore, Interaction SQS $= \|\hat{\mathbf{Y}}_{RC}\|^2 = 2991.20 - 1563.77 - 1139.92 = 287.51$, and Residual SQS $= \|\mathbf{Y} - \hat{\mathbf{Y}}\|^2$ = corrected total SQS–among cells SQS = 240.40. These may be summarized in an analysis-of-variance table (Table 11.5.2). Each row of the table corresponds to a subspace, usually labeled "Sources." We have indicated the subspace corresponding to each row. This is usually not done for

**TABLE 11.5.2 Analysis-of-Variance Table**

| Source and Subspace | df | SQS | Mean Square | $F$ | $p$-Value | Expected Mean Square |
|---|---|---|---|---|---|---|
| Hybrids $V_R$ | $3 - 1 = 2$ | 1563.77 | 781.88 | 39.03 | 0.0000056 | $\sigma^2 + (8/2)\sum_i \alpha_i^2$ |
| Fertilizer $V_C$ | $4 - 1 = 3$ | 1139.92 | 379.97 | 18.97 | 0.000075 | $\sigma^2 + (6/3)\sum_j \beta_j^2$ |
| $H \times F$ int. $V_{RC}$ | 6 | 287.51 | 47.92 | 2.39 | 0.094 | $\sigma^2 + (2/6)\sum_{ij} \alpha\beta_{ij}^2$ |
| Among cells $V_1$ | $4(3) - 1 = 11$ | 2991.20 | 271.93 | 13.97 | 0.000034 | $\sigma^2 + (2/11)\sum_{ij}(\mu - \mu)^2$ |
| Residual $V^{\perp}$ | $12(2 - 1) = 12$ | 240.40 | 20.04 | | | $\sigma^2$ |
| Corr. total $V_0^{\perp}$ | 23 | 3231.60 | | | | |

computer-produced analysis-of-variance (AOV) tables or for those given in elementary statistics book. The "Among cells" subspace is $V_1 = V \cap V_0^{\perp} = V_R + V_C + V_{RC}$. Thus, the subspace $V_0^{\perp}$ is decomposed into four orthogonal subspaces $V_R$, $V_C$, $V_{RC}$, and $V^{\perp}$, corresponding to the sources hybrid, fertilizer, H × F Interaction, and residual. The degrees of freedom for any subspace is the dimension of the subspace. The $F$-statistics are the ratios of the mean square for that source (or subspace) to $S^2 =$ residual mean square. The $p$-value is the probability, under the null hypothesis that the effects for that source are all zero, that the $F$-statistic would exceed the value obtained. For example, $P(F_{6,12} > 2.39) = 0.094$.

### S-Plus Code to Produce the Table

```
> Fert <- as.factor(rep(c(1,1,2,2,3,3,4,4),3))    Fert and Hyb are then vectors of "Factor
                                                      Levels," not numbers
> Hyb <- as.factor(rep(1:3, rep(8,3)))
> Fert
[1] 1 1 2 2 3 3 4 4 1 1 2 2 3 3 4 4 1 1 2 2 3 3 4 4
> Hyb
[1] 1 1 1 1 1 1 1 1 2 2 2 2 2 2 2 2 3 3 3 3 3 3 3 3
> y <- scan()
80.8  73.1  48.6  51.1  63.4  53.3  58.0  51.1  40.8  49.4
34.9  36.7  53.7  44.3  33.4  37.9  53.2  54.2  40.5  36.0
47.7  46.8  35.8  43.4
> a <- aov(y ~ Hyb * Fert)   Notice the order of the y's, the same as for Hyb and Fert.
> summary(a)
```

| | Df | Sum of Sq | Mean Sq | F Value | Pr(F) |
|---|---|---|---|---|---|
| Fert | 3 | 1139.915 | 379.9715 | 18.96736 | 0.00007548 |
| Hyb | 2 | 1563.766 | 781.8829 | 39.02991 | 0.00000560 |
| Fert:Hyb | 6 | 287.514 | 47.9190 | 2.39201 | 0.09359521 |
| Residuals | 12 | | 240.395 | 20.0329 | |

### SAS Code and Output

```
Data Corn;
Input Hybrid $ Fertilizer $ Yield @@;      # The @@ allowed more than one
                                           # observation per line.
Datalines;                                 # If the data had been available in a file, an
                                           # Infile statement would have been given
                                           # here.
H1 F1 80.8 H1 F1 73.1 H1 F2 48.6 H1 F2 51.1 H1 F3 63.4 H1 F3 53.3 H1 F4 58.0 H1 F4 51.1
H2 F1 40.8 H2 F1 49.4 H2 F2 34.9 H2 F2 36.7 H2 F3 53.7 H2 F3 44.3 H2 F4 33.4 H2 F4 37.9
H3 F1 53.2 H3 F1 54.2 H3 F2 40.5 H3 F2 36.0 H3 F3 47.7 H3 F3 46.8 H3 F4 35.8 H3 F4 43.4
;
Proc glm;
Class Hybrid Fertilizer;                   # Needed so that these two variables are
                                           # treated as factors. Similar to "as.factor" in S-Plus.
Model Yield = Hybrid Fertilizer Hybrid*Fertilizer;
Run;
```

**SAS Output**

The SAS System 14:13 Thursday, May 12, 2005 1

The GLM Procedure

Class Level Information

| Class | Levels | Values |
|---|---|---|
| Hybrid | 3 | H1 H2 H3 |
| Fertilizer | 4 | F1 F2 F3 F4 |

Number of Observations Read 24

Number of Observations Used 24

The SAS System 14:13 Thursday, May 12, 2005 2

The GLM Procedure

Dependent Variable: Yield

| Source | DF | Sum of Squares | Mean Square | F Value | Pr > F |
|---|---|---|---|---|---|
| Model | 11 | 2991.194583 | 271.926780 | 13.57 | <.0001 |
| Error | 12 | 240.395000 | 20.032917 | | |
| Corrected Total | 23 | 3231.589583 | | | |

| R-Square | Coeff Var | Root MSE | Yield Mean |
|---|---|---|---|
| 0.925611 | 9.196092 | 4.475815 | 48.67083 |

| Source | DF | Type I SS | Mean Square | F Value | Pr > F |
|---|---|---|---|---|---|
| Hybrid | 2 | 1563.765833 | 781.882917 | 39.03 | <.0001 |
| Fertilizer | 3 | 1139.914583 | 379.971528 | 18.97 | <.0001 |
| Hybrid*Fertilizer | 6 | 287.514167 | 47.919028 | 2.39 | 0.0936 |

| Source | DF | Type III SS | Mean Square | F Value | Pr > F |
|---|---|---|---|---|---|
| Hybrid | 2 | 1563.765833 | 781.882917 | 39.03 | <.0001 |
| Fertilizer | 3 | 1139.914583 | 379.971528 | 18.97 | <.0001 |
| Hybrid*Fertilizer | 6 | 287.514167 | 47.919028 | 2.39 | 0.0936 |

"R-square" is $R^2$ = Model SQS / Corrected total SQS $= 2991.19/3231.59 = 0.926$. type I and type III SS are the same here because the model is balanced. If, for example, just one observation was omitted, the two types of SQS would differ. We do not attempt to describe their precise meaning here.

□

Consider a subspace $V^*$ corresponding to a row of the AOV table. Let $\nu^*$, SQS$^*$, and MSQ$^*$ be the corresponding degrees of freedom, sum of squares, and mean square. Let $\mathbf{Y}^* = p(\mathbf{Y} \mid V^*)$, $\boldsymbol{\theta}^* = p(\theta \mid V^*)$. Then $||\mathbf{Y}^*||^2/\sigma^2 = \text{SQS}^*/\sigma^2 \sim \chi^2_{\nu^*}(\delta^*)$, where $\delta^* = ||\boldsymbol{\theta}^*||^2/\sigma^2$, so that $E(\text{MSQ}^*) = \sigma^2[\nu^* + \delta^*]/\nu^* = \sigma^2 + ||\boldsymbol{\theta}^*||^2/\nu^*$. If a formula is available for the computation of $||\mathbf{Y}^*||^2/\sigma^2$, then $\delta^*$ may be obtained simply by replacing $\mathbf{Y}$ by $\boldsymbol{\theta}$ in the same formula. This formula for $E(\text{MSQ}^*)$ depends only on the model assumption that $\text{Cov}(\boldsymbol{\varepsilon}) = \sigma^2\mathbf{I}_n$, not on the assumption that the $\varepsilon_i$ are normally distributed.

Since the SQS for the rows of the table corresponding to hybrids, fertilizer, $\text{H} \times \text{F}$, and residual are independent (under the normality of the $\varepsilon_{ijk}$), and (residual SQS)$/\sigma^2 \sim \chi^2_\nu$ for $\nu = (K-1)(4)(3) = 12$, the noncentrality parameter for the $F$-statistic $F^* = \text{MSQ}^*/\text{EMS}$ can be found from the formula

$\delta^* = ||\boldsymbol{\theta}^*||^2/\sigma^2 = \nu^*(EMS^* - \sigma^2)/\sigma^2$. For the hybrid row $\boldsymbol{\theta}^* = \boldsymbol{\theta}_H = \sum_i \alpha_i \mathbf{R}_i$, so that $\delta^* = 8(\Sigma\alpha_i^2)/\sigma^2$.

Usually, we don't know the values of these parameters. However, we have chosen the $\mu_{ij}$, and therefore know the $\alpha_i$, $\beta_j$, and $(\alpha\beta)_{ij}$, so that we can determine them in this case. Thus, $\Sigma\alpha_i^2 = 158$, so that E(MSQ for hybrids) $= 25 + 8(158)/2 = 657$. Compare this with its estimate 781.9 for these data. Since $\Sigma\beta_j^2 = 110$, $E$(MSQ for fertilizer) $= 25 + 6(110)/3 = 245$. Its estimate is 380.0. $\Sigma(\alpha\beta)_{ij}^2 = 52$, so that $E$(MSQ for H $\times$ F interaction) $= 25 + 2(52)/6 = 42.33$, with estimate 47.92. $\sum_{ij}(\mu_{ij} - \mu)^2 = 1014$, so that $E$(MSQ for cell means)$= 25 + 1014(2)/11 = 209.36$, with estimate 271.93. Of course, $E$(MSQ for residuals) $= \sigma^2 = 25$, with estimate 20.03.

Tests of hyptheses should usually be performed by moving up the table. That is, first test the null hypotheses that the 12 cell means are equal. This is the one-way AOV $F$-test for one-way AOV with 12 treatment levels. The appropriate $F$-statistic is given on the line entitled "Among Cells." In this case we certainly reject at any reasonable $\alpha$-level. Then test for the presence of interaction. In this case, since the observed $p$-value is 0.094, we cannot say with any certainty that any interaction is present at all, and if it is, it is not as large in magnitude as the hybrid and fertilizer effects, for which the observed $p$-values are extremely small. The various $F$-statistics are not independent, because the denominators are the same for all, although the numerators are independent under the normality assumption.

$100\gamma\%$ confidence intervals on differences in cell means $\eta_{iji'j'} = \mu_{ij} - \mu_{i'j'}$ are given by $I_{iji'j'} = [\hat{Y}_{ij} - \hat{Y}_{i'j'} \pm t_{(1+\gamma)/2} S\sqrt{1/2 + 1/2}]$, since each cell sample mean has variance $\sigma^2/2$. $S^2 =$ EMS. For $\gamma = 0.95$ we get $I_{iji'j'} = [\hat{Y}_{ij} - \hat{Y}_{i'j'} \pm 9.75]$ for these data. These are individual confidence intervals. The user may wish to use a simultaneous CI method instead. The Bonferroni, Tukey, and Scheffé methods are available for that purpose. We do not discuss them here.

Similarly, $100\gamma\%$ CIs on differences $\eta_{ii'} = \alpha_i - \alpha_{i'} = \overline{\mu}_{i+} - \overline{\mu}_{i'+}$ are given by $[\overline{Y}_{i+} - \overline{Y}_{i'+} \pm t_{(1+\gamma)/2} S\sqrt{1/8 + 1/8}] = [\overline{Y}_{i+} - \overline{Y}_{i'+} \pm 4.82]$ for $\gamma = 0.95$. In the same way, $100\gamma\%$ CIs on $\delta_{jj'} = \beta_j - \beta_{j'}$ are given by $[\overline{Y}_{+j} - \overline{Y}_{+j'} \pm t_{(1+\gamma)/2} S\sqrt{1/6 + 1/6}] = [\overline{Y}_{+j} - \overline{Y}_{+j'} \pm 5.63]$.

**Power**

The $F$-statistic for testing of $H_0 : \boldsymbol{\theta}^* \equiv p(\boldsymbol{\theta} \mid V^*) = \mathbf{0}$ versus $H_a : \boldsymbol{\theta}^* \neq \mathbf{0}$ is $F^* = \text{MSQ}^*/S^2 . F^* \sim F(\nu^*, \nu_R, \delta^*)$, where $\nu_R =$ (residual df) and $\delta^* = ||\boldsymbol{\theta}^*||^2/\sigma^2$. The $S$-Plus function "pf" may therefore be used to determine power. For our example, for H $\times$ F interaction, $\delta^* \equiv \delta_{HF} = ||\theta_{HF}||^2/\sigma^2 = 2(52)/25 = 4.16$. For $\alpha = 0.05$, the $F$-test for no $HxF$ interaction rejects for $F > F_{0.95}(6, 12) = 2.996$. The $S$-Plus command "$1 - \text{pf}(2.996, 6, 12, 4.16)$" produced the power 0.184. Similarly, the noncentrality parameters for the $F$-tests for hybrid and fertilizer effects are $\delta_H = 8(158)/25 = 50.56$ and $\delta_F = 26.4$. The test of $H_0$: no hybrid effects rejects for $F_H > F_{0.95}(2, 12) = 3.88$, so that the power is $1 - \text{pf}(3.885, 2, 12, 50.56) = 0.9999$. The test of $H_0$: (no fertilizer effects) rejects for $F_{\text{Fert}} > 3.49$, so that the power is $1 - \text{pf}(3.49, 3, 12, 26.4) = 0.9660$.

Approximations of the power may be obtained from the Pearson–Hartley charts (E. S. Pearson and H. O. Hartley, *Biometrika Tables for Statisticians*, vol. I, Cambridge University Press, 1954). These charts, first published in 1951, before the days of computers were truly remarkable. The charts are determined as functions of $\alpha$ (usually, 0.05), numerator and denominator degrees of freedom, $\nu_1$ and $\nu_2$, and the parameter $\phi = \sqrt{\delta^*/(v_1+1)}$. For H $\times$ F interaction, $\phi = 0.77$. Since the charts for $\nu_1 = 6$, $\nu_2 = 12$ do not give power for $\phi < 1$, we can only guess that the power is roughly 0.20, not far from the 0.184 produced from S-Plus. If there had been eight, rather than two observations per cell, $\delta_{\text{HF}}$ would have been four times as large, so that $\phi$ would have been $2(0.77) = 1.54$, $\nu_1 = 6$, $\nu_2 = 84$, power approximately 0.84. The S-Plus command "1 – pf(qf(.95, 6, 84), 6, 12, 4*4.16)" produced power 0.822.

### Problems for Section 11.5

**11.5.1** Consider the two-way factorial experiment with two levels of factor $A$, three levels of factor $B$, and two observations $Y_{ijk}$ per cell.

**(a)** Suppose that the $\mu_{ij}$ are as follows: $\mu_{11} = 27$, $\mu_{12} = 22$, $\mu_{23} = 17$, $\mu_{21} = 11$, $\mu_{22} = 10$, $\mu_{23} = 3$. Find $\mu$, the $\alpha_i$, the $\beta_j$, and the $(\alpha\beta)_{ij}$.

**(b)** Plot the $\mu_{ij}$ as in Figure 11.5.1.

**(c)** The following data table of $Y_{ijk}$ was generated by letting $Y_{ijk} = \mu_{ij} + \varepsilon_{ijk}$, with $\varepsilon_{ijk}$ independent N(0, 25).

| | $B_1$ | $B_2$ | $B_3$ |
|---|---|---|---|
| $\mathbf{A}_1$ | 32.6, 21.9 | 20.3, 28.7 | 15.5, 18.1 |
| $\mathbf{A}_2$ | 9.4, 22.3 | 1.0, 6.6 | 10.4, 4.6 |

Perform an anaysis of variance with a table similar to Table 11.5.2 and test the null hypotheses for $\alpha = 0.05$ (or determine a $p$-value) for (i) all cell means are equal, (ii) no $A \times B$ interaction, (iii) no $A$ effect, and (iv) no $B$ effect.

**(d)** Plot the $\overline{Y}_{ij}$ in the same way that you plotted the $\mu_{ij}$ in part (b).

**(e)** Give 95% CIs on $\mu_{11} - \mu_{23}$, on $\alpha_1 - \alpha_2 = 2\alpha_1$, on $\beta_1 - \beta_2$, and on $\beta_1 - \beta_3$. Give a lower bound on the confidence that all four CIs contain the corresponding parameters.

**(f)** For the parameters given in part (a) with $\sigma = 5$, give the power of the tests performed in part (c).

**(g)** For the $(\alpha\beta)_{ij}$ determined in part (a) with $\sigma = 5$, how many observations $K$ would be needed in each cell to have power at least 0.90 for $\alpha = 0.05$?

**11.5.2** For $\mathbf{Y} = (Y_{ijk})$ as in Problem 11.5.1, let $\mathbf{R}_1, \mathbf{R}_2, \mathbf{C}_1, \mathbf{C}_2, \mathbf{C}_3$ be the row and column indicators.

**(a)** Write these vectors with six cells, two observations per cell, similar to that in Table 11.5.2. Do that also for the cell indicators $\mathbf{C}_{12}$ and $\mathbf{C}_{23}$.

**(b)** Define $V_0$, $V_R$, $V_C$, $V_{RC}$ similar to the way they were defined for the corn example. For the $\mu_{ij}$ given in Problem 11.5.1, determine the vectors $\boldsymbol{\theta}_0$, $\boldsymbol{\theta}_R$, $\boldsymbol{\theta}_C$, $\boldsymbol{\theta}_{RC}$, $\boldsymbol{\theta}^* = \boldsymbol{\theta}_0 + \boldsymbol{\theta}_R + \boldsymbol{\theta}_C$, and $\boldsymbol{\theta} = \boldsymbol{\theta}^* + \boldsymbol{\theta}_{RC}$.

**11.5.3** Suppose that we perform an experiment designed to measure the relationship between the temperature ($T$) at which a metal is formed, the type of metal ($M$), and its strength $Y$. Temperature has four levels, $T_1$, $T_2$, $T_3$, $T_4$, and metal has three levels, $M_1$, $M_2$, $M_3$. The cost of the experiment is so high that only one observation, $Y_{ij}$, could be taken for each combination of temperature level $T_i$ and metal level $M_j$. Consider the model $Y_{ij} = \mu + \tau_i + m_j + \varepsilon_{ij}$, where the $\tau_i$ and $m_j$ are the temperature and machine effects, with $\Sigma\tau_i = 0$, $\Sigma m_j = 0$, and $\varepsilon_{ij} \sim \mathrm{N}(0, \sigma^2)$, all independent. The following $Y_{ij}$ were observed:

| | $M_1$ | $M_2$ | $M_3$ |
|---|---|---|---|
| $T_1$ | 27 | 15 | 15 |
| $T_2$ | 20 | 14 | 11 |
| $T_3$ | 12 | 6 | 12 |
| $T_4$ | 13 | 1 | 10 |

**(a)** Since the subspace $V$ spanned by the cell indicators is all of $R_{12}$, the subspace $V^\perp$ usually used to estimate $\sigma^2$ consists only of the zero vector. That is, $\dim(V^\perp) = 0$. However, under the additive model proposed, that the interaction terms $(\tau m)_{ij}$ are all zero, we can use what normally would be the interaction subspace as the residual (or error) subspace, so that the new denominator for the $F$-tests for temperature and metal effects becomes the residual MSQ, which was called the interaction MSQ for the model that included interaction. The full-model subspace is $V = V_0 + V_R + V_C$, and the error space is $V^\perp = V_{RC}$. For these data, determine the AOV table, omitting the last two rows corresponding to Table 11.5.2, and relabeling the rows with sources "H $\times$ F interaction" and "Among cells" as "Residual" and "Corrected total."

**(b)** Plot the data, with the four temperature levels on the $x$-axis.

**(c)** Find a 95% CI on $\eta_{12} = \alpha_1 - \alpha_2$.

**(d)** Suppose that the four temperatures were $T = 600, 700, 800, 900$. Define a multiple regression model for which just one term $\beta T$ corresponds to the temperature effect; that is, each $E(Y_{ij})$ is a linear function of $T$. Give the $12 \times 4$ design matrix $X$.

**(e)** Carry out the regression analysis for the model of part (d) and test $H_0$: The model of part (d) holds, given the full model stated in part (a),

$\alpha = 0.05$. It is possible to find formulas that are stated without the use of matrix algebra, so that computations could be done using a simple hand calculator, but a more general approach using matrix algebra would probably take less time.

**11.5.4** Consider factors $A$ and $B$ with $a \geq 2$ and $b \geq 2$ levels. Let $i$ index the levels of $A$. Let $j$ index the levels of $B$. Suppose that $K_{ij}$ observations $Y_{ijk}$ are taken on a variable $Y$ at level $i$ of $A$, level $j$ of $B$. Let these $Y_{ijk}$ be the entries of a table, with $i$ corresponding to row $i$, $j$ to column $j$, with cell $ij$ containing these $K_{ij}$ $Y_{ijk}$. Define $V_0$ as before. Let $\mathbf{R}_1, \ldots, \mathbf{R}_a$ and $\mathbf{C}_1, \ldots, \mathbf{C}_b$ be the indicators of the rows and columns. Let $V_R = \mathcal{L}(R_1, \ldots, R_a) \cap V_0^\perp$ and $V_C = \mathcal{L}(C_1, \ldots, C_b) \cap V_0^\perp$.

**(a)** Every vector in $V_R$ is of the form $\mathbf{v} = \Sigma a_i \mathbf{R}_i$. Give necessary and sufficient conditions on the $a_i$ such that $\mathbf{v} \in V_R$. Similarly, every vector in $V_C$ is of the form $\mathbf{w} = \Sigma b_j \mathbf{C}_j$. Give necessary and sufficient conditions on the $b_j$ such that $\mathbf{w} \in V_C$.

**(b)** What conditions must the $K_{ij}$ satisfy in order that $V_R \perp V_C$? Give examples of $K_{ij}$, not all equal, for the case that $a = 2, b = 3$ so that (1) $V_R \perp V_C$, and (2) $V_R(\text{not } \perp)V_C$.

**11.5.5** Give examples of $2 \times 3$ tables of data, with two observations per cell so that the following conditions are satisfied:

**(a)** $\text{SSA} > 0,\ \text{SSB} > 0,\ \text{SSAB} = 0,\ \text{SSE} = 12$.

**(b)** $\text{SSA} > 0,\ \text{SSB} = 0,\ \text{SSAB} = 8,\ \text{SSE} = 0$.

**(c)** $\text{SSA} = 300,\ \text{SSB} = 392,\ \text{SSAB} = 0,\ \text{SSE} = 108$.

CHAPTER TWELVE

# Frequency Data

## 12.1 INTRODUCTION

In this chapter we deal with count data. We considered such examples in Chapter Eight when observations were modeled as binomial or Poisson. The theory and statistical methods we discuss here can be considered to be the extension of the linear models and methods of Chapter Eleven to the case of Poisson, binomial, and multinomial distributions. The distribution theory is somewhat more difficult because it relies on limit theorems. Rather than attempting full proofs, we describe the results, make intuitive arguments, and demonstrate the approximations they provide by simulation. We begin with a discussion of the interval estimation of binomial and Poisson parameters, then discuss logistic regression, the extension to binomial data of simple and multiple linear regression.

## 12.2 CONFIDENCE INTERVALS ON BINOMIAL AND POISSON PARAMETERS

In Chapter Seven we discussed confidence intervals on the parameters $p$ for the binomial model and $\lambda$ for the Poisson model. These intervals rely on the normal approximations, which can be poor if $p$ is close to 0 or 1 or if the sample size $n$ is small. Better methods are available, which work, for example, for $n = 10$, and clearly $p$ is quite small. They rely on a rather fundamental inequality.

**The $F(X)$ Inequality** If $X$ has cdf $F$ and $0 < \alpha < 1$, then $P(F(X) \leq \alpha) \leq \alpha$.

*Proof:* Define $x_\alpha = \min\{x \mid F(x) \geq \alpha\}$. The fact that $F$ is continuous on the right assures us that $F(x_\alpha) \geq \alpha$. If $F$ is continuous at $x_\alpha$, then $P(F(X) \leq \alpha) = P(X \leq x_\alpha) = F(x_\alpha) = \alpha$. If $F(X_\alpha) > \alpha$, so that $F$ has a jump at $x_\alpha$ and $P(X = x_\alpha) > 0$, then $P(F(X) \leq \alpha) = P(X < x_\alpha) < \alpha$. □

---

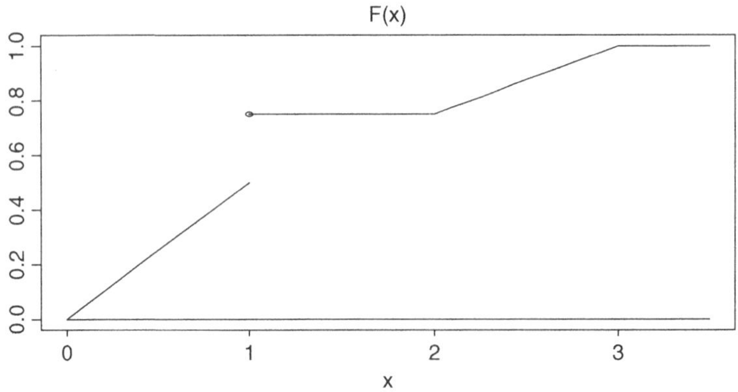


**FIGURE 12.2.1** CDF of a discrete–continuous mixture.

***Example 12.2.1*** Consider, for example, the cdf $F$, defined as follows: $F(x) = 0$ for $x \le 0$, $F(x) = x/2$ for $0 < x \le 1$, $F(x) = 3/4$ for $1 < x \le 2$, $F(x) = (x+1)/4$ for $2 < x \le 3$, $F(x) = 1$ for $x > 3$ (see Figure 12.2.1). For $0 < \alpha \le 1/2$, $x_\alpha = 2\alpha$. For $1/2 < \alpha \le 3/4$, $x_\alpha = 1$. For $3/4 < \alpha \le 1$, $x_\alpha = 4\alpha - 1$. Thus, for $1/2 < \alpha < 3/4$, $P(F(X) \le \alpha) = 1/2 < \alpha$. Otherwise, $P(F(X) \le \alpha) = \alpha$.

Define $\bar{F}(x) = P(X \ge x)$ for each $x$. Let $Y = -X$. The cdf of $Y$ is $G(y) = P(-X \le y) = P(X \ge -y) = \bar{F}(-y)$. Then, for any $\alpha$, $0 < \alpha < 1$, $P(\bar{F}(X) \le \alpha) = P(G(-X) \le \alpha) \le \alpha$ by the $F(X)$ inequality.

Now suppose that $X \sim \text{Binomial}(n, p)$ for $0 < p < 1$, and let its cdf be $F(x; p). F(x; p)$ is a continuous monotone decreasing function of $p$ for each $x$, $0 \le x \le n$, with limits 1 at $p = 0$ and 0 at $p = 1$. Let $\alpha_2$ be a chosen number, $0 < \alpha_2 < 1$ (usually small). Let $U(x)$ be the value of $p$ for which $F(x; p) = 1 - \alpha_2$. Then $P(p \ge U(X)) = P(F(X; p) \le F(X; U(X)) = P(F(X; p) \le \alpha_2) \le \alpha_2$ by the $F(X)$ inequality. Thus, $P(U(X) > p) \ge 1 - \alpha_2$, so that $U(X)$ is an upper $100(1 - \alpha_2)$ CI on $p$. For example, suppose that $n = 10$ and we observe $X = 0$. Choose $\alpha_2 = 0.05$. Since $F(0; p) = (1-p)^{10}$, $U(0) = 1 - \alpha^{1/10} = 0.2589$. If $X = 1$, $n = 10$, then $U(1)$ is the solution to $F(1; p) = \binom{n}{1} p^1 q^{n-1} + (1-p)^n = 0.05$. This can be solved numerically. We find $U(1) = 0.5069$. Later we introduce an easier method.

Let $\alpha_1$ be another chosen probability. $L(x)$ be value of $p$ for which $\bar{F}(x; p) = \alpha_1$. Then $P(p \le L(X)) = P(\bar{F}(X; p) \le \bar{F}(X; L(X))) = P(\bar{F}(X; p) \le \alpha_1) \le \alpha_1$. Hence, $P(L(X) < p) \ge 1 - \alpha_1$. For $n = 10$, $X = 9$, $L(9)$ is the solution to $\bar{F}(9; p) = \alpha_1$. For $n = 10$, $\alpha_1 = 0.05$, we find that $L(9) = 0.7411$. See Figure 12.2.2 for lower and upper 95% confidence limits for the case that $n = 20$ and $X$ observed to be 2 and 5. The values of $p$ for which each of these functions takes the value 0.05 are $L(2) = 0.0181$, $U(2) = 0.2826$, $L(5) = 0.1041$, $U(5) = 0.4556$. For example, $\bar{F}(5; 0.1041) = 0.05$. The interval $[L(5), U(5)]$ is a 90% CI on $p$. □

We do not derive it here, but the following method exploits the relationships between $F(x; p)$ as functions of $p$ for fixed $x$ and beta distributions, and between beta

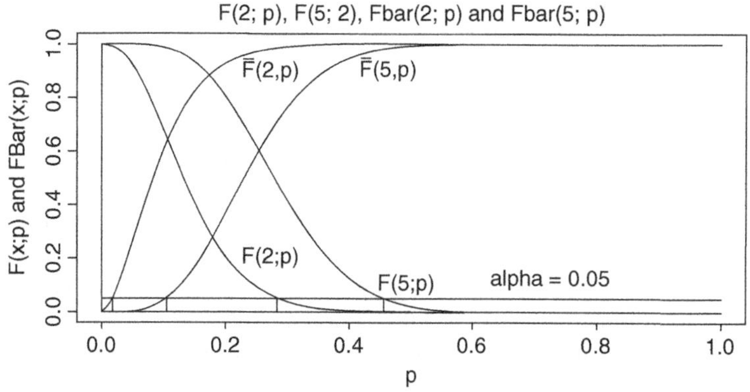


**FIGURE 12.2.2**

distributions and the $F$-distribution to provide simple formulas for $L(x)$ and $U(x)$. Let $\nu_1 = 2(n+1-x)$, $\nu_2 = 2x$, $L(x) = 1/[1 + (\nu_1/\nu_2)F_{1-\alpha_1}(\nu_1, \nu_2)]$. Let $\nu_1 = 2(x+1)$, $\nu_2 = 2(n-x)$, $W = (\nu_1/\nu_2)F_{1-\alpha_2}(\nu_1, \nu_2)$, $U(x) = W/(1+W)$. Then $P(L(X) \geq p; p) \leq \alpha_1$ and $P(U(X) \leq p; p) \leq \alpha_2$ for all $p$. Thus, $P(L(X) < p < U(X)) \geq 1 - \alpha_1 - \alpha_2$. That is, $[L(X), U(X)]$ is a $100(1-\alpha_1-\alpha_2)\%$ CI on $p$. We refer to this method as the *F-method*. The $F$ refers to the cdf., although the $F$-distribution is also involved.

### Confidence Intervals on $\Delta \equiv p_1 - p_2$

Suppose the $X_1 \sim \text{Binomial}(n_1, p_1)$, $X_2 \sim \text{Binomial}(n_2, p_2)$, with $X_1, X_2$ independent. These might, for example, be the numbers preferring $A$ to $B$ in sampling of voters before an election. The model would be appropriate if the number of possible voters is large compared to $n_1$ and $n_2$, and samples are taken independently. Since $Z = (\hat{p}_1 - \hat{p}_2)/\sqrt{\hat{\text{Var}}(\hat{p}_1 - \hat{p}_2)}$, where $\hat{\text{Var}}(\hat{p}_1 - \hat{p}_2) = \hat{p}_1(1-\hat{p}_1)/n_1 + \hat{p}_2(1-\hat{p}_2)/n_2$ is approximately distribution as N(0, 1) for large $n_1, n_2$, with $p_1, p_2$ not too close to 0 or 1, the interval $[(\hat{\Delta} = \hat{p}_1 - \hat{p}_2) \pm z_{1-\alpha/2}\sqrt{\hat{\text{Var}}(\hat{p}_1 - \hat{p}_2)}]$ is an approximate $100(1-\alpha)\%$ CI on $\Delta = p_1 - p_2$. Call this the *Wald interval*. However, the actual coverage probability is a function of $p_1, p_2, n_1, n_2$, and $\alpha$ which may be somewhat less than the nominal value, especially for smaller $n_1, n_2$, and $p_1, p_2$ near 0 or 1. An improvement is given by a method of Agresti and Coull (1998). Let $\tilde{p}_i = (X_i + 1)/(n_i + 2)$ for $i = 1, 2$. The interval is then the same but with the $\tilde{p}_i$ replacing the $\hat{p}_i$ and $n_i + 2$ replacing $n_i$ for each $i$. We pretend that there are two more observations for each $i$, one of the two being a "success." The random interval then has coverage probability closer to its nominal value. For example, if $n_1 = 30$, $n_2 = 40$, $X_1 = 13$, $X_2 = 23$, then $\hat{p}_1 = 13/30$, $\tilde{p}_1 = 14/32$, $\hat{p}_2 = 23/40$, $\tilde{p}_2 = 24/42$, and the Wald 95% interval $[-0.142 \pm 0.234]$ and Agresti–Coull 95% interval $[-0.134 \pm 0.228]$. This last interval was obtained by a method that has coverage probability closer to 0.95 than does the Wald method. For larger $n_1, n_2$, it makes little difference which method is used.

### Confidence Intervals on $\lambda$

Let $X \sim$ Poisson($\lambda$), for $\lambda > 0$. Let $F(x) = P(X \le x; \lambda)$ and $\bar{F}(x;\lambda) = P(X \ge x;\lambda)$. $F(x;\lambda)$ is a continuous monotone decreasing function of $\lambda$ for each $x$, and $\bar{F}(x;\lambda)$ is a continuous monotone increasing function of $\lambda$. Let $L(x)$ be the value of $\lambda$ for which $\bar{F}(x;\lambda) = \alpha_1$ and let $U(x)$ be the value of $\lambda$ for which $F(x;\lambda) = \alpha_2$. Then, as for the binomial distribution, $L(X)$ is a $100(1-\alpha_1)$ lower confidence limit on $\lambda$ and $U(X)$ is an upper $100(1-\alpha_2)\%$ confidence limit on $\lambda$. It follows that $[L(X), U(X)]$ is a $100(1-\alpha_1-\alpha_2)\%$ confidence interval on $\lambda$. For example, for observed $X = 0$, $F(0;\lambda) = e^{-\lambda}$. Setting this equal to $\alpha_2 = 0.05$, we get the upper 95% confidence limit $-\log(0.05) = 2.996$. For observed $X = 1$, $U(1)$ is the solution to $e^{-\lambda}(1+\lambda) = 0.05$. We can solve this numerically to get $U(1) = 4.744$.

We refer to this method as the *F-method*, as we did for CIs on $p$. Again the $F$ refers to the cdf. The $F$-distribution is not involved in this case. Fortunately, we can exploit the relationships between the Poisson distribution and the gamma and between the gamma and chi-square to give the following formulas for $L(x)$ and $U(x)$ (see Section 4.3): $L(x) = \chi^2(\alpha_1, 2x)/2$ (the $1-\alpha_1$ quantile of the chi-square distribution with $2x$ df). Let $U(x) = \chi^2(1-\alpha_2, 2(x+1))/2$. Then $P(L(X) < \lambda;\lambda) \ge 1-\alpha_1$ and $P(U(X) > \lambda;\lambda) \ge 1-\alpha_2$. For example, for $X = 1, \alpha_1 = 0.05$, $U(2) = \chi^2(0.95, 4)/2 = 9.488/2 = 4.744$, as we obtained by numerical methods.

For large $\lambda$ we can take advantage of the fact that $Z_\lambda = (X-\lambda)/\sqrt{\lambda}$ converges in distribution to N(0, 1) as $\lambda \to \infty$. Thus, we can take $I(X) = \{\lambda \mid |Z_\lambda| \le 1.96\} = \{\lambda \mid Q(\lambda) \equiv Z_\lambda^2 - 1.96^2 \le 0\} = \{\lambda \mid r_1 \le \lambda \le r_2\}$, where $r_1$ and $r_2$ are the roots of $Q(\lambda) = 0$. Or, for still larger $\lambda$ (as suggested by large $X$) we can replace $\lambda$ in the denominator of $Z_\lambda$ by $X$ to get the interval $[X \pm 1.96X^{1/2}]$. For example, for observed $X = 50$, to get a 95% interval, the $F$-method gives the interval $[\chi^2(0.025, 100)/2, \chi^2(0.975, 102)/2] = [37.11, 65.92]$. The $[r_1, r_2]$ method gives $[37.93, 65.91]$ and $[\bar{X} \pm 1.96\bar{X}^{1/2}] = [36.14, 63.86]$.

### Confidence Intervals on $R \equiv \lambda_1/\lambda_2$

Let $X_1$ and $X_2$ be independent, with $X_i \sim$ Poisson($\lambda_i$), $i = 1, 2$. For example, $X_1, X_2$ might be the number of murders, or the numbers of deaths from lung cancer, in a large city during two consecutive years. In the case of murders it might be necessary to consider incidents of murder, since there may be cases of multiple murder, causing the Poisson model to be suspect.

As shown in Example 5.2.1, conditionally on $Y = X_1 + X_2 = n$, $X_1 \sim$ Binomial($n, \theta$), where $\theta = \lambda_1/(\lambda_1 + \lambda_2)$. Let $R = \lambda_1/\lambda_2$. Then $\theta = R/(1+R)$ and $R = \theta/(1-\theta)$. That is, $\theta$ is the odds for a count for $X_1$, given $Y = n = X_1 + X_2$. We can use the lower and upper confidence limits $L(X_1)$ and $U(X_1)$ to establish $100(1-\alpha_1)\%$ and $100(1-\alpha_2)\%$ limits. If, for example, $U(X_1)$ is an upper 95% confidence interval on $\theta$, then since $R = \theta/(1-\theta)$ is a monotone increasing function of $\theta$, $U_R(X_1) = U(X_1)/(1-U(X_1))$ is a corresponding upper $100(1-\alpha_1)\%$ confidence limit on $R$.

***Example 12.2.2*** The number of cases of colon cancer reported among 15,761 men aged 40 to 55 living in county $A$, in which a nuclear reactor was located, over a five-year period, was 163. During the same time period in four adjacent counties there were 43,783 men of those ages, of which 427 were reported to have colon cancer. The rates per 1000 such men per year were $\rho_1 = \lambda_1/t_1$ and $\rho_2 = \lambda_2/t_2$, where $t_1 = (15{,}761/1000)/5$ and $t_2 = (43{,}783/1000)/5$. We would like lower and upper confidence limits on $\tau \equiv \rho_1/\rho_2 = (\lambda_1/\lambda_2)(t_2/t_1) = [\theta/(1-\theta)](t_2/t_1)$. Then $n = 510$, so that $L(X_1) = L(163) = 0.246$ is a lower 95% confidence limit on $\theta$, using the $F$-method. Similarly, $U(X_1) = U(163) = 0.308$ is an upper 95% confidence limit on $\theta$. It follows that $[0.246, 0.308]$, $[0.326, 0.445]$, and $(t_2/t_1)[0.326, 0.445] = [0.918, 1.253]$ are 90% confidence intervals on $\theta$, $R = \lambda_1/\lambda_2$, and $\tau = \rho_1/\rho_2$. The estimator $\hat{\tau} = (t_2/t_1)(X_1/X_2)$ is consistent for $\tau$ as $\lambda_1$ and $\lambda_2$ converge to $\infty$, with $\lambda_1/\lambda_2$ converging to $\tau$.

We could have estimated $\theta$ by $\hat{\theta} = X_1/(X_1 + X_2)$, then determined a 90% confidence limit on $\theta$ by $[\hat{\theta} \pm 1.645\sqrt{\hat{\theta}(1-\hat{\theta})/n}]$. In this case we get $[0.255, 0.317]$, rather close to the interval $[0.246, 0.308]$ found using the $F$-method. The resulting 90% CI on $\tau$ is $[0.965, 1.303]$, which is approximately the same as the interval above. In either case we would fail to reject $H_0\colon \rho = 1$ versus $H_a\colon \rho \neq 1$ at level $\alpha = 0.10$. □

## Problems for Section 12.2

**12.2.1** Let $X \sim \text{Binomial}(n = 8, \theta)$.

**(a)** For $X = 0$, find an upper 90% confidence limit on $\theta$. Do it in two ways: by using the quantiles of the $F$-distribution, and also as the solution to $F(0;\theta) = 0.10$.

**(b)** For $X = 8$, find a lower 90% confidence limit on $\theta$.

**(c)** For $X = 1$, find an upper 90% confidence limit on $\theta$ using the two methods of part (a).

**(d)** For $X = 4$, use the $F$-distribution method to find 95% lower and upper confidence limits on $\theta$.

**12.2.2** During 2005 there were 27 accidents at the corner of Washington and Lincoln in Adams City. After a change in the timing of traffic signals the number dropped to 17 during 2006. State a model and then find a 95% confidence interval on the ratio $\rho = \lambda_{2006}/\lambda_{2005}$. You will need to linearly interpolate in the $F$-table.

**12.2.3** Let $X$ have cdf $F(x)$, where $F(x) = x^2/2$ for $0 \le x < 1$ and $F(x) = 1/2 + x/4$ for $1 \le x < 2$. For each $\alpha$, $0 < \alpha < 1$, find $P(F(X) \le \alpha)$.

**12.2.4** Let $X$ denote the number of accidents among the employees of the Safe Bear Trap Company during 2007. Suppose that $X \sim \text{Poisson}(\lambda)$.

**(a)** Find 95% lower and upper confidence limits on $\lambda$ for $X = 10$.

**(b)** Use your calculator and numerical approximation to find an upper 95% confidence limit on $\lambda$ for $X = 2$. Verify your answer using chi-square quantiles.

**12.2.5** Let $X_1$ and $X_2$ be the numbers of deaths among babies born to mothers 21 or younger, and over 21, during the babies' first year. There were 1534 births and 7842 births to the younger and older mothers in Podunk City during 2006. Among these the numbers of deaths were 63 and 207.

**(a)** State a Poisson model.

**(b)** Use this model to define a parameter that would be of interest to study the effect of age on the likelihood of death. Then find a 95% confidence interval on this parameter. What do you conclude? Also find a point estimate of the parameter.

**12.2.6** Let $F(k; p)$ be the cdf of the binomial distribution for parameters $n$ and $p$. Show that for each fixed $k$, $F(k; p)$ is a decreasing function of $p$. *Hint:* Let $U_1, \ldots, U_n$ be a random sample from the Unif(0, 1) distribution. Let $p_1 < p_2$. Let $I_j = I[U_j \leq p_1]$ and $H_j = I[U_j \leq p_2]$. Compare $I_j$ and $H_j$. What are the distributions of $\Sigma I_j$ and of $\Sigma H_j$?

**12.2.7** Let $X \sim$ Binomial $(n, p)$. Show that $L(x)$, for $X = x$ observed, is a lower $100(1 - \alpha)\%$ confidence limit on $p$, for $L(x) = 1/[1 + (\nu_1/\nu_2)F_{1-\alpha_2}(\nu_1, \nu_2)]$, where $\nu_1 = 2(n + 1 - x)$, $\nu_2 = 2x$. *Hint:* Let $U_1, \ldots, U_n$ be a random sample from the Unif(0, 1) distribution and let $U_{(1)}, \ldots, U_{(n)}$ be the corresponding order statistics. Let $Y =$ (no. of $U_j \leq p$). From Section 5.3, $U_{(s)}$ has the Beta$(s, n + 1 - s)$ distribution. Also see Problem 9.5.4, relating the beta and $F$-distributions.

**12.2.8** Let $F(k; \lambda)$ be the cdf of $X \sim$ Poisson($\lambda$). Show that $F(k; \lambda)$ is a decreasing function of $\lambda$ for each $k$. *Hint:* Let $\lambda_1 < \lambda_2$. Define $U_j$ and $U_{(j)}$ as in Problem 12.2.6 for $j = 1, \ldots, n$. Let $Y_{1j} = -\log(1 - U_j)$ for each $j$. Let $X_1 =$ (no. $Y_j < \lambda_1$) and $X_2 =$ (no. $Y_j < \lambda_2$). Then from note 5 in Section 4.3, $X_1 \sim$ Poisson($\lambda_1$ and $X_2 \sim$ Poisson($\lambda_2$). Then $X_1 \leq X_2$ with probability 1.

## 12.3 LOGISTIC REGRESSION

Consider a study of two drugs $A$ and $B$ designed to decrease the probability of death within the next five years among men diagnosed to have a certain heart condition. The probability of death is a function of the drug $d$, the age $x$ of the man, and some other variables that we ignore for the present. Suppose that 800 volunteers took part in the study and that 400 were chosen randomly to receive drug $A$, the others to receive drug $B$. After five years the indicator of death $Y_i$ was determined. For reasons considered to be independent of the effects of the drugs, 13 of those taking drug $A$, and nine of those taking drug $B$ dropped out of the study, so that their $Y_i$ values were not determined. Thus, 778 observations $(d_i, x_i, Y_i)$ were made.

Let $p(d, x)$ be the probability of death of a man assigned to drug $d$, who was $x$ years old at the beginning of the study. Let $h(d, x) = p(d, x)/[1 - p(d, x)]$, the "odds for death" for a man using drug $d$, of age $x$. Let $\mu(d, x) = \log(h(d, x))$, the "log odds" for death. Suppose that $\mu(d, x) = \beta_0 + \beta_1 d_A + \beta_2 x$, where $d_A$ is the indicator for drug $A$. Suppose that $Y_i \sim \text{Bernoulli}(p(d_i, x_i))$ and are independent for $i = 1, \ldots, 778$. This is a logistic regression model. We are most interested in the coefficient $\beta_1$, since the odds ratio for death for the two drugs is $R \equiv h(A, x)/h(B, x) = \exp(\mu(A, x) - \mu(B, x)) = \exp(\beta_1)$. "exp" is the exponential function, so that $\exp(w) = e^w$ for any $w$. For example, if $\beta_1 = -0.5$, the odds for death of a man using drug $A$ are $\exp(\beta_1) = 0.6065$ times as great as they are for a man using drug $B$.

The inverse of the log-odds function $\mu = \log[p/(1 - p)]$ is easily determined to be $h(\mu) = e^\mu/(1 + e^\mu)$, so that $p(d, x) = h(\mu(d, x))$. Let $\mathbf{y}$ be the vector indicating deaths among the 778 men. Let $\mathbf{x}_0$ be the vector of 778 ones. Let $\mathbf{x}_a$ be the vector of ages and $\mathbf{d}_A$ be the vector indicating which of the men received drug $A$. Finally, let $\mathbf{p}$ and $\boldsymbol{\mu} = \log[\mathbf{p}/(\mathbf{1} - \mathbf{p})]$ be the vectors of probabilities and corresponding log odds. Then our model states that $\boldsymbol{\mu} = \beta_0\mathbf{x}_0 + \beta_1\mathbf{d}_A + \beta_2\mathbf{x}_a$. Let $\hat{\boldsymbol{\beta}}$ be the maximum likelihood estimator of $\boldsymbol{\beta} = (\beta_0, \beta_1, \beta_2)$. Let $\hat{\boldsymbol{\mu}} = \hat{\beta}_0\mathbf{x}_0 + \hat{\beta}_1\mathbf{d}_A + \hat{\beta}_2\mathbf{x}_a = \mathbf{X}\hat{\beta}$, where $\mathbf{X}$ is the $778 \times 3$ matrix $(\mathbf{x}_0, \mathbf{d}_A, \mathbf{x}_a)$. Let $\hat{\mathbf{p}} = h(\hat{\boldsymbol{\mu}})$, so that $\hat{\mathbf{p}}$ is the 778-component vector of estimates of probabilities of death. Similarly to the orthogonality conditions that $\hat{\mathbf{Y}}$ must satisfy for linear models, the MLE estimator $\hat{\boldsymbol{\beta}}$ must be chosen so that the inner product of $(\mathbf{y} - \hat{\mathbf{p}})$ with each of the explanatory vectors, $\mathbf{x}_0$, $\mathbf{d}_A$, $\mathbf{x}_a$ must be zero. Put another way, $\mathbf{X}^T(\mathbf{y} - \hat{\mathbf{p}}) = (0, 0, 0)^T$, equivalently $\mathbf{X}^T\hat{\mathbf{p}} = \mathbf{X}^T\mathbf{y}$. We do not provide a proof here (See Stapleton, 1995, Chap. 8).

The solution $\hat{\boldsymbol{\beta}}$ exists under rather general conditions. $\mathbf{X}$ must have rank 3 if the equation $\mathbf{X}^T\hat{\mathbf{p}} = \mathbf{X}^T\mathbf{y}$ is to have a unique solution. Instead, we demonstrate logistic regression analysis with simulations. Let $\boldsymbol{\beta} = (-5, 0.6, 0.07)$. The 778 ages were chosen by simulating a sample from the N(70, 25) distribution and rounding off to the nearest integer. The mean age was 69.96. The mean age for those receiving drug $A$ was 69.74. For drug $B$ it was 69.99. The probabilities of death within the next five years, depending on age and the drug used, were then determined. Their sum was 424.56. Then the vector $\mathbf{y}$, indicating the event of death for each of the patients, was chosen using the function "pbinom" in S-Plus. The components of $\mathbf{y}$ were independent Bernoulli random variables. Their sum was 428, meaning that 428 of the 778 men died. The age–drug–$\mathbf{y}$ information is summarized in Table 12.3.1.

Later we discuss a method that makes use of these frequencies directly rather than through the logistic model. The following command in S-Plus produced a logistic regression analysis: "druganalysis = glm($Y \sim d +$ ages, family = binomial)". $\mathbf{Y}$ was the $778 \times 2$ matrix with first column $\mathbf{y}$, second column $\mathbf{1}_{778} - \mathbf{y}$, where $\mathbf{1}_{778}$ is the vector of 778 1's. $\mathbf{d}$ is the vector indicating drug $A$, **ages** is the vector of ages.

TABLE 12.3.1

| | Drug $A$ | | | Drug $B$ | | |
|---|---|---|---|---|---|---|
| Age | Alive | Dead | % Dead | Alive | Dead | % Dead |
| 45–54 | 2 | 2 | 50.0 | 6 | 1 | 14.3 |
| 55–64 | 54 | 41 | 43.2 | 63 | 27 | 30.0 |
| 65–74 | 69 | 115 | 62.5 | 93 | 92 | 49.7 |
| 75–84 | 30 | 64 | 68.1 | 30 | 68 | 69.4 |
| 85–94 | 2 | 13 | 86.7 | 2 | 5 | 71.4 |

Part of the report produced by the command "summary(druganalysis)" was as follows:

| | Value | Std. Error | t-value |
|---|---|---|---|
| (Intercept) | 6.06 | 0.787 | −7.69 |
| d | 0.406 | 0.152 | 2.68 |
| ages | 0.067 | 0.0112 | 7.77 |

The column entitled "t-value" is the corresponding $\hat{\beta}$ divided by the estimate of its standard deviation, the standard error. The asymptotic theory states that when the sample size is large and the age distributions are "reasonable," the $t$-value is approximately distributed as N(0, 1) when the true $\beta$ is zero. 95% CIs can be determined by adding and subtracting 1.96 times the "std. error" to the $\hat{\beta}$. Thus, $[0.406 \pm 0.298]$ is a 95% CI on $\beta_1$, the "drug $A$ effect." Based on these data we can be quite sure that drug $A$ produces higher percentages of deaths within five years. That is, drug $B$ is better.

Figure 12.3.1 contains graphs of the functions $p(A, x)$ and $p(B, x)$, together with estimates based on three repetitions of the experiment. In all three cases the value of $\hat{\beta}_1$ is larger than the one given for our first sample. The simulation experiment was repeated 1000 times for the same parameters. The results are shown in Figure 12.3.2. The asymptotic theory, which will be described shortly, is well supported by the empirical evidence from the simulations. Let $\mathbf{X}_n$ be the design matrix corresponding to $n$ patients. Let $\mathbf{p}_n$ be the corresponding vector of probabilities of death. Then as $n \to \infty$, the distribution of $(\hat{\beta}_n - \beta)$ may be approximated by $N_3(0, \mathbf{M}_n^{-1})$, where $\mathbf{M}_n = \mathbf{X}_n^T \mathbf{D}_n \mathbf{X}_n$ and $\mathbf{D}_n = \text{diag}(\mathbf{p}_n(\mathbf{1}_n - \mathbf{p}_n))$. [The multiplication in $\mathbf{p}_n(\mathbf{1}_n - \mathbf{p}_n)$ is componentwise.) The approximation is good if the proportions of patients getting either drug is not close to zero or 1, and few or none of the ages are not "too far" from the other ages. That is, we should not expect good approximations if, for example, only 25 or 50 patients among 1000 receive drug $B$, or a patient is 120 years old. If $Y_i \sim \text{Binomial}(r_i, p_i)$, the $ii$ component of $D_n$ should be $r_i p_{ni}(1 - p_{ni})$. In vector language, $\mathbf{D}_n$ becomes $\mathbf{rp(1 - p)}$, where $\mathbf{r} = (r_1, \ldots, r_n)$.

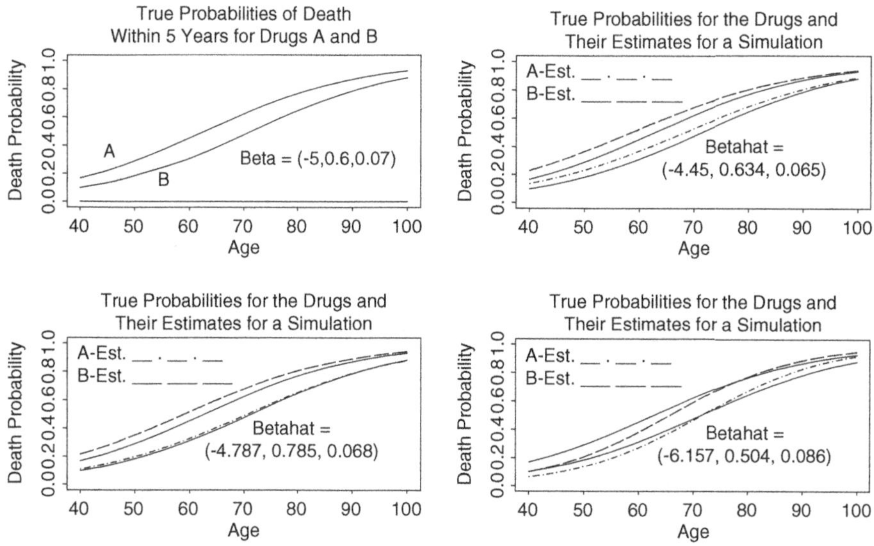


**FIGURE 12.3.1** $p(d, x)$ for Drugs $A$ and $B$ and estimates.

For these data we get $\mathbf{M}_n^{-1} = \mathbf{M}_{778}^{-1} = \begin{pmatrix} 1.3031 & -0.0.351 & -0.0192 \\ -0.0351 & 0.0575 & 0.000183 \\ -0.0192 & 0.000183 & 0.000289 \end{pmatrix}$.

Thus, the theory suggests that $\hat{\beta}_1$ should be approximately distributed as $N(0.6, 0.0575 = 0.240^2)$ and that $\hat{\beta}_2$ should be approximately distributed as $N(0.10, 0.000289 = 0.017^2)$. The simulations show that to be the case in very good approximation. The correlation among the $1000(\hat{\beta}_1, \hat{\beta}_2)$ pairs is quite close to the theoretical value $-0.000183/\sqrt{(0.0575)(0.000289)} = -0.045$.

## Estimation of $x_\eta$ and $p(d, x_0)$

For $0 < \eta < 1$, define $x_\eta$ to be the value of $x$ satisfying $p(A, x_\eta) = \eta$. Similarly, let $x_\eta^*$ satisfy $p(B, x_\eta^*) = \eta$. Since $\mu(d, x) = \beta_0 + \beta_1 d_A + \beta_2 x = \log(p(d, x)/(1 - p(d, x))$, $x_\eta$ is the $x$ solution to $\beta_0 + \beta_1 + \beta_2 x = \log(\eta/(1 - \eta)) \equiv \omega$ , and $x_\eta^*$ is the solution to $\beta_0 + \beta_2 x = \omega$. That is, $x_\eta = (\omega - \beta_0 - \beta_1)/\beta_2$ and $x_\eta^* = (\omega - \beta_0)/\beta_2$. Let $\hat{x}_\eta = (\omega - \hat{\beta}_0 - \hat{\beta}_1)/\hat{\beta}_2$ and $\hat{x}_\eta^* = (\omega - \hat{\beta}_0)/\hat{\beta}_2$. The $\delta$-method of Chapter Seven may be used to approximate the distribution of $(\hat{x}_\eta, \hat{x}_\eta^*)$. For large $n$, $(\hat{x}_\eta, \hat{x}_\eta^*)$ is approximately distributed as bivariate normal with mean vector $(x_\eta, x_\eta^*)$ and covariance matrix $\mathbf{C}\mathbf{M}^{-1}\mathbf{C}^T$, where $\mathbf{C}$ is the $2 \times 3$ matrix with rows $c_1$ and $c_2$, where $c_1 = (-1/\beta_2, -1/\beta_2, -(\omega - \beta_0 - \beta_1)/\beta_2^2)$, $c_2 = (-1/\beta_2, 0, -(\omega - \beta_0 - \beta_1)/\beta_2^2)$, and $\mathbf{D} = \text{diag}(\mathbf{p}(1 - \mathbf{p}))$, $\mathbf{M} = \mathbf{X}^T\mathbf{D}\mathbf{X}$. For $\beta_0 = -5$, $\beta_1 = 0.6$, $\beta_2 = 0.07$, $\eta = 0.8$, we find $\omega = \log(0.8/0.2) = 1.386$, $x_{0.6} = 62.86$, $x_{0.8}^* = 71.43$, $\mathbf{C}\mathbf{M}^{-1}\mathbf{C}^T = \begin{pmatrix} 3.233 & -0.156 \\ -0.156 & 2.256 \end{pmatrix}$. For 1000 simulations of the

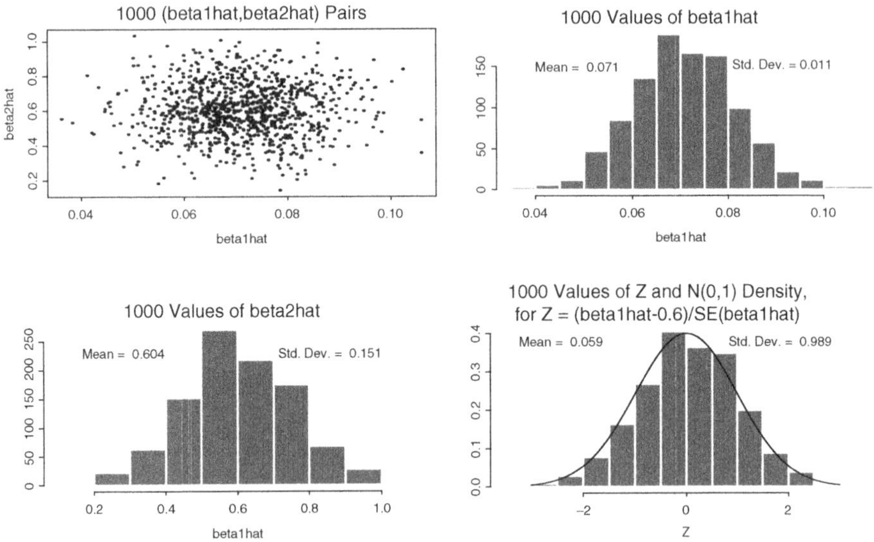


**FIGURE 12.3.2** 1000 Simulations of $(\hat{\beta}_1, \hat{\beta}_2)$.

experiment the corresponding mean vector was (62.68, 71.39), with sample covariance matrix $\begin{pmatrix} 3.502 & -0.130 \\ -0.130 & 2.299 \end{pmatrix}$. Of course, in practice the covariance matrix must be estimated by replacing the parameters by their estimates. Approximate 95% confidence intervals on $x_{0.8}$, for example, are then given by $[\hat{x}_{0.8} \pm 1.96\hat{\sigma}(\hat{x}_{0.8})]$. For the 1000 simulations the CI covered $x_{0.8}$ for 956 such intervals.

For a given age $x_0$ we might like to estimate $p(d, x_0)$ for $d = A$ or B. Again the δ-method may be used. Let $\hat{p}(d, x_0) = \exp(\hat{\mu}(d, x_0))/(1 + \exp(\hat{\mu}(d, x_0))$. Then in approximation for large $n$, $\hat{\mu}(d, x_0) \sim \mathrm{N}(\mu(d, x_0), v(d, x_0))$, where $v(d, x_0) = k(x_0)^{\mathrm{T}}\mathrm{M}^{-1}(1, 1, x_0)k(x_0)$ for $k(x_0)^{\mathrm{T}} = (1, 1, x_0)$ for $d = A$, and $k(x_0) = (1, 0, x_0)$ for $d = B$. Then, again using the δ-method, in approximation, $\hat{p}(d, x_0) \sim \mathrm{N}(p(d, x_0), w(x_0))$, where $w(x_0) = p(d, x_0)(1 - p(d, x_0))v(x_0)$. Thus, $I(x_0) \equiv [\hat{p}(d, x_0) \pm 1.96\sqrt{w(x_0)}]$ is an approximate 95% CI on $p(d, x_0)$.

### Goodness-of-Fit Statistics

The log-likelihood statistic is $G^2(\mathbf{y}, \hat{\mathbf{m}}) \equiv 2(\sum_{ij} y_{ij} \log(y_{ij}/\hat{m}_{ij})$, where $y_{i1} = y_i$, $y_{i2} = n_i - y_i$, $\hat{m}_{i1} = n_i\hat{p}_i$, $\hat{m}_{i2} = n_i(1 - \hat{p}_i)$. Similarly, Pearson's goodness-of-fit statistic is $\chi^2(\mathbf{y}, \hat{\mathbf{m}}) \equiv \sum_{ij}(y_{ij} - \hat{m})^2/\hat{m}_{ij}$. As the $n_i$ become large, these two statistics differ by small amounts. Each is asymptotically distributed as $\chi^2$ with $T-$ (no. of $\beta_j$'s) df. The S-Plus "summary" function after a model is fit using "glm" produces $G^2$ under the name *residual deviance*. For the drugs $A$ and $B$ example in this section, these statistics are not useful, since all the $n_i$ are 1's.

**Problems for Section 12.3**

**12.3.1** Suppose that $\text{logit}(p(x)) = -3 + 2x$ for $0 \le x \le 4$.

**(a)** Sketch the function $p(x)$.

**(b)** Find $x_{0.5}$ such that $p(x_{0.5}) = 0.5$. Find $x_{0.9}$ so that $p(x_{0.9}) = 0.9$.

**12.3.2** Suppose that $Y_1 \sim \text{Binomial}(50, p_1)$, $Y_2 \sim \text{Bimomial}(60, p_2)$, $Y_3 \sim \text{Binomial}(70, p_3)$ are independent, and that $\text{logit}(p_j) = \beta_0 + \beta_1 j$ for $j = 1, 2, 3$.

**(a)** For $Y_1 = 13, Y_2 = 43, Y_3 = 66$, find the ML estimates $(\hat{\beta}_0, \hat{\beta}_1)$ of $(\beta_0, \beta_1)$.

**(b)** Estimate the covariance matrix of $(\hat{\beta}_0, \hat{\beta}_1)$.

**(c)** Give an approximate 95% confidence interval on $\beta_1$.

**(d)** Find the $G^2$ and $\chi^2$ goodness-of-fit statistics.

**12.3.3** Let $x_i = (i-1)/2$ for $i = 1, 2, \ldots, 6$. Let $\mu(x) = \beta_0 + \beta_1 x$, $\beta_0 = -1.5$, $\beta_1 = 1.4$, $p(x) = e^{\mu(x)}/(1 + e^{\mu(x)})$. Let $n = (40, 60, 80, 80, 60, 40)$ be a vector of sample sizes. Therefore, the vector of probabilities is $\mathbf{p} = (0.1824, 0.3100, 0.4750, 0.6457, 0.7858, 0.8808)^{\mathrm{T}}$. Let $Y_i \sim \text{Binomial}(n_i, p(x_i))$ for $i = 1, \ldots, 6$, independent. **Y** was observed to be $\mathbf{Y} = (5, 18, 36, 47, 52, 31)^{\mathrm{T}}$. The vector of sample proportions of successes is therefore $(0.125, 0.3000, 0.4500, 0.5875, 0.8667, 0.7750)^{\mathrm{T}}$. Define **y** to be the $6 \times 2$ matrix with **Y** as the first column, **n** – **Y** as the second. The command "a = glm(**y** ~ **x**, family = binomial) and summary(a)" in S-Plus gives the following:

| Coefficients | Value | Std. Error | *t*-value |
|---|---|---|---|
| (Intercept) | −1.647165 | 0.2493343 | −6.606251 |
| *x* | 1.418378 | 0.1803626 | 7.864036 |
| Residual deviance: 7.333 on 4 degrees of freedom | | | |

The following commands were then used to produce $\hat{p}$, the vector of estimates of $p$ for this model:

```
phat <-  a$fitt
> phat
0.1615  0.2813  0.4431  0.6178  0.7667  0.8698
```

Let $\mathbf{m} = \mathbf{np}$ (component-wise multiplication). Thus, $\mathbf{m} = (7.297, 18.602, 38.002, 51.653, 47.150, 35.232)$. Multiplying "phat" by the vector of sample sizes, we get the estimate of $\mathbf{m}$: $\hat{\mathbf{m}} = (6.460, 16.879, 35.444, 49.427, 46.000, 34.790)$.

**(a)** Verify that the residual vector $(\mathbf{Y} - \hat{\mathbf{m}})$ is orthogonal to the vectors $\mathbf{1}_6$ and $\mathbf{x}$.

(b) How can the standard errors be determined from **n**, **phat**, and $\mathbf{X} = (\mathbf{1}_6, \mathbf{x})$? If you have matrix manipulation software, estimate the covariance matrix of $\hat{\boldsymbol{\beta}}$, using both the true $\boldsymbol{\beta}$ vector and its estimate $\hat{\boldsymbol{\beta}}$.

(c) Verify the value given for the residual deviance 7.333. Also find the Pearson chi-square version of the residual deviance. Find the corresponding $p$-values.

(d) Find a 95% CI on $x_{0.9}$, where $p(x_{0.9}) = 0.9$ for $\mathbf{Y}$ as given above. *Hint:* Replace $C$ in the discussion on the previous page by the vector $(-1/\hat{\beta}_0, -(\omega - \hat{\beta}_0)/\hat{\beta}_1^2)$.

## 12.4 TWO-WAY FREQUENCY TABLES

We begin with the smallest tables, $2 \times 2$ tables of frequencies. Suppose that a panel of 21 possible jurors has nine women, 12 men. From this panel a jury of six is selected "at random." Just one of the jurors selected is a woman. Is there reason to believe that the selection was not truly at random? Of course, there can be no "proof" that the selection was not random based on the selected sample alone. Neither can there be mathematical proof that the selection was not biased against women. We can clarify our thinking by considering the null hypothesis that the jury selection was at random versus the alternative that the selection process favors the choice of men, or that it favors the choice of women. Unless there is prior evidence of favoritism toward one gender, the alternative should be two-sided.

Let $X$ be the number of women chosen. If $H_0$ is true, $X$ has the hypergeometric distribution: $P(X = k) = f(k) = \binom{9}{k}\binom{12}{6-k}\Big/\binom{21}{6}$ for $k = 0, 1, \ldots, 6$. Let $K = \binom{21}{6} = 54{,}264$. Then $f(0) = 924/K = 0.017$, $f(1) = 7128/K = 0.131$, $f(2) = 17{,}820/K = 0.328$, $f(3) = 18{,}480/K = 0.341$, $f(4) = 18{,}480/K = 0.153$, $f(5) = 1512/K = 0.028$, $f(6) = 84/K = 0.002$.

A natural way to order the possible values of $X$ in order to decide when they are more "extreme" than the value $X = 1$ observed is to order them by their probabilities, those with smaller probabilities being more extreme. Thus, the observed $p$-value is $0.017 + 0.028 + 0.131 + 0.002 = 0.178$. There does not seem to be enough evidence to label the process as "biased," although there is some justification for collecting more data.

Suppose that data from 10 six-person jury panels has been collected, with a total of 223 people, of which 108 were women. Among 60 selected for jury duty, suppose that 19 were women. Let $W$ be the number of women selected. Under $H_0$, $W$ has the hypergeometric distribution with parameters 223, 60, 108, 115. For the alternative hypothesis of bias against the choice of women for juries, we should reject $H_0$ for small $X$. The observed $p$-value is therefore $P(W \le 19 | H_0$ true). Exact computations using S-Plus gave 0.00180. We could also make use of the normal approximation. Under $H_0$, $E(W) = 60p = 60(108/223) = 29.06$,

**TABLE 12.4.1 Cross-Classification of Respondents in 1972–1975 General Social Survey by Attitude to Treatment of Criminals by Courts and by Year of Survey**

| | Year of Survey | | | | |
|---|---|---|---|---|---|
| | 1972 | 1973 | 1974 | 1975 | Total |
| Too harshly | 105<br>(7.3%) | 68<br>(5.0%) | 42<br>(6.1%) | 61<br>(4.4%) | 276<br>(5.6%) |
| About right | 265<br>(18.5%) | 196<br>(14.5%) | 72<br>(10.4%) | 144<br>(10.4%) | 677<br>(13.9%) |
| Not harshly enough | 1066<br>(74.2%) | 1092<br>(80.5%) | 580<br>(83.6.0%) | 1174<br>(85.1%) | 3912<br>(80.4%) |
| Total | 1436 | 1356 | 694 | 1379 | 4865 |

$\text{Var}(W) = 60p(1-p)(223-6)/(223-1) = 11.003$, so that $P(W \leq 19) \doteq \Phi((19.5 - 29.05)/\sqrt{11.003}) = \Phi(-2.882) = 0.00198$. We have assumed that juries, in the absence of bias, are selected at random. In truth, defense lawyers and prosecutors can intervene, with and without cause, to have some eliminated from a jury. Thus, we must be careful when analyzing such data. This method for $2 \times 2$ tables is often called *Fisher's exact test* after Sir Ronald A. Fisher, knighted by the queen for his work in genetics and statistics.

### $r \times c$ Tables

***Example 12.4.1*** Consider Table 12.4.1, taken from the book by Shelby Haberman (1978, p. 120), which came originally from studies conducted by the National Opinion Research Center. From Haberman's book: "The question asked was: 'In general, do you think the courts in this area deal too harshly or not harshly enough with criminals?' The reduced 1974 sample results from an experimental change in the question used with one-half the sample." The percentages are by column, so that, for example, in 1972, $100(105/1613)\% = 6.5\%$ thought courts dealt too harshly with criminals. We have dropped the row "no answer or opinion."

There seems to be a drift over time in the direction of belief that courts are too lenient. We should try to decide whether the apparent drift is real or due to chance. Suppose that each of the three-component column vectors $\mathbf{Y}_j = (Y_{1j}, Y_{2j}, Y_{3j})$ has the multinomial distribution with parameters $\mathbf{p}_j = (p_{1j}, p_{2j}, p_{3j})$ and $n_j = Y_{+j} = \sum_{i=1}^{3} Y_{ij}$, the $j$th column sum, and that the $\mathbf{Y}_j$ are independent. The MLE of $\mathbf{p}_j$ under the full model is $\hat{\mathbf{p}}_j = \mathbf{Y}_j/n_j = (Y_{1j}, Y_{2j}, Y_{3j})/n_j$ for each $j$. Suppose that we wish to test $H_0$: $\mathbf{p}_1 = \mathbf{p}_2 = \mathbf{p}_3 = \mathbf{p}_4$ (homogeneity of probability vectors). Under $H_0$ the MLE for the common $\mathbf{p}_j$ is $\hat{\mathbf{p}}^0 = (Y_{1+}, Y_{2+}, Y_{3+})/Y_{++} = (\hat{p}_1^0, \hat{p}_2^0, \hat{p}_3^0)$. The log-likelihood statistic is therefore

$$G^2 = 2\sum_{j=1}^{4}\sum_{i=1}^{3} Y_{ij} \log \frac{Y_{ij}/n_j}{Y_{i+}/Y_{++}} = 2\sum_{j=1}^{4}\sum_{i=1}^{3} Y_{ij} \log \frac{Y_{ij}}{Y_{i+}Y_{j+}/Y_{++}}.$$

**TABLE 12.4.2 Cross-Classification of Respondents in 1972–1975 General Social Survey by Attitude to Treatment of Criminals by Courts and by Year of Survey**

| | Year of Survey | | | | |
|---|---|---|---|---|---|
| | 1972 | 1973 | 1974 | 1975 | Total |
| Too harshly | 105 | 68 | 42 | 61 | 276 |
| | 81.5 | 76.9 | 39.4 | 78.2 | 276 |
| | 23.5 | −8.9 | 2.6 | −17.2 | |
| | 2.61 | −1.02 | 0.42 | −1.95 | |
| About right | 265 | 196 | 72 | 144 | 677 |
| | 199.8 | 188.6 | 96.6 | 191.9 | 677 |
| | 65.2 | 7.3 | −24.6 | −47.9 | |
| | 4.61 | 0.53 | −2.50 | −3.46 | |
| Not harshly enough | 1066 | 1092 | 580 | 1174 | 3912 |
| | 1154.7 | 1090.4 | 558.1 | 1108.9 | 3912 |
| | −88.7 | 1.6 | 21.9 | 65.1 | |
| | −2.61 | 0.05 | 0.93 | 1.96 | |
| Total | 1436 | 1356 | 694 | 1379 | 4865 |

Let $\hat{m}_{ij} = n_j \hat{p}_i^0 = Y_{i+}Y_{j+}/Y_{++}$. Under $H_0$, $\hat{m}_{ij}$ is an unbiased estimator of $m_{ij} = n_j p_{ij}$ for each $i$. We will refer to $G^2$ as the *log-chi-square statistic*, in contrast to Pearson's chi-square statistic. $G^2$ may be expressed in the form $G^2(\hat{\mathbf{m}}, Y) = 2(\mathbf{Y}, \log(\mathbf{Y}/\hat{\mathbf{m}}))$, twice the inner product of the $3 \times 4$ table of $Y_{ij}$ with the table of $\log(Y_{ij}/\hat{m}_{ij})$. The Pearson chi-square statistic is

$$\chi^2 = \chi^2(\hat{\mathbf{m}}, Y) = \sum_{j=1}^{3}\sum_{i=1}^{4} \frac{(Y_{ij} - \hat{m}_{ij})^2}{\hat{m}_{ij}}.$$

As the $n_j \to \infty$, $G^2(\hat{\mathbf{m}}, \mathbf{Y})$ and $\chi^2(\hat{\mathbf{m}}, \mathbf{Y})$ take values that are relatively close, and each converges in distribution to the noncentral chi-square distribution with $(3 - 1)(4 - 1) = 6$ df. The noncentrality parameter is given by

$$G^2(\mathbf{m}^0, \mathbf{m}) = 2\sum_{j=1}^{3}\sum_{i=1}^{4} m_{ij} \log \frac{m_{ij}}{m_{i+}m_{j+}/m_{++}}.$$

Table 12.4.2 contains four quantities in each cell: $Y_{ij}$, $\hat{m}_{ij} = Y_{i+}Y_{+j}/Y_{++}$, $R_{ij} = Y_{ij} - \hat{m}_{ij}$ (the residuals), and "standardized residuals" $D_{ij} = R_{ij}/\sqrt{\hat{m}_{ij}}$. Under the null hypothesis of homogeneity the $D_{ij}$ are approximately distributed as standard normal. $\sum_{ij} D_{ij}^2 = \chi^2(\hat{m}, Y)$, Pearson's chi-square statistic, is 63.06. The $p$-value is $P(\chi_6^2 > 63.06) = 1.07 \times 10^{-11}$, indicating that the null hypothesis is almost certainly not true. Similarly, we obtain $G^2(\hat{\mathbf{m}}, \mathbf{Y}) = 62.69$. Notice that the $D_{ij}$ seem to indicate a trend toward "not harshly enough" over these four years. We investigate this by fitting a slightly more complex model. □

S-Plus may be used to fit the model of homogeneity of probability vectors as follows. Let $\mathbf{y} = (y_{11}, y_{21}, y_{31}, y_{12}, \ldots, y_{34})$ be the 12-component vector of observed frequencies. Let $\mathbf{R}_1, \mathbf{R}_2, \mathbf{R}_3, \mathbf{C}_1, \ldots, \mathbf{C}_4$ be the row and column indicators. Then the model of homogeneity of probability vectors holds if and only if $\boldsymbol{\mu} = \log(\mathbf{m}) = (\log(m_{11}), \ldots, \log(m_{34})) \in V = \mathcal{L}(\mathbf{R}_1, \mathbf{R}_2, \mathbf{R}_3, \mathbf{C}_1, \ldots, \mathbf{C}_4)$, *a* $[3 + (4 - 1) = 6]$-dimensional subspace of 12-space. Under the full model, the MLE $\hat{\mathbf{m}}$ of $\mathbf{m}$ is simply the table of frequencies $\mathbf{y}$.

We can obtain the MLE under the null hypothesis of homogeneity with the S-Plus command "crim = glm(y ~ R1 + R2 + C1 + C2 + C3, family = poisson)." "crim" is simply the name arbitrarily assigned to the results of the analysis, which is a list of objects that may be obtained using the command "attributes(crim)." Notice that $\mathbf{R}_3$ and $\mathbf{C}_4$ were omitted, since the vector of all 1's is included by default. The third component of "crim" is "crim[[3]]" or "crim$fitt," which we have called $\hat{\mathbf{m}}$. "summary(crim)" produces a short description of the results. For example, S-Plus produced the following. Some of the output is omitted.

```
> R2 = c(0,1,0,0,1,0,0,1,0,0,1,0)
> C2 = c(0,0,0,1,1,1,0,0,0,0,0,0)
> crim = glm(y ~ R1 + R2 + C1 + C2 + C3, family = poisson)
> attributes(crim)
$ names:
[1] "coefficients"  "residuals"  "fitted.values"  "effects"  "R"  "rank"
[7] "assign" "df.residual" "weights" "family" "linear.predictors" "deviance"
[13] "null.deviance" "call" "iter" "y" "contrasts" "terms" "formula" "control"
> summary(crim)
Call: glm(formula = y ~ R1 + R2 + C1 + C2 + C3, family = poisson)
Coefficients:
```

| | Value | Std. Error | t value |
|---|---|---|---|
| (Intercept) | 7.011 | 0.0278 | 251.8 |
| R1 | −2.651 | 0.0623 | −42.6 |
| R2 | −1.754 | 0.0416 | −42.1 |
| C1 | 0.041 | 0.0377 | 1.07 |
| C2 | −0.017 | 0.0382 | −0.44 |
| C3 | −0.687 | 0.0465 | −14.75 |

```
Residual Deviance: 62.69 on 6 degrees of freedom
> crim$ fitt
```

| | [,1] | [,2] | [,3] | [,4] |
|---|---|---|---|---|
| [1,] | 81.47 | 76.93 | 39.37 | 78.23 |
| [2,] | 199.83 | 188.70 | 96.58 | 191.90 |
| [3,] | 1154.70 | 1090.37 | 558.05 | 1108.87 |

**TABLE 12.4.3 Frequency Data**

| | | Raise Payroll Tax | No Opinion | Total |
|---|---|---|---|---|
| 21–35 | 67 (47.5%) | 53 (37.6%) | 21 (14.9%) | 141 |
| 36–50 | 59 (44.4%) | 50 (37.6%) | 24 (18.0%) | 133 |
| 51–65 | 43 (37.1%) | 55 (47.4%) | 18 (15.5%) | 116 |
| >65 | 35 (31.8%) | 53 (48.2%) | 22 (20.0%) | 110 |
| Total | 204 (40.8%) | 205 (41.0%) | 74 (14.8%) | 500 |

Since the homogeneity model does not fit well, we can try to improve the fit by adding a vector that will at least partially explain the deviations of $\hat{\mathbf{m}}$ from $\mathbf{y}$. Let $\mathbf{s} = (3, 0, -3, 1, 0, -1, -1, 0, 1, -3, 0, 3)$, the outer product, in vector form, of $(-1, 0, 1)$ and $(-3, -1, 1, 3)$. We then fit the model (in S-Plus language) $y \sim \text{R1} + \text{R2} + \text{C1} + \text{C2} + \text{C3} + \text{s}$. The fit was better, producing a log-chi-square value of 19.16 for $12 - (1 + 2 + 3 + 1) = 5$ df, with a corresponding $p$-value of 0.0018. This would still lead to rejection for the usual choices of $\alpha$, but the fit could very well be considered to be satisfactory. The real world is "messy," especially when it involves measurements on humans, so we cannot expect simple models to fit well. The $\hat{\mathbf{m}}$ (the "fitted values") and residual vector $\mathbf{y} - \hat{\mathbf{m}}$, in table form, were:

| $\hat{\mathbf{m}}$ | | | | $\mathbf{y} - \hat{\mathbf{m}}$ | | | |
|---|---|---|---|---|---|---|---|
| 112.3 | 82.2 | 32.4 | 49.2 | −7.3 | −14.2 | 9.6 | 11.8 |
| 231.6 | 196.7 | 90.0 | 158.8 | 33.4 | −0.7 | 18.0 | 14.8 |
| 1092.2 | 1077.1 | 571.7 | 1171.0 | −26.2 | 14.9 | 8.3 | 3.0 |

The residual vector $\mathbf{y} - \hat{\mathbf{m}}$ is necessarily orthogonal to each of the predicting vectors $\mathbf{1}, \mathbf{R}_1, \mathbf{R}_2, \mathbf{C}_1, \mathbf{C}_2, \mathbf{C}_3$ (so that row and column sums are zero), and therefore to $\mathbf{R}_3$ and $\mathbf{C}_4$ and to $\mathbf{s}$.

***Example 12.4.2*** To determine attitudes toward possible changes in the Social Security system, a study was planned around the following question: "The Social Security system will soon be short of money. Which of the following solutions do you favor? Select one answer: (1) Raise the retirement age; (2) Raise the payroll tax; or (3) No opinion." A telephone survey was conducted from a list of residential telephones for the state of Michikansas. The adult answering the telephone, or the first adult to come to the telephone when a child answered, was asked to provide (a) their answer to the question, and (b) to identify the age group to which they belonged: 21–35, 36–50, 51–65, and >65. Those who would not answer (a) were counted as "no opinion." Those who would not answer (b) (4%) were omitted.

The frequencies of response are given in Table 12.4.3. The percentages are row-wise, so that, for example, $100(67/141) = 47.5$. It seems that as people age, the percentage favoring the raising of the retirement age decreases, as one might expect. But can we be sure that these percentage changes aren't due to chance? The answer, of course, is that they could be, but are quite unlikely. Let's consider the problem formally. □

This example differs from the criminals example in that only the table total $n = 500$ is fixed, whereas in the criminal example, each column total was fixed. Let $Y_{ij}$ be the observed frequency in cell $ij$. Let $\mathbf{Y} = (Y_{ij})$ be the $4 \times 3$ table of frequencies. It seems reasonable to suppose that $\mathbf{Y} \sim$ Multinomial ($n = 500$, $\mathbf{p}$), where $\mathbf{p} = (p_{ij})$ is a probability table that has components which sum to 1. Call this the full model. Under this model the MLE of the $\mathbf{p}$ is $\hat{\mathbf{p}} = \mathbf{Y}/n$, the $4 \times 3$ matrix of sample proportions. Let $\mathbf{m} = (m_{ij}) = E(\mathbf{Y}) = (np_{ij})$.

Consider the null hypothesis of independence: $H_0$: $p_{ij} = p_{i+}p_{+j}$ for all $i$ and $j$. Under $H_0$ the MLE for $\mathbf{p}$ is $\hat{\mathbf{p}} = (\hat{p}_{ij} = (Y_{i+}/n)(Y_{j+}/n))$, and the MLE of $\mathbf{m} = (m_{ij})$ is therefore $\hat{\mathbf{m}} = (\hat{m}_{ij} = Y_{i+}Y_{+i}/n)$, the same as it was for the fixed column total model as applied to the criminal data. The goodness-of-fit statistics $G^2(\hat{\mathbf{m}}, \mathbf{Y})$ and $\chi^2(\hat{\mathbf{m}}, \mathbf{Y})$ remain the same, and their distributions converge to chi-square with $rc - (1 + r + c) = (r-1)(c-1)$ for the case of $r \times c$ tables, as they do for the model with fixed column totals. In this case these statistics are approximately distributed as chi-square with $(4-1)(3-1) = 6$ df. If the null hypothesis is not true, the distributions converge to noncentral chi-square with noncentrality parameter

$$G(\mathbf{m}^0, \mathbf{m}) = 2\sum_{j=1}^{3}\sum_{i=1}^{4} m_{ij} \log \frac{m_{ij}}{m_{i+}m_{j+}/m_{++}} \doteq \sum_{j=1}^{3}\sum_{i=1}^{4} \frac{(m_{ij} - m_{ij}^0)^2}{m_{ij}^0},$$

described for the criminal example. For these data we obtain $\hat{\mathbf{m}} = \begin{pmatrix} 57.5 & 59.5 & 24.0 & 54.3 \\ 56.1 & 22.6 & 47.3 & 49.0 \\ 19.7 & 44.9 & 46.4 & 18.7 \end{pmatrix}$, $G(\hat{\mathbf{m}}, \mathbf{Y}) = 8.86$, $\chi^2(\hat{\mathbf{m}}, \mathbf{Y}) = 8.79$, with corresponding $p$-values 0.182 and 0.186. These last two $p$-values are too large to reject $H_0$ for most choices of $\alpha$.

How good are these chi-square approximations? Some textbooks suggest that the approximations may be bad if some cells have expectations less than 5 under $H_0$. To investigate this, consider a $2 \times 3$ table. Let $\mathbf{Y} = (Y_{ij})$ have the multinomial distribution with parameters $n = 30$ and $\mathbf{p} = \begin{pmatrix} 0.14 & 0.21 & 0.35 \\ 0.06 & 0.09 & 0.15 \end{pmatrix}$. Then the null hypothesis of independence of rows and columns holds and $\mathbf{m} = n\mathbf{p} = \begin{pmatrix} 4.2 & 6.3 & 10.5 \\ 1.8 & 2.7 & 4.5 \end{pmatrix}$. Simulations were performed 10,000 times and the goodness-of-fit statistics $G(Y, \hat{\mathbf{m}})$ and $\chi^2(Y, \hat{\mathbf{m}})$ were determined for each (see Figure 12.4.1). The 0.95-quantile of the chi-square distribution is 5.991. $G^2$ exceeded 5.991 a total of 775 times, indicating that the true $\alpha$ is approximately 0.0775. On the other hand, $\chi^2$ exceeded 5.991 just 522 times, indicating that the null distribution of $\chi^2$ is better approximated by the chi-square distribution. For $n = 100$, the approximation is better, as shown by the histogram on the lower right. Suppose now that

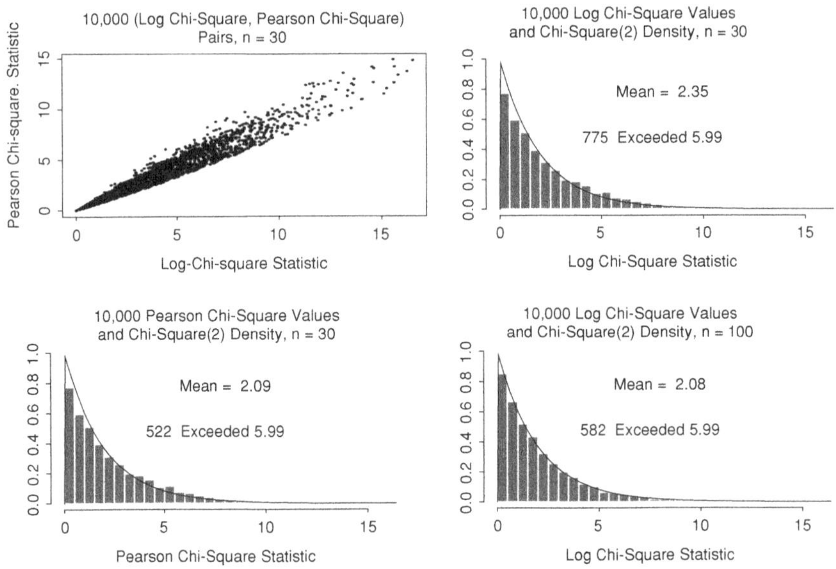


**FIGURE 12.4.1** Results for 10,000 simulations with small expected cell frequencies.

$\mathbf{p} = \begin{pmatrix} 0.17 & 0.25 & 0.28 \\ 0.03 & 0.05 & 0.22 \end{pmatrix}$. Row and column sums remain as they were, but rows and columns are now dependent. Again, 10,000 simulations were performed for $n = 30$ and $n = 100$. Results are summarized in Figure 12.4.2.

The power may be determined in approximation by evaluating the noncentrality parameter $\delta = G^2(m, m^0)$ [or $\chi^2$ $(\mathbf{m}, \mathbf{m}^0)$], where $\mathbf{m}^0$ is the table of expected values under the null hypothesis of independence. Since the marginal probabilities are as before, $\mathbf{m}^0$ is the $\mathbf{m}$ described above, the MLE of the table of expectations under the null hypothesis. We find that $\delta = 4.675$ (Pearson version) for $n = 50$, $\delta = 9.349$ for $n = 100$. Using the function "pf" (with any very large df for the denominator) in S-Plus we find that the power for $n = 50$ is 0.476, and for $n = 100$ the power is 0.787. The simulations indicate that the power is approximately 0.45 for $n = 50$, 0.80 for $n = 100$. The means are consistent with the fact that the expected value of a noncentral chi-square random variable is the sum of its two parameters.

## Problems for Section 12.4

**12.4.1** Consider a $2 \times 2$ table of frequencies $\mathbf{Y} = (Y_{ij})$. Suppose that $Y \sim$ multinomial with $n = \sum_{ij} Y_{ij}$ and $p = (p_{ij})$. Pearson's chi-square statistic for the test of $H_0$: rows and columns independent is $\chi^2(\mathbf{Y}, \hat{\mathbf{m}})$, where $\hat{\mathbf{m}} = (\hat{m}_{ij} = Y_{i+}Y_{+i}/n)$.

**(a)** Show that row and column sums for $\mathbf{Y}$ and $\hat{\mathbf{m}}$ are the same and therefore $(Y_{ij} - \hat{m}_{ij})^2 = (Y_{11} - \hat{m}_{11})^2$ for all $i$ and $j$.

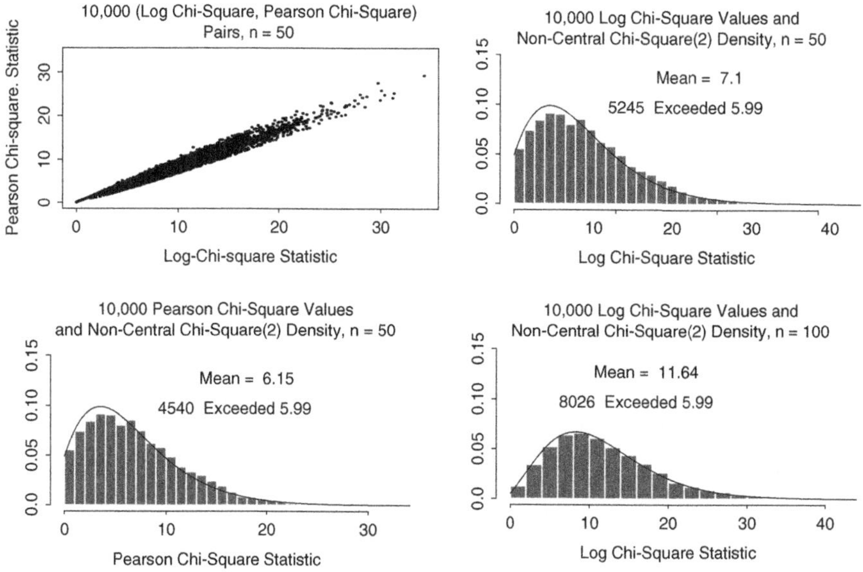


**FIGURE 12.4.2** Results of simulations for the case that rows and columns are not independent.

(**b**) Use part (a) to show that $\chi^2(\mathbf{Y}, \hat{\mathbf{m}}) = (Y_{11} - m_{11})^2 n^3/(Y_{1+}Y_{2+}Y_{+1}Y_{+2})$. Verify the formula for the table $\mathbf{Y} = \begin{pmatrix} 10 & 20 \\ 30 & 40 \end{pmatrix}$.

(**c**) Show that under the multinomial model the conditional distribution of $Y_{11}$, given $Y_{1+} = n_{1+}$ is Binomial $(n_1, p_1)$ and that the conditional distribution of $Y_{21}$ given $Y_2 = n_2$ is Binomial$(n_2, p_2)$, where $p_i = p_{i1}/(p_{i1} + p_{i2}), n_i = Y_{i1} + Y_{i2}$ for $i = 1, 2$. Also show that under these two conditions $Y_{11}$ and $Y_{21}$ are independent.

(**d**) Show that rows and columns are independent if and only if $p_1 = p_2$.

(**e**) Let $\hat{p}_i = Y_{i1}/n_i$ for $i = 1, 2$. Let $\hat{p} = X_{+1}/n$ and $\hat{q} = 1 - \hat{p}$, where $n = Y_{1+} + Y_{2+} = n_1 + n_2$. Let $Z = (\hat{p}_1 - \hat{p}_2)/\sqrt{\hat{p}\hat{q}\,(1/n_1 + 1/n_2)}$. Under $H_0$: $p_1 = p_2$, $Z$ is asymptotically distributed as N(0, 1) as $n_1$ and $n_2$ go to $\infty$. Show that $Z^2 = \chi^2(\mathbf{Y}, \hat{\mathbf{m}})$. Verify it for $\mathbf{Y}$ as in part (b).

**12.4.2** To investigate the effectiveness of a new surgical procedure for horses with broken legs, the owners of 21 such horses were asked to take part in the experiment, with the promise of free surgery. First, a subset $A$ of 11 of the integers $1, 2, \ldots, 21$ was chosen at random, then the $k$th horse to arrive was assigned to receive the new procedure if $k \in A$. The others received the old surgery. The owners were not told which surgery was used nor was the veterinarian who decided whether the surgery had been successful after one year. The evaluations were as follows:

| | Successful | Not Successful |
|---|---|---|
| New surgery | 9 | 2 |
| Old surgery | 4 | 6 |

**(a)** Test the null hypothesis that the two methods are equally effective, using the hypergeometric exact probability method. Give the exact $p$-value for the two-sided test.

**(b)** Give the $p$-value for the test based on the Pearson chi-square statistic.

**(c)** Determine the $Z$-statistic using the formula in Problem 12.4.1. Verify that $Z^2$ is the Pearson chi-square statistic obtained in part (b). Verify that the $p$-value is a better approximation to the $p$-value obtained in part (a) if the 1/2-correction is used in terms of the computing $Z$. *Hint:* First express $Z$ in terms of the frequencies $X_{ij}$. Use the fact that a hypergeometric random variable has variance $p(1-p)[(N-n)/(N-1)]$, where $N = 21, n = 11$, and $p$ is estimated by $13/21$.

**12.4.3** The data in Table 12.4.3 were taken from a paper of Dowdall (1974), as reported by Haberman (1978). Surveys were conducted in 1968 and 1969 by the Population Research Laboratory at Brown University. A total of 599 women between the ages of 15 and 64 were chosen randomly. They were asked to respond to the following question: "Do you think it is all right for a married woman to have a job instead of only taking care of the house and the children while her husband provides for the family?" The women were classified according to the reported national origin of their fathers. (There is little doubt that attitudes have changed considerably over the last 39 years.) Let $Y_{ij}$ be the observed frequency in cell $ij$. State a model.

**(a)** Show that for an $2 \times c$ table $\chi^2(\mathbf{Y}, \hat{\mathbf{m}}) = (Y_{++}^2/Y_{1+}Y_{2+})\sum_{j=1}^{c}(Y_{1j} - \hat{m}_{1j})^2/Y_{+j}$, where $\hat{m}_{1j} = Y_{1+}Y_{+j}/Y_{++}$.

**(b)** Verify the formula in part (a) for these data and test the null hypothesis of independence of attitude and ethnic origin for $\alpha = 0.05$. Conclusion?

**TABLE 12.4.3 Native-Born White Rhode Island Women Cross-Classified by Ethnic Origin and Attitude Toward Married Women Having Jobs**

| | Ethnic Origin | | | | | |
|---|---|---|---|---|---|---|
| Attitude | Italian | Northern Euoropean | Other Euoropean | English | Irish | French Canadian |
| Favorable | 78 | 56 | 43 | 53 | 43 | 36 |
| Unfavorable | 47 | 29 | 29 | 32 | 30 | 22 |
| Total | 125 | 85 | 72 | 85 | 73 | 58 |

**12.4.4** **(a)** Let $\mathbf{Y} \sim \mathbf{M}_k(n, \mathbf{p} = (p_1, \ldots, p_k))$, where $p$ is any probability vector. Show that the MLE for $\mathbf{p}$ is $\hat{\mathbf{p}} = \mathbf{Y}/n$. Write both $\mathbf{Y}$ and $\mathbf{p}$ as column vectors. Let $G(\mathbf{p}) = \text{diag}(\mathbf{p}) - \mathbf{p}\mathbf{p}^{\text{T}}$.

**(b)** Let $\mathbf{a} = (a_1, \ldots, a_k)^{\text{T}}$ be a vector of constants. Let $\eta = \mathbf{a}^T p = \Sigma a_i p_i$ and $\hat{\eta} = \mathbf{a}^{\text{T}}\hat{\mathbf{p}}$. Prove that $Z_n \equiv \sqrt{n}(\hat{\eta} - \eta)$ converges in distribution to $\text{N}(0, V(\mathbf{a}, \mathbf{p}))$, where $V(\mathbf{a}, \mathbf{p}) = \mathbf{a}^{\text{T}} G(\mathbf{p})\mathbf{a}$. *Hint:* $\mathbf{Y}$ can be expressed as the sum of $n$ independent generalized Bernoulli random vectors $\mathbf{U}_i$, each of which takes the $j$th unit vector $(0, \ldots 0, 1, 0, \ldots, 0)$, the indicator of the $j$th component, with probability $p_j$. Then $\hat{\eta}$ can be expressed as the sum of $n$ independent identically distributed random variables.

**(c)** Let $\mathbf{p} = (0.32, 0.32, 0.36)^{\text{T}}$ and $n = 100$. Find an approximation of $P(Y_3 < (Y_1 + Y_2)/2)$.

**(d)** For $\mathbf{p}$ and $n$ as in part (c), use a software package that provides bivariate normal probabilities to approximate $P(Y_3 < Y_1, Y_3 < Y_2)$. *Hint:* Consider the random variables $W_1 = Y_3 - Y_1$ and $W_2 = Y_3 - Y_2$. The distribution of the pair $(W_1, W_2)$ may be approximated by the bivariate normal distribution.

**12.4.5** Let $\mathbf{Y} = (Y_{ij})$ be a random matrix with $r$ rows and $c$ columns. Suppose that $\mathbf{Y} \sim \mathbf{M}_{rc}(n, \mathbf{p})$. Define $R(i_1, i_2, j_1, j_2) = (p_{i_1 j_2}/p_{i_1 j_1})/(p_{i_2 j_2}/p_{i_2 j_1}) = (p_{i_1 j_2}/p_{i_2 j_1})/(p_{i_1 j_1}/p_{i_2 j_2}$. These are the odds ratios.

**(a)** Show that independence of rows and columns is equivalent to $R = 1$ for all $i_1, i_2, j_1, j_2$.

**(b)** Let $\mu_{ij} = \log(p_{ij})$. Define $\mu$ be the mean of all $\mu_{ij}$, and let $\bar{\mu}_{1+}$ and $\bar{\mu}_{+j}$ be the corresponding row and column means. Define $\alpha_i = \bar{\mu}_{1+} - \mu$, $\beta_j = \bar{\mu}_{+j} - \mu$, $(\alpha\beta)_{ij} = \mu_{ij} - (\mu + \alpha_i + \beta_j)$. As for two-way analysis of variance, call these the $i$th row effect , the $j$th column effect, and the $(ij)$th interaction effect. Then $\mu_{ij} = \mu + \alpha_i + \beta_j + (\alpha\beta)_{ij}$. Show that $(\alpha\beta)_{ij} = 0$ for all $i$ and $j$ if and only if rows and columns are independent. *Hint:* Use part (a).

**12.4.6** The employees of the XYZ Company were classified as $A$, union members; $B$, managers; or $C$, upper-Level managers. The frequencies were $A$, 10,000; $B$, 2000; and $C$, 500. Simple random samples of 300, 200, and 100 were taken from these three subpopulations. All 600 were asked to assign a number 1, 2, 3, 4, or 5 as a rating of a new medical plan in comparison to an old plan: 1 = {new plan much better}, 2 = {new plan somewhat better}, 3 = {new plan and old plan about equal}, 4 = {new plan somewhat worse}, 5 = {new plan much worse}. The results are shown in Table 12.4.6.

**(a)** State a model and determine the approximate $p$-value for $H_0$: homogeneity of probability vectors.

**(b)** Determine the table of expected values $E(\mathbf{Y}) = \mathbf{m}$, then use this to find an approximate power of an $\alpha = 0.05$-level test for the case that

**TABLE 12.4.6 Medical Plan Data**

| | Rating | | | | | |
|---|---|---|---|---|---|---|
| | 1 | 2 | 3 | 4 | 5 | Total |
| Union | 39 | 52 | 81 | 93 | 35 | 300 |
| Managers | 27 | 46 | 64 | 40 | 23 | 200 |
| Upper-level managers | 20 | 22 | 35 | 14 | 9 | 100 |
| Total | 86 | 120 | 180 | 147 | 67 | 600 |

$\mathbf{p} = \begin{pmatrix} 0.12 & 0.18 & 0.25 & 0.29 & 0.16 \\ 0.15 & 0.25 & 0.25 & 0.20 & 0.15 \\ 0.17 & 0.27 & 0.30 & 0.16 & 0.10 \end{pmatrix}$. In 10,000 simulations, $H_0$ was rejected 8215 times. (Actually, samples of employees were taken with replacement for this simulation.) The data of the observed table $\mathbf{Y}$ above were generated using this $\mathbf{p}$.

**(c)** Determine the odds ratio (see Problem 12.4.5) $R(1, 2, 1, 2)$ and estimate it from the observed data table $\mathbf{Y}$.

**12.4.7** Let $\mathbf{X} = (X_{11}, X_{12}, X_{21}, X_{22}) \sim$ Multinomial $(n, p)$, where $\mathbf{p} = (p_{11}, p_{12}, p_{21}, p_{22})$. Define $R = (p_{11}/p_{12})/(p_{21}/p_{22}) = p_{11}p_{22}/p_{12}p_{21}$, the odds ratio. Define $\eta = \log(\hat{R})$, the log odds. Define $\hat{R} = (X_{11}/X_{12})/(X_{21}/X_{22}) = X_{11}X_{22}/(X_{12}X_{21})$, and $\hat{\eta} = \log(\hat{R})$.

**(a)** Use the $\delta$-method to show that for large $n$, $\text{Var}(\hat{\eta}) \doteq (1/n)(1/p_{11} + 1/p_{22} + 1/p_{12} + 1/p_{22})$.

**(b)** Use the fact that $\hat{\eta}$ is asymptotically normally distributed and Slutsky's theorem to derive a formula for a 95% confidence intervals on $\eta$ and $R$. For 1000 simulations with $n = 2000$, $\mathbf{p} = (0.1, 0.2, 0.3, 0.4)$, "95% CIs on $\eta = \log(0.04/0.06)$ contained $\eta$ 953 times.

**(c)** Apply the method in part (b) for $X_{11} = 221$, $X_{12} = 410$, $X_{21} = 613$, $X_{22} = 756$.

## 12.5 CHI-SQUARE GOODNESS-OF-FIT TESTS

The classical book *Mathematical Methods of Statistics*, by Harald Cramér (1951) contains a table of count data (p. 436). The counts are the numbers of $\alpha$-particles radiated from a disk in 2608 periods of 7.5 seconds. It was a classical experiment reported by Geiger and Rutherford. Their theory suggested that the observations $X_i$ should be values taken by 2608 independent Poisson random variables with common unknown parameter $\lambda$. The frequencies of counts $k$, labeled $Y_k$ here, estimates of expected frequencies $\hat{m}_k$ (to be explained), and contributions $\chi_k^2 = (Y_k - \hat{m}_k)^2/\hat{m}_k$ to the Pearson chi-square statistic are provided in Table 12.5.1.

**TABLE 12.5.1 Chi-Square Data**

| $k$ | $Y_k$ | $\hat{p}_k$ | $\hat{m}_k$ | $\chi_k^2$ |
|---|---|---|---|---|
| 0 | 57 | 0.021 | 54.4 | 0.13 |
| 1 | 203 | 0.081 | 210.5 | 0.26 |
| 2 | 383 | 0.157 | 407.3 | 1.45 |
| 3 | 525 | 0.202 | 525.4 | 0.00 |
| 4 | 532 | 0.195 | 508.4 | 1.09 |
| 5 | 408 | 0.150 | 393.6 | 0.53 |
| 6 | 273 | 0.097 | 253.9 | 1.44 |
| 7 | 139 | 0.053 | 140.4 | 0.01 |
| 8 | 45 | 0.026 | 67.9 | 7.73 |
| 9 | 27 | 0.011 | 29.2 | 0.17 |
| 10 | 10 | 0.005 | 11.3 | |
| 11 | 4 | 0.0014 | 4.0 | |
| 12 | 2 | 0.00048 | 1.25 | 0.15 |
| 13 | 0 | 0.00014 | 0.37 | |

To achieve a better approximation, the last four categories were combined, so that the total frequency of the last category was 16, with corresponding $\hat{m}_k = (11.3 + 4.0 + 1.3 + 0.4 + 0.1) = 17.1$, so that $\chi_k^2 = (16 - 17.1)^2/17.1 = 0.15$. The $\hat{m}_k$ were determined by first finding the MLE for $\lambda$, $\bar{X} = (1/n)\sum_{i=1}^{n} X_i = (1/n)\sum_{k=0}^{12} kY_k$, where $n = \sum_{k=0}^{12} Y_k = 2608$. Then the MLE for $p_k = P(X_i = k; \lambda) = e^{-\lambda}\lambda^k/k!$ was obtained by replacing $\lambda$ by $\hat{\lambda} = \bar{X}$ to obtain $\hat{p}_k$. Then $\hat{m}_k = n\hat{p}_k$. Under the null hypothesis that the $X_i$ constitute a random sample from some Poisson distribution, $\chi^2 = \sum_{k=0}^{11} \chi_k^2$ is approximately distributed as chi-square with $11 - 1 - 1 = 9$ degrees of freedom. We observe $\chi^2 = 12.65$, so that the observed $p$-value is 0.169. A curious feature of these data is that the observed frequency corresponding to $X = 8$ is so small, with observed frequency $Y_8 = 45$, though $\hat{m}_8 = 67$. The resulting $\chi_k^2$ is 7.73, a large part of the statistic $\chi^2 = 12.65$. To see that Pearson's goodness-of-fit chi-square statistic has a sampling distribution that is closely approximated by the chi-square distribution when samples of 2604 are taken from a Poisson distribution with $\lambda$ of this magnitude, simulations were conducted 10,000 times, each for $\lambda = 3.87$. Figure 12.5.1 shows that the approximation is excellent. For each the last category was $[X \geq 10]$, so that there were 11 categories, and therefore $11 - 2 = 9$ degrees of freedom.

So that the null hypothesis would not be true, the probabilities given by the Poisson distribution with $\lambda = 3.87$ were changed by adding 0.02 to $P(X = 0)$ and $P(X = 3)$ and subtracting 0.02 from $P(X = 1)$ and $P(X = 2)$. Call the resulting vector $\mathbf{p}$, and let $\mathbf{m} = 2608\mathbf{p}$. The resulting mean remained at 3.87, but the variance was 3.90, slightly larger. Figure 12.5.2 is a histogram for 10,000 Pearson chi-square statistics produced for $n = 2608$. The distribution of $\chi^2$ is then approximately noncentral chi-square with 9 df, noncentrality parameter $\delta = \sum_k (m_k - m_{k0})^2/m_{k0}$, where $m_{k0} = 2608p_k$, and $p_k$ is the Poisson probability for $\lambda = 3.86$. We find $\delta = 18.54$ and corresponding power 0.875. For 10,000 simulated values of $\chi^2$, 8550 exceeded the 0.95-quantile

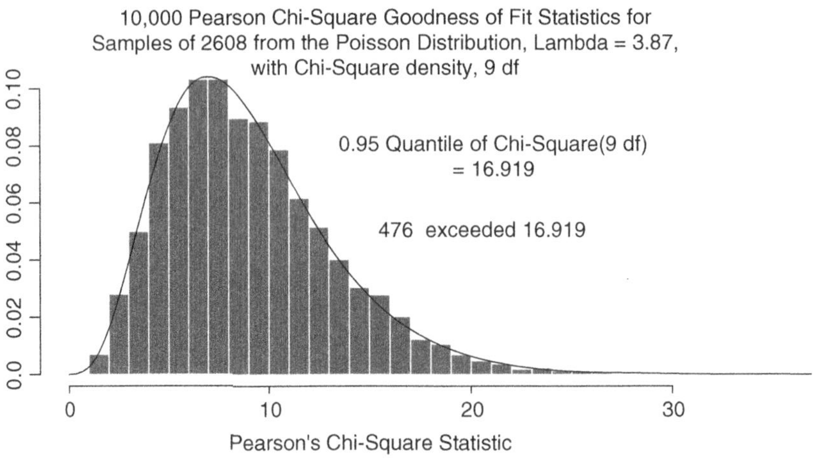


**FIGURE 12.5.1**

16.919 of the central chi-square distribution, 9 df, so that the agreement with the theory is excellent. For $n = 2608/2 = 1304$, $\delta$ is one-half as large, 9.27, with resulting approximate power 0.525. For $n = 2608/4 = 652$ and $n = 2608/8 = 326$, the ($\delta$, power) pairs are (4.63, 0.260) and (2.32, 0.140).

In general, suppose that a random sample of size $n$ is taken from a population and that each observation is classified as a member of some category $j$ for $j = 1, 2, \ldots, k$. Let $\mathbf{X} = (X_1, \ldots, X_n)$ be the vector of category numbers for the sample. We suppose that the $X_i$ are independent and that $P(X_i = j) = p_j$ for $i = 1, \ldots, n$ and $j = 1, \ldots, k$. Let $Y_j = (\text{no. } i \text{ for which } X_i = j) = \sum_{i=1}^{n} I[X_i = j]$.

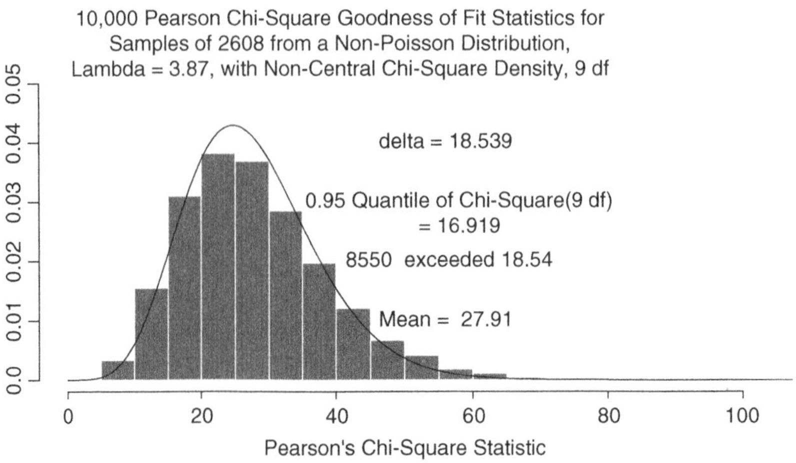


**FIGURE 12.5.2**

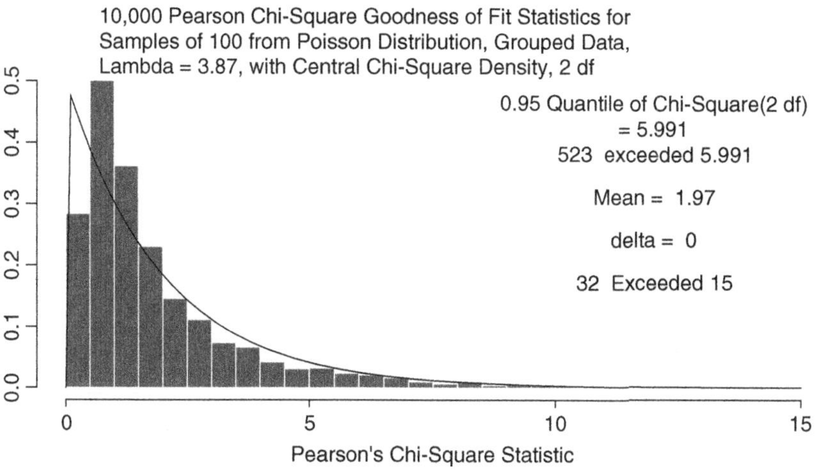


**FIGURE 12.5.3**

Then $\mathbf{Y} = (Y_1, \ldots, Y_k)$ has the multinomial distribution with parameters $n$ and $\mathbf{p} = (p_1, \ldots, p_k)$.

Let $g(j; \theta)$ for $j \in \{1, 2, \ldots, k\}$ be a probability mass function for each $\theta \in \Omega$ and suppose that we wish to test the null hypothesis $H_0$: [for some $\theta \in \Omega, \mathbf{p} = (g(1; \theta), \ldots, g(k; \theta))$]. For the $\alpha$-particle data above, the categories corresponded to the counts $X = 0, 1, \ldots, 9$ and $[X > 9]$, so that $k = 11$. The Poisson parameter was $\theta = \lambda$, $g(j; \lambda) = e^{-\lambda}\lambda^j/j!$ for $j = 0, \ldots, 9$, and $g(11; \lambda) = 1 - \sum_{j=0}^{9} g(j; \lambda)$.

Sometimes the number of observations $n$ may be too small to justify the use of categories for which the probabilities are relatively small. In that case, categories may be grouped, so that each has probability of sufficient size. The long-time rule was that the number expected in each category should be at least five. Simulations have shown that the chi-square approximation of the distribution of Pearson's chi-square statistic $\chi^2$ are often good even when several categories have expected counts of less than five. The estimator $\hat{\theta}$ of $\theta$ should be the MLE for the grouped data, not for the original ungrouped data. It has been shown that when the MLE for the ungrouped data is used, the distribution of $\chi^2$ under the null hypothesis is sometimes not well approximated by the chi-square distribution. The likelihood function corresponding to the grouped data is $L(\theta) = C \sum_{i=1}^{k} g(i; \theta)^{y_i}$, where $C$ does not depend on $\theta$. If $g(i; \theta)$ is not a simple function of $\theta$, as when it is a sum, there may not be a simple formula for the MLE corresponding to $L(\theta)$. However, with today's fast computational ability, that should cause no problem. For example, for samples of size 100 from the Poisson distribution with $\lambda = 3.87$, the observations were grouped into the categories $\{0, 1, 2, 3\}$, $\{4, 5, 6\}$, $\{7, 8, 9\}$, and $\{k | k \geq 10\}$. Figure 12.5.3 is a histogram of 10,000 values of Pearson's chi-square statistic, together with the density of the chi-square distribution for $4 - 1 - 1 = 2$ degrees of freedom. The fit is reasonably good.

To determine whether the noncentral chi-square distribution is a good approximation for the case that the distribution is not Poisson, the probabilities determined by

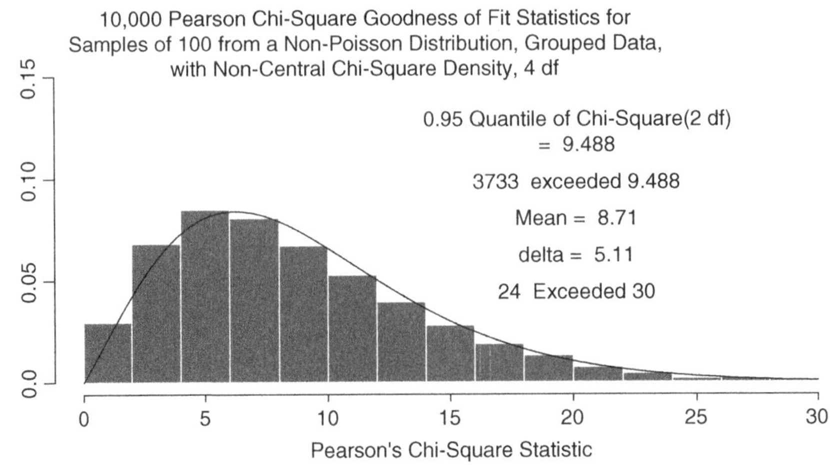


**FIGURE 12.5.4**

the Poisson (3.87) distribution were changed by adding 0.06 to the probabilities of 0 and 7, and by subtracting 0.06 from the probabilities of 2 and 6. This change did not alter the probability of the category $\{0, 1, 2, 3\}$, but did subtract and add 0.06 to the categories $\{4, 5, 6\}$, and $\{7, 8, 9\}$. These new probabilities and expectations $E_j$ for the four categories were then used to determine the MLE $\lambda^*$ of $\lambda$. That is, $\lambda^*$ maximized $\Pi_{j=1}^4 g(j;\lambda)^{E_j}$ as a function of $\lambda$, where $g(j;\lambda)$ is the Poisson probability of category $j$, expressed as a sum. Let $E_j^0$ be the expected frequency corresponding to $\lambda^*$. Let $\delta = \sum_{j=1}^4 (E_i - E_j^0)^2/E_j^0$. Theory states that $\chi^2(\mathbf{y}, \hat{\mathbf{E}}^0) = \sum_{j=1}^4 (y_i - \hat{E}_j^0)^2/E_j^0$ should be approximately distributed as noncentral chi-square with 2 df and noncentrality parameter $\delta$. $\hat{E}_j$ is the estimate of $E_j$ based on $\mathbf{y}$ rather than the vector $\mathbf{E}$. Figure 12.5.4 indicates that the approximation is quite good.

Use of the S-Plus command "1 - chisq(9.488, 2, 3.41)" produced 0.361, indicating that the power should be approximately 0.361. The actual power, as indicated by 10,000 simulations, seems to be a bit less, approximately 0.32 (see Figure 12.5.5). Simulations for other categories produced similar results.

### Testing for Normality

Suppose that $X_1, \ldots, X_n$ is a random sample from a distribution with cdf $F$ and we wish to test $H_0$: $F$ is the cdf of a normal distribution for some $\mu$ and $\sigma$. Samples of size 100 were taken from the N(50, 100) distribution. (The choices of $\mu = 50, \sigma = 10$ was arbitrary. The chi-square statistic would be the same for any choices.) For each sample the sample mean and standard deviation was determined, and the standardized observations $Z_i = (X_i - \bar{X})/S$ determined. These 100 $Z_i$'s were then categorized into the 10 intervals $(-\infty, -2.0], (-2.0, -1.5], (-1.5, -1.0], \ldots, (1.5, 2.0], (2.0, \infty)$, and their frequencies $Y_k$, $k = 1, \ldots, 10$ determined. For the $k$th interval $[a_k, b_k]$ the expected values $E_k^0 = 100[\Phi(b_k) - \Phi(a_k)]$ were then determined. The chi-square

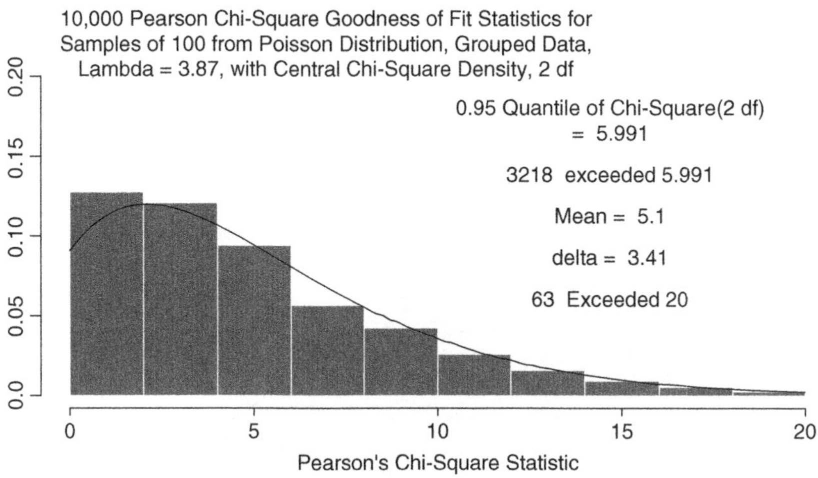


**FIGURE 12.5.5**

statistic $\chi^2 = \sum_{k=1}^{10}(Y_k - E_k^0)^2/E_k^0$ was then determined. This was done 100,000 times. The results are summarized in Figure 12.5.6. Theory suggests that $\chi^2$ should be approximately distributed as chi-square with $(10 - 1 - 2)$ degrees of freedom, since two parameters were estimated. Of the 100,000, 4753 exceeded the nominal cutoff point 14.067 for an $\alpha = 0.05$-level test. To study the power of the chi-square test for normality, samples were instead taken from the double-exponential distribution, again with $n = 100$. The chi-square statistic was determined for 100,000 samples (see

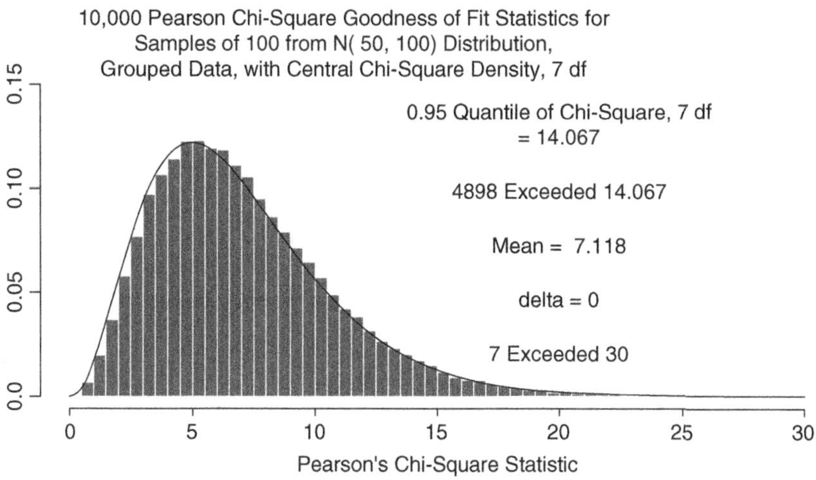


**FIGURE 12.5.6** Pearson chi-square statistics for chi-square test of normality.

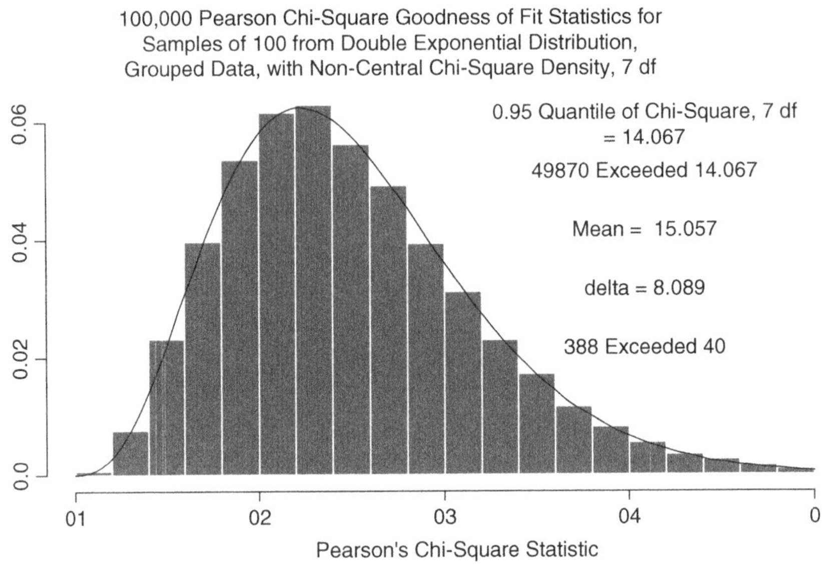


**FIGURE 12.5.7**

Figure 12.5.7). The $\alpha = 0.05$-level test led to rejection for 49,870 cases, so that the power is approximately 0.499.

We can find this in approximation by first finding the noncentrality parameter $\delta = \sum_{k=1}^{10}(E_k - E_k^0)^2/E_k^0$, where $E_k$ and $E_k^0$ are the expected frequencies for the double-exponential and normal distributions. We obtained $\delta = 8.089$. Thus, for samples from the double-exponential distribution $E(\chi^2) = \nu + \delta = 7 + 8.089 = 15.089$, very close to the mean 15.057 for the 100,000 samples. $P(\chi^2 > 14.067)$ for the case that $\chi^2$ has the noncentral chi-square distribution with parameters 7 and $\delta = 8.089$ is 0.507, very close to $49,870/10,0000$, as obtained for the simulations. This should convince a few skeptics. (The author was one until he was able to perform simulations.)

## Problems for Section 12.5

**12.5.1** A model for the number of hits $X$ that a certain major league baseball player gets in games in which he has four times at bat states that the $X \sim$ Binomial $(n, p)$ for some $p, 0 < p < 1$. In one season he played in 90 such games, with the following frequencies: 0 hits, 32; 1 hit, 26; 2 hits, 17; 3 hits, 12; 4 hits, 3.

**(a)** Test the null hypothesis that this model holds for $p = 0.320$. $\alpha = 0.05$.

**(b)** Test the hypothesis that this model holds for some $p$. $\alpha = 0.05$.

**(c)** If either of these test rejects $H_0$, can you suggest why the binomial model might be wrong?

**12.5.2** A random sample of 100 was taken from among the undergraduate students from a Midwest $U$. Their GPAs were as follows in order of size:

| | | | | | | | | | |
|---|---|---|---|---|---|---|---|---|---|
| 1.74 | 2.07 | 2.07 | 2.09 | 2.10 | 2.19 | 2.21 | 2.22 | 2.23 | 2.23 |
| 2.25 | 2.27 | 2.32 | 2.33 | 2.33 | 2.40 | 2.41 | 2.41 | 2.42 | 2.46 |
| 2.46 | 2.48 | 2.48 | 2.48 | 2.48 | 2.48 | 2.48 | 2.50 | 2.50 | 2.51 |
| 2.51 | 2.53 | 2.54 | 2.54 | 2.55 | 2.56 | 2.59 | 2.60 | 2.60 | 2.62 |
| 2.62 | 2.64 | 2.64 | 2.66 | 2.66 | 2.66 | 2.68 | 2.70 | 2.71 | 2.72 |
| 2.74 | 2.74 | 2.76 | 2.77 | 2.78 | 2.79 | 2.79 | 2.80 | 2.80 | 2.81 |
| 2.86 | 2.86 | 2.88 | 2.89 | 2.89 | 2.89 | 2.94 | 2.96 | 2.97 | 2.99 |
| 3.01 | 3.01 | 3.09 | 3.11 | 3.18 | 3.20 | 3.20 | 3.20 | 3.21 | 3.22 |
| 3.24 | 3.25 | 3.25 | 3.25 | 3.26 | 3.26 | 3.37 | 3.42 | 3.44 | 3.45 |
| 3.45 | 3.48 | 3.51 | 3.57 | 3.60 | 3.67 | 3.71 | 3.73 | 3.74 | 3.85 |

The sample mean and standard deviation are 2.797 and 0.448.

**(a)** Determine the Pearson $\chi^2$ statistic for $H_0$: the distribution of GPAs among undergraduates at Midwest $U$ is normal. Use the five intervals $(-\infty, -1.5), [-1.5, -0.5), [-0.5, 0.5], [0.5, 1.5], [1.5, \infty)$ for the standardized observations. Find the $p$-value.

**(b)** Present a histogram corresponding to these five intervals.

**12.5.3** The numbers of vehicles passing a point on I9999 freeway for each of the 120 one minute periods between 3:00 and 5:00 a.m. on July 13, 2006 were observed, then frequencies were determined as follows:

| | | | | | | | | |
|---|---|---|---|---|---|---|---|---|
| 0 | 1 | 2 | 3 | 4 | 5 | 6 | 7 | 8 |
| 6 | 20 | 22 | 32 | 11 | 16 | 9 | 3 | 1 |

**(a)** Test the simple null hypothesis that the 120 observations constitute a random sample from the Poisson distribution with $\lambda = 2.7$, using $\alpha = 0.05$. Use eight intervals with the last interval $[7, \infty)$.

**(b)** Repeat part (a) for the case that $H_0$: the 120 observations constitute a random sample from some Poisson distribution. Use the same intervals.

**(c)** Find an approximation for the power of the test in part (a) for the case that the observations were taken from the Poisson distribution with mean $\lambda = 3.0$, as they were.

**(d)** Under the model that this is a sample from a Poisson distribution, consider a test of $H_0$: $\lambda = 2.7$, $H_a$: $\lambda \neq 2.7$ based on $\bar{X}$. For $\alpha = 0.5$, for which values of $\bar{X}$ should $H_0$ be rejected? Give an approximation of the power for $\lambda = 3.0$. Why would you expect this to be larger than the power obtained in part (c)?

**12.5.4** Let $X = WU_1 + (1 - W)U_2$, where $W \sim \text{Bernoulli}(\theta)$ for $0 < \theta < 1$, $U_1 \sim \text{Unif}(0, 2)$, $U_2 \sim \text{Unif}(1, 3)$, and these three random variables are independent. $X$ has density $f(x; \theta) = \theta f_1(x) + (1 - \theta) f_2(x)$ and $f_1$ and $f_2$ are the densities of the uniform distributions on $[0, 2]$ and $[1, 3]$. Thus, $f(x; \theta)$ is $\theta/2$, $1/2$, and $(1 - \theta)/2$ on the intervals $[0, 1]$, $(1, 2)$, and $(2, 3)$. Suppose that a random sample $X_1, \ldots, X_n$ is taken from $f(x; \theta)$. Let $Y_1$, $Y_2$, and $Y_3$ be the numbers of these in these three intervals.

**(a)** Show that the MLE of $\theta$ based on $\mathbf{Y} = (Y_1, Y_2, Y_3)$ is $\hat{\theta} = Y_1/[(Y_1 + Y_3)]$.

**(b)** Determine the chi-square goodness-of-fit statistic $\chi^2$ for $n = 100$ and observed $Y = (17, 45, 38)$. Find the $p$-value corresponding to $\chi^2$. *Hint:* Recall that a chi-square random variable with 1 df is the square of a standard normal random variable.

**(c)** Let $\hat{p}_2 = Y_2/n$. Show that, in general, for any observed $\mathbf{Y} = \mathbf{y} = (y_1, y_2, y_3)$ with no zero components, $\chi^2 = z^2$, where $z = (\hat{p}_2 - 1/2)/\sqrt{1/(4n)}$. This is a test statistic for $H_0$: $p_2 = 1/2$.

**(d)** What is the approximate power for the $\alpha = 0.05$-level chi-square goodness-of-fit test for the case that the vector of probabilities for the three intervals is (0.2, 0.4, 0.4)? *Hint:* This can be found by using either the noncentral chi-square distribution or the $z$-statistic given in part (c). For 10,000 simulations the test rejected 5436 times.

**(e)** Give an approximation of the power for the case that the observations $X_1, \ldots, X_n$ constitute a random sample from the uniform distribution on $[0, 3]$. For 10,000 simulations the test rejected 9383 times.

**12.5.5** In a paper entitled "Sopra le Scoperte dei Dadi" (On a discovery concerning dice), written sometime in the period 1613–1623, Galileo Galilei, who was made famous by his then-controversial model of the solar system, in considering the totals on the throws of three dice, stated that "long experience has made dice-throwers consider 10 to be more advantageous than 9," even though both 9 and 10 may be partitioned in six ways. For example, $9 = 6 + 2 + 1 = 5 + 3 + 1 = 5 + 2 + 2 = 4 + 4 + 1 = 4 + 3 + 2 = 3 + 3 + 3$. If $T$ is the total, then $p_9 = P(T = 9) = 25/216$, while $p_{10} = P(T = 10) = 27/216$. In his book on the history of probability and statistics prior to 1750, Hald (2003) pointed out that there has been some discussion recently of whether such small differences could be detected in practice.

**(a)** Suppose that three dice are thrown $n$ times. Let $(X_9, X_{10}, X_0)$ be the frequencies of 9, 10, and neither. Let $\hat{p}_9 = X_9/n$ and $\hat{p}_{10} = X_{10}/n$. Show that Pearson's chi-square good-of-fit statistic for $H_0$: $p_9 = p_{10}$ versus $H_a$: $p_9 \neq p_{10}$ is $\chi^2 = n(\hat{p}_9 - \hat{p}_{10})^2/(\hat{p}_9 + \hat{p}_{10})$.

**(b)** Argue that asymptotically $\chi^2$ is distributed as noncentral chi-square with 1 df, noncentrality parameter $\delta = n(p_9 - p_{10})^2/(p_9 + p_{10})$.

**(c)** Show that a sample size of 10,782 is needed to have power 0.50 or more for $\alpha = 0.05$. *Hint:* The noncentral chi-square distribution with 1 df, noncentrality parameter $\delta$, is the distribution of $(Z + \delta)^2$ for $Z \sim \text{N}(0, 1)$, so probabilities may be obtained using a normal table.

**(d)** What is your conclusion about Galileo's statement?

**12.5.6** Let $X_1, \ldots, X_n$ be a random sample from a cdf $F$, where $F(0) = 0$, $F(1) = 1$, and $F$ is continuous. Suppose that we wish to test $H_0$: $F(x) = x, 0 \le x \le 1$ versus $H_a$: $H_0$ not true.

**(a)** Consider the Pearson $\chi^2$ statistic for the intervals $I_k = [(k-1)/4, k/4)$ for $k = 1, 2, 3, 4$. Show that Pearson's $\chi^2$-statistic for these intervals is $(4/n)\Sigma(Y_k - n/4)^2$, where $Y_k$ is the number of observations in $I_k$.

**(b)** Suppose that $F(x) = x^\theta$ for some $\theta > 0$. For $n = 100$, $\theta = 1.5$. Find an approximation for the power of the $\alpha = 0.05$-level test of $H_0$.

**(c)** Assuming the model $F(x) = x^\theta$ for $0 \le x \le 1$, find the likelihood ratio test of $H_0$: $\theta = 1$ versus $H_a$: $\theta \ne 1$. Find an approximation for the power of the $\alpha = 0.05$-level test for $n = 100$, $\theta = 1.5$.

CHAPTER THIRTEEN

# Miscellaneous Topics

## 13.1 INTRODUCTION

In this chapter we discuss four topics: survival analysis, bootstrapping, Bayesian analysis, and sample surveys. Many books have been written and entire courses have been devoted to each of these topics. Our discussion will serve only as brief introductions to each topic.

## 13.2 SURVIVAL ANALYSIS

Let $T$ be a random variable, the length of life of an object, animal, or person following some event. If the event is the birth of a human being, then $T$ could be his or her lifetime. Or $T$ could be the length of life of a cancer patient after diagnosis. Or $T$ could be the length of life of a 100-watt light bulb under continuous use. Let $F(t) = P(T \le t)$ be its cdf, and let $S(t) = P(T > t) = 1 - F(t)$ be its survival function. In this section we will be interested in estimating $S(t)$, particularly under random censoring, to be defined. Consider the conditional probability $G(t, \Delta) \equiv P(t < T \le t + \Delta \mid T > t) = [F(t + \Delta) - F(t)]/[1 - F(t)] = [S(t) - S(t + \Delta)]/S(t)$. If $F(t)$ has a derivative $f(t)$ for all $t > 0$, this is then, for $\Delta > 0$ and $\Delta$ small, in approximation, $\Delta f(t)/[1 - F(t)] = -\Delta \frac{d}{dt} \log(1 - F(t)) = \Delta \frac{d}{dt} \log(S(t))$. The function $\lambda(t) \equiv f(t)/[1 - F(t)] = f(t)/S(t) = -\frac{d}{dt} \log(1 - F(t)) = \frac{d}{dt} \log(S(t))$ is called the *instantaneous failure rate*, or simply the *failure rate*. By the fundamental theorem of calculus this implies that $\log(S(t)) = -\int_0^t \lambda(s)\,ds + \log(S(0)) = -\int_0^t \lambda(s)\,ds$, since $S(0) = 1$. The function $\Lambda(t) \equiv \int_0^t \lambda(s)\,ds$ for $t > 0$ is called the *cumulative failure rate*. Thus, $S(t) = e^{-\Lambda(t)}$.

*Models for Probability and Statistical Inference: Theory and Applications,* By James H. Stapleton

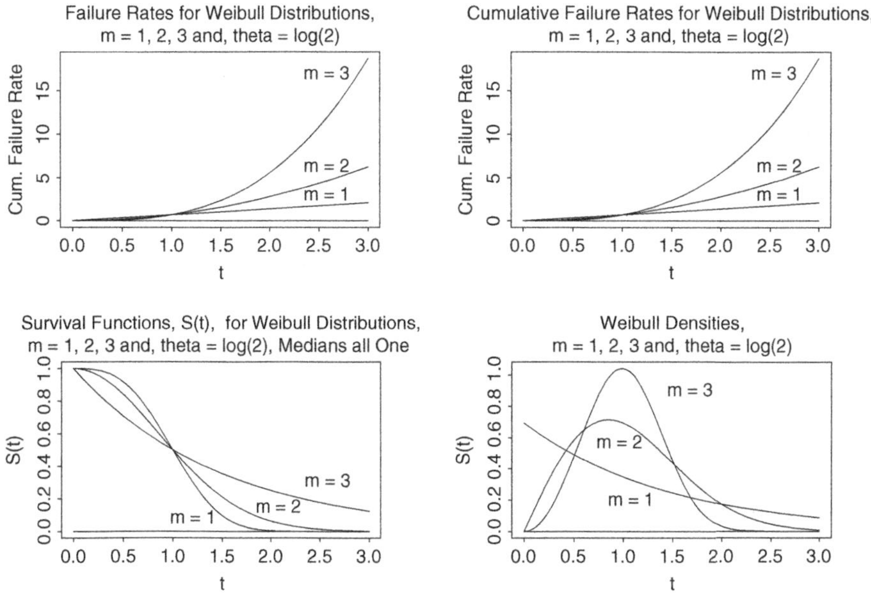


**FIGURE 13.2.1** Weibull distribution plots.

If the failure rate is constant, say $\lambda(t) = \theta > 0$ for $t > 0$, then $\Lambda(t) = \theta t$, so that $S(t) = e^{-\theta t}$. $T$ has the exponential distribution with mean $1/\theta$. If, for $\theta > 0, m > 0$, and $\Lambda(t) = (\theta t)^m$, it follows that $S(t) = e^{-(\theta t)^m}$, so that $T$ has the Weibull distribution with scale parameter $\theta$ and power parameter $m$. If $Y$ has the exponential distribution with mean 1, then $T = Y^{1/m}/\theta$ has the Weibull distribution with parameters $m$ and $\theta$. The Weibull distribution serves as a model for lifetime data for objects for which there is a *fatigue effect*, with larger values of $m$ corresponding to a propensity to wear out quickly. Human beings have lifetimes with failure rates $\lambda(t)$ on a yearly basis which are approximately 750/100,000 in the first year, drop to a minimum near $t = 10$ of approximately 22/100, 000, then gradually increase for the next 100 years, surpassing 10,000/100,000 at 85.

Figure 13.2.1 presents the functions $\lambda$, $\Lambda$, $S$, and $f$ for the Weibull distribution, for power parameters $m = 1, 2, 3$, with $\theta = \log(2)$, so that for each $m$, $S(1) = 1/2. E(T) = (1/\theta)\Gamma(1/m + 1) = 1.4427, 1.2786, 1.2883$ for $m = 1, 2, 3$. Their variances are $(1/\theta^2)[\Gamma(2/m + 1) - \Gamma(1/m + 1)^2] = 2.08, 0.45, 0.22$ for $m = 1, 2, 3$. As $m \to \infty$, $E(T) = (1/\theta)\Gamma(1/m + 1) \to 1$ and $\mathrm{Var}(T) \to 0$. For large $m$ most failures occur near 1.

Let $T_1, \ldots, T_n$ be a random sample from $F$. Let $F_n(t) = (1/n)$ (no. $T_i \le t) = (1/n)\sum_{i=1}^{n} I[T_i \le t]$ for each real number $t$, and let $S_n(t) = 1 - F_n(t)$ for each $t$. $F_n$ is the sample cdf (see Chapters Seven and Ten) and $S_n$ is the sample survival function. As discussed earlier, $E(F_n(t)) = F(t)$ and therefore $E(S_n(t)) = S(t)$ for each $t$. Also, $\mathrm{Var}(F_n(t)) = \mathrm{Var}(S_n(t)) = F(t)(1 - F(t))/n$. By the CLT both $F_n(t)$ and $S_n(t)$ are asymptotically normally distributed for each $t$ (after standardization)

**TABLE 13.2.1**

| | | | | | *i* | | | | |
|---|---|---|---|---|---|---|---|---|---|
| | 1 | 2 | 3 | 4 | 5 | 6 | 7 | 8 | 9 |
| $T_i$ | 0.050 | 0.991 | 0.768 | 0.742 | 0.738 | 0.306 | 0.089 | 0.594 | 0.977 |
| $C_i$ | 0.205 | 0.329 | 0.968 | 0.683 | 0.987 | 0.216 | 0.107 | 0.630 | 0.354 |
| $Y_i$ | 0.050 | 0.329 | 0.768 | 0.683 | 0.738 | 0.216 | 0.089 | 0.594 | 0.354 |
| $\Delta_i$ | 1 | 0 | 1 | 0 | 1 | 0 | 1 | 1 | 0 |

whenever $0 < F(t) < 1$, although the normal approximation provided may not be good for $F(t)$ close to 0 or 1. See Figure 10.3.1 for examples of sample cdf's. The interval $I(t) = [S_n(t) \pm 1.96\sqrt{S_n(t)(1 - S_n(t))/n}]$ is an approximate 95% confidence interval on $S(t)$, with a better approximation for $S(t)$ near 1/2, and, of course, large $n$. See Section 10.4 for a discussion of the use of the Kolmogorov statistic.

## Random Censoring

Suppose that cancer patients under study are diagnosed at random times between time 0 and time $A$. At time $A$ some patients will have died, so that their survival time has been observed. Others will still be alive, so that all we know is that their survival time exceeds a *censoring time*, the time from diagnosis until time $A$. Thus, we observe the minimum $Y$ of $T$, the patient's survival time, and $C$, the patient's censoring time. We also know whether $Y$ is the value of $T$ or the value of $C$. Define $\Delta$ to be the indicator of the event $[T \le C]$, equivalently that $Y = T$.

Suppose that $T_1, C_1, T_2, C_2, \ldots, T_n, C_n$ are independent random variables, with each $T_i \sim F$ and each $C_i \sim G$. Define $Y_i = \min(T_i, C_i)$ and $\Delta_i = I[T_i \le C_i]$ for each $i$. We observe the pair $(Y_i, \Delta_i)$, but not $T_i$ when $\Delta_i = 0$ and not $C_i$ when $\Delta_i = 1$. We consider the *Kaplan–Meier estimator* of $F$, or equivalently, the survival function $S(t) = 1 - F(t)$. The probability that an observation $Y_i$ is not censored is $p = P(\Delta_1 = 1) = P(T_1 \le C_1) = \int_0^\infty F(t)\,dG(t) = 1 - \int_0^\infty G(t)\,dF(t)$. These are *Steiltjes integrals*. If $G$ or $F$ has a corresponding density, $g$ or $f$, replace $dG(y)$ by $g(y)\,dy$ or $dF(y)$ by $f(y)\,dy$. For example, if $G(t) = t$ and $F(t) = t^2$ on [0, 1], then $p = 1/3 = 1 - 2/3$. For these same $F$ and $G$, nine pairs $(T_i, C_i)$ were generated, and rounded off to three decimal places (Table 13.2.1).

After ordering the pairs with respect to the $Y_i = \min(Y_i, C_i)$, smallest to largest, the pairs are (0.050, 1), (0.089, 1), (0.216, 0), (0.329, 0), (0.354, 0), (0.594, 1), (0.683, 0), (0.738, 1), (0.768, 1). Five of the nine observations were uncensored, four censored. Define $\hat{S}(t)$ for $t < 0.050$ to be 1 (see Figure 13.2.2). At $t = 0.050$, as we move to the right, $\hat{S}(t)$ should drop to 8/9. Since the next larger value of $Y_i$, 0.089, corresponds to $\Delta_i = 1$, an uncensored observation, $\hat{S}(t)$, drops to $(8/9)(7/8) = 7/9$ at 0.089. The next three values of $Y_i$ after ordering correspond to censored observations. As a result, define $\hat{S}(t)$ to remain at 7/9 for the interval [0.089, 0.594). Since 0.594 is an uncensored observation, drop $\hat{S}(t)$ to $\hat{S}(0.089)(3/4) = (7/9)(3/4)$ at $t = 0.594$,

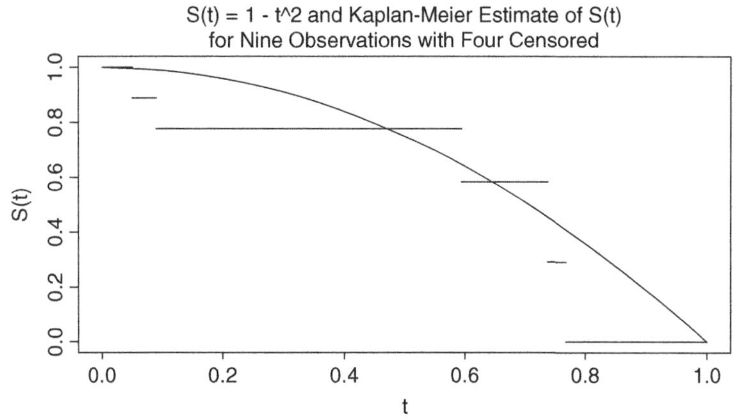


**FIGURE 13.2.2** Kaplan–meier estimate of $S(t)$ for nine observations.

since there are four values of $Y_i$ that are greater than or equal to 0.594. The next $Y_i$, 0.683, corresponds to a censored observation, so continue to define $\hat{S}(t)$ to be $\hat{S}(0.594)$ for $0.594 \leq t < 0.738$. At $t = 0.738$, drop $\hat{S}(t)$ to $\hat{S}(0.594)(1/2) = 0.292$. Finally, at $t = 0.768$, drop $\hat{S}(t)$ to zero.

Another sample of 50 $(T_i, C_i)$ pairs was taken from the same distributions $F$ and $G$. The $S$-Plus functions "Surv," "survfit," and "plot" were then used to estimate the survival function $S(t) = 1 - t^2$ for $0 \leq t \leq 1$ and to plot it (using the result of "survfit") (see Figure 13.2.3). $S(t)$ and the Kaplan–Meier estimate $\hat{S}(t)$, together with 95% confidence intervals $I(t)$ on $S(t)$, are plotted. Among the 50 observations, 18 were uncensored. $\hat{S}(t)$ was not plotted for $t > 0.873$ because there were no uncensored observations larger than 0.873.

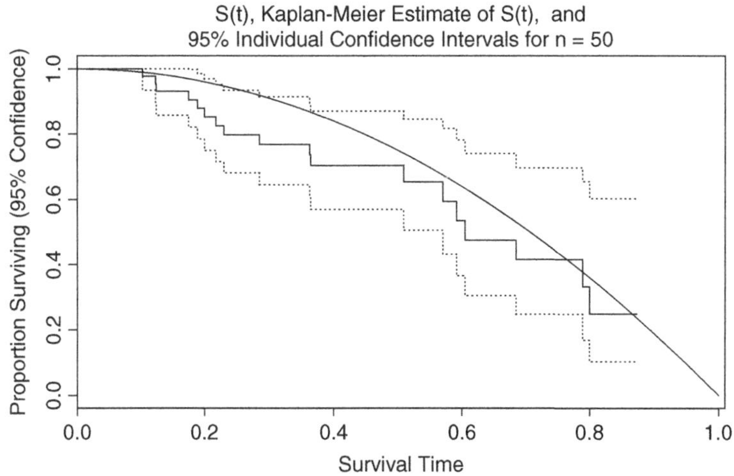


**FIGURE 13.2.3** Kaplan–Meier estimate and 95% confidence intervals.

The confidence bands seem to "miss" $S(t)$ for $t$ near $t = 0.25$. These are not simultaneous bands, so we should not be surprised that the bands on $S(t)$ do not contain $S(t)$ for some $t$. Formally, the KM estimator is defined as follows. We use the notation of S-Plus. Let $Y_i = \min(T_i, C_i)$, and $\Delta_i = I[T_i \le C_i]$ as before. Let $t_1 < t_2 < \cdots < t_m$ be the distinct values of $Y_i$ for which $\Delta_i = 1$. Define $n_j = \#[Y_i \ge t_j]$ and $D_j = \#[Y_i = t_j, \Delta_j = 1]$ for $j = 1, \ldots, m$. Let $C^* = \max(C_i)$. Define $\hat{S}(t) = 1$ for $t < t_1$; $\Pi_{t_j \le t}[(n_j - D_j)/n_j]$ for $t_1 \le t \le \max(t_m, C^*)$, undefined for $t > C^* \ge t_m$. Note that $\hat{S}(t) = 0$ for $t > t_m > C^*$, that $\hat{S}(t_{j+1}) = \hat{S}(t_j)[(n_{j+1} - d_{j+1})/n_{j+1}]$, that $\hat{S}(t)$ is "flat" between the $t_j$, and that $\hat{S}(t)$ is continuous on the right.

The KM estimator is the maximum likelihood estimator of $S(t)$ with respect to all pairs $(S, G)$ with the property that the sets of discontinuity of $F$ and of $G$ are disjoint. Thus, the KM estimator is the nonparametric MLE of $S(t)$ when it exists. It does not exist for $t$ that exceeds the largest $Y_i$ when the largest $Y_i$ corresponds to a censored observation.

## Problems for Section 13.2

**13.2.1** Let $X_1, X_2, \ldots, X_n$ be the times until failure of seven randomly sampled flashlight batteries, with survival function $S(t)$.

**(a)** For order statistics $X_{(i)}$ as follows, for $n = 7$, determine and plot the estimate $\hat{S}_7(t)$ of $S(t)$: $7, 11, 13, 15, 20, 23, 27$.

**(b)** Suppose that the observations 13, 15, and 23 in part (a) were censored with these as the censoring times. Determine the values of $\Delta_i$ and the Kaplan–Meier estimate $\hat{S}(t)$. Plot $\hat{S}(t)$ on the same axes as used in part (a).

**(c)** Plot $\hat{S}(t)$ for the case that $n = 10$ with observations $2^*, 5, 5, 7^*, 7, 7, 9, 10, 10, 13^*$. The censored observations are indicated with "*".

**13.2.2** Let $X$ have survival function $S(t)$, and failure rate $\lambda(t)$.

**(a)** Express $G(t, h) \equiv P(X > t + h \mid X > t)$ for $t > 0, h > 0$ in terms of $S$, $\lambda$, $t$, and $h$.

**(b)** What is this conditional probability in the case that $X$ has the exponential distribution with mean $1/\theta$? Show that $G(t, h)$ does not depend on $t, t > 0$.

**(c)** Determine $G(t, h)$ for the case that $X$ has the Weibull distribution with parameters $m$ and $\theta$.

**13.2.3** Let $T$ have failure rate $\lambda_T(t)$ and let $C$ have failure rate $\lambda_C(t)$. Suppose that $T$ and $C$ are independent. Show that $Y = \min(T, C)$ has failure rate $\lambda_T(t) + \lambda_C(t)$.

**13.2.4** Let $T$ and $C$ be independent, with exponential distributions with scale parameters $\theta_1$ and $\theta_2$ (means $1/\theta_1$ and $1/\theta_2$). Let $\Delta = I[T \le C]$ and $Y = \min(T, C)$. Show that:

**(a)** $P(T \le C) = E(\Delta) = \theta_1/(\theta_1 + \theta_2)$.

**(b)** $Y$ has the exponential distribution with scale parameter $\theta_1 + \theta_2$.

**(c)** $\Delta$ and $Y$ are independent. *Hint*: It is enough to show that $P(\Delta = 1, Y \le y) = P(0 \le T \le y, T < C) = P(\Delta = 1)P(Y \le y)$ for every $y > 0$.

**13.2.5** Consider the following 20 $(T_i, C_i)$ pairs.

$$
\begin{array}{ccccc}
(1.12, 2.98) & (0.09, 0.03) & (0.27, 0.41) & (0.25, 0.52) & (2.70, 0.10) \\
(0.59, 1.67) & (0.11, 1.07) & (1.42, 0.10) & (2.05, 1.47) & (1.97, 4.29) \\
(0.39, 2.64) & (0.11, 0.73) & (1.03, 2.06) & (0.50, 2.87) & (0.67, 0.17) \\
(3.49, 4.11) & (1.13, 2.60) & (0.54, 2.92) & (0.07, 0.36) & (0.56, 0.18).
\end{array}
$$

The $T_i$ and $C_i$ were generated independently so that $T_i \sim$ Exponential(mean 1), $C_i \sim$ Exponential(mean 2).

**(a)** Determine the $(\Delta_i, Y_i)$ pairs.

**(b)** Determine $P(\Delta_i = 1)$.

**(c)** For those with access to the S-Plus functions "surv" and "survfit," to the language R, or comparable "Proc's" in SAS, plot the Kaplan–Meier estimate of the survival function for $T$.

**13.2.6** Let $T \sim N(\mu_T, \sigma_T^2)$, $C \sim N(\mu_C, \sigma_C^2)$, with $T$ and $C$ independent.

**(a)** Show that $P(T \le C) = \Phi((\mu_T - \mu_C)/\sqrt{\sigma_T^2 + \sigma_C^2})$.

**(b)** Show that $Y = \min(T, C)$ has density $f(y) = (1/\sigma_T)\phi(-z_T)\Phi(-z_C) + (1/\sigma_C)\phi(-z_C)\Phi(-z_T)$, where $z_T = (y - \mu_T)/\sigma_T$, $z_C = (y - \mu_C)/\sigma_C$, and $\phi$ and $\Phi$ are the density and cdf of the standard normal distribution.

## 13.3 BOOTSTRAPPING

In 1979, Bradley Efron of Stanford University published a paper entitled "Bootstrap methods: another look at the jackknife." He followed that with a 1982 monograph *The Jackknife, the Bootstrap, and Other Resampling Methods*. The paper and the monograph were together a major impetus for much research in the following 28 years. Much of the success has been due to the rapid increase in computer power since then. We begin with a simple example.

Suppose that $Y = \theta + \varepsilon$, where $\varepsilon$, having cdf $F_\varepsilon$, is symmetrically distributed about zero. Suppose that $Y_1, \ldots, Y_n$ is a random sample with the distribution of $Y$ and that we wish to estimate $\theta$. If $\varepsilon$ is known to be normally distributed, the MLE for $\theta$ is $\bar{Y}$, and $\bar{Y}$ is known to be "best" in certain senses. Similarly, if $\varepsilon$ has a double-exponential (Laplace) distribution, the MLE is the sample median $M$, and again, $M$ has "nice" properties. However, we often or usually don't know the form of the distribution of $\varepsilon$.

**TABLE 13.3.1 Original Sample and Five Bootstrap Samples from It**

| | $X_1$ or $X_1^*$ | $X_2$ or $X_2^*$ | $X_3$ or $X_3^*$ | $X_4$ or $X_4^*$ | $X_5$ or $X_5^*$ | $X_6$ or $X_6^*$ | $X_7$ or $X_7^*$ | $X_8$ or $X_8^*$ | $X_9$ or $X_9^*$ | $X_{10}$ or $X_{10}^*$ | $\bar{X}_{0.3}$ or $\bar{X}_{0.3}^*$ | $\bar{\varepsilon}_{0.3}$ or $\bar{\varepsilon}_{0.3}^*$ |
|---|---|---|---|---|---|---|---|---|---|---|---|---|
| $X$'s | 7.36 | 6.63 | 9.37 | 7.17 | 7.96 | 7.23 | 7.57 | 8.24 | 6.83 | 8.62 | 7.53 | −0.47 |
| $X^*$'s | 7.23 | 8.62 | 8.82 | 8.24 | 7.17 | 7.23 | 8.62 | 6.83 | 7.57 | 8.24 | 7.82 | 0.29 |
| $X^*$'s | 7.36 | 7.17 | 6.83 | 7.96 | 7.17 | 9.37 | 6.63 | 8.24 | 8.24 | 6.63 | 7.41 | −0.12 |
| $X^*$'s | 7.23 | 9.37 | 7.57 | 7.23 | 9.37 | 7.17 | 6.83 | 8.62 | 6.63 | 6.83 | 7.30 | −0.23 |
| $X^*$'s | 7.57 | 8.24 | 8.62 | 7.17 | 6.63 | 7.17 | 9.37 | 8.62 | 7.23 | 7.23 | 7.57 | 0.04 |
| $X^*$'s | 6.83 | 6.83 | 7.17 | 6.63 | 8.62 | 7.57 | 6.83 | 7.96 | 7.23 | 6.63 | 7.02 | −0.51 |

Consider the α-trimmed mean $\bar{X}_\alpha$, the mean when the largest $k = [\alpha n]$ and smallest $k$ are omitted. For example, for a sample of $n = 50$, to determine the 0.3-trimmed mean, the smallest $k_{0.3} = 15$ and the largest 15 $X_i$'s are omitted and $\bar{X}_{0.3}$ is the mean of the middle 20 $X_i$'s. If $F_\varepsilon$ is known, we can estimate the distribution of the error $(\bar{\varepsilon} = \bar{X}_{0.3} - \theta$ for (say) $n = 50$ by taking a large number $B = 20{,}000$ of samples of $n$ ε's and determining $\bar{\varepsilon}_{0.3}$ for each. These $B$ $\bar{\varepsilon}_{0.3}$'s can then be used to estimate the cdf and density of $\bar{\varepsilon}_{0.3}$. The bias of $\bar{X}_{0.3}$ as an estimator of θ could be estimated by (mean of the $B$ $\bar{\varepsilon}_{0.3}$'s). Let $c_1$ and $c_2$ be the 0.025- and 0.975-quantiles of ε. Since $0.95 = P(c_1 \le \bar{X}_{0.3} - \theta \le c_2 \mid \theta) = P(\bar{X}_{0.3} - c_2 \le \theta \le \bar{X}_{0.3} - c_1 \mid \theta)$ for all θ, the interval $[\bar{X}_{0.3} - c_2, \bar{X}_{0.3} - c_1]$ is a 95% CI on θ. We could estimate $c_1$ and $c_2$ by the 500th and 19,500th order statistics among the 20,000 $\bar{\varepsilon}_{0.3}$'s.

The difficulty with this is that the distribution of ε is usually not known, and we have only the $n = 50 X_i$'s available. The bootstrap method provides one way out of the difficulty. Let $X_1^*, \ldots, X_n^*$ be a random sample (independent and identically distributed) from among $X_1, \ldots, X_n$. $\mathbf{X}^* = (X_1^*, \ldots, X_n^*)$ is called the *bootstrap sample*. Let $\bar{X}_{0.3}^*$ be the 0.3-trimmed mean for the sample $\mathbf{X}^*$. Let $\bar{\varepsilon}_{0.3}^* = \bar{X}_{0.3}^* - \bar{X}_{0.3}$. In theory we could consider all bootstrap samples. However, the number of such samples if different orderings are counted is $n^n$. Even for $n = 10$, we get 10 billion such samples. Usually, some large number $B$ of bootstrap samples is taken. Table 13.3.1 presents an original sample of 10 from the double-exponential distribution with $\theta = 8$ scale parameter 1, followed by five bootstrap samples. The last two columns contain the 0.3 trimmed mean and the corresponding values of $\bar{\varepsilon}_{0.3}$ or $\bar{\varepsilon}_{0.3}^*$.

In practice, we would take $B$ bootstrap samples with $B$ large rather than 5. Figure 13.3.1 presents the density of $\bar{\varepsilon}_{0.3}$, obtained by taking $N = 100{,}000$ samples of size $n = 100$ from the double-exponential distribution. The lower graph is an estimate of the density in the top graph obtained using $B = 20{,}000$ bootstrap samples of 100 from one sample from the double-exponential distribution.

The *percentile bootstrap method* for confidence intervals on θ works as follows. The $(\alpha/2)$- and $(1 - \alpha/2)$-quantiles $c_1$ and $c_2$ for $\bar{\varepsilon}_{0.3}$ are estimated by $\hat{c}_1$ and $\hat{c}_2$, the sample $(\alpha/2)$- and $(1 - \alpha/2)$-quantiles of $\bar{\varepsilon}_{0.3}$ for a large number $B$ of bootstrap samples, all the same size $n$ as the original sample. The CI on θ is then

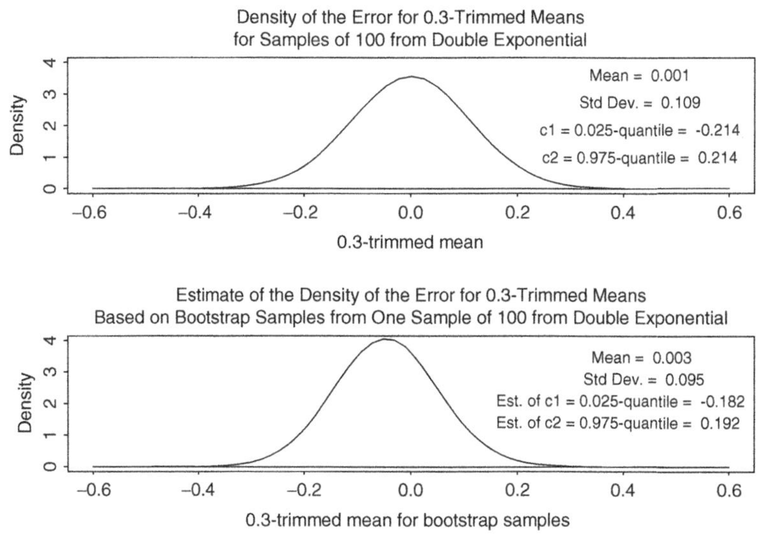


**FIGURE 13.3.1**

$[\bar{X}_{0.3} - \hat{c}_2, \bar{X}_{0.3} - \hat{c}_1]$. For $\alpha = 0.05$, this procedure was applied to 2000 original samples from the double-exponential distribution, each for $B = 1000$ bootstrap samples. Of the 2000 CIs, 1899 covered $\theta$, which in this case was chosen to be zero. Had $\theta$ been chosen to be another value, all CIs would have been $\theta$ larger, resulting in the same coverage percentage.

The same procedure was repeated for samples of 100 from the Cauchy distribution for 4000 original samples of 100. Of the CIs, 3855 covered $\theta$, 96.375%. Similarly, samples of 100 were taken from the contaminated normal distribution, with the cdf $F_\varepsilon(x) = 0.9\Phi(x) + 0.1\Phi(x/5)$, so that approximately $1/10$ of the $\varepsilon_i$ were taken randomly from the N(0, 25) distribution, the others from N(0, 1). Of 2000 such intervals, 94.95% covered $\theta$. The corresponding $t$ CIs covered $\theta$ for 94.5% of such samples. The mean lengths for the intervals were 0.702 for $t$-intervals, 0.481 for the bootstrap percentile method based on $\bar{X}_{0.3}$.

The lesson is that in cases for which the distribution sampled is thought to be symmetric about some $\theta$ but not necessarily normal, CIs based on the trimmed mean and bootstrapping will often provide shorter intervals than will the $t$-method. Of course, the $t$-method will not work at all if the distribution sampled has heavy tails. For 500 samples of $n = 100$ from the Cauchy distribution, the mean lengths of nominally 95% CIs were 14.8 for the $t$-method and 0.63 for the trimmed mean bootstrap method. These were based on 500 samples of 100. CIs covered $\theta$ for 97.0% and 93.2% of the samples.

## Another Example

Let $\mathbf{X} = (X_1, \ldots, X_n)$ be a random sample from a distribution $F$, with variance $\sigma^2$. Let $S_n^2$ be the sample variance. Of course, $S_n^2$ is an unbiased estimator of $\sigma^2$, and an argument based on the weak law of large numbers and Slutsky's theorem can be used to prove that the sequence $\{S_n^2\}$ is consistent for $\sigma^2$. If $F$ is a normal distribution, the fact that $(n-1)S_n^2/\sigma^2 \sim \chi_{n-1}^2$ can be used to show that $[S_n^2(n-1)/\chi_{n-1}^2(1-\alpha/2), S_n^2(n-1)/\chi_{n-1}^2(\alpha/2)]$ is a $100(1-\alpha)\%$ CI on $\sigma^2$. However, this interval, as discussed at the end of Section 9.3, is very sensitive to nonnormality. For example, only 784 of 1000 samples of 100 from the double-exponential distribution produced nominally 95% CIs intervals that covered the median $\theta$. Bootstrap sampling can "save the day."

We would like to know the $(\alpha/2)$- and $(1-\alpha/2)$-quantiles of the pivotal quantity $G = G(\mathbf{X}, F) \equiv (n-1)S^2/\sigma^2$. The bootstrap method replaces $G$ by $G^* = G(X^*, F_n)$, where $\mathbf{X}^*$ is bootstrap sample from $\mathbf{X}$ and $F_n$ is the sample cdf for $\mathbf{X}$. Thus, $G^* = (n-1)S^{2*}/S^2$, where $S^{2*}$ is the sample variance for $\mathbf{X}^*$ and $S^2$ is the sample variance for the original sample. We observe $G^*$ for B bootstrap samples and estimate the $(\alpha/2)$- and $(1-\alpha/2)$-quantiles $g_{\alpha/2}$ and $g_{1-\alpha/2}$ of $G$ by the corresponding quantiles $g_{\alpha/2}^*$ and $g_{1-\alpha/2}^*$ of the $B$ values of $G^*$.

For samples from the double-exponential, this procedure works reasonably well, with coverage probabilities a bit less than the nominal 0.95. For samples from the contaminated normal described above, it works less well (coverage probability approximately 0.85 for $n = 100$, 0.92 for $n = 400$), but far, far better than the normal theory method, for which the coverage probabilities are approximately 0.52 and 0.54 for $n = 100$ and $n = 400$ for the contaminated normal, 0.80 for both $n$ for the double-exponential. The lesson: If there is any suspicion at all that the distribution sampled is nonnormal, with heavy tails in particular, consider the bootstrapping method.

### The Bootstrap *t*-Method

Let $\mathbf{X} = (X_1, \ldots, X_n)$ be a random sample from a distribution $F$ with mean $\mu$, variance $\sigma^2$, both unknown. Let $T_n = (\bar{X}_n - \mu)/\sqrt{S^2/n}$, the $t$-statistic. If $F$ is a normal distribution, then $T \sim t_{n-1}$. In addition, $T_n$ converges in distribution to N(0, 1), so that even in the case of nonnormality but large $n$, $[\bar{X}_n \pm 1.96S/\sqrt{n})]$ is an approximate 95% CI on $\mu$. The approximation may not be very good for $n = 25$ (say) when $F$ has heavier tails than the normal, as it does, for example, for the exponential or contaminated normal distribution. The bootstrap method may be used to estimate the quantiles of the distribution of $T$ by estimates based on some large number $B$ of bootstrap samples, giving $B$ values of $T^* = (\bar{X}_n^* - \bar{X})/\sqrt{S^{2*}/n}$. If $c_1$ and $c_2$ are the $(\alpha/2)$- and $(1-\alpha/2)$-quantiles for $F$, and the estimates based on the bootstrap samples are $\hat{c}_1$ and $\hat{c}_2$, approximate $100(1-\alpha)\%$ CIs on $\mu$ are given by $[\bar{X}_n - \hat{c}_2 S/\sqrt{n}, \bar{X}_n - \hat{c}_1 S/\sqrt{n}]$. This method was used for $n = 50$, sampling from the contaminated normal distribution with $\mu = 0, \sigma_1 = 1, \sigma_2 = 5, \varepsilon = 0.2$, so that an observation came from N(0, 25) with probability 0.2. $B = 2000$ samples were taken. "95% CIs" covered $\mu = 0$ for 1917 cases when the usual $t$-method was used and 1969 times when the bootstrap $t$-method was used. One might guess that the bootstrap $t$-method produced longer intervals. They did, but only slightly, their mean lengths being 1.33 for the usual method, 1.37 for the bootstrap $t$-method.

When the same methods were used to determine CIs in the case that the distribution $F$ sampled was double-exponential with the same variance as for the contaminated normal, the two methods worked poorly, covering $\mu = 0$ in only 864 and 874 for 1000 samples, with mean lengths 1.34 and 1.35.

*The lesson*: Use the percentile method, not the $T$-method, if there is any fear at all that the tails of $F$ are heavy.

### The Bootstrap Method for Simple Linear Regression

Suppose that we observe $Y_i = \beta_0 + \beta_1 x_i + \varepsilon_i$ for $i = 1, \ldots, n$, with the $\varepsilon_i$ independent, each with distribution $F$. Suppose that we want a CI on $\beta_1$. The usual CI, based on the normality of $F$, is $[\hat{\beta}_1 \pm t_{1-\alpha/2}\sqrt{S^2/S_{xx}}]$. Is this appropriate when $F$ is not normal? If the $x_i$'s behave "appropriately" (don't have excessive outliers) and the tails of $F$ aren't too heavy, the distribution of $T = (\hat{\beta}_1 - \beta_1)/\sqrt{S^2/S_{xx}}$ is approximately $t_{n-2}$ for larger $n$, and the $t_{n-2}$ distribution is approximately N(0, 1), so this formula will produce CIs that cover $\beta_1$ with probability approximately $1 - \alpha$. However, if the tails of $F$ are heavy, the coverage probabilities may be smaller than "advertised" and/or the intervals may be excessively long. The $L_1$-*method*, which minimizes the sum of the absolute deviations rather than the sum of squares, may perform better. In S-Plus the function "l1reg" is available. Another robust method minimizes the median of the squared deviations, and a variety of other methods designed to overcome heavy tails for $F$ are available.

A problem with these methods is that although they are asymptotically unbiased, simple formulas for CIs based on these estimators are not available. The bootstrap method may be used as follows. If, for example, the $L_1$-method is used, let $\hat{\beta}_{L1}$ be the estimator of $\beta_1$. We would like to know the distribution of $D \equiv \hat{\beta}_{L1} - \beta_1$. Let $F_D$ be the cdf of D. If $F_D(c_1) = \alpha/2$ and $F_D(c_2) = 1 - \alpha/2$, then $c_1$ and $c_2$ may be estimated by *bootstrapping residuals* (or *resampling*). Let $R_i = Y_i - \hat{Y}_i$, where $\hat{Y}_i = \hat{\beta}_{L0} + \hat{\beta}_{L1}x_i$ is the $i$th predicted or fitted value. Let $\mathbf{X} = (\mathbf{x}_0, \mathbf{x})$, where $\mathbf{x}_0$ is the vector of all 1's. The vector $\mathbf{R} = (R_1, \ldots, R_n)$ has covariance matrix $(\mathbf{I}_n - \mathbf{H})\sigma^2$, where $\mathbf{H} = \mathbf{X}(\mathbf{X}^T\mathbf{X})^{-1}\mathbf{X}^T$ is projection onto the column space of $\mathbf{X} = (\mathbf{x}_0, \mathbf{x})$. Thus, the variances of the $R_i$ are proportional to the diagonal $\mathbf{d} = (d_1, \ldots, d_n)$ of $\mathbf{I} - \mathbf{H}$. For simple linear regression, $d_i = 1 - (1/n + (x_i - \bar{x})^2/S_{xx})$. The $R_i$ are not quite independent, but $W_i \equiv R_i/d_i^{1/2}$ has variance $\sigma^2$. Choose a random sample (with replacement) of $n$ from $(W_1, \ldots, W_n)$. Call the values obtained $(r_1^*, \ldots, r_n^*)$. Define $Y_i^* = \hat{Y}_i + r_i^*$. Perform a SLR on the pairs $(x_i, Y_i^*)$. Let $\hat{\beta}_{L1}^*$ be the resulting estimate of $\beta_1$, and let $D^* = \hat{\beta}_{L1} - \hat{B}_{L1}$. Do this $B$ times, to obtain $B$ copies of $D^*$. Estimate $c_1$ and $c_2$ by $\hat{c}_1$ and $\hat{c}_2$, the corresponding sample quantiles of the $D^*$. Then $[\hat{\beta}_{L1} - \hat{c}_2, \hat{\beta}_{L1} - \hat{c}_1]$ is an approximate $100(1 - \alpha)\%$ CI on $\beta_1$.

In the determination of $D^*$ it is enough to sample the $r_i^*$. Then $D^*$ is the slope of the simple linear regression line for the $(x_i, r_i^*)$ pairs (see Problem 13.3.3).

The method seems to provide "good" intervals in that their nominal confidence levels are close to their nominal values. For example, let $\mathbf{x} = (1, 1, 2, 2, \ldots, 10, 10)$, $\beta_0 = 50$, $\beta_1 = -4$. (These are completely arbitrary. The results would be the same for any choice of $\beta_0$, $\beta_1$), with contaminated normal errors, $\varepsilon_i$, with parameters $\sigma_1 =$

$1, \sigma_2 = 5, \varepsilon = 0.2$, as described a few paragraphs ago. $B$ was 200 for each. Four methods were used to estimate $\beta_1$: (1) the bootstrap $L_1$-method, (2) the bootstrap least squares method, (3) the bootstrap least median of squares method, and (4) the ordinary least squares method. Method 1 provided $142/150 = 0.946$ coverages of $\beta_1$; 2, $137/150 = 0.913$; 3, $139/150 = 0.927$; and 4, $140/150 = 0.933$. The method 1 CIs had a mean length 0.52, and the other three, 0.71, 0.70, and 0.74. Thus, method 1 wins "hands down." In practice, it would be better to choose $B$ to be 1000 or even 10,000. Our choice, $B = 200$, was suggested by the need to repeat this 150 times for each of our three distributional forms for the $\varepsilon_i$.

When the experiment was repeated for the double-exponential distribution for the $\varepsilon$'s, with the same variance, the coverage proportions for 150 repetitions were 0.947, 0.913, 0.927, and 0.933, with mean lengths 0.711, 0.729, 0.759, and 0.757, so that method 1 did well again. For the N(0, 5.80) distribution for 150 repetitions, the coverage proportions were 0.926, 0.920, 0.860, 0.920, with mean lengths 0.77, 0.75, 0.79, 0.79, so that methods 1, 2, and 4 do equally well, as might be expected. Thus, overall, when heavy tails are suspected, method 1 seems to be best. These comments are based on relatively little experimental evidence, of course.

### Problems for Section 13.3

**13.3.1** A random sample $\mathbf{X} = (X_1, X_2, X_3)$ is taken from a distribution that is symmetric about $\theta$, where $\theta$ is any real number. Let $M = \text{Median}(X_1, X_2, X_3)$ and $e_M = M - \theta$. Let $\mathbf{X}^* = (X_1^*, X_2^*, X_3^*)$ be a bootstrap sample from $\mathbf{X}$, let $M^* = \text{Median}(\mathbf{X}^*)$ and let $e_M^* = M^* - M$.

**(a)** For $\mathbf{X} = (3, 5, 2)$, find the bootstrap distribution of $e_M^*$.

**(b)** Replace $n = 3$ by $n = 25$, and suppose that $\mathbf{X}$ is observed to be

52.8 37.6 42.2 34.7 41.3 36.3 29.9 37.2 48.3 18.9 23.4 33.4 44.5
24.2 31.1 35.6 32.6 28.7 26.5 56.8 15.9 36.2 9.0 41.3 40.8

The sample mean and sample variance are 34.37 and 122.58. Use a computer to determine a histogram of $B = 1000$ values of $e_M^*$. The sample was taken from the double-exponential distribution with median $\theta = 35$, scale parameter 5. Figure 13.3.1 was determined by taking 10,000 samples of 25 from this distribution and determining $e = M - 35$ for each.

**(c)** Replace sample medians by sample means in part (b). Let $e^* = \bar{X}^* - \bar{X}$. Use mathematics to determine $E(e^* \mid \mathbf{X})$ and $\text{Var}(e^* \mid \mathbf{X})$. Find an approximation for $P(\mid e^* \mid > 0.20S \mid \mathbf{X})$ for $n = 100$. (The approximation, which should be good with high probability, should not depend on $\mathbf{X}$.) Also give an approximation for the unconditional probability $P(\mid e^* \mid > 0.20S)$.

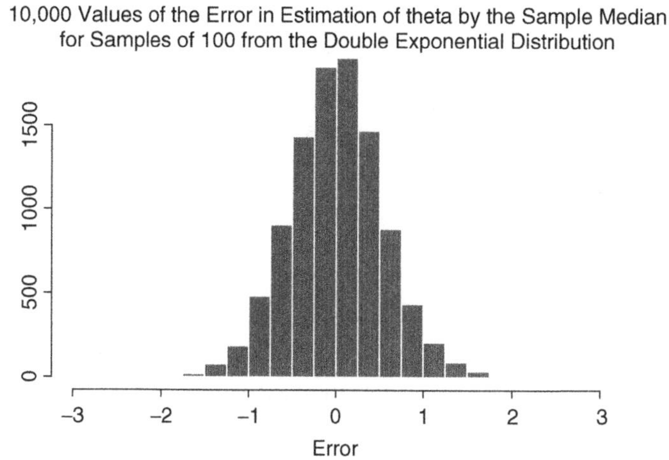


**FIGURE 13.3.1** Histogram of 10,000 estimates of $\theta$.

**13.3.2** Thirty $(x_i, y_i)$ pairs were determined from the model $Y_i = \beta_0 + \beta_1 x_i + \varepsilon_i$, where $\beta_0 = 3$, $\beta_1 = 2$, and the $\varepsilon_i$ were independent, each with the contaminated normal distribution, a (0.75, 0.25) mixture of N(0, 1), and N(0, 36) (see Figure 13.3.2 and Table 13.3.2).

**(a)** Find $\text{Var}(\hat{\beta}_1)$.

**(b)** Use the bootstrap-percentile method to give an approximate 95% CI on $\beta_1$. Use $B = 200$ if your computational method is slow. Otherwise, use $B = 1000$ or 10,000. Compare this interval to the one based on the $t$-method (nonbootstrap).

**(c)** For those who have software that produces the $L_1$-fit: Use "l1fit" in S-Plus. Use the bootstrap-$L_1$ method to produce both a 95% CI in $\beta_1$ and a 95% CI on $g(x_0) = \beta_0 + \beta_1 x_0$ for $x_0 = 5$.

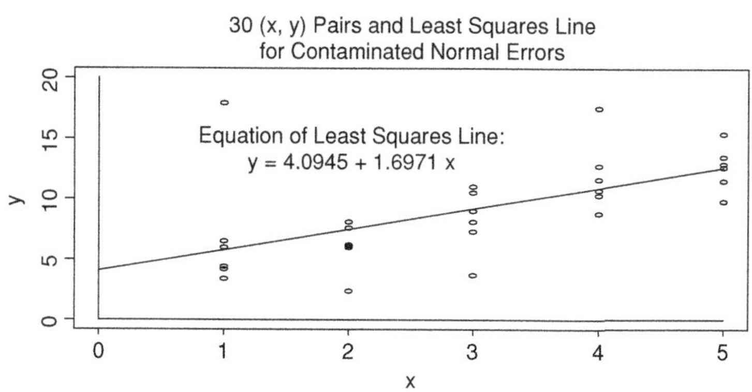


**FIGURE 13.3.2** Simple linear regression with contaminated errors.

**TABLE 13.3.2 Pair Data**

| $x_i$ | $y_i$ | $x_i$ | $y_i$ | $x_i$ | $y_i$ | $x_i$ | $y_i$ | $x_i$ | $y_i$ |
|---|---|---|---|---|---|---|---|---|---|
| 1 | 3.4 | 2 | 6.2 | 3 | 9.0 | 4 | 10.7 | 5 | 15.4 |
| | 17.9 | | 2.4 | | 10.5 | | 17.5 | | 12.9 |
| | 4.4 | | 8.1 | | 7.3 | | 10.3 | | 11.5 |
| | 4.2 | | 7.6 | | 3.7 | | 8.8 | | 12.6 |
| | 6.5 | | 6.1 | | 11.0 | | 11.6 | | 13.5 |
| | 6.0 | | 6.0 | | 8.1 | | 12.7 | | 9.8 |

**13.3.3** In the section "The Bootstrap Method for Simple Linear Regression," a two-sentence paragraph begins with "In the determination of $D^*$ it is enough to sample the $r_i^*$." Prove that.

**13.3.4** Suppose that $\mathbf{X} = (X_1, \ldots, X_n)$ is a random sample from the double-exponential distribution with density $f(x; \theta, \eta) = (1/\eta)e^{-|x-\theta|/\eta}$ for all $x$, with $\theta, \eta > 0$ unknown. Let $\hat{\theta} = \bar{X}_{0.3}$ be the 0.3-trimmed mean. Let $\hat{\eta} = (1/n)\sum |X_i - \bar{X}_{0.3}|$. Let $\mathbf{X}^* = (X_1^*, \ldots, X_n^*)$ be a random sample from the density $f(x; \bar{X}_{0.3}, \hat{\eta})$). Let $\hat{\theta}^* = \bar{X}_{0.3}^*$ and $\hat{\eta}^*$ be the values of these estimators for $\mathbf{X}^*$. For fixed $\mathbf{X}$ let $X^*(1), \ldots, X^*(B)$ be independent, each with the distribution of $\mathbf{X}^*$, and let $\hat{\theta}^*(j)$ be the estimator of $\theta$ corresponding to $X^*(j)$ for each $j$. Then the distribution of $e = \hat{\theta} - \theta$ may be estimated by the sampling distribution of $e^*(j) = \hat{\theta}^*(j) - \hat{\theta}$ for $j = 1, \ldots, B$. This sample distribution is the parametric bootstrap estimator of the distribution of $e$. If $c_1$ and $c_2$ are, for example, the 0.025- and 0.975-quantiles of the distribution of $e$, then $c_1$ and $c_2$ may be estimated by the corresponding quantiles of the $e^*(j)$. If $B = 10{,}000$, the $e^*(j)$ may be ordered and $\hat{c}_1$ and $\hat{c}_2$ may be taken to be the 250th and 9750th among the ordered $e^*(j)$. It follows that $[\hat{\theta} - \hat{c}_2, \hat{\theta} - \hat{c}_1]$ is an approximate 95% CI on $\theta$.

**(a)** Let $G(x; \theta, \eta)$ be the cdf of $e$. Show that $G(x; \theta, \eta) = G(x/\eta; 0, 1)$ for all $x, \theta, \eta > 0$.

**(b)** For $n = 100$, $\theta = 20$, $\eta = 5$, take a random sample from $f(x; \theta, \eta)$, determine $(\hat{\theta}, \hat{\eta})$, and estimate the variance of $\hat{\theta}$ by parametric bootstrap sampling. (You may take $\eta = 1$, then multiply through by $\hat{\eta}$ after determining the sample variance of the $\hat{\theta}^*$.) $\text{Var}(\bar{X}) = 2/100$.

**(c)** Determine 95% CIs on $\theta$ by parametric and nonparametric bootstrap sampling.

**(d)** Repeat parts (b) and (c) for a sample of 100 from the $N(\theta, 1)$ distribution.

## 13.4 BAYESIAN STATISTICS

Consider two discrete probability functions $f(k; \theta)$ corresponding to $\theta = 1, 2$:

| | | $k$ | |
|---|---|---|---|
| | 1 | 2 | 3 |
| $f(k, 1)$ | 0.6 | 0.3 | 0.1 |
| $f(k; 2)$ | 0.2 | 0.3 | 0.5 |

Suppose that we observe a random sample $\mathbf{X} = (X_1, X_2)$ for $\theta = 1$ or 2 but $\theta$ is unknown to us. To this point this is a familiar problem, and we might choose to estimate $\theta$ using the maximum likelihood principle. However, this time, suppose that we know that $P(\theta = 1) = 0.8$ and $P(\theta = 2) = 0.2$. Suppose also that we pay a penalty zero if we guess the value of $\theta$ correctly, 1 if we are incorrect. In more mathematical language, with $\hat{\theta} = \hat{\theta}(X_1, X_2)$ denoting our guess (estimator), we say that the *loss function* is $L(\hat{\theta}, \theta) = 0$ for $\hat{\theta} = \theta$, $L(\hat{\theta}, \theta) = 1$ for $\hat{\theta} \neq \theta$. We would like to determine the estimator $\hat{\theta}$ so that $E(L(\hat{\theta}, \theta))$ is minimum. Note that the loss $L(\hat{\theta}, \theta)$ is random for two reasons: (1) Given $\theta$, $\hat{\theta} = \hat{\theta}(X_1, X_2)$ is random, and (2) $\theta$ is random.

For this simple loss function $E(L(\hat{\theta}, \theta)) = P(\hat{\theta} \neq \theta) = \sum_x P(\hat{\theta} \neq \theta \mid X = \mathrm{x})P(\mathrm{X} = \mathrm{x})$. Only the first term following the summation depends on the choice of the function $\hat{\theta}(x)$. Therefore, for each possible $\mathbf{x}$, we should choose $\hat{\theta}(\mathbf{x})$ to minimize $P(\hat{\theta}(\mathbf{X}) \neq \theta \mid \mathbf{X} = \mathbf{x})$, or equivalently, maximize $P(\hat{\theta}(\mathbf{X}) = \theta \mid \mathbf{X} = \mathbf{x})$.

The following table contains values of $P(\theta = 1 \mid \mathbf{X} = \mathbf{x})$, for each possible $\mathbf{x}$.

| | | $x_2$ | |
|---|---|---|---|
| $x_1$ | 1 | 2 | 3 |
| 1 | 0.288/0.296 | 0.144/0.156 | 0.048/0.068 |
| 2 | 0.144/0.156 | 0.720/0.900 | 0.024/0.054 |
| 3 | 0.048/0.068 | 0.024/0.054 | 0.008/0.058 |

For example, $P(\theta = 1 \mid X_1 = 1, X_2 = 2) = (0.8)(0.6)(0.3)/[(0.8)(0.6)(0.3) + (0.2)(0.2)(0.3)] = 0.144/0.156$. It is helpful to use a tree diagram beginning at the base with the probabilities $P(\theta = 1) = 0.8$ and $P(\theta = 2) = 0.2$. All these conditional probabilities are greater than 1/2 except for the three cases $\mathbf{x} = (2, 3)$, $(3, 2)$, and $(3, 3)$. Therefore, the expected loss is smallest for $\hat{\theta} = 2$ for each of these $\mathbf{x}$, and $\hat{\theta}(\mathbf{x}) = 1$ otherwise.

The probabilities $P(\theta = 1) = 0.8$ and $P(\theta = 2) = 0.2$ are called the *prior probabilities*. That is, $\theta$ has the prior probabilities 0.8 and 0.2 for $\theta = 1$ and 2. The probabilities given in the table are the *posterior probabilities* of $\theta = 1$ for each the nine possible values of $\mathbf{x} = (x_1, x_2)$. We say, for example, that the posterior distribution of $\theta$ for $\mathbf{X} = (2, 3)$ assigns probability $0.720/0.900 = 0.80$ to $\theta = 1$, probability $0.180/0.900 = 0.2$ to $\theta = 2$. For each possible $\mathbf{x}$ we choose $\hat{\theta}(\mathbf{x})$ to maximize $P(\hat{\theta}(\mathbf{X}) = \theta \mid \mathbf{X} = \mathbf{x})$, or equivalently, minimize $P(\hat{\theta}(\mathbf{X}) \neq \theta \mid \mathbf{X} = \mathbf{x})$.

For this choice of the estimator $\hat{\theta}(\mathbf{X})$, the expected loss is $E(L(\hat{\theta}, \theta) = 0.8 [2(0.3)(0.1) + (0.1)(0.1)] + 0.2[(1 - 2(0.3)(0.5) - 0.5^2] = 0.056 + 0.090 = 0.146$. If we change the estimator a bit by redefining $\hat{\theta}(2, 2) = 2$, then $E(L(\hat{\theta}, \theta)) = 0.8[2(0.3)(0.1) + 0.1(0.1) + 0.3^2] + 0.2[1 - 2(0.3)(0.5) - 0.5^2 - 0.3^2] = 0.136 + 0.072 = 0.208$. Since $\theta$ takes only two values, we can recast the problem into the language of testing hypotheses. Suppose that we wish to test $H_0$: $\theta = 1$ versus $H_a$: $\theta = 2$. We seek a critical region $C$, a subset of the sample space, so that $\hat{\theta}(\mathbf{x}) = 2$ for $\mathbf{x} \in C$, $\hat{\theta}(\mathbf{x}) = 1$ otherwise. Let the power function corresponding to $C$ be $\gamma(\theta) = P(X \in C \mid \theta)$. Then $E(L(\hat{\theta}, \theta)) = P(\theta = 1)\gamma(1) + P(\theta = 2)[1 - \gamma(2)]$. We have shown that this is minimum for $C = \{(2, 3), (3, 2), (3, 3)\}$.

Let's briefly review what we did. We wanted to choose the estimator $\hat{\theta}(\mathbf{X})$ in order to minimize $E(L(\hat{\theta}, \theta))$. This expectation may be expressed in two ways:

$$E(L(\hat{\theta}, \theta)) = \begin{cases} \sum_{t \in \omega} \rho(t) P(\theta = t), & \text{where } \rho(t) = E(L(\hat{\theta}, \theta) \mid \theta = t), \\ \sum_{\mathbf{x}} \tau(\mathbf{x}) P(\mathbf{X} = \mathbf{x}; \theta), & \text{where } \tau(\mathbf{x}) = E(L(\hat{\theta}, \theta) \mid \mathbf{X} = \mathbf{x}). \end{cases}$$

We determined the estimator $\hat{\theta}(\mathbf{X})$ by using the second expression, choosing $\hat{\theta}(\mathbf{x})$ to minimize $\tau(\mathbf{x})$ for each $\mathbf{x}$. The function $\rho(t)$ on the parameter space $\Omega$ in the first expression is the *risk function.*

The approach considered here, in which we impose a probability distribution on the parameter $\theta$, is called *Bayesian*. The estimator $\hat{\theta}(\mathbf{X})$ is called the *Bayesian estimator* with respect to the loss function $L$ and the prior distribution (0.8, 0.2) on the parameter space $\Omega = \{1, 2\}$.

In real applications, how could we know the probability distribution of $\theta$? Usually, we do not. However, there are certainly cases in which we do know it, or think we know it, at least in approximation. For example, we may have observed that among college basketball players, the probability $\theta$ that a player can make a free-throw is $\theta$, and that $\theta$ in rough approximation is uniformly distributed on the interval [0.5, 0.9]. Suppose that we observe the results of 10 free-throws for a "randomly chosen" player. We don't even know whether he is a guard or center. Guards are usually better free-throw shooters. Given $\theta$, we suppose that the number of free-throws he makes is $X \sim \text{Binomial}(10, \theta)$. If we observe $X = 8$, what is the posterior distribution of $\theta$? What should our estimate $\hat{\theta}(8)$ be? We will consider such problems shortly.

In some cases, "personal beliefs" in the form of probability distributions are imposed on the parameter space $\Omega$, so that two different reasonable people determine different posterior distribution *s* on $\Omega$. Because of its mathematical tractability, the most commonly used loss function is squared error. In that case, $\tau(\mathbf{x}) = E(L(\hat{\theta}, \theta) \mid \mathbf{X} = \mathbf{x}) = E((\hat{\theta}(\mathbf{X}) - \theta)^2 \mid \mathbf{X} = \mathbf{x}) = \sum_{t \in \Omega} (\hat{\theta}(\mathbf{x}) - t)^2 P(\theta = t \mid \mathbf{X} = \mathbf{x})$ when $\theta$ has a discrete distribution for each $\mathbf{x}$. For $\mathbf{X} = \mathbf{x}$ this is minimum for $\hat{\theta}(x) = E(\theta \mid \mathbf{X} = \mathbf{x})$, the mean of the posterior distribution of $\theta$ given $\mathbf{X} = \mathbf{x}$. Thus, the Bayesian estimator of $\theta$ for the squared error loss function is the mean of the posterior distribution of $\theta$ given $\mathbf{X} = \mathbf{x}$. These arguments hold if the distribution of $\mathbf{X}$ or $\theta$ is continuous. We need only replace summations by integrals.

Bayesian *confidence intervals* may be determined as follows. Let $L(\mathbf{X})$ and $U(\mathbf{X})$ be the $\alpha_1$- and $(1-\alpha_2)$-quantiles of the posterior distribution of $\theta$ for given $\mathbf{X}$. Thus, $P(L(\mathbf{X}) \leq \theta \leq U(\mathbf{X})) = 1 - \alpha_1 - \alpha_2$. (For simplicity we assume the continuity of these posterior distributions, which is usually the case.)

***Example 13.4.1*** Let $\mathbf{X} = (X_1, \ldots, X_n)$ be a random sample from the Bernoulli($\theta$) distribution, and suppose that $\theta \sim \text{Beta}(\alpha, \beta)$. Then $(\theta, \mathbf{X})$ has the joint mass function $\times$ density $f(t, \mathbf{x}) = f_\theta(t)\Pi_{i=1}^n t^{xi}(1-t)^{1-x_i} = C(\alpha, \beta)t^{\alpha-1}(1-t)^{\beta-1}t^w(1-t)^{n-w} = C(\alpha, \beta)t^{\alpha+w-1}(1-t)^{\beta+n-w-1}$ for $< 0 < t < 1$, each $x_i$ is zero or 1, $w = \sum x_i$, and $C(\alpha, \beta) = \Gamma(\alpha+\beta)/\Gamma(\alpha)\Gamma(\beta)$. This means that $P(\theta \in B, \mathbf{X} = \mathbf{x}) = \int_B f(t, \mathbf{x})\,dt$ for any (Borel) subset $B$ of $[0, 1]$ and vector $\mathbf{x}$ of $n$ zeros and 1's. The conditional density of $\theta$ given $\mathbf{X} = \mathbf{x}$ is $f_\theta(t \mid \mathbf{X} = \mathbf{x}) = f(t, \mathbf{x})/f_X(\mathbf{x})$, which is of the form of the Beta($\alpha + w, \beta + n - w$) density, with constant term $C = \Gamma(\alpha+\beta+n)/\Gamma(\alpha+w)\Gamma(\beta+n-w)$. It follows that $f_X(\mathbf{x}) = [\Gamma(\alpha+\beta)/\Gamma(\alpha)\Gamma(\beta)]\,[\Gamma(\alpha+w)\Gamma(\beta+n-w)/\Gamma(\alpha+\beta+n)]$. The mean of this posterior distribution of $\theta$ is therefore $\hat{\theta} = (\alpha+w)/(\alpha+\beta+n)$. The Bayesian estimator of $\theta$ for the squared error loss function and this Beta($\alpha, \beta$) prior is $\hat{\theta}(W_n) = (\alpha + W_n)/(\alpha+\beta+n) = [(\alpha/n)/(\alpha/n + \beta/n + 1)] + [(W_n/n)/(\alpha/n + \beta/n + 1)]$, where $W_n = \Sigma X_i$. The random variable $W_n$, conditionally on $\theta = t$, has the Binomial($n, t$) distribution. For each $\theta = t$, as $n \to \infty$, $\hat{\theta}(W_n)$ converges in probability to $t$. That is, as $n \to \infty$, more and more weight is put on the sample proportion $W_n/n$, less on the sequence of constants $(\alpha/n)/(\alpha/n + \beta/n + 1)$.The usual estimator $W_n/n$ and the Bayesian estimators differ little if $n$ is large relative to $\alpha$ and $\beta$. □

***Example 13.4.2*** Given $\theta > 0$, suppose that $\mathbf{X} = (X_1, \ldots, X_n)$ is a random sample from the exponential distribution with scale parameter $\theta$, having density $f(x;\theta) = \theta e^{-\theta x}$ for $x > 0$. Suppose that $\theta$ has the prior distribution $\Gamma(\alpha, 1/\eta)$, with density $g(u;\eta,\alpha) = [\eta^\alpha/\Gamma(\alpha)]u^{\alpha-1}e^{-\eta u}$ for $u > 0$, $\eta > 0$, $\alpha \geq 1$. It follows that $(\mathbf{X}, \theta)$ has the density $f((\mathbf{x}, u);\eta,\alpha) = \Pi_{i=1}^n f(x_i;u)g(t;\eta,\alpha) = [u^n\eta^\alpha/\Gamma(\alpha)]u^{\alpha-1}e^{-u(\sum x_i+\eta)}$ for $t > 0$ and all $x_i > 0$. To determine the conditional density $f_\theta(u \mid \mathbf{X} = \mathbf{x};\eta,\alpha)$ of $\theta$ given $\mathbf{x}$, we need to divide this by the marginal density for $\mathbf{X}$. However, from the form of the joint density of $(\mathbf{X}, \theta)$ and the fact that the marginal density of $\mathbf{X}$ depends on $\mathbf{x}$, not on $u$, it should be clear that $f_\theta(u \mid \mathbf{X} = \mathbf{x};\eta,\alpha)$, the posterior density of $\theta$, is the density of the $\Gamma(\alpha+n, 1/(t+\eta))$ distribution, where $t = \Sigma x_i$, so that $f_\theta(u \mid \mathbf{X} = \mathbf{x};\eta,\alpha) = Cu^{n+\alpha-1}e^{-u(t+\eta)}$, where $C = (t+\eta)^{n+\alpha}\Gamma(n+\alpha) = [\eta^\alpha/\Gamma(\alpha)]/f_X(\mathbf{x};\eta,\alpha)$. Thus, conditionally on $\Sigma X_i = t$, $\theta$ has the $\Gamma(n+\alpha, 1/(t+\eta))$ distribution. It follows that the marginal density of $\mathbf{X}$ is $f_{\mathbf{X}}(\mathbf{x}) = [\eta^\alpha/\Gamma(\alpha)]/[\Gamma(n+\alpha)(t+\eta)^{n+\alpha}]$ for all $x_i > 0$.

The random interval $[L(\mathbf{X}), U(\mathbf{X})]$ is a 95% Bayesian CI if $L(\mathbf{X})$ and $U(\mathbf{X})$ are the 0.025- and 0.975-quantiles of the $\Gamma(n+\alpha, 1/(\Sigma X_i + \eta))$ distribution. These are given by $L(\mathbf{X}) = \chi^2_{\nu\alpha/2}/[2(\Sigma X_i + \eta)]$, $U(\mathbf{X}) = \chi^2_{\nu 1-\alpha/2}/[2(\Sigma X_i + \eta)]$, where $\nu = 2(n+\alpha)$. For the squared error loss function the Bayesian estimator $\hat{\theta}$ of $\theta$ is the mean of the posterior distribution. Thus, $\hat{\theta} = \hat{\theta}(\mathbf{X}) = (n+\alpha)/(\Sigma X_i + \eta) =$

$(1+\alpha/n)/(\bar{X}+\eta/n)$. The method of moments estimator and the MLE are both $1/\bar{X}_n$. As $n \to \infty$, both the Bayes estimator $\hat{\theta} = \hat{\theta}_n$ and the MLE converge to $\theta$ in probability. □

As indicated by Examples 13.4.1 and 13.4.2, the posterior distribution of a parameter $\theta$, given a sample $\mathbf{X} = \mathbf{x}$, depends on $\mathbf{x}$ only through a sufficient statistic T($\mathbf{X}$). For example, the mean $\hat{\theta}(\mathbf{X}) = E(\theta \mid \mathbf{X})$ of the posterior distribution of $\theta$ may be expressed as a function $m(T(\mathbf{X}))$. The proof makes use of Neyman factorization.

Consider the case that both $\mathbf{X}$, given $\theta$, and $\theta$ have discrete distributions. The joint mass function of $(\theta, \mathbf{X})$ is $f(u, \mathbf{x}) = g(T(\mathbf{x}), u)h(\mathbf{x})P(\theta = u)$ for all $u \in \Omega$, the parameter space, and all $\mathbf{x}$. [We have used the argument $u$ rather than $t$ so as not to confuse it with possible values of $T(\mathbf{x})$.] The marginal mass function of $\mathbf{X}$ is $f_{\mathbf{X}}(\mathbf{x}) = \sum_u f(u, \mathbf{x})P(\theta = u) = h(\mathbf{x}) \sum_u g(T(\mathbf{x}), u)P(\theta = u)$, so that the posterior mass function of $\theta$ given $\mathbf{x}$ is $f_\theta(u \mid \mathbf{X} = \mathbf{x}) = g(T(\mathbf{x}), u)P(\theta = u)/\sum_u g(T(\mathbf{x}), u)P(\theta = u)$, which is the same for all $\mathbf{x}$ for which $T(\mathbf{x}) = t$. Thus, the posterior distribution of $\theta$ is the same for all $\mathbf{x}$ for which $T(\mathbf{x})$ is constant. The argument is the same for the case that $\theta$ has density $f_\theta(u)$ if the sum is replaced by an integral. For Examples 13.4.1 and 13.4.2, the sum $\Sigma X_i$ is sufficient in each case, as it is for the next example.

***Example 13.4.3*** (Normally Distributed Random Sample and Normal Prior) Suppose that $\mathbf{X} = (X_1, \ldots, X_n)$ is a random sample from the $N(\theta, \sigma^2)$ distribution, with $\theta$ unknown, $\sigma^2$ known. Let $\bar{X}$ be the sample mean. Suppose that $\theta$ has the prior distribution $N(\theta_0, \tau^2)$, with both parameters known. Then the conditional distribution of $\theta$, given $\mathbf{X} = \mathbf{x}$, is $N([R/(1+R)](\bar{X} - \theta_0) + \theta_0, \tau^2/(1+R))$, where $R = n\tau^2/\sigma^2$ (see Problem 13.4.3). Thus, as $R$ converges to zero, $E(\theta \mid \mathbf{X})$ approaches $\theta_0$. Similarly, as $R$ converges to $\infty$, as it does for fixed $\sigma$ and $\tau$ as $n \to \infty$, $E(\theta \mid \mathbf{X})$ converges to $\bar{X}$. The estimator $\hat{\theta} = E(\theta \mid X) = [R/(1+R)](\bar{X} - \theta_0) + \theta_0 = [R/(1+R)]\bar{X} + [1/(1+R)]\theta_0$, a convex combination of $\bar{X}$ and $\theta_0$, is the Bayesian estimator of $\theta$ for squared error loss function and prior distribution $N(\theta_0, \tau^2)$. Notice that $E(\hat{\theta} \mid \theta) = [R/(1+R)](\theta - \theta_0) + \theta_0$, so that bias$(\hat{\theta} \mid \theta) = (\theta_0 - \theta)/(1+R)$. If $\theta$ is close to $\theta_0$, the bias is small. As $n \to \infty$, so that $R \to \infty$, the bias converges to zero for all $\theta$. Since $E(\hat{\theta} - \theta \mid \mathbf{X}) = 0$ for each $\mathbf{X} = \mathbf{x}$, it follows that $E(\hat{\theta} - \theta) = E(E(\hat{\theta} - \theta \mid \mathbf{X})) = 0$, so that the mean bias is zero. Let $D = \hat{\theta} - \theta$. Then $\mathrm{Var}(D) = (\sigma^2/n)[R/(1+R)]$ (see Problem 13.5.5). This will be small if $n$ is large or if $R$ is small, as it is when $\tau$ is small; that is, $\theta$ is usually close to $\theta_0$ and therefore $\hat{\theta}$ is usually close to $\theta_0$. □

**Multinomial Distribution with Dirichlet Prior Distribution**

Suppose that we randomly sample $n$ consumers of breakfast cereal from a population of 1 million and wish to estimate the proportions $\theta_1, \theta_2, \theta_3$ in the population who prefer each of three brands. If $X_1, X_2, X_3$ are the corresponding frequencies in the sample, $\mathbf{X} = (X_1, X_2, X_3) \sim \text{Multinomial}(n, \boldsymbol{\theta} = (\theta_1, \theta_2, \theta_3))$. From past studies we have a rough idea concerning the parameter $\theta$. Suppose that $\theta$ has the *Dirichlet prior distribution* with parameter $\alpha = (\alpha_1, \alpha_2, \alpha_3)$ with each $\alpha_i > 0$.That is, $(\theta_1, \theta_2)$ has the joint density $f(\mathbf{t} = (t_1, t_2) \mid \alpha = \alpha_1, \alpha_2, \alpha_3) = [\Gamma(\alpha_1 + \alpha_2 + \alpha_3)/\Gamma(\alpha_1)\Gamma(\alpha_2)\Gamma(\alpha_3)]t_1^{\alpha_1} t_2^{\alpha_2} t_3^{\alpha_3}$, where $t_3 = 1 - t_1 - t_2$, for all $0 < t_1, t_2 < 1$ with $t_1 + t_2 < 1$. This is the natural generalization of the beta distribution. For example, if $\mathbf{T} = (T_1, T_2)$ has this density, $E(T_i) = \alpha_i/\Sigma\alpha_j$ for each $i$, and for $T_3 = 1 - T_1 - T_2$, $E(T_3) = \alpha_3/\Sigma\alpha_j$. The marginal distribution of $T_i$ is Beta$(\Sigma\alpha_j - \alpha_i)$. We may simulate observations from the Dirichlet distribution with parameter vector $\alpha$ as follows. Let $W_i \sim \Gamma(\alpha_i, 1)$ independently for $i = 1, 2, 3$. Let $W = W_1 + W_2 + W_3$. Then $(W_1/W, W_2/W, W_3/W)$ has the Dirichlet distribution with parameter vector $\alpha$.

For observed $\mathbf{X} = \mathbf{x} = (x_1, x_2, x_3)$, $\theta$ has this posterior distribution, also Dirichlet, with parameter vector $(\alpha_1 + x_1, \alpha_2 + x_2, \alpha_3 + x_3)$. This posterior distribution therefore has posterior mean vector with components $(\alpha_i + x_i)/(\Sigma\alpha_j + n)$. Thus, the Bayes estimator of $\theta$ is $\hat{\theta} = ((\alpha_1 + X_1)/(\Sigma\alpha_j + n), ((\alpha_2 + X_2)/(\Sigma\alpha_j + n), (\alpha_3 + X_3)/(\Sigma\alpha_j + n)$. As $n$ increases, the "new evidence," the $X_i$'s, overwhelms the "old" evidence (the $\alpha_i$'s), and the Bayes estimator $\hat{\theta}$ differs in smaller and smaller amounts from the non-Bayesian unbiased estimator $(X_1/n, X_2/n, X_3/n)$.

### Example: Simulation Using S-Plus

```
V = rgamma(3,c(2,3,4))
> V
[1] 1.2971   3.3172   3.0376   # Observations from Γ(2,1), Γ(3, 1) and Γ(4, 1)
> theta = V/sum(V)
> theta
[1] 0.1695   0.4335   0.3970   # An Observation on θ
x = rmultnom(1, 100, theta)
> x
   1      2      3
  16     46     38              #An Observation of X ~ Multinomial(100, θ)
> thetahat = (c(16,46,38)+c(2,3,4))/(100 + 9)
> thetahat
[1]      0.1651      0.4495      0.3853      #Bayesian Estimate of θ
```

All of this may be generalized in an obvious way to the case that 3 is replaced by $k \geq 3$. The Bayesian approach has become a more popular approach as computing power has increased, making it possible to approximate posterior distributions, even though mathematical methods are too complex. We will go no further in this book but refer readers to books and papers on the Bayesian method. See, for example, the February 2004 issue of *Statistical Science*, No. 1, Vol. 19.

### Problems for Section 13.4

**13.4.1** Consider the discrete example at the beginning of Section 13.4. Suppose that the prior distribution of $\theta$ puts masses 0.3 and 0.7 on $\theta = 1$ and

$\theta = 2$ and that a random sample $\mathbf{X} = (X_1, X_2)$ is observed from one of these distributions.

**(a)** Find the posterior probabilities $P(\theta = 1 \mid \mathbf{X} = \mathbf{x})$ for each possible $\mathbf{x} = (x_1, x_2)$.

**(b)** Determine $\hat{\theta} = \hat{\theta}(X)$ so that $E(L(\hat{\theta}, \theta)) = P(\hat{\theta} \neq \theta)$ is minimized. Determine $P(\hat{\theta} \neq \theta)$.

**(c)** Let $f(k; \theta)$ remain the same for $\theta = 1, 2$. Define $f(k; 3) = 0.3, 0.4, 0.3$ for $k = 1, 2, 3$. Let $\theta$ take one of the values 1, 2, 3 with prior probabilities 0.5, 0.4, 0.1. For a random sample $\mathbf{X} = (X_1, X_2)$ from one of these distributions and loss function $L(\hat{\theta}, \theta) = I[\hat{\theta} \neq \theta]$, find the Bayes estimator $\hat{\theta}(X_1, X_2)$ and $P(\hat{\theta} \neq \theta)$.

**13.4.2** Let $X \sim \text{Uniform}(0, \theta)$, and let $\theta$ have the uniform prior on $(0, 1)$. For squared error loss, find the Bayes estimator $\hat{\theta}(X)$.

**13.4.3** Prove the statement beginning "Then the conditional distribution ..." in Example 13.4.3.

**13.4.4** For the model of Example 13.4.3 give a formula for a 95% Bayesian CI on the mean $\theta$. Determine it for the case that $\sigma = 5$, $\tau = 3$, $n = 4$, $\bar{X} = 43$, $\theta_0 = 40$.

**13.4.5** Let $\mathbf{X} = (X_1, \ldots, X_n)$ be a random sample from the geometric distribution with parameter $\theta$, $0 < \theta < 1$. Suppose that $\theta$ has the uniform distribution on $[0, 1]$.

**(a)** Show that $T = \sum_{i=1}^{n} X_i$ is sufficient for $\theta$.

**(b)** What is the joint "density" of $(\theta, T)$? [This is a density with respect to the length (Lebesgue) measure on the cross product of the positive real line and counting measure on the set of positive integers.]

**(c)** Show that the conditional distribution of $\theta$, given $\mathbf{X} = \mathbf{x}$, is Beta($\alpha = n + 1, \beta = t - n + 1$), where $t = \Sigma x_i$, and that $E(\theta \mid \mathbf{X} = \mathbf{x}) = \alpha/(\alpha + \beta) = (n + 1)/(t + 2) = (1 + 1/n)/(\bar{X} + 2/n)$.

**(d)** One observation on $\theta$ was made. Then for this $\theta$ the vector $\mathbf{X} = (X_1, \ldots, X_{60})$ was observed as follows:

| | | | | | | | | | | | | | | | | | | | |
|---|---|---|---|---|---|---|---|---|---|---|---|---|---|---|---|---|---|---|---|
| 3 | 2 | 3 | 6 | 5 | 4 | 1 | 1 | 2 | 2 | 3 | 3 | 1 | 1 | 4 | 2 | 1 | 8 | 10 | 4 |
| 3 | 3 | 4 | 13 | 3 | 1 | 7 | 1 | 13 | 1 | 4 | 2 | 2 | 4 | 5 | 1 | 1 | 2 | 3 | 2 |
| 6 | 6 | 5 | 3 | 1 | 1 | 2 | 1 | 8 | 2 | 1 | 1 | 9 | 2 | 5 | 4 | 1 | 1 | 3 | 1 |

The $X_i$ and their frequencies were

| 0 | 1 | 2 | 3 | 4 | 5 | 6 | 7 | 8 | 9 | 12 |
|---|---|---|---|---|---|---|---|---|---|---|
| 18 | 11 | 10 | 7 | 4 | 3 | 1 | 2 | 1 | 1 | 2 |

What is the posterior expected value of $\theta$ given these $X_i$'s? (See the "Answers" section at the back of the book for the value of $\theta$.)

**13.4.6** Let $\mathbf{X} = (X_1, \ldots, X_n)$ be a random sample from the Poisson distribution with mean $\lambda$. Suppose that $\lambda$ has the prior distribution $\Gamma(\alpha, 1)$, having density $g(u; \alpha) = [1/\Gamma(\alpha)]u^{\alpha-1}e^{-u}$ for $u > 0$.

**(a)** Show that the conditional distribution of $\lambda$, given $T = t$, is $\Gamma(t + \alpha, 1/(n + 1))$. *Hints*: First use the fact that $T = \sum_{i=1}^{n} X_i$ is a sufficient statistic for $\lambda$. The form of the joint density of $\lambda$, $T$ then indicates that the conditional distribution of $\lambda$, given $T = t$, is gamma.

**(b)** Show that the Bayes estimator of $\lambda$ for squared error loss is $\hat{\lambda} = \hat{\lambda}(\mathbf{X}) = (T + \alpha)/(n + 1) = (\bar{X} + \alpha/n)/(1 + 1/n)$.

**13.4.7** Let $X_1 \sim \mathrm{N}(\mu_1, \sigma_1^2)$, and conditionally on $X_1 = x_1$, let $X_2 \sim \mathrm{N}(\mu_2 + \rho\sigma_2 z_1, (1 - \rho^2)\sigma_2^2)$, where $z_1 = (x_1 - \mu_1)/\sigma_1$. It was shown in Section 5.3 that this implies that $(X_1, X_2)$ has the bivariate normal distribution with parameters $\mu_1, \mu_2, \sigma_1^2, \sigma_2^2, \rho$, equivalently that the standardized random variables $Z_1$, $Z_2$ have the bivariate normal distribution with parameters 0, 0, 1, 1, $\rho$.

**(a)** Show that the conditional distribution of $Z_1$, given $Z_2 = z_2$, is $\mathrm{N}(\rho z_2, (1 - \rho^2))$.

**(b)** Use the results of Section 5.3 and part (a) to prove the statement referred to in Problem 13.4.3.

## 13.5 SAMPLING

The author was once asked by the local bus company, CATA (Capital Area Transportation Authority, www.CATA.org) to help it estimate the number of bus riders on certain days. The company receives grants from the federal government and must periodically submit reports on its ridership. For the next several pages we use this as an example. The data provided, although of the same magnitude, is fictitious. The numbers of routes and the numbers of "trips" will be approximations, for illustrative purposes only. For each of approximately 40 routes, "trips" were run, usually consisting of round trips from a point $A$ to other points, say $B$, $C$, $D$, $E$, and return. At the Web site the times at which a trip passes these points are available. Riders get on the bus at varying places and get off after short or long rides. The numbers of trips $M_i$ made for route $\mathrm{R}_i$ on any weekday varies with $i$, $i = 1, \ldots, 40$. Let $y_{ij}$ be the number of riders on trip $j$ for route $i$. Our problem was to estimate $\tau = \sum_{ij} y_{ij}$ for a particular day, say Wednesday, July 12, 2006.

One easy solution was simply to perform a census. Have each bus driver record the number of customers for each of her or his trips. Experience had shown, however, that the counts, say $x_{ij}$, by the drivers were often quite inaccurate. Usually, $x_{ij} < y_{ij}$. There was also the danger that a driver who was distracted by the need to count would be more prone to have an accident. The amounts of money deposited in the fare boxes could not be used because people used many means to pay fares: transfer coupons, monthly passes, and so on. The determination of $y_{ij}$ for trip $j$ on route

$i$ was to be made by a trained "counter," who would ride the entire trip with a "clicker," one click for each person getting on the bus. The total number of trips was $N = \Sigma M_i$, which was approximately 1000. The simplest method of sampling was to take a simple random sample from the $N$ trips, then use the "expansion estimator" of $\tau$. For a simple random sample of $n$ trips (sampling without replacement) the numbers of riders $W_1, \ldots, W_n$ would be recorded, and the sample mean $\bar{W}$ and sample variance $S_W^2$ determined. The expansion estimator is then $\hat{\tau} = N\bar{W}$. If we let $\mu = \tau/N$, then $E(\hat{\tau}) = NE(\bar{W}) = N\mu = \tau$, so that $\hat{\tau}$ is an unbiased estimator of $\tau$, with $\text{Var}(\hat{\tau}) = N^2\text{Var}(\bar{W}) = N^2(\sigma^2/n)\,[(N-n)/(N-1)]$ where $\sigma^2$ is the variance of all $N$ of the $y_{ij}$.

WARNING: In many sampling texts $\sigma^2$ is replaced by the symbol $S^2$, called the *variance of the population*, although it uses $(N-1)$ as the divisor. We have chosen to use $\sigma^2$ as the population variance, with $N$ divisor because the notation is more consistent with previous notation in this book. Since $S^2 = \sigma^2 N/(N-1)$, there is little difference even for moderate $N$. Rather than use the traditional sampling notation, which is somewhat contrary to that used for most statistical literature, we stick to the notation used throughout this book. Greek letters will denote parameters. Capital letters near the end of the alphabet will denote random variables.

Since $\bar{W}$ is approximately normally distributed for large $n$, and the sample variance $s^2$ among the $W_i$ will be close to $\sigma^2$ with high probability, it follows that $Z = (\bar{W} - \mu)/\sqrt{(S^2/n)[(N-n)/(N-1)]}$ is approximately distributed as N(0, 1). Therefore, $[\bar{W} \pm z_{1-\alpha/2}\sqrt{(s^2/n)[(N-n)(N-1)]}]$ is an approximate $100(1-\alpha)$ CI on $\mu$. To obtain a CI on $\tau$, we need only multiply this interval by $N$.

How large should $n$ be? Let's say that CATA needed to estimate $\tau$ within 1500 riders with probability 0.95. Equivalently, it needed $1.96\sqrt{N^2(\sigma^2/n)[(N-n)/(N-1)]} \leq 1500$. By omitting the $-1$ in $N-1$ for simplicity and squaring both sides, we arrive at $(\sigma^2/n)[(N-n)/N] \leq [1500/(1.96)N)]^2$, or equivalently, $n \geq n_w/(1 + n_w/N)$, where $n_w = [1.96\sigma/(1500/N)]^2$ is the sample size needed for with replacement sampling. The subscript $w$ means "with." There is one big thing wrong with this formula, however. We seldom know $\sigma$. In the case of CATA there was some data from previous sampling, based on a sample of about 200. These data indicated that $\sigma$ was about 12. From this we determine that $n_w = 245.9$, $n = 197$. That meant, assuming that each counter could take eight trips on that day, that 25 counters would be needed.

Other sampling methods were considered. Stratified sampling was suggested because some routes, known to CATA, tended to have large numbers of riders, while others had just a few riders. Thus, with the help of management the routes were "stratified" (divided) into "heavy," "average," and "light," with seven routes in the heavy stratum, 18 routes in the average stratum, 13 routes in the light stratum. Let $N_k$ be the numbers of trips made for stratum $k$, $k = 1, 2, 3$. These were $N_1 = 223$, $N_2 = 372$, $N_3 = 405$. Thus, $\Sigma M_k = N = 1000$. Let $\mu_1$, $\mu_2$, and $\mu_3$ be the mean of the numbers of riders for all trips made during that day for each of the strata. Let $\tau_k = N_k\mu_k$ for $k = 1, 2, 3$

be the corresponding totals, and let $\sigma_1^2, \sigma_2^2, \sigma_3^2$ be the corresponding variances. Simple random samples of sizes $n_1, n_2, n_3$ of the trips were then chosen. Let $\bar{Y}_k$ and $s_k^2$ for $k = 1, 2, 3$ be the corresponding sample means and variances of the numbers of riders on the trips sampled. Then $\hat{\tau}_{\text{str}} = \sum_{k=1}^{3} N_k \bar{Y}_k$ and $\hat{\mu}_{\text{str}} = \hat{\tau}_{\text{str}}/N$ are unbiased estimators of $\tau$ and $\mu = \tau/N$, with variance $\text{Var}(\hat{\tau}_{\text{str}}) = \sum_{k=1}^{3} N_k^2 \text{Var}(\bar{Y}_k)$, where $\text{Var}(\bar{Y}_k) = (\sigma_k^2/n_k)[(N_k - n_k)/(N_k - 1)]$, $\text{Var}(\hat{\mu}_{\text{str}}) = N^2 \text{Var}(\hat{\tau}_{\text{str}})$. These variances may be estimated "unbiasedly" by replacing each $\sigma_k^2$ by $s_k^2$.

***Example 13.5.1*** The histograms of Figure 13.5.1 indicate the combined population and the three subpopulations, the strata heavy, average, and light. The population total is $\tau = 42{,}282$, with population variance $\sigma^2 = 1973.54$. Of course, these strata distributions would not be known in practice. They could be estimated by using the samples. Suppose that sample sizes $n_1 = 80$, $n_2 = 60$, $n_3 = 40$ were chosen. It should seem intuitively clear that larger sample sizes should correspond to larger $M_i$ and $\sigma_i^2$. We will investigate that shortly. For these sample sizes $\text{Var}(\hat{\tau}_{\text{str}}) = 7{,}669{,}140$, so that $\text{SD}(\hat{\tau}_{\text{str}}) = 2769.32$. $\text{Var}(\hat{\mu}_{\text{str}}) = \text{Var}(\hat{\tau}_{\text{str}})/N^2 = 7.669140$, $\text{SD}(\hat{\mu}_{\text{str}}) = 2.7693$. Again, these variances and standard deviations would not be known in practice, but could be estimated.

Samples were taken and the following statistics determined: $\hat{Y}_1 = 67.32$, $s_1^2 = 2376.6$, $\bar{Y}_2 = 50.33$, $s_2^2 = 2257.5$, $\bar{Y}_3 = 16.88$, $s_3^2 = 124.1$. Therefore, $\hat{\tau}_{\text{str}} = N_1\bar{Y}_1 + N_2 Y_2 + N_3 \bar{Y}_3 = 46{,}624$, $\hat{\mu}_{\text{str}} = 4.6624$. We estimate $\text{Var}(\hat{\tau}_{\text{str}})$ by $S^2(\hat{\tau}_{\text{str}}) = \sum_{k=1}^{3} N_k^2 S^2(\bar{Y}_k) = 7{,}283{,}793$, where $S^2(\bar{Y}_k) = (S_k^2/n_k)[(N_k - n_k)/(N_k - 1)]$. We

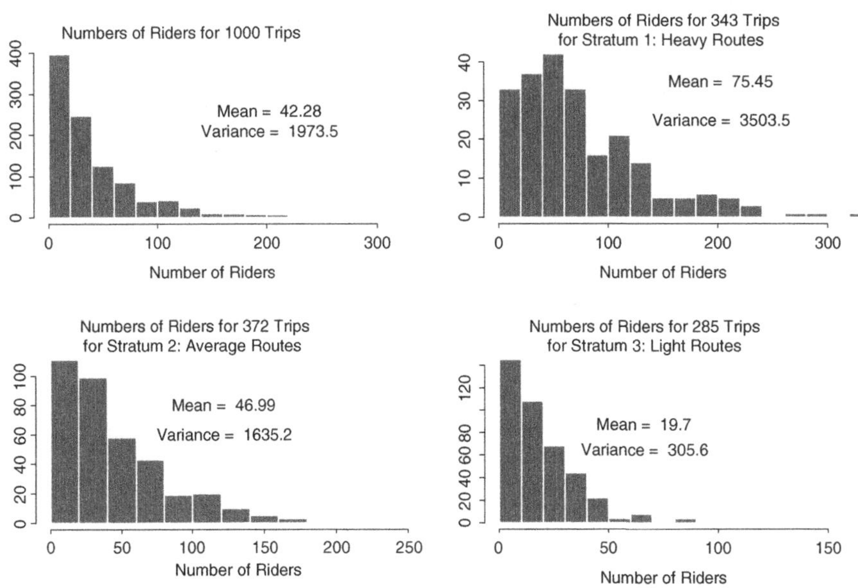


**FIGURE 13.5.1** Histograms for the number of riders for the entire population of trips and for the three strata.

were somewhat lucky, since $\text{Var}(\hat{\tau}_{\text{str}}) = 7{,}669{,}140$. We therefore estimate $\text{Var}(\hat{\mu}_{\text{str}})$ to be $S^2(\hat{\mu}_{\text{str}}) = S^2(\bar{Y}_k)/N^2 = 7.6691$. A 95% CI on $\tau$ is given by $[\hat{\tau}_{\text{strr}} \pm 1.96S(\hat{\tau}_{\text{str}})] = [46{,}624 \pm 5428]$.

What would the variance of the expansion estimator $\hat{\tau} = N\bar{Y}$ of $\tau$ have been had we simply taken a simple random sample of $n = 80 + 60 + 40 = 180$ from the population of trips? The variance of this population was $\sigma^2 = 1973.54$, so $\text{Var}(\hat{\tau}) = N^2(\sigma^2/n)[(N - n)/(N - 1)] = 8{,}999{,}552$, $\text{SD}(\hat{\tau}_{\text{str}}) = 2999.9$, a bit larger than $\text{Var}(\hat{\tau}_{\text{str}})$ and $\text{SD}(\hat{\tau}_{\text{str}})$. The optimum choice for the $\text{n}_k$ for fixed total sample size $n = \Sigma n_k$ is provided by the following analysis. In good approximation we can replace $N_k - 1$ by $N_k$, so that this is in good approximation $\sum_{k=1}^{3} N_k^2(\sigma_k^2/n_k) - \sum_{k=1}^{3} N_k\sigma_k^2$. The second term does not depend on the $n_k$. Letting $G(n_1, n_2, n_3, \lambda) = \sum_{k=1}^{3} N_k^2(\sigma_k^2/n_k) + \lambda(\Sigma n_k - n)$, then taking partial derivatives with respect to $n_1, n_2, n_3$ and the Lagrangian multiplier $\lambda$, we find that the optimum choice for $n_k$ is n $(N_k\sigma_k/M)$, where $M = \Sigma N_k\sigma_k$. For these strata we get $n_1 = 90.62$, $n_2 = 67.14$, $n_3 = 22.24$. We could therefore use sample sizes 90, 67, 22, resulting in $\text{Var}(\hat{\tau}_{\text{str}}) = 7{,}209{,}416$, smaller than for the sample sizes 80, 60, 40. 10,000 simulations, each for these optimum sample sizes, provided 10,000 CIs on $\tau$. 9399 of these covered $\tau$, with mean CI length 8759.8. 10,000 simulations, each for these optimum sample sizes, provided 10,000 CIs on $\tau$. 9399 of these covered $\tau$, with mean CI length 8759.8.

For the optimum choice of the $n_k$, $\text{Var}(\hat{\tau}_{\text{str}}) = (1/n)[(\Sigma N_k\sigma_k)^2 - \Sigma N_k\sigma_k^2]$. Of course, in practice the $\sigma_k^2$ are unknown. Good guesses concerning the values of the $N_k\sigma_k$ and choices for the $n_k$ based on these are better than arbitrary choices.

Insight into the value of stratified sampling is provided by the following analysis. From Example 11.4.1 on one-way analysis of variance, we have the identity $\sigma^2 = (1/N)[\Sigma(\mu_k - \mu)^2 N_k + \Sigma N_k\sigma_k^2]$. For optimum allocation for stratified sampling, we therefore have, for the same total sample sizes $n$, ignoring the finite correction factors,

$$\frac{\text{Var}(\bar{Y})}{\text{Var}(\hat{\tau}_{\text{str}})} = \frac{\Sigma p_k\sigma_k^2 + \Sigma p_k\delta_k^2}{(\Sigma p_k\sigma_k)^2}, \tag{13.5.1}$$

where $p_k = N_k/N$ and $\delta_k = \mu_k - \mu$. Since $\Sigma p_k(\sigma_k - \bar{\sigma})^2 = \Sigma p_k\sigma_k^2 - \Sigma p_k\sigma_k)^2$, the ratio of the first term in the numerator to the denominator in (13.5.1) is at least (1), and is (1) only if all $\sigma_k$ are equal. The second term in the numerator is the variance of the $\mu_k$ (with respect to the probabilities $p_k$), being largest when the $\mu_k$ differ by greater amounts. Thus, it is advantageous to choose strata whose means differ as much as possible. We don't know the $\mu_k$, of course, but usually do have some educated guesses as to the values of the $\sigma_k$, from previous studies, or merely intuition. For the bus trips it was clear from knowledge of bus use that the more heavily used trips would have means that differed greatly from those for lightly used trips.

For this population we get $\mathbf{p} = (p_1, p_2, p_3) = (0.223, 0.372, 0.405)$ and the ratio $\text{Var}(\bar{\text{Y}})/\text{Var}(\hat{\tau}_{\text{str}}) = (1513.34 + 479.72)/35.32^2 = 1993.06/1247.67 = 1.597$. Taking the correction factors into account, we get the ratio 8999552/5435861 = 1.65. □

There were more complicated factors in designing the survey which we will not discuss. For example, if more counters were needed at some times of the day than others because the trips were chosen randomly, more people would have to be hired. It was therefore necessary to "block" the trips, so that the number of counters needed in blocks of time were approximately equal.

It should be clear that the number $k$ of strata can be any positive integer $k \geq 2$ and that the formulas for expected values and variances remain the same, with $k$ replacing 3. Although we have concentrated on the estimation of $\tau$, we can determine an estimator $\hat{\mu} = \hat{\tau}/N$ for any estimator $\hat{\tau}$ of $\tau$, with corresponding $\text{Var}(\hat{\mu}) = \text{Var}(\hat{\tau})/N^2$. For bias $b(\tau) = E(\hat{\tau}) - \tau$, $\hat{\mu}$ has bias $(E(\hat{\tau}) - \tau)/N$ as an estimator of $\mu$.

### Ratio Estimation

For each of the 1000 trips, the driver was asked to record $x_{ij}$, his or her estimate of the number of riders for route $i$, trip $j$. As stated earlier, there tends to be significant differences between the $x_{ij}$ and the $y_{ij}$, but it is possible to take advantage of this information. The idea is to express $\tau = \tau_y = \Sigma y_{ij}$ in the form $\tau_y = R\tau_x$, where $\text{R} = \tau_y/\tau_x$ and $\tau_x = \Sigma x_{ij}$. Since $\tau_x$ is known, we can estimate the ratio $R$ from a sample, obtaining $\hat{R}$, then define $\hat{\tau}_R = \hat{R}\tau_x$. It often turns out that the estimator is "approximately unbiased" and that its variance is much smaller than for the estimators already discussed. Figure 13.5.2 is a scatter diagram of the $(x, y)$ pairs in our fictional population. For these data the $y$'s are the same as for the population considered earlier, so that $\tau_y = 42{,}282$. The 1000 $x_i$'s were all reported, so that $\tau_x = 37{,}993$ and $R = 1.113$. Since the population $y_i$'s were not known.

Suppose that a simple random sample of $n$ is taken from the population and that $(X_i, Y_i)$ is recorded for $i = 1, \ldots, n$. Let $\bar{X}$ and $\bar{Y}$ be the sample mean and variance. Let $\hat{R} = \bar{Y}/\bar{X}$. Let $D_j = Y_j - RX_j$, $\bar{D} = (1/n)\sum_{j=1}^{n} D_i$, $d_i = y_i - Rx_i$. Then $\hat{R} - R = (\bar{Y} - R\bar{X})/\bar{X} = \bar{D}/\bar{X}$. For large $n$, $\bar{X}$ will be close to $\mu_x = \tau_x/N$ with high

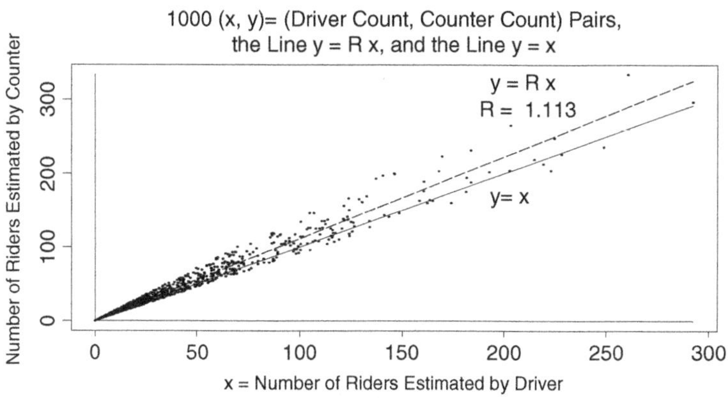


**FIGURE 13.5.2** 1000 (driver count, counter count) pairs.

probability. Therefore, $\hat{R} - R$ and $\bar{D}/\mu_x$ have approximately the same distributions for large $n$. Since $E(\bar{D}) = (1/N)\Sigma d_i = 0$, it follows that for large $n$, $E(\bar{D}) \doteq 0$, so that $\hat{R}$ is approximately unbiased for $R$ for large $n$. Also, $\text{Var}(\hat{R}) \doteq \text{Var}(\bar{D}/\mu_x) = (1/\mu_x^2)(\sigma_d^2/n)[(N-n)/(N-1)]$, where $\sigma_d^2 = (1/N)\sum_{i=1}^{N} d_i^2$. Notice that for our example the $d_i$'s, being the vertical distances from the points to the straight line with slope $R$, are relatively small in absolute value. We find that $\sigma_d^2 = 61.05$. It follows that $E(\hat{\tau}_R) = E(\hat{R}\tau_x) \doteq R\tau_x = \tau_y$ and $\text{Var}(\hat{\tau}_R) \doteq N^2(\sigma_d^2/n)[(N-n)/(N-1)]$. For our example, we find that $\text{Var}(\hat{R}) \doteq 0.000193$ and $\text{Var}(\hat{\tau}_R) \doteq 278{,}280$, so that $\text{SD}(\hat{\tau}_R) = 527.5$, considerably smaller than $\text{Var}(\hat{\tau}_{\text{str}}) = 7{,}669{,}140$, $\text{SD}(\hat{\tau}_{\text{str}}) = 2769.3$.

By the central limit theorem for sampling without replacement $\bar{D}/\mu_x$ will therefore be distributed approximately as $N(\tau_y, \text{Var}(\hat{\tau}_R))$. $\text{Var}(\hat{\tau}_R)$ may be estimated by replacing $\sigma_d^2$ by $S_d^2 = (1/n)\Sigma(Y_j - \hat{R}x_j)^2$, so that $S^2(\hat{\tau}_R) = N^2(S_d^2/n)[(N-n)/(N-1)]$. It follows that the interval $[\hat{\tau}_R \pm z_{1-\alpha/2}S(\hat{\tau}_R)]$ is an approximate $100[1-\alpha]\%$ CI on $\tau_y$. 10,000 simple random samples of 180 were taken from the trip population and a 95% CI on $\tau_y$ determined for each. 9535 of the intervals contained $\tau_y$. Their mean length was 2076.3. The analogous simulation was conducted using stratified sampling and the estimator $\hat{\tau}_{\text{str}}$. For sample sizes 80, 60, 40 9478 of the 10,000 intervals covered $\tau_y$ with mean length 9391.5. For the optimum sample sizes 90, 67, 22 9478 of the intervals covered $\tau_y$, with mean length 8711.5. Thus, the ratio estimator was far better.

We could, of course, combine stratification with ratio estimation. Optimal sample sizes were determined by the standard deviations of the $d_i$'s replacing the $\sigma_k$ in the formula $n_k = nN_k\sigma_k/M$, $M = \Sigma\sigma_k N_k$. The optimal sample sizes were determined to be 71, 72, 37. 10,000 simulations provided 10,000 CIs on $\tau_y$. Of these, 9531 covered $\tau_y$. Their mean length was 1820.6, the smallest obtained for all the methods considered.

### Medical Records

The author was once asked by the state government to testify at an administrative hearing concerning charges that the state government had made against a medical laboratory for lab tests. A random sample of approximately 200 bills had been taken from a population of several thousand bills that the laboratory had sent the state. For each bill, medical professionals had had to evaluate what the bill should have been, $y$, when the actual amount billed was $x$. For many bills, $y = x$. However, there were many for which either $y = 0$ or $y$ was much less than $x$. The cost of hiring the professionals had ruled out a complete census. The total $\tau_x$ of the amounts billed was known, and $(x, y)$ measurements were available for the bills sampled, so that the ratio estimation of $\tau_y$ was used. The amount overpaid $\Delta = \tau_y - \tau_x$ could therefore be estimated both by point and interval estimates. Due to some mistakes in the selections of the bills sampled, there was much heated discussion among lawyers on both sides. Ultimately, the laboratory was forced to return approximately \$300,000 to the state.

### Cluster Sampling of Bus Trips

Although it was not implemented, cluster sampling was considered. That is, from among the 1000 trips it might have saved labor costs if trips were chosen randomly in clusters, with clusters consisting of consecutive trips on the same route by the same counter. He or she would then have less wasted time moving from one route to another or waiting between trips on the same route. Random sampling of clusters usually creates larger variances than they do for simple random sampling of the same number of units because the $y$ values for sampling units within the same cluster tend to be positively correlated. However, cluster sampling is often less expensive. The sampling of households, for example, is often less expensive when all the households on the same block are included in the sample, if any are.

The author once helped a telephone company estimate the value of its telephone poles. For poles chosen for a sample, pole "evaluators" had to drive to each pole, often several miles, then briefly "score" the pole. It seemed very reasonable to ask the evaluator to score several poles whenever he was at any given location. In another consultation the author and a colleague were asked by the state elections bureau to estimate the standard error of an estimator of the proportion of the signatures on roughly 20,000 petition sheets, each with up to 20 signatures, which were "valid." Since the probabilities of good signatures tended to vary from sheet to sheet, the sampling of all the signatures on a sheet was cluster sampling. That had to be taken into consideration in estimating the standard error.

### Problems for Section 13.5

**13.5.1** Consider a population of 5 units, with $(x, y)$ measurements $(1, 3)$, $(2, 7)$, $(4, 10)$, $(6, 18)$, $(7, 22)$.

**(a)** Find the population parameters $\tau_x, \tau_y, \mu_x, \mu_y, R, \sigma_y^2, d_i$ for $i = 1, \ldots, 5, \sigma_d^2$.

**(b)** For a simple random sample of $n = 2$, use formulas to find $\text{Var}(\bar{Y})$ and $\text{Var}(\hat{\tau})$, where $\hat{\tau} = N\bar{Y}$.

**(c)** For $n = 2$, find the sampling distribution of $\hat{\tau}$.

**(d)** Use the result of part (c) to show that $E(\hat{\tau}) = \tau_y$ and $\text{Var}(\hat{\tau})$ has the value obtained in part (b).

**(e)** Find the sampling distribution of $\hat{\tau}_R = \tau_x \hat{R}$. Determine $E(\hat{\tau}_R)$ and $\text{Var}(\hat{\tau}_R)$. Compare these to $\tau_y$ and the approximate value given by $\tau_x^2(\sigma_d^2/n)(N - n)/(N - 1)$.

**13.5.2** Consider a population with two strata as follows:

Stratum 1: $y_1 = 7, y_2 = 3, y_3 = 1, y_4 = 5$

Stratum 2: $y_5 = 10, y_6 = 6, y_7 = 12, y_8 = 9, y_9 = 8, y_{10} = 9.$

**(a)** Find the parameters $\mu_1, \mu_2, \sigma_1^2, \sigma_2^2, \tau_y$.

**(b)** Find the sampling distribution of $\hat{\tau}_{\text{str}}$ for $n_1 = 2, n_2 = 3$.
*Hint:* The possible estimates are $54, 56, \ldots, 84$, with probabilities $(2, 2, 4, 4, 10, 8, 12, 10, 16, 10, 12, 8, 10, 4, h(82), h(84), h(86))/120$. You find $h(80), h(82), h(84)$.

**(c)** Find $E(\hat{\tau}_{\text{str}})$ by (i) using the sampling distribution of $\hat{\tau}_{\text{str}}$ found in part (b), and (ii) using an appropriate formula. Do the same thing for $\text{Var}(\hat{\tau}_{\text{str}})$.

**(d)** Find the variance of $\hat{\tau}$, the expansion estimator of $\tau_y$ for $n = 5$.

**13.5.3** Consider a population with four strata with the following standard deviations $\sigma_i$ and sizes $N_i$: $\sigma_1 = 5, N_1 = 30, \sigma_2 = 3, N_2 = 20, \sigma_3 = 4, N_3 = 70, \sigma_4 = 8, N_4 = 30$.

**(a)** For stratified sampling with $n_1 = 10, n_2 = 6, n_3 = 15, n_4 = 9$, determine $\text{Var}(\hat{\tau}_{\text{str}})$.

**(b)** For the same total sample size $n$, find the optimum sample sizes for the strata. What is $\text{Var}(\hat{\tau}_{\text{str}})$ for these sample sizes?

**(c)** Suppose that the strata means are $\mu_1 = 40, \mu_2 = 50, \mu_3 = 60, \mu_4 = 44$. Find the population mean $\mu$ and variance $\sigma^2$. Determine the variance of the expansion estimator of $\tau$ for simple random sampling with $n = 40$ from the population of $N = 150$.

**13.5.4** Suppose that a population of seven $(x_i, y_i)$ measurements were $(7, 3), (2, 12), (5, 6), (9, 3)(1, 8), (4, 7), (3, 10)$. Let $\hat{\tau}$ and $\hat{\tau}_R$ the expansion and ratio estimators of $\tau$ for simple random sampling with $n = 3$ from the population.

**(a)** Plot the points.

**(b)** Find $E(\hat{\tau})$ and $\text{Var}(\hat{\tau})$. Find an approximation for $\text{Var}(\hat{\tau}_R)$. Exact computations based on all 35 possible samples determined that $E(\hat{\tau}_R) = 56.56$ and $\text{Var}(\hat{\tau}_R) = 886.73$.

**(c)** What can you conclude in general concerning the use of ratio estimators?

**13.5.5** A small town had 2000 voters. The town was divided into three districts, having 400, 900, and 700 voters. The town had two candidates for mayor, $A$ and $B$. In an effort to determine the proportion $p_A$ of voters who would vote for $A$, simple random samples of sizes 60, 80, and 70 were taken from the three districts. Among the voters sampled, the numbers favoring $A$ were 37, 46, and 31.

**(a)** State a model concerning these three observations $X_1, X_2, X_3$. The model should enable you to answer part (b).

**(b)** Determine a 95% confidence interval on $p_A$.

**(c)** Give a 95% CI on the difference $p_A - p_B \equiv \Delta$. Careful: $\hat{p}_A$ and $\hat{p}_B$ are not independent.

**(d)** What sample sizes might have been better, keeping the same total sample size $n = 210$?

**13.5.6** Consider a population of $N = 8$, consisting of four clusters, each of size 2, with corresponding measurements $y_{12}, y_{12}, y_{21}, \ldots, y_{41}, y_{42}$: cluster 1: 3, 5; cluster 2: 9, 7; cluster 3: 8, 10; cluster 4: 2, 4. Samples of size 4 are to be taken by choosing a simple random sample of two clusters, then determining the corresponding $y_{ij}$. Let $Y_k = (Y_{k1}, Y_{k2})$ for $k = 1, 2$ be the pair of measurements for the $k$th cluster chosen. Let $\mu$ be the population mean among all eight $y_{ij}$.

**(a)** Give a formula for an unbiased estimator $\hat{\mu}_{\text{clus}}$ of the population mean $\mu$.

**(b)** Let $\bar{Y}$ be the mean of a simple random sample of four from this population. Find $\text{Var}(\hat{\mu}_{\text{clus}})$ and $\text{Var}(\bar{Y})$.

**(c)** Repeat parts (a) and (b) for the case that the clusters are: cluster 1: 3, 10; cluster 2: 9, 2; cluster 3: 8, 5; cluster 4: 7, 4. The values are the same as in part (a), but they are clustered differently.

**(d)** Can you draw any conclusions from this example as to when cluster sampling may be better than simple random sampling?

**(e)** Another sampling method would be to take one observation randomly from each cluster (stratified sampling). For which of the populations in parts (a) and (c) would this be the best sampling method?

# References


Agresti, A. (1990). *Categorical Data Analysis*. New York: Wiley.

Agresti, A., and B. Coull (1998). Approximate is better than "exact" for interval estimation of binomial proportions. *Amer. Statist.* **52**, 119–126.

Bartlett, M. S. (1947). The use of transformations. *Biometrics* **3**, 39–52.

Bernoulli, D. (1778). Dijudicatio maxime probablis plurium observationum discrepantium atque verismillima induction inde formanda. *Acta Academiae Scientuarum Imperalis Petropolitanae for 1777, pars prior*, 3–23. Reprinted in translation in Kendall (1961).

Bertrand, J. (1889). *Calcul des probabilités*. Paris: Gauthier-Villars. Second ed., 1907. Reprinted by Chelsea, New York, 1972.

Bickel, P., and K. Doksum (1977). *Mathematical Statistics*. San Francisco, CA: Holden-Day.

Bickel, P., E. Hammel, and J. O'Connell (1975). Sex bias in graduate admissions. *Science* **187**, 398–403.

Blackwell, D. (1947). Conditional expectation and unbiased sequential estimation. *Ann. Math. Statist.* **18**, 105–110.

Brown, T.W. (1984). Poisson approximations and the definition of the Poisson process. *Amer. Math. Monthly* **91**, 116–123.

Casella, G., and R. Berger (2002). *Statistical Inference*, 2nd ed. Pacific Grove, CA: Duxbury Press.

Chebychev, P. L. (1867). Des valeurs moyennes. *J. Math. Pures Appl.* Ser. 2 ***12***. *Ouvres*, Vol. 1. New York: Chelsea, 1962.

Cramér, H. (1951). *Mathematical Methods of Statistics*. Princeton, NJ: Princeton University Press.

Doob, J. (1953). *Stochastic Processes*. New York: Wiley.

Durrett, R. (2004). *Probability: Theory and Examples*. Pacific Grove, CA: Duxbury Press.

Edgeworth, F. (1908–1909). On the probable error of frequency-constants. *J. Roy. Statist. Soc.* **71**, 381–397, 499–512, 651–678; 72, 81–90.

Efron, B. (1979). Bootstrap methods: another look at the jackknife. *Ann. Statist.* **1**, 1–26.

Efron, B. (1982). *The Jackknife, the Bootstrap, and Other Resampling Methods*. CBMS-NSF 38. Philadelphia, PA: Society for Industrial and Applied Mathematics.



*Models for Probability and Statistical Inference: Theory and Applications*, By James H. Stapleton

Fabian, V., and J. F. Hannan (1985). *Introduction to Probability and Mathematical Statistics*. New York: Wiley.

Feller, W. (1950). *An Introduction to Probability Theory and Its Applications*. New York: Wiley.

Ferguson, T. (1996). *A Course in Large Sample Theory*. New York, Chapman & Hall.

Fisher, R. A. (1912). On an absolute criterion for fitting frequency curves. *Messenger Math.* **41**, 155–160. Reprinted in Fisher (1974).

Fisher, R. A. (1922). On the mathematical foundations of theoretical statistics. *Philos. Trans. Roy. Soc. London Ser. A* **222**, 309–368. Reprinted as paper 10 in Fisher (1950) and as paper 19 in Fisher (1974).

Fisher, R. A. (1934). Two new properties of mathematical likelihood. *Proc. Roy. Soc. Ser. A* **144**, 285–307.

Fisher, R. A. (1950). *Contributions to Mathematical Statistics*. New York: Wiley.

Fisher, R. A. (1974). *The Collected Papers of R. A. Fisher*. Edited by J. H. Bennett. Adelaide, Australia: University of Adelaide Press.

Galton, F. (1889). *Natural Inheritance*. London: Macmillan.

Galton, F. (1890). Kinship and correlation. *North Amer. Rev.* **150**, 419–431.

Gosset, W. (1908). On the probable error of the mean. *Biometrika* **5**, 315. [Published under the name "A Student."]

Hajek, J. (1960). Limiting distributions in simple random sampling from a finite population. *Publ. Math. Inst. Hung. Acad. Sci. Ser. A* **5**(3), 361–374.

Hald, A. (2003). *History of Probability and Statistics and Their Applications*. New York: Wiley.

Huber, P. J. (1981a). Robust estimation of a location parameter. *Ann. Math. Statist.* **35**, 73–101.

Huber, P. J. (1981b). *Robust Statistics*. New York: Wiley.

Kennedy, W. J., and J. E. Gentle (1980). *Statistical Computing*. New York: Marcel Dekker.

Khintchine, A. Y. (1929). Sur la loi des grands nombres. *C. R. Acad. Sci. Paris* **191**.

Le Cam, L. (1953). On the asymptotic properties of maximum likelihood estimates and related Bayes estimates. *Univ. Calif. Publ. Statist.* **1**, 277–330.

Le Cam, L. (1960). An approximation of the Poisson binomial distribution. *Pacific J. Math.* **10**, 1181–1197.

Legendre, A.-M. (1805). *Nouvelles méthods pour la détermination des orbites des comètes*. Paris: Courcier.

Lehmann, E. (1975). *Nonparametrics: Statistical Methods Based on Ranks*. San Francisco, CA: Holden-Day.

Lehmann, E., and G. Casella (1998). *On the Theory of Point Estimation*. New York: Springer Verlag.

Lilliefors, H. (June 1967), On the Kolmogorov-Smirnov test for normality with mean and variance unknown. *J. Am. Stat. Assoc.* **62**, 399–402.

Mallows, C. L. (1964). Choosing variables in linear regression: a graphical aid. Presented at the Central Regional Meeting of the Institute of Mathematical Statistics, Manhattan, KS.

Mann, H., and D. Whitney (1947). On a test of whether one of two random variables is stochastically larger than the other. *Ann. Math. Statist.* **18**, 50–60.

Neyman, J. (1935). Sur un teorèma concèrnente le cosidétte statische sufficiènti. *Ital. Statist.* **6**, 320–334.

Neyman, J., and E. Pearson (1933). On the problem of the most efficient tests of statistical hypotheses. *Philos. Trans. Roy. Soc. London Ser. A.* **231**, 289–337.

Pearson, E., and H. O. Hartley (1966). *Biometrika Tables for Statisticians*, 3rd ed. Cambridge: Cambridge University Press.

Poisson, S. (1837). *Reserches sur la probabilité de jugements en matière criminelle et en matriere civile, précédés des règles, généralés du calcul des probabilités.* Paris: Bachelier.

Rao, C. R. (1945). Information and the accuracy attainable in the estimation of statistical parameters. *Bull. Calcutta Math. Soc.* **37**, 81–91.

Rao, C. R. (1947). Minimum variance and the estimation of several parameters. *Cambridge Philos Soc.* **43**, 280–283.

Rao, C. R. (1952). *Advanced Statistical Methods in Biometric Research.* New York: Wiley.

Scheffé, H. (1953). A method for judging all contrasts in the analysis of variance, *Biometrika* **40**, 87–104.

Scheffé , H. (1959). *The Analysis of Variance.* New York: John Wiley.

Simpson, E. H. (1951). The interpretation of interaction in contingency tables. *J. Royal Statist. Soc. London Ser. B* **13**, 238–241.

Stapleton, J. (1995). *Linear Statistical Models.* New York: Wiley.

Stigler, S. (1986). *The History of Statistics: The Measurement of Uncertainty Before 1990.* Cambridge, MA: Harvard University Press.

Stigler, S. (1989). Galton's account of the invention of correlation. *Statist. Sci.* **4**, 73–86.

Stigler, S. (1999). *Statistics on the Table: The History of Statistical Concepts and Methods.* Cambridge, MA: Harvard University Press.

Stirling, J. (1730). *Methodes Differentialis.* London.

Wald, A. (1949). Note on the consistency of the maximum likelihood estimate. *Ann. Math. Stat.* **20**, 595–601.

Welch, R. (1947). The generalization of Student's problem when several different population variances are involved. *Biometrika* **34**, 28.

Wilcoxon, F. (1945). Individual comparisons by ranking methods. *Biometrics* **1**, 80–83.

Wilks, S. (1938). The large sample distribution of the likelihood ratio for testing hypotheses. *Ann. Math. Statist.* **9**, 60–62.

Xia, Y. (2002). Interval estimation for the difference of two binomial proportions in non-adaptive and adaptive designs. Ph.D. dissertation, Michigan State University, East Lansing, MI.

# Appendix

## TABLE 1

**Cumulative Binomial Tables for $B(x; n, p) = \sum_{k=0}^{x} \binom{n}{k} p^k (1-p)^{n-k}$**

For example, if $X \sim \text{Binomial}(10, 1/3)$, then $P(X \le 3) = 0.559$, and $P(X = 3) = P(X \le 3) - P(X \le 2) = 0.559 - 0.299 = 0.260 = \binom{10}{3}(1/3)^3(2/3)^7 = b(3; 10, 1/3). P(X = 7; n = 10, p = 2/3) = b(3; n = 10, p = 1/3) = 0.260$.

Linear interpolation can provide rough approximations. For example, for $n = 10$, since $(0.632 - 0.6)/(0.70 - 0.6) = 0.32$, $P(X \le 7; p = 0.632) \doteq 0.68P(X \le 7; p = 0.6) + 0.32P(X \le 7; p = 0.7)) = 0.764$. More exact computation gives $P(X \le 7; p = 0.632) = 0.780$.

Let $\Phi$ be the standard normal cdf. The normal approximation for $P(X \le 13; n = 25, p = 0.4)$ is $\Phi((13.5 - 25)(0.4)/\sqrt{25(0.4)(0.6)}) = \Phi(1.429) = 0.923$. The table value is 0.922.

$n = 5$

| | | $p$ | | | | | | | | | | | |
|---|---|---|---|---|---|---|---|---|---|---|---|---|---|
| | | .02 | .10 | .20 | .25 | .30 | 1/3 | .40 | .50 | .60 | .70 | .80 | .90 |
| | 0 | .904 | .590 | .328 | .237 | .168 | .132 | .078 | .031 | .010 | .002 | .000 | .000 |
| | 1 | .996 | .919 | .737 | .633 | .528 | .461 | .337 | .187 | .087 | .031 | .007 | .0001 |
| $x$ | 2 | 1.000 | .991 | .942 | .896 | .837 | .790 | .683 | .500 | .317 | .163 | .058 | .009 |
| | 3 | 1.000 | 1.000 | .993 | .984 | .969 | .955 | .913 | .812 | .663 | .472 | .263 | .081 |
| | 4 | 1.000 | 1.000 | 1.000 | .999 | .998 | .966 | .990 | .969 | .922 | .832 | .672 | .410 |

**$n = 10$**

| $x$ \ $p$ | .02 | .10 | .20 | .25 | .30 | 1/3 | .40 | .50 | .60 | .70 | .80 | .90 |
|---|---|---|---|---|---|---|---|---|---|---|---|---|
| 0 | .817 | .349 | .107 | .056 | .028 | .017 | .006 | .001 | .000 | .000 | .000 | .000 |
| 1 | .984 | .736 | .376 | .244 | .149 | .104 | .046 | .011 | .002 | .000 | .000 | .000 |
| 2 | .999 | .930 | .678 | .526 | .383 | .299 | .167 | .055 | .012 | .002 | .000 | .000 |
| 3 | 1.000 | .987 | .879 | .776 | .650 | .559 | .382 | .172 | .055 | .011 | .001 | .000 |
| 4 | 1.000 | 1.000 | .967 | .922 | .850 | .787 | .633 | .377 | .166 | .047 | .006 | .000 |
| 5 | 1.000 | 1.000 | .994 | .980 | .953 | .923 | .834 | .623 | .367 | .150 | .033 | .002 |
| 6 | 1.000 | 1.000 | 0.999 | .996 | .989 | .980 | .945 | .828 | .618 | .350 | .121 | .013 |
| 7 | 1.000 | 1.000 | 1.000 | 1.000 | .998 | .997 | .988 | .945 | .833 | .617 | .322 | .070 |
| 8 | 1.000 | 1.000 | 1.000 | 1.000 | 1.000 | 1.000 | .998 | .989 | .954 | .851 | .624 | .264 |
| 9 | 1.000 | 1.000 | 1.000 | 1.000 | 1.000 | 1.000 | 1.000 | .999 | .994 | .972 | .893 | .651 |

**$n = 15$**

| $x$ \ $p$ | .02 | .10 | .20 | .25 | .30 | 1/3 | .40 | .50 | .60 | .70 | .80 | .90 |
|---|---|---|---|---|---|---|---|---|---|---|---|---|
| 0 | .739 | .206 | .035 | .002 | .000 | .000 | .000 | .000 | .000 | .000 | .000 | .000 |
| 1 | .965 | .549 | .167 | .080 | .035 | .019 | .005 | .000 | .000 | .000 | .000 | .000 |
| 2 | .997 | .816 | .398 | .236 | .127 | .079 | .027 | .004 | .000 | .000 | .000 | .000 |
| 3 | 1.000 | .944 | .648 | .461 | .297 | .209 | .091 | .018 | .002 | .000 | .000 | .000 |
| 4 | 1.000 | .987 | .836 | .686 | .515 | .404 | .217 | .059 | .009 | .001 | .000 | .000 |
| 5 | 1.000 | .998 | .939 | .852 | .722 | .618 | .403 | .151 | .034 | .004 | .000 | .000 |
| 6 | 1.000 | 1.000 | .982 | .943 | .869 | .797 | .610 | .304 | .095 | .015 | .001 | .000 |
| 7 | 1.000 | 1.000 | .996 | .983 | .950 | .912 | .787 | .500 | .213 | .050 | .004 | .000 |
| 8 | 1.000 | 1.000 | .999 | .996 | .985 | .969 | .905 | .696 | .390 | .131 | .018 | .000 |
| 9 | 1.000 | 1.000 | 1.000 | .999 | .996 | .991 | .966 | .849 | .597 | .278 | .061 | .002 |
| 10 | 1.000 | 1.000 | 1.000 | 1.000 | .999 | .998 | .991 | .941 | .783 | .485 | .164 | .013 |
| 11 | 1.000 | 1.000 | 1.000 | 1.000 | 1.000 | 1.000 | .998 | .982 | .909 | .703 | .352 | .056 |
| 12 | 1.000 | 1.000 | 1.000 | 1.000 | 1.000 | 1.000 | 1.000 | .996 | .973 | .873 | .682 | .184 |
| 13 | 1.000 | 1.000 | 1.000 | 1.000 | 1.000 | 1.000 | 1.000 | 1.000 | .995 | .965 | .833 | .451 |
| 14 | 1.000 | 1.000 | 1.000 | 1.000 | 1.000 | 1.000 | 1.000 | 1.000 | 1.000 | .995 | .965 | .794 |

***n* = 20**

| | | p | | | | | | | | | | | |
|---|---|---|---|---|---|---|---|---|---|---|---|---|---|
| | | .02 | .10 | .20 | .25 | .30 | 1/3 | .40 | .50 | .60 | .70 | .80 | .90 |
| | 0 | .668 | .122 | .012 | .003 | .001 | .000 | .000 | .000 | .000 | .000 | .000 | .000 |
| | 1 | .940 | .392 | .069 | .024 | .008 | .003 | .001 | .000 | .000 | .000 | .000 | .000 |
| | 2 | .993 | .677 | .206 | .091 | .035 | .018 | .004 | .000 | .000 | .000 | .000 | .000 |
| | 3 | .999 | .867 | .411 | .225 | .107 | .060 | .016 | .001 | .000 | .000 | .000 | .000 |
| | 4 | 1.000 | .957 | .630 | .415 | .238 | .152 | .051 | .006 | .000 | .000 | .000 | .000 |
| | 5 | 1.000 | .989 | .804 | .617 | .416 | .297 | .126 | .021 | .002 | .000 | .000 | .000 |
| | 6 | 1.000 | .998 | .913 | .786 | .608 | .479 | .250 | .058 | .006 | .000 | .000 | .000 |
| | 7 | 1.000 | 1.000 | .968 | .898 | .772 | .661 | .416 | .132 | .021 | .001 | .000 | .000 |
| | 8 | 1.000 | 1.000 | .990 | .959 | .887 | .809 | .596 | .252 | .057 | .005 | .000 | .000 |
| | 9 | 1.000 | 1.000 | .997 | .986 | .952 | .908 | .755 | .412 | .128 | .017 | .001 | .000 |
| *x* | 10 | 1.000 | 1.000 | 1.000 | .996 | .983 | .962 | .872 | .588 | .245 | .048 | .003 | .000 |
| | 11 | 1.000 | 1.000 | 1.000 | .999 | .995 | .987 | .943 | .748 | .404 | .113 | .010 | .000 |
| | 12 | 1.000 | 1.000 | 1.000 | 1.000 | .999 | .996 | .979 | .868 | .584 | .228 | .032 | .000 |
| | 13 | 1.000 | 1.000 | 1.000 | 1.000 | 1.000 | .999 | .994 | .942 | .750 | .392 | .087 | .002 |
| | 14 | 1.000 | 1.000 | 1.000 | 1.000 | 1.000 | 1.000 | .999 | .979 | .874 | .584 | .196 | .011 |
| | 15 | 1.000 | 1.000 | 1.000 | 1.000 | 1.000 | 1.000 | 1.000 | .994 | .949 | .762 | .370 | .043 |
| | 16 | 1.000 | 1.000 | 1.000 | 1.000 | 1.000 | 1.000 | 1.000 | .999 | .984 | .893 | .589 | .133 |
| | 17 | 1.000 | 1.000 | 1.000 | 1.000 | 1.000 | 1.000 | 1.000 | 1.000 | .996 | .965 | .794 | .323 |
| | 18 | 1.000 | 1.000 | 1.000 | 1.000 | 1.000 | 1.000 | 1.000 | 1.000 | .999 | .992 | .931 | .608 |
| | 19 | 1.000 | 1.000 | 1.000 | 1.000 | 1.000 | 1.000 | 1.000 | 1.000 | 1.000 | .999 | .988 | .878 |

***n* = 25**

| | | | | | | *p* | | | | | | |
|---|---|---|---|---|---|---|---|---|---|---|---|---|
| | .02 | .10 | .20 | .25 | .30 | 1/3 | .40 | .50 | .60 | .70 | .80 | .90 |
| 0 | .603 | .072 | .004 | .001 | .000 | .000 | .000 | .000 | .000 | .000 | .000 | .000 |
| 1 | .911 | .271 | .027 | .007 | .002 | .001 | .000 | .000 | .000 | .000 | .000 | .000 |
| 2 | .987 | .537 | .098 | .032 | .009 | .004 | .000 | .000 | .000 | .000 | .000 | .000 |
| 3 | .999 | .764 | .234 | .096 | .033 | .015 | .002 | .000 | .000 | .000 | .000 | .000 |
| 4 | 1.000 | .902 | .421 | .214 | .090 | .046 | .009 | .000 | .000 | .000 | .000 | .000 |
| 5 | 1.000 | .967 | .617 | .378 | .193 | .112 | .029 | .002 | .000 | .000 | .000 | .000 |
| 6 | 1.000 | .991 | .780 | .561 | .341 | .222 | .074 | .007 | .000 | .000 | .000 | .000 |
| 7 | 1.000 | .998 | .891 | .727 | .521 | .370 | .154 | .022 | .001 | .000 | .000 | .000 |
| 8 | 1.000 | 1.000 | .953 | .851 | .677 | .538 | .274 | .054 | .004 | .000 | .000 | .000 |
| 9 | 1.000 | 1.000 | .983 | .929 | .811 | .696 | .425 | .115 | .013 | .000 | .000 | .000 |
| 10 | 1.000 | 1.000 | .994 | .970 | .902 | .822 | .586 | .212 | .034 | .002 | .000 | .000 |
| 11 | 1.000 | 1.000 | .998 | .989 | .956 | .908 | .732 | .345 | .078 | .006 | .000 | .000 |
| *x* 12 | 1.000 | 1.000 | 1.000 | .997 | .983 | .958 | .846 | .500 | .154 | .017 | .000 | .000 |
| 13 | 1.000 | 1.000 | 1.000 | .999 | .994 | .984 | .922 | .655 | .268 | .044 | .002 | .000 |
| 14 | 1.000 | 1.000 | 1.000 | 1.000 | .998 | .994 | .966 | .788 | .414 | .098 | .006 | .000 |
| 15 | 1.000 | 1.000 | 1.000 | 1.000 | 1.000 | .998 | .987 | .885 | .575 | .189 | .017 | .000 |
| 16 | 1.000 | 1.000 | 1.000 | 1.000 | 1.000 | 1.000 | .996 | .946 | .726 | .323 | .047 | .000 |
| 17 | 1.000 | 1.000 | 1.000 | 1.000 | 1.000 | 1.000 | .999 | .978 | .846 | .488 | .109 | .002 |
| 18 | 1.000 | 1.000 | 1.000 | 1.000 | 1.000 | 1.000 | 1.000 | .993 | .926 | .659 | .220 | .009 |
| 19 | 1.000 | 1.000 | 1.000 | 1.000 | 1.000 | 1.000 | 1.000 | .998 | .971 | .807 | .383 | .033 |
| 20 | 1.000 | 1.000 | 1.000 | 1.000 | 1.000 | 1.000 | 1.000 | 1.000 | .991 | .910 | .579 | .098 |
| 21 | 1.000 | 1.000 | 1.000 | 1.000 | 1.000 | 1.000 | 1.000 | 1.000 | .998 | .967 | .766 | .236 |
| 22 | 1.000 | 1.000 | 1.000 | 1.000 | 1.000 | 1.000 | 1.000 | 1.000 | 1.000 | .991 | .902 | .463 |
| 23 | 1.000 | 1.000 | 1.000 | 1.000 | 1.000 | 1.000 | 1.000 | 1.000 | 1.000 | .998 | .973 | .729 |
| 24 | 1.000 | 1.000 | 1.000 | 1.000 | 1.000 | 1.000 | 1.000 | 1.000 | 1.000 | 1.000 | .996 | .928 |

## TABLE 2

**Cumulative Poisson Tables for $F(x; n, p) = \sum_{k=0}^{x} e^{-\lambda}\lambda^k/k!$**

*Approximations*: The tables may be interpolated linearly to provide good approximations. For example, since $(3.73 - 3.6)/(3.8 - 3.6) = 0.65$, so $P(X \leq 5; \lambda = 3.73) \doteq 0.35P(X \leq 5; \lambda = 3.6) + 0.35P(X \leq 5; \lambda = 3.8) = (0.35)(0.844) + (0.65)(0.816) = 0.8287$, while more exact computation gives 0.8302.

Let $\Phi$ be the standard normal cdf. Then $P(X \leq x; \lambda) \doteq \Phi((x + 1/2 - \lambda)/\sqrt{\lambda})$ for integers $x$ for $\lambda$ "large." For example, for $\lambda = 16.63$, $P(X \leq 20) \doteq \Phi((20.5 - 16.63)/\sqrt{16.63}) = 0.8203$. More exact computations gives 0.8302, while linear interpolation using this table gives 0.8283.

| | λ | | | | | | | | | |
|---|---|---|---|---|---|---|---|---|---|---|
| x | .20 | .40 | .60 | .80 | 1.00 | 1.20 | 1.40 | 1.60 | 1.80 | 2.00 |
| 0 | .819 | .670 | .549 | .449 | .368 | .301 | .247 | .202 | .165 | .135 |
| 1 | .982 | .938 | .878 | .809 | .736 | .663 | .592 | .525 | .463 | .406 |
| 2 | .999 | .992 | .977 | .953 | .920 | .879 | .833 | .783 | .731 | .677 |
| 3 | 1.000 | .999 | .997 | .991 | .981 | .966 | .946 | .921 | .891 | .857 |
| 4 | 1.000 | 1.000 | 1.000 | .999 | .996 | .992 | .986 | .976 | .964 | .947 |
| 5 | 1.000 | 1.000 | 1.000 | 1.000 | .999 | .998 | .997 | .994 | .990 | .983 |
| 6 | 1.000 | 1.000 | 1.000 | 1.000 | 1.000 | 1.000 | .999 | .999 | .997 | .995 |
| 7 | 1.000 | 1.000 | 1.000 | 1.000 | 1.000 | 1.000 | 1.000 | 1.000 | .999 | .999 |
| 8 | 1.000 | 1.000 | 1.000 | 1.000 | 1.000 | 1.000 | 1.000 | 1.000 | 1.000 | 1.000 |

| | λ | | | | | | | | | |
|---|---|---|---|---|---|---|---|---|---|---|
| x | 2.2 | 2.4 | 2.6 | 2.8 | 3.0 | 3.2 | 3.4 | 3.6 | 3.8 | 4.0 |
| 0 | .111 | .091 | .074 | .061 | .050 | .041 | .033 | .027 | .022 | .018 |
| 1 | .355 | .308 | .267 | .231 | .199 | .171 | .147 | .126 | .107 | .092 |
| 2 | .623 | .570 | .518 | .469 | .423 | .380 | .340 | .303 | .269 | .238 |
| 3 | .819 | .779 | .736 | .692 | .647 | .603 | .558 | .515 | .473 | .433 |
| 4 | .928 | .904 | .877 | .848 | .815 | .781 | .744 | .706 | .668 | .629 |
| 5 | .975 | .964 | .951 | .935 | .916 | .895 | .871 | .844 | .816 | .785 |
| 6 | .993 | .988 | .983 | .976 | .966 | .955 | .942 | .927 | .909 | .889 |
| 7 | .998 | .997 | .995 | .992 | .988 | .983 | .977 | .969 | .960 | .949 |
| 8 | 1.000 | .999 | .999 | .998 | .996 | .994 | .992 | .988 | .984 | .979 |
| 9 | 1.000 | 1.000 | 1.000 | .999 | .999 | .998 | .997 | .996 | .994 | .992 |
| 10 | 1.000 | 1.000 | 1.000 | 1.000 | 1.000 | 1.000 | .999 | .999 | .998 | .997 |
| 11 | 1.000 | 1.000 | 1.000 | 1.000 | 1.000 | 1.000 | 1.000 | 1.000 | .999 | .999 |

| | λ | | | | | | | | | |
|---|---|---|---|---|---|---|---|---|---|---|
| x | 4.2 | 4.4 | 4.6 | 4.8 | 5.0 | 5.2 | 5.4 | 5.6 | 5.8 | 6.0 |
| 0 | .015 | .012 | .010 | .080 | .007 | .006 | .005 | .004 | .003 | .002 |
| 1 | .078 | .066 | .056 | .048 | .040 | .034 | .029 | .024 | .021 | .017 |
| 2 | .210 | .185 | .163 | .143 | .125 | .109 | .095 | .082 | .072 | .062 |
| 3 | .395 | .359 | .326 | .294 | .265 | .238 | .213 | .191 | .170 | .151 |
| 4 | .590 | .551 | .513 | .476 | .440 | .406 | .373 | .342 | .313 | .285 |
| 5 | .753 | .720 | .686 | .651 | .616 | .581 | .546 | .512 | .478 | .446 |
| 6 | .867 | .844 | .818 | .791 | .762 | .732 | .702 | .670 | .638 | .606 |
| 7 | .936 | .921 | .905 | .887 | .867 | .845 | .822 | .797 | .771 | .744 |
| 8 | .972 | .964 | .955 | .944 | .932 | .918 | .903 | .886 | .867 | .847 |
| 9 | .989 | .985 | .980 | .975 | .968 | .960 | .951 | .941 | .929 | .916 |
| 10 | .996 | .994 | .992 | .990 | .986 | .982 | .977 | .972 | .965 | .957 |
| 11 | .999 | .998 | .997 | .996 | .995 | .993 | .990 | .988 | .984 | .980 |
| 12 | 1.000 | .999 | .999 | .999 | .998 | .997 | .996 | .995 | .993 | .991 |
| 13 | 1.000 | 1.000 | 1.000 | 1.000 | .999 | .999 | .999 | .998 | .997 | .996 |
| 14 | 1.000 | 1.000 | 1.000 | 1.000 | 1.000 | 1.000 | 1.000 | .999 | .999 | .999 |
| 15 | 1.000 | 1.000 | 1.000 | 1.000 | 1.000 | 1.000 | 1.000 | 1.000 | 1.000 | .999 |

| | λ | | | | | | | | | |
|---|---|---|---|---|---|---|---|---|---|---|
| x | 6.2 | 6.4 | 6.6 | 6.8 | 7.0 | 7.2 | 7.4 | 7.6 | 7.8 | 8.0 |
| 0 | .002 | .002 | .001 | .001 | .001 | .001 | .001 | .000 | .000 | .000 |
| 1 | .015 | .012 | .010 | .009 | .007 | .006 | .005 | .004 | .004 | .003 |
| 2 | .054 | .046 | .040 | .034 | .030 | .025 | .022 | .019 | .016 | .014 |
| 3 | .134 | .119 | .105 | .093 | .082 | .072 | .063 | .055 | .048 | .042 |
| 4 | .259 | .235 | .213 | .192 | .173 | .156 | .140 | .125 | .112 | .100 |
| 5 | .414 | .384 | .355 | .327 | .301 | .276 | .253 | .231 | .210 | .191 |
| 6 | .574 | .542 | .511 | .480 | .450 | .420 | .392 | .365 | .338 | .313 |
| 7 | .716 | .687 | .658 | .628 | .599 | .569 | .539 | .510 | .481 | .453 |
| 8 | .826 | .803 | .780 | .755 | .729 | .703 | .676 | .648 | .620 | .593 |
| 9 | .902 | .886 | .869 | .850 | .830 | .810 | .788 | .765 | .741 | .717 |
| 10 | .949 | .939 | .927 | .915 | .901 | .887 | .871 | .854 | .835 | .816 |
| 11 | .975 | .969 | .963 | .955 | .947 | .937 | .926 | .915 | .902 | .888 |
| 12 | .989 | .986 | .982 | .978 | .973 | .967 | .961 | .954 | .945 | .936 |
| 13 | .995 | .994 | .992 | .990 | .987 | .984 | .980 | .976 | .971 | .966 |
| 14 | .998 | .997 | .997 | .996 | .994 | .993 | .991 | .989 | .986 | .983 |
| 15 | .999 | .999 | .999 | .998 | .998 | .997 | .996 | .995 | .993 | .992 |
| 16 | 1.000 | 1.000 | .999 | .999 | .999 | .999 | .998 | .998 | .997 | .996 |
| 17 | 1.000 | 1.000 | 1.000 | 1.000 | 1.000 | 1.000 | .999 | .999 | .999 | .998 |
| 18 | 1.000 | 1.000 | 1.000 | 1.000 | 1.000 | 1.000 | 1.000 | 1.000 | 1.000 | .999 |

| | | λ | | | | |
|---|---|---|---|---|---|---|
| | | 8.0 | 8.2 | 8.4 | 8.6 | 8.8 |
| | 0 | .000 | .000 | .000 | .000 | .000 |
| | 1 | .003 | .003 | .002 | .002 | .001 |
| | 2 | .014 | .012 | .010 | .009 | .007 |
| | 3 | .042 | .037 | .037 | .028 | .024 |
| | 4 | .100 | .089 | .079 | .070 | .062 |
| | 5 | .191 | .174 | .157 | .142 | .128 |
| | 6 | .313 | .290 | .267 | .246 | .226 |
| | 7 | .453 | .425 | .399 | .373 | .348 |
| | 8 | .593 | .565 | .537 | .509 | .482 |
| | 9 | .717 | .692 | .666 | .640 | .614 |
| | 10 | .816 | .796 | .774 | .752 | .729 |
| $x$ | 11 | .888 | .873 | .857 | .840 | .822 |
| | 12 | .936 | .926 | .915 | .903 | .890 |
| | 13 | .966 | .960 | .952 | .945 | .936 |
| | 14 | .983 | .979 | .975 | .970 | .965 |
| | 15 | .992 | .990 | .987 | .985 | .982 |
| | 16 | .996 | .995 | .994 | .993 | .991 |
| | 17 | .998 | .998 | .997 | .997 | .996 |
| | 18 | .999 | .999 | .999 | .999 | .998 |
| | 19 | 1.000 | 1.000 | 1.000 | .999 | .999 |
| | 20 | 1.000 | 1.000 | 1.000 | 1.000 | 1.000 |

| | λ | | | | | | | | | |
|---|---|---|---|---|---|---|---|---|---|---|
| $x$ | 9 | 10 | 11 | 12 | 13 | 14 | 15 | 16 | 17 | 18 |
| 1 | .001 | .000 | .000 | .000 | .000 | .000 | .000 | .000 | .000 | .000 |
| 2 | .006 | .003 | .001 | .001 | .000 | .000 | .000 | .000 | .000 | .000 |
| 3 | .021 | .010 | .005 | .002 | .001 | .000 | .000 | .000 | .000 | .000 |
| 4 | .055 | .029 | .015 | .008 | .004 | .002 | .001 | .000 | .000 | .000 |
| 5 | .116 | .067 | .038 | .020 | .011 | .006 | .003 | .001 | .001 | .000 |
| 6 | .207 | .130 | .079 | .046 | .026 | .014 | .008 | .004 | .002 | .001 |
| 7 | .324 | .220 | .143 | .090 | .054 | .032 | .018 | .010 | .005 | .003 |
| 8 | .456 | .333 | .232 | .155 | .100 | .062 | .037 | .022 | .013 | .007 |
| 9 | .587 | .458 | .341 | .242 | .166 | .109 | .070 | .043 | .026 | .015 |
| 10 | .706 | .583 | .460 | .347 | .252 | .176 | .118 | .077 | .049 | .030 |
| 11 | .803 | .697 | .579 | .462 | .353 | .260 | .185 | .127 | .085 | .055 |
| 12 | .876 | .792 | .689 | .576 | .463 | .358 | .268 | .193 | .135 | .092 |
| 13 | .926 | .864 | .781 | .682 | .573 | .464 | .363 | .275 | .201 | .143 |
| 14 | .959 | .917 | .854 | .772 | .675 | .570 | .466 | .368 | .281 | .208 |
| 15 | .978 | .951 | .907 | .844 | .764 | .669 | .568 | .467 | .371 | .287 |
| 16 | .989 | .973 | .944 | .899 | .835 | .756 | .664 | .566 | .468 | .375 |
| 17 | .995 | .986 | .968 | .937 | .890 | .827 | .749 | .659 | .564 | .469 |
| 18 | .998 | .993 | .982 | .963 | .930 | .883 | .819 | .742 | .655 | .562 |
| 19 | .999 | .997 | .991 | .979 | .957 | .923 | .875 | .812 | .736 | .651 |
| 20 | 1.000 | .998 | .995 | .988 | .975 | .952 | .917 | .868 | .805 | .731 |
| 21 | 1.000 | .999 | .998 | .994 | .986 | .971 | .947 | .911 | .861 | .799 |
| 22 | 1.000 | 1.000 | .999 | .997 | .992 | .983 | .967 | .942 | .905 | .855 |
| 23 | 1.000 | 1.000 | 1.000 | .999 | .996 | .991 | .981 | .963 | .937 | .899 |
| 24 | 1.000 | 1.000 | 1.000 | .999 | .998 | .995 | .989 | .978 | .959 | .932 |
| 25 | 1.000 | 1.000 | 1.000 | 1.000 | .999 | .997 | .994 | .987 | .975 | .955 |
| 26 | 1.000 | 1.000 | 1.000 | 1.000 | 1.000 | .999 | .997 | .993 | .985 | .972 |
| 27 | 1.000 | 1.000 | 1.000 | 1.000 | 1.000 | .999 | .998 | .996 | .991 | .983 |
| 28 | 1.000 | 1.000 | 1.000 | 1.000 | 1.000 | 1.000 | .999 | .998 | .995 | .990 |
| 29 | 1.000 | 1.000 | 1.000 | 1.000 | 1.000 | 1.000 | 1.000 | .999 | .997 | .994 |
| 30 | 1.000 | 1.000 | 1.000 | 1.000 | 1.000 | 1.000 | 1.000 | .999 | .999 | .997 |

## TABLE 3

## Standard Normal Cumulative Density Function $\Phi(z)$

| z | 0.00 | ,01 | 0.02 | 0.03 | 0.04 | 0.05 | 0.06 | 0.07 | 0.08 | 0.09 |
|---|---|---|---|---|---|---|---|---|---|---|
| 0.0 | 0.5000 | 0.5040 | 0.5080 | 0.5120 | 0.5160 | 0.5199 | 0.5239 | 0.5279 | 0.5319 | 0.5359 |
| 0.1 | 0.5398 | 0.5438 | 0.5478 | 0.5517 | 0.5557 | 0.5596 | 0.5636 | 0.5675 | 0.5714 | 0.5753 |
| 0.2 | 0.5793 | 0.5832 | 0.5871 | 0.5910 | 0.5948 | 0.5987 | 0.6026 | 0.6064 | 0.6103 | 0.6141 |
| 0.3 | 0.6179 | 0.6217 | 0.6255 | 0.6293 | 0.6331 | 0.6368 | 0.6406 | 0.6443 | 0.6480 | 0.6517 |
| 0.4 | 0.6554 | 0.6591 | 0.6628 | 0.6664 | 0.6700 | 0.6736 | 0.6772 | 0.6808 | 0.6844 | 0.6879 |
| 0.5 | 0.6915 | 0.6950 | 0.6985 | 0.7019 | 0.7054 | 0.7088 | 0.7123 | 0.7157 | 0.7190 | 0.7224 |
| 0.6 | 0.7257 | 0.7291 | 0.7324 | 0.7357 | 0.7389 | 0.7422 | 0.7454 | 0.7486 | 0.7517 | 0.7549 |
| 0.7 | 0.7580 | 0.7611 | 0.7642 | 0.7673 | 0.7704 | 0.7734 | 0.7764 | 0.7794 | 0.7823 | 0.7852 |
| 0.8 | 0.7881 | 0.7910 | 0.7939 | 0.7967 | 0.7995 | 0.8023 | 0.8051 | 0.8078 | 0.8106 | 0.8133 |
| 0.9 | 0.8159 | 0.8186 | 0.8212 | 0.8238 | 0.8264 | 0.8289 | 0.8315 | 0.8340 | 0.8365 | 0.8389 |
| 1.0 | 0.8413 | 0.8438 | 0.8461 | 0.8485 | 0.8508 | 0.8531 | 0.8554 | 0.8577 | 0.8599 | 0.8621 |
| 1.1 | 0.8643 | 0.8665 | 0.8686 | 0.8708 | 0.8729 | 0.8749 | 0.8770 | 0.8790 | 0.8810 | 0.8830 |
| 1.2 | 0.8849 | 0.8869 | 0.8888 | 0.8907 | 0.8925 | 0.8944 | 0.8962 | 0.8980 | 0.8997 | 0.9015 |
| 1.3 | 0.9032 | 0.9049 | 0.9066 | 0.9082 | 0.9099 | 0.9115 | 0.9131 | 0.9147 | 0.9162 | 0.9177 |
| 1.4 | 0.9192 | 0.9207 | 0.9222 | 0.9236 | 0.9251 | 0.9265 | 0.9279 | 0.9292 | 0.9306 | 0.9319 |
| 1.5 | 0.9332 | 0.9345 | 0.9357 | 0.9370 | 0.9382 | 0.9394 | 0.9406 | 0.9418 | 0.9429 | 0.9441 |
| 1.6 | 0.9452 | 0.9463 | 0.9474 | 0.9484 | 0.9495 | 0.9505 | 0.9515 | 0.9525 | 0.9535 | 0.9545 |
| 1.7 | 0.9554 | 0.9564 | 0.9573 | 0.9582 | 0.9591 | 0.9599 | 0.9608 | 0.9616 | 0.9625 | 0.9633 |
| 1.8 | 0.9641 | 0.9649 | 0.9656 | 0.9664 | 0.9671 | 0.9678 | 0.9686 | 0.9693 | 0.9699 | 0.9706 |
| 1.9 | 0.9713 | 0.9719 | 0.9726 | 0.9732 | 0.9738 | 0.9744 | 0.9750 | 0.9756 | 0.9761 | 0.9767 |
| 2.0 | 0.9772 | 0.9778 | 0.9783 | 0.9788 | 0.9793 | 0.9798 | 0.9803 | 0.9808 | 0.9812 | 0.9817 |
| 2.1 | 0.9821 | 0.9826 | 0.9830 | 0.9834 | 0.9838 | 0.9842 | 0.9846 | 0.9850 | 0.9854 | 0.9857 |
| 2.2 | 0.9861 | 0.9864 | 0.9868 | 0.9871 | 0.9875 | 0.9878 | 0.9881 | 0.9884 | 0.9887 | 0.9890 |
| 2.3 | 0.9893 | 0.9896 | 0.9898 | 0.9901 | 0.9904 | 0.9906 | 0.9909 | 0.9911 | 0.9913 | 0.9916 |
| 2.4 | 0.9918 | 0.9920 | 0.9922 | 0.9925 | 0.9927 | 0.9929 | 0.9931 | 0.9932 | 0.9934 | 0.9936 |
| 2.5 | 0.9938 | 0.9940 | 0.9941 | 0.9943 | 0.9945 | 0.9946 | 0.9948 | 0.9949 | 0.9951 | 0.9952 |
| 2.6 | 0.9953 | 0.9955 | 0.9956 | 0.9957 | 0.9959 | 0.9960 | 0.9961 | 0.9962 | 0.9963 | 0.9964 |
| 2.7 | 0.9965 | 0.9966 | 0.9967 | 0.9968 | 0.9969 | 0.9970 | 0.9971 | 0.9972 | 0.9973 | 0.9974 |
| 2.8 | 0.9974 | 0.9975 | 0.9976 | 0.9977 | 0.9977 | 0.9978 | 0.9979 | 0.9979 | 0.9980 | 0.9981 |
| 2.9 | 0.9981 | 0.9982 | 0.9982 | 0.9983 | 0.9984 | 0.9984 | 0.9985 | 0.9985 | 0.9986 | 0.9986 |

## The Standard Normal $10^6*[1 - \Phi(z)]$

| z | 0.0 | 0.1 | 0.2 | 0.3 | 0.4 | 0.5 | 0.6 | 0.7 | 0.8 | 0.9 |
|---|---|---|---|---|---|---|---|---|---|---|
| 2.5 | 6209.67 | 6036.56 | 5867.74 | 5703.13 | 5542.62 | 5386.15 | 5233.61 | 5084.93 | 4940.02 | 4798.80 |
| 2.6 | 4661.19 | 4527.11 | 4396.49 | 4269.24 | 4145.30 | 4024.59 | 3907.03 | 3792.56 | 3681.11 | 3572.60 |
| 2.7 | 3466.97 | 3364.16 | 3264.10 | 3166.72 | 3071.96 | 2979.76 | 2890.07 | 2802.82 | 2717.95 | 2635.40 |
| 2.8 | 2555.13 | 2477.08 | 2401.18 | 2327.40 | 2255.68 | 2185.96 | 2118.21 | 2052.36 | 1988.38 | 1926.21 |
| 2.9 | 1865.81 | 1807.14 | 1750.16 | 1694.81 | 1641.06 | 1588.87 | 1538.20 | 1489.00 | 1441.24 | 1394.89 |
| 3.0 | 1349.90 | 1306.24 | 1263.87 | 1222.77 | 1182.89 | 1144.21 | 1106.69 | 1070.29 | 1035.00 | 1000.78 |
| 3.1 | 967.60 | 935.44 | 904.26 | 874.03 | 844.74 | 816.35 | 788.85 | 762.20 | 736.38 | 711.36 |
| 3.2 | 687.14 | 663.68 | 640.95 | 618.95 | 597.65 | 577.03 | 557.06 | 537.74 | 519.04 | 500.94 |
| 3.3 | 483.42 | 466.48 | 450.09 | 434.23 | 418.89 | 404.06 | 389.71 | 375.84 | 362.43 | 349.46 |
| 3.4 | 336.93 | 324.81 | 313.11 | 301.79 | 290.86 | 280.29 | 270.09 | 260.23 | 250.71 | 241.51 |
| 3.5 | 232.63 | 224.05 | 215.77 | 207.78 | 200.06 | 192.62 | 185.43 | 178.49 | 171.80 | 165.34 |
| 3.6 | 159.11 | 153.10 | 147.30 | 141.71 | 136.32 | 131.12 | 126.11 | 121.28 | 116.62 | 112.13 |
| 3.7 | 107.80 | 103.63 | 99.61 | 95.74 | 92.01 | 88.42 | 84.96 | 81.62 | 78.41 | 75.32 |
| 3.8 | 72.35 | 69.48 | 66.73 | 64.07 | 61.52 | 59.06 | 56.69 | 54.42 | 52.23 | 50.12 |
| 3.9 | 48.10 | 46.15 | 44.27 | 42.47 | 40.74 | 39.08 | 37.48 | 35.94 | 34.46 | 33.04 |
| 4.0 | 31.67 | 30.36 | 29.10 | 27.89 | 26.73 | 25.61 | 24.54 | 23.51 | 22.52 | 21.57 |
| 4.1 | 20.66 | 19.78 | 18.94 | 18.14 | 17.37 | 16.62 | 15.91 | 15.23 | 14.58 | 13.95 |
| 4.2 | 13.35 | 12.77 | 12.22 | 11.69 | 11.18 | 10.69 | 10.22 | 9.77 | 9.35 | 8.93 |
| 4.3 | 8.54 | 8.16 | 7.80 | 7.46 | 7.12 | 6.81 | 6.50 | 6.21 | 5.93 | 5.67 |
| 4.4 | 5.41 | 5.17 | 4.94 | 4.71 | 4.50 | 4.29 | 4.10 | 3.91 | 3.73 | 3.56 |

| Φ(z) | z |
|---|---|
| 0.800 | 0.84162 |
| 0.900 | 1.28156 |
| 0.950 | 1.64485 |
| 0.975 | 1.95996 |
| 0.990 | 2.32635 |

| z | 1 - Φ(z) |
|---|---|
| 4.5 | 3.3977 E-6 |
| 4.6 | 2.1125 E-6 |
| 4.7 | 1.3008 E-6 |
| 4.8 | 7.9333 E-7 |
| 4.9 | 4.7918 E-7 |

| z | 1 - Φ(z) |
|---|---|
| 5.0 | 2.8665 E-7 |
| 6.0 | 9.8659 E-10 |
| 7.0 | 1.2798 E-12 |
| 8.0 | 6.2210 E-16 |
| 9.0 | 1.1286 E-19 |

# TABLE 4

## Student's-$t$ $\gamma$-Quantiles for $\nu$ Degrees of Freedom

| $\nu$ \ $\gamma$ | 0.550 | 0.600 | 0.650 | 0.700 | 0.750 | 0.800 | 0.850 | 0.900 | 0.950 | 0.975 | 0.990 | 0.995 | 0.999 |
|---|---|---|---|---|---|---|---|---|---|---|---|---|---|
| 1 | 0.158 | 0.325 | 0.510 | 0.727 | 1.000 | 1.376 | 1.963 | 3.078 | 6.314 | 12.71 | 31.82 | 63.66 | 318.3 |
| 2 | 0.142 | 0.289 | 0.445 | 0.617 | 0.817 | 1.061 | 1.386 | 1.886 | 2.920 | 4.303 | 6.965 | 9.925 | 22.33 |
| 3 | 0.137 | 0.277 | 0.424 | 0.584 | 0.765 | 0.979 | 1.250 | 1.638 | 2.353 | 3.182 | 4.541 | 5.841 | 10.21 |
| 4 | 0.134 | 0.271 | 0.414 | 0.569 | 0.741 | 0.941 | 1.190 | 1.533 | 2.132 | 2.776 | 3.747 | 4.604 | 7.173 |
| 5 | 0.132 | 0.267 | 0.408 | 0.559 | 0.727 | 0.920 | 1.156 | 1.476 | 2.015 | 2.571 | 3.365 | 4.032 | 5.893 |
| 6 | 0.131 | 0.265 | 0.404 | 0.553 | 0.718 | 0.906 | 1.134 | 1.440 | 1.943 | 2.447 | 3.143 | 3.707 | 5.208 |
| 7 | 0.130 | 0.263 | 0.402 | 0.549 | 0.711 | 0.896 | 1.119 | 1.415 | 1.895 | 2.365 | 2.998 | 3.500 | 4.785 |
| 8 | 0.130 | 0.262 | 0.400 | 0.546 | 0.706 | 0.889 | 1.108 | 1.397 | 1.860 | 2.306 | 2.897 | 3.355 | 4.501 |
| 9 | 0.129 | 0.261 | 0.398 | 0.544 | 0.703 | 0.883 | 1.100 | 1.383 | 1.833 | 2.262 | 2.821 | 3.250 | 4.297 |
| 10 | 0.129 | 0.260 | 0.397 | 0.542 | 0.700 | 0.879 | 1.093 | 1.372 | 1.813 | 2.228 | 2.764 | 3.169 | 4.144 |
| 11 | 0.129 | 0.260 | 0.396 | 0.540 | 0.697 | 0.876 | 1.088 | 1.363 | 1.796 | 2.201 | 2.718 | 3.106 | 4.025 |
| 12 | 0.128 | 0.259 | 0.395 | 0.539 | 0.696 | 0.873 | 1.083 | 1.356 | 1.782 | 2.179 | 2.681 | 3.055 | 3.930 |
| 13 | 0.128 | 0.259 | 0.394 | 0.538 | 0.694 | 0.870 | 1.080 | 1.350 | 1.771 | 2.160 | 2.650 | 3.012 | 3.852 |
| 14 | 0.128 | 0.258 | 0.393 | 0.537 | 0.692 | 0.868 | 1.076 | 1.345 | 1.761 | 2.145 | 2.625 | 2.977 | 3.787 |
| 15 | 0.128 | 0.258 | 0.393 | 0.536 | 0.691 | 0.866 | 1.074 | 1.341 | 1.753 | 2.131 | 2.603 | 2.947 | 3.733 |
| 16 | 0.128 | 0.258 | 0.392 | 0.535 | 0.690 | 0.865 | 1.071 | 1.337 | 1.746 | 2.120 | 2.584 | 2.921 | 3.686 |
| 17 | 0.128 | 0.257 | 0.392 | 0.534 | 0.689 | 0.863 | 1.069 | 1.333 | 1.740 | 2.110 | 2.567 | 2.898 | 3.646 |
| 18 | 0.127 | 0.257 | 0.392 | 0.534 | 0.688 | 0.862 | 1.067 | 1.330 | 1.734 | 2.101 | 2.552 | 2.878 | 3.611 |
| 19 | 0.127 | 0.257 | 0.391 | 0.533 | 0.688 | 0.861 | 1.066 | 1.328 | 1.729 | 2.093 | 2.540 | 2.861 | 3.579 |
| 20 | 0.127 | 0.257 | 0.391 | 0.533 | 0.687 | 0.860 | 1.064 | 1.325 | 1.725 | 2.086 | 2.528 | 2.845 | 3.552 |
| 21 | 0.127 | 0.257 | 0.391 | 0.533 | 0.686 | 0.859 | 1.063 | 1.323 | 1.721 | 2.080 | 2.518 | 2.831 | 3.527 |
| 22 | 0.127 | 0.256 | 0.390 | 0.532 | 0.686 | 0.858 | 1.061 | 1.321 | 1.717 | 2.074 | 2.508 | 2.819 | 3.505 |
| 23 | 0.127 | 0.256 | 0.390 | 0.532 | 0.685 | 0.858 | 1.060 | 1.320 | 1.714 | 2.069 | 2.500 | 2.807 | 3.485 |
| 24 | 0.127 | 0.256 | 0.390 | 0.531 | 0.685 | 0.857 | 1.059 | 1.318 | 1.711 | 2.064 | 2.492 | 2.797 | 3.467 |
| 25 | 0.127 | 0.256 | 0.390 | 0.531 | 0.684 | 0.856 | 1.058 | 1.316 | 1.708 | 2.060 | 2.485 | 2.787 | 3.450 |
| 26 | 0.127 | 0.256 | 0.390 | 0.531 | 0.684 | 0.856 | 1.058 | 1.315 | 1.706 | 2.056 | 2.479 | 2.779 | 3.435 |
| 27 | 0.127 | 0.256 | 0.389 | 0.531 | 0.684 | 0.855 | 1.057 | 1.314 | 1.703 | 2.052 | 2.473 | 2.771 | 3.421 |
| 28 | 0.127 | 0.256 | 0.389 | 0.530 | 0.683 | 0.855 | 1.056 | 1.313 | 1.701 | 2.048 | 2.467 | 2.763 | 3.408 |
| 29 | 0.127 | 0.256 | 0.389 | 0.530 | 0.683 | 0.854 | 1.055 | 1.311 | 1.699 | 2.045 | 2.462 | 2.756 | 3.396 |
| 30 | 0.127 | 0.256 | 0.389 | 0.530 | 0.683 | 0.854 | 1.055 | 1.310 | 1.697 | 2.042 | 2.457 | 2.750 | 3.385 |
| 35 | 0.127 | 0.255 | 0.389 | 0.529 | 0.682 | 0.852 | 1.052 | 1.306 | 1.690 | 2.030 | 2.438 | 2.724 | 3.340 |
| 40 | 0.127 | 0.255 | 0.388 | 0.529 | 0.681 | 0.851 | 1.050 | 1.303 | 1.684 | 2.021 | 2.423 | 2.705 | 3.307 |
| 50 | 0.126 | 0.255 | 0.388 | 0.528 | 0.679 | 0.849 | 1.047 | 1.299 | 1.676 | 2.009 | 2.403 | 2.678 | 3.261 |
| 60 | 0.126 | 0.255 | 0.387 | 0.527 | 0.679 | 0.848 | 1.046 | 1.296 | 1.671 | 2.000 | 2.390 | 2.660 | 3.232 |
| 90 | 0.126 | 0.254 | 0.387 | 0.526 | 0.677 | 0.846 | 1.042 | 1.291 | 1.662 | 1.987 | 2.369 | 2.632 | 3.183 |
| 120 | 0.126 | 0.254 | 0.386 | 0.526 | 0.677 | 0.845 | 1.041 | 1.289 | 1.658 | 1.980 | 2.358 | 2.617 | 3.160 |
| Infinity | 0.1256 | 0.2533 | 0.3853 | 0.5244 | 0.6745 | 0.8416 | 1.0364 | 1.2816 | 1.6449 | 1.9560 | 2.3263 | 2.576 | 3.090 |

## Student's-$t$ $\gamma$-Quantiles for $\gamma = 0.05/(2k)$, $\nu$ df

| $\nu$ \ k | 2 | 3 | 4 | 5 | 6 | 7 | 8 | 9 | 10 |
|---|---|---|---|---|---|---|---|---|---|
| 1 | 25.452 | 38.189 | 50.923 | 63.657 | 76.390 | 89.123 | 101.856 | 114.589 | 127.321 |
| 2 | 6.205 | 7.649 | 8.860 | 9.925 | 10.886 | 11.769 | 12.590 | 13.360 | 14.089 |
| 3 | 4.177 | 4.857 | 5.392 | 5.841 | 6.232 | 6.580 | 6.895 | 7.185 | 7.453 |
| 4 | 3.495 | 3.961 | 4.315 | 4.604 | 4.851 | 5.068 | 5.261 | 5.437 | 5.598 |
| 5 | 3.163 | 3.534 | 3.810 | 4.032 | 4.219 | 4.382 | 4.526 | 4.655 | 4.773 |
| 6 | 2.969 | 3.288 | 3.521 | 3.707 | 3.863 | 3.997 | 4.115 | 4.221 | 4.317 |
| 7 | 2.841 | 3.128 | 3.335 | 3.500 | 3.636 | 3.753 | 3.855 | 3.947 | 4.029 |
| 8 | 2.752 | 3.016 | 3.206 | 3.355 | 3.479 | 3.584 | 3.677 | 3.759 | 3.833 |
| 9 | 2.685 | 2.933 | 3.111 | 3.250 | 3.364 | 3.462 | 3.547 | 3.622 | 3.690 |
| 10 | 2.634 | 2.870 | 3.038 | 3.169 | 3.277 | 3.368 | 3.448 | 3.518 | 3.581 |
| 11 | 2.593 | 2.820 | 2.981 | 3.106 | 3.208 | 3.295 | 3.370 | 3.437 | 3.497 |
| 12 | 2.560 | 2.780 | 2.935 | 3.055 | 3.153 | 3.236 | 3.308 | 3.371 | 3.428 |
| 13 | 2.533 | 2.746 | 2.896 | 3.012 | 3.107 | 3.187 | 3.257 | 3.318 | 3.373 |
| 14 | 2.510 | 2.718 | 2.864 | 2.977 | 3.069 | 3.146 | 3.214 | 3.273 | 3.326 |
| 15 | 2.490 | 2.694 | 2.837 | 2.947 | 3.036 | 3.112 | 3.177 | 3.235 | 3.286 |
| 16 | 2.473 | 2.673 | 2.813 | 2.921 | 3.008 | 3.082 | 3.146 | 3.202 | 3.252 |
| 17 | 2.458 | 2.655 | 2.793 | 2.898 | 2.984 | 3.056 | 3.119 | 3.174 | 3.222 |
| 18 | 2.445 | 2.639 | 2.775 | 2.878 | 2.963 | 3.034 | 3.095 | 3.149 | 3.197 |
| 19 | 2.433 | 2.625 | 2.759 | 2.861 | 2.944 | 3.014 | 3.074 | 3.127 | 3.174 |
| 20 | 2.423 | 2.613 | 2.744 | 2.845 | 2.927 | 2.996 | 3.055 | 3.107 | 3.153 |
| 21 | 2.414 | 2.601 | 2.732 | 2.831 | 2.912 | 2.980 | 3.038 | 3.090 | 3.135 |
| 22 | 2.406 | 2.591 | 2.720 | 2.819 | 2.899 | 2.966 | 3.023 | 3.074 | 3.119 |
| 23 | 2.398 | 2.582 | 2.710 | 2.807 | 2.886 | 2.953 | 3.010 | 3.060 | 3.104 |
| 24 | 2.391 | 2.574 | 2.700 | 2.797 | 2.875 | 2.941 | 2.997 | 3.047 | 3.091 |
| 25 | 2.385 | 2.566 | 2.692 | 2.787 | 2.865 | 2.930 | 2.986 | 3.035 | 3.078 |
| 26 | 2.379 | 2.559 | 2.684 | 2.779 | 2.856 | 2.920 | 2.975 | 3.024 | 3.067 |
| 27 | 2.373 | 2.553 | 2.676 | 2.771 | 2.847 | 2.911 | 2.966 | 3.014 | 3.057 |
| 28 | 2.369 | 2.547 | 2.670 | 2.763 | 2.839 | 2.902 | 2.957 | 3.005 | 3.047 |
| 29 | 2.364 | 2.541 | 2.663 | 2.756 | 2.832 | 2.895 | 2.949 | 2.996 | 3.038 |
| 30 | 2.360 | 2.536 | 2.657 | 2.750 | 2.825 | 2.887 | 2.941 | 2.988 | 3.030 |
| 35 | 2.342 | 2.515 | 2.633 | 2.724 | 2.797 | 2.858 | 2.910 | 2.955 | 2.996 |
| 40 | 2.329 | 2.499 | 2.616 | 2.705 | 2.776 | 2.836 | 2.887 | 2.931 | 2.971 |
| 50 | 2.311 | 2.477 | 2.591 | 2.678 | 2.747 | 2.805 | 2.855 | 2.898 | 2.937 |
| 60 | 2.299 | 2.463 | 2.575 | 2.660 | 2.729 | 2.786 | 2.834 | 2.877 | 2.915 |
| 90 | 2.280 | 2.440 | 2.549 | 2.632 | 2.698 | 2.753 | 2.800 | 2.841 | 2.878 |
| 120 | 2.270 | 2.428 | 2.536 | 2.617 | 2.683 | 2.737 | 2.784 | 2.824 | 2.860 |

## Student's-$t$ $\gamma$-Quantiles for $\gamma = 1 - 0.05/[k(k-1)]$, $\nu$ df

| $\nu$ \ k | 3 | 4 | 5 | 6 | 7 | 8 |
|---|---|---|---|---|---|---|
| 1 | 38.189 | 76.390 | 127.321 | 190.984 | 267.379 | 356.506 |
| 2 | 7.649 | 10.886 | 14.089 | 17.277 | 20.457 | 23.633 |
| 3 | 4.857 | 6.232 | 7.453 | 8.575 | 9.624 | 10.617 |
| 4 | 3.961 | 4.851 | 5.598 | 6.254 | 6.847 | 7.392 |
| 5 | 3.534 | 4.219 | 4.773 | 5.247 | 5.667 | 6.045 |
| 6 | 3.288 | 3.863 | 4.317 | 4.698 | 5.030 | 5.326 |
| 7 | 3.128 | 3.636 | 4.029 | 4.355 | 4.636 | 4.884 |
| 8 | 3.016 | 3.479 | 3.833 | 4.122 | 4.370 | 4.587 |
| 9 | 2.933 | 3.364 | 3.690 | 3.954 | 4.179 | 4.374 |
| 10 | 2.870 | 3.277 | 3.581 | 3.827 | 4.035 | 4.215 |
| 11 | 2.820 | 3.208 | 3.497 | 3.728 | 3.923 | 4.091 |
| 12 | 2.780 | 3.153 | 3.428 | 3.649 | 3.833 | 3.993 |
| 13 | 2.746 | 3.107 | 3.373 | 3.584 | 3.760 | 3.912 |
| 14 | 2.718 | 3.069 | 3.326 | 3.530 | 3.699 | 3.845 |
| 15 | 2.694 | 3.036 | 3.286 | 3.484 | 3.648 | 3.788 |
| 16 | 2.673 | 3.008 | 3.252 | 3.444 | 3.604 | 3.740 |
| 17 | 2.655 | 2.984 | 3.222 | 3.410 | 3.565 | 3.698 |
| 18 | 2.639 | 2.963 | 3.197 | 3.380 | 3.532 | 3.661 |
| 19 | 2.625 | 2.944 | 3.174 | 3.354 | 3.503 | 3.629 |
| 20 | 2.613 | 2.927 | 3.153 | 3.331 | 3.477 | 3.601 |
| 21 | 2.601 | 2.912 | 3.135 | 3.310 | 3.453 | 3.575 |
| 22 | 2.591 | 2.899 | 3.119 | 3.291 | 3.432 | 3.552 |
| 23 | 2.582 | 2.886 | 3.104 | 3.274 | 3.413 | 3.531 |
| 24 | 2.574 | 2.875 | 3.091 | 3.258 | 3.396 | 3.513 |
| 25 | 2.566 | 2.865 | 3.078 | 3.244 | 3.380 | 3.496 |
| 26 | 2.559 | 2.856 | 3.067 | 3.231 | 3.366 | 3.480 |
| 27 | 2.553 | 2.847 | 3.057 | 3.219 | 3.353 | 3.465 |
| 28 | 2.547 | 2.839 | 3.047 | 3.208 | 3.340 | 3.452 |
| 29 | 2.541 | 2.832 | 3.038 | 3.198 | 3.329 | 3.440 |
| 30 | 2.536 | 2.825 | 3.030 | 3.189 | 3.319 | 3.428 |
| 35 | 2.515 | 2.797 | 2.996 | 3.150 | 3.276 | 3.382 |
| 40 | 2.499 | 2.776 | 2.971 | 3.122 | 3.244 | 3.347 |
| 50 | 2.477 | 2.747 | 2.937 | 3.083 | 3.201 | 3.300 |
| 60 | 2.463 | 2.729 | 2.915 | 3.057 | 3.173 | 3.270 |
| 90 | 2.440 | 2.698 | 2.878 | 3.016 | 3.127 | 3.220 |
| 120 | 2.428 | 2.683 | 2.860 | 2.995 | 3.104 | 3.195 |
| Infinity | 2.394 | 2.638 | 2.807 | 2.935 | 3.038 | 3.124 |

## TABLE 5 Chi-Square Quantiles

| $\nu$ = d.f. | 0.10 | 0.20 | 0.30 | 0.40 | 0.50 | 0.60 | 0.70 | 0.80 | 0.90 | 0.95 | 0.975 | 0.990 | 0.995 | 0.999 |
|---|---|---|---|---|---|---|---|---|---|---|---|---|---|---|
| 1 | 0.02 | 0.06 | 0.15 | 0.27 | 0.45 | 0.71 | 1.07 | 1.64 | 2.71 | 3.84 | 5.02 | 6.63 | 7.88 | 10.83 |
| 2 | 0.21 | 0.45 | 0.71 | 1.02 | 1.39 | 1.83 | 2.41 | 3.22 | 4.61 | 5.99 | 7.38 | 9.21 | 10.60 | 13.82 |
| 3 | 0.58 | 1.01 | 1.42 | 1.87 | 2.37 | 2.95 | 3.66 | 4.64 | 6.25 | 7.81 | 9.35 | 11.34 | 12.84 | 16.27 |
| 4 | 1.06 | 1.65 | 2.19 | 2.75 | 3.36 | 4.04 | 4.88 | 5.99 | 7.78 | 9.49 | 11.14 | 13.28 | 14.86 | 18.47 |
| 5 | 1.61 | 2.34 | 3.00 | 3.66 | 4.35 | 5.13 | 6.06 | 7.29 | 9.24 | 11.07 | 12.83 | 15.09 | 16.75 | 20.52 |
| 6 | 2.20 | 3.07 | 3.83 | 4.57 | 5.35 | 6.21 | 7.23 | 8.56 | 10.64 | 12.59 | 14.45 | 16.81 | 18.55 | 22.46 |
| 7 | 2.83 | 3.82 | 4.67 | 5.49 | 6.35 | 7.28 | 8.38 | 9.80 | 12.02 | 14.07 | 16.01 | 18.48 | 20.28 | 24.32 |
| 8 | 3.49 | 4.59 | 5.53 | 6.42 | 7.34 | 8.35 | 9.52 | 11.03 | 13.36 | 15.51 | 17.53 | 20.09 | 21.95 | 26.12 |
| 9 | 4.17 | 5.38 | 6.39 | 7.36 | 8.34 | 9.41 | 10.66 | 12.24 | 14.68 | 16.92 | 19.02 | 21.67 | 23.59 | 27.88 |
| 10 | 4.87 | 6.18 | 7.27 | 8.30 | 9.34 | 10.47 | 11.78 | 13.44 | 15.99 | 18.31 | 20.48 | 23.21 | 25.19 | 29.59 |
| 11 | 5.58 | 6.99 | 8.15 | 9.24 | 10.34 | 11.53 | 12.90 | 14.63 | 17.28 | 19.68 | 21.92 | 24.72 | 26.76 | 31.26 |
| 12 | 6.30 | 7.81 | 9.03 | 10.18 | 11.34 | 12.58 | 14.01 | 15.81 | 18.55 | 21.03 | 23.34 | 26.22 | 28.30 | 32.91 |
| 13 | 7.04 | 8.63 | 9.93 | 11.13 | 12.34 | 13.64 | 15.12 | 16.98 | 19.81 | 22.36 | 24.74 | 27.69 | 29.82 | 34.53 |
| 14 | 7.79 | 9.47 | 10.82 | 12.08 | 13.34 | 14.69 | 16.22 | 18.15 | 21.06 | 23.68 | 26.12 | 29.14 | 31.32 | 36.12 |
| 15 | 8.55 | 10.31 | 11.72 | 13.03 | 14.34 | 15.73 | 17.32 | 19.31 | 22.31 | 25.00 | 27.49 | 30.58 | 32.80 | 37.70 |
| 16 | 9.31 | 11.15 | 12.62 | 13.98 | 15.34 | 16.78 | 18.42 | 20.47 | 23.54 | 26.30 | 28.85 | 32.00 | 34.27 | 39.25 |
| 17 | 10.09 | 12.00 | 13.53 | 14.94 | 16.34 | 17.82 | 19.51 | 21.61 | 24.77 | 27.59 | 30.19 | 33.41 | 35.72 | 40.79 |
| 18 | 10.86 | 12.86 | 14.44 | 15.89 | 17.34 | 18.87 | 20.60 | 22.76 | 25.99 | 28.87 | 31.53 | 34.81 | 37.16 | 42.31 |
| 19 | 11.65 | 13.72 | 15.35 | 16.85 | 18.34 | 19.91 | 21.69 | 23.90 | 27.20 | 30.14 | 32.85 | 36.19 | 38.58 | 43.82 |
| 20 | 12.44 | 14.58 | 16.27 | 17.81 | 19.34 | 20.95 | 22.77 | 25.04 | 28.41 | 31.41 | 34.17 | 37.57 | 40.00 | 45.31 |
| 21 | 13.24 | 15.44 | 17.18 | 18.77 | 20.34 | 21.99 | 23.86 | 26.17 | 29.62 | 32.67 | 35.48 | 38.93 | 41.40 | 46.80 |
| 22 | 14.04 | 16.31 | 18.10 | 19.73 | 21.34 | 23.03 | 24.94 | 27.30 | 30.81 | 33.92 | 36.78 | 40.29 | 42.80 | 48.27 |
| 23 | 14.85 | 17.19 | 19.02 | 20.69 | 22.34 | 24.07 | 26.02 | 28.43 | 32.01 | 35.17 | 38.08 | 41.64 | 44.18 | 49.73 |
| 24 | 15.66 | 18.06 | 19.94 | 21.65 | 23.34 | 25.11 | 27.10 | 29.55 | 33.20 | 36.42 | 39.36 | 42.98 | 45.56 | 51.18 |
| 25 | 16.47 | 18.94 | 20.87 | 22.62 | 24.34 | 26.14 | 28.17 | 30.68 | 34.38 | 37.65 | 40.65 | 44.31 | 46.93 | 52.62 |

| $\nu$ = d.f. | 0.10 | 0.20 | 0.30 | 0.40 | 0.50 | 0.60 | 0.70 | 0.80 | 0.90 | 0.95 | 0.975 | 0.990 | 0.995 | 0.999 |
|---|---|---|---|---|---|---|---|---|---|---|---|---|---|---|
| 26 | 17.29 | 19.82 | 21.79 | 23.58 | 25.34 | 27.18 | 29.25 | 31.79 | 35.56 | 38.89 | 41.92 | 45.64 | 48.29 | 54.05 |
| 27 | 18.11 | 20.70 | 22.72 | 24.54 | 26.34 | 28.21 | 30.32 | 32.91 | 36.74 | 40.11 | 43.19 | 46.96 | 49.64 | 55.48 |
| 28 | 18.94 | 21.59 | 23.65 | 25.51 | 27.34 | 29.25 | 31.39 | 34.03 | 37.92 | 41.34 | 44.46 | 48.28 | 50.99 | 56.89 |
| 29 | 19.77 | 22.48 | 24.58 | 26.48 | 28.34 | 30.28 | 32.46 | 35.14 | 39.09 | 42.56 | 45.72 | 49.59 | 52.34 | 58.30 |
| 30 | 20.60 | 23.36 | 25.51 | 27.44 | 29.34 | 31.32 | 33.53 | 36.25 | 40.26 | 43.77 | 46.98 | 50.89 | 53.67 | 59.70 |
| 31 | 21.43 | 24.26 | 26.44 | 28.41 | 30.34 | 32.35 | 34.60 | 37.36 | 41.42 | 44.99 | 48.23 | 52.19 | 55.00 | 61.10 |
| 32 | 22.27 | 25.15 | 27.37 | 29.38 | 31.34 | 33.38 | 35.66 | 38.47 | 42.58 | 46.19 | 49.48 | 53.49 | 56.33 | 62.49 |
| 33 | 23.11 | 26.04 | 28.31 | 30.34 | 32.34 | 34.41 | 36.73 | 39.57 | 43.75 | 47.40 | 50.73 | 54.78 | 57.65 | 63.87 |
| 34 | 23.95 | 26.94 | 29.24 | 31.31 | 33.34 | 35.44 | 37.80 | 40.68 | 44.90 | 48.60 | 51.97 | 56.06 | 58.96 | 65.25 |
| 35 | 24.80 | 27.84 | 30.18 | 32.28 | 34.34 | 36.47 | 38.86 | 41.78 | 46.06 | 49.80 | 53.20 | 57.34 | 60.27 | 66.62 |
| 40 | 29.05 | 32.34 | 34.87 | 37.13 | 39.34 | 41.62 | 44.16 | 47.27 | 51.81 | 55.76 | 59.34 | 63.69 | 66.77 | 73.40 |
| 50 | 37.69 | 41.45 | 44.31 | 46.86 | 49.33 | 51.89 | 54.72 | 58.16 | 63.17 | 67.50 | 71.42 | 76.15 | 79.49 | 86.66 |
| 60 | 46.46 | 50.64 | 53.81 | 56.62 | 59.33 | 62.13 | 65.23 | 68.97 | 74.40 | 79.08 | 83.30 | 88.38 | 91.95 | 99.61 |
| 80 | 64.28 | 69.21 | 72.92 | 76.19 | 79.33 | 82.57 | 86.12 | 90.41 | 96.58 | 101.9 | 106.6 | 112.3 | 116.3 | 124.8 |
| 100 | 82.36 | 87.95 | 92.13 | 95.81 | 99.33 | 102.95 | 106.91 | 111.67 | 118.50 | 124.3 | 129.6 | 135.8 | 140.2 | 149.5 |
| 200 | 174.84 | 183.00 | 189.05 | 194.32 | 199.33 | 204.43 | 209.99 | 216.61 | 226.02 | 233.99 | 241.06 | 249.45 | 255.26 | 267.54 |

**The Cube Root Approximation**

For large n the $\gamma$th quantile is given in good approximation by $u = \nu[a + z\, b]^3$, where $a = 1 - 2/(9\nu)$, $b = (2/(9\nu))$, and $\Phi(z) = \gamma$. For example, for $\gamma = 0.99$, $\nu = 100$, $a = 0.99778$, $b = 0.04714$, $z = 2.32635$, and $u = 135.82$.

# TABLE 6

## 0.90-Quantiles of the $F$-Distribution

| $\nu_2$ \ $\nu_1$ | 1 | 2 | 3 | 4 | 5 | 6 | 7 | 8 | 9 | 10 | 12 | 15 | 20 | 24 | 30 | 40 | 60 | 120 | Inf. |
|---|---|---|---|---|---|---|---|---|---|---|---|---|---|---|---|---|---|---|---|
| 1 | 39.9 | 49.5 | 53.6 | 55.8 | 57.2 | 58.2 | 58.9 | 59.4 | 59.9 | 60.2 | 60.7 | 61.2 | 61.7 | 62.0 | 62.3 | 62.5 | 62.8 | 63.1 | 63.3 |
| 2 | 8.53 | 9.00 | 9.16 | 9.24 | 9.29 | 9.33 | 9.35 | 9.37 | 9.38 | 9.39 | 9.41 | 9.42 | 9.44 | 9.45 | 9.46 | 9.47 | 9.47 | 9.48 | 9.49 |
| 3 | 5.54 | 5.46 | 5.39 | 5.34 | 5.31 | 5.28 | 5.27 | 5.25 | 5.24 | 5.23 | 5.22 | 5.20 | 5.18 | 5.18 | 5.17 | 5.16 | 5.15 | 5.14 | 5.13 |
| 4 | 4.54 | 4.32 | 4.19 | 4.11 | 4.05 | 4.01 | 3.98 | 3.95 | 3.94 | 3.92 | 3.90 | 3.87 | 3.84 | 3.83 | 3.82 | 3.80 | 3.79 | 3.78 | 3.76 |
| 5 | 4.06 | 3.78 | 3.62 | 3.52 | 3.45 | 3.40 | 3.37 | 3.34 | 3.32 | 3.30 | 3.27 | 3.24 | 3.21 | 3.19 | 3.17 | 3.16 | 3.14 | 3.12 | 3.11 |
| 6 | 3.78 | 3.46 | 3.29 | 3.18 | 3.11 | 3.05 | 3.01 | 2.98 | 2.96 | 2.94 | 2.90 | 2.87 | 2.84 | 2.82 | 2.80 | 2.78 | 2.76 | 2.74 | 2.72 |
| 7 | 3.59 | 3.26 | 3.07 | 2.96 | 2.88 | 2.83 | 2.78 | 2.75 | 2.72 | 2.70 | 2.67 | 2.63 | 2.59 | 2.58 | 2.56 | 2.54 | 2.51 | 2.49 | 2.47 |
| 8 | 3.46 | 3.11 | 2.92 | 2.81 | 2.73 | 2.67 | 2.62 | 2.59 | 2.56 | 2.54 | 2.50 | 2.46 | 2.42 | 2.40 | 2.38 | 2.36 | 2.34 | 2.32 | 2.29 |
| 9 | 3.36 | 3.01 | 2.81 | 2.69 | 2.61 | 2.55 | 2.51 | 2.47 | 2.44 | 2.42 | 2.38 | 2.34 | 2.30 | 2.28 | 2.25 | 2.23 | 2.21 | 2.18 | 2.16 |
| 10 | 3.29 | 2.92 | 2.73 | 2.61 | 2.52 | 2.46 | 2.41 | 2.38 | 2.35 | 2.32 | 2.28 | 2.24 | 2.20 | 2.18 | 2.16 | 2.13 | 2.11 | 2.08 | 2.06 |
| 11 | 3.23 | 2.86 | 2.66 | 2.54 | 2.45 | 2.39 | 2.34 | 2.30 | 2.27 | 2.25 | 2.21 | 2.17 | 2.12 | 2.10 | 2.08 | 2.05 | 2.03 | 2.00 | 1.97 |
| 12 | 3.18 | 2.81 | 2.61 | 2.48 | 2.39 | 2.33 | 2.28 | 2.24 | 2.21 | 2.19 | 2.15 | 2.10 | 2.06 | 2.04 | 2.01 | 1.99 | 1.96 | 1.93 | 1.90 |
| 13 | 3.14 | 2.76 | 2.56 | 2.43 | 2.35 | 2.28 | 2.23 | 2.20 | 2.16 | 2.14 | 2.10 | 2.05 | 2.01 | 1.98 | 1.96 | 1.93 | 1.90 | 1.88 | 1.85 |
| 14 | 3.10 | 2.73 | 2.52 | 2.39 | 2.31 | 2.24 | 2.19 | 2.15 | 2.12 | 2.10 | 2.05 | 2.01 | 1.96 | 1.94 | 1.91 | 1.89 | 1.86 | 1.83 | 1.80 |
| 15 | 3.07 | 2.70 | 2.49 | 2.36 | 2.27 | 2.21 | 2.16 | 2.12 | 2.09 | 2.06 | 2.02 | 1.97 | 1.92 | 1.90 | 1.87 | 1.85 | 1.82 | 1.79 | 1.76 |
| 16 | 3.05 | 2.67 | 2.46 | 2.33 | 2.24 | 2.18 | 2.13 | 2.09 | 2.06 | 2.03 | 1.99 | 1.94 | 1.89 | 1.87 | 1.84 | 1.81 | 1.78 | 1.75 | 1.72 |
| 17 | 3.03 | 2.64 | 2.44 | 2.31 | 2.22 | 2.15 | 2.10 | 2.06 | 2.03 | 2.00 | 1.96 | 1.91 | 1.86 | 1.84 | 1.81 | 1.78 | 1.75 | 1.72 | 1.69 |
| 18 | 3.01 | 2.62 | 2.42 | 2.29 | 2.20 | 2.13 | 2.08 | 2.04 | 2.00 | 1.98 | 1.93 | 1.89 | 1.84 | 1.81 | 1.78 | 1.75 | 1.72 | 1.69 | 1.66 |
| 19 | 2.99 | 2.61 | 2.40 | 2.27 | 2.18 | 2.11 | 2.06 | 2.02 | 1.98 | 1.96 | 1.91 | 1.86 | 1.81 | 1.79 | 1.76 | 1.73 | 1.70 | 1.67 | 1.63 |
| 20 | 2.97 | 2.59 | 2.38 | 2.25 | 2.16 | 2.09 | 2.04 | 2.00 | 1.96 | 1.94 | 1.89 | 1.84 | 1.79 | 1.77 | 1.74 | 1.71 | 1.68 | 1.64 | 1.61 |
| 21 | 2.96 | 2.57 | 2.36 | 2.23 | 2.14 | 2.08 | 2.02 | 1.98 | 1.95 | 1.92 | 1.87 | 1.83 | 1.78 | 1.75 | 1.72 | 1.69 | 1.66 | 1.62 | 1.59 |
| 22 | 2.95 | 2.56 | 2.35 | 2.22 | 2.13 | 2.06 | 2.01 | 1.97 | 1.93 | 1.90 | 1.86 | 1.81 | 1.76 | 1.73 | 1.70 | 1.67 | 1.64 | 1.60 | 1.57 |
| 23 | 2.94 | 2.55 | 2.34 | 2.21 | 2.11 | 2.05 | 1.99 | 1.95 | 1.92 | 1.89 | 1.84 | 1.80 | 1.74 | 1.72 | 1.69 | 1.66 | 1.62 | 1.59 | 1.55 |
| 24 | 2.93 | 2.54 | 2.33 | 2.19 | 2.10 | 2.04 | 1.98 | 1.94 | 1.91 | 1.88 | 1.83 | 1.78 | 1.73 | 1.70 | 1.67 | 1.64 | 1.61 | 1.57 | 1.53 |
| 25 | 2.92 | 2.53 | 2.32 | 2.18 | 2.09 | 2.02 | 1.97 | 1.93 | 1.89 | 1.87 | 1.82 | 1.77 | 1.72 | 1.69 | 1.66 | 1.63 | 1.59 | 1.56 | 1.52 |
| 30 | 2.88 | 2.49 | 2.28 | 2.14 | 2.05 | 1.98 | 1.93 | 1.88 | 1.85 | 1.82 | 1.77 | 1.72 | 1.67 | 1.64 | 1.61 | 1.57 | 1.54 | 1.50 | 1.46 |
| 40 | 2.84 | 2.44 | 2.23 | 2.09 | 2.00 | 1.93 | 1.87 | 1.83 | 1.79 | 1.76 | 1.71 | 1.66 | 1.61 | 1.57 | 1.54 | 1.51 | 1.47 | 1.42 | 1.38 |
| 60 | 2.79 | 2.39 | 2.18 | 2.04 | 1.95 | 1.87 | 1.82 | 1.77 | 1.74 | 1.71 | 1.66 | 1.60 | 1.54 | 1.51 | 1.48 | 1.44 | 1.40 | 1.35 | 1.29 |
| 120 | 2.75 | 2.35 | 2.13 | 1.99 | 1.90 | 1.82 | 1.77 | 1.72 | 1.68 | 1.65 | 1.60 | 1.55 | 1.48 | 1.45 | 1.41 | 1.37 | 1.32 | 1.26 | 1.19 |
| Inf. | 2.71 | 2.30 | 2.09 | 1.95 | 1.85 | 1.78 | 1.72 | 1.67 | 1.63 | 1.60 | 1.55 | 1.49 | 1.42 | 1.39 | 1.34 | 1.30 | 1.24 | 1.17 | 1.00 |

## 0.95-Quantiles of the $F$-Distribution

| $\nu_2$ \ $\nu_1$ | 1 | 2 | 3 | 4 | 5 | 6 | 7 | 8 | 9 | 10 | 12 | 15 | 20 | 24 | 30 | 40 | 60 | 120 | Inf. |
|---|---|---|---|---|---|---|---|---|---|---|---|---|---|---|---|---|---|---|---|
| 1 | 162 | 200 | 216 | 225 | 230 | 234 | 237 | 239 | 241 | 242 | 244 | 246 | 248 | 249 | 250 | 251 | 252 | 253 | 254 |
| 2 | 18.5 | 19.0 | 19.2 | 19.3 | 19.3 | 19.3 | 19.4 | 19.4 | 19.4 | 19.4 | 19.4 | 19.4 | 19.5 | 19.5 | 19.5 | 19.5 | 19.5 | 19.5 | 19.5 |
| 3 | 10.1 | 9.55 | 9.28 | 9.12 | 9.01 | 8.94 | 8.89 | 8.85 | 8.81 | 8.79 | 8.74 | 8.70 | 8.66 | 8.64 | 8.62 | 8.59 | 8.57 | 8.55 | 8.53 |
| 4 | 7.71 | 6.94 | 6.59 | 6.39 | 6.26 | 6.16 | 6.09 | 6.04 | 6.00 | 5.96 | 5.91 | 5.86 | 5.80 | 5.77 | 5.75 | 5.72 | 5.69 | 5.66 | 5.63 |
| 5 | 6.61 | 5.79 | 5.41 | 5.19 | 5.05 | 4.95 | 4.88 | 4.82 | 4.77 | 4.74 | 4.68 | 4.62 | 4.56 | 4.53 | 4.50 | 4.46 | 4.43 | 4.40 | 4.37 |
| 6 | 5.99 | 5.14 | 4.76 | 4.53 | 4.39 | 4.28 | 4.21 | 4.15 | 4.10 | 4.06 | 4.00 | 3.94 | 3.87 | 3.84 | 3.81 | 3.77 | 3.74 | 3.70 | 3.67 |
| 7 | 5.59 | 4.74 | 4.35 | 4.12 | 3.97 | 3.87 | 3.79 | 3.73 | 3.68 | 3.64 | 3.57 | 3.51 | 3.44 | 3.41 | 3.38 | 3.34 | 3.30 | 3.27 | 3.23 |
| 8 | 5.32 | 4.46 | 4.07 | 3.84 | 3.69 | 3.58 | 3.50 | 3.44 | 3.39 | 3.35 | 3.28 | 3.22 | 3.15 | 3.12 | 3.08 | 3.04 | 3.01 | 2.97 | 2.93 |
| 9 | 5.12 | 4.26 | 3.86 | 3.63 | 3.48 | 3.37 | 3.29 | 3.23 | 3.18 | 3.14 | 3.07 | 3.01 | 2.94 | 2.90 | 2.86 | 2.83 | 2.79 | 2.75 | 2.71 |
| 10 | 4.96 | 4.10 | 3.71 | 3.48 | 3.33 | 3.22 | 3.14 | 3.07 | 3.02 | 2.98 | 2.91 | 2.85 | 2.77 | 2.74 | 2.70 | 2.66 | 2.62 | 2.58 | 2.54 |
| 11 | 4.84 | 3.98 | 3.59 | 3.36 | 3.20 | 3.09 | 3.01 | 2.95 | 2.90 | 2.85 | 2.79 | 2.72 | 2.65 | 2.61 | 2.57 | 2.53 | 2.49 | 2.45 | 2.40 |
| 12 | 4.75 | 3.89 | 3.49 | 3.26 | 3.11 | 3.00 | 2.91 | 2.85 | 2.80 | 2.75 | 2.69 | 2.62 | 2.54 | 2.51 | 2.47 | 2.43 | 2.38 | 2.34 | 2.30 |
| 13 | 4.67 | 3.81 | 3.41 | 3.18 | 3.03 | 2.92 | 2.83 | 2.77 | 2.71 | 2.67 | 2.60 | 2.53 | 2.46 | 2.42 | 2.38 | 2.34 | 2.30 | 2.25 | 2.21 |
| 14 | 4.60 | 3.74 | 3.34 | 3.11 | 2.96 | 2.85 | 2.76 | 2.70 | 2.65 | 2.60 | 2.53 | 2.46 | 2.39 | 2.35 | 2.31 | 2.27 | 2.22 | 2.18 | 2.13 |
| 15 | 4.54 | 3.68 | 3.29 | 3.06 | 2.90 | 2.79 | 2.71 | 2.64 | 2.59 | 2.54 | 2.48 | 2.40 | 2.33 | 2.29 | 2.25 | 2.20 | 2.16 | 2.11 | 2.07 |
| 16 | 4.49 | 3.63 | 3.24 | 3.01 | 2.85 | 2.74 | 2.66 | 2.59 | 2.54 | 2.49 | 2.42 | 2.35 | 2.28 | 2.24 | 2.19 | 2.15 | 2.11 | 2.06 | 2.01 |
| 17 | 4.45 | 3.59 | 3.20 | 2.96 | 2.81 | 2.70 | 2.61 | 2.55 | 2.49 | 2.45 | 2.38 | 2.31 | 2.23 | 2.19 | 2.15 | 2.10 | 2.06 | 2.01 | 1.96 |
| 18 | 4.41 | 3.55 | 3.16 | 2.93 | 2.77 | 2.66 | 2.58 | 2.51 | 2.46 | 2.41 | 2.34 | 2.27 | 2.19 | 2.15 | 2.11 | 2.06 | 2.02 | 1.97 | 1.92 |
| 19 | 4.38 | 3.52 | 3.13 | 2.90 | 2.74 | 2.63 | 2.54 | 2.48 | 2.42 | 2.38 | 2.31 | 2.23 | 2.16 | 2.11 | 2.07 | 2.03 | 1.98 | 1.93 | 1.88 |
| 20 | 4.35 | 3.49 | 3.10 | 2.87 | 2.71 | 2.60 | 2.51 | 2.45 | 2.39 | 2.35 | 2.28 | 2.20 | 2.12 | 2.08 | 2.04 | 1.99 | 1.95 | 1.90 | 1.84 |
| 21 | 4.32 | 3.47 | 3.07 | 2.84 | 2.68 | 2.57 | 2.49 | 2.42 | 2.37 | 2.32 | 2.25 | 2.18 | 2.10 | 2.05 | 2.01 | 1.96 | 1.92 | 1.87 | 1.81 |
| 22 | 4.30 | 3.44 | 3.05 | 2.82 | 2.66 | 2.55 | 2.46 | 2.40 | 2.34 | 2.30 | 2.23 | 2.15 | 2.07 | 2.03 | 1.98 | 1.94 | 1.89 | 1.84 | 1.78 |
| 23 | 4.28 | 3.42 | 3.03 | 2.80 | 2.64 | 2.53 | 2.44 | 2.37 | 2.32 | 2.27 | 2.20 | 2.13 | 2.05 | 2.01 | 1.96 | 1.91 | 1.86 | 1.81 | 1.76 |
| 24 | 4.26 | 3.40 | 3.01 | 2.78 | 2.62 | 2.51 | 2.42 | 2.36 | 2.30 | 2.25 | 2.18 | 2.11 | 2.03 | 1.98 | 1.94 | 1.89 | 1.84 | 1.79 | 1.73 |
| 25 | 4.24 | 3.39 | 2.99 | 2.76 | 2.60 | 2.49 | 2.40 | 2.34 | 2.28 | 2.24 | 2.16 | 2.09 | 2.01 | 1.96 | 1.92 | 1.87 | 1.82 | 1.77 | 1.71 |
| 30 | 4.17 | 3.32 | 2.92 | 2.69 | 2.53 | 2.42 | 2.33 | 2.27 | 2.21 | 2.16 | 2.09 | 2.01 | 1.93 | 1.89 | 1.84 | 1.79 | 1.74 | 1.68 | 1.62 |
| 40 | 4.08 | 3.23 | 2.84 | 2.61 | 2.45 | 2.34 | 2.25 | 2.18 | 2.12 | 2.08 | 2.00 | 1.92 | 1.84 | 1.79 | 1.74 | 1.69 | 1.64 | 1.58 | 1.51 |
| 60 | 4.00 | 3.15 | 2.76 | 2.53 | 2.37 | 2.25 | 2.17 | 2.10 | 2.04 | 1.99 | 1.92 | 1.84 | 1.75 | 1.70 | 1.65 | 1.59 | 1.53 | 1.47 | 1.39 |
| 120 | 3.92 | 3.07 | 2.68 | 2.45 | 2.29 | 2.18 | 2.09 | 2.02 | 1.96 | 1.91 | 1.83 | 1.75 | 1.66 | 1.61 | 1.55 | 1.50 | 1.43 | 1.35 | 1.25 |
| Inf. | 3.84 | 3.00 | 2.60 | 2.37 | 2.21 | 2.10 | 2.01 | 1.94 | 1.88 | 1.83 | 1.75 | 1.67 | 1.57 | 1.52 | 1.46 | 1.39 | 1.32 | 1.22 | 1.00 |

## 0.975-Quantiles of the $F$-Distribution

| $\nu_2$ \ $\nu_1$ | 1 | 2 | 3 | 4 | 5 | 6 | 7 | 8 | 9 | 10 | 12 | 15 | 20 | 24 | 30 | 40 | 60 | 120 | Inf. |
|---|---|---|---|---|---|---|---|---|---|---|---|---|---|---|---|---|---|---|---|
| 1 | 648 | 800 | 864 | 900 | 922 | 937 | 948 | 957 | 963 | 969 | 977 | 985 | 993 | 249 | 250 | 251 | 252 | 253 | 254 |
| 2 | 38.5 | 39.0 | 39.2 | 39.3 | 39.3 | 39.3 | 39.4 | 39.4 | 39.4 | 39.4 | 39.4 | 39.4 | 39.5 | 19.5 | 19.5 | 19.5 | 19.5 | 19.5 | 19.5 |
| 3 | 17.4 | 16.0 | 15.4 | 15.1 | 14.9 | 14.7 | 14.6 | 14.5 | 14.5 | 14.4 | 14.3 | 14.3 | 14.2 | 8.64 | 8.62 | 8.59 | 8.57 | 8.55 | 8.53 |
| 4 | 12.2 | 10.7 | 9.98 | 9.60 | 9.36 | 9.20 | 9.07 | 8.98 | 8.90 | 8.84 | 8.75 | 8.66 | 8.56 | 5.77 | 5.75 | 5.72 | 5.69 | 5.66 | 5.63 |
| 5 | 10.0 | 8.43 | 7.76 | 7.39 | 7.15 | 6.98 | 6.85 | 6.76 | 6.68 | 6.62 | 6.52 | 6.43 | 6.33 | 4.53 | 4.50 | 4.46 | 4.43 | 4.40 | 4.37 |
| 6 | 8.81 | 7.26 | 6.60 | 6.23 | 5.99 | 5.82 | 5.70 | 5.60 | 5.52 | 5.46 | 5.37 | 5.27 | 5.17 | 3.84 | 3.81 | 3.77 | 3.74 | 3.70 | 3.67 |
| 7 | 8.07 | 6.54 | 5.89 | 5.52 | 5.29 | 5.12 | 4.99 | 4.90 | 4.82 | 4.76 | 4.67 | 4.57 | 4.47 | 3.41 | 3.38 | 3.34 | 3.30 | 3.27 | 3.23 |
| 8 | 7.57 | 6.06 | 5.42 | 5.05 | 4.82 | 4.65 | 4.53 | 4.43 | 4.36 | 4.30 | 4.20 | 4.10 | 4.00 | 3.12 | 3.08 | 3.04 | 3.01 | 2.97 | 2.93 |
| 9 | 7.21 | 5.71 | 5.08 | 4.72 | 4.48 | 4.32 | 4.20 | 4.10 | 4.03 | 3.96 | 3.87 | 3.77 | 3.67 | 2.90 | 2.86 | 2.83 | 2.79 | 2.75 | 2.71 |
| 10 | 6.94 | 5.46 | 4.83 | 4.47 | 4.24 | 4.07 | 3.95 | 3.85 | 3.78 | 3.72 | 3.62 | 3.52 | 3.42 | 2.74 | 2.70 | 2.66 | 2.62 | 2.58 | 2.54 |
| 11 | 6.72 | 5.26 | 4.63 | 4.28 | 4.04 | 3.88 | 3.76 | 3.66 | 3.59 | 3.53 | 3.43 | 3.33 | 3.23 | 2.61 | 2.57 | 2.53 | 2.49 | 2.45 | 2.40 |
| 12 | 6.55 | 5.10 | 4.47 | 4.12 | 3.89 | 3.73 | 3.61 | 3.51 | 3.44 | 3.37 | 3.28 | 3.18 | 3.07 | 2.51 | 2.47 | 2.43 | 2.38 | 2.34 | 2.30 |
| 13 | 6.41 | 4.97 | 4.35 | 4.00 | 3.77 | 3.60 | 3.48 | 3.39 | 3.31 | 3.25 | 3.15 | 3.05 | 2.95 | 2.42 | 2.38 | 2.34 | 2.30 | 2.25 | 2.21 |
| 14 | 6.30 | 4.86 | 4.24 | 3.89 | 3.66 | 3.50 | 3.38 | 3.29 | 3.21 | 3.15 | 3.05 | 2.95 | 2.84 | 2.35 | 2.31 | 2.27 | 2.22 | 2.18 | 2.13 |
| 15 | 6.20 | 4.77 | 4.15 | 3.80 | 3.58 | 3.41 | 3.29 | 3.20 | 3.12 | 3.06 | 2.96 | 2.86 | 2.76 | 2.29 | 2.25 | 2.20 | 2.16 | 2.11 | 2.07 |
| 16 | 6.12 | 4.69 | 4.08 | 3.73 | 3.50 | 3.34 | 3.22 | 3.12 | 3.05 | 2.99 | 2.89 | 2.79 | 2.68 | 2.24 | 2.19 | 2.15 | 2.11 | 2.06 | 2.01 |
| 17 | 6.04 | 4.62 | 4.01 | 3.66 | 3.44 | 3.28 | 3.16 | 3.06 | 2.98 | 2.92 | 2.82 | 2.72 | 2.62 | 2.19 | 2.15 | 2.10 | 2.06 | 2.01 | 1.96 |
| 18 | 5.98 | 4.56 | 3.95 | 3.61 | 3.38 | 3.22 | 3.10 | 3.01 | 2.93 | 2.87 | 2.77 | 2.67 | 2.56 | 2.15 | 2.11 | 2.06 | 2.02 | 1.97 | 1.92 |
| 19 | 5.92 | 4.51 | 3.90 | 3.56 | 3.33 | 3.17 | 3.05 | 2.96 | 2.88 | 2.82 | 2.72 | 2.62 | 2.51 | 2.11 | 2.07 | 2.03 | 1.98 | 1.93 | 1.88 |
| 20 | 5.87 | 4.46 | 3.86 | 3.51 | 3.29 | 3.13 | 3.01 | 2.91 | 2.84 | 2.77 | 2.68 | 2.57 | 2.46 | 2.08 | 2.04 | 1.99 | 1.95 | 1.90 | 1.84 |
| 21 | 5.83 | 4.42 | 3.82 | 3.48 | 3.25 | 3.09 | 2.97 | 2.87 | 2.80 | 2.73 | 2.64 | 2.53 | 2.42 | 2.05 | 2.01 | 1.96 | 1.92 | 1.87 | 1.81 |
| 22 | 5.79 | 4.38 | 3.78 | 3.44 | 3.22 | 3.05 | 2.93 | 2.84 | 2.76 | 2.70 | 2.60 | 2.50 | 2.39 | 2.03 | 1.98 | 1.94 | 1.89 | 1.84 | 1.78 |
| 23 | 5.75 | 4.35 | 3.75 | 3.41 | 3.18 | 3.02 | 2.90 | 2.81 | 2.73 | 2.67 | 2.57 | 2.47 | 2.36 | 2.01 | 1.96 | 1.91 | 1.86 | 1.81 | 1.76 |
| 24 | 5.72 | 4.32 | 3.72 | 3.38 | 3.15 | 2.99 | 2.87 | 2.78 | 2.70 | 2.64 | 2.54 | 2.44 | 2.33 | 1.98 | 1.94 | 1.89 | 1.84 | 1.79 | 1.73 |
| 25 | 5.69 | 4.29 | 3.69 | 3.35 | 3.13 | 2.97 | 2.85 | 2.75 | 2.68 | 2.61 | 2.51 | 2.41 | 2.30 | 1.96 | 1.92 | 1.87 | 1.82 | 1.77 | 1.71 |
| 30 | 5.57 | 4.18 | 3.59 | 3.25 | 3.03 | 2.87 | 2.75 | 2.65 | 2.57 | 2.51 | 2.41 | 2.31 | 2.20 | 1.89 | 1.84 | 1.79 | 1.74 | 1.68 | 1.62 |
| 40 | 5.42 | 4.05 | 3.46 | 3.13 | 2.90 | 2.74 | 2.62 | 2.53 | 2.45 | 2.39 | 2.29 | 2.18 | 2.07 | 1.79 | 1.74 | 1.69 | 1.64 | 1.58 | 1.51 |
| 60 | 5.29 | 3.93 | 3.34 | 3.01 | 2.79 | 2.63 | 2.51 | 2.41 | 2.33 | 2.27 | 2.17 | 2.06 | 1.94 | 1.70 | 1.65 | 1.59 | 1.53 | 1.47 | 1.39 |
| 120 | 5.15 | 3.80 | 3.23 | 2.89 | 2.67 | 2.52 | 2.39 | 2.30 | 2.22 | 2.16 | 2.05 | 1.94 | 1.82 | 1.61 | 1.55 | 1.50 | 1.43 | 1.35 | 1.25 |
| Inf. | 5.02 | 3.69 | 3.12 | 2.79 | 2.57 | 2.41 | 2.29 | 2.19 | 2.11 | 2.05 | 1.94 | 1.83 | 1.71 | 1.52 | 1.46 | 1.39 | 1.32 | 1.22 | 1.00 |

## 0.99-Quantiles of the $F$-Distribution

$\nu_1$ across the top, $\nu_2$ down the side.

| $\nu_2$ \ $\nu_1$ | 1 | 2 | 3 | 4 | 5 | 6 | 7 | 8 | 9 | 10 | 12 | 15 | 20 | 24 | 30 | 40 | 60 | 120 | Inf. |
|---|---|---|---|---|---|---|---|---|---|---|---|---|---|---|---|---|---|---|---|
| 1 | 4052 | 5000 | 5403 | 5625 | 5764 | 5859 | 5928 | 5981 | 6022 | 6056 | 6106 | 6157 | 6209 | 6235 | 6261 | 6287 | 6313 | 6339 | 6366 |
| 2 | 98.5 | 99.0 | 99.2 | 99.3 | 99.3 | 99.3 | 99.4 | 99.4 | 99.4 | 99.4 | 99.4 | 99.4 | 99.5 | 99.5 | 99.5 | 99.5 | 99.5 | 99.5 | 99.5 |
| 3 | 34.1 | 30.8 | 29.5 | 28.7 | 28.2 | 27.9 | 27.7 | 27.5 | 27.4 | 27.2 | 27.1 | 26.9 | 26.7 | 26.6 | 26.5 | 26.4 | 26.3 | 26.2 | 26.1 |
| 4 | 21.2 | 18.0 | 16.7 | 16.0 | 15.5 | 15.2 | 15.0 | 14.8 | 14.7 | 14.6 | 14.4 | 14.2 | 14.0 | 13.9 | 13.8 | 13.8 | 13.7 | 13.6 | 13.5 |
| 5 | 16.3 | 13.3 | 12.1 | 11.4 | 11.0 | 10.7 | 10.5 | 10.3 | 10.2 | 10.1 | 9.89 | 9.72 | 9.55 | 9.47 | 9.38 | 9.29 | 9.20 | 9.11 | 9.46 |
| 6 | 13.8 | 10.9 | 9.78 | 9.15 | 8.75 | 8.47 | 8.26 | 8.10 | 7.98 | 7.87 | 7.72 | 7.56 | 7.40 | 7.31 | 7.23 | 7.14 | 7.06 | 6.97 | 6.88 |
| 7 | 12.3 | 9.55 | 8.45 | 7.85 | 7.46 | 7.19 | 6.99 | 6.84 | 6.72 | 6.62 | 6.47 | 6.31 | 6.16 | 6.07 | 5.99 | 5.91 | 5.82 | 5.74 | 5.65 |
| 8 | 11.3 | 8.65 | 7.59 | 7.01 | 6.63 | 6.37 | 6.18 | 6.03 | 5.91 | 5.81 | 5.67 | 5.52 | 5.36 | 5.28 | 5.20 | 5.12 | 5.03 | 4.95 | 4.86 |
| 9 | 10.6 | 8.02 | 6.99 | 6.42 | 6.06 | 5.80 | 5.61 | 5.47 | 5.35 | 5.26 | 5.11 | 4.96 | 4.81 | 4.73 | 4.65 | 4.57 | 4.48 | 4.40 | 4.31 |
| 10 | 10.0 | 7.56 | 6.55 | 5.99 | 5.64 | 5.39 | 5.20 | 5.06 | 4.94 | 4.85 | 4.71 | 4.56 | 4.41 | 4.33 | 4.25 | 4.17 | 4.08 | 4.00 | 3.91 |
| 11 | 9.65 | 7.21 | 6.22 | 5.67 | 5.32 | 5.07 | 4.89 | 4.74 | 4.63 | 4.54 | 4.40 | 4.25 | 4.10 | 4.02 | 3.94 | 3.86 | 3.78 | 3.69 | 3.60 |
| 12 | 9.33 | 6.93 | 5.95 | 5.41 | 5.06 | 4.82 | 4.64 | 4.50 | 4.39 | 4.30 | 4.16 | 4.01 | 3.86 | 3.78 | 3.70 | 3.62 | 3.54 | 3.45 | 3.36 |
| 13 | 9.07 | 6.70 | 5.74 | 5.21 | 4.86 | 4.62 | 4.44 | 4.30 | 4.19 | 4.10 | 3.96 | 3.82 | 3.66 | 3.59 | 3.51 | 3.43 | 3.34 | 3.25 | 3.17 |
| 14 | 8.86 | 6.51 | 5.56 | 5.04 | 4.69 | 4.46 | 4.28 | 4.14 | 4.03 | 3.94 | 3.80 | 3.66 | 3.51 | 3.43 | 3.35 | 3.27 | 3.18 | 3.09 | 3.00 |
| 15 | 8.68 | 6.36 | 5.42 | 4.89 | 4.56 | 4.32 | 4.14 | 4.00 | 3.89 | 3.80 | 3.67 | 3.52 | 3.37 | 3.29 | 3.21 | 3.13 | 3.05 | 2.96 | 2.87 |
| 16 | 8.53 | 6.23 | 5.29 | 4.77 | 4.44 | 4.20 | 4.03 | 3.89 | 3.78 | 3.69 | 3.55 | 3.41 | 3.26 | 3.18 | 3.10 | 3.02 | 2.93 | 2.84 | 2.75 |
| 17 | 8.40 | 6.11 | 5.18 | 4.67 | 4.34 | 4.10 | 3.93 | 3.79 | 3.68 | 3.59 | 3.46 | 3.31 | 3.16 | 3.08 | 3.00 | 2.92 | 2.83 | 2.75 | 2.65 |
| 18 | 8.29 | 6.01 | 5.09 | 4.58 | 4.25 | 4.01 | 3.84 | 3.71 | 3.60 | 3.51 | 3.37 | 3.23 | 3.08 | 3.00 | 2.92 | 2.84 | 2.75 | 2.66 | 2.57 |
| 19 | 8.18 | 5.93 | 5.01 | 4.50 | 4.17 | 3.94 | 3.77 | 3.63 | 3.52 | 3.43 | 3.30 | 3.15 | 3.00 | 2.92 | 2.84 | 2.76 | 2.67 | 2.58 | 2.49 |
| 20 | 8.10 | 5.85 | 4.94 | 4.43 | 4.10 | 3.87 | 3.70 | 3.56 | 3.46 | 3.37 | 3.23 | 3.09 | 2.94 | 2.86 | 2.78 | 2.69 | 2.61 | 2.52 | 2.42 |
| 21 | 8.02 | 5.78 | 4.87 | 4.37 | 4.04 | 3.81 | 3.64 | 3.51 | 3.40 | 3.31 | 3.17 | 3.03 | 2.88 | 2.80 | 2.72 | 2.64 | 2.55 | 2.46 | 2.36 |
| 22 | 7.95 | 5.72 | 4.82 | 4.31 | 3.99 | 3.76 | 3.59 | 3.45 | 3.35 | 3.26 | 3.12 | 2.98 | 2.83 | 2.75 | 2.67 | 2.58 | 2.50 | 2.40 | 2.31 |
| 23 | 7.88 | 5.66 | 4.76 | 4.26 | 3.94 | 3.71 | 3.54 | 3.41 | 3.30 | 3.21 | 3.07 | 2.93 | 2.78 | 2.70 | 2.62 | 2.54 | 2.45 | 2.35 | 2.26 |
| 24 | 7.82 | 5.61 | 4.72 | 4.22 | 3.90 | 3.67 | 3.50 | 3.36 | 3.26 | 3.17 | 3.03 | 2.89 | 2.74 | 2.66 | 2.58 | 2.49 | 2.40 | 2.31 | 2.21 |
| 25 | 7.77 | 5.57 | 4.68 | 4.18 | 3.85 | 3.63 | 3.46 | 3.32 | 3.22 | 3.13 | 2.99 | 2.85 | 2.70 | 2.62 | 2.54 | 2.45 | 2.36 | 2.27 | 2.17 |
| 30 | 7.56 | 5.39 | 4.51 | 4.02 | 3.70 | 3.47 | 3.30 | 3.17 | 3.07 | 2.98 | 2.84 | 2.70 | 2.55 | 2.47 | 2.39 | 2.30 | 2.21 | 2.11 | 2.01 |
| 40 | 7.31 | 5.18 | 4.31 | 3.83 | 3.51 | 3.29 | 3.12 | 2.99 | 2.89 | 2.80 | 2.66 | 2.52 | 2.37 | 2.29 | 2.20 | 2.11 | 2.02 | 1.92 | 1.80 |
| 60 | 7.08 | 4.98 | 4.13 | 3.65 | 3.34 | 3.12 | 2.95 | 2.82 | 2.72 | 2.63 | 2.50 | 2.35 | 2.20 | 2.12 | 2.03 | 1.94 | 1.84 | 1.73 | 1.60 |
| 120 | 6.85 | 4.79 | 3.95 | 3.48 | 3.17 | 2.96 | 2.79 | 2.66 | 2.56 | 2.47 | 2.34 | 2.19 | 2.03 | 1.95 | 1.86 | 1.76 | 1.66 | 1.53 | 1.38 |
| Inf. | 6.64 | 4.61 | 3.78 | 3.32 | 3.02 | 2.80 | 2.64 | 2.51 | 2.41 | 2.32 | 2.19 | 2.04 | 1.88 | 1.79 | 1.70 | 1.59 | 1.48 | 1.33 | 1.00 |

## TABLE 7

### Wilcoxon, Mann–Whitney, Cumulative Distribution Function

For $mX_i$'s, $nY_i$'s, $W_{XY} = \sum_{ij} I[X_i \le Y_j] = W - n(n+1)$, where $W$ is the sum of the ranks of the $Y_j$'s. The table entries are $F(k; m, n) = P(W_{XY} \le k)$ for the case that the ranks of the $Y_j$'s are a simple random sample of the integers $1, 2, \ldots, N = m + n$. Since $F(k; m, n) = F(k; n, m)$ and $F(k; m, n) = P(W_{XY} \ge mn - k) = 1 - P(W_{XY} \le mn - k - 1)$, $F(k; m, n)$ is given for only some combinations of $k$, $m$, $n$. For example, for $m = 4$, $n = 3$, $P(W_{XY} \le 9) = P(W_{XY} \ge 3) = 1 - P(W_{XY} \le 2) = 1 - 0.114 = 0.886 = P(W \le 15)$. Even for relatively small $m$ and $n$ the normal approximation $P(W_{XY} \le k) \doteq \Phi((k + 1/2 - mn/2)/\sqrt{(N+1)mn/12})$ provides good approximations. For $m = 7$, $n = 6$, the normal approximation gives 0.133, whereas the table value is 0.147.

To shorten the tables probabilities are given only for some $k$ when $n \ge 6$.

$n = 2$

| $k$ \ $m$ | 2 | 3 | 4 | 5 | 6 | 7 | 8 | 9 | 10 |
|---|---|---|---|---|---|---|---|---|---|
| 0 | .167 | .100 | .067 | .048 | .036 | .028 | .022 | .018 | .015 |
| 1 | .333 | .200 | .133 | .095 | .071 | .056 | .044 | .036 | .030 |
| 2 | .667 | .4 | .267 | .190 | .143 | .111 | .089 | .073 | .061 |
| 3 | .833 | .6 | .400 | .286 | .214 | .167 | .133 | .109 | .091 |
| 4 | 1.00 | .8 | .600 | .429 | .321 | .250 | .200 | .164 | .136 |
| 5 | 1.00 | .9 | .733 | .571 | .429 | .333 | .267 | .218 | .182 |
| 6 | 1.00 | 1.00 | .867 | .714 | .571 | .444 | .356 | .291 | .242 |
| 7 | 1.00 | 1.00 | .933 | .810 | .679 | .556 | .444 | .364 | .303 |
| 8 | 1.00 | 1.00 | 1.00 | .905 | .786 | .667 | .556 | .454 | .379 |
| 9 | 1.00 | 1.00 | 1.00 | .952 | .857 | .750 | .644 | .546 | .454 |
| 10 | 1.00 | 1.00 | 1.00 | 1.00 | .929 | .833 | .733 | .636 | .546 |

$n = 3$

| $k$ \ $m$ | 3 | 4 | 5 | 6 | 7 | 8 | 9 | 10 | 11 |
|---|---|---|---|---|---|---|---|---|---|
| 0 | .050 | .029 | .018 | .012 | .008 | .006 | .005 | .003 | .003 |
| 1 | .100 | .057 | .036 | .024 | .017 | .012 | .009 | .007 | .005 |
| 2 | .200 | .114 | .071 | .048 | .033 | .024 | .018 | .014 | .011 |
| 3 | .350 | .200 | .125 | .083 | .058 | .042 | .032 | .024 | .019 |
| 4 | .500 | .314 | .196 | .131 | .092 | .067 | 050 | 038 | 030 |
| 5 | .650 | .429 | .286 | .190 | .133 | .097 | .073 | .056 | .044 |
| 6 | .800 | .571 | .393 | .274 | .192 | .139 | .105 | .080 | .063 |
| 7 | .90 | .686 | .500 | .357 | .258 | .188 | .141 | .108 | .085 |
| 8 | .95 | .800 | .607 | .452 | .333 | .248 | .186 | .143 | .113 |
| 9 | 1.00 | .886 | .714 | .548 | .417 | .315 | .241 | .185 | .146 |

**$n = 3$**

| | | | | | $m$ | | | | |
|---|---|---|---|---|---|---|---|---|---|
| | 4 | 5 | 6 | 7 | 8 | 9 | 10 | 11 | 12 |
| 10 | 1.00 | .943 | .804 | .643 | .500 | .388 | .300 | .234 | .184 |
| 11 | 1.00 | .971 | .875 | .726 | .583 | .461 | .364 | .287 | .228 |
| 12 | 1.00 | 1.00 | .929 | .810 | .667 | .539 | .432 | .346 | .277 |
| $k$ 13 | 1.00 | 1.00 | .964 | .869 | .742 | .612 | .500 | .406 | .330 |
| 14 | 1.00 | 1.00 | .982 | .917 | .808 | .685 | .568 | .469 | .385 |
| 15 | 1.00 | 1.00 | 1.00 | .952 | .867 | .752 | .636 | .531 | .442 |
| 16 | 1.00 | 1.00 | 1.00 | .976 | .908 | .812 | .700 | .594 | .500 |

**$n = 4$**

| | | | | | $m$ | | | | |
|---|---|---|---|---|---|---|---|---|---|
| | 4 | 5 | 6 | 7 | 8 | 9 | 10 | 11 | 12 |
| 0 | .014 | .008 | .005 | .003 | .002 | .001 | .001 | .001 | .001 |
| 1 | .029 | .016 | .010 | .006 | .004 | .003 | .002 | .001 | .001 |
| 2 | .057 | .032 | .019 | .012 | .080 | .006 | .004 | .003 | .002 |
| 3 | .100 | .056 | .033 | .021 | .014 | .010 | .007 | .005 | .004 |
| 4 | .171 | .095 | .057 | .036 | .024 | .017 | .012 | .009 | .007 |
| 5 | .243 | .143 | .086 | .055 | .036 | .025 | .018 | .013 | .010 |
| 6 | .343 | .206 | .129 | .082 | .055 | .038 | .027 | .020 | .015 |
| 7 | .443 | .278 | .176 | .115 | .077 | .053 | .038 | .028 | .021 |
| 8 | .557 | .365 | .238 | .158 | .107 | .074 | .053 | .039 | .029 |
| 9 | .657 | .452 | .305 | .206 | .141 | .099 | .071 | .052 | .039 |
| 10 | .757 | .548 | .381 | .264 | .184 | .130 | .094 | .069 | .052 |
| 11 | .829 | .635 | .457 | .324 | .230 | .165 | .120 | .089 | .066 |
| $k$ 12 | .900 | .722 | .543 | .394 | .285 | .207 | .152 | .113 | .085 |
| 13 | .943 | .794 | .619 | .464 | .341 | .252 | .187 | .140 | .106 |
| 14 | .971 | .857 | .695 | .536 | .404 | .302 | .227 | .171 | .131 |
| 15 | .986 | .905 | .762 | .606 | .467 | .355 | .270 | .206 | .158 |
| 16 | 1.00 | .944 | .824 | .676 | .533 | .413 | .318 | .245 | .190 |
| 17 | 1.00 | .968 | .871 | .736 | .596 | .470 | .367 | .286 | .223 |
| 18 | 1.00 | .984 | .914 | .794 | .659 | .530 | .420 | .330 | .260 |
| 19 | 1.00 | .992 | .943 | .842 | .715 | .587 | .473 | .377 | .299 |
| 20 | 1.00 | 1.00 | .967 | .885 | .770 | .645 | .527 | .426 | .342 |
| 21 | 1.00 | 1.00 | .981 | .918 | .816 | .698 | .580 | .475 | .385 |
| 22 | 1.00 | 1.00 | .990 | .945 | .859 | .748 | .633 | .525 | .431 |
| 23 | 1.00 | 1.00 | .995 | .964 | .893 | .793 | .682 | .574 | .476 |
| 24 | 1.00 | 1.00 | 1.00 | .979 | .923 | .835 | .730 | .623 | .524 |

$n = 5$

| | m: 5 | 6 | 7 | 8 | 9 | 10 | 11 | 12 | 13 |
|---|---|---|---|---|---|---|---|---|---|
| 0 | .004 | .002 | .001 | .001 | .000 | .000 | .000 | .000 | .000 |
| 1 | .008 | .004 | .003 | .002 | .001 | .001 | .000 | .000 | .000 |
| 2 | .016 | .009 | .005 | .003 | .002 | .001 | .001 | .001 | .000 |
| 3 | .028 | .015 | .009 | .005 | .003 | .002 | .002 | .001 | .001 |
| 4 | .048 | .026 | .015 | .009 | .006 | .004 | .003 | .002 | .001 |
| 5 | .075 | .041 | .024 | .015 | .009 | .006 | .004 | .003 | .002 |
| 6 | .111 | .063 | .037 | .023 | .014 | .010 | .007 | .005 | .003 |
| 7 | .155 | .089 | .053 | .033 | .021 | .014 | .010 | .007 | .005 |
| 8 | .210 | .123 | .074 | .047 | .030 | .020 | .014 | .010 | .007 |
| 9 | .274 | .165 | .101 | .064 | .041 | .028 | .019 | .013 | .010 |
| 10 | .345 | .214 | .134 | .085 | .056 | .038 | .026 | .018 | .013 |
| 11 | .421 | .268 | .172 | .111 | .073 | .050 | .034 | .024 | .018 |
| 12 | .500 | .331 | .216 | .142 | .095 | .065 | .045 | .032 | .023 |
| 13 | .579 | .396 | .265 | .177 | .120 | .082 | .057 | .041 | .030 |
| 14 | .655 | .465 | .319 | .218 | .149 | .103 | .073 | .052 | .038 |
| 15 | .726 | .535 | .378 | .262 | .182 | .127 | .090 | .065 | .047 |
| k 16 | .790 | .604 | .438 | .311 | .219 | .155 | .111 | .080 | .059 |
| 17 | .845 | .669 | .500 | .362 | .259 | .185 | .134 | .097 | .072 |
| 18 | 889 | .732 | .562 | .416 | .303 | .220 | .160 | .117 | .087 |
| 19 | .925 | .786 | .622 | .472 | .350 | .257 | .189 | .139 | .104 |
| 20 | .952 | .835 | .681 | .528 | .399 | .297 | .220 | .164 | .123 |
| 21 | .972 | .877 | .735 | .584 | .449 | .339 | .255 | .191 | .144 |
| 22 | .984 | .911 | .784 | .638 | .500 | .384 | .292 | .221 | .168 |
| 23 | .992 | .937 | .828 | .689 | .551 | .430 | .331 | .253 | .194 |
| 24 | .996 | .959 | .866 | .738 | .601 | .477 | .371 | .287 | .222 |
| 25 | 1.00 | .974 | .899 | .782 | .650 | .523 | .413 | .323 | .251 |
| 26 | 1.00 | .985 | .926 | .823 | .697 | .570 | .457 | .361 | .283 |
| 27 | 1.00 | .991 | .947 | .858 | .741 | .616 | .500 | .399 | .317 |
| 28 | 1.00 | .996 | .963 | .889 | .781 | .661 | .543 | .439 | .351 |
| 29 | 1.00 | .998 | .976 | .915 | .818 | .703 | .587 | .480 | .387 |
| 30 | 1.00 | 1.00 | .985 | .936 | .851 | .743 | .629 | .520 | .424 |
| 31 | 1.00 | 1.00 | .991 | .953 | .880 | .780 | .669 | .561 | .462 |
| 32 | 1.00 | 1.00 | .995 | .967 | .905 | .815 | .708 | .601 | .500 |

$n = 6$

| | | m | | | | | | | | |
|---|---|---|---|---|---|---|---|---|---|---|
| | | 6 | 7 | 8 | 9 | 10 | 11 | 12 | 13 | 14 |
| | 0 | .001 | .001 | .000 | .000 | .000 | .000 | .000 | .000 | .000 |
| | 1 | .002 | .001 | .001 | .000 | .000 | .000 | .000 | .000 | .000 |
| | 2 | .004 | .002 | .001 | .001 | .000 | .000 | .000 | .000 | .000 |
| | 3 | .008 | .004 | .002 | .001 | .001 | .001 | .000 | .000 | .000 |
| | 4 | .013 | .007 | .004 | .001 | .001 | .001 | .000 | .000 | .000 |
| | 5 | .021 | .011 | .006 | .004 | .002 | .002 | .001 | .001 | .000 |
| | 6 | .032 | 017 | .100 | .006 | .004 | .002 | .002 | .001 | .001 |
| | 7 | .047 | .026 | .015 | .009 | .005 | .004 | .002 | .002 | .001 |
| | 8 | .066 | .037 | .021 | .013 | .008 | .005 | .003 | .002 | .002 |
| | 9 | .090 | .051 | .030 | .018 | .011 | .007 | .005 | .003 | .002 |
| | 10 | .120 | .069 | .041 | .025 | .016 | .010 | .007 | .005 | .003 |
| | 11 | .155 | .090 | .054 | .033 | .021 | .014 | .009 | .006 | .004 |
| | 12 | .197 | .117 | .071 | .044 | .028 | .018 | .012 | .008 | .006 |
| | 13 | .242 | .147 | .091 | .057 | .036 | .024 | .016 | .011 | .008 |
| | 14 | .294 | .183 | .114 | .072 | .047 | .031 | .021 | .014 | .010 |
| | 15 | .335 | .223 | .141 | .091 | .059 | .039 | .026 | .018 | .013 |
| | 16 | .409 | .267 | .172 | .112 | .074 | .049 | .033 | .023 | .016 |
| | 17 | .469 | .314 | .207 | .136 | .090 | .061 | .042 | .029 | .020 |
| | 18 | .531 | .365 | .245 | .164 | .110 | .074 | .051 | .036 | .025 |
| | 19 | .591 | .418 | .286 | .194 | .132 | .090 | .062 | .044 | .031 |
| | 20 | .650 | .473 | .331 | .228 | .157 | .108 | .075 | .053 | .038 |
| k | 21 | .706 | .527 | .377 | .264 | .184 | .128 | .090 | .064 | .046 |
| | 22 | .758 | .582 | .426 | .303 | .214 | .151 | .106 | .076 | .055 |
| | 23 | .803 | .635 | .475 | .344 | .246 | .175 | .125 | .090 | .065 |
| | 24 | .845 | .686 | .525 | .388 | .281 | .202 | .145 | .105 | .076 |
| | 25 | .880 | .733 | .574 | .432 | .318 | .231 | .168 | .122 | .089 |
| | 26 | .910 | .777 | .623 | .477 | .356 | .262 | .192 | .141 | .104 |
| | 27 | .934 | .817 | .669 | .523 | .396 | .295 | .219 | .161 | .120 |
| | 28 | .953 | .853 | .714 | .568 | .437 | .330 | .247 | .184 | .137 |
| | 29 | .968 | .883 | .755 | .612 | .479 | .366 | .277 | .208 | .156 |
| | 30 | .979 | .910 | .793 | .656 | .521 | .404 | .308 | .234 | .177 |
| | 31 | .987 | .931 | .828 | .697 | .563 | .442 | .341 | .261 | .199 |
| | 32 | .992 | .949 | .859 | .736 | .604 | .481 | .375 | .289 | .222 |
| | 33 | .996 | .963 | .886 | .772 | .644 | .519 | .410 | .319 | .247 |
| | 34 | .998 | .974 | .909 | .806 | .682 | .558 | .446 | .351 | .273 |
| | 35 | .999 | .983 | .929 | .836 | .719 | .596 | .482 | .383 | .301 |
| | 36 | 1.00 | .989 | .946 | .864 | .754 | .634 | .518 | .416 | .329 |
| | 37 | 1.00 | .993 | .959 | .888 | .786 | .670 | .554 | .449 | .359 |
| | 38 | 1.00 | .996 | .970 | .909 | .816 | .705 | .590 | .483 | .390 |
| | 39 | 1.00 | .998 | .979 | .928 | .843 | .738 | .625 | .517 | .421 |
| | 40 | 1.00 | .999 | .985 | .943 | .868 | .769 | .659 | .551 | .452 |
| | 41 | 1.00 | .999 | .990 | .956 | .890 | .798 | .692 | .584 | .484 |
| | 42 | 1.00 | 1.000 | .994 | .967 | .910 | .825 | .723 | .617 | .516 |

**$n = 7$**

| | *m* | | | | | | | | |
|---|---|---|---|---|---|---|---|---|---|
| *k* | 7 | 8 | 9 | 10 | 11 | 12 | 13 | 14 | 15 |
| 2 | .001 | .001 | .000 | .000 | .000 | .000 | .000 | .000 | .000 |
| 3 | .002 | .001 | .001 | .000 | .000 | .000 | .000 | .000 | .000 |
| 4 | .003 | .002 | .001 | .001 | .000 | .000 | .000 | .000 | .000 |
| 5 | .013 | .007 | .004 | .002 | .001 | .001 | .001 | .000 | .000 |
| 6 | .019 | .010 | .006 | .003 | .002 | .001 | .001 | .001 | .000 |
| 7 | .027 | .014 | .008 | .005 | .003 | .002 | .001 | .001 | .001 |
| 8 | .036 | .020 | .011 | .007 | .004 | .003 | .002 | .001 | .001 |
| 9 | .049 | .027 | .016 | .009 | .006 | .004 | .002 | .002 | .001 |
| 10 | .064 | .036 | .021 | .012 | .008 | .005 | .003 | .002 | .001 |
| 11 | .082 | .047 | .027 | .017 | .010 | .006 | .004 | .003 | .002 |
| 12 | .104 | .060 | .036 | .022 | .013 | .009 | .006 | .004 | .003 |
| 13 | .130 | .076 | .045 | .028 | .017 | .011 | .007 | .005 | .003 |
| 14 | .159 | .095 | .057 | .035 | .022 | .014 | .009 | .006 | .004 |
| 15 | .191 | .116 | .071 | .044 | .028 | .018 | .012 | .008 | .005 |
| 16 | .228 | .140 | .087 | .054 | .035 | .022 | .015 | .010 | .007 |
| 17 | .267 | .168 | .105 | .067 | .043 | .028 | .018 | .012 | .009 |
| 18 | .310 | .198 | .126 | .081 | .052 | .034 | .023 | .015 | .011 |
| 19 | .355 | .232 | .150 | .097 | .063 | .042 | .028 | .019 | .013 |
| 20 | .402 | .268 | .176 | .115 | .075 | .050 | .034 | .023 | .016 |
| 21 | .451 | .306 | .204 | .135 | 090 | 060 | 041 | .028 | .019 |
| 22 | .500 | .347 | .235 | .157 | .105 | .071 | .048 | .033 | .023 |
| 23 | .549 | .389 | .268 | .182 | .123 | .084 | .057 | .040 | .028 |
| 24 | .598 | .433 | .303 | .209 | .143 | .098 | .067 | .047 | .033 |
| 25 | .645 | .478 | .340 | .237 | .164 | .113 | .079 | .055 | .039 |
| 26 | .690 | .522 | .379 | .268 | .187 | .131 | .091 | .064 | .046 |
| 27 | .733 | .567 | .419 | .300 | .213 | .150 | .105 | .074 | .053 |
| 28 | .772 | .611 | .459 | .335 | .239 | .170 | .121 | .086 | .061 |
| 29 | .809 | .653 | .500 | .370 | .268 | .192 | .137 | .098 | .071 |
| 30 | .841 | .694 | .541 | .406 | .298 | .216 | .156 | .112 | .081 |
| 31 | .870 | .732 | .581 | .443 | .329 | .241 | .175 | .127 | .093 |
| 32 | .896 | .768 | .621 | .481 | .362 | .268 | .196 | .144 | .105 |

**$n = 8$**

| $k$ | $m$ = 8 | 9 | 10 | 11 |
|---|---|---|---|---|
| 3 | .001 | .000 | .000 | .000 |
| 4 | .001 | .000 | .000 | .000 |
| 5 | .001 | .001 | .000 | .000 |
| 6 | .002 | .001 | .001 | .000 |
| 7 | .003 | .002 | .001 | .001 |
| 8 | .005 | .003 | .002 | .001 |
| 9 | .007 | .004 | .002 | .001 |
| 10 | .010 | .006 | .003 | .002 |
| 11 | .014 | .008 | .004 | .002 |
| 12 | .019 | .010 | .006 | .003 |
| 13 | .025 | .014 | .008 | .005 |
| 14 | .032 | .018 | .010 | .006 |
| 15 | .041 | .023 | .013 | .008 |
| 16 | .052 | .030 | .017 | .010 |
| 17 | .065 | .037 | .022 | .013 |
| 18 | .080 | .046 | .027 | .016 |
| 19 | .097 | .057 | .034 | .020 |
| 20 | .117 | .069 | .042 | .025 |
| 21 | .139 | .084 | .051 | .031 |
| 22 | .164 | .100 | .061 | .038 |
| 23 | .191 | .118 | .073 | .045 |
| 24 | .221 | .138 | .086 | .054 |
| 25 | .253 | .161 | .102 | .064 |
| 26 | .287 | .185 | .118 | .076 |
| 27 | .323 | .212 | .137 | .089 |
| 28 | .360 | .240 | .158 | .103 |
| 29 | .399 | .271 | .180 | .119 |

**$n = 9$**

| $k$ | $m$ = 9 | 10 | 11 |
|---|---|---|---|
| 6 | .001 | .000 | .000 |
| 7 | .001 | .000 | .000 |
| 8 | .001 | .001 | .000 |
| 9 | .002 | .001 | .001 |
| 10 | .003 | .001 | .001 |
| 11 | .004 | .002 | .001 |
| 12 | .005 | .003 | .002 |
| 13 | .007 | .004 | .002 |
| 14 | .009 | .005 | .003 |
| 15 | .012 | .007 | .004 |
| 16 | .016 | .009 | .005 |
| 17 | .020 | .011 | .006 |
| 18 | .025 | .014 | .008 |
| 19 | .031 | .017 | .010 |
| 20 | .039 | .022 | .013 |
| 21 | .047 | .027 | .016 |
| 22 | .057 | .033 | .019 |
| 23 | .068 | .039 | .023 |
| 24 | .081 | .047 | .028 |
| 25 | .095 | .056 | .034 |
| 26 | .111 | .067 | .040 |
| 27 | .129 | .078 | .048 |
| 28 | .149 | .091 | .056 |
| 29 | .170 | .106 | .065 |
| 30 | .193 | .121 | .076 |
| 31 | .218 | .139 | .088 |
| 32 | .245 | .158 | .101 |

## TABLE 8

### Wilcoxon Signed Rank Statistic $W^+$

The table values are $P(W^+ \leq k)$ for the case that $W^+$ has the distribution of $\sum_{j=1}^{N} j V_j$, where the $V_j$ are Bernoulli(1/2), independent.

$W^+$ is symmetric around $\mu_N = N(N+1)/4$. That is, $P(W^+ \geq [N(N+1)/4] + u) = P(W^+ \leq [N(N+1)/4] - u)$ for each $u$. For larger $N$ the approximation $P(X \leq k) \doteq \Phi((k + 1/2 - \mu_N)/\sigma_N)$, for $\sigma_N^2 = N(N+1)(2N+1)/24$ works well.

For example, for $N = 9$, $P(X \leq 26) \doteq \Phi((26.5 - 22.5)/8.44) = \Phi(0.474) = 0.682$. The table value is 0.674.

| $k$ \ $N$ | 2 | 3 | 4 | 5 | 6 | 7 | 8 | 9 | 10 |
|---|---|---|---|---|---|---|---|---|---|
| 0 | .25 | .125 | .062 | .031 | .016 | .008 | .004 | .002 | .001 |
| 1 | .50 | .250 | .125 | .062 | .03 | .016 | .008 | .004 | .002 |
| 2 | .75 | .375 | .188 | .094 | .047 | .023 | .012 | .006 | .003 |
| 3 | 1.00 | .625 | .312 | .156 | .078 | .039 | .020 | .010 | .005 |
| 4 | 1.00 | .750 | .438 | .219 | .109 | .055 | .027 | .014 | .007 |
| 5 | 1.00 | .875 | .562 | .312 | .156 | .078 | .039 | .020 | .010 |
| 6 | 1.00 | 1.000 | .688 | .406 | .219 | .109 | .055 | .027 | .014 |
| 7 | 1.00 | 1.000 | .812 | .500 | .281 | .148 | .074 | .037 | .019 |
| 8 | 1.00 | 1.000 | .875 | .594 | .344 | .188 | .098 | .049 | .024 |
| 9 | 1.00 | 1.000 | .938 | .688 | .422 | .234 | .125 | .064 | .032 |
| 10 | 1.00 | 1.000 | 1.000 | .781 | .500 | .289 | .156 | .082 | .042 |
| 11 | 1.00 | 1.000 | 1.000 | .844 | .578 | .344 | .191 | .102 | .053 |
| 12 | 1.00 | 1.000 | 1.000 | .906 | .656 | .406 | .230 | .125 | .065 |
| 13 | 1.00 | 1.000 | 1.000 | .938 | .719 | .469 | .273 | .150 | .080 |
| 14 | 1.00 | 1.000 | 1.000 | .969 | .781 | .531 | .320 | .180 | .097 |
| 15 | 1.00 | 1.000 | 1.000 | 1.000 | .844 | .594 | .371 | .213 | .116 |
| 16 | 1.00 | 1.000 | 1.000 | 1.000 | .891 | .656 | .422 | .248 | .138 |
| 17 | 1.00 | 1.000 | 1.000 | 1.000 | .922 | .711 | .473 | .285 | .161 |
| 18 | 1.00 | 1.000 | 1.000 | 1.000 | .953 | .766 | .527 | .326 | .188 |
| 19 | 1.00 | 1.000 | 1.000 | 1.000 | .969 | .812 | .578 | .367 | .216 |
| 20 | 1.00 | 1.000 | 1.000 | 1.000 | .984 | .852 | .629 | .410 | .246 |
| 21 | 1.00 | 1.000 | 1.000 | 1.000 | 1.000 | .891 | .680 | .455 | .278 |
| 22 | 1.00 | 1.000 | 1.000 | 1.000 | 1.000 | .922 | .727 | .500 | .312 |
| 23 | 1.00 | 1.000 | 1.000 | 1.000 | 1.000 | .945 | .770 | .545 | .348 |
| 24 | 1.00 | 1.000 | 1.000 | 1.000 | 1.000 | .961 | .809 | .590 | .385 |
| 25 | 1.00 | 1.000 | 1.000 | 1.000 | 1.000 | .977 | .844 | .633 | .423 |
| 26 | 1.00 | 1.000 | 1.000 | 1.000 | 1.000 | .984 | .875 | .674 | .461 |
| 27 | 1.00 | 1.000 | 1.000 | 1.000 | 1.000 | .992 | .902 | .715 | .500 |
| 28 | 1.00 | 1.000 | 1.000 | 1.000 | 1.000 | 1.000 | .926 | .752 | .539 |

# TABLE 9 Pearson and Hartley Charts for the Power of the $F$-Test

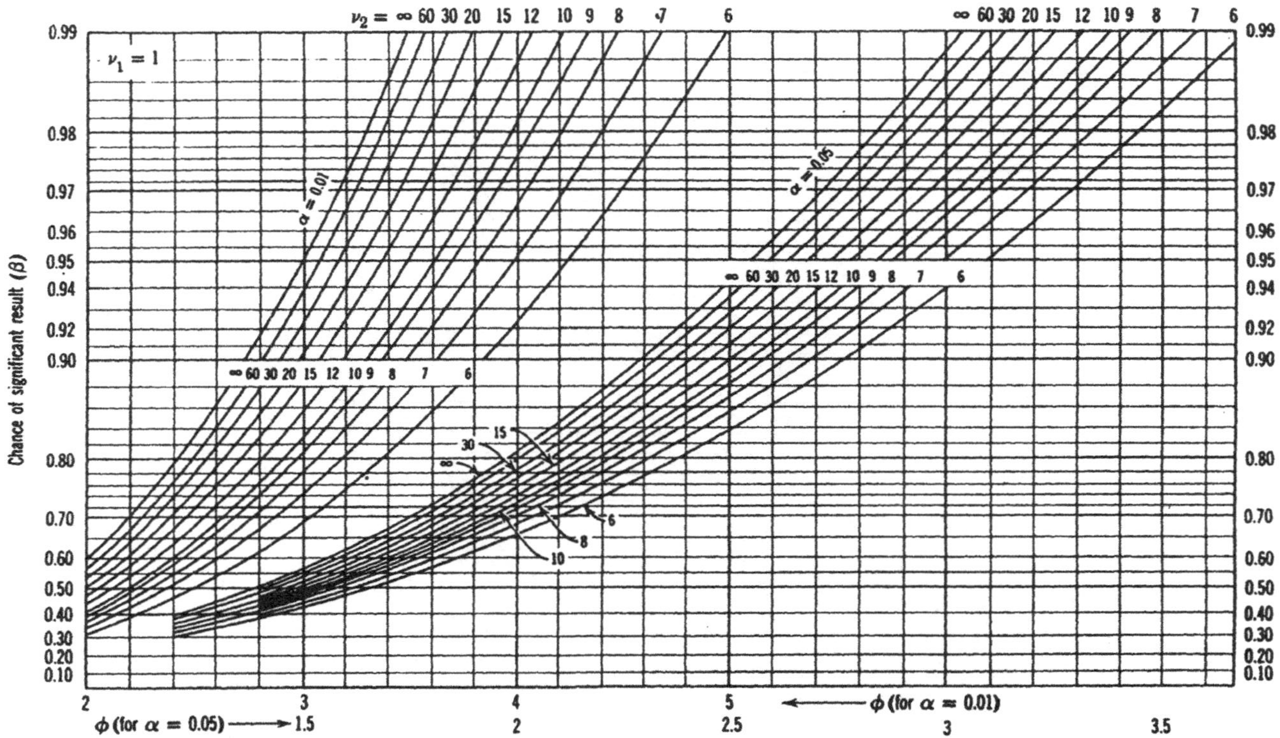


*Source:* E. S. Pearson and H. O. Hartley, *Biometrika*, Vol. 38, pp. 115–122 (1951). Reproduced with permission of the authors and the editor.

# TABLE 9 Pearson and Hartley Charts for the Power of the $F$-Test

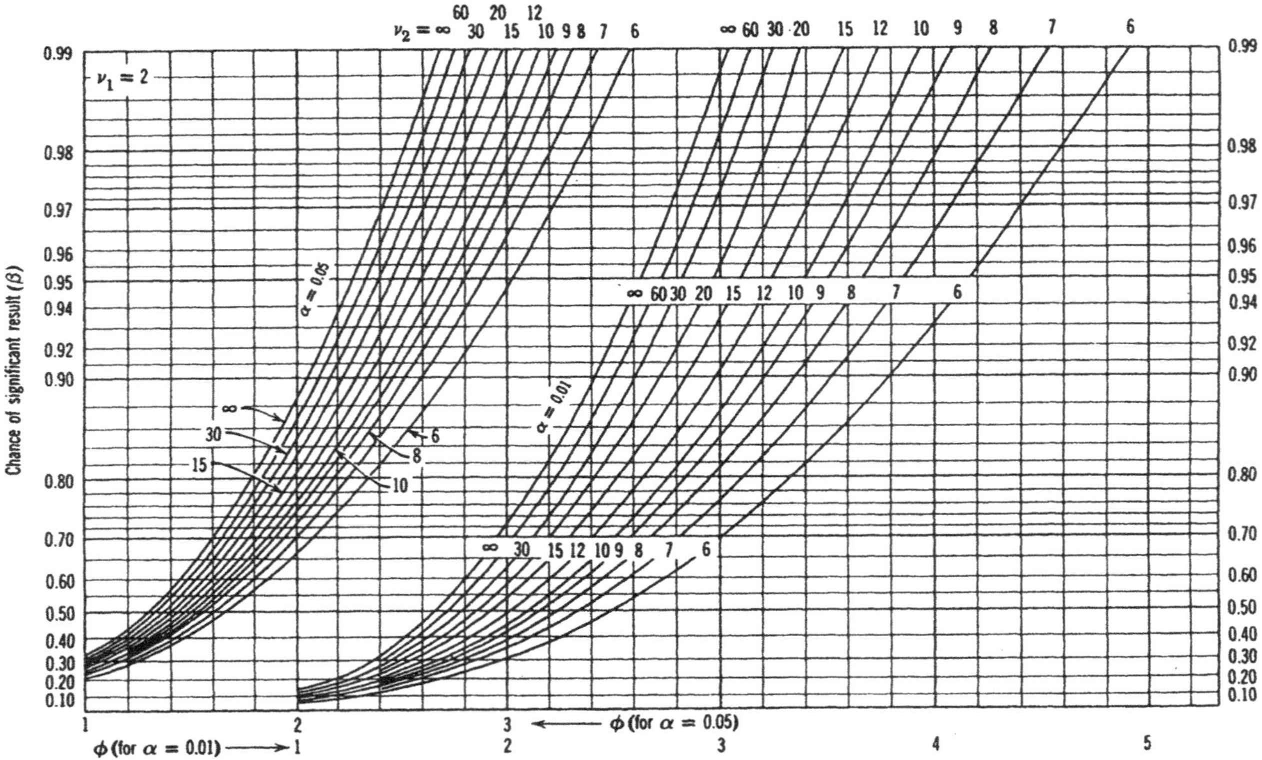


*Source:* E. S. Pearson and H. O. Hartley, *Biometrika*, Vol. 38 pp. 115–122 (1951). Reproduced with permission of the authors and the editor.

## TABLE 9 Pearson and Hartley Charts for the Power of the $F$-Test

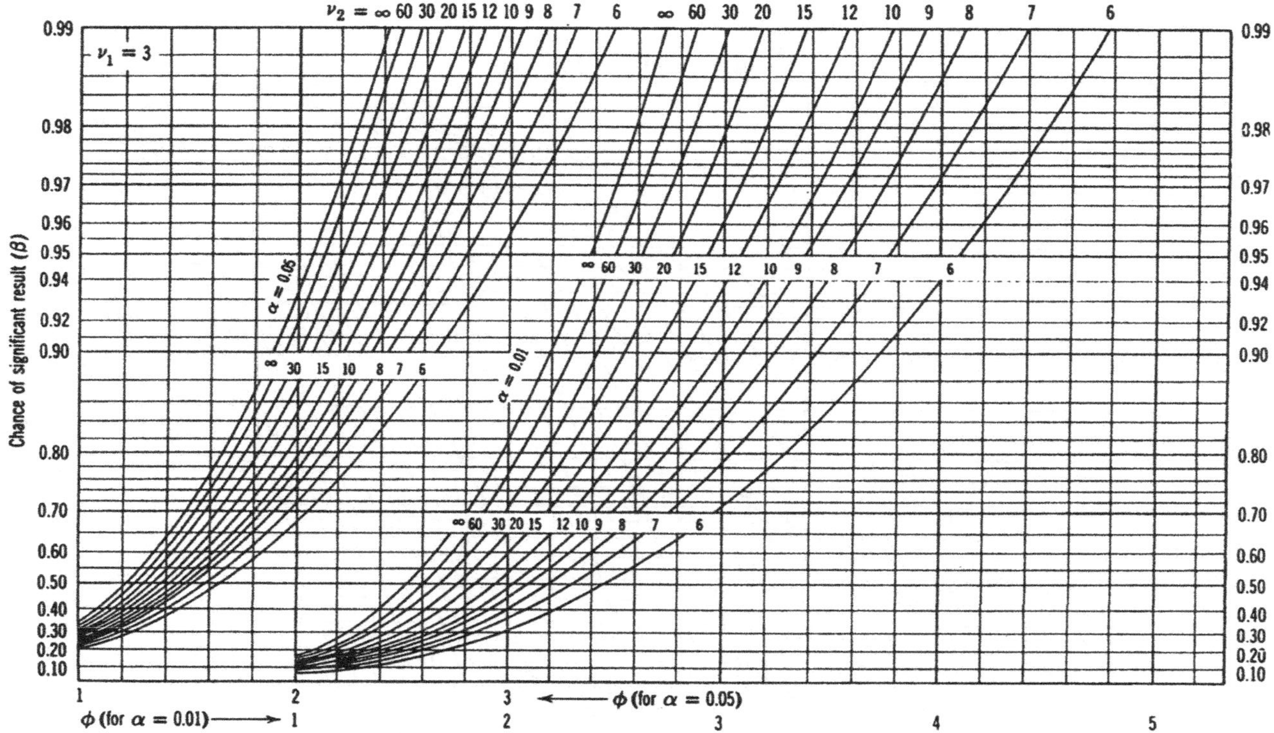


*Source:* E. S. Pearson and H. O. Hartley, *Biometrika*, Vol. 38, pp. 115–122 (1951). Reproduced with permission of the authors and the editor.

# TABLE 9 Pearson and Hartley Charts for the Power of the $F$-Test

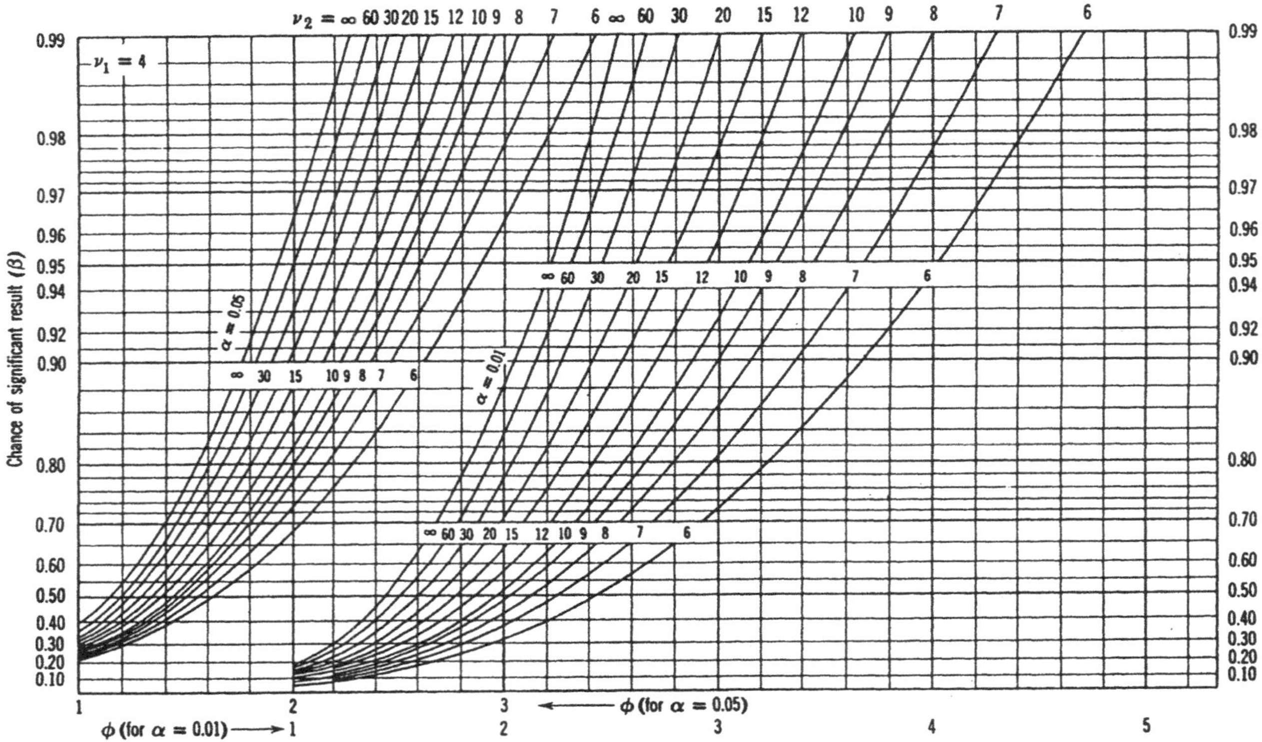


*Source:* E. S. Pearson and H. O. Hartley, *Biometrika* Vol. 38, pp. 115–122 (1951). Reproduced with permission of the authors and the editor.

# Answers to Selected Problems

## Section 1.2

**1.2.1** 0.05, 0.55, 0.45, 0.65, 0.90, 0.90.

**1.2.2** (**a**) $P$((red, red)) $= 12/35$, $P$((red,white)) $= 8/35$, $P$((white, red)) $= 9/35$, $P$((white, white)) $= 6/35$. (**b**) 6/35, 17/35, 12/35.

**1.2.3** Approximate answers: 1/64974, 1/4165, 1/536, 1/47.3, 1/2.37.

**1.2.5** (**b**) 5/16.

**1.2.7** (**a**) $1 - |7 - k|/36$ for $k = 1, 2, \ldots, 6$. (**b**) $(j/6)^2 - (j - 1)/6]^2 = (2j - 1)/36$ for $j = 1, 2, \ldots, 6$.

**1.2.8** (**a**) $1 - 55/144 = 89/144 \doteq 0.6181$ (**b**) $1 - 811,008/1,712,304 \doteq 0.5264$.

**1.2.9** Method 1: $1/3$, $1/3$, $1/3$, $3/5$; method 2: $2/9$, $1/3$, $2/3$, $1/2$.

**1.2.10** $5/8 = 0.625$.

**1.2.11** $1456/8008 = 2/11 \doteq 0.1818$.

**1.2.12** (**a**) 2/27. (**b**) 7/27.

**1.2.13** (**c**) 0.15 and 0.85.

**1.2.14** (**b**) $P$(at least one match) $\doteq 1 - e^{-1} \doteq 0.632$.

## Section 1.3

**1.3.1** 0.6, 0.7, 4/7, 4/6, 1/3.

**1.3.2**

| | Independent | mutually exclusive |
|---|---|---|
| $A, B$ | Yes | No |
| $A, C$ | No | Yes |
| $A, D$ | No | No |
| $B, C$ | No | No |
| $B, D$ | Yes | No |
| $C, D$ | No | No |

**1.3.3** (**a**) $P(\text{first has number } j, \text{ second has number } k) = (1/5)(1/j)$ for $1 \leq k \leq j, j = 1, 2, \ldots, 5$. Therefore, $P(\text{first has number } j \mid \text{second has number } k) = [(1/5)(1/j)]/[(1/5)(1/k) + (1/5)(1/(k+1) + \cdots + (1/5)(1/5)) = (1/j)/[1/k + 1/(k+1) + \cdots + (1/5)]$ for $1 \leq k \leq j$ and $j = 1, 2, \ldots, 5$. (**b**) Replace 5 by $N$.

**1.3.4** Let $A_1$ = [same on tosses 1 and 2], $A_2$ = [same on tosses 1 and 3], $A_3$ = [same on tosses 2 and 3]. Then $P(A_i) = 1/2$ for $i = 1, 2, 3$, and $P(A_i A_j) = P(\text{all three tosses are the same}) = 1/8$ for $i \neq j$, so these three events are pairwise independent, but $P(A_1 A_2 A_3) = P(\text{all three tosses are the same}) = 1/8$, so these three events are not independent.

**1.3.5** Solve the inequality $1 - (35/36)^n \geq 0.90$. Thus, $n$ must be 82 or more.

**1.3.6** (**a**) 0.8624. (**b**) place component 1 in parallel with 4, 2 in parallel with 3. (**c**) The answer is as in (**b**). It's easier to prove it by expressing the reliabilities in terms of the $q_i = 1 - p_i$ for $i = 1, 2, 3, 4$.

**1.3.7** The conditional probability that she is lying, given that she says she got $n$ heads in $n$ tosses, is greater than $1/2$ when $n$ is 7 or more. If we wish this probability to exceed $p$, take $n > \log(99/p)/\log(2)$.

**1.3.8** (**a**) 0.004099. (**b**) 0.05343 (**c**) $k$ must be 4 or more. If $k = 4$, the conditional probability is 0.9139.

**1.3.9** (**a**) 0.3292. It happened in 32 of 80 trials. The proportion was 32/80 = 0.40. (**b**) $P(Y = 0) = 0.5685$, $P(Y = 1) = 0.3449$, $P(Y = 2) = 0.0784$, $P(Y = 3) = 0.0079$, $P(Y = 4) = 0.0003$. $s(0) = 11/20$, $s(1) = 7/20$, $s(2) = 1/20$, $s(3) = 0$, $s(4) = 0$.

**1.3.10** (**a**) $\sum_{k=2}^{12}(6 - |7 - k|)/36^2) = 0.1127$. (**b**) 0.01427.

**1.3.11** In July 1978 there were 5 Saturdays, 5 Sundays, 5 Mondays, and 4 each of the other 4 days. Suppose that Patrick has probability 5/31 of having his birthday on each of Saturday, Sunday, Monday, and probability 4/31 of having his birthday on each of the other days. Suppose that the same is true for Hildegarde. The outcomes are the pairs $(d_1, d_2)$ of birthdays for Patrick and Hildegarde. Thus, the sample space $S$ has 49 outcomes with probabilities the products of the probabilities for Patrick and Hildegarde. For example, the pair (Monday, Wednesday) has probability (5/31)(4/31). The probability they were born on the same day of the week is therefore 0.1449, slightly more than 1/7.

**1.3.12** (**b**) $k_{\max}$ = (greatest integer $\le ne^{-1} \doteq 0.3679n$. The probability is then approximately $e^{-1}$.

## Section 1.4

**1.4.1** (**a**) $f_X(x) = 9/36, 12/36, 10/36, 4/36, 1/36$ for $x = 2, 3, 4, 5, 6$. $f_M(m) = 9/36, 16/36, 11/36$ for $m = 1, 2, 3$.

**1.4.2** $f(k) = 0.336, 0.452, 0.188, 0.024$ for $k = 0, 1, 2, 3$.

**1.4.3** (**a**) $p_1p_2/(1 - q_1q_2)$ for $q_1 = 1 - p_1, q_2 = 1 - p_2$. (**b**) Show that $P([W = w]) = (q_1q_2)^{w-1}[1 - q_1q_2]$ and that $P(A \cap [W = w]) = (q_1q_2)^{w-1}p_1p_2$. (**c**) $p/(2 - p)$, $1/3$, and $1/11$. (**d**) $f(y) = (y - 1)p^2q^{y-2}$ for $y = 2, 3, \ldots$.

**1.4.4** (**c**) $f_X(x) = 1/126, 20/126, 60/126, 40/126, 5/126$ for $x = 0, 1, 2, 3, 4$. $f_Y(y) = 6/126, 45/126, 60/126, 15/126$ for $y = 0, 1, 2, 3$.

**1.4.5** 0.4744.

**1.4.7** $f_M(m) = (4m^3 - 6m^2 + 4m - 1)/6^4$ for $m = 1, 2, 3, 4, 5, 6$.

**1.4.8** (**a**) $f_T(t) = 0.10584, 0.3024, 0.342, 0.1912, 0.0528$ for $t = 0, 1, 2, 3, 4, 5$. (**b**) 0.3132.

**1.4.10** (**a**) 4914. (**b**) 164.

**1.4.11** (**b**) $f_2(x_2) = 137/300, 77/300, 47/300, 27/300, 12/137$ for $x_2 = 1, 2, 3, 4, 5$. (**c**) 15/77. (**d**) $f_Y(y) = (1/300)(60, 30, 50, 35, 47, 27, 27, 12, 12)$ for $y = 2, 3, \ldots, 10$.

**1.4.12** (**a**) 0.0649. (**b**) Multinomial(10, (1/8, 3/8, 3/8, 1/8)). (**c**) $B(10, 1/2)$.

## Section 1.5

**1.5.1** (**a**) $f_X(1) = 0.2$, $f_X(2) = 0.4$, $f_X(3) = 0.3$, $f_X(4) = 0.1$, $f_Y(0) = 0.4$, $f_Y(1) = 0.5$, $f_Y(2) = 0.1$. (**b**) $E(X) = 2.3$. (**c**) $E(Y) = 0.7$. (**d**) $E(XY) = 1.9$.

**1.5.2** (**a**) $E(M) = 161/36$.

**1.5.3** (**a**) $f_R(0) = 3/15$, $f_R(1) = 9/15$, $f_R(20) = 3/15$, $f_W(0) = 6/15$, $f_W(1) = 8/15$, $f_W(2) = 1/15$. (**b**) 1, 2/3, 2/5. (**c**) $B = 2 - R - W$. (**d**) $f_W(1) = 5/15$, $f_W(2) = 10/15$, $E(W) = 5/3$.

**1.5.4** (**a**) $K = 1/15$. (**b**) $E(X) = E(Y) = 9/5$, no, no.
(**c**) $f_W(2) = 1/15$, $f_W(3) = 4/15$, $f_W(4) = 10/15$, $E(W) = 18/5$.

**1.5.6** For $T$ = (no. correct answers), $E(T) = 35$.

**1.5.7** $E(X_0) = 405/256$, $E(X_1) = 540/256$, $E(X_2) = 270/256$, $E(X_3) = 60/256$, $E(X_4) = 5/256$. $\sum_{k=0}^{4} E(X_k) = 5$.

**1.5.8** Bake 3. $E(X) = 2.8$.

**1.5.9** 180 and 33.

**1.5.10** $P(B) = 601/625$.

**1.5.11** (**a**) $f_T(0) = 0.024$, $f_T(1) = 0.188$, $f_T(2) = 0.452$, $f_T(3) = 0.336$, $E(T) = 2.1$.

## Section 1.6

**1.6.1** (**a**) $E(X) = 4$, $\text{Var}(X) = 2$.

**1.6.4** 5 and 1.8.

**1.6.6** $E(T) = 95$, $\text{SD}(T) = \sqrt{290}$.

**1.6.7** (**a**) $E(X) = 7/3$, $E(Y) = 2/3$, $\text{Var}(X) = 5/9$, $\text{Var}(Y) = 5/9$. (**b**) $f_T(1) = 3/36$, $f_T(2) = 8/36$, $f_T(3) = 14/36$, $f_T(4) = 8/36$, $f(5) = 3/36$. (**c**) $E(T) = 3$, $\text{Var}(T) = 10/9$.

**1.6.8** (**a**) 53/60 and 7/60. (**b**) 7/240.

**1.6.9** (**a**) For $N = 6$, $E(T) = 14.7$, $\text{Var}(T) = 38.99$. For $N = 20$, $E(T) = 71.95$, $\text{Var}(T) = 566.51$.

**1.6.10** (**a**) \$855.47.

## Section 1.7

**1.7.1** (**a**) $(X, Y)$ takes the values (1,2), (1,3), (1,4), (2,3), 2,4), (3,4), each with probability 1/6. (**b**) $E(X) = 5/3$, $E(Y) = 10/3$, $\text{Var}(X) = \text{Var}(Y) = 5/9$, $\text{Cov}(X, Y) = 5/18$, $\rho(X, Y) = 1/2$. (**c**) $\hat{Y} = 10/3 + (1/2)(X - 5/3)$. (**d**) $\text{Var}(\hat{Y}) = 5/36$, $\text{Var}(Y - \hat{Y}) = 15/36$.

**1.7.2** (**a**) $\mu = 4$, $\sigma^2 = 5$. (**b**) $E(T) = 8$, $\text{Var}(T) = 20/3$, $E(\bar{X}) = 2$, $\text{Var}(\bar{X}) = 20/12$.

**1.7.3** $\sqrt{k/n}$.

**1.7.4** $\rho(X, Y) = 4\theta - 1$ for $0 \le \theta \le 1/2$.

**1.7.5** (**a**) $\text{Var}(T) = n\sigma^2(1 + (n-1)\rho)$, $\text{Var}(T/n) = (\sigma^2/n)(1 + (n-1)\rho)$. (**b**) $\rho \geq -1/(n-1)$. (**c**) $\rho(X_i, X_j) = \sigma_A^2/(\sigma_A^2 + \sigma_\epsilon^2)$.

**1.7.6** (**a**) $g(x) = \overline{y} + b(x - \overline{x})$, where $b = S_{xy}/S_{xx}$. (**b**) $g(x) = 4 - 2(x-2)$, proportion 4/7. $\rho(X, Y) = -2/\sqrt{7}$.

## Section 2.2

**2.2.1** (**a**) 0.2060. (**b**) 0.3125. (**c**) 0.2752. (**d**) 1/2, (**e**) 2.

**2.2.2** (**a**) 2. (**b**) 0. (**c**) 3, 4. (**d**) 40. (**e**) $1/3 \leq p < 1/2$.

**2.2.3** (**a**) 0.7216. (**b**) 0.5948. (**c**) 0.1700, 0.2186, 0.2061. (**d**) 0.1311.

## Section 2.3

**2.3.1** (**a**) For example, $P(X = 2) = 0.2398$; $b(2; 5, 0.6) = 0.230$. (**b**) $E(X) = 3.0$, $\text{Var}(X) = 0.8571$.

**2.3.2** (**b**) $\hat{R}$ may be expressed in terms of $X$, $N$, and $n$. $\text{Var}(\hat{R})$ may be expressed in terms of $N$, $np = R/N$, and $q = 1 - p$. For $N = 400$, $n = 100$, $X = 43$, $\hat{\text{Var}}(\hat{R}) = 2.949$, and the resulting 95% confidence interval is [168.63, 175.37]. (Of course, $R$ must be an integer.)

**2.3.3** The hypergeometric probabilities $f_X(x)$ for (10, 15) are 0.1978, 0.4565, 0.2935, 0.0522 for $x = 0, 1, 2, 3$. The corresponding binomial probabilities are 0.216, 0.432, 0.288, 0.064.

## Section 2.4

**2.4.1** (**b**) $P(Y_3 = 5) = 54/1024$. (**c**) $P(Y_3 \leq 4) = 13/256$. (**e**) $E(Y_3) = 12$, $\text{Var}(Y_3) = 36$.

**2.4.2** (**a**) 0.4845. (**b**) 0.0804.

**2.4.4** (**a**) Negative binomial with parameters $r_1 + r_2$ and $p$.
(**b**) $\binom{x-1}{r_1-1}\binom{w-1-x}{r_2-1} \Big/ \binom{w-1}{r_1+r_2-1}$ for $x = r_1, r_1+1, \ldots, w - r_2$.
(**c**) $f_M(m) = q^{2m}(1 - q^2)$ for $m = 1, 2, \ldots$, where $q = 1 - p$.

**2.4.5** (**a**) $P(X = Y) = p_1p_2/(1 - q_1q_2)$. (**b**) $P(X > Y) = p_2q_1/(1 - q_1q_2)$.

**2.4.6** $p = \mu/(\sigma^2 + \mu)$, $r = \mu^2/(\sigma^2 + \mu)$, so $(\hat{p}, \hat{r}) = (0.2024, 3.023)$. If $r$ were known to be an integer, we would replace 3.023 by 3.

**2.4.7** $M = [(r-1)/p] + 1$, with maxima at both $M$ and $M - 1$ if $(r-1)/p$ is an integer. For the four examples: 1, 10 and 11, 45 and 46, 120 and 121.

## Section 2.5

**2.5.1** $P(X_n = 2)$ is 0.2277, 0.2248, 0.2242 for the three binomial distributions. It is 0.2240 for the Poisson distribution with $\lambda = 3.0$. $P(X_n = 3)$ is 0.4113, 0.4209, 0.4227 for the three binomial distributions. It is 0.4232 for the Poisson distribution with $\lambda = 3.0$.

**2.5.2** (**a**) $P(X \leq 2) = e^{-1.6}(1 + 1.6 + 1.6^2/2) = 3.88e^{-1.6} = 0.7834$. $P(X \geq 3) = 1 - 0.7834 = 0.2166$. (**b**) $1 - 0.7834^4 = 0.6234$. (**c**) 292 and 292.

**2.5.4** $e^{-\lambda d}$.

**2.5.5** Given $N = n$, $X_1 \sim \text{Binomial}(n, p)$ with $p = \lambda_1/(\lambda_1 + \lambda_2)$.

**2.5.6** (**a**) $e^{-2.8}$. (**b**) $1 - e^{-3.8}(1 + 3.8 + 3.8^2/2)$.
(**c**) $7[0.4(7) + 0.8(10) + 1.5(7)] = 149.1$, so $W \sim \text{Poisson}(149.1)$.

**2.5.7** (**a**)

| $k$: | 0 | 1 | 2 | 3 | 4 | 5 |
|---|---|---|---|---|---|---|
| $P(Y = k)$ | 0.69768 | 0.25116 | 0.04521 | 0.00542 | 0.00049 | 0.00004 |

(**b**) $\sum p_i^2 = 0.0204$. (**c**) Taking $A = \{1, 5\}$, we get $G(A) = 0.0124$.

**2.5.8**

| $k$: | 0 | 1 | 2 | 3 | 4 | 5 | 6 | 7 |
|---|---|---|---|---|---|---|---|---|
| $P(X = k)$ | 0.13506 | 0.27067 | 0.27094 | 0.18063 | 0.09022 | 0.03602 | 0.01197 | 0.00344 |
| $P(Y = k)$ | 0.13534 | 0.27067 | 0.27067 | 0.18045 | 0.09022 | 0.03609 | 0.01203 | 0.00341 |

Let $A = \{k | P(X = k) > P(Y = k)\}$. Then $A = \{2, 3\}$, and for this $A$, $G(A) = 0.00045$. The upper bound given by Le Cam's theorem is $\sum p_i^2 = 0.0040$, which is larger than 0.00045.

## Section 3.2

**3.2.1** (**a**) $C = 1$. (**b**) $1/4$. (**c**) $F(u) = 1/2 - (1 - u)^2/2$ for $0 \leq u < 1$, $1/2 + (u - 1)^2/2$ for $1 \leq u \leq 2$, 0 for $u < 0$, 1 for $u > 2$.
(**d**) $F$ is continuous at all points except at $x = 0, x = 2$, so the derivative of $F$ exists at all points except at $x = 0, x = 6$, and has the values $f(x)$ at $x \neq 0, 6$.
(**e**) For example, redefine $f(0) = 78$, $f(3/2) = 17$. $F$ remains the same, and then has derivatives equal to $f(x)$ except for $x = 0, 3/2, 2, 6$
(**f**) $x_{0.6} = 1 + \sqrt{0.2}$.

**3.2.2** (**a**) $C = 2/9$. (**b**) $F(u) = 0$ for $u < 0$, $u^3/3$ for $0 \leq u < 1$, $1/3 + (2/9)(u - 1)$ for $1 \leq u < 4$, 1 for $u \geq 4$. (**c**) 183/216. (**d**) All $u$ except for $u = 1, 4$. Yes.

**3.2.3** $F_Y(y) = 0$ for $y < 0$, $y^{1/2}$ for $0 \leq y < 1$, 1 for $y \geq 1$. $f_Y(y) = 0$ for $y < 0$, $(1/2)y^{-1/2}$ for $0 < y < 1$, 0 for $y > 1$. We can define $f$ in any way that we wish for $y = 0$ and $y = 1$. (Of course, the values assigned cannot be less than zero.)

**3.2.4** (**a**) $C = 0.7/8$. (**b**) $0.2 + 5.6/16$. (**c**) For all $x$ except for $x = 0, 2, 4$, with derivative $Cx$ for $0 < x < 4$, 0 for $x < 0$ and $x > 4$. (**d**) $G(y) = 0.1I[y \geq 0] + 0.2I[y \geq 1] + (C/2)(y - 1)$ for $1 \leq y < 17$, 0 for $y < 1$, 1 for $y \geq 17$.

**3.2.5** (**a**) $C = 1.4/16$. (**b**) $F(3) - F(1) = 0.2 + C(9/2 - 1/2)$. (**c**) All $u$ except for $u = 0, 2, 4$. (**d**) $F_Y(y) = F((y-1)^{1/2}) = 0.1I(y \geq 1) + 0.2I(v \geq 5) + C(y-1)/2$ for $0 \leq y < 17$, 1 for $y \geq 17$.

## Section 3.3

**3.3.1** $E(X) = 0$, $\text{Var}(X) = E(X^2) = 2/3$.

**3.3.2** $E(X) = 13/12$, $\text{Var}(X) = E(X^2) - [E(X)]^2 = 5/3 - [13/12]^2 = 71/144$.

**3.3.6** $B = 10 + \sqrt{50c/s} = 10 + \sqrt{20}$.

**3.3.7** $G(D; B) = (s-2)D - (c-2)B$ for $D < B$, $(s-c)B[(s-2)-(c-2)]B$ for $D \geq B$. Thus, $G$ and $H$ are the same as before after replacement of $s$ by $s-2$ and $c$ by $c-2$.
$H(B) = (s-2)\int_0^B [1 - F(x)]dx$. Choose $B = x$, where $x$ satisfies $F(x) = 1 - (c-2)/(s-2)$.

## Section 3.4

**3.4.1** $Y$ has the mass function $f(0) = 3/8$, $f(1) = 1/2$, $f(2) = 1/8$.

**3.4.2** (**a**) $g(y) = 1$ for $0 \leq y \leq 1$, Unif(0, 1). (**b**) 1/2 by both methods. (**c**) $H(U) = F^{-1}(U) = \sqrt{1-2U}$ for $0 \leq U \leq 1/2$, and $\sqrt{2U-1}$ for $1/2 < U \leq 1$.

**3.4.3** (**a**) $H(X) = F(X) = X^2$. (**b**) $H(X) = G^{-1}(H(X)) = 1/(1-X^2)$ for all $X$. (**c**) $H(X) = G^{-1}(H(X)) = \sqrt{2X^2} = -\sqrt{2}X$ for $-1 \leq X \leq 0$, $8X^2 - 3$ for $0 < X \leq 1$.

**3.4.4** (**a**) $H(X) = F(X) = X/(1+X)$ for all $X$.
(**b**) $K(U) = F^{-1}(U) = U/(1-U)$. (**c**) $Y$ has (surprisingly!) the same cdf as $X$ and the same density $f(y) = (1+y)^{-2}$ for $y \geq 0$.

**3.4.5** (**a**) $X$ has cdf $F(x) = (1/\pi)\tan^{-1}(x/d)$, so that the density is $f(x) = (1/(d\pi))1/[1 + (x/d)^2]$ for all $x$. (**b**) The integral $\int_d^\infty xf(x)dx > (1/(2d\pi))\int_d^\infty x/(x/d)^2 dx$ does not converge, so $E(X)$ does not exist.

## Section 3.5

**3.5.1** (**a**) $C = 1$. (**b**) $f_X(x) = xe^{-x}$ for $x > 0$, $f_Y(y) = e^{-y}$ for $y > 0$. (**c**) $P(X > bY) = 1/b$ for $b > 1$. Therefore, $P(W \leq w) = P(Y/X \leq w) = P(X \geq (1/w)Y) = w$ for $0 < w \leq 1$. Thus, $W \sim$ Unif(0, 1). (d) $E(XY) = 3$.

**3.5.2** (**a**) $f_X(x) = f_Y(x) = x + 1/2$ for $0 \leq x \leq 1$. (**b**) 7/24.

**3.5.3** (**a**) $F(u_1, u_2) = F(u_2)^n - [F(u_2) - F(u_1)]^n = [1 - e^{-u_1}]^n - [e^{-u_1} - e^{-u_2}]^n$ for $0 < u_1 < u_2 < \infty$. Taking partial derivatives with respect to $u_1$ and $u_2$, we get the joint density $f(u_1, u_2) = n(n-1)[e^{-u_1} - e^{-u_2}]^{n-2}e^{-u_1-u_2}$ for $0 < u_1 < u_2 < \infty$.

(**b**) We can get the marginal density of $X_{(1)}$ by integrating over $u_2 > u_1$. It is easier to find the cdf of $X_{(1)}$ first: $F_1(u_1) = 1 - P(X_{(1)} > u_1) = P(\text{all} X_i > u_1) = [e^{-u_1}]^n = e^{-nu_1}$, so that $X_{(1)}$ has density $f_1(u_1) = ne^{-nu_1}$ for $u_1 > 0$.

**3.5.4** $f_M(m) = 2m^2$ for $m \in [0, 1]$, $= m^2/6$ for $m \in (1, 2]$, and $= m/3$ for $m \in (2, 3]$.

## Section 3.6

**3.6.1** (**a**) $Y$ has cdf $F(y) = y[1 - \log(y)]$ for $0 < y < 1$, density $f(y) = -\log(y)$ for $0 < y < 1$. $E(Y) = E(U_1)E(U_2) = 1/4$, $E(Y^2) = E(U_1^2)E(U_2^2) = (1/3)^2 = 1/9$, $\text{Var}(Y) = 7/144$. $E(Y^2)$ and $E(Y)$ can be obtained using the density of $Y$, making the change of variable $w = -\log(y)$, so $y = e^{-w}$.

**3.6.2** (**a**) $f(x_1, x_2) = 2e^{-x_1 - x_2}$ for $0 < x_1 < x_2 < \infty$.
(**b**) Integrating over the region for which $x_2 - x_1 > d$, we find that $P(X_{(2)} - X_{(1)} \leq d) = 1 - e^{-d}$ for $d > 0$, so $D$ has density $f_D(d) = e^{-d}$ for $d > 0$.
(**c**) The Jacobian of the transformation from $(X_{(1)}, X_{(2)})$ to $(Y_1, Y_2)$ is 1, so $Y = (Y_1, Y_2)$ has joint density $f_Y(y_1, y_2) = 2e^{-2y_1 - d}$ for $0 < y_1 < \infty$ and $0 < d < \infty$. Since this is a product, $Y_1$ and $D = Y_2$ are independent, with the obvious densities, exponential distributions with means 1/2 and 1.

**3.6.3** (**a**) $f_Y(y_1, y_2) = (3/4)(y_1 + y_2)$ over the triangle with vertices (0, 0), (0, 2), (1, 1). $Y_1$ has density $g(y_1) = (9/8)y_1^2$ for $0 < y_1 < 1$, $(3/8)[2 - (y_1 - 1)^2$ for $1 \leq y_1 \leq 2$.

## Section 4.2

**4.2.1** (**a**) 0.1093. (**b**) 0.7396. (**c**) 0.0349.

**4.2.2** (**a**) 0.1093. (**b**) 0.9356. (**c**) 0.9651. (**d**) 58.42. (**e**) $d = 69.60$.

**4.2.3** $f_Y(y) = 1\sqrt{2\pi y}e^{-y}$ for $y > 0$.

**4.2.4** (**a**) 0.9123. (**b**) 0.9906.

**4.2.7** $E(Y) = \exp(\sigma^2/2 + \mu)$, $\text{Var}(Y) = (\exp(2\sigma^2) - \exp(\sigma^2))\exp(2\mu)$. $\exp(u) \equiv e^u$ for any real number $u$.

## Section 4.3

**4.3.1** $r(2.5) = 1.0337$, $r(4.5) = 1.0187$, $r(6) = 1.0168$.

**4.3.4** (**a**) $E(1/X)$ exists for $\alpha > 1$, with value $\alpha - 1$. (**b**) No.

**4.3.5** (**c**) $P(R \leq r) = (1 + r)^{-\beta}$ for $r > 0$. For $\beta = 2$, $P(R \leq 3) = 15/16$.

**4.3.6** (**a**) $f_X(x) = [1/2^{\nu/2}\Gamma(\nu/2)]x^{\nu/2-1}e^{-x/2}$ for $x > 0$, $\nu$ and $2\nu$. (**b**) 278.79.

**4.3.8** (**c**) 0.90, 0.6838, 0.9487. (**d**) $7/11, 28/1452$.

**4.3.12** (**a**) $E(Y) = \theta\Gamma(1/m+1)$, $\text{Var}(Y) = \theta^2[\Gamma(2/m+1) - (\Gamma(1/m+1))^2]$. (**c**) 6.6025. (**d**) $H(u; m, \theta) = \theta[-\log(1-U)]^{1/m}$. $1-U$ could be replaced by $U$.

**4.3.13** (**c**) (1.874, 2.896).

## Section 5.1

**5.1.1** $P(R = r | T = t)$ is as follows:

| $t$ | $r$: 0 | 1 | 2 | 3 |
|---|---|---|---|---|
| 1 | 3/7 | 4/7 | 0 | 0 |
| 2 | 6/42 | 24/42 | 12/42 | 0 |
| 3 | 1/35 | 12/35 | 18/35 | 4/35 |

**5.1.2** $P(X_1 = x_1, X_2 = x_2) = (1/6)^2(5/6)^{x_2-2}$ for $1 \le x_1 < x_2 < \infty$, so $P(X_1 = x_1 | X_2 = k) = 1/(k-1)$ for $1 \le x_1 < k < \infty$. This is the uniform distribution on $1, 2, \ldots, k-1$.

**5.1.3** Let $\mathbf{X} = (X_0, X_1, X_2, X_3)$, $\mathbf{x} = (x_0, x_1, x_2, x_3)$. $X$ has the multinomial distribution with $n = 5$, $\mathbf{p} = (1/8, 3/8, 3/8, 1/8)$. $X_3$ has the binomial distribution with $n = 5$, $p_3$. A few computations show that conditionally on $X_3 = x_3$, $(X_0, X_1, X_2)$ has the multinomial distribution with $n = 5 - x_3$, probability vector $(1/7, 3/7, 3/7)$.

**5.1.7** (**c**) Show that $\mathbf{p}\text{P} = \mathbf{p}$.

## Section 5.2

**5.2.1** (**a**) $C = 1/36$. (**b**) $g(1) = 20/9$, $g(2) = 26/12$, $g(3) = 32/15$, $E[g(X)] = 13/6 = E(Y)$.

**5.2.2** $g(t_2) = t_2 + \mu_3$. $E(g(T_2)) = E(T_2) + \mu_3 = \mu_1 + \mu_2 + \mu_3 = E(T_3)$.

**5.2.3** (**a**) $g(1) = 7/2$, $g(2) = 13/2$, $g(3) = 9$, $g(4) = 11$, $g(5) = 12.5$. For example, $g(3) = 3(1+2+4+5)/4$. (**b**) $E(Y) = (1/5)(g(1) + g(2) + g(3) + g(4) + g(5)) = 172/4 = 43$.

## Section 5.3

**5.3.1** (**a**) $f_X(x) = (C/2)x^2$ for $0 < x < 1$, so $C = 6$. (**c**) $f_Y(y|X = x)(2y)/x^2$ for $0 < y < x \le 1$. (**d**) $g(x) = 2x$ for $0 < x < 1$, $E(g(X)) = 1/2 = E(Y)$.

**5.3.2** (**a**) $f_Y(y \mid X = x) = xe^{-xy}$ for $y > 0$, $f_X(x) = e^{-x}$ for $x > 0$, $f_Y(y) = 1/(1+y)^2$ for all $y > 0$, $g(x) = 1/x$, $E(Y)$ does not exist.

**5.3.3** $P(X=k) = \int_0^\infty \binom{n}{k} p^k (1-p)^{n-k} dp = \dfrac{\Gamma(k+1)\Gamma(n-k+1)}{\Gamma(n+2)} \binom{n}{k} = 1/(n+1)$ for $k = 0, 1, \ldots, n$. Uniform distribution on $0, 1, \ldots, n$.

**5.3.4** $f_Y(y) = (2/A^2)(A - y)$ for $0 \le y \le A$.

**5.3.5** $f_W(w|X = x) = e^{-w+x}$, for $0 < x < w < \infty$, $f_W(w) = we^{-w}$ for $0 < w < \infty$.

**5.3.8** (**a**) 0.9367 and 0.9993. (**b**) $g(x) = 165 + 0.3429(x - 178)$, $h(y) = 178 + 0.4667(y - 165)$. (**c**) 5.76 and 30.24. (**d**) 0.662.

## Section 6.2

**6.2.1** (**a**) $M(t) = (1/t)(e^t - 1) = 1 + t/2 + t^2/3! + \cdots$, so the derivative at $t = 0$ is $1/2 = E(X)$. (**b**) $[(1/t)(e^t - 1)]^2$.

**6.2.2** (**a**) $M(t) = 1/(1 - t)^\alpha$ for $t < 1$. The mgf of $Y$ is $1/(1 - t)^{\alpha_1+\alpha_2}$, the mgf of the Γ-distribution with power parameter $\alpha_1 + \alpha_2$.

**6.2.3** $\nu_{2k} = \sigma^{2k}(2k)!/[k!2^k]$ for $k$ any positive integer. $\nu_{2k+1} = 0$ for $k = 0, 1, 2, \ldots$.

**6.2.4** (**a**) $P(Z > x) = P(e^{Zt} > e^{xt}) \le E(e^{Zt})/e^{xt}) = M(t)/e^{xt}$. This in a minimum for $t = x$, with value $e^{-t^2/2}$, so this is an upper bound. (**b**) The probability is less than $2e^{-K^2n/2}$.

**6.2.5** (**a**) $M(t) = e^t p/(1 - e^t q)$ for $e^t q < 1$. (**b**) $M(t)^r = e^{tr} p^r (1 - qe^t)^{-r}$ for $e^t q < 1$. (**c**) Let $M_r(t)$ be the mgf. $M_r(t)$ is as given in (**b**). To show it, use the fact that $\binom{k-1}{r-1} = \binom{-r}{k-1} = (-1)^{k-r}$. (See 2.4.2.) Then the infinite sum, after factoring out $e^{tr} p^r$ and letting $m = k - r$, is the binomial expansion of $(1 - qe^t)^{-r}$.
(**d**) As in Section 2.4, let $W_r = X - r$. ($W_r$ is the number of failures before the $r$th success.) The mgf of $W_r$ is $M_W(t) = p^r(1 - qe^t)^{-r}$. Then $M_W(t) = rp^r qe^t(1 - qe^t)^{-r-1}$ and $M''_W(t) = r(r-1)p^r q^2 e^{2t}(1 - qe^t)^{-r-2} + M_W(t)$, so $M_W(0) = rq/p$ and $M''_W(0) = r(r+1)q^2/p^2 + rq/p$, so that $W_r$ has mean $q/p$ and variance $rq/p + rq^2/p^2 = rq/p^2$. Thus, $E(X) = r + rq/p = r/p$ and $\text{Var}(X) = \text{Var}(W_r) = rq/p^2$.

## Section 6.3

**6.3.1** Let $\varepsilon > 0$. We must show that $P(|m_n - 0| > \varepsilon) = P(m_n > \varepsilon)$ converges to zero. But $P(m_n > \varepsilon) = P(\text{all} X_i$ are greater than $\varepsilon) = (e^{-\varepsilon})^n$, which converges to zero because $e^{-\varepsilon} < 1$.

**6.3.3** (**a**) Determine $E(\hat{p}_n)$ and $\text{Var}(\hat{p}_n)$. (**b**) Yes, because $g(u) = u^2$ is continuous. (**c**) For all $p$ except for $p = 1/2$, where $g$ is discontinuous.

**6.3.4** For any $\varepsilon > 0$, $P(|X_n - C| \le \varepsilon) \ge P(C - \varepsilon < X_n \le C + \varepsilon) = F_n(C + \varepsilon) - F_n(C - \varepsilon)$, where $F_n$ is the cdf of $X_n$. Since $\{X_n\}$ converges in distribution to $C$, $F_n(C + \varepsilon)$ converges to 1. $F_n(C - \varepsilon)$ converges to zero. Thus, $P(|X_n - C| \le \varepsilon)$ converges to 1.

**6.3.5** Let $\varepsilon > 0$. We must show that $P(|W_n - W| > \varepsilon)$ converges to zero. But $|W_n - W| = |X_n - X + Y_n - Y| \leq |X_n - X| + |Y_n - Y|$. Therefore $P(|W_n - W| > \varepsilon) \leq P(|X_n - X| + |Y_n - Y| > \varepsilon) \leq P((|X_n - X| > \varepsilon/2) + P(|Y_n - Y| > \varepsilon/2)$. Each of these last two probabilities converges to zero.

**6.3.6** (**a**) $M_n(t) = (1/n)(1 - e^{t(n+1)/n})/(1 - e^{t/n})$. The numerator converges to $1 - e^t$ as $n \to \infty$. The denominator is $n[1 - (1 + t/n + (t/n)^2/2 + \cdots)$, which converges to $-t$. Therefore, $M_n(t)$ converges to $(e^t - 1)/t$ for all $t$, which is the mgf of the uniform distribution on [0, 1].

**6.3.7** The mgf of the uniform distribution on $[-1/n, 1/n]$ is $(e^{t/n} - e^{-t/n})(n/2t) = (n/2t)[(1 + t/n + (t/n)^2 + \cdots) - (1 - t/n + (t/n)^2/n - \cdots)] = (n/2t)[2(t/n) + 2(t/n)^3/3! + \cdots]$, which converges to 1 for all $t$. This is the MGF of the distribution with point mass 1 at zero.

**6.3.8** The mgf of $X_n$ is $M_n(t) = (q_n + p_n e^t)^n$. Show that $\log(M_n(t))$ converges to $\lambda(e^t - 1)$.

## Section 6.4

**6.4.1** (**a**) 0.1714, 0.3759, 0.8286, 0.9431. (**b**) 0.2045, 0.1145 The binomial probabilities are 0.2051 and 0.1172.

**6.4.2** (**a**) 0.9545. (**b**) 0.9545.

**6.4.3** 0.0168.

**6.4.4** $Z_n = (S_n - np)/(npq)^{1/2}$ has mgf $M_n(t) = e^{-npt/(npq)^{1/2}}(q + pe^{t/\sqrt{npq}})$. Show that the log of this converges to $t^2/2$.

**6.4.5** (**a**) 0.1783. (**b**) 0.1763.

**6.4.6** 0.1580.

**6.4.7** (**a**) 0.2056569 using the normal approximation of the Poisson distribution, 0.1990 using the normal approximation of the gamma distribution.

**6.4.8** (**a**) 14.5 and 32.73. (**b**) 0.3314 and 0.3379. (**c**) 0.1915 and 0.6179.

**6.4.9** (**a**) $\log(M_n(t)) = -\lambda_n^{1/2} + \lambda_n(e^{t\lambda_n^{1/2}} - 1)$. (**b**) 0.8789.

**6.4.10** (**a**) See Problem 6.2.2. (**d**) 0.1587 and 0.0786.

**6.4.11** $n = 801$.

**6.4.12** (**a**) 0.943. (**b**) Solving a quadratic equation we get (no. reservations) $= s = 429$. (**c**) 423.

**6.4.13** (**a**) $Y_n$ has the mgf $M_n(t) = [(1 - q_n e^{tp_n})p_n]^{-1}$. The term in brackets converges to $(1 - t)^{-1}$, the mgf of the exponential distribution with mean 1. Thus, $\{Y_n\}$ converges in distribution to the exponential distribution with mean 1. (**c**) $\{W_n{}^*\}$ converges in distribution to the $\Gamma(r, 1)$ distribution.

## Section 7.2

**7.2.1** (**a**) $\text{MSE}(T_4) = \text{Var}(T_4) = \theta^2/[n(n+2)]$. (**b**) $c_n = (n+2)/(n+1)$, $r_{T_5}(\theta) = \theta^2/(n+1)^2$, $r_{T_5}(\theta)/r_{T_2}(\theta) = (n+2)/[2(n+1)] < 1$ for all $n$ and $r_{T_5}(\theta)/r_{T_4}(\theta) = n(n+2)/(n+1)^2 < 1$ for all $n$.

**7.2.2** (**a**) Let the components of $\mathbf{a}_0$ be $a_j = (1/\sigma_j^2)/C$, where $C = \sum_{i=1}^n 1/\sigma_1^2$ for $j = 1, \ldots, n$. (**b**) For $\mathbf{a} = \mathbf{a}_0$, $\text{Var}(T) = 1/C$.

**7.2.3** $p(1-p)/n$.

**7.2.4** Since the means for these three distribution s are 2, 2.5, and 3, we might let $\hat{\theta} = 2\overline{X} - 3$, so that $\hat{\theta}$ is an unbiased estimator of $\theta$, and $\text{MSE}(\hat{\theta}) = 4\text{Var}(\overline{X}) = 4v(\theta)/2$, where $v(\theta)$ is the variance of one observation from the mass function $f_\theta$. Thus, $v(1) = 1$, $v(2) = 1.05$, $v(3) = 1$. Thus, $\text{MSE}(\hat{\theta}) = 2$, 2.10, and 2 for $\theta = 1, 2, 3$. A problem with this estimator is that it takes values that are not possible values of $\theta$. Another estimator could be the value $\hat{\theta}^*$ of $\theta$ that maximizes $P_\theta(X_1 = x_1, X_2 = x_2)$ for each $(x_1, x_2)$, so that, for example, for $x_1 = 1, x_2 = 3$, $\hat{\theta}^* = 1$. $\hat{\theta}^*$ is called the maximum likelihood estimator.

**7.2.5** (**a**) Bias $= \lambda/n - 1)$ for $n > 1$, $\text{MSE}(\hat{\lambda}) = \lambda^2(n+2)/[n-1)(n-2)]$ for $n > 2$. (**b**) For $C_n = (n-1)/n$, $\text{MSE} = \lambda^2/(n-2)$.

**7.2.7** (**a**) Let $T(k) = 1$ for $k = 0$, 0 for $k > 0$. Then $T(X)$ is the indicator of the event $[X = 0]$.

**7.2.9** (**a**) $\text{Var}(T) = n/[(n+1)^2(n+2)]$. (**b**) $\text{Var}(T)/\text{Var}(\hat{\theta}) = 2n/(n+1)$.

## Section 7.3

**7.3.1** (**a**) $\hat{\theta}_2 = \overline{X} + \sqrt{3}\hat{\sigma}$, $\hat{\theta}_2 = \overline{X} - \sqrt{3}\hat{\sigma}$. (**b**) No, it doesn't make sense.

**7.3.2** $1/(\overline{X} + 1)$.

**7.3.3** $(\hat{r}, \hat{\theta}) = (\overline{X}^2/(\hat{\sigma}^2 - \overline{X}), \overline{X}/(\hat{\sigma}^2 - \overline{X}))$. We get (3.3355, 0.6148).

**7.3.4** $\hat{\lambda} = 1/\hat{\sigma}$, $\hat{\eta} = \overline{X} - \hat{\sigma}$.

**7.3.5** $r_{\hat{\theta}}(\theta) = 0.08$ for without-replacement sampling for each $\theta$. $r_{\hat{\theta}}(\theta) = 0.06$ for without-replacement sampling.

**7.3.6** $\hat{p} = 1 - \hat{\sigma}^2/\overline{X}$, $\hat{m} = \overline{X}^2/(\overline{X} - \hat{\sigma}^2)$.

**7.3.7** (**a**) The MME is $e^{-\overline{X}}$.

## Section 7.4

**7.4.1** (**a**) $\hat{R}(0) = 0$, $\hat{R}(1) = 2$ or 3, $\hat{R}(2) = 5$ or 6, $\hat{R}(3) = 8$.

**7.4.2** Both $1/\overline{X}$.

**7.4.3** $(X_1 + X_2)/(n_1 + n^2)$.

**7.4.6** $\hat{\eta} = \min(X_1, \ldots, X_n)$, $\hat{\lambda} = 1/(\overline{X} - \hat{\eta})$.

**7.4.7** (**a**) The MME is $\overline{X}/(\overline{X} - 1)$; the MLE is $1/\overline{Y}$, where $Y_i = \log(X_i)$.

**7.4.8** (**a**) $\hat{\theta} = (X_1 + 2X_2)/2n$; yes. (**b**) $((1 - \hat{\theta})^2, 2\hat{\theta}(1 - \hat{\theta}), \hat{\theta}^2)$; no.

**7.4.9** (**a**) $\hat{\beta} = \sum x_i Y_i / \Sigma x_i^2$. (**b**) 3 and 1. (**c**) $(\overline{X} \sum_{i=1}^{n} (X_1 - \overline{X})^2)$. (**e**) $\sigma^2 / \sum x_i^2$. (**f**) $\sum a_i x_i = 1$.

**7.4.10** (**a**) $\hat{\theta}(0, 0) = 3$, $\hat{\theta}(0, 1) = 3$, $\hat{\theta}(1, 2) = 1$, $\hat{\theta}(1, 0) = 3$, $\hat{\theta}(1, 1) = 2$, $\hat{\theta}(1, 2) = 1$, $\hat{\theta}(2, 0) = 1$, $\hat{\theta}(2, 1) = 1$, $\hat{\theta}(2, 2) = 1$. $C(1) = 0.84$, $C(2) = 0.64$, $C(3) = 0.72$. (**c**) For example, $\hat{\theta}(0, 1, 2) = 1$, $\hat{\theta}(0, 1, 0) = 3$. $C(1) = 0.756$, $C(2) = 0.896$, $C(3) = 0.648$. (**d**) MLE is 3.

## Section 7.6

**7.6.1** $E(h_1(X)) = 1/9$, $E(h_2(X)) = 0.11438$, $\text{Var}(h_1(X)) = 5.0805 \times 10^{-5}$, $E(1/X) = 0.11157$, $\text{Var}(1/X) = 5.1733 \times 10^{-5}$.

**7.6.2** $E(h_2(X_1, X_2)) = \theta_1\theta_2 + (1/2)\sigma_{12}$. $\text{Var}(h_1(X_1, X_2)) = \theta_1^2\text{Var}(X_1) + \theta_2^2\text{Var}(X_2) + (1/2)\sigma_{12}$.

**7.6.3** $\text{Var}(h_2(\overline{X})) = 7.615 \times 10^{-6}$, $P(|h(\hat{X}) - h(\theta_0)| \le 1) \doteq 0.610$, where $\theta_0 = 30(\pi/180) = 0.5236$.

**7.6.7** (**a**) 0.843 and 0.166. (**b**) $-0.343 \pm 0.281$.

**7.6.8** (**a**) $E(\hat{\theta}_1) \doteq \theta + \theta(1 + \theta)/(\theta + 2)n$ (using quadratic approximation), $\text{Var}(\hat{\theta}_1) \doteq \theta(\theta + 1)^2/n(\theta + 2)$ (using linear approximation). (**b**) $E(\hat{\theta}_2) = [n/(n - 1)]\theta$ (exact), $E(\hat{\theta}_2) \doteq \theta + \theta/n$, $\text{Var}(\hat{\theta}_2) = n^2\theta^2/(n - 1)^2(n - 2)$ (exact), $\text{Var}(\hat{\theta}_2) \doteq \theta^2/n$.
(**c**) Normal approximations: 0.363 and 0.383. [The expectations and variances given in part (a) and (b) were used.]

## Section 7.7

**7.7.1** (**a**) $U = X/0.1054$. (**b**) The 0.05- and 0.95-quantiles of the chi-square distribution with 20 df are 10.85 and 31.41. Therefore, the 0.10-and 0.95-quantiles for the gamma distribution with $\alpha = 10$ are $10.85/2 = 5.42$ and $31.41/2 = 15.70$. The 90% CI on $\theta$ is $[T/15.70, T/5.42] = [6.00, 17.38]$.

**7.7.2** Let $T = \min(X_1, \ldots, X_n)$. $D = T - \eta$ has cdf $G(x) = 1 - e^{-nx}$ for $x > 0$. $u = -(1/n)\log(0.05)$. $L = T - u$, so that the lower limit is 12.96.

**7.7.3** (**a**) 423. (**b**) 349.

**7.7.4** (**a**) 492. (**b**) 0.110. (**c**) $0.094 + 0.0215$.

**7.7.5** 1691.

**7.7.6** $[6 \pm 2.35]$.

**7.7.7** (**a**) [0.967, 1.433]. *Hint*: $X_1 \sim \text{Poisson}(\lambda_1 = 8.791\theta_1)$, $X_2 \sim \text{Poisson}(\lambda_2 = 62.547\ \theta_2)$, where $\boldsymbol{\theta}_1$ and $\boldsymbol{\theta}_2$ are the rates per 1000. First find a CI on $p = \lambda_1/(\lambda_1 + \lambda_2) = \rho/(1 + \rho)$, where $\rho = \lambda_1/\lambda_2$, then use this interval to get an interval on $R = \theta_1/\theta_2$.
(**b**) [7.74, 11.15] and [ 7.14, 8.30].

**7.7.8** $[-0.85, 6.65]$.

**7.7.10** (**a**) $R = 4$, since $P(X \geq 3|R = 4) = 0.119$, $P(X \geq 3|R = 5) = 0.357$, and $P(X \geq 3|R) < 0.119$ for $R < 4$.
(**b**) 65, using the finite correction factor for the variance of $X$, the number of red balls chosen.

**7.7.11** (**e**) $[-0.054 \pm 0.055]$.

**7.7.12** [0.1604, 0.1764], the solutions of a quadratic equation in $\theta$.

## Section 7.8

**7.8.1** (**a**) $\theta^2/n$. (**b**) $\hat{\theta} = 1/\overline{X}$. (**c**) $C_n = n/(n-1)$. (**d**) $d(\theta) = \theta$. (**e**) $[\hat{\theta}_n \pm 1.96\hat{\theta}_n n^{1/2}] = \hat{\theta}_n[1 \pm 1.96n^{1/2}]$.

**7.8.2** (**a**) $\theta/n$. (**b**) $\hat{\theta} = \overline{X}$. (**c**) $C(n) = 1$. (**d**) $d(\theta) = \theta^{1/2}$. (**e**) $[\overline{X} \pm 1.96n^{1/2}\overline{X}^{1/2}]$.

**7.8.3** (**a**) Cramer–Rao bound: $1/[n\theta^2(1-\theta)^2]$.

**7.8.4** (**a**) $1/n\alpha^2$. (**b**) $n/[(n-1)^2\alpha^2]$, (**c**) $C(\alpha) = \sqrt{\alpha}$ and $[(\hat{\alpha}_n \pm 1.96\sqrt{\hat{\alpha}_n/n}]$.

**7.8.5** (**a**) $\text{Var}(\hat{\tau}_2) \geq 2\tau^2/n$. (**b**) $\hat{\tau} = (1/n)\sum_{i=1}^n (X_i - \mu)^2$. (c) $C_n = [n/(n-1)]$, $\text{Var}(\tau^2) = 2\sigma^2/(n-1)$. (**d**) $d(\eta) = 2^{1/2}\eta$.
(**e**) $[\hat{\tau} \pm 1.96\sqrt{2\hat{\tau}/n}]$ [$n$ or $\hat{\tau}$ could be replaced by $(n-1)$ or $\hat{\tau}_2$].

**7.8.6** (**a**) $L(\theta) = C\theta^{2x_2+x_1}(1-\theta)^{2x_0+x_1}$, where $C$ does not depend on $\theta$. (**b**) $\hat{\theta} = 0.70$, $p(\hat{\theta}) = (0.09, 0.42, 0.49)$. $(X_0, X_1, X_2)/1000 = (0.094, 0.412, 0.494)$. Good fit. (**d**) $B(\theta, n) = \theta(1-\theta)/(2n)$, where $n = 1000$. (**e**) $[\hat{\theta} \pm 1.96(\hat{\theta}(1-\hat{\theta})/2n)^{1/2}]$.

**7.8.7** (**a**) Let $X_{i+}$ and $X_{+j}$ be the row and column totals for $i = 0, 1$ and $j = 0, 1$. Then $\hat{\theta}_1 = X_{1+}/n, \hat{\theta}_2 = X_{+1}/n$. The pair $(\hat{\theta}_1, \hat{\theta}_2)$, is the MLE for $(\theta_1, \theta_2), \hat{\theta} = (0.12, 0.18, 0.28, 0.42)$. (**b**) A 95% CI on $\theta_1$ is $[\hat{\theta}_1 \pm 1.96\sqrt{\hat{\theta}_1(1 - \hat{\theta}_1/n}] = [0.672, 0.728]$.

## Section 7.9

**7.9.1** (**a**) Yes. (**b**) No.

**7.9.2** (**a**) Yes. (**b**) $g(t, \theta) = \theta(1 - \theta)^{t-1}$, for $t = 1, 2, \ldots, 0 < \theta, < 1, h(\mathbf{x}) = 1$ for all $x_i$ positive integers, 0 otherwise.

**7.9.4** (**b**) $\hat{\theta}^* = \exp(-\overline{X})$. (**c**) $\text{Var}(\hat{\theta}) = \exp(-\lambda)(1 - \exp(-\lambda))/n$, $\text{Var}(\hat{\theta}^*) = \exp(n\lambda(\exp(-2/n) - 1)) - \exp(2n\lambda(\exp(-1/n) - 1))$. (**d**) Yes. (**e**) $\hat{\theta}^*$.

**7.9.7** $T = Y_1 + 2Y_2, \hat{\theta} = T/2n$.

**7.9.9** (**a**) $(M \equiv \min(X_1, \ldots, X_n), T = \Sigma X_i)$ is one such pair. (**b**) $(M, \overline{X} - M)$, where $\overline{X} = T/n$.

## Section 8.1

**8.1.1** (**a**) $\gamma(1) = 0.07, \gamma(2) = 0.33, \gamma(3) = 0.65$. (**b**) 0.33. (**c**) $\gamma(1) = 0.16, \gamma(2) = 0.49, \gamma(3) = 0.81$. Whether this second test is better or not depends on the relative importance of the two types of error: type I error, made when $H_0$ is true but we reject, and type II error, made when $H_a$ is true but we do not reject $H_0$. (**d**) Yes. For example, the test that rejects for $T_3 \leq 2$, is uniformly better.

**8.1.2** (**a**) Reject for $X \leq 74$. (**b**) $\gamma(p) = \Phi((74.5 - 100p)/\sqrt{100p(1 - p)})$. (**c**) 0.977, 0.0.837, 0.454, 0.085, 0.002.

**8.1.3** (**a**) Reject for $T \leq 61$. (**b**) power $\doteq 0.846$.

**8.1.4** (**a**) $\Omega = \{A, B, C\}$. (**b**) $A \times A$, where $A = \{1, 2, 3\}$.
(**c**) One good choice is: critical region $= \{(3, 3), (3, 2), (2, 3), (2, 2)\}$. For this region $\gamma(A) = 9/36, \gamma(B) = 16/36, \gamma(C) = 25/36$. (**c**) For this critical region, $\gamma(A) = 3/15, \gamma(B) = 6/15, \gamma(C) = 10/15$. (**d**) If we reject whenever $T = X_1 + X_2 + X_3 \leq 6$, except for the case that $X_1 = 1, X_2 = 2, X_3 = 2$, then $\gamma(A) = 30/120 = 0.25, \gamma(B) = 72/120 = 0.6$, and $\gamma(C) = 96/120 = 0.8$.
(**d**) The mass functions for $T = X_1 + X_2 + X_3$ for random samples of three taken without replacement are

| $\theta$ | $t$: 3 | 4 | 5 | 6 | 7 | 8 | 9 |
|---|---|---|---|---|---|---|---|
| $A$ | 0.05 | 0.30 | 0.30 | 0.30 | 0.05 | 0 | 0 |
| $B$ | 0 | 0.10 | 0.20 | 0.40 | 0.20 | 0.10 | 0 |
| $C$ | 0 | 0 | 0.05 | 0.30 | 0.30 | 0.30 | 0.05 |

If you want $\alpha = 0.25$, for example, you might reject for $T = 7, 8, 9$ and also with probability 2/3 when $T = 6$. Then the power for $\theta = B$ would be $(6/7)(0.40 + 0.20 + 0.10) = 0.643$. What would it be for $\theta = C$?

**8.1.5** (**a**) $H_0: \mu \geq 2.0$, $H_a: \mu < 2.0$. (**b**) Take $n = 54$. Reject $H_0$ if $\bar{X}_n < 1.9888$. (**c**) For $\mu = 1.99$, power $= 0.430$.

**8.1.6** (**a**) Let $M = \max(X_1, \ldots, X_{10})$. Reject for $M < m = 20(0.05)^{1/10} = 14.823$. $\gamma(\theta) = (m/\theta)^{10}$ for all $\theta$, (**b**) Reject for $M < 14.823$ and for $M > 20$. Then $\gamma(\theta) = (m/20)^{10} + (1 - (20/\theta)^{10})$, $\gamma(22) = 0.634$, $\gamma(25) = 0.898$, $\gamma(30) = 0.984$.

**8.1.7** (**a**) Reject for $X \leq 3$ and for $X \geq 12$. Then $\alpha = 0.0776$.
(**b**) $\gamma(p) = P(X \leq 3 \mid p) + P(X \geq 12 \mid p)$. For $p = 0.1, 0.2, 0.3, 0.4, 0.5$, these are 0.764, 0.236, 0.0775, 0.270, 0.655.
(**c**) In order that $\alpha \doteq 0.078$, we can let $k_1 = 21$, $k_2 = 38$, in which case $\alpha = 0.082$. These were found using the exact binomial distribution. Normal approximation using the 1/2-correction should give approximately the same values.
(**d**) $\gamma(p) \doteq \Phi((21.5 - 100p)/\sqrt{100p(1-p)}) + 1 - \Phi((37.5 - 100p)/\sqrt{100p(1-p)})$. For $p = 0.15, 0.20, \ldots, 0.45$, we get 0.966, 0.646, 0.211, 0.083, 0.302, 0.695, 0.934, using the normal approximation.
(**e**) $P(X \leq 21 \mid p = 0.25) = 0.209$, very close to $\gamma(0.25) = 0.211$, since $P(X \geq 38 \mid p = 0.25) \doteq 0.002$.

## Section 8.2

**8.2.1** (**a**) Reject for $\Sigma X_i \geq k$ or some $k$, or equivalently, $\overline{X} \geq k' = k/n$. In order to have $\alpha = 0.05$, take $k = n\mu_0 + 1.645\sigma\sqrt{n}$, equivalently $\overline{X} \geq \mu_0 + 1.645\sigma/\sqrt{n}$.
(**b**) $\gamma(\mu) = 1 - \Phi(1.645 + (\mu_0 - \mu)/(\sigma/\sqrt{n})) = \Phi(-1.645 + (\mu - \mu_0)/(\sigma/\sqrt{n}))$.
(**c**) Yes. Why?
(**d**) The test that rejects for $\overline{X} \leq \mu_0 - 1.645\sigma/\sqrt{n}$ has $\alpha = 0.05$, with power function $\gamma^*(\mu) = \Phi(-1.645 - (\mu - \mu_0)/(\sigma/\sqrt{n}))$, which is larger than $\gamma(\mu)$ for $\mu < \mu_0$ since $1 - \Phi(1.645 + (\mu_0 - \mu)/\sigma/\sqrt{n}) = \Phi(-1.645 + (\mu - \mu_0)/\sigma/\sqrt{n}) < \Phi(-1.645 - (\mu - \mu_0)/(\sigma/\sqrt{n}))$ for $\mu < \mu_0$.
(**e**) Let $Q = \Sigma(X_i - \mu)^2$. Reject for $Q > k$. To find $k$, do as follows. $P(Q/\sigma^2 > k/\sigma^2)$ must be 0.05 when $\sigma^2 = 100$. The 0.95-quantile of the $\chi^2$-distribution with 25 df is 37.65. Reject for $Q/100 > 37.65$, equivalently for $\hat{\sigma}^2 = Q/25 > (100)(37.65)/25 = 150.6$ then $\gamma(140) = P(Q/100 > 35.65|\sigma^2 = 140) = P(\chi^2 = Q/140 > 37.65(100/140) = 26.89)$, where $\chi^2$ is distributed as chi-square with 25 df. From the chi-square tables for 25 df, we find that the power is approximately 0.36.

**8.2.2** Reject for $r(\mathbf{X}) = r(X_1, X_2, X_3) > 3.0$. This is equivalent to defining the critical region to be $C = \{(x_1, x_2, x_3)|x_1 + x_2 + x_3 \geq 7\}$. $\gamma(0) = 0.076$, $\gamma(1) = 0.81$.

**8.2.3** (**a**) Reject for $T = X_1 + X_2 + X_3 \geq 10$. Then $\alpha = 0.084$. (**b**) $\gamma(2.5) = 0.224$, $\gamma(3) = 0.603$, $\gamma(3.5) = 0.865$. (**c**) Yes. Why? (**d**) 0.0499, 0.3665, 0.8069.

**8.2.5** (**a**) Let $M = \max(X_1, \ldots, X_5)$. Reject for $M > 19.796$. (**b**) $\gamma(\theta) = 1 - (19.796/\theta)^5$ for $\theta \geq 19.796$, 0 for $\theta < 19.796$. (**c**) Yes, since the NP lemma gives the same test for any alternative $\theta_a > \theta_0 = 20$. (**d**) Reject for $M \leq c$, where $c$ is chosen so that $P(M \leq c|\theta = 20) = 0.05$. Thus, $(c/20)^5 = 0.05$, so that $c = 20(0.05^{1/5}) = 10.986$. Power $= P(M \leq c)|\theta = 15) = P(M/15 \leq c/15|\theta = 15) = (c/15)^5 = (0.05)(20/15)^5 = 0.2107$.

**8.2.6** (**a**) Let $Y_1, Y_2, Y_3$ be the frequencies of 1, 2, 3. Reject for $V = \log(0.6/0.5)Y_1 + \log(0.3/0.3)Y_2 + \log(0.1/0.2)Y_3 \geq 0.7203$. (**b**) No (a close call). (**c**) 119.

## Section 8.3

**8.3.1** (**a**) For example, $\Lambda(0, 1) = 1/6$, $\Lambda(2, 3) = 6$. (**b**) Reject for $(X_1, X_2) = (1, 1), (1, 2), (2, 1), (2, 2)$. (**c**) $\gamma(1) = \gamma(2) = 0.09$, $\gamma(3) = \gamma(4) = 0.49$.

**8.3.2** (**a**) $\Lambda(\mathbf{x}) = (\max(\mathbf{x})/\theta_0)^n$ for $\max(x) \leq \theta_0$, 0 for $\max(\mathbf{x}) > \theta_0$. (**b**) $c = \theta_0\alpha^{1/n}$, do not reject. (**c**) $\gamma(\theta) = 1$ for $\theta \leq c$, $\alpha(\theta_0/\theta)^n$ for $c \leq \theta \leq \theta_0$, $1 - (\theta_0/\theta)^n[1 - \alpha]$ for $\theta > \theta_0$. (**d**) Same test, same power function as in (b) and (c).

**8.3.4** (**c**) $C_n = 2.216$. Do not reject $H_0$.

**8.3.5** (**b**) $\hat{\theta} = 0.7025$, $C_n = C_{400} = 3.238$. Do not reject.

**8.3.6** (**b**) $C_{141} = 3.858$. Reject $H_0$.

## Section 8.4

**8.4.1** (**a**) 0.0060, 0.0464, 0.1673. (**b**) 0.0.0015, 0.1792, 0.5406, 0.8692 using normal approximation; 0.0012, 0.1795, 0.5433, 0.8689 using exact binomial.

**8.4.2** (**a**) $T_2 \geq 10$, $\alpha = 0.084$. (**b**) 0.020. (**c**) 0.758. (**d**) $T_{100} \geq 325$, $\alpha = 0.0786$. (**e**) $p$-value $= 0.0499$. (**f**) 0.6190.

**8.4.3** (**a**) 0.012. (**b**) [80.864, 87.136]; reject. (**c**) 0.295.

**8.4.4** (**a**) Reject for $M \leq c = \alpha^{1/n}\theta_0$. (**b**) $C = [M, \alpha^{-1/n}M]$. (**c**) 0.566. (**d**) 0.202.

**8.4.5** (**a**) .04, .08, .21, .43, .67, .91, 1.00 for $T = 0, 1, \ldots, 6$.

**8.4.6** (**a**) 11.15. (**b**) 0.165.

## Section 9.2

**9.2.1** $f(x_1, x_2) = Ce^{-(1/2)}Q(x_1, x_2)$, where $C = [1/(2\pi)][1/((1 - \rho^2)\sigma_1\sigma_2)$.

**9.2.2** $D = Y - X \sim \mathrm{N}(13, 49.09)$. (**a**) $P(D > 0) = 0.9682$. (**b**) $P(|D| > 5) = 0.8783$.

**9.2.3** (**a**) $\rho(\mathbf{X}) = \begin{pmatrix} 1 & 1/2 & -1/8 \\ 1/2 & 1 & -5/12 \\ -1/8 & 5/12 & 1 \end{pmatrix}$. (**b**) Multivariate normal with covariance matrix, $A\Sigma A^T = \begin{pmatrix} 47 & 0 \\ 0 & 208 \end{pmatrix}$, mean vector (6, 88). (**c**) 0. (**d**) 0.068.

**9.2.4** Let $X = \sigma_X Z_1 + \mu_X$ and $Y = aZ_1 + bZ_2 + \mu_y$, where $a = \rho\sigma_Y$, $b = \sigma_y(1-\rho^2)^{1/2}$.

**9.2.5** (**b**) 0. (**c**) No in both cases, because neither vector is of the form $\mu + u$, where $u$ is in the column space of $\Sigma$.

**9.2.6** 0.0598.

## Section 9.3

**9.3.1** $Z^2$ has density $g(u) = u^{-1/2}e^{-u/2}/(2\pi)^{1/2}$ for $u > 0$. This is $\Gamma(\alpha = 1/2, \theta = 2)$.

**9.3.2** (**a**) By definition the sum of the squares of two independent standard normal rv's has the chi-square distribution with two df's. From what we know of the $\Gamma(\alpha, \theta)$ distribution, the sum of two independent $\Gamma(1/2, 2)$ rv's has the $\Gamma(1, 2)$ distribution. The density is therefore $(1/(2\Gamma(2)))u^{1-1}e^{-u/2} = (1/2)e^{-u/2}$ for $u > 0$. This is the exponential density with mean 2.

**9.3.3** (**a**) $K = 1/9$. 3 df, with noncentrality parameter $\delta = 14/3$. (**b**) $C_1 = 1/9$, $C_2 = 1/18$, $\delta = 4.5$, 2 df.

**9.3.4** (**a**) There are an infinity of choices. Perhaps the simplest are $\mathbf{a}_3 = (0, 0, 1, -1)^T/\sqrt{2}$ and $\mathbf{a}_4 = (1, 1, -1, -1)^T/2$. (**b**) $b_1 = 10$, $b_2 = -\sqrt{2}$, $b_3 = -\sqrt{2}$, $b_4 = -6$. (**c**) 140 and 40. (**d**) $\mathbf{U} \sim N_4((50, 0, 0, 0)^T, 25I_4)||\mathbf{U}||^2/25 \sim \chi_4^2(100)$.

**9.3.5** 30.404 and 31.403.

**9.3.6** (0.0668)(0.2277) = 0.01521.

## Section 9.4

**9.4.1** $k_1 = 1/3$, $k_2 = 2$.

**9.4.2** The solution can be found by writing $f_T(t) = \int_0^\infty \phi(z) f_V(tz) t\,dz$, where $f_v$ is the density of $V = \sqrt{W/\nu}$ and $W$ has the chi-square density for $\nu$ df. Students are advised not to spend much time on this.

**9.4.4** (**b**) $W > 1$ if and only if $X_1 > 0$ and $X_2 > 0$. Therefore, $P(W > 1) = 1/4$.

**9.4.6** (**a**) $[26.84 \pm 2.25]$. (**b**) $t = 2.491$, $p$-value approximately 0.457. (**c**) $[1.293 \pm 3.040]$ (**d**) $t = 0.746$, $p$-value $= 0.467$.
(**e**) The lazy-$t$ test has noncentrality parameter $\theta_1 = \Delta/[(2\sigma^2)/8]^{1/2}$. The two-sample $t$-test has noncentrality parameter $\theta_2 = \Delta/[\sigma^2(1/9 + 1/8)]^{1/2}$. Their ratio is $\theta_2/\theta_1 = [2(1/8)/(1/9 + 1/8)]^{1/2} = (18/17)^{1/2} > 1$, so that the two-sample noncentrality parameter is larger in magnitude.
(**f**) $\hat{\Delta} = 0.9$, so the 90% CI on $\Delta$ is $[0.9 \pm 3.360]$, $t = 0.9/1.774 = 0.507$, $p$-value $= 0.628$.

**9.4.7** (**b**) $K = 1/\sqrt{1/n + 1/m}$, $(n - 1)$, $\theta = (K/\sigma)(\Delta - \Delta_0)$, where $\Delta = \mu_2 - \mu_1$.

## Section 9.5

**9.5.1** (**d**) $P(S_p^2 > 36.784, R < 3.438) = P(16S_p^2/25 > (16/25)36.784)P(R < 3.438)$. $W = 16S_p^2/25$ has the chi-square distribution with $n + m - 2 = 16$ df, and $R$ has the $F(8, 8)$ distribution. Therefore, using the chi-square and $F$-tables, we find that $(0.10)(0.95) = 0.095$.

**9.5.2** (**a**) $K = \sqrt{n/2}$, $\nu = 2(n - 1)$, $\theta = 0$. (**b**) $F(1, n - 1)$.

**9.5.3** (**a**) Reject for $F > 2.5876$, $\gamma(\tau) = P(F = S_1^2/S_2^2 > 2.5876) = P(F/\tau > 2.5876/\tau)$. $Q = F\tau$ has the central $F$-distribution with $m - 1$ and $n - 1$ df, so $\gamma(\tau)$ is the area to the right of $2.5876/\tau$ under the $F$-density.

**9.5.4** $S_1^2 = 11.5$, $S_2^2 = 67.33$, $F = S_2^2/S_1^2 = 5.855$, $F_{0.015}(3, 4) = 0.0662$, $F_{0.975}(3, 4) = 9.979$. 95% CI $= [5.855/9.979 = 0.587, 5.855/0.0662 = 88.427]$.

**9.5.5** (**a**) $\alpha = \nu_1/2$, $\beta = \nu_2/2$. (**b**) $F_{0.95}(10, 12) = 2.7534$, so the 0.95-quantile of the Beta(5, 6) distribution is 0.69646.

**9.5.6** (**a**) Generate independent rv's $V_1$, $V_2$ which have central chi-square distributions with 10 and 15 df's. (How could you do that?) For each such pair $(V_1, V_2)$, compute $F = (V_1/10)/(V_2/15)$. Repeat this (say) 10,000 times. Order these 10,000 $F$-values. Choose the 6000th one as an estimate of $x_{0.6}$.
(**b**) You would need approximately $[(1.96)(0.5)/0.005]^2 = 36,864$ $F$-values. For each $F$ you need 10 + 15 = 25 pseudostandard normal rv's. Total $= 230,400$.

## Section 10.2

**10.2.1** For example, $P(W_{XY} = 4) = P(W = 7) = 2/10$.

**10.2.2** (**a**) $W = 87$, $p$-value $= 0.0061$, normal approximation: $z = 2.604$, $p$-value $= 0.0046$. (**b**) 26. (**c**) $s = 11, t = 44$, $[8, 51]$. (**d**) $[6.28, 48.47]$. (**e**) 0.676. (**f**) 0.0.779. (**g**) $W^* = 84.5$, $z = 2.387$, $p$-value $= 0.0085$.

**10.2.3** (**a**) Let $f_0$ and $f_a$ be the probability mass function of $W$ under $H_0$ and $H_a$. Then $f_0(3) = f_0(4) = f_0(8) = f_0(9) = 0.1$, $f_0(5) = f_0(6) = f_0(7) = 0.2$, and $f_a(3) = f_a(4) = 0.021/40$, $f_a(5) = 2/40$, $f_a(6) = f_a(7) = 7/40$, $f_a(8) = 6/40$, $f_a(9) = 16/40$. (**b**) $E(W|H_0) = 6$, $E(W|H_a) = 7.5$, $\text{Var}(W|H_0) = 3.0$, $\text{Var}(W|H_a) = 2.5$. (**c**) Reject for $W = 8$ or 9, power $= 0.55$. (**e**) No, since another test has power 0.85.

**10.2.4** $4/126$.

**10.2.5** $z = -1.308$, $p$-value $= 0.191$.

**10.2.6** (**a**) For example, $P(W^* = 3.5) = 2/15$, $P(W^* = 5) = 1/15$, $P(W^* = 5.5) = 2/15$ (**b**) $E(W^*) = 7$, $\text{Var}(W^*) = 4.4$.

**10.2.7** (**a**) $12/792 = 0.0151$. (**b**) 0.0174.

## Section 10.3

**10.3.2** (**b**) 0.0766. (**c**) 0.0213.

**10.3.4** (**a**) $[X_{(137)}, X_{(184)}]$. (**b**) Take $t - s \doteq 1450$.

**10.3.5** 5 and 7.5.

**10.3.6** (**a**) $10/128 = 0.0781$. (**b**) 0.0754.

**10.3.7** (**a**) Reject for $Y \geq 217$, $W^+ \geq 43,906$, $\hat{D} \geq 0.8225$. (**b**) Powers: 0.477, 0.618, 0.639.

**10.3.9** For Wilcoxon: 0.0090. For sign: 0.0665 (using normal approximation with 1/2-correction).

**10.3.10** (**a**) $W^+$ takes the values $0, 1, \ldots, 6$ with probabilities $(1, 1, 1, 5, 4, 7, 8)/27$. (**b**) $E(W^+) = 117/27$, $\text{Var}(W^+) = 577/27$.

**10.3.11** $6/512$.

## Section 10.4

**10.4.1** (**b**) $D^+ = 0.3416$, $D^- = 0.0645$, $D = 0.3416$.

**10.4.2** (**a**) [0.030, 0.210], [0.529, 0.791], [0.790, 0.970]. (**b**) [320, 475] or [318, 468]. (**c**) 0.0071 for normal approximation, 0.0057 for exact binomial. (**d**) $D = 0.1749$, $p$-value $= 0.094$.

**10.4.3** (**a**) 0.65. (**b**) 0.0234. (**c**) 0.0088.

## Section 11.2

**11.2.1** (**a**) $b = 3$. (**c**) 414.

**11.2.4** (**a**) $b = (5, 3)^T$.

**11.2.6** (**a**) $\hat{y} = b_0 + b_1 x$, where $b_0 = -12,856.38, b_1 = 9.7175$. (**b**) Total SSQS = 7,118,267,246, Regr. SSQS = 3,032,350,922, Error SSQS = 4,085,916,324, $R^2 =$ 0.4260, $R = 0.6797$. (**c**) $y = 3.8525(10^{-3})x^{1.8925}$.

**11.2.7** (**e**) The second row of $P_V$ is $(1, 3, -1, 1)(1/4)$, $\hat{y} = (6, 4, 3, 1)^T$.

**11.2.8** (**a**) $a_1 = (1/3)(1, 1, 0, 1)^{\mathrm{T}}$. (**b**) $\hat{W}_1 = 5$.

## Section 11.3

**11.3.1** (**a**) $\hat{\beta} = 2$, 95%, CI : [0.591, 3.408]. (**b**) Let $(L(x_0, Y, x), U(x_0, Y, x)) = x_0 I$, where $I$ is as given above. The graphs are straight lines through (0, 0) with slopes given by the first and second terms of $I$.

**11.3.2** $T = \sqrt{n-2}r/\sqrt{1-r^2}$.

**11.3.3** (**a**) $\hat{\beta} = (3, 2)^{\mathrm{T}}, \hat{y} = (5, 7, 6, 11, 13)^{\mathrm{T}}, S^2 = 4/3$. (**b**) $[-2.021, 10.02]$. (**c**) $\mathbf{a} = (47, 33, -42, 5, -9)^{\mathrm{T}}/38 = (1.237, 0.868, -1.105, 0.132, -0.237)^{\mathrm{T}}, \mathbf{h} = (-1, 1, 0, 1, -1)^{\mathrm{T}}$ is an example.

**11.3.4** (**a**) $\mathbf{Y} = (Y_{11}, Y_{12}, Y_{21}, Y_{22}, Y_{23}, Y_{31}, Y_{32})^{\mathrm{T}}, \mathbf{x}_1 = (1, 1, 0, 0, 0, 0, 0)^{\mathrm{T}}, \mathbf{x}_2 = (0, 0, 1, 1, 1, 0, 0)^{\mathrm{T}}, \mathbf{x}_3 = (0, 0, 0, 0, 0, 1, 1)^{\mathrm{T}}, \varepsilon_{ij} = Y_{ij} - \mu_I, \varepsilon = (\varepsilon_{11}, \ldots, \varepsilon_{32})$. Then $\mathbf{Y} = \mu_1\mathbf{x}_1 + \mu_2\mathbf{x}_2 + \mu_3\mathbf{x}_3 + \varepsilon$. (**b**) $\hat{\mu}_1 = 7, \hat{\mu}_2 = 10, \hat{\mu}_3 = 13$. (**c**) $S^2 = 18/4 = 4.5$. (**d**) Let $h_{23} = S^2(1/2 + 1/3) = 3.75$. Then $I_{13} = [-3 \pm 2.776\sqrt{h_{23}}] = [-3 \pm 5.38]$. (**e**) $R^2 = 1 - \mathrm{ESS/TSS}$. For the data, $R^2 = 1 - 18/54 = 36/54 = 2/3. R = 0$ if the $\overline{Y}_i$ are all equal.

**11.3.5** $\hat{\beta} = (5, 2)^{\mathrm{T}}, \mathrm{Var}(\hat{\eta}) = (2/3)\sigma^2, S^2 = \mathrm{ESS}/(4-2) = 3/2, \mathrm{CI} = [3 \pm 4.303\sqrt{(2/3)S^2}] = [3 \pm 4.303]$.

**11.3.7** (**a**) $\hat{\beta} = (58.2028, 16.1198, -16.9435)^{\mathrm{T}}$. (**b**) $[-1.7189, 0.0715]$.

## Section 11.4

**11.4.1** (**a**) $= (3, 2, 4)^{\mathrm{T}}$ (**b**) $S^2 = 4/2 = 2$, (**c**) $[3 \pm 12.30]$. (**d**) $F = 17.2$. Reject. (**e**) $||\theta - \theta_0||^2 = 7.693, \delta = 1.923$.

**11.4.2** (**a**)

| | df | Sum of Squares | Mean Squares |
|---|---|---|---|
| Treatments | 3 | 148 | 49.333 |
| Residuals | 10 | 56 | 5.600 |
| Corr. total | 13 | 204 | |

(**b**) From the table, $F = 8.809$ with observed $p$-value 0.0037, so we reject for any $\alpha \geq 0.0037$. (**c**) $[-0.043, 7.377]$.

**11.4.3** (**a**) $\boldsymbol{\beta} = \mathbf{M}^{-1}\mathbf{X}^{\mathrm{T}}\boldsymbol{\theta}$, so $C\boldsymbol{\beta} = \mathbf{C}\mathbf{M}^{-1}\mathbf{X}^{\mathrm{T}}\boldsymbol{\theta} = \mathbf{A}^{\mathrm{T}}\boldsymbol{\theta}$, where $\mathbf{A} = \mathbf{X}\mathbf{M}^{-1}\mathbf{C}^{\mathrm{T}}$. Let $V_1 =$ (column space of $\mathbf{A}$). Let $V_0 = V \cup V_1^{\perp}$. Then $\mathbf{C}\boldsymbol{\beta} = \mathbf{0}$ is equivalent to $\boldsymbol{\theta} \in V_0$.
(**b**) Write the projection matrix with respect to $V_1$ in terms of $\mathbf{A}$, then in terms of $\mathbf{X}$ and $\mathbf{C}$. Then use this to write $||\hat{Y}_1||^2$ in terms of $\mathbf{C}$, $\mathbf{X}$, and $\hat{\boldsymbol{\beta}}$.

**11.4.5** (**a**) $F = [(n-k)/(n-1)][R^2/(1-R^2)]$. (**b**) $P(R^2 \geq 0.2525) = 0.1000$.

**11.4.6** $n_0 = 21$.

**11.4.7** (**a**) Express $\mathrm{ESS}_1$ and $\mathrm{ESS}_2$, the error sums of squares for the two lines, in terms of $\mathrm{ESS}_1 = S_{1yy} - (S_{1xy})^2/S_{1xx}$, and a similar expression for the second line. Then $\mathrm{SSE}_{\mathrm{FM}} = \mathrm{ESS}_1 + \mathrm{ESS}_2$. Under $H_0$ there is only one straight line. Let $S_{xy}$, $S_{xx}$, $S_{yy}$ be defined similarly for all $n = n_1 + n_2$ pairs. Then $\mathrm{ESS}_{H_0} = S_{yy} - S_{xy}^2/S_{xx}$. Then $F = [(\mathrm{ESS}_{H_0} - \mathrm{ESS}_{\mathrm{FM}})/2]/\mathrm{ESS}_{\mathrm{FM}}/(n-4)$ This can be written in other ways.
(**b**) $F = 3.33$, (observed $p$-value for 2 and 3 df) = 0.173.

**11.4.8** (**a**) $g(t) = \beta_0 + \beta_1 t$ for $50 \leq t \leq 10$ and $\beta_0 + 100\beta_1 + \beta_2(t-100)$ for $100 < t \leq 150$. (**b**) Let $\mathbf{X}$ have column vectors $x_0 = (1, 1, 1, 1, 1)^{\mathrm{T}}$, $\mathbf{x}_1 = (50, 80, 100, 100, 100)^{\mathrm{T}}$, $\mathbf{x}_2 = (0, 0, 10, 30, 50)^{\mathrm{T}}$, $\mathbf{y} = (144, 190, 270, 345, 381)^{\mathrm{T}}$. $\mathrm{ESS}_{H_0} = ||\mathbf{y} - \hat{Y}||^2 = 741.94$, $||\hat{Y} - \hat{Y}_0||^2 = \mathrm{ESS}_{H_0} - \mathrm{ESS}_{\mathrm{FM}} = 741.94 - 422 = 319.94$, $F = (319.94/(422/2) = 1.516$ for 1 and 2 df. Observed $p$-value $= 0.343$.

## Section 11.5

**11.5.1** (**c**)

| Source, Subspace | df | SQS | Mean SQ | $F$ | $p$-value |
|---|---|---|---|---|---|
| $A$ | 1 | 571.32 | 571.32 | 16.20 | 0.0069 |
| $B$ | 2 | 196.16 | 98.08 | 2.78 | 0.1398 |
| $A \times B$ | 2 | 73.62 | 36.81 | 1.04 | 0.4083 |
| Among cells | 5 | 841.10 | 168.22 | 13.97 | 0.0030 |
| Residual | 6 | 211.61 | 35.27 | | |
| Corr. total | 11 | 1052.71 | | | |

(**e**) $[19.75 \pm 14.53]$, $[13.16 \pm 8.39]$, $[7.40 \pm 10.28]$, $[9.40 \pm 10.28]$, 80%.
(**f**) For equality of the cell means: $\delta = 30.56$, power $= 0.804$. For $A \times B$ Interaction: $\delta = 0.32$, Power $= 0.065$. For $A$ effects: $\delta = 23.52$, power $= 0.979$. For $B$ Effects: $\delta = 6.72$, power $= 0.420$.
(**g**) 80 observations per cell.

**11.5.3** (**b**)

| Source | df | Sum of Squarea | Mean Squarea | $F$ | $\Pr(F)$ |
|---|---|---|---|---|---|
| Temperature | 3 | 222 | 74.0 | 6.17 | 0.029 |
| Machine | 2 | 168 | 84.0 | 7.00 | 0.027 |
| Residual | 6 | 72 | 12.0 | | |
| Corr. Total | 11 | 462 | | | |

(**c**) $[4 \pm 6.92]$. (**d**) $Y_{ij} = \mu + \beta T + m_i + \varepsilon_{ij}$, where the $m_i$ sum to zero, or $Y_{ij} = \beta T + m_i + \varepsilon_{ij}$, where the $m_i$ do not necessarily sum to zero. For each case, suppose that the $\varepsilon_{ij}$ are independent, each $N(0, \sigma^2)$. The two models are equivalent. (**e**) The $F$-statistic is $((172.5 - 72)/2)/(72/6) = 4.1875$. $p$-value $= 0.073$.

## Section 12.2

**12.2.1** (**a**) 0.2501. (**b**) 0.7499. (**c**) 0.406. (**d**) $L(4) = 0.1929$, $U(4) = 0.8071$.

**12.2.2** On $\rho$: [0.835, 3.098].

**12.2.4** (**a**) $L(10) = 5.42$, $U(10) = 16.96$. (**b**) $U(2) = 6.296$.

**12.2.5** (**b**) CI[1.154, 3.484] on $\rho$, point estimate 1.556.

## Section 12.3

**12.3.1** (**b**) $x_{0.5} = 1.5$, $x_{0.9} = 2.599$.

**12.3.2** (**a**) $\hat{\boldsymbol{\beta}} = (-2.191, 1.502)^T$. (**b**) $\widehat{\mathrm{Cov}}(\hat{\boldsymbol{\beta}}) = \begin{pmatrix} 0.242 & -0.115 \\ -0.115 & 0.063 \end{pmatrix}$. (**c**) [1.010, 1.994] (**d**), $G^2 = 0.294$, $\chi^2 = 0.295$.

**12.3.3** (**b**) The diagonal terms of $\mathrm{Cov}(\hat{\boldsymbol{\beta}})$ are 0.595 and 0.0234. The diagonal terms of $\widehat{\mathrm{Cov}}(\hat{\boldsymbol{\beta}})$ are 0.0622 and 0.0325. (**c**) $G^2 = 7.333$, $\chi^2 = 7.349$, $p$-values $= 0.1186$ and 0.1193. (**d**) [2.261, 3.156].

## Section 12.4

**12.4.2** (**a**) 0.0805. (**b**) $\chi^2 = 3.884$, $p$-value $= 0.0487$. (**c**) $Z = 1.9708$, $p$-value without 1/2-correction $= 0.0487$. For 1/2-correction it is 0.090.

**12.4.3** (**b**) $\chi^2 = 1.0124$.

**12.4.4** (**c**) 0.024. (**d**) 0.155.

**12.4.6** (**b**) $\chi^2 = 18.71$, $p$-value $= 0.0235$. (**b**) Power $\doteq 0.819$. (**c**) 1.278 and 1.111.

## Section 12.5

**12.5.1** (**a**) $\chi^2 = 18.28$, combining the last two categories, $p$-value $= 0.00386$. (**b**) $\chi^2 = 17.64$, combining last two categories, $p$-value 0.00015.

**12.5.2** (**a**) $\chi^2 = 4.400$, $p$-value $= 0.111$.

**12.5.3** (**a**) $\chi^2 = 16.47$, $p$-value $= 0.021$. (**b**) $\chi^2 = 7.788$, $p$-value $= 0.1859$. (**c**) $\delta = 4.017$, power $\doteq 0.239$. (**d**) 0.515.

**12.5.4** (**b**) $\chi^2 = 1.0$, $p$-value $= 0.317$ (**d**) Approximate power 0.516. (**e**) $\delta = 11.11$, power approximately 0.915.

**12.5.6** (**b**) 0.816. (**c**) 0.980.

## Section 13.2

**13.2.1** (**a**) For example, $S_7(t) = 2/7$ for $20 \le t < 23$. (**b**) For example, $\hat{S}(t) = 10/21$ for $20 \le t < 27$. (**c**) For example, $\hat{S}(t) = 5/9$ for $7 \le t < 9$, undefined for $t > 13$.

**13.2.2** (**a**) $\exp[-\int_t^{t+h} \lambda(u)du]$. (**b**) $G(t, h) = \exp(-h\theta)$. (**c**) $\exp(-\theta^m[(t+h)^m - t^m])$.

**13.2.5** (**a**) For example, $(Y_1, \Delta_1) = (1.12, 1)$. $\Delta = 1$ for 14 pairs. (**b**) 2/3.

## Section 13.3

**13.3.1** (**a**) For example, $e_M^*$ takes the value $-1$ with probability 7/27. (**b**) 0.317 and 0.317.

**13.3.2** (**a**) 0.1625. (**b**) [0.833, 2.561] and [0.886, 2.499] for $B = 10,000$.
(**c**) [1.525, 2.625], [12.05, 14.50] for bootstrap samples of $B = 1000$.

**13.3.4** (**c**) 0.0127 for $B = 1000$ parametric bootstrap samples. (**d**) [18.330, 21.12], again for $B = 1000$.

## Section 13.4

**13.4.1** (**a**) For example, $P(\theta = 1|X = (1, 1)) = 41/42$, $P(\theta = 1|X = (1, 2)) = 7/8$. (**b**) $\hat{\theta}(x_1, x_2) = 2$ for $(x_1, x_2) = (2, 3), (3, 2), (3, 3), (2, 2)$, 1 otherwise. $P(\hat{\theta} = \theta) = 0.7152$. (**c**) $\hat{\theta}(x_1, x_2) = 1$ for $(x_1, x_2) = (1, 1), (1, 2), (2, 1), (2, 2)$, $=$ 2 otherwise. $P(\hat{\theta} \ne \theta) = 0.705$.

**13.4.2** $\theta(X) = (1 - X)/(-\log(X))$.

**13.4.3** $[(R/(1+R)(\overline{X} - \theta_0) + \theta_0 \pm 1.96\sqrt{\tau^2/(1+R)}]$ for $R = n\tau^2/\sigma^2$, [38.01, 45.53].

**13.4.5** (**d**) $\hat{\theta} = 0.296$. The observations were generated with $\theta = 0.3$.

## Section 13.5

**13.5.1** (**b**) Var$\overline{Y} = 18.45$, Var$(\hat{\tau}) = 461.45$. (**d**)$E(\hat{R}) = 3.0025$, $R = 3$, Var$(\hat{R}) = 0.0385$, Var$(\hat{\tau}_R) = 15.42$. By formula we get 36.

**13.5.2** (**c**) Var$(\hat{\tau}_{str}) = 50\,2/3$. (**d**) Var$(\hat{\tau}) = 1000/9$.

**13.5.3** (**a**) 10,794.5. (**b**) 8, 3, 16, 13, and 9639.4. (**c**) $\sigma^2 = 98.18$, Var$(\hat{\tau}) = 40,772.0$.

**13.5.4** (**b**) 49 and 105.78. Approximation: Var$(\hat{\tau}_R) = 320.62$.

**13.5.5** (**b**) $[0.472, 0.602]$. (**c**) $[-0.055, 0.203]$. (**d**) 41, 95, 74.

**13.5.6** (**b**) 2.17, 1.07. (**c**) $1/12$, 1.07. (**e**) The first.

# Index

# WILEY SERIES IN PROBABILITY AND STATISTICS

ESTABLISHED BY WALTER A. SHEWHART AND SAMUEL S. WILKS



The ***Wiley Series in Probability and Statistics*** is well established and authoritative. It covers many topics of current research interest in both pure and applied statistics and probability theory. Written by leading statisticians and institutions, the titles span both state-of-the-art developments in the field and classical methods.

Reflecting the wide range of current research in statistics, the series encompasses applied, methodological and theoretical statistics, ranging from applications and new techniques made possible by advances in computerized practice to rigorous treatment of theoretical approaches.

This series provides essential and invaluable reading for all statisticians, whether in academia, industry, government, or research.

† ABRAHAM and LEDOLTER · Statistical Methods for Forecasting
AGRESTI · Analysis of Ordinal Categorical Data
AGRESTI · An Introduction to Categorical Data Analysis, *Second Edition*
AGRESTI · Categorical Data Analysis, *Second Edition*
ALTMAN, GILL, and McDONALD · Numerical Issues in Statistical Computing for the Social Scientist
AMARATUNGA and CABRERA · Exploration and Analysis of DNA Microarray and Protein Array Data
ANDĚL · Mathematics of Chance
ANDERSON · An Introduction to Multivariate Statistical Analysis, *Third Edition*
* ANDERSON · The Statistical Analysis of Time Series
ANDERSON, AUQUIER, HAUCK, OAKES, VANDAELE, and WEISBERG · Statistical Methods for Comparative Studies
ANDERSON and LOYNES · The Teaching of Practical Statistics
ARMITAGE and DAVID (editors) · Advances in Biometry
ARNOLD, BALAKRISHNAN, and NAGARAJA · Records
* ARTHANARI and DODGE · Mathematical Programming in Statistics
* BAILEY · The Elements of Stochastic Processes with Applications to the Natural Sciences
BALAKRISHNAN and KOUTRAS · Runs and Scans with Applications
BALAKRISHNAN and NG · Precedence-Type Tests and Applications
BARNETT · Comparative Statistical Inference, *Third Edition*
BARNETT · Environmental Statistics
BARNETT and LEWIS · Outliers in Statistical Data, *Third Edition*
BARTOSZYNSKI and NIEWIADOMSKA-BUGAJ · Probability and Statistical Inference
BASILEVSKY · Statistical Factor Analysis and Related Methods: Theory and Applications
BASU and RIGDON · Statistical Methods for the Reliability of Repairable Systems
BATES and WATTS · Nonlinear Regression Analysis and Its Applications

*Now available in a lower priced paperback edition in the Wiley Classics Library.
†Now available in a lower priced paperback edition in the Wiley–Interscience Paperback Series.

BECHHOFER, SANTNER, and GOLDSMAN · Design and Analysis of Experiments for Statistical Selection, Screening, and Multiple Comparisons
BELSLEY · Conditioning Diagnostics: Collinearity and Weak Data in Regression
† BELSLEY, KUH, and WELSCH · Regression Diagnostics: Identifying Influential Data and Sources of Collinearity
BENDAT and PIERSOL · Random Data: Analysis and Measurement Procedures, *Third Edition*
BERRY, CHALONER, and GEWEKE · Bayesian Analysis in Statistics and Econometrics: Essays in Honor of Arnold Zellner
BERNARDO and SMITH · Bayesian Theory
BHAT and MILLER · Elements of Applied Stochastic Processes, *Third Edition*
BHATTACHARYA and WAYMIRE · Stochastic Processes with Applications
BILLINGSLEY · Convergence of Probability Measures, *Second Edition*
BILLINGSLEY · Probability and Measure, *Third Edition*
BIRKES and DODGE · Alternative Methods of Regression
BISWAS, DATTA, FINE, and SEGAL · Statistical Advances in the Biomedical Sciences: Clinical Trials, Epidemiology, Survival Analysis, and Bioinformatics
BLISCHKE AND MURTHY (editors) · Case Studies in Reliability and Maintenance
BLISCHKE AND MURTHY · Reliability: Modeling, Prediction, and Optimization
BLOOMFIELD · Fourier Analysis of Time Series: An Introduction, *Second Edition*
BOLLEN · Structural Equations with Latent Variables
BOLLEN and CURRAN · Latent Curve Models: A Structural Equation Perspective
BOROVKOV · Ergodicity and Stability of Stochastic Processes
BOULEAU · Numerical Methods for Stochastic Processes
BOX · Bayesian Inference in Statistical Analysis
BOX · R. A. Fisher, the Life of a Scientist
BOX and DRAPER · Response Surfaces, Mixtures, and Ridge Analyses, *Second Edition*
* BOX and DRAPER · Evolutionary Operation: A Statistical Method for Process Improvement
BOX and FRIENDS · Improving Almost Anything, *Revised Edition*
BOX, HUNTER, and HUNTER · Statistics for Experimenters: Design, Innovation, and Discovery, *Second Editon*
BOX and LUCEÑO · Statistical Control by Monitoring and Feedback Adjustment
BRANDIMARTE · Numerical Methods in Finance: A MATLAB-Based Introduction
† BROWN and HOLLANDER · Statistics: A Biomedical Introduction
BRUNNER, DOMHOF, and LANGER · Nonparametric Analysis of Longitudinal Data in Factorial Experiments
BUCKLEW · Large Deviation Techniques in Decision, Simulation, and Estimation
CAIROLI and DALANG · Sequential Stochastic Optimization
CASTILLO, HADI, BALAKRISHNAN, and SARABIA · Extreme Value and Related Models with Applications in Engineering and Science
CHAN · Time Series: Applications to Finance
CHARALAMBIDES · Combinatorial Methods in Discrete Distributions
CHATTERJEE and HADI · Regression Analysis by Example, *Fourth Edition*
CHATTERJEE and HADI · Sensitivity Analysis in Linear Regression
CHERNICK · Bootstrap Methods: A Guide for Practitioners and Researchers, *Second Edition*
CHERNICK and FRIIS · Introductory Biostatistics for the Health Sciences
CHILÈS and DELFINER · Geostatistics: Modeling Spatial Uncertainty
CHOW and LIU · Design and Analysis of Clinical Trials: Concepts and Methodologies, *Second Edition*
CLARKE and DISNEY · Probability and Random Processes: A First Course with Applications, *Second Edition*
* COCHRAN and COX · Experimental Designs, *Second Edition*

*Now available in a lower priced paperback edition in the Wiley Classics Library.
†Now available in a lower priced paperback edition in the Wiley–Interscience Paperback Series.

CONGDON · Applied Bayesian Modelling
CONGDON · Bayesian Models for Categorical Data
CONGDON · Bayesian Statistical Modelling
CONOVER · Practical Nonparametric Statistics, *Third Edition*
COOK · Regression Graphics
COOK and WEISBERG · Applied Regression Including Computing and Graphics
COOK and WEISBERG · An Introduction to Regression Graphics
CORNELL · Experiments with Mixtures, Designs, Models, and the Analysis of Mixture Data, *Third Edition*
COVER and THOMAS · Elements of Information Theory
COX · A Handbook of Introductory Statistical Methods
* COX · Planning of Experiments
CRESSIE · Statistics for Spatial Data, *Revised Edition*
CSÖRGŐ and HORVÁTH · Limit Theorems in Change Point Analysis
DANIEL · Applications of Statistics to Industrial Experimentation
DANIEL · Biostatistics: A Foundation for Analysis in the Health Sciences, *Eighth Edition*
* DANIEL · Fitting Equations to Data: Computer Analysis of Multifactor Data, *Second Edition*
DASU and JOHNSON · Exploratory Data Mining and Data Cleaning
DAVID and NAGARAJA · Order Statistics, *Third Edition*
* DEGROOT, FIENBERG, and KADANE · Statistics and the Law
DEL CASTILLO · Statistical Process Adjustment for Quality Control
DEMARIS · Regression with Social Data: Modeling Continuous and Limited Response Variables
DEMIDENKO · Mixed Models: Theory and Applications
DENISON, HOLMES, MALLICK and SMITH · Bayesian Methods for Nonlinear Classification and Regression
DETTE and STUDDEN · The Theory of Canonical Moments with Applications in Statistics, Probability, and Analysis
DEY and MUKERJEE · Fractional Factorial Plans
DILLON and GOLDSTEIN · Multivariate Analysis: Methods and Applications
DODGE · Alternative Methods of Regression
* DODGE and ROMIG · Sampling Inspection Tables, *Second Edition*
* DOOB · Stochastic Processes
DOWDY, WEARDEN, and CHILKO · Statistics for Research, *Third Edition*
DRAPER and SMITH · Applied Regression Analysis, *Third Edition*
DRYDEN and MARDIA · Statistical Shape Analysis
DUDEWICZ and MISHRA · Modern Mathematical Statistics
DUNN and CLARK · Basic Statistics: A Primer for the Biomedical Sciences, *Third Edition*
DUPUIS and ELLIS · A Weak Convergence Approach to the Theory of Large Deviations
EDLER and KITSOS · Recent Advances in Quantitative Methods in Cancer and Human Health Risk Assessment
* ELANDT-JOHNSON and JOHNSON · Survival Models and Data Analysis
ENDERS · Applied Econometric Time Series
† ETHIER and KURTZ · Markov Processes: Characterization and Convergence
EVANS, HASTINGS, and PEACOCK · Statistical Distributions, *Third Edition*
FELLER · An Introduction to Probability Theory and Its Applications, Volume I, *Third Edition*, Revised; Volume II, *Second Edition*
FISHER and VAN BELLE · Biostatistics: A Methodology for the Health Sciences
FITZMAURICE, LAIRD, and WARE · Applied Longitudinal Analysis
* FLEISS · The Design and Analysis of Clinical Experiments
FLEISS · Statistical Methods for Rates and Proportions, *Third Edition*

*Now available in a lower priced paperback edition in the Wiley Classics Library.
†Now available in a lower priced paperback edition in the Wiley–Interscience Paperback Series.

† FLEMING and HARRINGTON · Counting Processes and Survival Analysis
FULLER · Introduction to Statistical Time Series, *Second Edition*
† FULLER · Measurement Error Models
GALLANT · Nonlinear Statistical Models
GEISSER · Modes of Parametric Statistical Inference
GELMAN and MENG · Applied Bayesian Modeling and Causal Inference from Incomplete-Data Perspectives
GEWEKE · Contemporary Bayesian Econometrics and Statistics
GHOSH, MUKHOPADHYAY, and SEN · Sequential Estimation
GIESBRECHT and GUMPERTZ · Planning, Construction, and Statistical Analysis of Comparative Experiments
GIFI · Nonlinear Multivariate Analysis
GIVENS and HOETING · Computational Statistics
GLASSERMAN and YAO · Monotone Structure in Discrete-Event Systems
GNANADESIKAN · Methods for Statistical Data Analysis of Multivariate Observations, *Second Edition*
GOLDSTEIN and LEWIS · Assessment: Problems, Development, and Statistical Issues
GREENWOOD and NIKULIN · A Guide to Chi-Squared Testing
GROSS and HARRIS · Fundamentals of Queueing Theory, *Third Edition*
* HAHN and SHAPIRO · Statistical Models in Engineering
HAHN and MEEKER · Statistical Intervals: A Guide for Practitioners
HALD · A History of Probability and Statistics and their Applications Before 1750
HALD · A History of Mathematical Statistics from 1750 to 1930
† HAMPEL · Robust Statistics: The Approach Based on Influence Functions
HANNAN and DEISTLER · The Statistical Theory of Linear Systems
HEIBERGER · Computation for the Analysis of Designed Experiments
HEDAYAT and SINHA · Design and Inference in Finite Population Sampling
HEDEKER and GIBBONS · Longitudinal Data Analysis
HELLER · MACSYMA for Statisticians
HINKELMANN and KEMPTHORNE · Design and Analysis of Experiments, Volume 1: Introduction to Experimental Design, *Second Edition*
HINKELMANN and KEMPTHORNE · Design and Analysis of Experiments, Volume 2: Advanced Experimental Design
HOAGLIN, MOSTELLER, and TUKEY · Exploratory Approach to Analysis of Variance
* HOAGLIN, MOSTELLER, and TUKEY · Exploring Data Tables, Trends and Shapes
* HOAGLIN, MOSTELLER, and TUKEY · Understanding Robust and Exploratory Data Analysis
HOCHBERG and TAMHANE · Multiple Comparison Procedures
HOCKING · Methods and Applications of Linear Models: Regression and the Analysis of Variance, *Second Edition*
HOEL · Introduction to Mathematical Statistics, *Fifth Edition*
HOGG and KLUGMAN · Loss Distributions
HOLLANDER and WOLFE · Nonparametric Statistical Methods, *Second Edition*
HOSMER and LEMESHOW · Applied Logistic Regression, *Second Edition*
HOSMER and LEMESHOW · Applied Survival Analysis: Regression Modeling of Time to Event Data
† HUBER · Robust Statistics
HUBERTY · Applied Discriminant Analysis
HUBERTY and OLEJNIK · Applied MANOVA and Discriminant Analysis, *Second Edition*
HUNT and KENNEDY · Financial Derivatives in Theory and Practice, *Revised Edition*

*Now available in a lower priced paperback edition in the Wiley Classics Library.
†Now available in a lower priced paperback edition in the Wiley–Interscience Paperback Series.

HURD and MIAMEE · Periodically Correlated Random Sequences: Spectral Theory and Practice
HUSKOVA, BERAN, and DUPAC · Collected Works of Jaroslav Hajek—with Commentary
HUZURBAZAR · Flowgraph Models for Multistate Time-to-Event Data
IMAN and CONOVER · A Modern Approach to Statistics
† JACKSON · A User's Guide to Principle Components
JOHN · Statistical Methods in Engineering and Quality Assurance
JOHNSON · Multivariate Statistical Simulation
JOHNSON and BALAKRISHNAN · Advances in the Theory and Practice of Statistics: A Volume in Honor of Samuel Kotz
JOHNSON and BHATTACHARYYA · Statistics: Principles and Methods, *Fifth Edition*
JOHNSON and KOTZ · Distributions in Statistics
JOHNSON and KOTZ (editors) · Leading Personalities in Statistical Sciences: From the Seventeenth Century to the Present
JOHNSON, KOTZ, and BALAKRISHNAN · Continuous Univariate Distributions, Volume 1, *Second Edition*
JOHNSON, KOTZ, and BALAKRISHNAN · Continuous Univariate Distributions, Volume 2, *Second Edition*
JOHNSON, KOTZ, and BALAKRISHNAN · Discrete Multivariate Distributions
JOHNSON, KEMP, and KOTZ · Univariate Discrete Distributions, *Third Edition*
JUDGE, GRIFFITHS, HILL, LÜTKEPOHL, and LEE · The Theory and Practice of Econometrics, *Second Edition*
JUREČKOVÁ and SEN · Robust Statistical Procedures: Aymptotics and Interrelations
JUREK and MASON · Operator-Limit Distributions in Probability Theory
KADANE · Bayesian Methods and Ethics in a Clinical Trial Design
KADANE AND SCHUM · A Probabilistic Analysis of the Sacco and Vanzetti Evidence
KALBFLEISCH and PRENTICE · The Statistical Analysis of Failure Time Data, *Second Edition*
KARIYA and KURATA · Generalized Least Squares
KASS and VOS · Geometrical Foundations of Asymptotic Inference
† KAUFMAN and ROUSSEEUW · Finding Groups in Data: An Introduction to Cluster Analysis
KEDEM and FOKIANOS · Regression Models for Time Series Analysis
KENDALL, BARDEN, CARNE, and LE · Shape and Shape Theory
KHURI · Advanced Calculus with Applications in Statistics, *Second Edition*
KHURI, MATHEW, and SINHA · Statistical Tests for Mixed Linear Models
KLEIBER and KOTZ · Statistical Size Distributions in Economics and Actuarial Sciences
KLUGMAN, PANJER, and WILLMOT · Loss Models: From Data to Decisions, *Second Edition*
KLUGMAN, PANJER, and WILLMOT · Solutions Manual to Accompany Loss Models: From Data to Decisions, *Second Edition*
KOTZ, BALAKRISHNAN, and JOHNSON · Continuous Multivariate Distributions, Volume 1, *Second Edition*
KOVALENKO, KUZNETZOV, and PEGG · Mathematical Theory of Reliability of Time-Dependent Systems with Practical Applications
KOWALSKI and TU · Modern Applied U-Statistics
KVAM and VIDAKOVIC · Nonparametric Statistics with Applications to Science and Engineering
LACHIN · Biostatistical Methods: The Assessment of Relative Risks
LAD · Operational Subjective Statistical Methods: A Mathematical, Philosophical, and Historical Introduction
LAMPERTI · Probability: A Survey of the Mathematical Theory, *Second Edition*

*Now available in a lower priced paperback edition in the Wiley Classics Library.
†Now available in a lower priced paperback edition in the Wiley–Interscience Paperback Series.

LANGE, RYAN, BILLARD, BRILLINGER, CONQUEST, and GREENHOUSE · Case Studies in Biometry
LARSON · Introduction to Probability Theory and Statistical Inference, *Third Edition*
LAWLESS · Statistical Models and Methods for Lifetime Data, *Second Edition*
LAWSON · Statistical Methods in Spatial Epidemiology
LE · Applied Categorical Data Analysis
LE · Applied Survival Analysis
LEE and WANG · Statistical Methods for Survival Data Analysis, *Third Edition*
LEPAGE and BILLARD · Exploring the Limits of Bootstrap
LEYLAND and GOLDSTEIN (editors) · Multilevel Modelling of Health Statistics
LIAO · Statistical Group Comparison
LINDVALL · Lectures on the Coupling Method
LIN · Introductory Stochastic Analysis for Finance and Insurance
LINHART and ZUCCHINI · Model Selection
LITTLE and RUBIN · Statistical Analysis with Missing Data, *Second Edition*
LLOYD · The Statistical Analysis of Categorical Data
LOWEN and TEICH · Fractal-Based Point Processes
MAGNUS and NEUDECKER · Matrix Differential Calculus with Applications in Statistics and Econometrics, *Revised Edition*
MALLER and ZHOU · Survival Analysis with Long Term Survivors
MALLOWS · Design, Data, and Analysis by Some Friends of Cuthbert Daniel
MANN, SCHAFER, and SINGPURWALLA · Methods for Statistical Analysis of Reliability and Life Data
MANTON, WOODBURY, and TOLLEY · Statistical Applications Using Fuzzy Sets
MARCHETTE · Random Graphs for Statistical Pattern Recognition
MARDIA and JUPP · Directional Statistics
MASON, GUNST, and HESS · Statistical Design and Analysis of Experiments with Applications to Engineering and Science, *Second Edition*
McCULLOCH and SEARLE · Generalized, Linear, and Mixed Models
McFADDEN · Management of Data in Clinical Trials, *Second Edition*
* McLACHLAN · Discriminant Analysis and Statistical Pattern Recognition
McLACHLAN, DO, and AMBROISE · Analyzing Microarray Gene Expression Data
McLACHLAN and KRISHNAN · The EM Algorithm and Extensions, *Second Edition*
McLACHLAN and PEEL · Finite Mixture Models
McNEIL · Epidemiological Research Methods
MEEKER and ESCOBAR · Statistical Methods for Reliability Data
MEERSCHAERT and SCHEFFLER · Limit Distributions for Sums of Independent Random Vectors: Heavy Tails in Theory and Practice
MICKEY, DUNN, and CLARK · Applied Statistics: Analysis of Variance and Regression, *Third Edition*
* MILLER · Survival Analysis, *Second Edition*
MONTGOMERY, PECK, and VINING · Introduction to Linear Regression Analysis, *Fourth Edition*
MORGENTHALER and TUKEY · Configural Polysampling: A Route to Practical Robustness
MUIRHEAD · Aspects of Multivariate Statistical Theory
MULLER and STOYAN · Comparison Methods for Stochastic Models and Risks
MURRAY · X-STAT 2.0 Statistical Experimentation, Design Data Analysis, and Nonlinear Optimization
MURTHY, XIE, and JIANG · Weibull Models
MYERS and MONTGOMERY · Response Surface Methodology: Process and Product Optimization Using Designed Experiments, *Second Edition*
MYERS, MONTGOMERY, and VINING · Generalized Linear Models. With Applications in Engineering and the Sciences

*Now available in a lower priced paperback edition in the Wiley Classics Library.
†Now available in a lower priced paperback edition in the Wiley–Interscience Paperback Series.

† NELSON · Accelerated Testing, Statistical Models, Test Plans, and Data Analyses
† NELSON · Applied Life Data Analysis
NEWMAN · Biostatistical Methods in Epidemiology
OCHI · Applied Probability and Stochastic Processes in Engineering and Physical Sciences
OKABE, BOOTS, SUGIHARA, and CHIU · Spatial Tesselations: Concepts and Applications of Voronoi Diagrams, *Second Edition*
OLIVER and SMITH · Influence Diagrams, Belief Nets and Decision Analysis
PALTA · Quantitative Methods in Population Health: Extensions of Ordinary Regressions
PANJER · Operational Risk: Modeling and Analytics
PANKRATZ · Forecasting with Dynamic Regression Models
PANKRATZ · Forecasting with Univariate Box-Jenkins Models: Concepts and Cases
* PARZEN · Modern Probability Theory and Its Applications
PEÑA, TIAO, and TSAY · A Course in Time Series Analysis
PIANTADOSI · Clinical Trials: A Methodologic Perspective
PORT · Theoretical Probability for Applications
POURAHMADI · Foundations of Time Series Analysis and Prediction Theory
POWELL · Approximate Dynamic Programming: Solving the Curses of Dimensionality
PRESS · Bayesian Statistics: Principles, Models, and Applications
PRESS · Subjective and Objective Bayesian Statistics, *Second Edition*
PRESS and TANUR · The Subjectivity of Scientists and the Bayesian Approach
PUKELSHEIM · Optimal Experimental Design
PURI, VILAPLANA, and WERTZ · New Perspectives in Theoretical and Applied Statistics
† PUTERMAN · Markov Decision Processes: Discrete Stochastic Dynamic Programming
QIU · Image Processing and Jump Regression Analysis
* RAO · Linear Statistical Inference and Its Applications, *Second Edition*
RAUSAND and HØYLAND · System Reliability Theory: Models, Statistical Methods, and Applications, *Second Edition*
RENCHER · Linear Models in Statistics
RENCHER · Methods of Multivariate Analysis, *Second Edition*
RENCHER · Multivariate Statistical Inference with Applications
* RIPLEY · Spatial Statistics
* RIPLEY · Stochastic Simulation
ROBINSON · Practical Strategies for Experimenting
ROHATGI and SALEH · An Introduction to Probability and Statistics, *Second Edition*
ROLSKI, SCHMIDLI, SCHMIDT, and TEUGELS · Stochastic Processes for Insurance and Finance
ROSENBERGER and LACHIN · Randomization in Clinical Trials: Theory and Practice
ROSS · Introduction to Probability and Statistics for Engineers and Scientists
ROSSI, ALLENBY, and McCULLOCH · Bayesian Statistics and Marketing
† ROUSSEEUW and LEROY · Robust Regression and Outlier Detection
* RUBIN · Multiple Imputation for Nonresponse in Surveys
RUBINSTEIN · Simulation and the Monte Carlo Method
RUBINSTEIN and MELAMED · Modern Simulation and Modeling
RYAN · Modern Engineering Statistics
RYAN · Modern Experimental Design
RYAN · Modern Regression Methods
RYAN · Statistical Methods for Quality Improvement, *Second Edition*
SALEH · Theory of Preliminary Test and Stein-Type Estimation with Applications
* SCHEFFE · The Analysis of Variance
SCHIMEK · Smoothing and Regression: Approaches, Computation, and Application
SCHOTT · Matrix Analysis for Statistics, *Second Edition*
SCHOUTENS · Levy Processes in Finance: Pricing Financial Derivatives

*Now available in a lower priced paperback edition in the Wiley Classics Library.
†Now available in a lower priced paperback edition in the Wiley–Interscience Paperback Series.

SCHUSS · Theory and Applications of Stochastic Differential Equations
SCOTT · Multivariate Density Estimation: Theory, Practice, and Visualization
† SEARLE · Linear Models for Unbalanced Data
† SEARLE · Matrix Algebra Useful for Statistics
† SEARLE, CASELLA, and McCULLOCH · Variance Components
SEARLE and WILLETT · Matrix Algebra for Applied Economics
SEBER · A Matrix Handbook For Statisticians
† SEBER · Multivariate Observations
SEBER and LEE · Linear Regression Analysis, *Second Edition*
† SEBER and WILD · Nonlinear Regression
SENNOTT · Stochastic Dynamic Programming and the Control of Queueing Systems
* SERFLING · Approximation Theorems of Mathematical Statistics
SHAFER and VOVK · Probability and Finance: It's Only a Game!
SILVAPULLE and SEN · Constrained Statistical Inference: Inequality, Order, and Shape Restrictions
SMALL and McLEISH · Hilbert Space Methods in Probability and Statistical Inference
SRIVASTAVA · Methods of Multivariate Statistics
STAPLETON · Linear Statistical Models
STAPLETON · Models for Probability and Statistical Inference: Theory and Applications
STAUDTE and SHEATHER · Robust Estimation and Testing
STOYAN, KENDALL, and MECKE · Stochastic Geometry and Its Applications, *Second Edition*
STOYAN and STOYAN · Fractals, Random Shapes and Point Fields: Methods of Geometrical Statistics
STREET and BURGESS · The Construction of Optimal Stated Choice Experiments: Theory and Methods
STYAN · The Collected Papers of T. W. Anderson: 1943–1985
SUTTON, ABRAMS, JONES, SHELDON, and SONG · Methods for Meta-Analysis in Medical Research
TAKEZAWA · Introduction to Nonparametric Regression
TANAKA · Time Series Analysis: Nonstationary and Noninvertible Distribution Theory
THOMPSON · Empirical Model Building
THOMPSON · Sampling, *Second Edition*
THOMPSON · Simulation: A Modeler's Approach
THOMPSON and SEBER · Adaptive Sampling
THOMPSON, WILLIAMS, and FINDLAY · Models for Investors in Real World Markets
TIAO, BISGAARD, HILL, PEÑA, and STIGLER (editors) · Box on Quality and Discovery: with Design, Control, and Robustness
TIERNEY · LISP-STAT: An Object-Oriented Environment for Statistical Computing and Dynamic Graphics
TSAY · Analysis of Financial Time Series, *Second Edition*
UPTON and FINGLETON · Spatial Data Analysis by Example, Volume II: Categorical and Directional Data
VAN BELLE · Statistical Rules of Thumb
VAN BELLE, FISHER, HEAGERTY, and LUMLEY · Biostatistics: A Methodology for the Health Sciences, *Second Edition*
VESTRUP · The Theory of Measures and Integration
VIDAKOVIC · Statistical Modeling by Wavelets
VINOD and REAGLE · Preparing for the Worst: Incorporating Downside Risk in Stock Market Investments
WALLER and GOTWAY · Applied Spatial Statistics for Public Health Data
WEERAHANDI · Generalized Inference in Repeated Measures: Exact Methods in MANOVA and Mixed Models
WEISBERG · Applied Linear Regression, *Third Edition*

*Now available in a lower priced paperback edition in the Wiley Classics Library.
†Now available in a lower priced paperback edition in the Wiley–Interscience Paperback Series.

WELSH · Aspects of Statistical Inference
WESTFALL and YOUNG · Resampling-Based Multiple Testing: Examples and Methods for *p*-Value Adjustment
WHITTAKER · Graphical Models in Applied Multivariate Statistics
WINKER · Optimization Heuristics in Economics: Applications of Threshold Accepting
WONNACOTT and WONNACOTT · Econometrics, *Second Edition*
WOODING · Planning Pharmaceutical Clinical Trials: Basic Statistical Principles
WOODWORTH · Biostatistics: A Bayesian Introduction
WOOLSON and CLARKE · Statistical Methods for the Analysis of Biomedical Data, *Second Edition*
WU and HAMADA · Experiments: Planning, Analysis, and Parameter Design Optimization
WU and ZHANG · Nonparametric Regression Methods for Longitudinal Data Analysis
YANG · The Construction Theory of Denumerable Markov Processes
YOUNG, VALERO-MORA, and FRIENDLY · Visual Statistics: Seeing Data with Dynamic Interactive Graphics
ZELTERMAN · Discrete Distributions—Applications in the Health Sciences
* ZELLNER · An Introduction to Bayesian Inference in Econometrics
ZHOU, OBUCHOWSKI, and McCLISH · Statistical Methods in Diagnostic Medicine

*Now available in a lower priced paperback edition in the Wiley Classics Library.
†Now available in a lower priced paperback edition in the Wiley–Interscience Paperback Series.

CPSIA information can be obtained at www.ICGtesting.com
Printed in the USA
BVOW11*0108250614

357275BV00007B/79/P

9 780470 073728